Student Solutions Manual

JOHN POLKING

DAVID ARNOLD

Differential Equations With Boundary Value Problems

SECOND EDITION

POLKING BOGGESS ARNOLD

Upper Saddle River, NJ 07458

Editor-in-Chief: Sally Yagan
Supplement Editor: Jennifer Urban
Executive Managing Editor: Kathleen Schiaparelli
Assistant Managing Editor: Karen Bosch
Production Editor: Jenelle J. Woodrup
Supplement Cover Manager: Paul Gourhan
Supplement Cover Design: Joanne Alexandris
Manufacturing Buyer: Ilene Kahn
Manufacturing Manager: Alexis Heydt-Long

Pearson Prentice Hall
Pearson Education, Inc.
Upper Saddle River, NJ 07458

Printed in the United States of America

10 9 8 7 6 5 4 3 2

ISBN 0-13-143739-9

Pearson Education Ltd., *London*
Pearson Education Australia Pty. Ltd., *Sydney*
Pearson Education Singapore, Pte. Ltd.
Pearson Education North Asia Ltd., *Hong Kong*
Pearson Education Canada, Inc., *Toronto*
Pearson Educación de Mexico, S.A. de C.V.
Pearson Education—Japan, *Tokyo*
Pearson Education Malaysia, Pte. Ltd.

Contents

Chapter 1. Introduction

Section 1.1. Introduction to Differential Equations

1. Let $y(t)$ be the number of bacteria at time t. The rate of change of the number of bacteria is $y'(t)$. Since this rate of change is given to be proportional to $y(t)$, the resulting differential equation is $y'(t) = ky(t)$. Note that k is a positive constant since $y'(t)$ must be positive (i.e. the number of bacteria is growing).

3. Let $y(t)$ be the number of ferrets at time t. The rate of change of the number of ferrets is $y'(t)$. Since this rate of change is given to be proportional to the product of $y(t)$ and the difference between the maximum population and $y(t)$ (i.e. $100 - y(t)$), the resulting differential equation is $y'(t) = ky(t)(100 - y(t))$. Note that k is a positive constant since $y'(t)$ must be positive (i.e. the number of ferrets is growing provided $y(t) < 100$).

5. Let $y(t)$ be the quantity of material at time t. The rate of change of the material is $y'(t)$. Since this rate of change (decay) is given to be inversely proportional to $y(t)$, the resulting differential equation is $y'(t) = -k/y(t)$. Note that k is a positive constant since $y'(t)$ must be negative (i.e. the quantity of material is decreasing).

7. Let $y(t)$ be the temperature of the thermometer at time t. The rate of change of the temperature is $y'(t)$. Since this rate of change is given to be proportional to the difference between the thermometer's temperature and that of the surrounding room (i.e. $77 - y(t)$), the resulting differential equation is $y'(t) = k(77 - y(t))$. Note that k is a positive constant since $y'(t)$ must be positive (i.e. the thermometer is warming) and since $77 - y(t) > 0$ (i.e. the thermometer is cooler than the surrounding room).

9. Let $x(t)$ be the position (displacement) of the particle at time t. The force on the particle is given to be proportional to the square of the particle's velocity, i.e. $(x'(t))^2$. As a first guess, one might surmise that the force is given by $F = -k(x'(t))^2$, where k is a positive constant. However, closer inspection reveals that this will have the force pointing to the left, regardless of whether the velocity is positive or negative. We can work around this difficulty by letting the force equal $F = -kx'(t)|x'(t)|$. The reader will recognize that the force is positive when $x'(t) < 0$, while the force is negative when $x'(t) > 0$, thus insuring that the force is always opposite the particle's motion. Newton's law states $F = ma$ where m is the mass of the object and $a = x''(t)$ is its acceleration. Therefore, $F = ma$ becomes $-k(x'(t))|x'(t)| = mx''(t)$, which is the differential equation governing the motion of this particle.

11. Let $V(t)$ be the voltage drop across the inductor and $I(t)$ be the current at time t. The rate of change of the current is $I'(t)$. Since the voltage drop is proportional to the rate of change of I, we obtain the differential equation $V(t) = kI'(t)$, where k is a constant.

Section 1.2. The Derivative

1. $$\begin{aligned} D_x(3x-5) &= 3D_x x - D_x 5 \\ &= 3(1) - 0 \\ &= 3 \end{aligned}$$

3. $$\begin{aligned} D_x(3\sin 5x) &= 3D_x \sin 5x \\ &= 3(\cos 5x)D_x(5x) \\ &= 15\cos 5x \end{aligned}$$

5. $$\begin{aligned} D_x(e^{3x}) &= e^{3x}D_x(3x) \\ &= 3e^{3x} \end{aligned}$$

7. $$\begin{aligned} D_x \ln|5x| &= \tfrac{1}{5x}D_x(5x) \\ &= \tfrac{1}{x} \end{aligned}$$

9. $$\begin{aligned} D_x x\ln x &= (D_x x)\ln x + xD_x \ln x \\ &= (1)\ln x + x\left(\tfrac{1}{x}\right) \\ &= 1 + \ln x \end{aligned}$$

11. $$\begin{aligned} D_x\left(\tfrac{x^2}{\ln x}\right) &= \tfrac{(D_x x^2)\ln x - x^2 D_x \ln x}{[\ln x]^2} \\ &= \tfrac{2x\ln x - x^2\left(\frac{1}{x}\right)}{[\ln x]^2} \\ &= \tfrac{2x\ln x - x}{[\ln x]^2} \end{aligned}$$

13. If $L(x) = f(x_0) + f'(x_0)(x - x_0)$, then

$$\begin{aligned} R(x) &= f(x) - L(x) \\ &= f(x) - f(x_0) - f'(x_0)(x - x_0). \end{aligned}$$

Thus,

$$\begin{aligned} \lim_{x\to x_0}\frac{R(x)}{x - x_0} &= \lim_{x\to x_0}\left[\frac{f(x) - f(x_0)}{x - x_0} - f'(x_0)\right] \\ &= f'(x_0) - f'(x_0) \\ &= 0. \end{aligned}$$

15. Given that $f(x) = \cos x$, the derivative is $f'(x) = -\sin x$. At $x_0 = \pi/4$, $f'(\pi/4) = -\sqrt{2}/2$. Thus, the linearization is

$$\begin{aligned} L(x) &= f(\pi/4) + f'(\pi/4)(x - \pi/4) \\ &= \frac{\sqrt{2}}{2} - \frac{\sqrt{2}}{2}\left(x - \frac{\pi}{4}\right). \end{aligned}$$

The graph of f, together with its linear approximation at $x_0 = \pi/4$, is shown in the following figure.

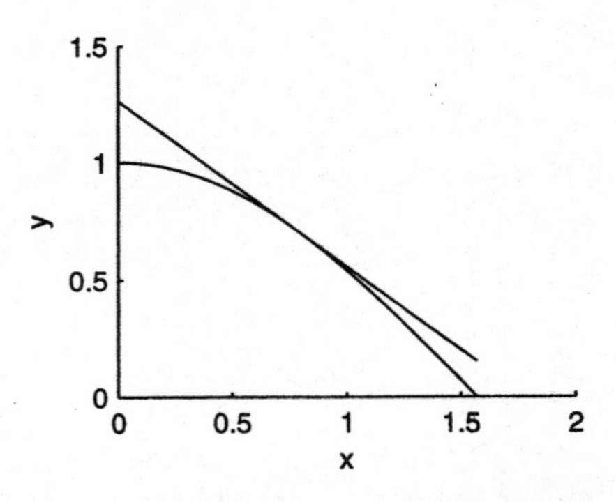

17. Given that $f(x) = \ln(1 + x)$, the derivative is $f'(x) = 1/(1 + x)$. At $x_0 = 0$, $f'(0) = 1$. Thus, the linearization is

$$L(x) = f(0) + f'(0)(x - 0) = x.$$

The graph of f, together with its linear approximation at $x_0 = 0$, is shown in the following figure.

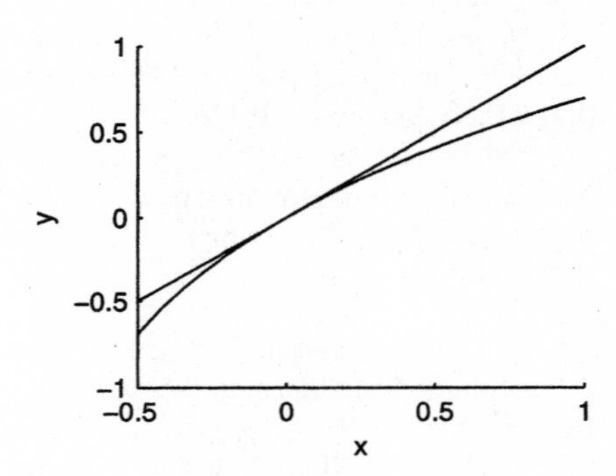

19. Given that $f(x) = \sin 2x$, the derivative is $f'(x) = 2\cos 2x$. At $x_0 = \pi/8$, $f'(\pi/8) = \sqrt{2}$. Thus, the linearization is

$$\begin{aligned} L(x) &= f(\pi/8) + f'(\pi/8)(x - \pi/8) \\ &= \frac{\sqrt{2}}{2} + \sqrt{2}\left(x - \frac{\pi}{8}\right). \end{aligned}$$

The graph of $y = x - \pi/8$, together with the graph of the remainder $R(x) = f(x) - L(x)$, is shown in the following figure.

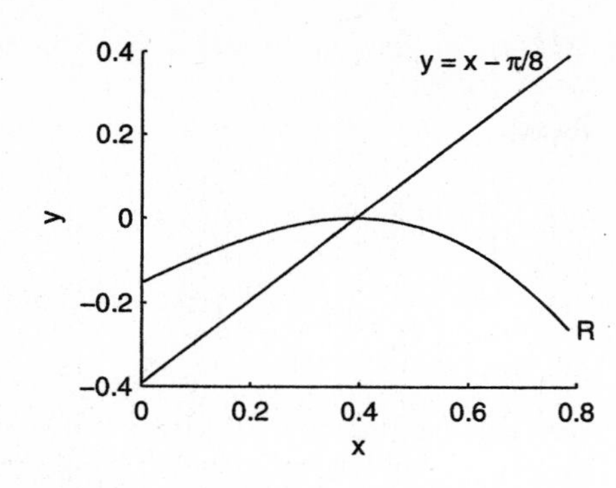

Note that both graphs approach zero as $x \to \pi/8$, but the graph of R approaches zero at a more rapid rate.

21. Given that $f(x) = xe^{x-1}$, the derivative is $f'(x) = (x+1)e^{x-1}$. At $x_0 = 1$, $f'(1) = 2$. Thus, the linearization is

$$L(x) = f(1) + f'(1)(x-1) = 1 + 2(x-1).$$

The graph of $y = x - 1$, together with the graph of the remainder $R(x) = f(x) - L(x)$, is shown in the following figure.

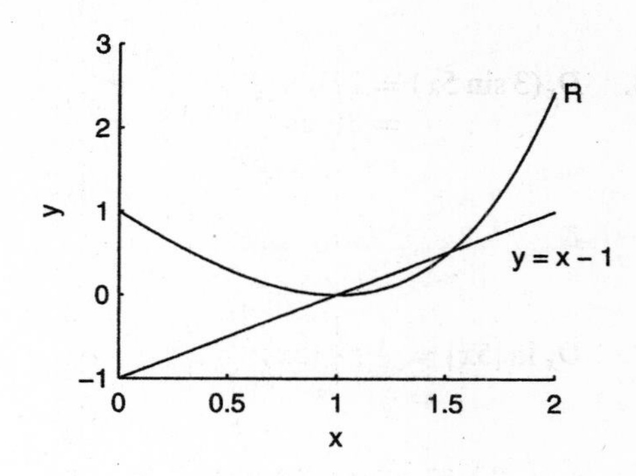

Note that both graphs approach zero as $x \to 1$, but the graph of R approaches zero at a more rapid rate.

Section 1.3. Integration

1. $y' = 2t + 3$. Integrate to obtain $y = t^2 + 3t + C$.

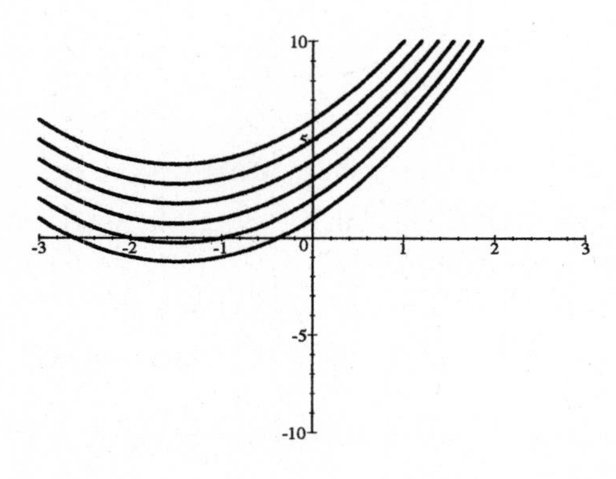

3. $y' = \sin 2t + 2\cos 3t$. Integrate to obtain $y = (-1/2)\cos 2t + (2/3)\sin 3t + C$.

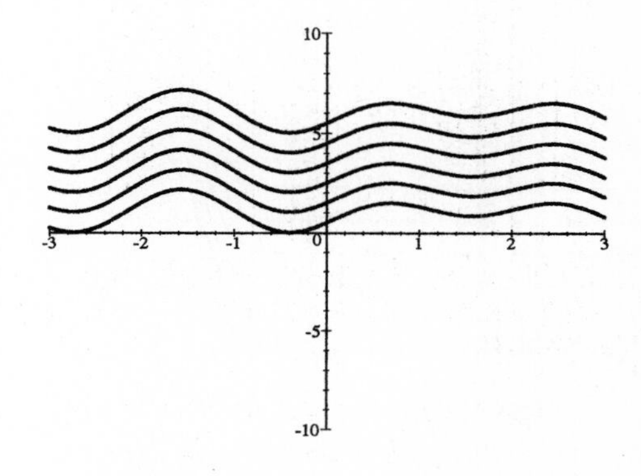

5. $y' = t/(1+t^2)$. Use $u = 1 + t^2$, $du = 2t\,dt$ and get $dy = (1/2)du/u$. Integrate to obtain

$$y = (1/2)\ln u + C = (1/2)\ln(1+t^2) + C.$$

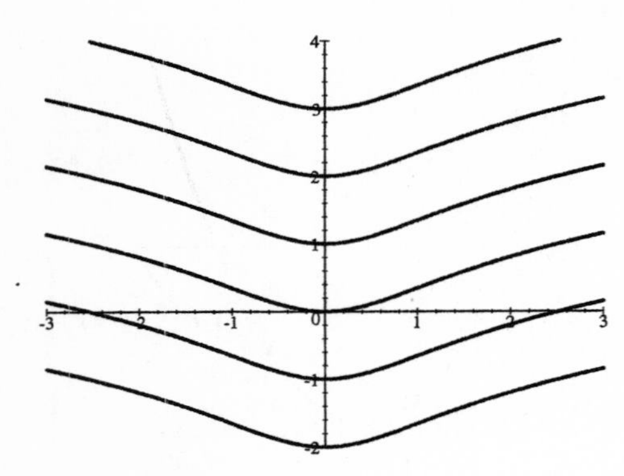

7. $y' = t^2 e^{3t}$. Integrate by parts with $u = t^2$ and $dv = e^{3t}$ to obtain

$$y = \frac{t^2 e^{3t}}{3} - \int \frac{e^{3t} 2t}{3} dt$$

Integrate by parts once more and obtain

$$y = \frac{t^2 e^{3t}}{3} - \frac{2te^{3t}}{9} + \frac{2e^{3t}}{27} + C$$

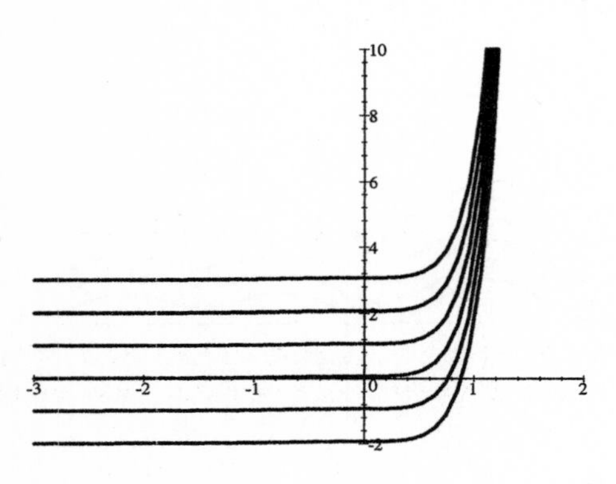

9. $y' = e^{-2\omega}\sin\omega$. Integrate by parts with $u = e^{-2\omega}$ and $dv = \sin\omega$ to obtain

$$\int e^{-2\omega}\sin\omega\, d\omega = -e^{-2\omega}\cos\omega - 2\int \cos\omega e^{-2\omega}\, d\omega.$$

Integrate by parts again with $u = e^{-2\omega}$ and $dv = \cos\omega$, to obtain

$$\int e^{-2\omega}\sin\omega\, d\omega = -e^{-2\omega}\cos\omega - 2\sin\omega e^{-2\omega} - 4\int \sin\omega e^{-2\omega}\, d\omega$$

Then add the integral on the right to the integral on the left, which then becomes $5\int \sin\omega e^{-2\omega}\, d\omega$; divide by the 5 and obtain the answer:

$$y = \left(-e^{-2\omega}\cos\omega - 2\sin\omega e^{-2\omega}\right)/5 + C$$

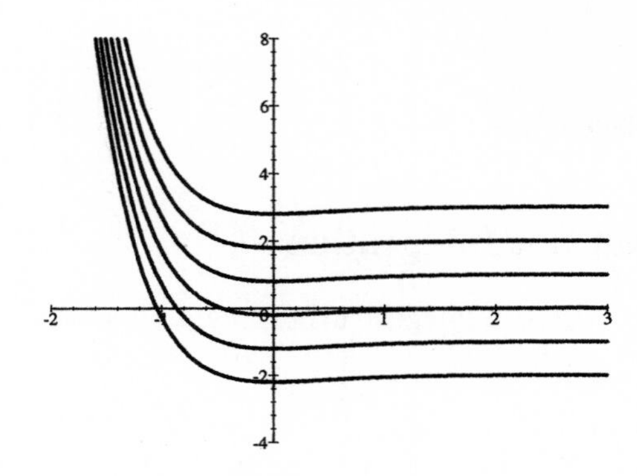

11. $x' = s^2 e^{-s}$. Integrate by parts with $u = s^2$ and $dv = e^{-s}$ and obtain

$$x = -s^2 e^{-s} + 2\int e^{-s} s\, ds.$$

Integrate by parts again with $u = s$ and $dv = e^{-s}$ to obtain the answer:

$$x = -s^2 e^{-s} - 2se^{-s} - 2e^{-s} + C.$$

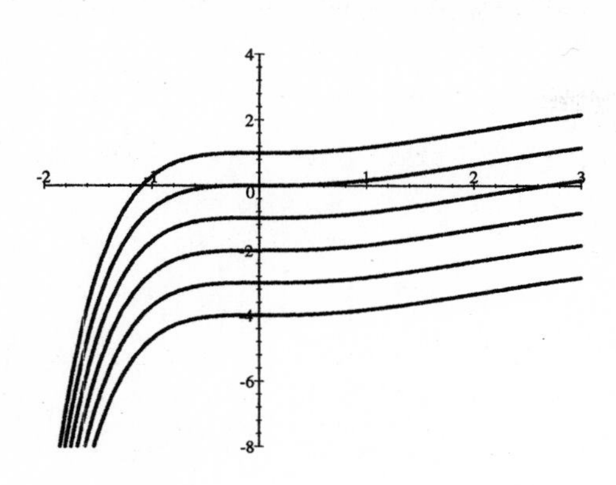

13. Use partial fractions to write

$$r' = \left[\frac{1}{u} + \frac{1}{1-u}\right].$$

Then integrate to obtain

$$r = \ln u - \ln(1-u) = \ln\left(\frac{u}{1-u}\right).$$

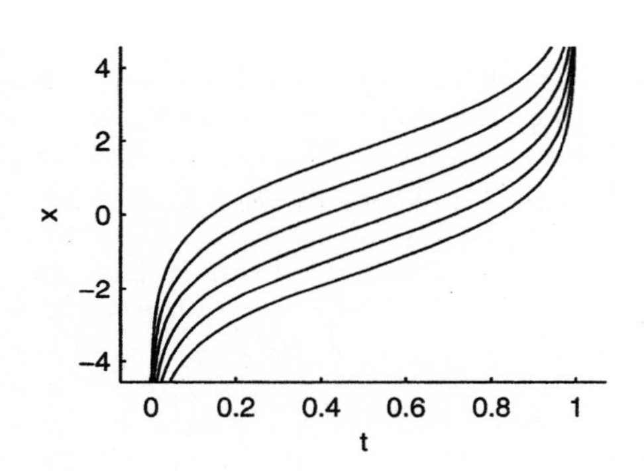

15. $y' = 4t - 6$. Integrate y' to obtain $y = 2t^2 - 6t + C$; the initial condition $y(0) = 1$ gives $1 = C$; so

$$y(t) = 2t^2 - 6t + 1.$$

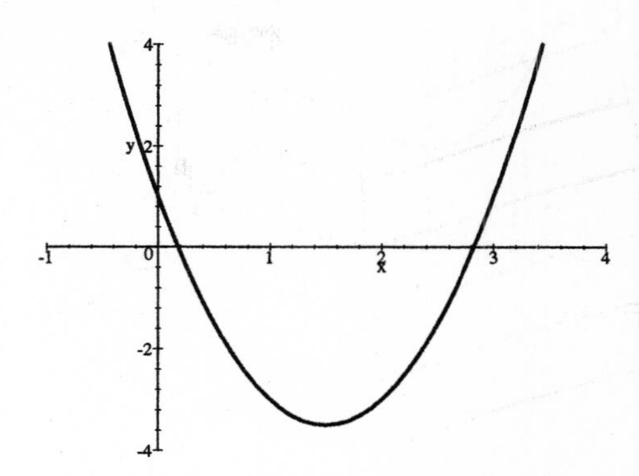

17. $x'(t) = te^{-t^2}$. Integrate to obtain $x(t) = (-1/2)e^{-t^2} + C$; the initial condition $x(0) = 1$ gives $1 = (-1/2) + C$; so $C = 3/2$ and $x(t) = (-1/2)e^{-t^2} + (3/2)$.

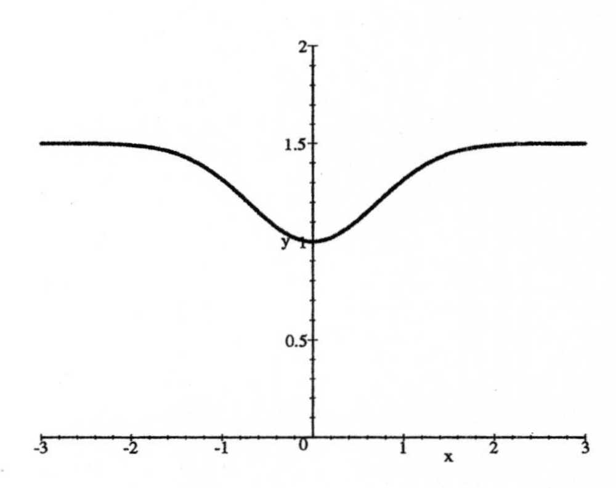

19. $s'(r) = r^2 \cos 2r$. Integrate by parts twice with dv being the trig - term (cos $2r$ and then sin $2r$ to obtain

$$s(r) = \frac{r^2 \sin 2r}{2} + \frac{r \cos 2r}{2} - \frac{\sin 2r}{4} + C.$$

The initial condition, $s(0) = 1$ gives $1 = C$ so

$$s(r) = \frac{r^2 \sin 2r}{2} + \frac{r \cos 2r}{2} - \frac{\sin 2r}{4} + 1$$

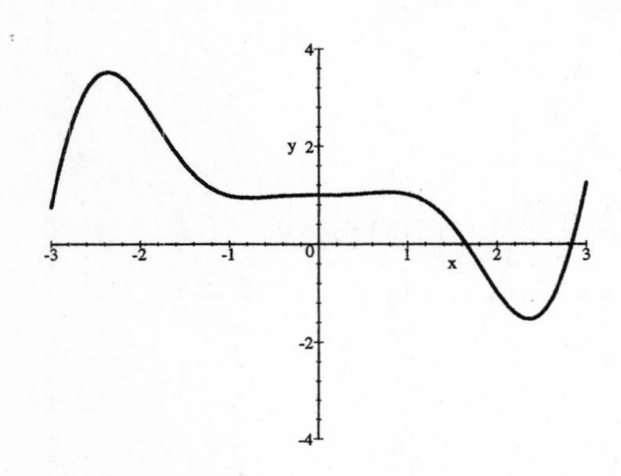

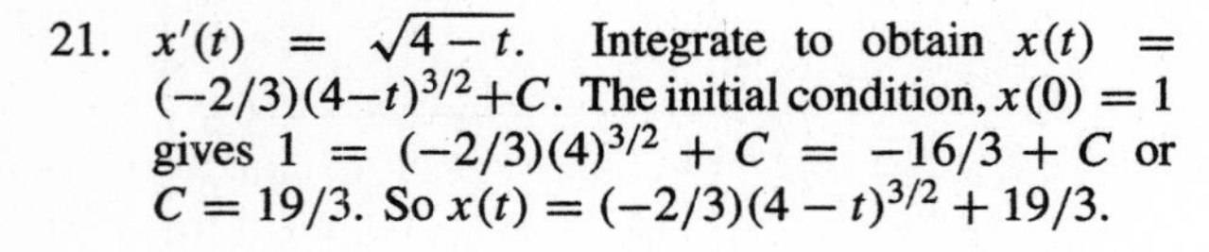

21. $x'(t) = \sqrt{4-t}$. Integrate to obtain $x(t) = (-2/3)(4-t)^{3/2}+C$. The initial condition, $x(0) = 1$ gives $1 = (-2/3)(4)^{3/2} + C = -16/3 + C$ or $C = 19/3$. So $x(t) = (-2/3)(4-t)^{3/2} + 19/3$.

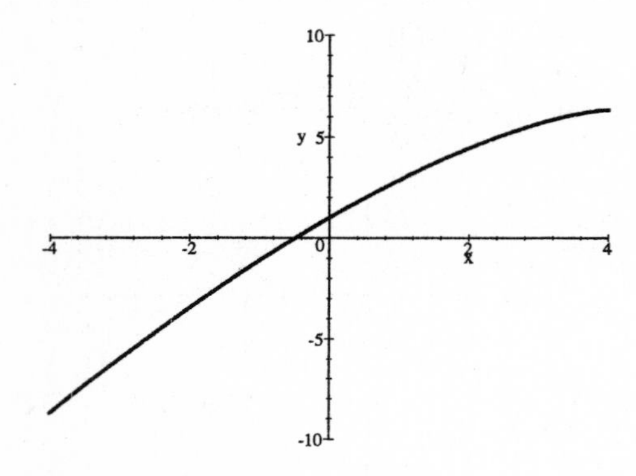

23. $y'(t) = \frac{t+1}{t(t+4)}$. Partial fractions gives

$$y'(t) = \frac{1/4}{t} + \frac{3/4}{t+4}$$

Integrating, we obtain

$$y(t) = (1/4)\ln|t| + (3/4)\ln|t+4| + C$$

The initial condition, $y(-1) = 0$ gives

$$0 = (1/4)\ln 1 + (3/4)\ln 3 + C = (3/4)\ln 3 + C$$

or $C = -(3/4)\ln 3$. So

$$y(t) = (1/4)\ln|t| + (3/4)\ln|t+4| - (3/4)\ln 3$$

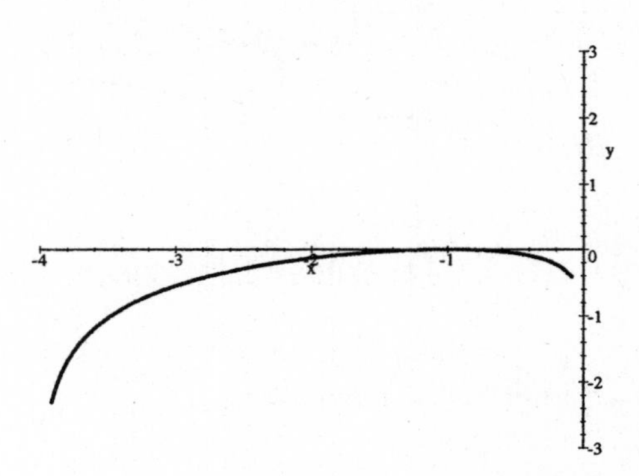

25. Let $s(t)$ be the height of the ball at time t seconds. If $g = -9.8$ is the gravitational constant, then $s''(t) = g$. Integrating we obtain, $s' = gt + v_0$, where v_0 is a constant. The initial condition $s'(0) = 50$ gives $v_0 = 50$; so $s'(t) = gt + 50$. The velocity at $t = 3$ seconds is $s'(3) = 3g + 50 = 20.6$ meters/second. Integrating s', gives $s(t) = gt^2/2 + 50t + s_0$. The initial condition, $s(0) = 3$ gives $s_0 = 3$, so $s(t) = gt^2/2 + 50t + 3$. The height at $t = 3$ seconds is $s(3) = (9/2)g + 153 = 108.9$ meters.

27. Let $s(t)$ be the height of the ball at time t seconds. If $g = -9.8$ is the gravitational constant, then $s''(t) = g$. Integrating we obtain, $s' = gt + v_0$, where v_0 is a constant. The initial condition $s'(0) = 120$ gives $v_0 = 120$, so $s'(t) = gt + 120$. The maximum height occurs when the velocity reaches zero, i.e. when $gt + 120 = 0$, or $t = -120/g = 12.24$ seconds. Integrating s' gives $s(t) = gt^2/2 + 120t + s_0$. The initial condition $s(0) = 6$ gives $s_0 = 6$, so $s(t) = gt^2/2 + 120t + 6$. When $t = 12.24$, the maximum height is $s(12.24) = 740.69$ meters.

Chapter 2. First-Order Equations

Section 2.1. Differential Equations and Solutions

1. $\phi(t, y, y') = t^2y' + (1+t)y = 0$ must be solved for y'. We get

$$y' = -\frac{(1+t)y}{t^2}.$$

3. $y'(t) = -Cte^{-(1/2)t^2}$ and $-ty(t) = -tCe^{-(1/2)t^2}$, so $y' = -ty$.

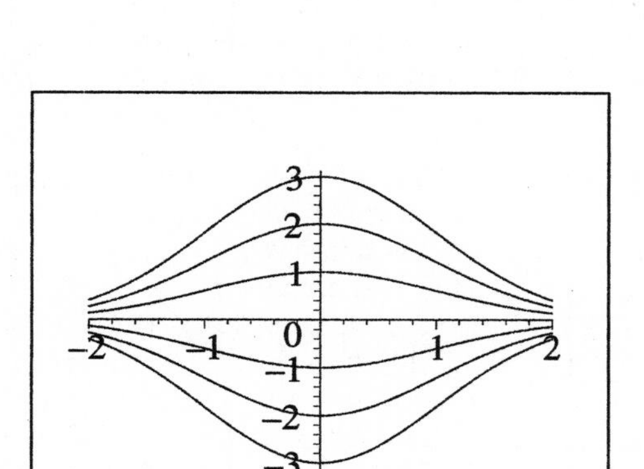

5. If $y(t) = (4/5)\cos t + (8/5)\sin t + Ce^{-(1/2)t}$, then

$$\begin{aligned}
&y(t)' + (1/2)y(t) \\
&= [-(4/5)\sin t + (8/5)\cos t - (C/2)e^{-(1/2)t}] \\
&\quad + (1/2)[(4/5)\cos t + (8/5)\sin t + Ce^{-(1/2)t}] \\
&= 2\cos t.
\end{aligned}$$

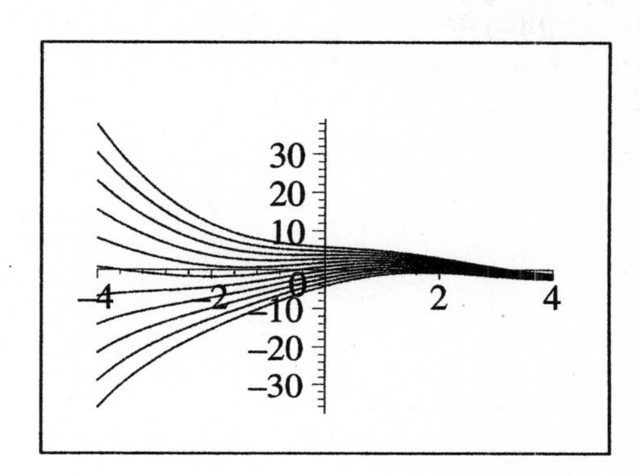

7. For $y(t) = 0$, $y'(t) = 0$ and $y(t)(4 - y(t)) = 0(4-0) = 0$.

9. (a) If $t^2 - 4y^2 = c^2$, then

$$\begin{aligned}
\frac{d}{dt}(t^2 - 4y^2) &= \frac{d}{dt}t^2 \\
2t - 8yy' &= 0 \\
t - 4yy' &= 0.
\end{aligned}$$

(b) If $y(t) = \pm\sqrt{t^2 - C^2}/2$, then $y' = \pm t/(2\sqrt{t^2 - C^2})$, and

$$\begin{aligned}
&t - 4yy' \\
&= t - 4[\pm\sqrt{t^2 - C^2}/2][\pm t/(2\sqrt{t^2 - C^2})] \\
&= t - t \\
&= 0.
\end{aligned}$$

(c) The interval of existence is either $-\infty < t < -C$ or $C < t < \infty$.

(d)

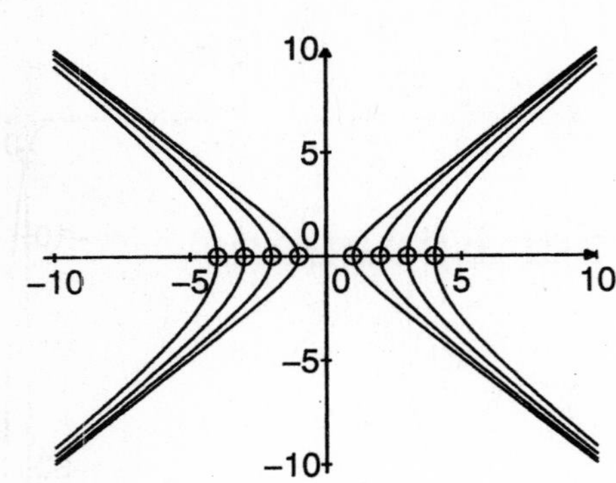

11. See Exercise 6. The interval of existence is $(-\infty, \ln(5)/4)$.

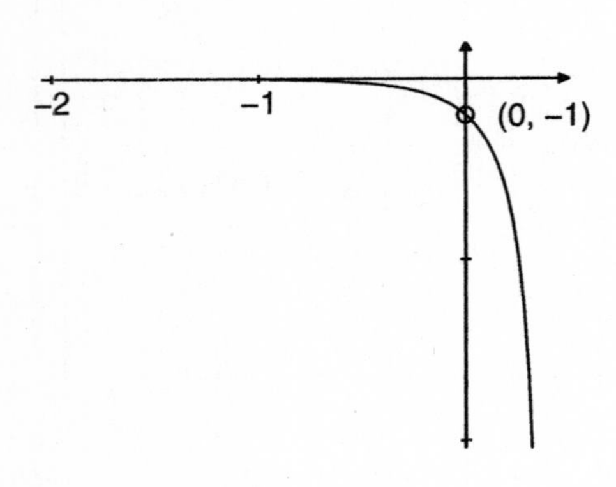

13. $y(t) = \dfrac{1}{3}t^2 + \dfrac{5}{3t}$. The interval of existence is $(0, \infty)$.

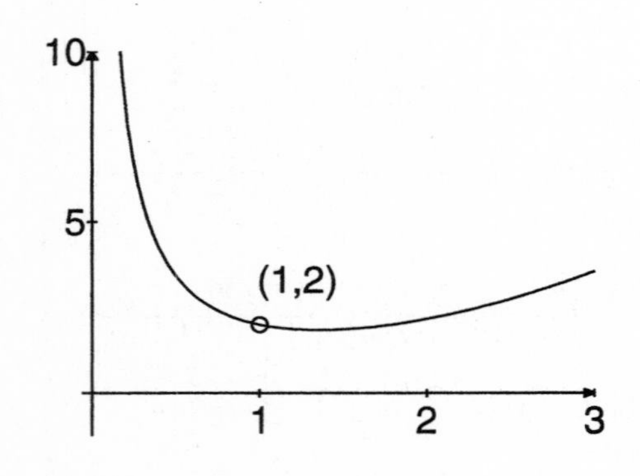

15. $y(t) = 2/(-1 + e^{-2t/3})$. The interval of existence is $(-\ln(3)/2, \infty)$.

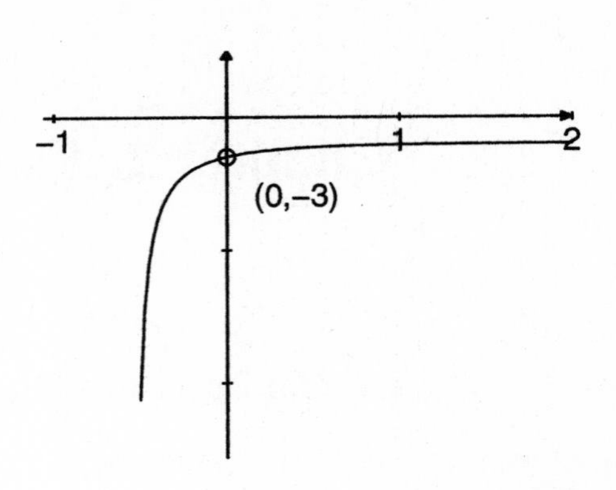

17.

y ' = y + t

19.

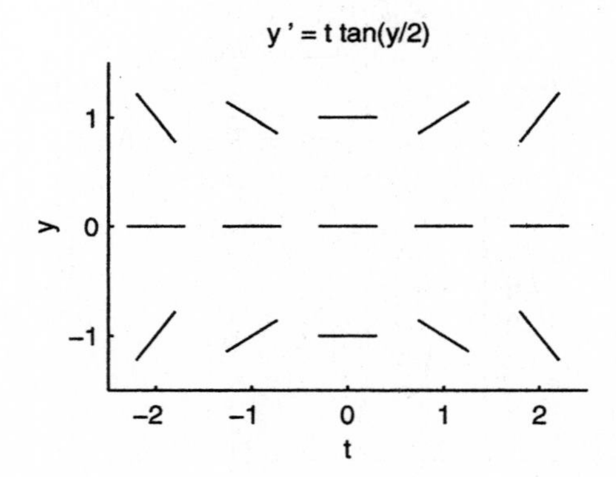

21.

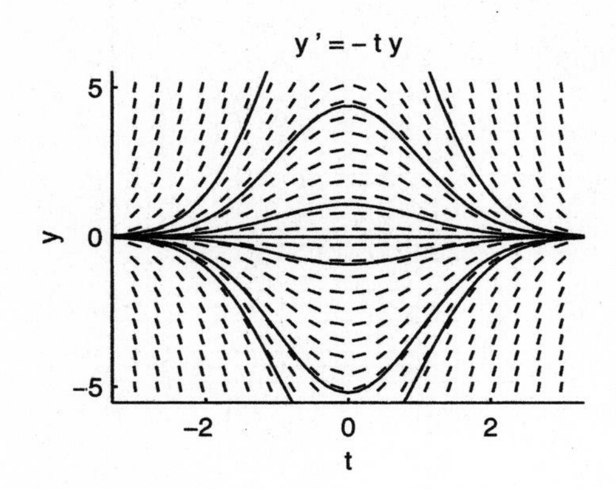

27.

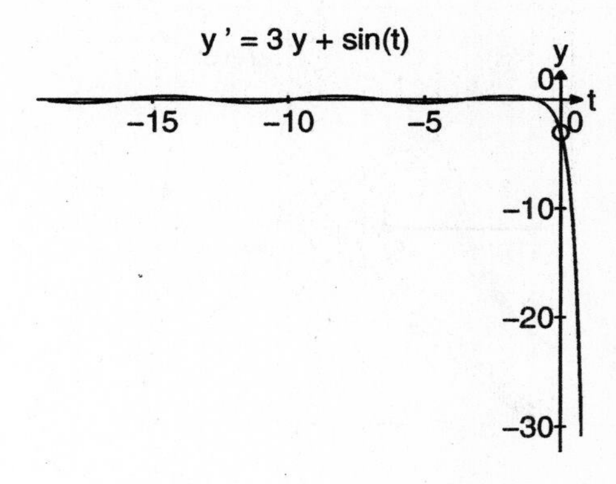

23.

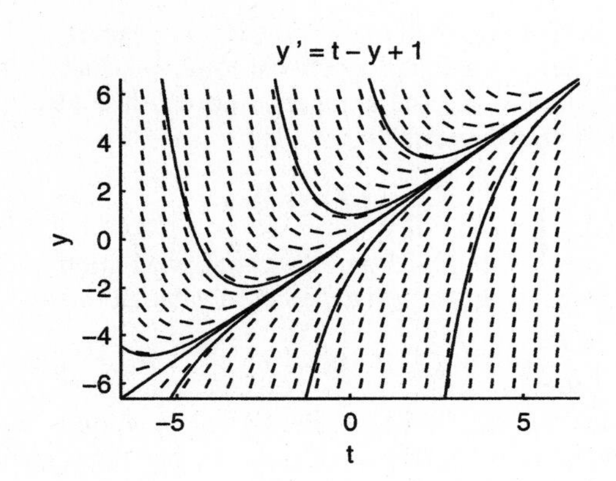

29.

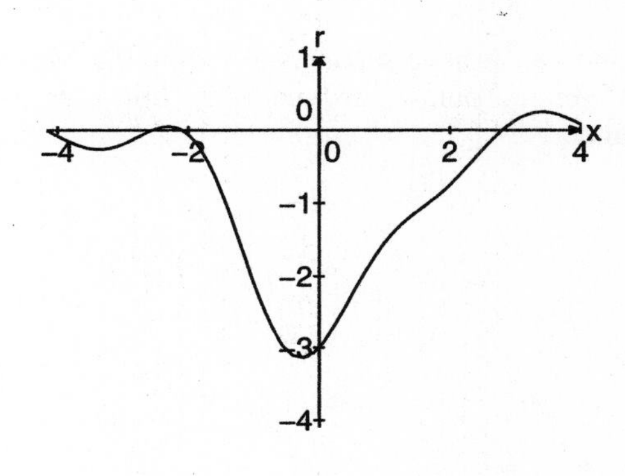

25.

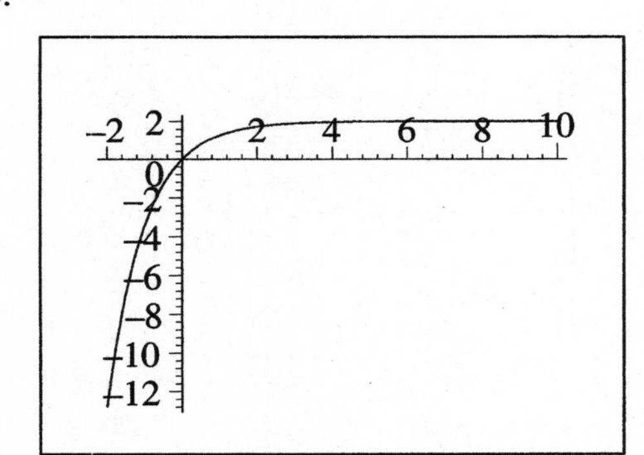

31.

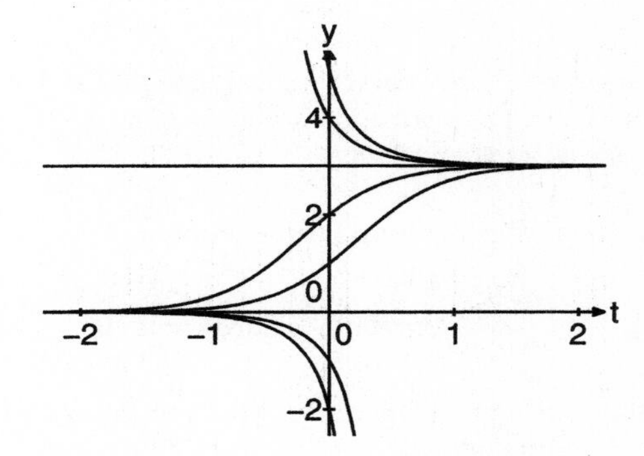

33.

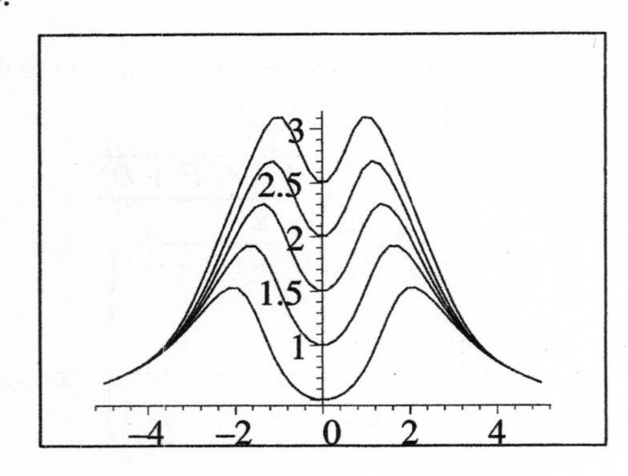

35. We must solve the initial value problem

$$\frac{dP}{dt} = 0.44P, \qquad P(0) = 1.5.$$

Using our numerical solver, we input the equation and initial condition, arriving at the following solution curve.

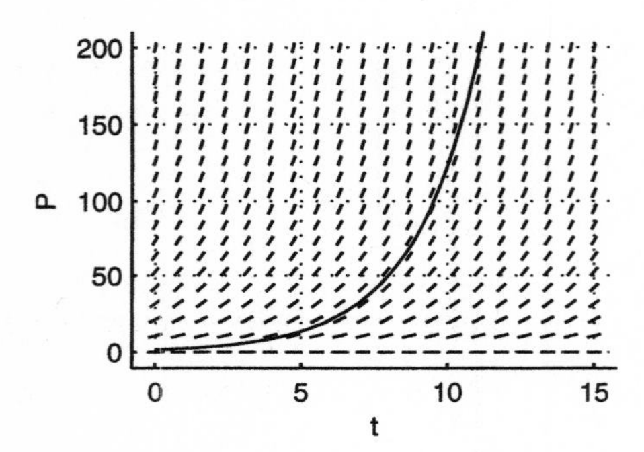

Using the solution curve, we estimate that $P(10) \approx 124$. Thus, there are approximately 124 mg of bacteria present after 10 days.

37. We must solve the initial value problem

$$\frac{dc}{dt} = -0.055c, \qquad c(0) = 0.10.$$

Using our numerical solver, we input the equation and initial condition, arriving at the following solution curve.

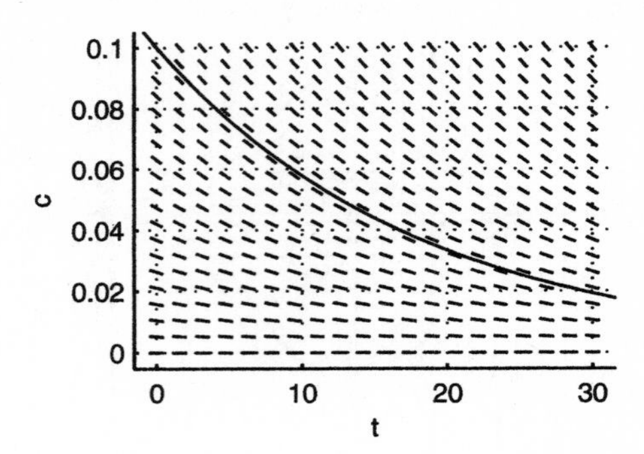

Use the solution curve to estimate that it takes a little more than 29 days for the concentration level to dip below 0.02.

39. The rate at which the population is changing with respect to time is proportional to the product of the population and the number of critters less than the "carrying capacity" (100). Thus,

$$\frac{dP}{dt} = kP(100 - P).$$

With $k = 0.00125$ and an initial population of 20 critters, we must solve the initial value problem

$$\frac{dP}{dt} = 0.00125P(100 - P), \qquad P(0) = 20.$$

Note that the right-hand side of this equation is positive if the number of critters is less than the carrying capacity (100). Thus, we have growth. Using our numerical solver, we input the equation and the initial condition, arriving at the following solution curve.

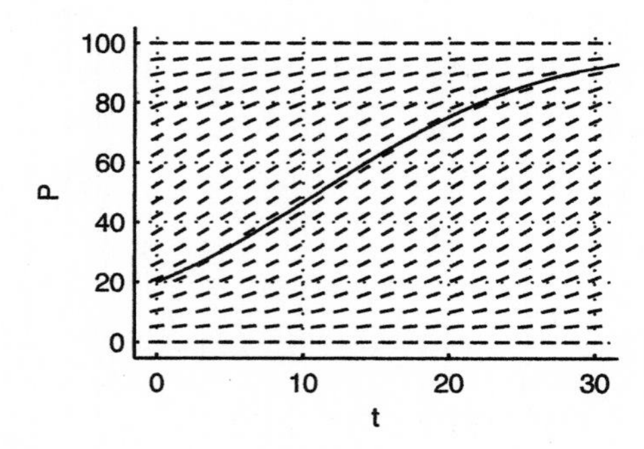

Use the solution curve to estimate that there are about 91 critters in the environment at the end of 30 days.

Section 2.2. Solutions to Separable Equations

1. Separate the variables and integrate.

$$\frac{dy}{dx} = xy$$
$$\frac{dy}{y} = x\,dx$$
$$\ln|y| = \frac{1}{2}x^2 + C$$
$$|y(x)| = e^{x^2/2+C}$$
$$y(x) = \pm e^C \cdot e^{x^2/2}$$
$$= Ae^{x^2/2},$$

Where the constant $A = \pm e^C$ is arbitrary.

3. Separate the variables and integrate.

$$\frac{dy}{dx} = e^{x-y}$$
$$e^y dy = e^x dx$$
$$e^y = e^x + C$$
$$y(x) = \ln(e^x + C)$$

5. Separate the variables and integrate.

$$\frac{dy}{dx} = y(x+1)$$
$$\frac{1}{y}\,dy = (x+1)\,dx$$
$$\ln|y| = \frac{1}{2}x^2 + x + C$$
$$|y| = e^{(1/2)x^2+x+C}$$
$$y(x) = \pm e^C e^{(1/2)x^2+x}$$

Letting $D = \pm e^C$, $y = De^{(1/2)x^2+x}$.

7. Separate the variables and integrate,

$$\frac{dy}{dx} = \frac{x}{y+2},$$
$$(y+2)\,dy = x\,dx,$$
$$\frac{1}{2}y^2 + 2y = \frac{1}{2}x^2 + C,$$
$$y^2 + 4y - (x^2 + D) = 0,$$

With D replacing $2C$ in the last step. We can use the quadratic formula to solve for y.

$$y(x) = \frac{-4 \pm \sqrt{16 + 4(x^2 + D)}}{2}$$
$$y(x) = -2 \pm \sqrt{x^2 + (D+4)}$$

If we replace $D + 4$ with another arbitrary constant E, then $y(x) = -2 \pm \sqrt{x^2 + E}$.

9. First a little algebra.

$$x^2y' = y\ln y - y'$$
$$(x^2+1)y' = y\ln y$$

Separate the variables and integrate.

$$\frac{1}{y\ln y}\,dy = \frac{1}{x^2+1}\,dx$$
$$\frac{1}{u}\,du = \frac{1}{x^2+1}\,dx,$$

where $u = \ln y$ and $du = \frac{1}{y}\,dy$. Hence, $\ln|u| = \tan^{-1}x + C$. Solve for u:

$$|u| = e^{\tan^{-1}x + C}$$
$$u = \pm e^C e^{\tan^{-1}x}$$

Let $D = \pm e^C$, replace u with $\ln y$, and solve for y.

$$\ln y = De^{\tan^{-1}x}$$
$$y(x) = e^{De^{\tan^{-1}x}}$$

11.

$$(y^3 - 2)\frac{dy}{dx} = x$$
$$(y^3 - 2)dy = x\,dx$$
$$\frac{y^4}{4} - 2y = \frac{x^2}{2} + C.$$

The solution is given implicitly by the equation $y^4 - 8y - 2x^2 = A$, where we have set $A = 4C$.

13.

$$
\begin{aligned}
\frac{dy}{dx} &= \frac{y}{x} \\
\frac{dy}{y} &= \frac{dx}{x} \\
\lambda|y| &= \ln|x| + C \\
|y(x)| &= e^{\ln|x|+C} = e^C\,|x| \\
y(x) &= Ax.
\end{aligned}
$$

The initial condition $y(1) = -2$ gives $A = -2$. The solution is $y(x) = -2x$. The solution is defined for all x, but the differential equation is not defined at $x = 0$ so the interval of existence is $(0, \infty)$.

15.

$$
\begin{aligned}
\frac{dy}{dx} &= \frac{\sin x}{y} \\
y\,dy &= \sin x\,dx \\
\frac{1}{2}y^2 &= -\cos x + C_1 \\
y^2 &= -2\cos x + C \quad (C = 2C_1) \\
y(x) &= \pm\sqrt{C - 2\cos x}
\end{aligned}
$$

Using the initial condition we notice that we need the plus sign, and $1 = y(\pi/2) = \sqrt{C}$. Thus $C = 1$ and the solution is

$$y(x) = \sqrt{1 - 2\cos x}.$$

The interval of existence will be the interval containing $\pi/2$ where $2\cos x < 1$. This is $\pi/3 < x < 5\pi/3$.

17.

$$
\begin{aligned}
\frac{dy}{dt} &= 1 + y^2 \\
\frac{dy}{1+y^2} &= dt \\
\tan^{-1}(y) &= t + C \\
y(t) &= \tan(t + C)
\end{aligned}
$$

For the initial condition we have $1 = y(0) = \tan C$, so $C = \pi/4$ and the solution is $y(t) = \tan(t + \pi/4)$. Since the tangent is continuous on the interval $(-\pi/2, \pi/2)$, the solution $y(t) = \tan(t + \pi/4)$ is continuous on the interval $(-3\pi/4, \pi/4)$.

19. With $y(0) = 1$, we get the solution $y(x) = \sqrt{1 + x^2}$, with interval of existence $(-\infty, \infty)$. This solution is plotted with the solid curve in the following figure. With $y(0) = -1$, we get the solution $y(x) = -\sqrt{1 + x^2}$, with interval of existence $(-\infty, \infty)$. This solution is plotted with the dashed curve in the following figure.

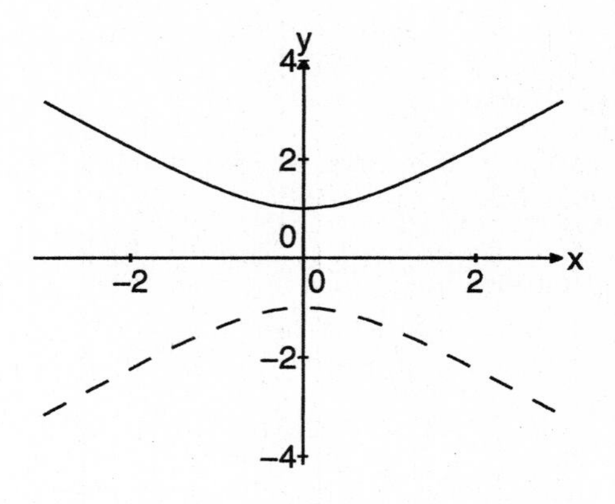

21. With $y(0) = 3$ the solution is $y(t) = 2 + e^{-t}$, on $(-\infty, \infty)$. This solution is plotted is the solid curve in the next figure. With $y(0) = 1$ the solution is $y(t) = 2 - e^{-t}$, on $(-\infty, \infty)$. This solution is plotted is the dashed curve in the next figure.

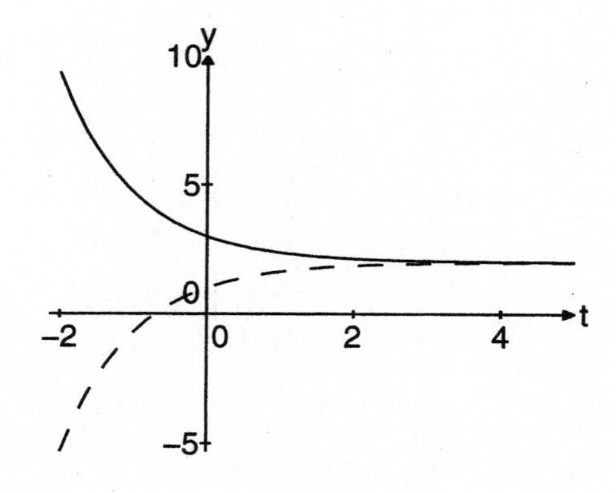

23. We have $N(t) = N_0 e^{-\lambda t}$, and

$$\begin{aligned} N(t + T_{1/2}) &= N_0 e^{-\lambda(t+T_{1/2})} \\ &= N_0 e^{-\lambda t} \cdot e^{-\lambda T_{1/2}} \\ &= N(t) \cdot e^{-\lambda T_{1/2}} \\ &= N(t) \cdot \frac{1}{2} \end{aligned}$$

if $e^{-\lambda T_{1/2}} = 1/2$, or $T_{1/2} = \ln 2/\lambda$.

25. We have $80 = N(4) = 100e^{-4\lambda}$. Hence $\lambda = \ln(100/80)/4 = 0.0558$. Then $T_{1/2} = \ln 2/\lambda = 12.4251$ hours.

27. Using $T_{1/2} = 8.04$ days, we have $\lambda = \ln 2/T_{1/2} = 0.0862$. Then $N(20) = 500e^{-20\lambda} = 89.1537$mg.

29. (a) If $N = N_0 e^{-\lambda t}$, then substituting $T_\lambda = 1/\lambda$,

$$\begin{aligned} N &= N_0 e^{-\lambda T_\lambda} \\ &= N_0 e^{-\lambda(1/\lambda)} \\ &= N_0 e^{-1}. \end{aligned}$$

Therefore, after a period of one time constant $T_\lambda = 1/\lambda$, the material remaining is $N_0 e^{-1}$. Thus, the amount of radioactive substance has decreased to e^{-1} of its original value N_0.

(b) If the half-life is 12 hours, then

$$\begin{aligned} \frac{1}{2} N_0 &= N_0 e^{-\lambda(12)} \\ e^{-12\lambda} &= \frac{1}{2} \\ -12\lambda &= \ln \frac{1}{2} \\ \lambda &= \frac{\ln \frac{1}{2}}{-12}. \end{aligned}$$

Hence, the time constant is

$$T_\lambda = \frac{1}{\lambda} = \frac{-12}{\ln \frac{1}{2}} \approx 17.3\,\text{hr}.$$

(c) If 1000 mg of the substance is present initially, then the amount of substance remaining as a function of time is given by $N = 1000e^{(t \ln(1/2))/12}$. The graph over four time periods ($[0, 4T_\lambda]$) follows.

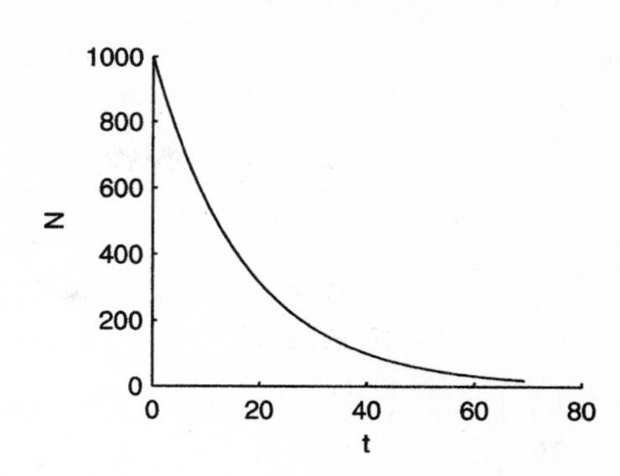

31. The half-life is related to the *decay constant* by

$$T_{1/2} = \frac{\ln 2}{\lambda_{226}}.$$

The *decay rate* is related to the number of atoms present by

$$R = \lambda_{226} N.$$

Substituting,

$$T_{1/2} = \frac{N \ln 2}{R}.$$

Calculate the number of atoms present in the 1g sample.

$$\begin{aligned} N &= 1\text{g} \times \frac{1 \text{ mol}}{226 \text{ g}} \times \frac{6.02 \times 10^{23} \text{ atoms}}{\text{mole}} \\ &= 266 \times 10^{21} \text{ atoms}. \end{aligned}$$

Now,

$$T_{1/2} = \frac{(2.66 \times 10^{21} \text{ atoms})(\ln 2)}{3.7 \times 10^{10} \text{ atom/s}} = 4.99 \times 10^{10}\text{s}.$$

In years, $T_{1/2} \approx 1582$ yr. The dedicated reader might check this result in the CRC Table.

33. Let $t = 0$ correspond to midnight. Thus, $T(0) = 31$°C. Because the temperature of the surrounding medium is $A = 21$°C, we can use $T = A + (T_0 - A)e^{-kt}$ and write

$$T = 21 + (31 - 21)e^{-kt} = 21 + 10e^{-kt}.$$

At $t = 1$, $T = 29$°C, which can be used to calculate k.

$$\begin{aligned} 29 &= 21 + 10e^{-k(1)} \\ k &= -\ln(0.8) \\ k &\approx 0.2231 \end{aligned}$$

Thus, $T = 21 + 10e^{-0.2231t}$. To find the time of death, enter "normal" body temperature, $T = 37°C$ and solve for t.

$$37 = 21 + 10e^{-0.2231t}$$
$$t = \frac{\ln 1.6}{-0.2231}$$
$$t \approx -2.1 \text{ hrs}$$

Thus, the murder occurred at approximately 9:54 PM.

35. The same differential equation and solution hold as in the previous problem:

$$y(t) = 70 - Ce^{-kt}$$

We let $t = 0$ correspond to when the beer was discovered, so $y(0) = 50$. This means $C = 20$. We also have $y(10) = 60$ or

$$60 = 70 - 20e^{-10k}$$

Therefore, $k = (-1/10)\ln(1/2) \approx .0693$. We want to find the time T when $y(T) = 40$, which gives the equation

$$70 - 20e^{-kT} = 40$$

Since we know k, we can solve this equation for T to obtain

$$T = (-1/k)\ln(3/2) \approx -5.85$$

or about 5.85 minutes before the beer was discovered on the counter.

37. The tangent line at the point (x, y) is $\hat{y} - y = y'(x)(\hat{x} - x)$ (the variables for the tangent line have the hats). The $\hat{x}$ intercept is $\hat{x}_{\text{int}} = -y/y' + x$. Since (x, y) bisects the tangent line, we have $\hat{x}_{\text{int}} = 2x$. Therefore

$$2x = \frac{-y}{y'} + x$$

or

$$\frac{y'}{y} = \frac{-1}{x}$$

This separable differential equation is easily solved to obtain $y(x) = C/x$, where C is an arbitrary constant.

39. Let ϕ be the angle from the radius to the tangent.

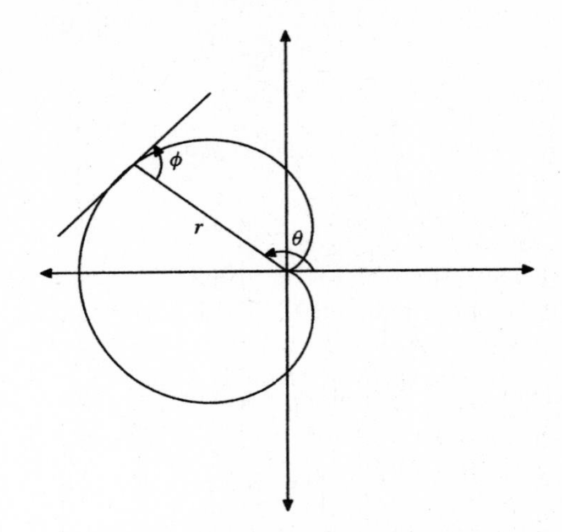

From geometry, $\tan\phi = r\,d\theta/dr$. Since $\theta = 2\phi$, we obtain

$$\frac{dr}{d\theta} = r\cot\phi = r\cot(\theta/2)$$

which can be separated as $dr/r = \cot(\theta/2)d\theta$. This can be solved for r as $r(\theta) = C\sin^2(\theta/2)$, where C is a constant.

41. Center the football at the origin with equation

$$z^2 + x^2 + \frac{y^2}{4} = 4.$$

The top half of the football is the graph of the function

$$z = \sqrt{4 - x^2 - y^2/4}.$$

The (x, y)-components of the path of a rain drop form a curve in the (x, y)- plane which must always point in the direction of the gradient of the function z (the path of steepest descent). The gradient of z is given by

$$\nabla z = \frac{-xi - (y/4)j}{\sqrt{4 - x^2 - y^2/4}}$$

where i and j are the unit vectors parallel to the x and y-axis. Since the path traced by the drop, $y = y(x)$, must point in the direction of ∇z, we must have

$$\frac{dy}{dx} = \text{slope of the gradient} = \frac{z_y}{z_x} = \frac{y}{4x}.$$

This differential equation can be separated and solved as $x = Cy^4$. C can be solved from the initial position of the drop (x_0, y_0) to be $C = x_0/y_0^4$. The final answer is given by inserting $x = Cy^4$ into the expression for z:

$$(x, y, z) = (Cy^4, y, \sqrt{4 - C^2y^8 - y^2/4}),$$

(here, y is the independent variable).

43. Let the unknown curve forming the outside of the bowl be given by $y = y(x)$ (the bowl is then formed by revolving this curve around the y-axis). We can also write this equation as $x = x(y)$ (reversing the roles of the independent and dependent variables). As in the analysis of the last problem, the rate of change of volume, dV/dt is the cross sectional area multiplied by the rate of change in height, dy/dt. The cross sectional area is $\pi x^2 = \pi x(y)^2$. Thus

$$\frac{dV}{dt} = \pi x^2 \frac{dy}{dt}$$

From Torricelli's law:

$$\frac{dV}{dt} = -\pi a^2 v = -\pi a^2 \sqrt{2gy}$$

Since $dy/dt = C$ (a negative constant), we obtain

$$C\pi x^2 = -\pi a^2 \sqrt{2gy}$$

Solving for y, we obtain $y = Kx^4$ where K is a constant.

Section 2.3. Models of Motion

1. We need $gt = c/5$. or $t = c/5g = 612{,}240$ seconds. The distance traveled will be $s = gt^2/2 = 1.84 \times 10^{14}$ meters.

3. The depth of the well satisfies $d = 4.9t^2$, where t is the amount of time it takes the stone to hit the water. It also satisfies $d = 340s$, where $s = 8 - t$ is the amount of time it takes for the noise of the splash to reach the ear. Thus we must solve the quadratic equation $4.9t^2 = 340(8 - t)$. The solution is $t = 7.2438$sec. The depth is $d = 340(8 - t) = 257.1$m.

5. The distance dropped in time t is $4.9t^2$. If T is the time taken for the first half of the trip, then $4.9(T+1)^2 = 2 \times 4.9T^2$, or $4.9(T^2 - 2T - 1) = 0$. Solving we find that $T = 1 + \sqrt{2} = 2.4142$s. So the body fell $2 \times 4.9T^2 = 57.12$m, and it took $T + 1 = 3.4142$s.

7. The velocities must be changed to ft/s, so $v_0 = 60$ mi/h $= 60 \times 5280/3600 = 88$ft/s, and $v = 30$ mi/h $= 44$ft/s. Then $a = (v^2 - v_0^2)/2(x - x_0) = -5.8\text{ft/s}^2$.

9. The resistance force has the form $R = -rv$. When $v = 0.2$, $R = -1$ so $r = 5$. The terminal velocity is $v_{\text{term}} = -mg/r = -0.196$m/s.

11. Without air resistance, $v_0 = \sqrt{2 \times 13.5g} = 16.2665$m/s. With air resistance, v_0 is defined by

$$\int_{v_0}^{0} \frac{v\,dv}{v + mg/r} = -\frac{r}{m}\int_{1.5}^{15} dy.$$

Hence,

$$-v_0 + (mg/r)\ln(v_0 + mg/r) - (mg/r)\ln(mg/r)$$
$$= -13.5\frac{r}{m} \quad \text{or}$$
$$-v_0 + 49\ln(v_0 + 49) - 49\ln(49) = -2.7$$

This is an implicit equation for v_0. Solving on a calculator or a computer yields $v_0 = 18.1142$m/s.

13. Following the lead of Exercise 11, we find that

$$v\,dv = (-g + R(v)/m)\,dy = \left(-9.8 - 0.5v^2\right)dy$$

Hence if y_1 is the maximum height we have

$$\int_{230}^{0} \frac{v\,dv}{v^2 + 19.6} = -0.5\int_0^{y_1} dy.$$

Hence

$$y_1 = \log(v^2 + 19.6)\Big|_0^{230}$$
$$= 7.9010.$$

15. Let $x(t)$ be the distance from the mass to the center of the Earth. The force of gravity is kx (proportional to the distance from the center of the Earth). Since the force of gravity at the surface (when $x = R$) is $-mg$, we must have $k = -mg/R$. Newton's law becomes
$$m\frac{d^2x}{dt^2} = \frac{-mgx}{R}$$
Using the reduction of order technique as given in the hint, we obtain
$$v\frac{dv}{dx} = \frac{-gx}{R}$$
which can be separated with a solution given by $v = \sqrt{C - gx^2/R}$. The constant C can be evaluated from the initial condition, $v(x = R) = 0$, to be $C = gR$. When $x = 0$ (the center of the Earth), we obtain $v = \sqrt{C} = \sqrt{gR}$ or approximately 4.93 miles per second.

17. The force acting on the chain is the force of gravity applied to the piece of the chain that hangs off the table. This force is $mgx(t)$ where m is the mass density of the chain. Newton's law gives
$$mx''(t) = mgx(t)$$
Using the hint, $x''(t) = dv/dt = v(dv/dx)$ and so this equation becomes
$$v\frac{dv}{dx} = gx$$
Separating this equation and integrating gives $v^2 = gx^2 + C$. Since $v = 0$ when $x = 2$ (initial velocity is zero), we obtain $C = -4g$. Therefore
$$v = \sqrt{g(x^2 - 4)}.$$
Since $dx/dt = v = \sqrt{g(x^2 - 4)}$, we can separate this equation and integrate to obtain
$$\ln(x + \sqrt{x^2 - 4}) = \sqrt{g}t + K$$
where K is a constant. From the initial condition that $x = 2$ when $t = 0$, we obtain $K = \ln 2$. Inserting $x = 10$ and solving for t, we obtain
$$t = \frac{1}{\sqrt{g}}\ln\left(\frac{10 + \sqrt{96}}{2}\right) \approx .405 \text{ seconds}$$

19. Let x be the height of the parachuter and let v be his velocity. The resistance force is proportional to v and to e^{-ax}. Hence it is given by $R(x, v) = -ke^{ax}v$, where k is a positive constant. Newton's second law gives us $mx'' = -mg - ke^{-ax}x'$, or $mx'' + ke^{ax}x' + mg = 0$.

Section 2.4. Linear Equations

1. Compare $y' = -y + 2$ with $y' = a(t)y + f(t)$ and note that $a(t) = -1$. Consequently, an integrating factor is found with
$$u(t) = e^{\int -a(t)\,dt} = e^{\int 1\,dt} = e^t.$$
Multiply both sides of our equation by this integrating factor and check that the left-hand side of the resulting equation is the derivative of a product.
$$e^t(y' + y) = 2e^t$$
$$(e^t y)' = 2e^t$$
Integrate and solve for y.
$$e^t y = 2e^t + C$$
$$y(t) = 2 + Ce^{-t}$$

3. Compare $y' + (2/x)y = (\cos x)/x^2$ with $y' = a(x)x + f(x)$ and note that $a(x) = -2/x$. Consequently, an integrating factor is found with

$$u(x) = e^{\int -a(x)\,dx} = e^{\int 2/x\,dx} = e^{2\ln|x|} = |x|^2 = x^2.$$

Multiply both sides of our equation by the integrating factor and note that the left-hand side of the resulting equation is the derivative of a product.

$$x^2\left(y' + \frac{2}{x}y\right) = \cos x$$
$$x^2y' + 2xy = \cos x$$
$$(x^2y)' = \cos x$$

Integrate and solve for x.

$$x^2y = \sin x + C$$
$$y(x) = \frac{\sin x + C}{x^2}$$

5. Compare $x' - 2x/(t+1) = (t+1)^2$ with $x' = a(t)x + f(t)$ and note that $a(t) = -2/(t+1)$. Consequently, an integrating factor is found with

$$u(t) = e^{\int -a(t)\,dt} = e^{\int -2/(t+1)\,dt}$$
$$= e^{-2\ln|t+1|} = |t+1|^{-2} = (t+1)^{-2}.$$

Multiply both sides of our equation by the integrating factor and note that the left-hand side of the resulting equation is the derivative of a product.

$$(t+1)^{-2}\left(x' - \frac{2}{t+1}x\right) = 1$$
$$\left((t+1)^{-2}x\right)' = 1$$

Integrate and solve for x.

$$(t+1)^{-2}x = t + C$$
$$x(t) = t(t+1)^2 + C(t+1)^2$$

7. Divide both sides by $1+x$ and solve for y'.

$$y' = -\frac{1}{1+x}y + \frac{\cos x}{1+x}$$

Compare this result with $y' = a(x)y + f(x)$ and note that $a(x) = -1/(1+x)$. Consequently, an integrating factor is found with

$$u(x) = e^{\int -a(x)\,dx} = e^{\int 1/(1+x)\,dx} = e^{\ln|1+x|} = |1+x|.$$

If $1 + x > 0$, then $|1+x| = 1+x$. If $1+x < 0$, then $|1+x| = -(1+x)$. In either case, if we multiply both sides of our equation by either integrating factor, we arrive at

$$(1+x)y' + y = \cos x.$$

Check that the left-hand side of this result is the derivative of a product, integrate, and solve for y.

$$((1+x)y)' = \cos x$$
$$(1+x)y = \sin x + C$$
$$y(x) = \frac{\sin x + C}{1+x}$$

9. Divide both side of this equation by L and solve for di/dt.

$$\frac{di}{dt} = -\frac{R}{L}i + \frac{E}{L}$$

Compare this with $i' = a(t)i + f(t)$ and note that $a(t) = -R/L$. Consequently, an integrating factor is found with

$$u(t) = e^{\int -a(t)\,dx} = e^{\int R/L\,dt} = e^{Rt/L}$$

Multiply both sides of our equation by this integrating factor and note that the resulting left-hand side is the derivative of a product.

$$e^{Rt/L}\left(\frac{di}{dt} + \frac{R}{L}i\right) = \frac{E}{L}e^{Rt/L}$$
$$\left(e^{Rt/L}i\right)' = \frac{E}{L}e^{Rt/L}$$

Integrate and solve for i.

$$e^{Rt/L}i = \frac{E}{R}e^{Rt/L} + C$$
$$i(t) = \frac{E}{R} + Ce^{-Rt/L}$$

11. Compare $y' = \cos x - y\sec x$ with $y' = a(x)y + f(x)$ and note that $a(x) = -\sec x$. Consequently, an integrating factor is found with

$$u(x) = e^{\int -a(x)\,dx} = e^{\int \sec x\,dx}$$
$$= e^{\ln|\sec x + \tan x|} = |\sec x + \tan x|.$$

If $\sec x + \tan x > 0$, then $|\sec x + \tan x| = \sec x + \tan x$. If $\sec x + \tan x < 0$, then $|\sec x + \tan x| = -(\sec x + \tan x)$. In either case, when we multiply both sides of the differential equation by this integrating factor, we arrive at

$$(\sec x + \tan x)(y' + y\sec x) = \cos x(\sec x + \tan x),$$

or

$$(\sec x + \tan x)y' + (\sec^2 x + \sec x \tan x)y = 1 + \sin x$$

Again, check that the left-hand side of this equation is the derivative of a product, then integrate and solve for y.

$$((\sec x + \tan x)y)' = 1 + \sin x$$
$$(\sec x + \tan x)y = x - \cos x + C$$
$$y = \frac{x - \cos x + C}{\sec x + \tan x}$$

13. (a) Compare $y' + y\cos x = \cos x$ with $y' = a(x)y + f(x)$ and note that $a(x) = -\cos x$. Consequently, an integrating factor is found with

$$u(x) = e^{-\int a(x)\,dx} = e^{\int \cos x\,dx} = e^{\sin x}.$$

Multiply both sides of the differential equation by the integrating factor and check that the resulting left-hand side is the derivative of a product.

$$e^{\sin x}(y' + y\cos x) = e^{\sin x}\cos x$$
$$\left(e^{\sin x}y\right)' = e^{\sin x}\cos x$$

Integrate and solve for y.

$$e^{\sin x}y = e^{\sin x} + C$$
$$y(x) = 1 + Ce^{-\sin x}$$

(b) Separate the variables and integrate.

$$\frac{dy}{dx} = \cos x(1 - y)$$
$$\frac{dy}{1 - y} = \cos x\,dx$$
$$-\ln|1 - y| = \sin x + C.$$

Take the exponential of each side.

$$|1 - y| = e^{-\sin x - C}$$
$$1 - y = \pm e^{-C}e^{-\sin x}$$

If we let $A = \pm e^{-C}$, then

$$y(x) = 1 - Ae^{-\sin x},$$

where A is any real number, except zero. However, when we separated the variables above by dividing by $y - 1$, this was a valid operation only if $y \neq 1$. This hints at another solution. Note that $y = 1$ easily checks in the original equation. Consequently,

$$y(x) = 1 - Ae^{-\sin x},$$

where A is any real number. Note that this will produce the same solutions a $y = 1 + Ce^{-\sin x}$, C any real number, the solution found in part (a).

15. Solve for y'.

$$y' = -\frac{3x}{x^2 + 1}y + \frac{6x}{x^2 + 1}$$

Compare this with $y' = a(x)y + f(x)$ and note that $a(x) = -3x/(x^2 + 1)$. Consequently, an integrating factor is found with

$$u(x) = e^{\int -a(x)\,dx} = e^{\int 3x/(x^2+1)\,dx}$$
$$= e^{(3/2)\ln(x^2+1)} = (x^2 + 1)^{3/2}.$$

Multiply both sides of our equation by the integrating factor and note that the left-hand side of the resulting equation is the derivative of a product.

$$(x^2 + 1)^{3/2}y' + 3x(x^2 + 1)^{1/2}y = 6x(x^2 + 1)^{1/2}$$
$$\left((x^2 + 1)^{3/2}y\right)' = 6x(x^2 + 1)^{1/2}$$

Integrate and solve for y.

$$(x^2 + 1)^{3/2}y = 2(x^2 + 1)^{3/2} + C$$
$$y = 2 + C(x^2 + 1)^{-3/2}$$

The initial condition gives

$$-1 = y(0) = 2 + C(0^2 + 1)^{-3/2} = 2 + C.$$

Therefore, $C = -3$ and $y(x) = 2 - 3(x^2 + 1)^{-3/2}$.

17. Compare $x' + x\cos t = (1/2)\sin 2t$ with $x' = a(t)x + f(t)$ and note that $a(t) = -\cos t$. Consequently, an integrating factor is found with

$$u(t) = e^{\int -a(t)\,dt} = e^{\int \cos t\,dt} = e^{\sin t}.$$

Multiply both sides of our equation by the integrating factor and note that the left-hand side of the resulting equation is the derivative of a product.

$$\begin{aligned} e^{\sin t}x' + e^{\sin t}(\cos t)x &= \frac{1}{2}e^{\sin t}\sin 2t \\ \left(e^{\sin t}x\right)' &= \frac{1}{2}e^{\sin t}\sin 2t \end{aligned}$$

Use $\sin 2t = 2\sin t\cos t$.

$$\left(e^{\sin t}x\right)' = e^{\sin t}\sin t\cos t$$

Let $u = \sin t$ and $dv = e^{\sin t}\cos t\,dt$. Then,

$$\begin{aligned} \int e^{\sin t}\cos t\sin t\,dt &= \int u\,dv \\ &= uv - \int v\,du \\ &= (\sin t)e^{\sin t} - \int e^{\sin t}\cos t\,dt \\ &= (\sin t)e^{\sin t} - e^{\sin t} + C \end{aligned}$$

Therefore,

$$\begin{aligned} e^{\sin t}x &= e^{\sin t}\sin t - e^{\sin t} + C \\ x(t) &= \sin t - 1 + Ce^{-\sin t} \end{aligned}$$

The initial condition gives

$$1 = x(0) = \sin(0) - 1 + Ce^{-\sin(0)} = -1 + C.$$

Consequently, $C = 2$ and $x(t) = \sin t - 1 + 2e^{-\sin t}$.

19. Solve for y'.

$$y' = \frac{1}{2x+3}y + (2x+3)^{-1/2}$$

Compare this with $y' = a(x)y + f(x)$ and note that $a(x) = 1/(2x+3)$ and $f(x) = (2x+3)^{-1/2}$. It is important to note that a is continuous everywhere except $x = -3/2$, but f is continuous only on $(-3/2, +\infty)$, facts that will heavily influence our interval of existence.

An integrating factor is found with

$$\begin{aligned} u(x) &= e^{\int -a(x)\,dx} = e^{\int -1/(2x+3)\,dx} \\ &= e^{-(1/2)\ln|2x+3|} = |2x+3|^{-1/2}. \end{aligned}$$

However, we will assume that $x > -3/2$ (a domain where both a and f are defined), so $u(x) = (2x+3)^{-1/2}$. Multiply both sides of our equation by the integrating factor and note that the left-hand side of the resulting equation is the derivative of a product.

$$\begin{aligned} (2x+3)^{-1/2}y' - (2x+3)^{-3/2}y &= (2x+3)^{-1} \\ \left((2x+3)^{-1/2}y\right)' &= (2x+3)^{-1} \end{aligned}$$

Integrate and solve for y.

$$(2x+3)^{-1/2}y = \frac{1}{2}\ln(2x+3) + C,$$

or

$$y = \frac{1}{2}(2x+3)^{1/2}\ln(2x+3) + C(2x+3)^{1/2}$$

The initial condition provides

$$0 = y(-1) = C.$$

Consequently, $y = (1/2)(2x+3)^{1/2}\ln(2x+3)$. The interval of existence is $(-3/2, +\infty)$ and the solution curve is shown in the following figure.

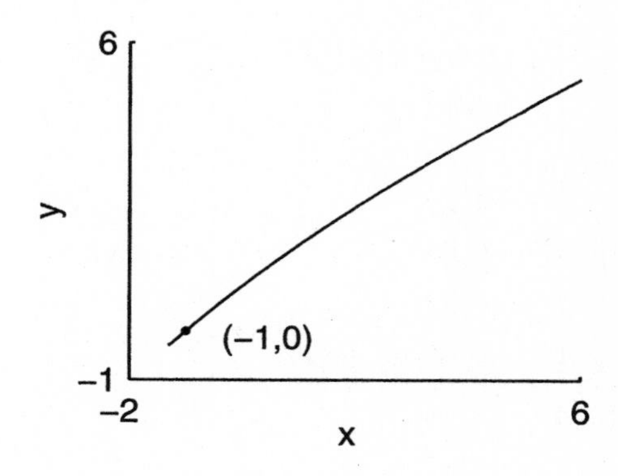

21. Solve for x.

$$x' = -\frac{1}{1+t}x + \frac{\cos t}{1+t}$$

Compare this result with $x' = a(t)x + f(t)$ and note that $a(t) = -1/(1+t)$ and $f(t) = \cos t/(1+t)$, neither of which are continuous at $t = -1$. An integrating factor is found with

$$u(t) = e^{\int -a(t)\,dt} = e^{\int 1/(t+1)} = e^{\ln|1+t|} = |1+t|.$$

However, the initial condition dictates that our solution pass through the point $(-\pi/2, 0)$. Because of the discontinuity at $t = -1$, our solution must remain to the left of $t = -1$. Consequently, with $t < -1, u(t) = -(1+t)$. However, multiplying our equation by $u(t)$ produces

$$\begin{aligned}(1+t)x' + x &= \cos t,\\ ((1+t)x)' &= \cos t,\\ (1+t)x &= \sin t + C.\end{aligned}$$

Use the initial condition.

$$\left(1 - \frac{\pi}{2}\right)(0) = \sin\left(-\frac{\pi}{2}\right) + C$$

Consequently, $C = 1$ and

$$x = \frac{1+\sin t}{1+t}.$$

The interval of existence is maximally extended to $(-\infty, -1)$, as shown in the following figure.

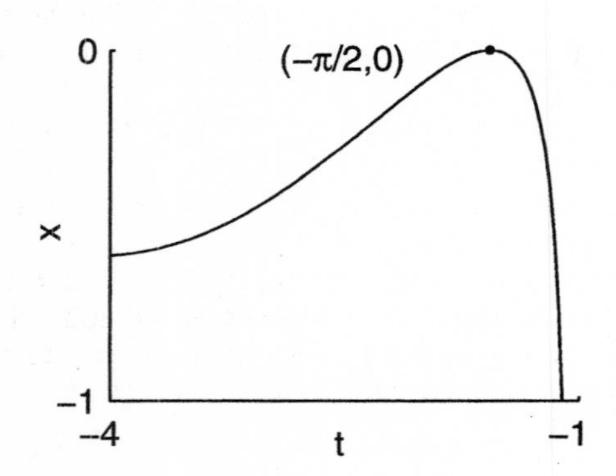

23. In this case $n = 2$, so we set $z = y^{-1}$. Then

$$\begin{aligned}\frac{dz}{dx} &= \frac{dz}{dy}\frac{dy}{dx}\\ &= -y^{-2}\left[xy^2 - \frac{y}{x}\right]\\ &= -x + \frac{1}{xy}\\ &= -x + \frac{z}{x}.\end{aligned}$$

This is a linear equation for z. The integrating factor is $1/x$, so we have

$$\begin{aligned}\frac{1}{x}\left[\frac{dz}{dx} - \frac{z}{x}\right] &= \frac{1}{x}(-x)\\ \left[\frac{z}{x}\right]' &= -1\\ \frac{z}{x} &= -x + C\\ z(x) &= x(C - x)\end{aligned}$$

Since $z = 1/y$, our solution is $y(x) = \dfrac{1}{x(C-x)}$.

25. Solve for y'.

$$y' = -\frac{1}{x}y + x^3y^3$$

Compare this with $y' = a(x)y + f(x)y^n$ and note that this has the form of Bernoulli's equation with $n = 3$. Let $z = y^{1-3} = y^{-2}$. Then

$$\frac{dz}{dx} = \frac{dz}{dy}\frac{dy}{dx} = -2y^{-3}\frac{dy}{dx}.$$

Multiply the equation by $-2y^{-3}$.

$$-2y^{-3}\frac{dy}{dx} = \frac{2}{x}y^{-2} - 2x^3$$

Replace $-2y^{-3}(dy/dx)$ with dz/dx and y^{-2} with z.

$$\frac{dz}{dx} = \frac{2}{x}z - 2x^3$$

This equation is linear with integrating factor

$$u(x) = e^{\int -a(x)\,dx} = e^{\int -2/x\,dx} = e^{-2\ln|x|} = x^{-2}.$$

Multiply by the integrating factor and integrate.

$$\begin{aligned} x^{-2}z' - 2x^{-3}z &= -2x \\ \left(x^{-2}z\right)' &= -2x \\ x^{-2}z &= -x^2 + C \\ z &= -x^4 + Cx^2 \end{aligned}$$

Replace z with y^{-2} and solve for y.

$$\begin{aligned} y^{-2} &= -x^4 + Cx^2 \\ y &= \pm 1/\sqrt{Cx^2 - x^4} \end{aligned}$$

27. (a) Since $y = y_1 + z$, $y^2 = y_1^2 + 2y_1 z + z^2$. Hence

$$\begin{aligned} z' &= y' - y_1' \\ &= -[\psi y^2 + \phi y + \chi] + [\psi y_1^2 + \phi y_1 + \chi] \\ &= \psi[y_1^2 - y^2] + \phi[y_1 - y] \\ &= -\psi[2y_1 z + z^2] - \phi z \\ &= -(2y_1\psi + \phi)z - \psi z^2 \end{aligned}$$

(b) Since $y_1 = 1/t$ is a solution, we set $z = y + 1/t$. Then $y = z - 1/t$, and $y^2 = z^2 - 2z/t + 1/t^2$, so

$$\begin{aligned} z' &= y' - \frac{1}{t^2} \\ &= -\frac{1}{t^2} - \frac{y}{t} + y^2 - \frac{1}{t^2} \\ &= -\frac{3z}{t} + z^2. \end{aligned}$$

This is a Bernoulli equation with $n = 2$. Thus we set $w = 1/z$. Differentiating, we get

$$\begin{aligned} w' &= -\frac{1}{t^2}z' \\ &= -\frac{1}{t^2}\left[-\frac{3z}{t} + z^2\right] \\ &= \frac{3}{tz} - 1 \\ &= \frac{3w}{t} - 1. \end{aligned}$$

This is a linear equation, and t^{-3} is an integrating factor.

$$\begin{aligned} t^{-3}\left[w' - \frac{3w}{t}\right] &= -t^{-3} \\ \left[t^{-3}w\right] &= -t^{-3} \\ t^{-3}w(t) &= \frac{1}{2}t^{-2} + C \\ w(t) &= \frac{1}{2}t + Ct^3 \end{aligned}$$

Now it is a matter of unravelling the changes of variable. First

$$z(t) = \frac{1}{w(t)} = \frac{2}{t + 2Ct^3} = \frac{2}{t + Bt^3},$$

Where we have set $B = 2C$. Then

$$y(t) = z(t) - \frac{1}{t} = \frac{2}{t + Bt^3} - \frac{1}{t}.$$

This looks a little better if we use partial fractions to write

$$\begin{aligned} y(t) &= \frac{2}{t + Bt^3} - \frac{1}{t} \\ &= \frac{2}{t} - \frac{2Bt}{1 + Bt^2} - \frac{1}{t} \\ &= \frac{1}{t} - \frac{2Bt}{1 + Bt^2}. \end{aligned}$$

29. Newton's law of cooling says the rate of change of temperature is equal to k times the difference between the current temperature and the ambient temperature. In this case the ambient temperature is decreasing from 0°C, and at a constant rate of 1°C per hour. Hence the model equation is $T' = -k(T + t)$, where we are taking $t = 0$ to be midnight. This is a linear equation. The solution is $T(t) = -t + 1/k + Ce^{-kt}$. Since $T(0) = 31$, the constant evaluates to $C = 31 - 1/k$. The solution is $T(t) = -t + 1/k + (31 - 1/k)e^{-kt}$.

We need to compute the time t_0 when $T(t_0) = 37$, using $k \approx 0.2231$ from Exercise 35 of Section 2. This is a nonlinear equation, but using a calculator or a computer we can find that $t_0 \approx -0.8022$. Since $t = 0$ corresponds to midnight, this means that the time of death is approximately 11:12 PM.

31. The homogeneous equation, $y' = -2y$ has solution $y_h(t) = e^{-2t}$. We look for a particular solution in

the form $y_p(t) = v(t)y_h(t)$, where v is an unknown function. Since

$$\begin{aligned} y_p' &= v'y_h + vy_h' \\ &= v'y_h - 2vy_h \\ &= v'y_h - 2y_p, \end{aligned}$$

and $y_p' = -2y_p + 5$, we have $v' = 5/y_h = 5e^{2t}$. Integrating we see that $v(t) = 5e^{2t}/2$, and

$$y_p(t) = v(t)y_h(t) = \frac{5}{2}e^{2t} \cdot e^{-2t} = \frac{5}{2}.$$

The general solution is

$$y(t) = y_p(t) + Cy_h(t) = \frac{5}{2} + Ce^{-2t}.$$

33. The homogeneous equation, $y' = -y/t$, has solution $y_h(t) = 1/t$. We look for a particular solution in the form $y_p(t) = v(t)y_h(t)$, where v is an unknown function. Since

$$\begin{aligned} y_p' &= v'y_h + vy_h' \\ &= v'y_h - 2vy_h \\ &= v'y_h - y_p/t, \end{aligned}$$

and $y_p' = -y_p/t + 4t$, we have $v' = 4t/y_h = 4t^2$. Integrating we get $v(t) = 4t^3/3$, and

$$y_p(t) = v(t)y_h(t) = \frac{4}{3}t^2.$$

The general solution is

$$y(t) = y_p(t) + Cy_h(t) = \frac{4}{3}t^2 + C/t.$$

35. The homogeneous equation, $y' = -2xy$ has solution $y_h(x) = e^{-x^2}$. We look for a particular solution in the form $y_p(x) = v(x)y_h(x)$, where v is an unknown function. Since

$$\begin{aligned} y_p' &= v'y_h + vy_h' \\ &= v'y_h - 2xy_p, \end{aligned}$$

and $y_p' = -2xy_p + 4x$, we have $v' = 4x/y_h = 4xe^{x^2}$. Hence $v(x) = 2e^{x^2}$, and

$$y_p(x) = v(x)y_h(x) = 2.$$

The general solution is

$$y(t) = y_p(t) + Cy_h(t) = 2 + Ce^{x^2}.$$

37. The homogeneous equation $y' = -y/2$ has solution $y_h(t) = e^{-t/2}$. We look for a particular solution of the form $y_p(t) = v(t)y_h(t)$, where v is an unknown function. Since

$$\begin{aligned} y_p' &= v'y_h + vy_h' \\ &= v'y_h - vy_h/2 \\ &= v'y_h - y_p/2, \end{aligned}$$

and $y_p' = -y_p/2 + t$, we have $v' = t/y_h(t) = te^{t/2}$. Integrating we find that $v(t) = (2t-4)e^{t/2}$, and $y_p(t) = v(t)y_h(t) = (2t-4)$. The general solution is $y(t) = y_p(t) + Cy_h(t) = (2t-4) + Ce^{-t/2}$. From $y(0) = 1$ we compute that $C = 5$, so the solution is

$$y(t) = (2t-4) + 5e^{-t/2}.$$

39. The homogeneous equation $y' = -2xy$ has solution $y_h(x) = e^{-x^2}$. We look for a particular solution of the form $y_p(x) = v(x)y_h(x)$, where v is an unknown function. Since

$$\begin{aligned} y_p' &= v'y_h + vy_h' \\ &= v'y_h - 2xvy_h \\ &= v'y_h - 2xy_p, \end{aligned}$$

and $y_p' = -2xy_p + 2x^3$, we see that $v' = 2x^3e^{x^2}$. Integrating we get $v(x) = (x^2-1)e^{x^2}$, and $y_p(x) = x^2 - 1$. The general solution is $y(x) = y_p(x) + Cy_h(x) = x^2 - 1 + Ce^{-x^2}$. Since $y(0) = -1$, we have $C = 0$, and the solution is

$$y(x) = x^2 - 1.$$

41. The homogeneous equation $x' = -4tx/(1+t^2)$ has solution $x_h(t) = (1+t^2)^{-2}$. We look for a particular solution of the form $x_p(t) = v(t)x_h(t)$, where v is an unknown function. Since

$$\begin{aligned} x_p' &= v'x_h + vx_h' \\ &= v'x_h - 4tx_p/(1+t^2), \end{aligned}$$

and $x_p' = -4tx_p/(1+t^2) + t/(1+t^2)$, so

$$v' = \frac{t}{1+t^2} \cdot \frac{1}{x_h(t)} = t(1+t^2).$$

Integrating, we get $v(t) = t^2/2 + t^4/4$. Thus

$$x_p(t) = v(t)x_h(t) = \frac{2t^2+t^4}{4(1+t^2)^2}.$$

The general solution is

$$x(t) = x_p(t) + Cx_h(t) = \frac{4C+2t^2+t^4}{4(1+t^2)^2}.$$

The initial condition $x(0) = 1$ implies that $C = 1$, so the solution is

$$y(t) = \frac{4+2t^2+t^4}{4(1+t^2)^2}.$$

43. (a) The solution of the homogeneous equation $T' + kT = 0$ is $T_h = Fe^{-kt}$, with F an arbitrary constant.

(b) We guess that $T_p = C\cos\omega t + D\sin\omega t$ is a particular solution. Substituting T_p and $T_p' = -C\omega\sin\omega t + D\omega\cos\omega t$ in the left side of $T' + kT = kA\sin\omega t$, then gathering coefficients of $\cos\omega t$ and $\sin\omega t$, we obtain

$$T_p' + kT_p = (-\omega C + kD)\sin\omega t + (kC + \omega D)\cos\omega t. \tag{4.1}$$

Comparing this with the right side of $T_p' + kT_p = kA\sin\omega t$, we see that

$$-\omega C + kD = kA \quad \text{and} \quad kC + \omega D = 0.$$

(c) Solving these equations simultaneously (for example, multiply the first equation by k, the second by ω, then add the equations to eliminate C) provides

$$C = -\frac{\omega kA}{k^2+\omega^2} \quad \text{and} \quad D = \frac{k^2A}{k^2+\omega^2}.$$

Substituting these results in $T_p = C\cos\omega t + D\sin\omega t$ provides the particular solution

$$T_p = -\frac{\omega kA}{k^2+\omega^2}\cos\omega t + \frac{k^2A}{k^2+\omega^2}\sin\omega t.$$

Hence, the general solution is

$$\begin{aligned} T &= T_h + T_p \\ &= Fe^{-kt} + \frac{kA}{k^2+\omega^2}\left[k\sin\omega t - \omega\cos\omega t\right]. \end{aligned}$$

Section 2.5. Mixing Problems

1. (a) Let $S(t)$ denote the amount of sugar in the tank, measured in pounds. The rate in is 3 gal/min × 0.2 lb/gal = 0.6 lb/min. The rate out is 3 gal/min × $S/100$ lb/gal = $3S/100$ lb/min. Hence

$$\begin{aligned} \frac{dS}{dt} &= \text{rate in} - \text{rate out} \\ &= 0.6 - 3S/100 \end{aligned}$$

This linear equation can be solved using the integrating factor $u(t) = e^{3t/100}$ to get the general solution $S(t) = 20 + Ce^{-3t/100}$. Since $S(0) = 0$, the constant $C = -20$ and the solution is $S(t) = 20(1 - e^{-3t/100})$. $S(20) = 10(1 - e^{-0.6}) \approx 9.038$lb.

(b) $S(t) = 15$ when $e^{-3t/100} = 1 - 15/20 = 1/4$. Taking logarithms this translates to $t = (100\ln 4)/3 \approx 46.2098$m.

(c) As $t \to \infty$ $S(t) \to 20$.

3. (a) Let $x(t)$ represent the number of pounds of salt in the tank at time t. The rate at which the salt in the tank is changing with respect to time is equal to the rate at which salt enters the tank

minus the rate at which salt leaves the tank, i.e.,

$$\frac{dx}{dt} = \text{rate in} - \text{rate out}.$$

In order that the units match in this equation, dx/dt, the rate In, and the rate Out must each be measured in pounds per minute (lb/min). Solution enters the tank at 5 gal/min, but the concentration of this solution is 1/4 lb/gal. Consequently,

$$\text{rate in} = 5\ \text{gal/min} \times \frac{1}{4}\ \text{lb/gal} = \frac{5}{4}\ \text{lb/min}.$$

Solution leaves the tank at 5 gal/min, but at what concentration? Assuming perfect mixing, the concentration of salt in the solution is found by dividing the amount of salt by the volume of solution, $c(t) = x(t)/100$. Consequently,

$$\text{rate out} = 5\ \text{gal/min} \times \frac{x(t)}{100}\ \text{lb/gal} = \frac{1}{20}x(t)\ \text{lb/min}.$$

As there are 2 lb of salt present in the solution initially, $x(0) = 2$ and

$$\frac{dx}{dt} = \frac{5}{4} - \frac{1}{20}x, \quad x(0) = 2.$$

Multiply by the integrating factor, $e^{(1/20)t}$, and integrate.

$$\begin{aligned}\left(e^{(1/20)t}x\right)' &= \frac{5}{4}e^{(1/20)t}\\ e^{(1/20)t}x &= 25e^{(1/20)t} + C\\ x &= 25 + Ce^{-(1/20)t}\end{aligned}$$

The initial condition $x(0) = 2$ gives $C = -23$ and

$$x(t) = 25 - 23e^{-(1/20)t}.$$

Thus, the concentration at time t is given by

$$c(t) = \frac{x(t)}{100} = \frac{25 - 23e^{-(1/20)t}}{100},$$

and the eventual concentration can be found by taking the limit as $t \to +\infty$.

$$\lim_{t\to+\infty} \frac{25 - 23e^{-(1/20)t}}{100} = \frac{1}{4}\ \text{lb/gal}$$

Note that this answer is quite reasonable as the concentration of solution entering the tank is also $1/4$ lb/gal.

(b) We found it convenient to manipulate our original differential equation before using our solver. The key idea is simple: we want to sketch the concentration $c(t)$, not the salt content $x(t)$. However,

$$c(t) = \frac{x(t)}{100} \quad \text{or} \quad x(t) = 100c(t).$$

Consequently, $x'(t) = 100c'(t)$. Substituting these into our balance equation gives

$$\begin{aligned}x' &= \frac{5}{4} - \frac{1}{20}x,\\ 100c' &= \frac{5}{4} - \frac{1}{20}(100c),\\ c' &= \frac{1}{80} - \frac{1}{20}c,\end{aligned}$$

with $c(0) = x(0)/100 = 2/100 = 0.02$. The numerical solution of this ODE is presented in the following figure. Note how the concentration approaches 0.25 lb/gal.

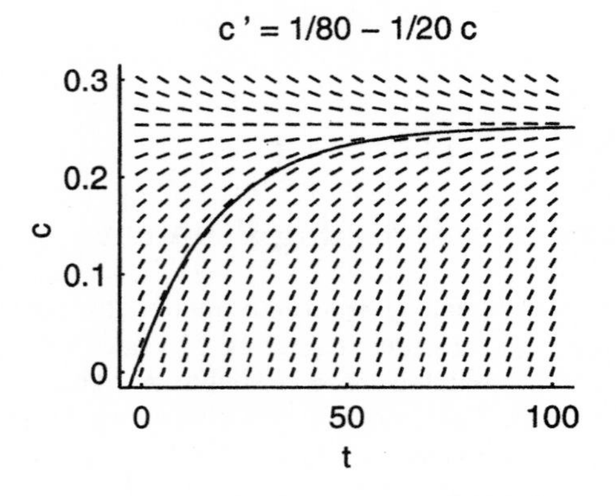

5. The volume is increasing at the rate of 2 gal/min, so the volume at time t is $V(t) = 20 + 2t$. The tank is full when $V(t) = 50$, or when $t = 15$ min. If $x(t)$ is the amount of salt in the tank at time t, then the concentration is $x(t)/V(t)$. The rate in is 4 gal/min

$\cdot$ 0.5 lb/gal = 2 lb/min. The rate out is 2 gal/min $\cdot x/V$ lb/gal. Hence the model equation is

$$x' = 2 - 2x/V = 2 - \frac{x}{10+t}.$$

This linear equation can be solved using the integrating factor $u(t) = 10 + t$, giving the general solution $x(t) = 10 + t + C/(10+t)$. The initial condition $x(0) = 0$ enables us to compute that $C = -100$, so the solution is $x(t) = 10 + t - 100/(10+t)$. At $t = 15$, when the tank is full, we have $x(15) = 21$ lb.

7. (a) The volume of liquid in the tank is increasing by 2 gal/min. Hence the volume is $V(t) = 100 + 2t$ gal. Let $x(t)$ be the amount of pollutant in the tank, measured in lbs. The rate in during this initial period is 6 gal/min $\cdot$ 0.5 lb/gal = 3 lb/gal. The rate out is 8 gl/min $\cdot x/V = 4x/(50+t)$. Hence the model equation is

$$x' = 3 - 4x/(50+t).$$

This linear equation can be solved using the integrating factor $u(t) = (50+t)^4$. The general solution is $x(t) = 3(50+t)/5 + C(50+t)^{-4}$. The initial condition $x(0) = 0$ allows us to compute the constant to be $C = -1.875 \times 10^8$. Hence the solution is

$$x(t) = \frac{3t}{5} + 30 - \frac{1.875 \times 10^8}{(50+t)^4}.$$

After 10 minutes the tank contains $x(10) = 21.5324$ lb of salt.

(b) Now the volume is decreasing at the rate of 4 gal/min from the initial volume of 120 gal. Hence if we start with $t = 0$ at the 10 minute mark, the volume is $V(t) = 120 - 4t$ gal. Now the rate in is 0, and the rate out is 8 gal/min $\cdot x/V = 2x/(30-t)$. Hence the model equation is

$$x' = -\frac{2x}{30-t}.$$

This homogeneous linear equation can solved by separating variables to find the general solution $x(t) = A(30-t)^2$. At $t = 0$ we have $x(0) = 21.5342$, from which we find that $A = 21.5342/900$, and the solution is

$$x(t) = \frac{21.5342}{900}(30-t)^2.$$

We are asked to find when this is one-half of 21.5342. This happens when $(30-t)^2 = 450$ or at $t = 8.7868$ min.

9. (a) The rate at which pollutant enters the lake is

$$\text{rate in} = p \text{ km}^3/\text{yr}.$$

The rate at which the pollutant leaves the lake is found by multiplying the flow rate by the concentration of pollutant in the lake.

$$\begin{aligned}\text{rate out} &= (r+p) \text{ km}^3/\text{yr} \times \frac{x(t)}{V} \text{ km}^3/\text{km}^3 \\ &= \frac{r+p}{V} x(t) \text{ km}^3/\text{yr}\end{aligned}$$

Consequently,

$$\frac{dx}{dt} = p - \frac{r+p}{V} x.$$

But $c(t) = x(t)/V$, so $Vc'(t) = x'(t)$ and

$$\begin{aligned}Vc' &= p - \frac{r+p}{V}(Vc), \\ c' + \frac{r+p}{V} c &= \frac{p}{V}.\end{aligned}$$

(b) With $r = 50$ and $p = 2$, the equation becomes

$$\begin{aligned}c' &= \frac{2}{100} - \frac{50+2}{100} c, \\ c' &= 0.02 - 0.52c.\end{aligned}$$

This is linear and solved in the usual manner.

$$\begin{aligned}\left(e^{0.52t} c\right)' &= 0.02 e^{0.052t} \\ e^{0.52t} c &= \frac{0.02}{0.52} e^{0.052t} + K \\ c &= \frac{1}{26} + K e^{-0.52t}\end{aligned}$$

The initial concentration is zero, so $c(0) = 0$ produces $K = -1/26$ and

$$c = \frac{1}{26} - \frac{1}{26} e^{-0.52t}.$$

The question asks when the concentration reaches 2%, or when $c(t) = 0.02$. Thus,

$$0.02 = \frac{1}{26}\left(1 - e^{-0.52t}\right),$$
$$e^{-0.52t} = 0.48,$$
$$t = -\frac{\ln 0.48}{0.52},$$
$$t \approx 1.41 \text{ years}.$$

11. (a) The concentrations are plotted in the following figure. In steady-state the concentration varies periodically.

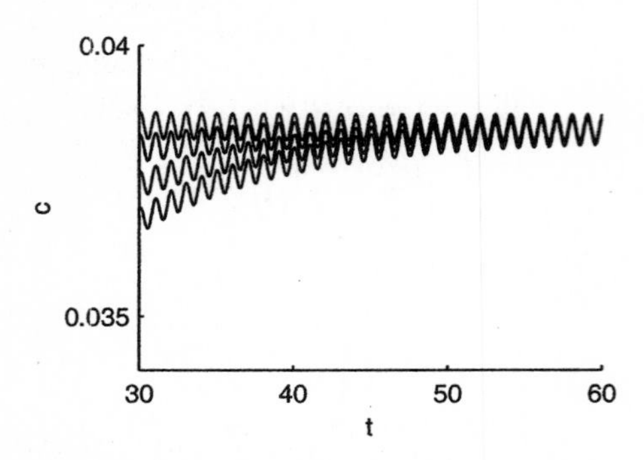

(b) The following figure shows one year of the oscillation, and indicates that the maximum concentration occurs early in February. This is four months after the time of the minimum flow. Thus there is a shift of phase between the cause and the effect.

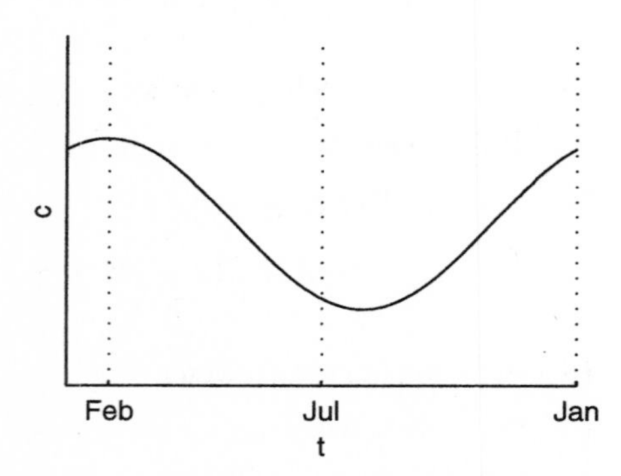

13. (a) Let $x(t)$ be the amount of pollutant (measured in km^3) in Lake Happy Times. The rate in for Lake Happy Times is $2\,\text{km}^3/\text{yr}$. The rate out is $52\,\text{km}^3/\text{yr} \times x/100 = 0.52x\,\text{km}^3/\text{yr}$. Hence the model equation is

$$x' = 2 - 0.52x.$$

This linear equation can be solved using the integrating factor $u(t) = e^{0.52t}$. With the initial condition $x(0) = 0$ we find the solution $x(t) = 2[1 - e^{-0.52t}]/0.52$.

Let $y(t)$ be the amount of pollutant (measured in km^3) in Lake Sad Times. The rate into lake Sad Times is the same as the rate out of Lake Happy Times, or $0.52x\,\text{km}^3/\text{yr}$. The rate out is $52\,\text{km}^3/\text{yr} \times y/100 = 0.52y\,\text{km}^3/\text{yr}$. Hence the model equation is

$$y' = 0.52(x - y) = 2[1 - e^{-0.52t}] - 0.52y.$$

This linear equation can also be solved using the integrating factor $u(t) = e^{0.52t}$. With the initial condition $y(0) = 0$ we find the solution

$$\begin{aligned} y(t) &= 2[1 - e^{-0.52t}]/0.52 - 2te^{-0.52t} \\ &= x(t) - 2te^{-0.52t}. \end{aligned}$$

After 3 months, when $t = 1/4$, we have $x(1/4) = 0.4689\text{km}^3$ and $y(1/4) = 0.0298\text{km}^3$.

(b) If the factory is shut down, then the flow of pollutant at the rate of $2\,\text{km}^3/\text{yr}$ is stopped. This means that the flow between the lakes and that out of Lake Sad Times will be reduced to $50\,\text{km}^3/\text{yr}$ in order to maintain the volumes. We will start time over at this point and we have the initial conditions $x(0) = x_1 = 0.4689\text{km}^3$, and $y(0) = y_1 = 0.0298\text{km}^3$.

Now there is no flow of pollutant into Lake Happy Times, and the rate out is $x/2\,\text{km}^3/\text{yr}$. Hence the model equation is $x' = -x/2$. The solution is $x(t) = x_1e^{-t/2}$.

The rate into Lake Sad times is $x/2\,\text{km}^3/\text{yr}$, and the rate out is $y/2\,\text{km}^3/\text{yr}$. The model equation is $y' = (x - y)/2 = x_1e^{-t/2}/2 - y/2$. this time we use the integrating factor $u(t) = e^{t/2}$ and find the solution

$$y(t) = [x_1t/2 + y_1]e^{-t/2}.$$

The plot of the solution over 10 years is shown in the following figure. It is perhaps a little surprising to see that the level of pollution in Lake Sad Times continues to rise for some time after the factory is closed.

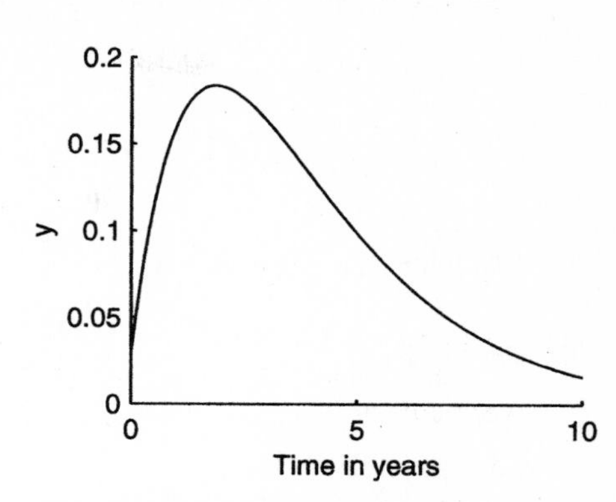

Using a computer or a calculator, we find that $y(t) = y_1/2$ when $t = 10.18$ yrs.

Section 2.6. Exact Differential Equations

1. $dF = 2ydx + (2x + 2y)dy$

3. $dF = \dfrac{xdx + ydy}{\sqrt{x^2 + y^2}}$

5.
$$dF = \frac{1}{x^2 + y^2}(x^2ydx + y^3dx - ydx + x^3dy + xy^2dy + xdy)$$

7.
$$dF = \left(\frac{2x}{x^2 + y^2} + 1/y\right)dx + \left(\frac{2y}{x^2 + y^2} - \frac{x}{y^2}\right)dy$$

9. With $P = 2x + y$ and $Q = x - 6y$, we see that
$$\frac{\partial P}{\partial y} = 1 = \frac{\partial Q}{\partial x}$$
so the equation is exact. We solve by setting
$$F(x, y) = \int P(x, y)\, dx = \int (2x + y)\, dx$$
$$= x^2 + xy + \phi(y).$$
To find ϕ, we differentiate
$$Q(x, y) = \frac{\partial F}{\partial y} = x + \phi'(y).$$
Hence $\phi' = -6y$, and we can take $\phi(y) = -3y^2$. Hence the solution is $F(x, y) = x^2 + xy - 3y^2 = C$.

11. With $P = 1 + \dfrac{y}{x}$ and $Q = -\dfrac{1}{x}$, we compute
$$\frac{\partial P}{\partial y} = \frac{1}{x} \neq \frac{\partial Q}{\partial x} = \frac{1}{x^2}.$$
Hence the equation is not exact.

13. Exact $x^3 + xy - y^3 = C$

15. Exact $u^2/2 + vu - v^2/2 = C$

17. Not exact

19. Exact $x\sin 2t - t^2 = C$

21. Not exact.

23. $x^2y^2/2 - \ln x + \ln y = C$

25. $x - (1/2)\ln(x^2+y^2) = C$

27. $\mu(x) = \dfrac{1}{x}$. $F(x,y) = xy - \ln x - \dfrac{y^2}{2} = C$.

29. $\mu(y) = 1/y^2$. $F(x,y) = \dfrac{yx+x^2}{y} = C$

31. $x+y$ and $x-y$ are homogeneous of degree one.

33. $x - \sqrt{x^2+y^2}$ and $-y$ are homogenous of degree one.

35. $x^2 - Cx = y^2$

37. $F(x,y) = xy + (3/2)x^2 = C$.

39. $y(x) = \dfrac{x+Cx^4}{1-2Cx^3}$

41. (a) First,

$$\frac{dy}{dx} = \frac{dy/dt}{dx/dt} = \frac{v_0\sin\theta - \omega}{v_0\cos\theta}.$$

However, $\cos\theta = x/\sqrt{x^2+y^2}$ and $\sin\theta = y/\sqrt{x^2+y^2}$, so

$$\frac{dy}{dx} = \frac{\dfrac{v_0 y}{\sqrt{x^2+y^2}} - \omega}{\dfrac{v_0 x}{\sqrt{x^2+y^2}}} = \frac{v_0 y - \omega\sqrt{x^2+y^2}}{v_0 x}.$$

Divide top and bottom by v_0 and replace ω/v_0 with k.

$$\frac{dy}{dx} = \frac{y - \dfrac{\omega}{v_0}\sqrt{x^2+y^2}}{x} = \frac{y - k\sqrt{x^2+y^2}}{x}.$$

(b) Write the equation

$$\frac{dy}{dx} = \frac{y - k\sqrt{x^2+y^2}}{x}$$

in the form

$$(y - k\sqrt{x^2+y^2})\,dx - x\,dy = 0.$$

Both terms are homogeneous (degree 1), so make substitutions $y = xv$ and $dy = x\,dv + v\,dx$.

$$(xv - k\sqrt{x^2+x^2v^2})\,dx - x(x\,dv + v\,dx) = 0$$

After cancelling the common factor x and combining terms,

$$k\sqrt{1+v^2}\,dx - x\,dv = 0.$$

Separate variables and integrate.

$$\frac{k\,dx}{x} - \frac{dv}{\sqrt{1+v^2}} = 0$$

$$k\ln x - \ln(\sqrt{1+v^2}+v) = C$$

Note the initial condition $(x,y) = (a,0)$. Because $y = xv$, v must also equal zero at this point. Thus, $(x,v) = (a,0)$ and

$$k\ln a - \ln(\sqrt{1-0^2}+0) = C$$
$$C = k\ln a.$$

Therefore,

$$k\ln x - \ln(\sqrt{1+v^2}+v) = k\ln a.$$

Taking the exponential of both sides,

$$e^{\ln x^k - \ln(\sqrt{1+v^2}+v)} = e^{\ln a^k}$$

$$\frac{x^k}{v+\sqrt{1+v^2}} = a^k$$

$$\left(\frac{x}{a}\right)^k = v + \sqrt{1+v^2}.$$

Solve for v.

$$\left(\frac{x}{a}\right)^k - v = \sqrt{1+v^2}$$

$$\left(\frac{x}{a}\right)^{2k} - 2v\left(\frac{x}{a}\right)^k + v^2 = 1 + v^2$$

$$\left(\frac{x}{a}\right)^{2k} - 1 = 2v\left(\frac{x}{a}\right)^k$$

$$\frac{1}{2}\left[\left(\frac{x}{a}\right)^k - \left(\frac{x}{a}\right)^{-k}\right] = v$$

Finally, recall that $y = xv$, so

$$\frac{y}{x} = \frac{1}{2}\left[\left(\frac{x}{a}\right)^{k} - \left(\frac{x}{a}\right)^{-k}\right]$$
$$y = \frac{a}{2}\left[\left(\frac{x}{a}\right)^{1+k} - \left(\frac{x}{a}\right)^{1-k}\right].$$

(c) The following three graphs show the cases where $a = 1$, and $k = 1/2,\ 1,\ 3/2$. When $0 < k < 1$, the wind speed is less than that of the goose and the goose flies home. When $k = 1$ the two speeds are equal, and try as he might, the goose can't get home. Instead he approaches a point due north of the nest. When $k > 1$ the wind speed is greater, so the goose loses ground and keeps getting further from the nest.

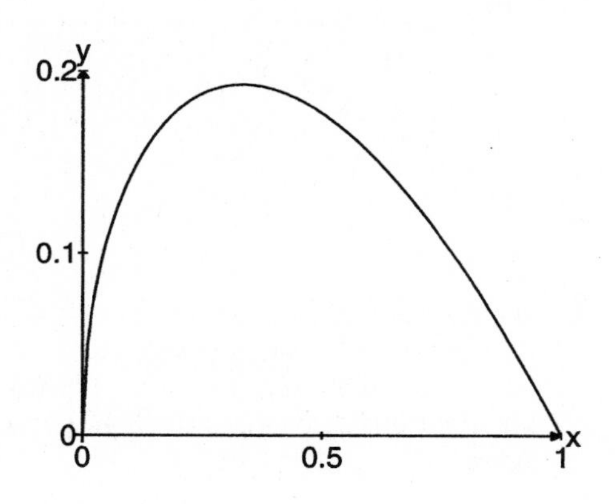

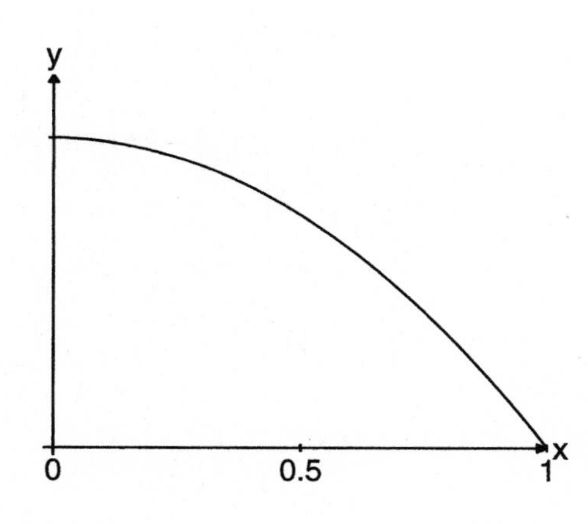

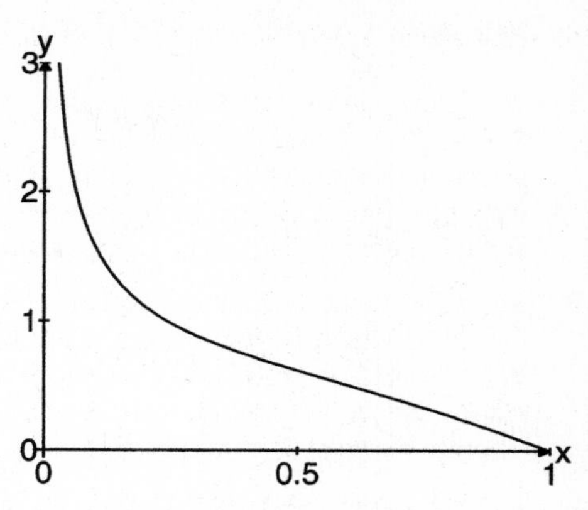

43. (a) The curves are defined by the equation $F(x, y) = x/(x^2 + y^2) = c$. Hence the orthogonal family must satisfy

$$\frac{dy}{dx} = \frac{\partial F}{\partial y} \Big/ \frac{\partial F}{\partial x} = \frac{2xy}{x^2 - y^2}.$$

(b) The differential equation is homogeneous. Solving in the usual way we find that the orthogonal family is defined implicitly by

$$G(x, y) = \frac{y}{x^2 + y^2} = C.$$

The original curves are the solid curves in the following figure, and the orthogonal family is dashed.

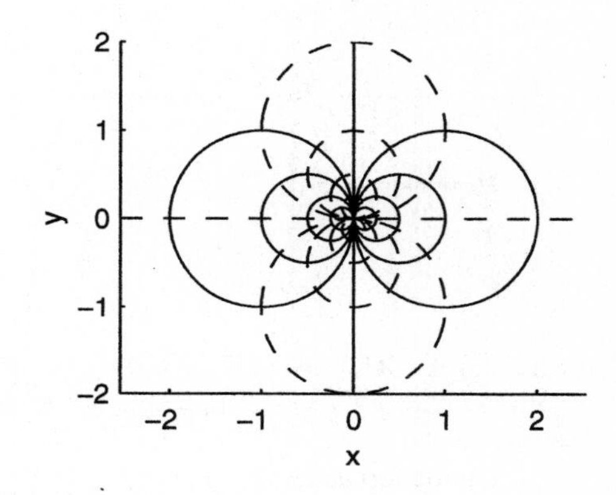

45. $\arctan(y/x) - y^4/4 = C$

47. $\arctan(y/x) - (1/4)(y^2 + x^2)^2 = C$

49. $Cxy = \dfrac{x + y}{x - y}$

Section 2.7. Existence and Uniqueness of Solutions

1. The right hand side of the equation is $f(t, y) = 4 + y^2$. f is continuous in the whole plane. Its partial derivative $\partial f/\partial y = 2y$ is also continuous on the whole plane. Hence the hypotheses are satisfied and the theorem guarantees a unique solution.

3. The right hand side of the equation is $f(t, y) = t \tan^{-1} y$, which is continuous in the whole plane. $\partial f/\partial y = t/(1 + y^2)$ is also continuous in the whole plane. Hence the hypotheses are satisfied and the theorem guarantees a unique solution.

5. The right hand side of the equation is $f(t, x) = t/(x + 1)$, which is continuous in the whole plane, except where $x = -1$. $\partial f/\partial x = -t/(x + 1)^2$ is also continuous in the whole plane, except where $x = -1$. Hence the hypotheses are satisfied in a rectangle containing the initial point $(0, 0)$, so the theorem guarantees a unique solution.

7. The equation is linear. The general solution is $y(t) = t \sin t + Ct$. Several solutions are plotted in the following figure.

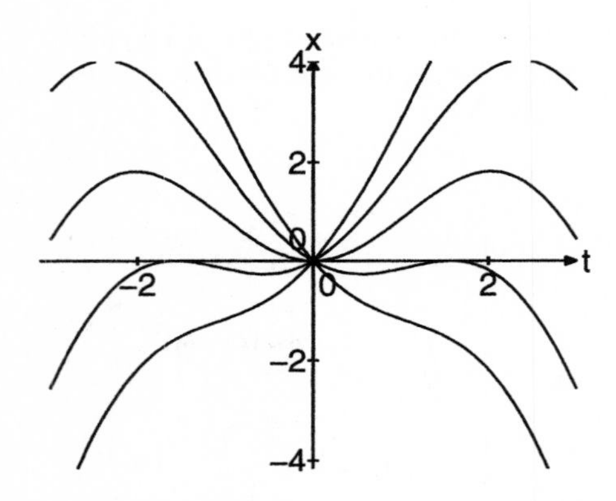

Since every solution satisfies $y(0) = 0$, there is no solution with $y(0) = -3$. If we put the equation into normal form

$$\frac{dy}{dt} = \frac{1}{t} y + t \cos t,$$

we see that the right hand side $f(t, y)$ fails to be continuous at $t = 0$. Consequently the hypotheses of the existence theorem are not satisfied.

9. The y-derivative of the right hand side $f(t, y) = 3y^{2/3}$ is $2y^{-1/3}$, which is not continuous at $y = 0$. Hence the hypotheses of Theorem 7.16 are not satisfied.

11. The exact solution is $y(t) = -1 + \sqrt{t^2 - 3}$. The interval of solution is $(\sqrt{3}, \infty)$. The solver has trouble near $\sqrt{3}$. The point where the difficulty arises is circled in the following figure.

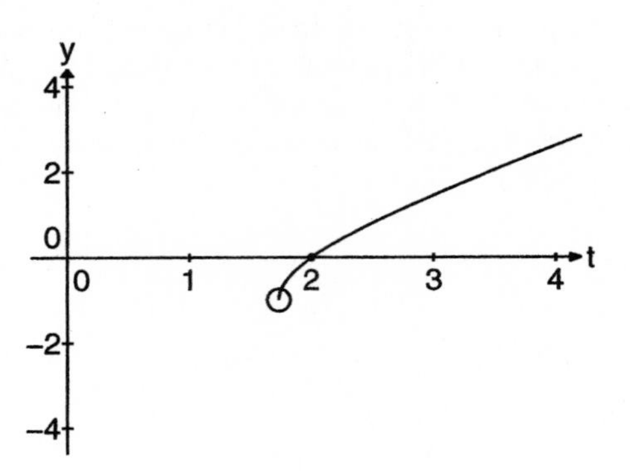

13. The exact solution is $y(t) = -1 + \sqrt{4 + 2\ln(1 - t)}$. The interval of existence is $(-\infty, 1 - e^{-2})$. The solver has trouble near $1 - e^{-2} \approx 0.8647$. The point where the difficulty arises is circled in the following figure.

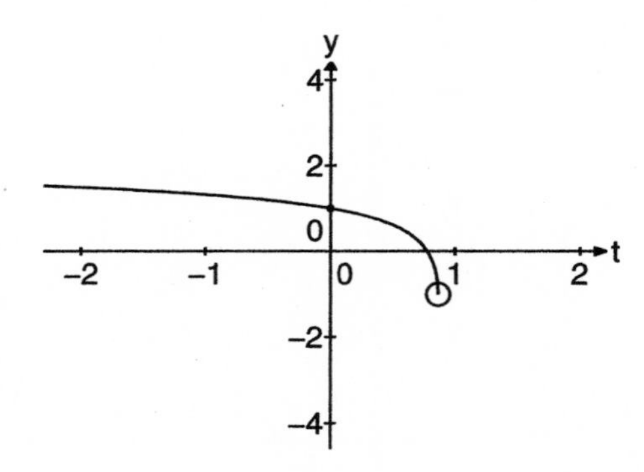

15. The solution is defined implicitly by the equation $y^3/3+y^2-3y=2t^3/3$. The solver has trouble near $(t_1, 1)$, where $t_1 = -(5/2)^{1/3} \approx -1.3572$, and also near $(t_2, -3)$, where $t_2 = (27/2)^{1/3} \approx 2.3811$. The points where the difficulty arises are circled in the following figure.

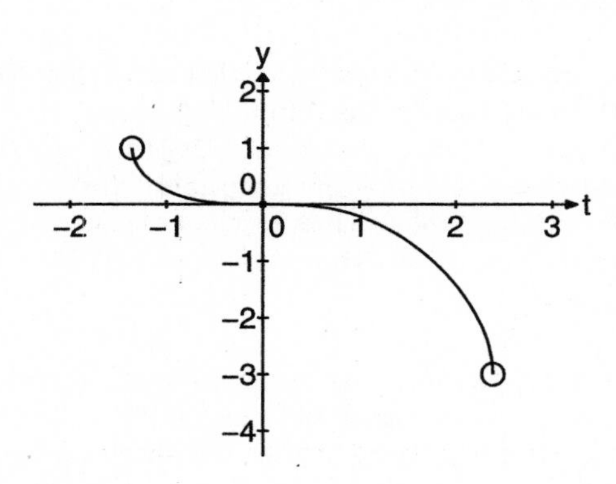

17. The computed solution is shown in the following figure.

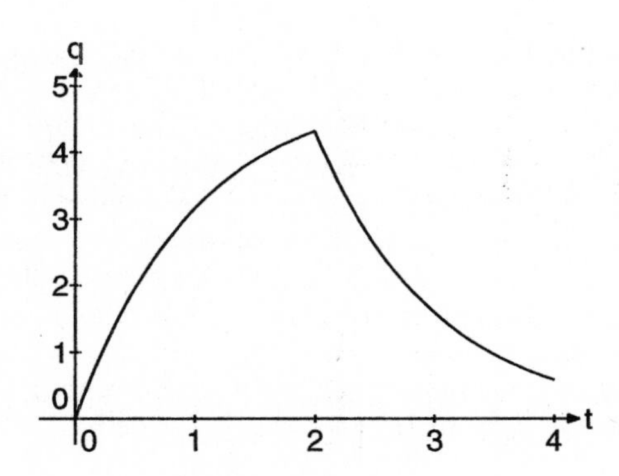

The exact solution is

$$q(t) = \begin{cases} 5 - 5e^{-t}, & \text{if } 0 < t < 2, \\ 5(1 - e^{-2})e^{2-t}, & \text{if } t \geq 2 \end{cases}$$

Hence $q(4) = 5(1 - e^{-2})e^{-2} \approx 0.5851$.

19. The computed solution is shown in the following figure.

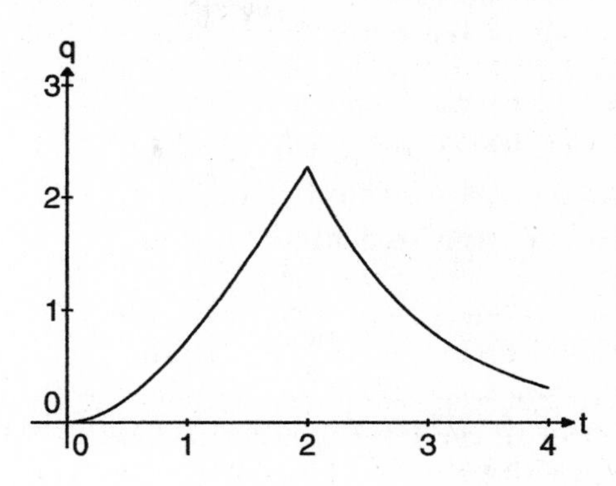

The exact solution is

$$q(t) = \begin{cases} 2(t - 1 + e^{-t}), & \text{if } 0 < t < 2, \\ 2(1 + e^{-2})e^{2-t}, & \text{if } t \geq 2 \end{cases}$$

Hence $q(4) = 2(1 + e^{-2})e^{-2} \approx 0.3073$.

21. (a) If

$$y(t) = \begin{cases} 0, & \text{if } t \leq t_0 \\ (t - t_0)^3, & \text{if } t > t_0, \end{cases}$$

then

$$\begin{aligned} y'(t_0^+) &= \lim_{t \searrow t_0} \frac{y(t) - y(t_0)}{t - t_0} \\ &= \lim_{t \to t_0^+} \frac{(t - t_0)^3 - 0}{t - t_0} \\ &= \lim_{t \to t_0^+} (t - t_0)^2 \\ &= 0. \end{aligned}$$

On the other hand,

$$\begin{aligned} y'(t_0^-) &= \lim_{t \nearrow t_0} \frac{y(t) - y(t_0)}{t - t_0} \\ &= \lim_{t \to t_0^-} \frac{0 - 0}{t - t_0} \\ &= 0. \end{aligned}$$

Therefore, $y'(t_0) = 0$, since both the left and right-hand derivatives equal zero.

(b) The right hand side of the equation, $f(t, y) = 3y^{2/3}$, is continuous, but $\partial f/\partial y = 2y^{-1/3}$ is not continuous where $y = 0$. Hence the hypotheses of Theorem 7.16 are not satisfied.

23. If

$$y(t) = \begin{cases} 0, & t < 0 \\ t^4, & t \geq 0, \end{cases}$$

then it is easily seen that

$$y'(t) = \begin{cases} 0, & t < 0 \\ 4t^3, & t > 0. \end{cases}$$

It remains to check the existence of $y'(0)$. First, the left-derivative.

$$y'_-(0) = \lim_{t \to 0^-} \frac{y(t) - y(0)}{t - 0} = \lim_{t \to 0^-} \frac{0 - 0}{t} = 0.$$

Secondly,

$$y'_+(0) = \lim_{t \to 0^+} \frac{y(t) - y(0)}{t - 0} = \lim_{t \to 0^+} \frac{t^4}{t} = \lim_{t \to 0^+} t^3 = 0.$$

Thus, $y'(0) = 0$ and we can write

$$y'(t) = \begin{cases} 0, & t < 0 \\ 4t^3, & t \geq 0. \end{cases}$$

Now,

$$ty'(t) = \begin{cases} 0, & t < 0 \\ 4t^4, & t \geq 0, \end{cases}$$

and

$$4y(t) = \begin{cases} 0, & t < 0 \\ 4t^4, & t \geq 0, \end{cases}$$

so $y = y(t)$ is a solution of $ty' = 4y$. Finally, $y(0) = 0$. In a similar manner, it is not difficult to show that

$$\omega(t) = \begin{cases} 0, & t < 0 \\ 5t^4, & t \geq 0 \end{cases}$$

is also a solution of the initial value problem $ty' = 4y$, $y(0) = 0$. At first glance, it would appear that we have contradicted uniqueness. However, if $ty' = 4y$ is written in normal form,

$$y' = \frac{4y}{t},$$

then

$$\frac{\partial}{\partial y}\left(\frac{4y}{t}\right) = \frac{4}{t}$$

is *not* continuous on any rectangular region containing the vertical axis (where $t = 0$), so the hypotheses of the Uniqueness Theorem are not satisfied. There is no contradiction of uniqueness.

25. The equation $x' = f(t, x)$ satisfies the hypotheses of the uniqueness theorem. Notice that $x_1(0) = x_2(0) = 0$. If they were both solutions $x' = f(t, x)$ near $t = 0$, then by the uniqueness theorem they would have to be equal everywhere. Since they are not, they cannot both be solutions of the differential equation.

27. Notice that $x_1(t) = 0$ is a solution to the same differential equation with initial value $x_1(0) = 0 < 1 = x(0)$. The right hand side of the differential equation, $f(t, x) = x \cos^2 t$ and $\partial f/\partial x = \cos^2 t$ are both continuous on the whole plane. Consequently the uniqueness theorem applies, so the solution curves for x and x_1 cannot cross. Hence we must have $x(t) > x_1(t) = 0$ for all t.

29. Notice that the right hand side of the equation is $f(t, y) = (y^2 - 1)e^{ty}$ and f is continuous on the whole plane. Its partial derivative $\partial f/\partial y = 2ye^{ty} + t(y^2 - 1)e^{ty}$ is also continuous on the whole plane. Thus the hypotheses of the uniqueness theorem are satisfied. By direct substitution we discover that $y_1(t) = -1$ and $y_2(t) = 1$ are both solutions to the differential equation. If y is a solution and satisfies $y(1) = 0$, then $y_1(1) < y(1) < y_2(1)$. By the uniqueness theorem we must have $y_1(t) < y(t) < y_2(t)$ for all t for which y is defined. Hence $-1 < y(t) < 1$ for all t for which y is defined.

31. Notice that $x_1(t) = t^2$ is a solution to the same differential equation with initial value $x_1(0) = 0 < 1 = x(0)$. The right hand side of the differential equation, $f(t, x) = x - t^2 + 2t$ and $\partial f/\partial x = 1$ are both continuous on the whole plane. Consequently the uniqueness theorem applies, so the solution curves for x and x_1 cannot cross. Hence we must have $t^2 = x_1(t) < x(t)$ for all t.

Section 2.8. Dependence of Solutions on Initial Conditions

1. $x(0) = 0.8009$
3. $x(0) = 0.9596$
5. $x(0) = 0.7275$
7. $x(0) = 0.7290106$
9. $x(0) = -3.2314$
11. $x(0) = -3.2320923$
13. Ten! :-)
15. The only adjustment from the previous exercise is that we now want $|x_0 - y_0| < 0.01$. This leads to

$$1 - e^{\sin t} - 0.01e^{|t|} \le y(t) \le 1 - e^{\sin t} + 0.01e^{|t|}$$

and this image.

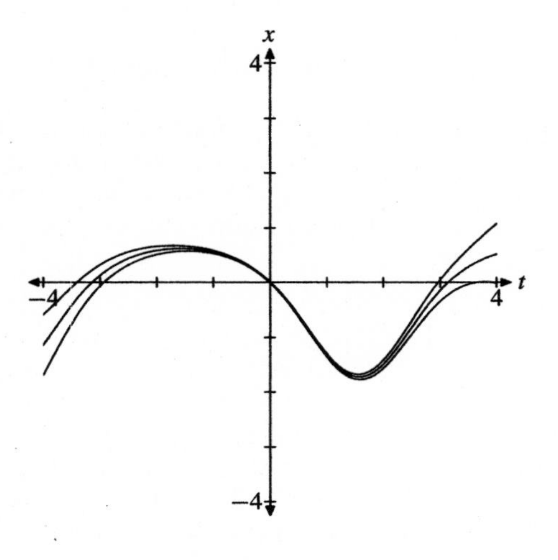

17. (a) The right hand side of the equation is $f(t, x) = -2x + \sin t$, and $\partial f/\partial x = -2$. Hence $M = \max(|\partial f/\partial x|) = 2$, and Theorem 7.15 predicts that $|y(t) - x(t)| \le |y(0) - x(0)|e^{M|t|} = |y(0) - x(0)|e^{2|t|}$.

    (b) The equation is linear, and we find that $x(t) = [2\sin t - \cos t]/5$, and $y(t) = [2\sin t - \cos t]/5 - e^{-2t}/10$. Hence

$$\begin{aligned} x(t) - y(t) &= e^{-2t}/10 = |x(0) - y(0)|e^{-2t} \\ &\le |y(0) - x(0)|e^{2|t|}. \end{aligned}$$

    (c) Since $e^{-2t} = e^{2|t|}$ for $t < 0$, we see that the maximum predicted error is achieved for all $t < 0$.

Section 2.9. Autonomous Equations and Stability

1. Note that $P' = 0.05P - 1000$ is autonomous, having form $P' = f(P)$. Solving the equation $0 = f(P) = 0.05P - 1000$, we find the equilibrium point $P = 20000$. Thus, $P(t) = 20000$ is an unstable equilibrium solution, as shown in the following figure.

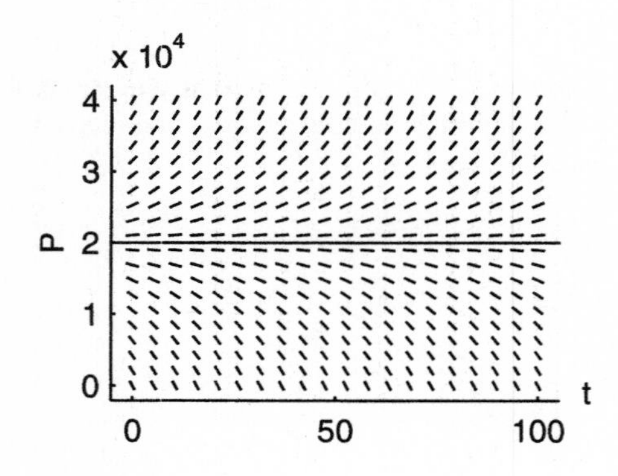

3. Note that $x' = t^2 - x^2$ is not autonomous, having form $x' = f(t, x)$, where $f(t, x) = t^2 - x^2$. The explicit dependence of the right-hand side of this differential equation on the independent variable t causes the equation to be non-autonomous.

5. The equation is autonomous. The point $q = 2$ is an unstable equilibrium point, as the following figure shows. In addition every solution of $\sin q = 0$ is an equilibrium point. These are the points $k\pi$, where k is any integer, positive of negative. The stability of the equilibrium points alternates between asymptotic stable and unstable, as is seen in the figure.

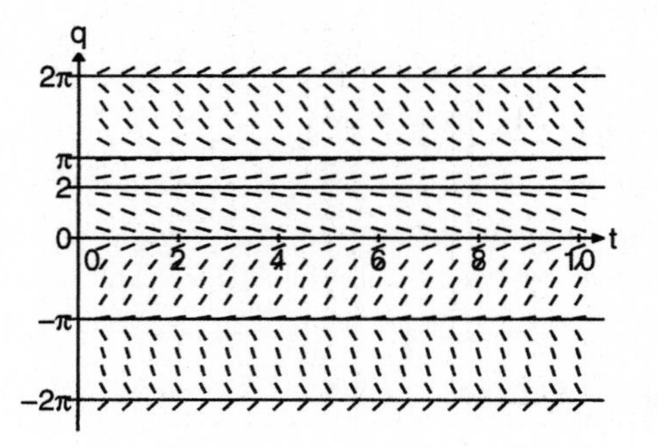

7. Note that the graph of $f(y)$ intercepts the y-axis at $y = 3$. Consequently, $y = 3$ is an equilibrium point ($f(3) = 0$) and $y(t) = 3$ is an equilibrium solution, shown in the following figure. The solution $y = 3$ is unstable.

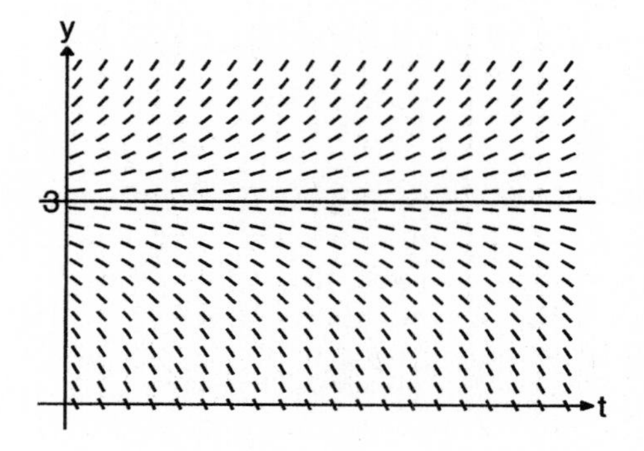

9. Since $f(y)$ has zeros at $y = -1$ and $y = 1$, these are equilibrium points. Correspondingly, $y(t) = -1$ and $y(t) = 1$ are equilibrium solutions, and are plot-

ted in the following figure. Both are unstable.

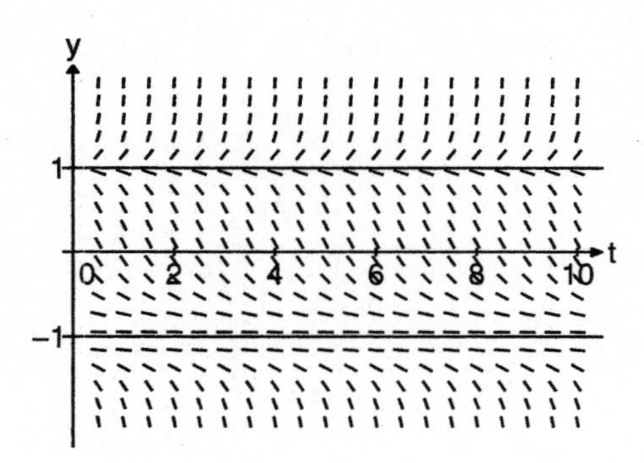

11. Because the differential equation $y' = f(y)$ is autonomous, the slope at any point (t, x) in the direction field does not depend on t, only on y, as shown in the following figure.

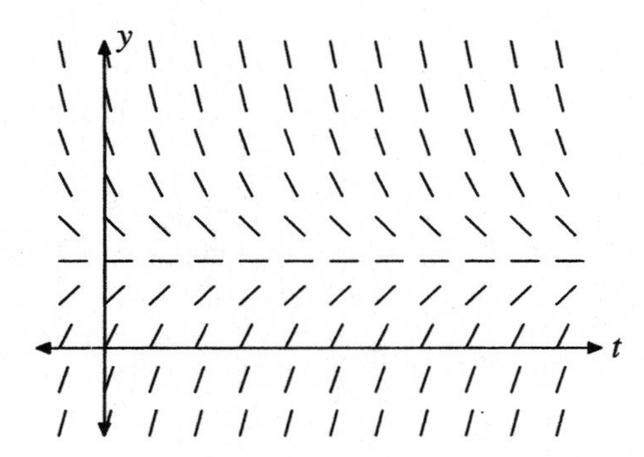

The equilibrium point is asymptotically stable.

13. The key thing to note is the fact that y' and $f(y)$ are equal. Consequently, the value of $f(y)$ is the slope of the direction line positioned at (t, y).

    - At $y = 3$, $f(y) = 0$ and the slope is zero. Thus $y = 3$ is an equilibrium point. This is shown in the following figure.
    - To the right of $y = 3$, note that the graph of f dips below the y-axis. Therefore, as y increases beyond 3, the slope becomes increasingly negative. This is also shown in the following figure.
    - To the left of $y = 3$, note that the graph of f rise above the y-axis. Therefore, as y decreases below 3, the slope becomes increasingly positive. This is also shown in the following figure. In particular, this means that the equilibrium point $y = 3$ is asymptotically stable.

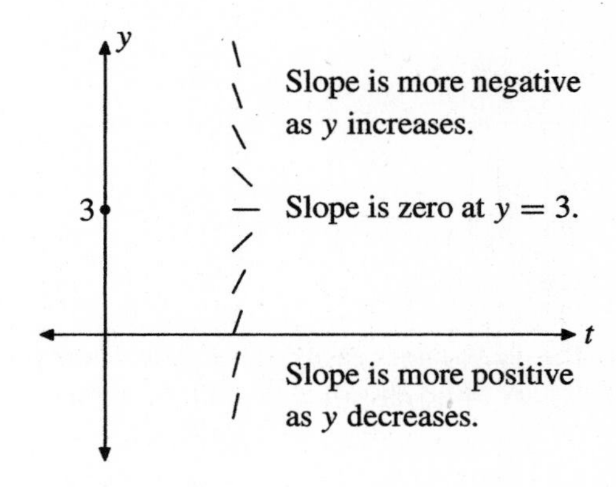

Finally, because the equation $y' = f(y)$ is autonomous, the slope of a direction line positioned at (t, y) depends only on y and not on t. Consequently, the rest of the direction field is easily completed, as shown in the next figure.

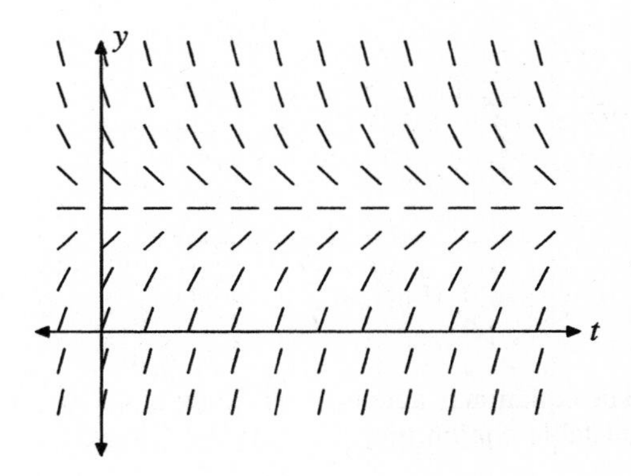

15. (i) In this case, $f(y) = 2 - y$, whose graph is shown

in the following figure.

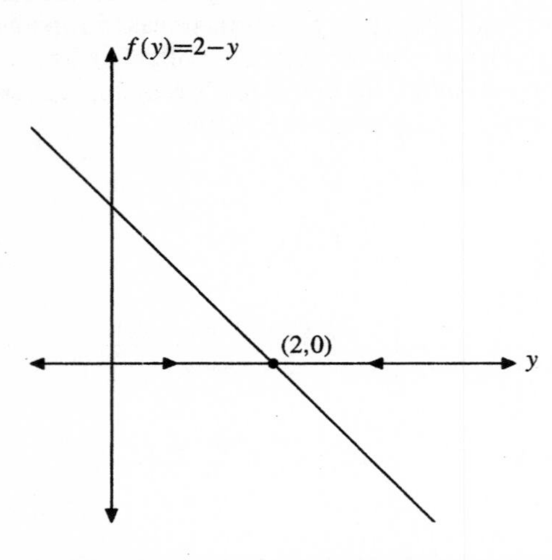

(ii) The phase line is easily captured from the previous figure, and is shown in the following figure.

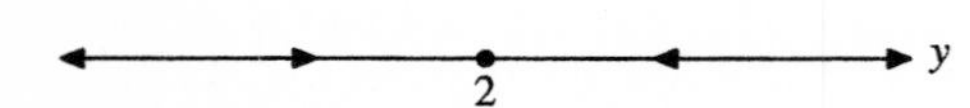

(iii) The phase line in the second figure indicates that solutions increase if $y < 2$ and decrease if $y > 2$. This allows us to easily construct the phase portrait shown in the ty plane in the next figure. Note the stable equilibrium solution, $y(t) = 2$.

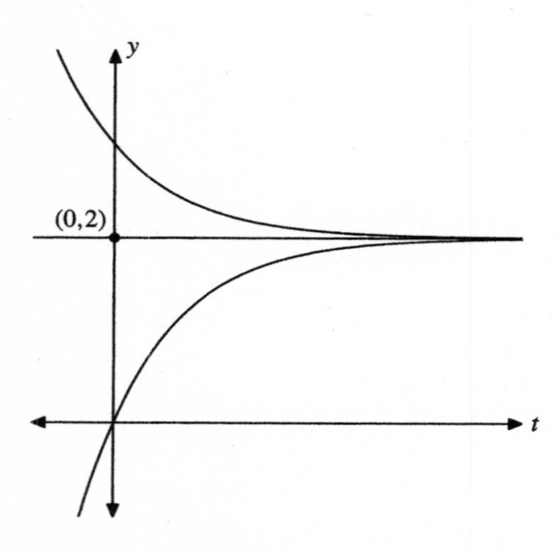

17. (i) In this case, $f(y) = (y+1)(y-4)$, whose graph is shown in the next figure.

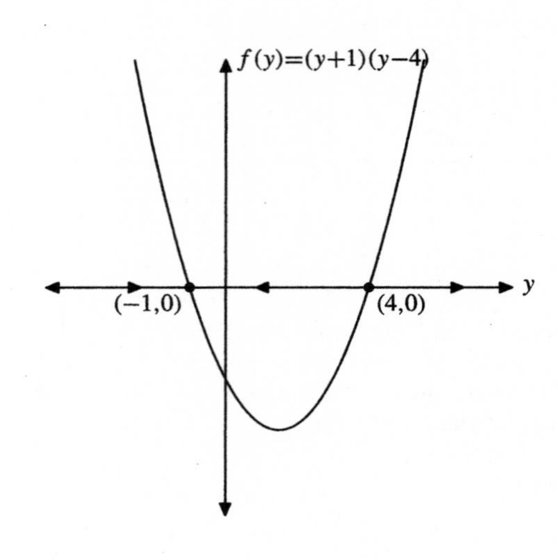

(ii) The phase line is easily captured from the previous figure, and is shown in the next figure.

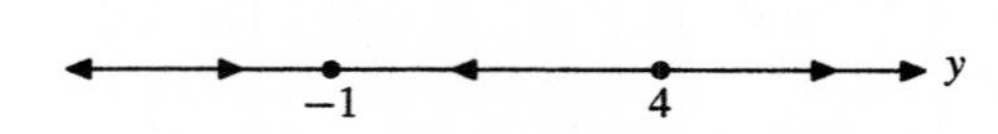

(iii) The phase line in the second figure indicates that solutions increase if $y < -1$, decrease for $-1 < y < 4$, and increase if $y > 4$. This allows us to easily construct the phase portrait shown in the ty plane in the next figure. Note the unstable equilibrium solution, $y(t) = 4$, and the stable equilibrium solution, $y(t) = -1$.

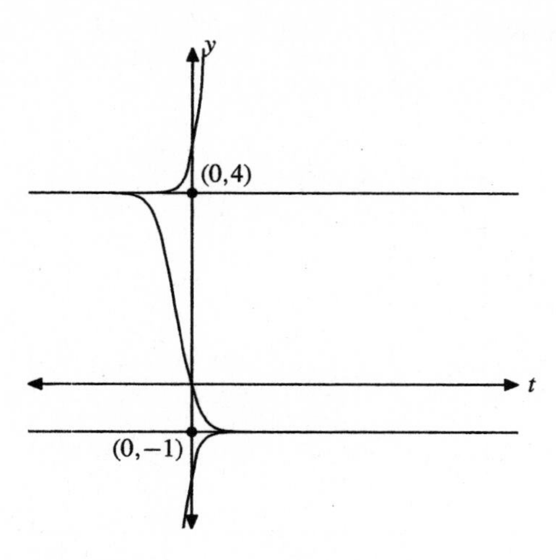

19. (i) In this case, $f(y) = 9y - y^3$ factors as $f(y) = y(y+3)(y-3)$, whose graph is shown in the next figure.

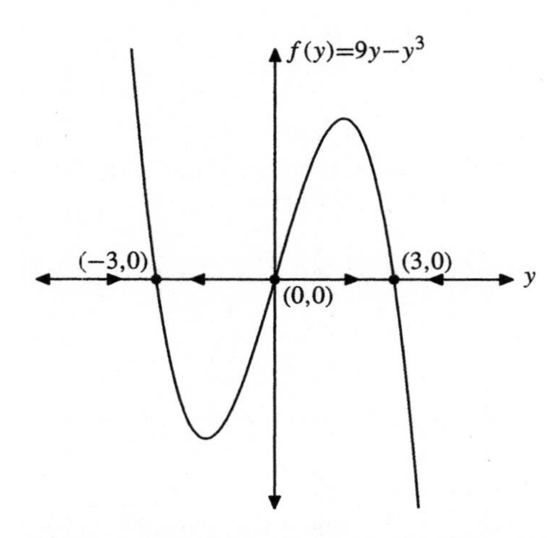

(ii) The phase line is easily captured from the previous figure, and is shown in the next figure.

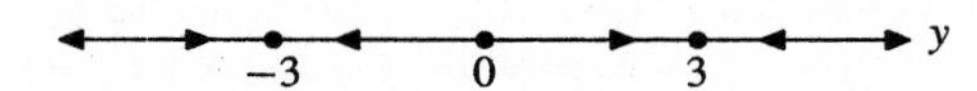

(iii) The phase line in the second figure indicates that solutions increase if $y < -3$, decrease for $-3 < y < 0$, increase if $0 < y < 3$, and decrease for $y > 3$. This allows us to easily construct the phase portrait shown in the ty plane in the next figure. Note the stable equilibrium solution, $y(t) = -3$, the unstable equilibrium solution, $y(t) = 0$, and the stable equilibrium solution, $y(t) = 3$.

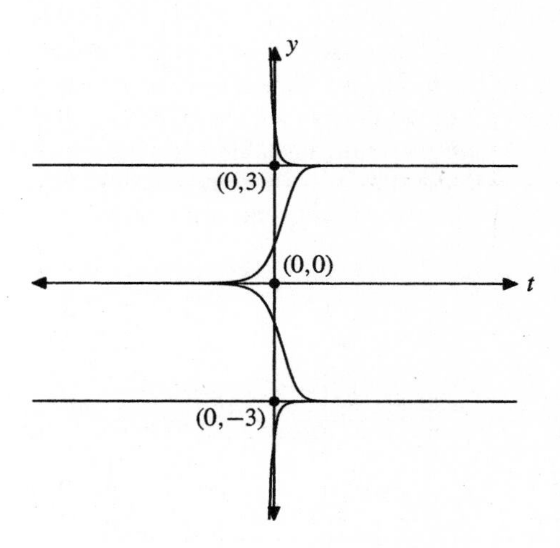

21. Due to the periodic nature of this equation, we sketch only a few regions. You can easily use the periodicity to produce more regions.

(i) In this case, $f(y) = \sin y$, whose graph is shown in the next figure.

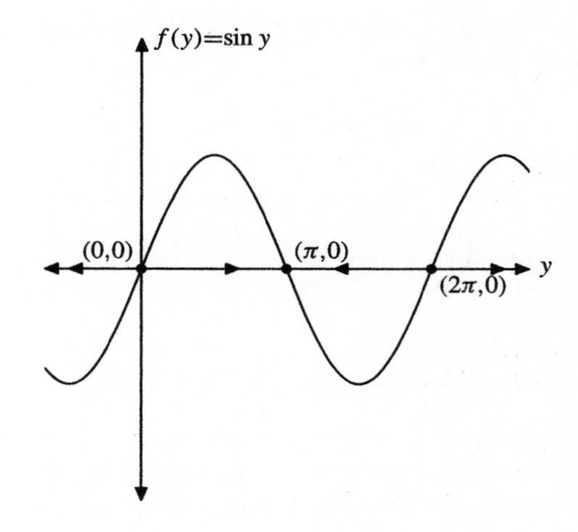

(ii) The phase line is easily captured from the previous figure, and is shown in the next figure.

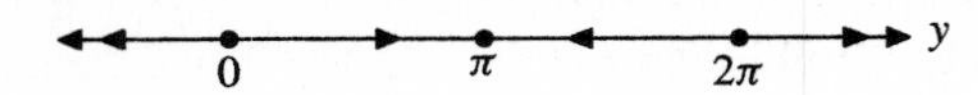

(iii) The phase line in the second figure indicates that solutions decrease if $-\pi < y < 0$, increase for $0 < y < \pi$, decrease if $\pi < y < 2\pi$, and increase for $2\pi < y < 3\pi$. This allows us to easily construct the phase portrait shown in the ty plane in the next figure. Note the unstable equilibrium solution, $y(t) = 0$, the stable equilibrium solution, $y(t) = \pi$, and the unstable equilibrium solution, $y(t) = 2\pi$.

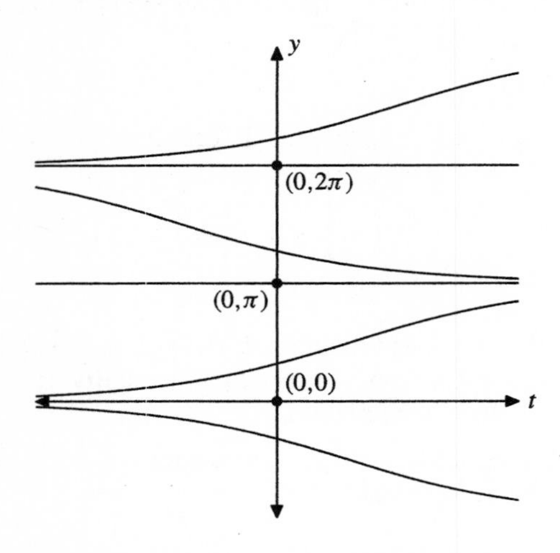

23. The equation is linear, so multiply by the integrating factor and integrate.

$$\begin{aligned}(e^t y)' &= 6e^t \\ e^t y &= 6e^t + C \\ y(t) &= 6 + Ce^{-t}\end{aligned}$$

The initial condition $y(0) = 2$ produces $C = -4$ and $y(t) = 6 - 4e^{-t}$. Now, e^{-t} approaches zero as $t \to +\infty$, so

$$\lim_{t\to+\infty} y(t) = \lim_{t\to+\infty} \left(6 - 4e^{-t}\right) = 6.$$

Compare $y' = f(y)$ with $y' = 6 - y$. Then $f(y) = 6 - y$, whose graph is shown in the first figure below. The phase line on the y-axis in this figure shows that $y = 6$ is a stable equilibrium point, so a trajectory with initial condition $y(0) = 2$ should approach the stable equilibrium solution $y(t) = 6$, as shown in the second figure. This agrees nicely with the analytical solution.

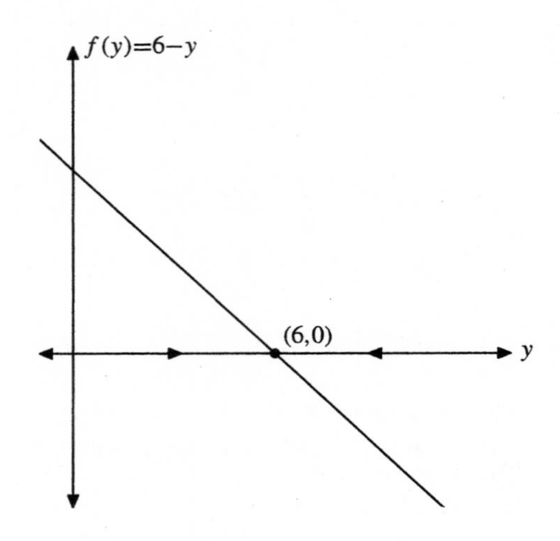

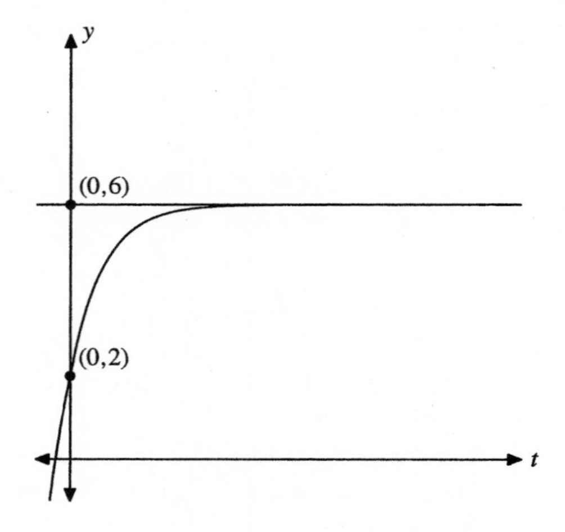

25. The equation has the form $y' = f(y)$, where $f(y) = (1 + y)(5 - y)$. The graph of f is in the next figure. We have also indicated the direction of the solutions on the y-axis. This shows that $y = -1$ is an unstable equilibrium point, and $y = 5$ is an asymptotically

stable equilibrium point. Therefore, a solution starting with $y(0) = 2$ will increase and approach $y = 5$ as t increases.

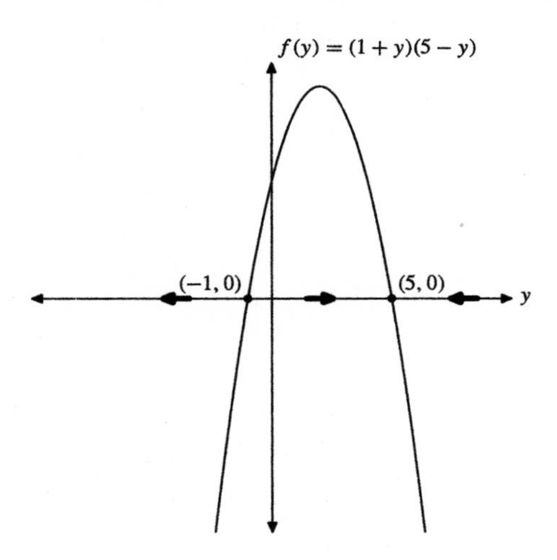

To find the exact solution, we separate variables and use partial fractions to get

$$\frac{1}{6}\left[\frac{1}{1+y} + \frac{1}{5-y}\right] dy = dt$$

Integrate,

$$\begin{aligned}
\frac{1}{6}\ln|1+y| - \frac{1}{6}\ln|5-y| &= t + C,\\
\ln|1+y| - \ln|5-y| &= 6t + 6C,\\
\ln\left|\frac{y+1}{y-5}\right| &= 6t + 6C\\
\frac{y+1}{y-5} &= Ae^{6t},
\end{aligned}$$

where $A = \pm e^{6C}$. Using the initial condition $y(0) = 2$ we see that $A = -1$, so

$$\frac{y+1}{y-5} = -e^{6t}.$$

Solving for y, we find that

$$y(t) = \frac{5e^{6t} - 1}{1 + e^{6t}} = \frac{5 - e^{-6t}}{1 + e^{-6t}}.$$

From this we see that $y(t) \to 5$ as $t \to \infty$, agreeing with what we discovered earlier.

27. We have the equation $x' = f(x) = 4 - x^2$. The equilibrium points are at $x = \pm 2$, where $f(x) = 0$. We have $f'(x) = -2x$. Since $f'(-2) = 4 > 0$, $x = -2$ is unstable. Since $f'(2) = -4 < 0$, $x = 2$ is asymptotically stable.

29. (a) $f(x) = x^2$, $f(x) = x^3$, or $f(x) = x^4$.

    (b) $f(x) = -x^3$, $f(x) = -x^5$, or $f(x) = -x^7$.

31. Let $x(t)$ represent the amount of salt in the tank at time t. The rate at which solution enters the tank is given by

$$\text{Rate In} = 2\,\text{gal/min} \times 3\,\text{lb/gal} = 6\,\text{lb/min}.$$

The rate at which solution leaves the tank is

$$\text{Rate Out} = 2\,\text{gal/min} \times \frac{x}{100}\,\text{lb/gal} = \frac{1}{50}x\,\text{lb/min}.$$

Consequently,

$$\frac{dx}{dt} = 6 - \frac{1}{50}x.$$

Let $c(t)$ represent the concentration of salt in the solution at time t. Thus, $c(t) = x(t)/100$ and $100c' = x'$.

$$\begin{aligned}
100c' &= 6 - \frac{1}{50}(100c)\\
c' &= \frac{6}{100} - \frac{1}{50}c
\end{aligned}$$

Let $f(c) = 6/100 - (1/50)c$. Setting $f(c) = 0$ produces the equilibrium point $c = 3$, as shown in the following figure.

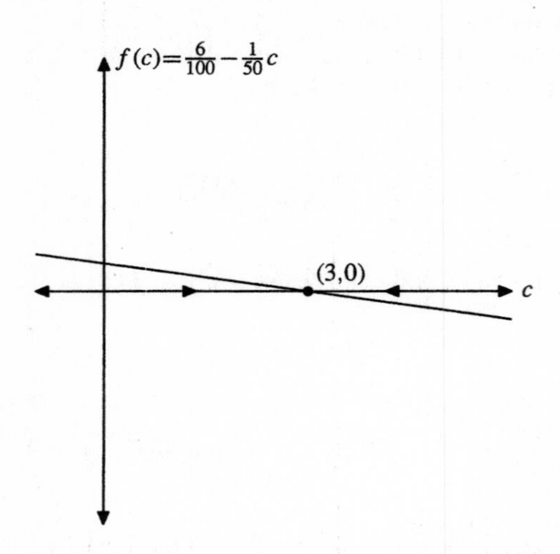

The phase line on the c-axis in this figure shows that $c = 3$ is a stable equilibrium point so a trajectory with initial condition $c(0) = 0$ (the initial concentration of salt is zero) should approach the stable equilibrium solution $c(t) = 3$.

Chapter 3. Modeling and Applications

Section 3.1. Modeling Population Growth

1. The equation of the Malthusian model is $P(t) = Ce^{rt}$. Apply the initial condition $P(0) = 100$. Then $100 = Ce^0$, or $C = 100$. Next apply the condition $P(1) = 300$. Then $300 = 100e^r$. Solving gives the reproductive rate $r = \ln 3 \approx 1.0986$. The population after five days is $P(5) = 100e^{5\ln 3} = 24300$.

3. Recall the equation of the Malthusian model $P(t) = Ce^{rt}$. The constant C is the initial population $P(0)$, so the tripling condition is stated as $P(10) = P(0)e^{10r} = 3P(0)$. Thus, $r = (1/10)\ln 3 \approx 0.1099$. The population doubles precisely when $P(t) = 2P(0) = P(0)e^{(t\ln 3)/10}$, that is, when $(t\ln 3)/10 = \ln 2$. Solving, one obtains $t = (10\ln 2)/\ln 3 \approx 6.3093$.

5. The modified Malthusian model will take the form $P' = rP - h$. Multiplying by the integrating factor e^{-rt} and putting the terms involving P on the left gives the equation $P'e^{-rt} - re^{-rt}P = -he^{-rt}$. Solve to obtain $Pe^{-rt} = (h/r)e^{-rt} + C$, where C is a constant. This constant has the value $P(0) - (h/r)$, obtained by substituting $t = 0$. Multiply by e^{rt} and get the solution $P(t) = (h/r) + (P(0) - (h/r))e^{rt}$. The long-time activity of the population depends entirely on the sign of $P(0) - (h/r)$. If it is negative, then the population will die out. If it is positive, the population will grow exponentially. If it is zero, then the population will remain at a constant h/r.

7. The reproduction rate r is calculated as before by solving $P(4) = 2P(0) = P(0)e^{4r}$. Thus, $r = (1/4)\ln 2 \approx 0.1733$. Using the result from problem 3.1.5, the population is given by $P(t) = (h/r) + (P(0) - (h/r))e^{rt}$. In order for the population to not explode, the coefficient $P(0) - (h/r)$ must not be positive. Hence the harvesting rate must be at least $rP(0)$, which here is $(10000)(1/4)\ln 2 = 2500\ln 2 \approx 1732.9$.

9. (a) For simplicity of notation, we will let $\sum$ represent $\sum_{i=1}^{n}$. Start with

$$S = \sum [y_i - (mx_i + b)]^2 \qquad (1.1)$$

and differentiate with respect to m to obtain

$$\frac{\partial S}{\partial m} = 2\sum [y_i - (mx_i + b)]\,(-x_i).$$

Set this result equal to zero and divide both sides of the resulting equation by -2 to obtain the result

$$0 = \sum \left[x_i y_i - mx_i^2 - bx_i\right].$$

Familiar properties of summation allow us to write this in the form

$$m\sum x_i^2 + b\sum x_i = \sum x_i y_i. \qquad (1.2)$$

Next, differentiate S, as defined in equation (1.1), with respect to b to obtain

$$\frac{\partial S}{\partial b} = 2\sum [y_i - (mx_i + b)]\,(-1).$$

Set this result equal to zero then divide both sides of the resulting equation to obtain

$$0 = \sum [y_i - mx_i - b].$$

Familiar properties of summation (recall $\sum_{i=1}^{n} b = bn$) allows to write this in the form

$$m\sum x_i + bn = \sum y_i. \qquad (1.3)$$

Let's eliminate b from equations (1.2) and (1.3). Multiply equation (1.2) by n, then multiply equation (1.3) by $-\sum x_i$ to obtain

$$mn\sum x_i^2 + bn\sum x_i = n\sum x_i y_i$$

$$-m\left(\sum x_i\right)^2 - bn\sum x_i = -\sum x_i \sum y_i.$$

Adding,

$$mn\sum x_i^2 - m\left(\sum x_i\right)^2 = n\sum x_i y_i - \sum x_i \sum y_i.$$

Solving this for m,

$$m = \frac{n\sum x_i y_i - \sum x_i \sum y_i}{n\sum x_i^2 - \left(\sum x_i\right)^2}. \tag{1.4}$$

Finally, solve equation (1.3) for b to obtain

$$b = \frac{\sum y_i - m\sum x_i}{n}. \tag{1.5}$$

(b) Now, compare

$$y = mx + b$$

with equation

$$\ln P = \ln C + rt$$

and you will note that we need to associate t_i with x_i, $\ln P_i$ with y_i, r with m, and $\ln C$ with b. With these changes, equation (1.4) becomes

$$r = \frac{n\sum t_i \ln P_i - \sum t_i \sum \ln P_i}{n\sum t_i^2 - \left(\sum t_i\right)^2}, \tag{1.6}$$

and equation (1.5) becomes

$$\ln C = \frac{\sum \ln P_i - r\sum t_i}{n}. \tag{1.7}$$

(c) It's helpful to arrange the data in tabular form.

t_i	P_i	$\ln P_i$	t_i^2	$t_i \ln P_i$
0	10	2.3026	0	0.0000
1	25	3.2189	1	3.2189
2	61	4.1109	4	8.2217
3	144	4.9698	9	14.9094
4	360	5.8861	16	23.5444
10	600	20.4883	30	49.8944

Substitute the appropriate totals from the table into equation (1.6).

$$r = \frac{5(49.8944) - 10(20.4883)}{5(30) - 10^2} \approx 0.8918$$

Substitute this value of r and the appropriate totals from the table into equation (1.7).

$$\ln C = \frac{20.4883 - 0.8918(10)}{5} \approx 2.3141$$

Thus, $C \approx 10.1154$.

(d) Substituting the values of C and r in $P = Ce^{rt}$, $P = 10.1154e^{0.8918t}$. The plot of the data and the exponential fit follows.

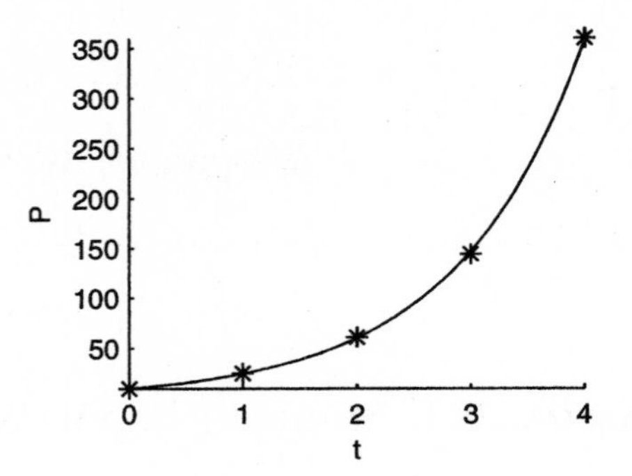

11. (a) Let $\omega = \alpha P$ and $s = \beta t$. Then solving for P and t and substituting into the logistic equation, one obtains

$$\frac{\beta}{\alpha}\frac{d\omega}{ds} = r\frac{\omega}{\alpha}\left(1 - \frac{\omega}{\alpha K}\right).$$

Isolating $d\omega/ds$ and simplifying, one obtains

$$\frac{d\omega}{ds} = \frac{r}{\beta}\omega - \frac{r}{\alpha\beta K}\omega^2.$$

This is as desired.

(b) If $\beta = r$ and $\alpha = 1/K$, then we have $d\omega/ds = \omega - \omega^2$.

(c) Separating variables and using partial fractions, one obtains

$$\left(\frac{1}{\omega} - \frac{1}{1-\omega}\right) d\omega = ds.$$

Thus, $\ln|\omega| - \ln|1-\omega| = s + c$. Exponentiating and combining all constants into a constant a gives

$$\frac{\omega}{1-\omega} = ae^s.$$

Solving for ω gives the desired form

$$\omega = \frac{1}{1-(-a)e^{-s}}.$$

(d) Note that $P = \omega K$ and $s = rt$. Thus,

$$P = \frac{K}{1 + ae^{-rt}}.$$

The initial condition $P(t_0) = P_0$ gives $P_0 + P_0 ae^{-rt_0} = K$, so

$$a = \frac{K - P_0}{P_0} e^{rt_0}.$$

Using this in the equation for P, one obtains

$$P = \frac{KP_0}{P_0 + (K - P_0)e^{-r(t-t_0)}}.$$

This is the solution (1.13) given in the text.

13. The carrying capacity $K = 20,000$, and the initial condition $P_0 = 1000$, and it is given that $P(8) = 1200$. Using equation (1.13), one obtains

$$1200 = \frac{(20000)(1000)}{1000 + (19000)e^{-8r}}.$$

Solving we get

$$r = (-1/8)\ln(188000/((12)(19000)) \approx 0.0241.$$

Now, find t so that $P(t) = 3K/4 = 15000$. This gives $e^{-rt} = 1/57$. That is, $t = (\ln 57)/r \approx 167.6713$.

15. The data is shown in the following figure plotted as points. Using this we estimate $K = 9990$. To find r we use the solution to the logistic model

$$P(t) = \frac{KP_0}{P_0 + (K - P_0)e^{-rt}},$$

and solve for

$$r = \frac{1}{t}\ln\left(\frac{P(t)(K - P_0)}{P_0(K - P(t))}\right).$$

If we use $t = 60$, with $P(60) = 5510$, we get $r = 0.08$. The logistic solution with these parameters is the solid curve in the figure.

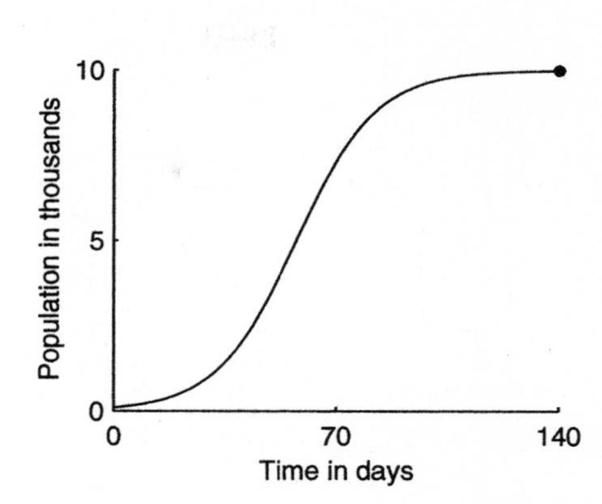

The fit seems to be quite good. However, this method depends on the choice of middle point. If that is an outlier the fit might not be so good as it was in this case.

17. (a) In the following figure, the five curves are given by initial conditions $P(0) = 410$, $P(0) = 414$, $P(0) = 415$, $P(0) = 420$, and $P(0) = 450$. Notice that the solution curve with initial condition $P(0) = 415$ never reaches zero, although the curve with initial condition $P(0) = 414$ does. Thus, the critical population is between 414 and 415.

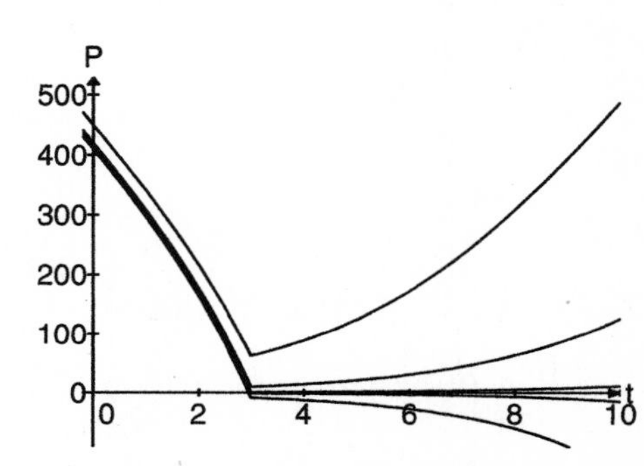

(b) The critical initial population will be the initial condition for a solution curve which has $P = 0$ when $t = 3$. Writing the differential equation for $t < 3$ gives

$$\frac{dP}{dt} = 0.38P - 0.00038P^2 - 200.$$

Separate variables.

$$\frac{dP}{-0.00038P^2 + 0.38P - 200} = dt.$$

Integrating this, one obtains

$$t + C = \frac{2}{\sqrt{.1596}} \tan^{-1}\left(\frac{-0.00076P + 0.38}{\sqrt{.1596}}\right).$$

Using conditions $P(3) = 0$, one obtains

$$3 + C = \frac{2}{\sqrt{.1596}} \tan^{-1}\left(\frac{0.38}{\sqrt{.1596}}\right)$$

or $C = 0.8067$. Now setting $t = 0$ and solving for $P = P_0$, one obtains $P_0 = 414.5557$.

19. (a) The harvesting rate is $0.01P$, so the modified model is

$$P' = 0.1P(1 - P/10) - 0.01P.$$

(b) The model equation simplifies to $P' = 0.01P(9 - P)$. This is another logistic equation. Thus $P = 0$ is an unstable equilibrium point, and $P = 9$ is an asymptotically stable equilibrium point, representing the carrying capacity.

(c) The population tends to the carrying capacity $P = 9$.

21. Refer to the answer to Exercise 19. The units of γP are individuals/ unit time. It represents the number of individuals harvested in a unit of time. For this reason it is called the yield. Notice that for values of $\gamma < r$ the model is another logistic equation with carrying capacity $K_\gamma = (r - \gamma)K/r$. For populations at the carrying capacity the yield is $\gamma K_\gamma = \gamma(r - \gamma)K/r$. This function of γ is maximized when $\gamma = r/2$. The carrying capacity for this value of γ is $K/2$ and the yield is $rK/4$.

Section 3.2. Models and the Real World

1. We have $P_0 = P(1790) = 3929827$, and $P(1800) = 5305925 = P_0e^{10r}$. Solving the last equation for r we get $r = \ln(5305925/3929827)/10 \approx 0.03$.

3. Answers will vary depending on the data chosen.

Section 3.3. Personal Finance

1. (a) Let $P(t)$ represent the balance in the account after t years. Because the initial deposit at time $t = 0$ is $P = \$1200$,

$$P' = 0.05P, \quad P(0) = 1200.$$

Separate the variables and solve for P, then use the initial condition to produce

$$P(t) = 1200e^{0.05t}.$$

Thus, the amount in the account after 10 years is

$$P(10) = 1200e^{0.05(10)} \approx \$1978.50.$$

(b) Solve $P(t) = 5000$, or

$$\begin{aligned} 1200e^{0.05t} &= 5000 \\ e^{(1/20)t} &= \frac{25}{6} \\ t &= 20\ln\frac{25}{6} \\ t &\approx 28.5\,\text{yr} \end{aligned}$$

3. Let $P(t)$ represent the balance in the account t years after the initial investment. Let r represent the annual rate, d the yearly deposit, and P_0 the initial investment. Then,

$$P' = rP + d, \quad P(0) = P_0.$$

This equation is linear, with integrating factor e^{-rt}. Consequently,

$$\begin{aligned} \left(e^{-rt}P\right)' &= de^{-rt}, \\ e^{-rt}P &= -\frac{d}{r}e^{-rt} + C, \\ P &= -\frac{d}{r} + Ce^{rt}. \end{aligned}$$

Use $P(0) = P_0$ to produce $C = P_0 + d/r$ and

$$P(t) = -\frac{d}{r} + \left(P_0 + \frac{d}{r}\right)e^{rt}.$$

Thus, the amount in the account after 10 years is

$$\begin{aligned} P(10) &= -\frac{1200}{0.06} + \left(5000 + \frac{1200}{0.06}\right)e^{0.06(10)} \\ &\approx \$25,553. \end{aligned}$$

5. Let $P(t)$ represent the balance in the account after t years. Let r represent the annual rate, w the yearly withdrawal, and P_0 the amount of the inheritance. Then

$$P' = rP - w \quad P(0) = P_0.$$

The equation is linear with integrating factor e^{-rt}. Consequently,

$$\begin{aligned} \left(e^{-rt}P\right)' &= -we^{-rt}, \\ e^{-rt}P &= \frac{w}{r}e^{-rt} + C, \\ P &= \frac{w}{r} + Ce^{rt}. \end{aligned}$$

Use $P(0) = P_0$ to produce $C = P_0 - w/r$ and

$$P(t) = \frac{w}{r} + \left(P_0 - \frac{w}{r}\right)e^{rt}.$$

Now, to find when the funds are depleted, set $P(t) = 0$.

$$\begin{aligned} 0 &= \frac{w}{r} + \left(P_0 - \frac{w}{r}\right)e^{rt}, \\ e^{rt} &= \frac{w/r}{w/r - P_0}, \\ t &= \frac{1}{r}\ln\frac{w/r}{w/r - P_0}. \end{aligned}$$

Thus, the account will be depleted in

$$t = \frac{1}{0.05}\ln\frac{8000/0.05}{8000/0.05 - 50000} \approx 7.5\,\text{years}.$$

7. Let $P(t)$ represent the loan balance after t years. Let r represent the annual rate, w the annual payment, and P_0 the amount of the loan. Then

$$P' = rP - w \quad P(0) = P_0.$$

The equation is linear with integrating factor e^{-rt}. Consequently,

$$\begin{aligned} \left(e^{-rt}P\right)' &= -we^{-rt}, \\ e^{-rt}P &= \frac{w}{r}e^{-rt} + C, \\ P &= \frac{w}{r} + Ce^{rt}. \end{aligned}$$

Use $P(0) = P_0$ to produce $C = P_0 - w/r$ and

$$P(t) = \frac{w}{r} + \left(P_0 - \frac{w}{r}\right)e^{rt}.$$

Now, the loan is exhausted at the end of 30 years. Consequently, $P(30) = 0$, so

$$\begin{aligned} 0 &= \frac{w}{r} + \left(P_0 - \frac{w}{r}\right)e^{r(30)}, \\ \frac{w}{r}\left(e^{30r} - 1\right) &= P_0e^{30r}, \\ w &= \frac{rP_0}{1 - e^{-30r}}, \\ w &= \frac{0.08(100000)}{1 - e^{-30(0.08)}} \\ w &\approx \$8,798.15 \end{aligned}$$

9. Let $S(t)$ represent José's salary after t years. Consequently,

$$S' = 0.01S, \quad S(0) = 28,$$

where S is measured in *thousands* of dollars. Therefore, José's salary is given by

$$S(t) = 28e^{0.01t}.$$

Let $P(t)$ represent the balance of the account after t years. Let ρ represent the fixed percentage of José's salary that is deposited in the account on an annual basis. Thus,

$$P' = 0.06P + 28\rho e^{0.01t}, \quad P(0) = 2.5,$$

where $P(t)$ is also measured in *thousands* of dollars. Multiply by the integrating factor, $e^{-0.06t}$, and integrate.

$$\begin{aligned} \left(e^{-0.06t}P\right)' &= 28\rho e^{-0.05t} \\ e^{-0.06t}P &= \frac{28}{-0.05}\rho e^{-0.05t} + C \\ P &= -560\rho e^{0.01t} + Ce^{0.06t} \end{aligned}$$

The initial condition $P(0) = 2.5$ produces $C = 2.5 + 560\rho$ and

$$P = -560\rho e^{0.01t} + (2.5 + 560\rho)e^{0.06t}.$$

Now, José wants $50,000 in the account at the end of 20 years, so $P(20) = 50$ and

$$\begin{aligned} 50 &= -560\rho e^{0.01(20)} + (2.5 + 560\rho)e^{0.06(20)}, \\ \rho &= \frac{50 - 2.5e^{1.2}}{560(e^{1.2} - e^{0.2})}, \\ \rho &\approx 0.035, \end{aligned}$$

or 3.5%.

11. (a) Let $P(n)$ represent the balance at the end of n compounding periods, I the annual interest rate, m the number of compounding periods per year, and P_0 the initial investment. Thus,

$$P(n+1) = \left(1 + \frac{I}{m}\right)P(n), \quad P(0) = P_0.$$

(b) Compare

$$P(n+1) = \left(1 + \frac{I}{m}\right)P(n), \quad P(0) = P_0.$$

with

$$a(n+1) = ra(n), \quad a(0) = a_0,$$

and note that $r = 1 + I/m$ and $a_0 = P_0$. Consequently,

$$a(n) = a_0 r^n$$

becomes

$$P(n) = P_0\left(1 + \frac{I}{m}\right)^n.$$

13. (a) The series $1 + r + r^2 + \cdots + r^{n-1}$ is geometric, finite, with sum $(1-r^n)/(1-r)$. Consequently,

$$\begin{aligned} a(n) &= a_0 r^n + b\left(1 + r + r^2 + \cdots + r^{n-1}\right), \\ a(n) &= a_0 r^n + b\left(\frac{1 - r^n}{1 - r}\right), \\ a(n) &= a_0 r^n + \frac{b}{1-r} - \frac{b}{1-r}r^n, \\ a(n) &= \left(a_0 - \frac{b}{1-r}\right)r^n + \frac{b}{1-r}. \end{aligned}$$

(b) Comparing

$$P(n+1) = \left(1 + \frac{I}{m}\right)P(n) + d, \quad P(0) = P_0.$$

with

$$a(n+1) = ra(n) + b, \quad a(0) = a_0$$

shows us that $r = 1 + I/m$, $b = d$, and $a_0 = P_0$. Consequently, the solution

$$a(n) = \left(a_0 - \frac{b}{1-r}\right)r^n + \frac{b}{1-r}$$

becomes

$$\begin{aligned} P(n) &= \left(P_0 - \frac{d}{1 - (1 + I/m)}\right)\left(1 + \frac{I}{m}\right)^n \\ &\quad + \frac{d}{1 - (1 + I/m)}, \\ P(n) &= \left(P_0 + \frac{md}{I}\right)\left(1 + \frac{I}{m}\right)^n - \frac{md}{I}. \end{aligned}$$

15. (a) Let $P(t)$ represent the balance of the loan after t years, I the annual interest rate, w the yearly payment, and P_0 the amount of the loan. Then,

$$P' = rP - w, \quad P(0) = P_0.$$

Multiply by the integrating factor, e^{-rt}, and integrate.

$$\begin{aligned}
\left(e^{-rt}P\right)' &= -we^{-rt} \\
e^{-rt}P &= \frac{w}{r}e^{-rt} + C \\
P &= \frac{w}{r} + Ce^{-rt}
\end{aligned}$$

The initial condition $P(0) = P_0$ gives $C = P_0 - w/r$ and

$$P = \frac{w}{r} + \left(P_0 - \frac{w}{r}\right)e^{-rt}.$$

Because the loan expires in 5 years, $P(5) = 0$ and

$$\begin{aligned}
0 &= \frac{w}{r} + \left(P_0 - \frac{w}{r}\right)e^{r(5)}, \\
w &= \frac{P_0 r}{1 - e^{-5r}}, \\
w &= \frac{(12000)(0.08)}{1 - e^{-5(0.08)}}, \\
w &\approx \$2,911.91,
\end{aligned}$$

or \$242.66 per month.

(b) Let $P(n)$ represent the balance on the loan after n compounding periods, I the annual rate, m the number of compounding periods per year, w the monthly payment, and P_0 the initial amount of the loan. Then, in a manner similar to that in Exercise 13, the progress of the loan is modeled by the first order difference equation

$$P(n+1) = \left(1 + \frac{I}{m}\right)P(n) - w, \quad P(0) = P_0,$$

with solution

$$P(n) = \left(P_0 - \frac{mw}{I}\right)\left(1 + \frac{I}{m}\right)^n + \frac{mw}{I}.$$

There are 60 compounding periods in 5 years, so $P(60) = 0$ and

$$\begin{aligned}
0 &= \left(P_0 - \frac{mw}{I}\right)\left(1 + \frac{I}{m}\right)^{60} + \frac{mw}{I}, \\
w &= \frac{P_0 I(1 + I/m)^{60}}{m((1 + I/m)^{60} - 1)}, \\
w &= \frac{P_0 I}{m(1 - (1 + I/m)^{-60})}, \\
w &= \frac{(12000)(0.08)}{12(1 - (1 + 0.08/12)^{-60})}, \\
w &\approx \$243.32.
\end{aligned}$$

Section 3.4. Electrical Circuits

1. The model equation is $Q' + Q/2 = 5$. The solution is $Q(t) = 10 - 10e^{-t/2}$.

3. The model equation is $Q' + Q/2 = 5\sin 2t$. The solution is

$$Q(t) = (-40\cos 2t + 10\sin 2t + 40e^{-t/2})/17.$$

5. The model equation is $Q' + Q/2 = 5 - t/20$. The solution is

$$Q(t) = 51(1 - e^{-t/2})/5 - t/10.$$

7. The model equation is $I' + I/10 = 1$. The solution is $I(t) = 10(1 - e^{-t/10})$.

9. The model equation is $I' + I/10 = 5\sin 2\pi t$. The solution is

$$I(t) = \frac{50(-20\pi\cos 2\pi t + 20\pi e^{-t/10} + \sin 2\pi t)}{1 + 400\pi^2}.$$

11. The model equation is $I' + I/10 = 10 - 2t$. The

solution is

$$I(t) = 300 - 20t - 300e^{-t/10}.$$

13. $Q(t) = EC(1 - e^{-t/RC})$

15. The model equation, $Q' + Q/20 = (1/2)e^{-t/100}$ with $Q(0) = 0$, has solution

$$Q(t) = 25(e^{-t/100} - e^{-t/20})/2.$$

We find the maximum by setting the derivative

$$Q'(t) = 25\left[\frac{1}{20}e^{-t/20} - \frac{1}{100}e^{-t/100}\right] \Big/ 2 = 0,$$

and solving for t. It has a maximum at $t = 25 \ln 5 \approx 40.24$. The maximum value is $Q(25 \ln 5) = 6.69$.

17. (a) 23.87

(b) The model equation, $Q' + Q/10 = 10$ with $Q(0) = 0$, has solution $Q = 100(1 - e^{-t/10})$ for $0 \le t \le 5$. Note that $Q(5) = Q_5 = 100(1 - e^{-1/2})$. Then for the next 5 seconds, the model equation is $Q' = -Q/10$ with $Q(5) = Q_5$. The solution to this equation is $Q = Q_5 e^{-(t-5)/10}$. At $t = 10$ seconds, the amount of charge is $Q = Q_5 e^{-1/2} \approx 23.87$.

19. The nine solutions are plotted below.

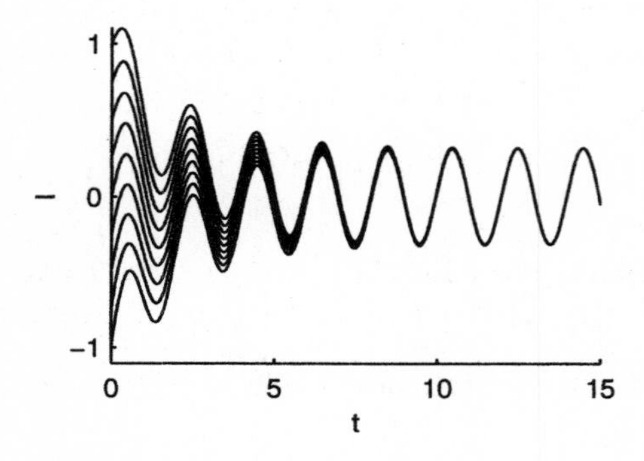

The frequency is of the force is $2\pi/2 = \pi$ Hz. This also appears to be the frequency of the steady-state response.

21. If we use $I = Q'$, we get the differential equation

$$Q' + \frac{1}{RC}Q = \frac{E}{C}\sin \omega t.$$

This linear equation can be solved using the integrating factor $e^{t/RC}$. We get the solution

$$Q(t) = \frac{EC}{1 + R^2C^2\omega^2}(\sin \omega t - RC\omega \cos \omega t) + Ae^{-t/RC},$$

where A is an arbitrary constant. The term $Ae^{-t/RC}$ dies out as t increases, so it is the transient term. The first term does not die out and is therefore the steady-state solution.

23. By the chain rule

$$\frac{dI}{dt} = \frac{dI}{dQ}\frac{dQ}{dt} = I\frac{dI}{dt}.$$

Hence the equation becomes

$$LI\frac{dI}{dQ} + Q/C = 0.$$

This separable equation has the solution $I = \sqrt{k - Q^2/LC}$ where k is a constant. Since $I = dQ/dt$, the equation now becomes

$$\frac{dQ}{dt} = \sqrt{k - Q^2/LC}.$$

This is another separable equation, and

$$\frac{dQ}{\sqrt{k - Q^2/LC}} = dt$$

The integral on the left involves an arcsine and the solution is

$$Q(t) = k_1\sqrt{LC}\sin\left(\frac{t}{\sqrt{LC}} + k_2\right).$$

Since $I = dQ/dt$, we also obtain

$$I(t) = k_1 \cos\left(\frac{t}{\sqrt{LC}} + k_2\right),$$

where k_1 and k_2 are constants.

Chapter 4. Second-Order Equations

Section 4.1. Definitions and Examples

1. Compare

$$y'' + 3y' + 5y = 3\cos 2t$$

with

$$y'' + p(t)y' + q(t)y = g(t),$$

and note that $p(t) = 3$, $q(t) = 5$, and $g(t) = 3\cos 2t$. Hence, the equation is linear and inhomogeneous.

3. Expand $t^2y'' + (1 - y)y' = \cos 2t$ to obtain

$$t^2y'' + y' - yy' = \cos 2t.$$

Note that the term yy' is nonlinear. Hence, this equation is nonlinear.

5. In

$$t^2y'' + 4yy' = 0,$$

note that the term $4yy'$ is nonlinear. Hence, this equation is nonlinear.

7. In

$$y'' + 3y' + 4\sin y = 0$$

note that the term $4\sin y$ is nonlinear. Hence, this equation is nonlinear.

9. Use $k = (mg)/x$ to determine the spring constant.

$$k = \frac{2\,\text{kg} \times 9.8\,\text{m/s}^2}{0.5\,\text{m}} = 39.2\,\text{N/m}.$$

Using the model

$$my'' + \mu y' + ky = F(t),$$

we note that there is no damping ($\mu = 0$) and there is no driving force ($F(t) = 0$), so the equation becomes

$$my'' + ky = 0.$$

With $m = 2$ kg and $k = 39.2$ N/m, this becomes

$$2y'' + 39.2y = 0.$$

Because the initial displacement was 12 cm, $y(0) = 0.12$ (assuming an orientation where y is positive in the downward direction). Because the mass is released from rest, the initial velocity is zero and $y'(0) = 0$.

11. The period of the driving force is 2 seconds. Thus, the circular frequency is

$$\omega = \frac{2\pi}{T} = \frac{2\pi}{2} = \pi \text{ rad/s}.$$

Because the amplitude is $A = 0.5$ m, and the spring is initially displaced 0.5 m upward (remember, upward is negative) by the driving force, the driving force can be described with

$$F(t) = -0.5\cos \pi t$$

Now, $m = 2$ kg, $k = 39.2$ N/m, and the damping force is given by $R(v) = -0.05v$. This makes the damping constant $\mu = 0.05$. Thus, the equation

$$my'' + \mu y' + ky = F(t)$$

becomes

$$2y'' + 0.05y' + 39.2y = -0.5\cos \pi t$$

From Exercise 9, the initial conditions are $y(0) = 0.12$ and $y'(0) = 0$.

13. If $y_1(t) = e^{3t}$, then

$$\begin{aligned} y_1'' - y_1' - 6y_1 &= (e^{3t})'' - (e^{3t})' - 6(e^{3t}) \\ &= 9e^{3t} - 3e^{3t} - 6e^{3t} \\ &= 0. \end{aligned}$$

If $y_2(t) = e^{-2t}$, then

$$\begin{aligned} y_2'' - y_2' - 6y_2 &= (e^{-2t})'' - (e^{-2t})' - 6(e^{-2t}) \\ &= 4e^{-2t} + 2e^{-2t} - 6e^{-2t} \\ &= 0. \end{aligned}$$

Finally, if $y(t) = C_1e^{3t} + C_2e^{-2t}$, then

$$\begin{aligned} & y'' - y' - 6y \\ &= (C_1e^{3t} + C_2e^{-2t})'' - (C_1e^{3t} + C_2e^{-2t})' \\ &\quad - 6(C_1e^{3t} + C_2e^{-2t}) \\ &= 9C_1e^{3t} + 4C_2e^{-2t} - 3C_1e^{3t} + 2C_2e^{-2t} \\ &\quad - 6C_1e^{3t} - 6C_2e^{-2t} \\ &= 0. \end{aligned}$$

15. If $y_1(t) = e^t \cos t$, then

$$\begin{aligned} y_1'(t) &= e^t(-\sin t) + e^t \cos t \\ &= e^t(\cos t - \sin t), \\ y_1''(t) &= e^t(-\sin t - \cos t) + e^t(\cos t - \sin t) \\ &= -2e^t \sin t, \end{aligned}$$

and

$$\begin{aligned} & y_1'' - 2y_1' + 2y_1 \\ &= -2e^t \sin t - 2e^t(\cos t - \sin t) + 2e^t \cos t \\ &= 0. \end{aligned}$$

If $y_2(t) = e^t \sin t$, then

$$\begin{aligned} y_2'(t) &= e^t \cos t + e^t \sin t \\ &= e^t(\cos t + \sin t), \\ y_2''(t) &= e^t(-\sin t + \cos t) + e^t(\cos t + \sin t) \\ &= 2e^t \cos t, \end{aligned}$$

and

$$\begin{aligned} & y_2'' - 2y_2' + 2y_2 \\ &= 2e^t \cos t - 2e^t(\cos t + \sin t) + 2e^t \sin t \\ &= 0. \end{aligned}$$

Finally, if $y(t) = C_1e^t \cos t + C_2e^t \sin t$, or if $y(t) = e^t(C_1 \cos t + C_2 \sin t)$, then

$$\begin{aligned} y'(t) &= e^t(-C_1 \sin t + C_2 \cos t) \\ &\quad + e^t(C_1 \cos t + C_2 \sin t) \\ &= e^t((C_1 + C_2)\cos t + (-C_1 + C_2)\sin t), \\ y''(t) &= e^t((C_1 + C_2)(-\sin t) + (-C_1 + C_2)\cos t) \\ &\quad + e^t((C_1 + C_2)\cos t + (-C_1 + C_2)\sin t) \\ &= 2e^t(C_2 \cos t - C_1 \sin t), \end{aligned}$$

and

$$\begin{aligned} & y'' - 2y' + 2y \\ &= 2e^t(C_2 \cos t - C_1 \sin t) \\ &\quad - 2e^t((C_1 + C_2)\cos t + (-C_1 + C_2)\sin t) \\ &\quad + 2e^t(C_1 \cos t + C_2 \sin t) \\ &= 0. \end{aligned}$$

17. We leave it to our readers to first check that $y_1(t) = e^{-t}$ and $y_2(t) = e^{2t}$ are solutions of $y'' - y' - 2y = 0$. Next note that

$$\frac{y_1(t)}{y_2(t)} = \frac{e^{-t}}{e^{2t}} = e^{-3t} \neq c,$$

where c is some constant. Therefore $y_1(t)$ is not a constant multiple of $y_2(t)$ and the solutions are linearly independent. Further,

$$W(t) = \begin{vmatrix} y_1(t) & y_2(t) \\ y_1'(t) & y_2'(t) \end{vmatrix} = \begin{vmatrix} e^{-t} & e^{2t} \\ -e^{-t} & 2e^{2t} \end{vmatrix} = 3e^t,$$

which is never equal to zero. Therefore, the solutions $y_1(t)$ and $y_2(t)$ are linearly independent and form a fundamental set of solutions.

19. We leave it to our readers to first check that $y_1(t) = e^{-2t} \cos 3t$ and $y_2(t) = e^{-2t} \sin 3t$ are solutions of $y'' + 4y' + 13y = 0$. Next note that

$$\frac{y_1(t)}{y_2(t)} = \frac{e^{-2t} \cos 3t}{e^{-2t} \sin 3t} = \cot 3t \neq c,$$

where c is some constant. Therefore $y_1(t)$ is not a constant multiple of $y_2(t)$ and the solutions are lin-

early independent. Further,

$$\begin{aligned} W(t) &= \begin{vmatrix} y_1(t) & y_2(t) \\ y_1'(t) & y_2'(t) \end{vmatrix} = y_1(t)y_2'(t) - y_1'(t)y_2(t) \\ &= e^{-2t}\cos 3t \times e^{-2t}(3\cos 3t - 2\sin 3t) \\ &\quad - e^{-2t}\sin 3t \times e^{-2t}(-3\sin 3t - 2\cos 3t) \\ &= e^{-4t}(3\cos^2 3t - 2\cos 3t\sin 3t) \\ &= 3e^{-4t}, \end{aligned}$$

which is never equal to zero. Therefore, the solutions $y_1(t)$ and $y_2(t)$ are linearly independent and form a fundamental set of solutions.

21. If $y_1(t) = t^2$ and $y_2(t) = t|t|$, then

$$\frac{y_1(t)}{y_2(t)} = \frac{t^2}{t|t|} = \begin{cases} -1, & t < 0, \\ 1, & t > 0. \end{cases}$$

Thus, $y_1(t)$ is not a constant multiple of $y_2(t)$ on $(-\infty, +\infty)$. However, the Wronskian

$$\begin{aligned} W(t) &= \begin{vmatrix} y_1(t) & y_2(t) \\ y_1'(t) & y_2'(t) \end{vmatrix} \\ &= \begin{vmatrix} t^2 & t|t| \\ 2t & 2|t| \end{vmatrix} \\ &= 2t^2|t| - 2t^2|t| \\ &= 0. \end{aligned}$$

Thus, the Wronskian is identically zero on $(-\infty, +\infty)$, seemingly contradicting Proposition 1.27, that is, until one realizes that the hypothesis that requires that y_1 and y_2 are solutions of the differential equation $y'' + p(t)y' + q(t)y = 0$ is unsatisfied.

23. If $y_1(t) = \cos 4t$, then

$$y_1'' + 16y_1 = -16\cos 4t + 16\cos 4t = 0,$$

and if $y_2(t) = \sin 4t$, then

$$y_2'' + 16y_2 = -16\sin 4t + 16\sin 4t = 0.$$

Furthermore,

$$\frac{y_1(t)}{y_2(t)} = \frac{\cos 4t}{\sin 4t} = \cot 4t,$$

which is nonconstant. Thus, y_1 is not a constant multiple of y_2 and the solutions $y_1(t) = \cos 4t$ and $y_2(t) = \sin 4t$ form a fundamental set of solutions. Thus, the general solution of $y'' + 16y = 0$ is

$$y(t) = C_1\cos 4t + C_2\sin 4t,$$

and its derivative is

$$y'(t) = -4C_1\sin 4t + 4C_2\cos 4t.$$

The initial conditions, $y(0) = 2$ and $y'(0) = -1$ lead to the equations

$$\begin{aligned} 2 &= C_1 \\ -1 &= 4C_2 \end{aligned}$$

and the constants $C_1 = 2$ and $C_2 = -1/4$. Thus, the solution of the initial value problem is

$$y(t) = 2\cos 4t - \frac{1}{4}\sin 4t.$$

25. If $y_1(t) = e^{-4t}$, then

$$y_1'' + 8y_1' + 16y_1 = 16e^{-4t} + 8(-4e^{-4t}) + 16e^{-4t} = 0.$$

Thus, y_1 is a solution. If $y_2(t) = te^{-4t}$, then

$$\begin{aligned} y_2'(t) &= e^{-4t}(1 - 4t), \quad \text{and} \\ y_2''(t) &= e^{-4t}(-8 + 16t). \end{aligned}$$

Thus,

$$\begin{aligned} &y_2'' + 8y_2' + 16y_2 \\ &\quad = e^{-4t}(-8 + 16t) + 8e^{-4t}(1 - 4t) + 16te^{-4t} \\ &\quad = 0. \end{aligned}$$

Thus, y_2 is a solution. Because

$$\frac{y_1(t)}{y_2(t)} = \frac{e^{-4t}}{te^{-4t}} = \frac{1}{t}$$

is nonconstant, y_1 and y_2 are independent and form a fundamental set of solutions. Thus, the general solution is

$$y(t) = C_1e^{-4t} + C_2te^{-4t}.$$

Substituting the initial condition $y(0) = 2$ provides $C_1 = 2$. After a bit of work, the derivative of the general solution is

$$y'(t) = e^{-4t}((C_2 - 4C_1) - 4C_2t).$$

Substituting the initial condition $y'(0) = -1$ gives

$$-1 = C_2 - 4C_1,$$

and the fact that $C_1 = 2$ provides us with

$$\begin{aligned} -1 &= C_2 - 8 \\ C_2 &= 7. \end{aligned}$$

Thus, the solution is

$$y(t) = 2e^{-4t} + 7te^{-4t}.$$

27. If $y_1(t) = t$, then

$$t^2y_1'' - 2ty_1' + 2y_1 = t^2(0) - 2t(1) + 2t = 0.$$

Thus, $y_1(t) = t$ is a solution. Let $y_2(t) = v(t)y_1(t) = v(t)t$, where v is to be determined. Then,

$$\begin{aligned} y_2 &= tv \\ y_2' &= tv' + v \\ y_2'' &= tv'' + 2v'. \end{aligned}$$

Substituting these into the differential equation,

$$\begin{aligned} 0 &= t^2y_2'' - 2ty_2' + 2y_2 \\ &= t^2(tv'' + 2v') - 2t(tv' + v) + 2tv \\ &= t^3v''. \end{aligned}$$

Because t^3 is not identically equal to zero, $v'' = 0$ and

$$\begin{aligned} v' &= k_1 \\ v &= k_1t + k_2. \end{aligned}$$

Choose $k_1 = 1$ and $k_2 = 0$, which gives $v(t) = t$ and $y_2(t) = v(t)y_1(t) = tt = t^2$. Checking, we see that

$$t^2y_2'' - 2ty_2' + 2y_2 = t^2(2) - 2t(2t) + 2(t^2) = 0,$$

making $y_2(t) = t^2$ a solution. Note that

$$\frac{y_1(t)}{y_2(t)} = \frac{t}{t^2} = \frac{1}{t}$$

is nonconstant, so y_1 and y_2 are independent and form a fundamental set of solutions. Thus, the general solution is

$$y(t) = C_1t + C_2t^2.$$

29. If $y_1(t) = t$, then

$$t^2y_1'' - 3ty_1' + 3y_1 = t^2(0) - 3t(1) + 3t = 0,$$

making $y_1(t) = t$ a solution. Let $y_2(t) = v(t)y_1(t) = tv(t)$, with v to be determined. Thus,

$$\begin{aligned} y_2 &= tv \\ y_2' &= tv' + v \\ y_2'' &= 2v' + tv''. \end{aligned}$$

Substituting these into the differential equation,

$$\begin{aligned} 0 &= t^2y_2'' - 3ty_2' + 3y_2 \\ &= t^2(2v' + tv'') - 3t(tv' + v) + 3tv \\ &= t^2(tv'' - v'). \end{aligned}$$

Because t^2 is not identically zero,

$$\begin{aligned} tv'' - v' &= 0 \\ \frac{v''}{v'} &= \frac{1}{t} \\ \ln v' &= \ln t \\ v' &= t \\ v &= \frac{1}{2}t^2. \end{aligned}$$

Thus,

$$y_2(t) = v(t)y_1(t) = \frac{1}{2}t^2(t) = \frac{1}{2}t^3.$$

Readers should check that this is a solution of the differential equation. Further,

$$\frac{y_1(t)}{y_2(t)} = \frac{t}{(1/2)t^3} = \frac{2}{t^2}$$

is nonconstant, so y_1 and y_2 are independent and form a fundamental set of solutions. Thus, the general solution is

$$y(t) = C_1t + C_2t^3,$$

where we have absorbed the $1/2$ into the arbitrary constant C_2.

Section 4.2. Second Order Equations and Systems

1. First, solve the differential equation $y''+2y'-3y=0$ for the highest derivative of y present in the equation.

$$y''=-2y'+3y$$

Next, set $v=y'$. Then

$$v'=y''=-2y'+3y=-2v+3y.$$

Thus, we now have the following system of first order equations.

$$\begin{aligned} y' &= v \\ v' &= -2v+3y \end{aligned}$$

3. First, solve the differential equation $y''+3y'+4y=2\cos 2t$ for the highest derivative of y present in the equation.

$$y''=-3y'-4y+2\cos 2t$$

Next, set $v=y'$. Then

$$\begin{aligned} v' = y'' &= -3y'-4y+2\cos 2t \\ &= -3v-4y+2\cos 2t. \end{aligned}$$

Thus, we now have the following system of first order equations.

$$\begin{aligned} y' &= v \\ v' &= -3v-4y+2\cos 2t \end{aligned}$$

5. First, solve the differential equation $y''+\mu(t^2-1)y'+y=0$ for the highest derivative of y present in the equation.

$$y''=-\mu(t^2-1)y'-y$$

Next, set $v=y'$. Then

$$v'=y''=-\mu(t^2-1)y'-y=-\mu(t^2-1)v-y.$$

Thus, we now have the following system of first order equations.

$$\begin{aligned} y' &= v \\ v' &= -\mu(t^2-1)v-y \end{aligned}$$

7. First, solve the differential equation $LQ''+RQ'+(1/C)Q=E(t)$ for the highest derivative of Q present in the equation.

$$Q''=-\frac{R}{L}Q'-\frac{1}{LC}Q+\frac{1}{L}E(t)$$

Next, set $I=Q'$. Then

$$\begin{aligned} I' = Q'' &= -\frac{R}{L}Q'-\frac{1}{LC}Q+\frac{1}{L}E(t) \\ &= -\frac{R}{L}I-\frac{1}{LC}Q+\frac{1}{L}E(t). \end{aligned}$$

Thus, we now have the following system of first order equations.

$$\begin{aligned} Q' &= I \\ I' &= -\frac{R}{L}I-\frac{1}{LC}Q+\frac{1}{L}E(t) \end{aligned}$$

9. Our solver requires that we first change the second order equation $my''+\mu y'+ky=0$ into a system of two first order equations. If we let $y'=v$, then

$$v'=y''=-\frac{\mu}{m}y'-\frac{k}{m}y=-\frac{\mu}{m}v-\frac{k}{m}y,$$

which leads to the following system of first order equations.

$$\begin{aligned} y' &= v \\ v' &= -\frac{\mu}{m}v-\frac{k}{m}y \end{aligned}$$

Now, if $m=1\,\text{kg}$, $\mu=0\,\text{kg/s}$, $k=4\,\text{kg/s}^2$, $y(0)=-2\,\text{m}$, and $y'(0)=-2\,\text{m/s}$, then

$$\begin{aligned} y' &= v \\ v' &= -4y, \end{aligned}$$

and

$$\begin{aligned} y(0) &= -2 \\ v(0) &= y'(0) = -2. \end{aligned}$$

Entering this system in our solver, we generate plots of position versus time and velocity versus time.

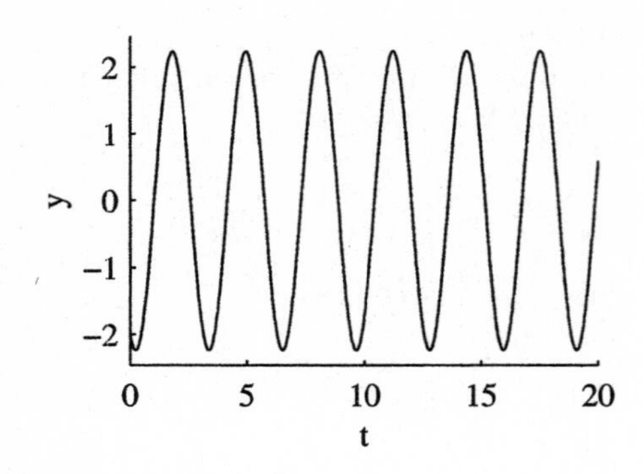

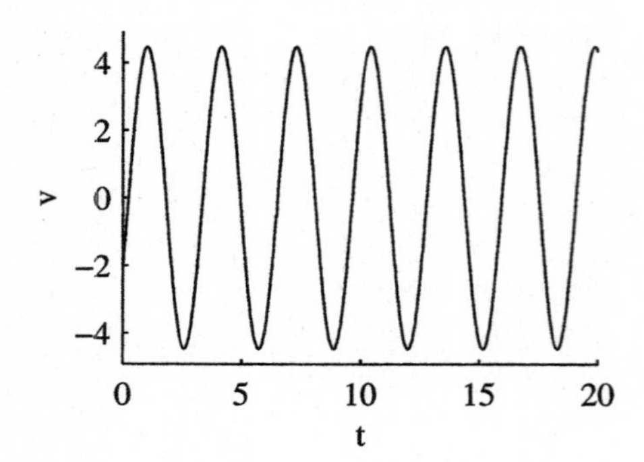

Our solver also generates a combined plot of both position and velocity versus time, and a plot of velocity versus position in the phase plane.

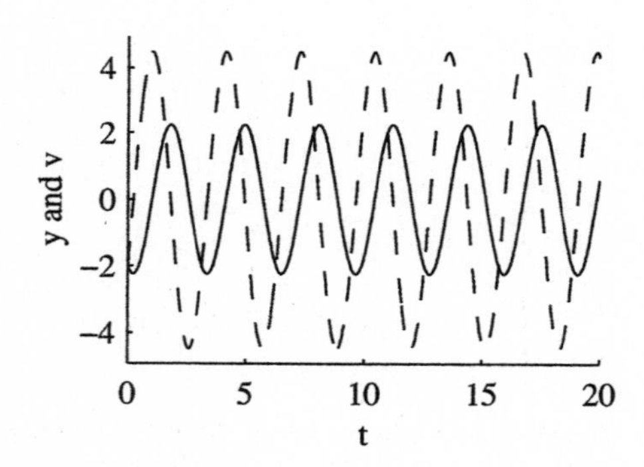

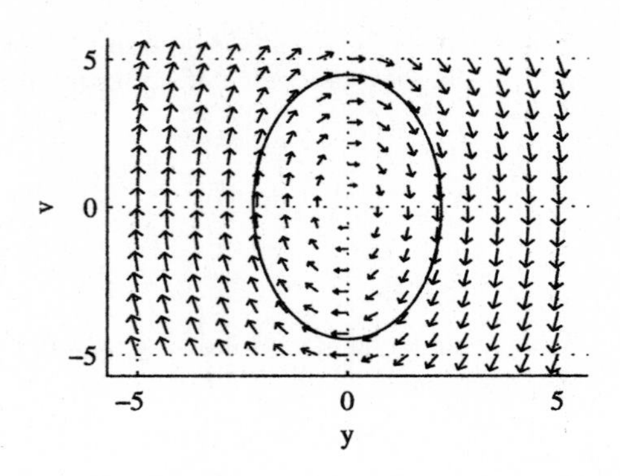

11. Our solver requires that we first change the second order equation $my'' + \mu y' + ky = 0$ into a system of two first order equations. If we let $y' = v$, then

$$v' = y'' = -\frac{\mu}{m}y' - \frac{k}{m}y = -\frac{\mu}{m}v - \frac{k}{m}y,$$

which leads to the following system of first order equations.

$$\begin{aligned} y' &= v \\ v' &= -\frac{\mu}{m}v - \frac{k}{m}y \end{aligned}$$

Now, if $m = 1\,\text{kg}$, $\mu = 2\,\text{kg/s}$, $k = 1\,\text{kg/s}^2$, $y(0) = -3\,\text{m}$, and $y'(0) = -2\,\text{m/s}$, then

$$\begin{aligned} y' &= v \\ v' &= -2v - y, \end{aligned}$$

and

$$\begin{aligned} y(0) &= -3 \\ v(0) &= y'(0) = -2. \end{aligned}$$

Entering this system in our solver, we generate plots of position versus time and velocity versus time.

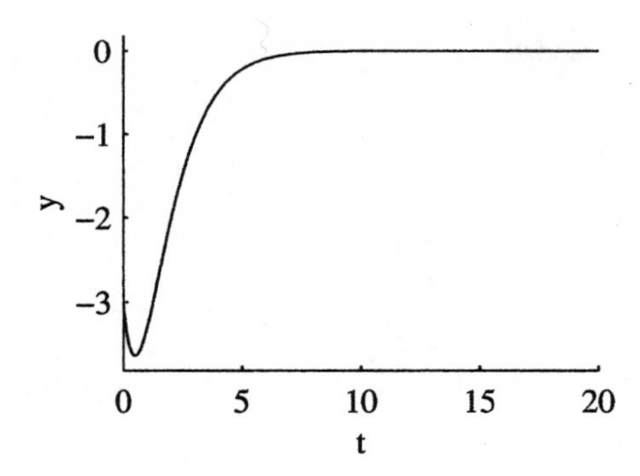

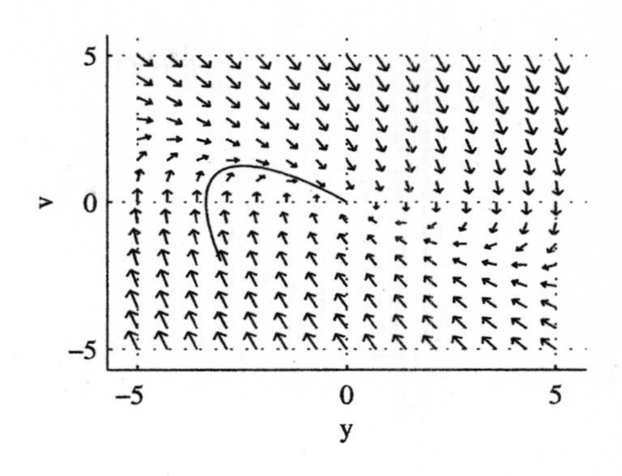

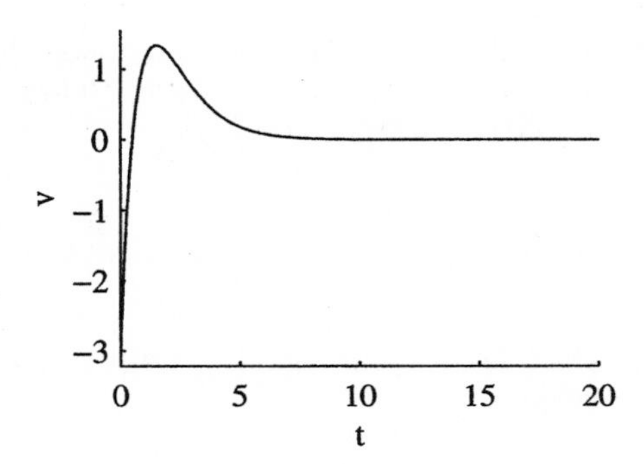

Our solver also generates a combined plot of both position and velocity versus time, and a plot of velocity versus position in the phase plane.

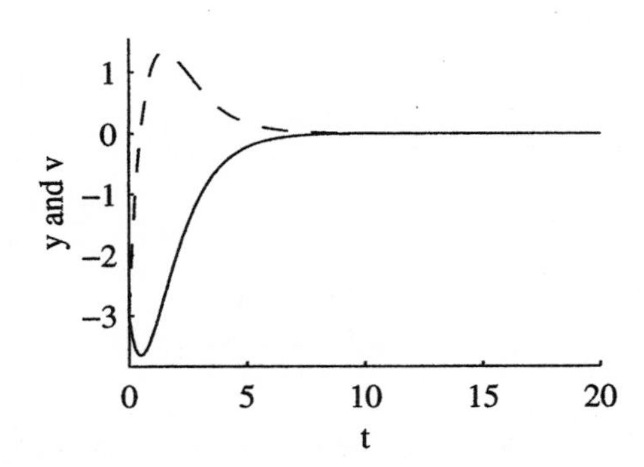

13. Our solver requires that we first change the second order equation $my'' + \mu y' + ky = 0$ into a system of two first order equations. If we let $y' = v$, then

$$v' = y'' = -\frac{\mu}{m}y' - \frac{k}{m}y = -\frac{\mu}{m}v - \frac{k}{m}y,$$

which leads to the following system of first order equations.

$$\begin{aligned} y' &= v \\ v' &= -\frac{\mu}{m}v - \frac{k}{m}y \end{aligned}$$

Now, if $m = 1\,\text{kg}$, $\mu = 0.5\,\text{kg/s}$, $k = 4\,\text{kg/s}^2$, $y(0) = 2\,\text{m}$, and $y'(0) = 0\,\text{m/s}$, then

$$\begin{aligned} y' &= v \\ v' &= -0.5v - 4y, \end{aligned}$$

and

$$\begin{aligned} y(0) &= 2 \\ v(0) &= y'(0) = 0. \end{aligned}$$

Entering this system in our solver, we generate plots of position versus time and velocity versus time.

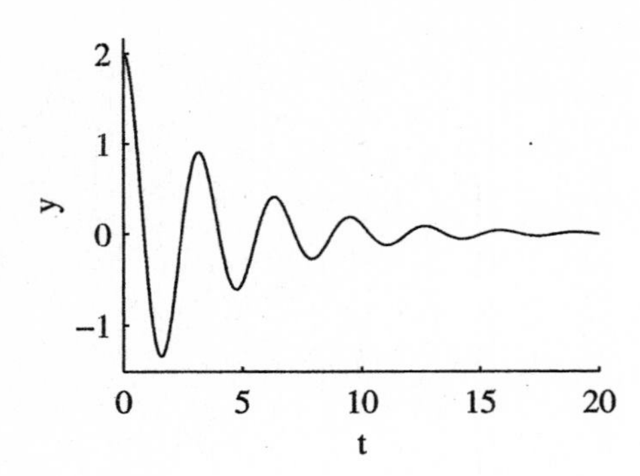

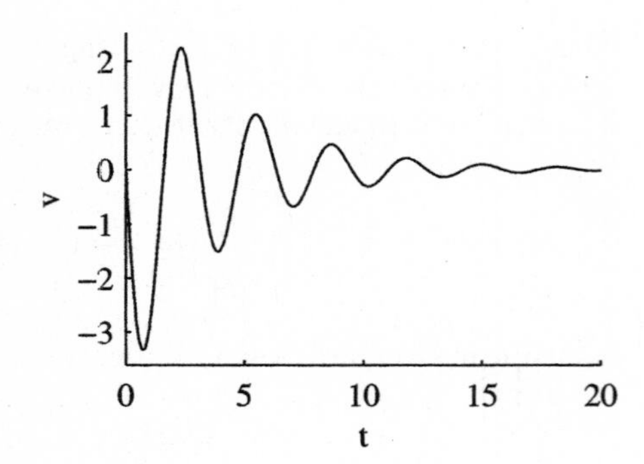

Our solver also generates a combined plot of both position and velocity versus time, and a plot of velocity versus position in the phase plane.

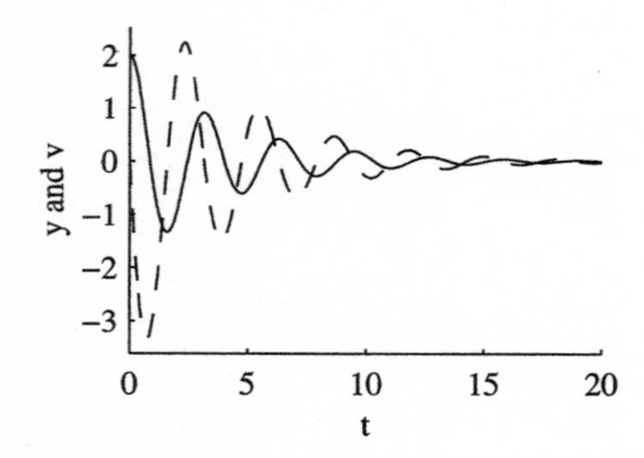

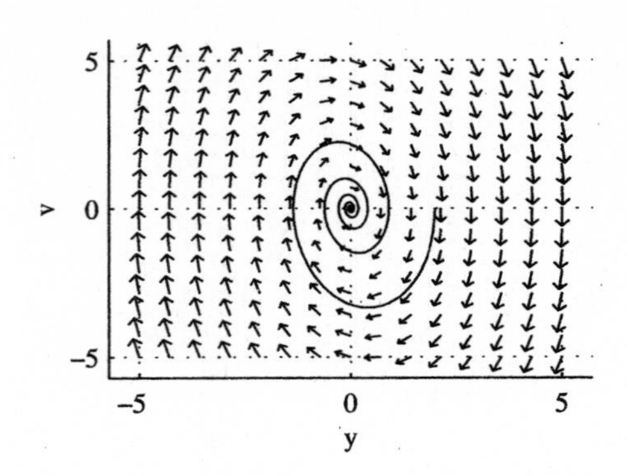

15. Our solver requires that we first change the second order equation $my'' + \mu y' + ky = 0$ into a system of two first order equations. If we let $y' = v$, then

$$v' = y'' = -\frac{\mu}{m}y' - \frac{k}{m}y = -\frac{\mu}{m}v - \frac{k}{m}y,$$

which leads to the following system of first order equations.

$$\begin{aligned} y' &= v \\ v' &= -\frac{\mu}{m}v - \frac{k}{m}y \end{aligned}$$

Now, if $m = 1\,\text{kg}$, $\mu = 3\,\text{kg/s}$, $k = 1\,\text{kg/s}^2$, $y(0) = -1\,\text{m}$, and $y'(0) = -5\,\text{m/s}$, then

$$\begin{aligned} y' &= v \\ v' &= -3v - y, \end{aligned}$$

and

$$\begin{aligned} y(0) &= -1 \\ v(0) &= y'(0) = -5. \end{aligned}$$

Entering this system in our solver, we generate plots of position versus time and velocity versus time.

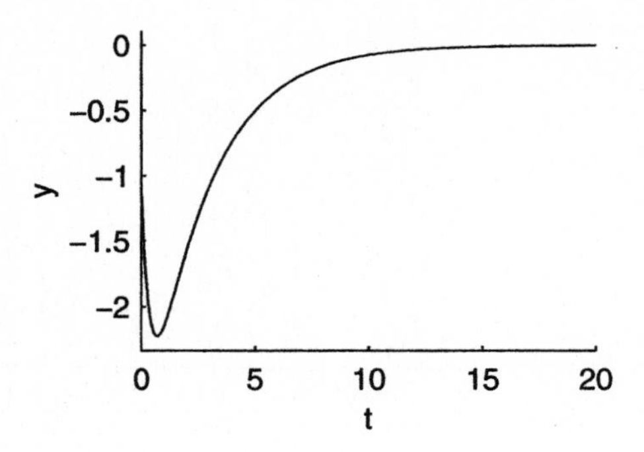

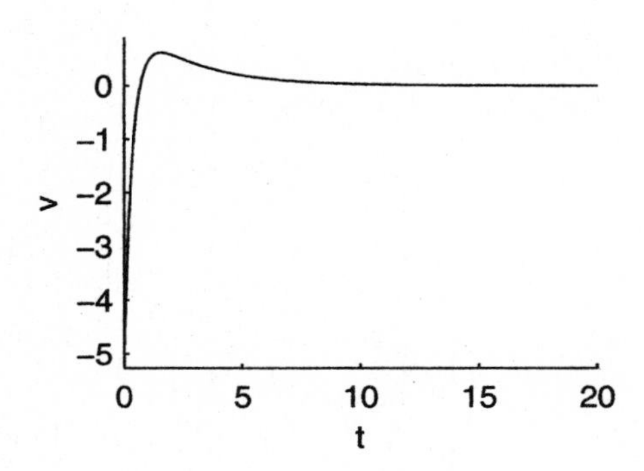

Our solver also generates a combined plot of both position and velocity versus time, and a plot of velocity versus position in the phase plane.

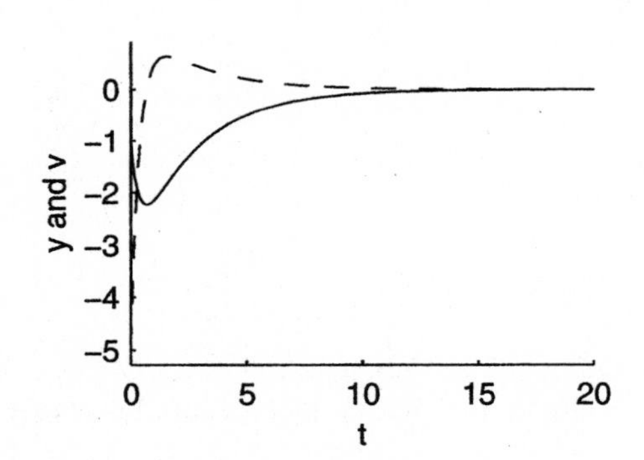

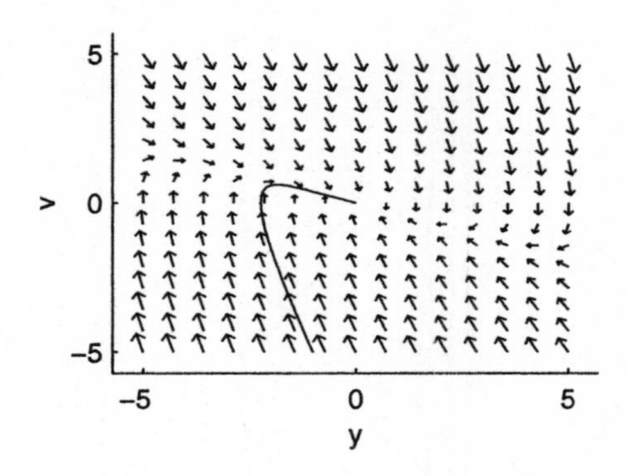

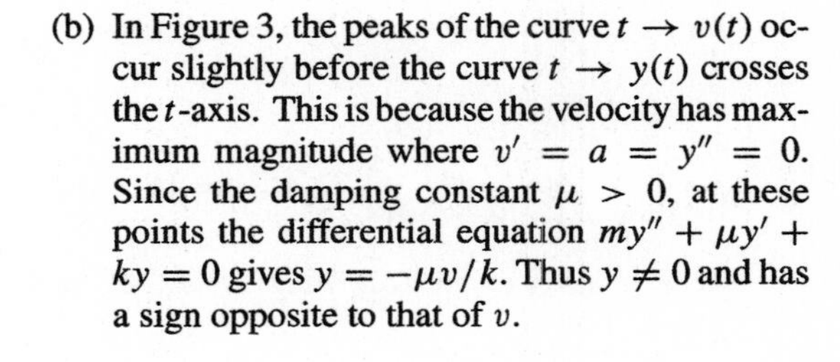

17. (a) In Figure 3, the peaks of the curve $t \to y(t)$ occur where $y' = v = 0$, and these are the points where the curve $t \to v(t)$ crosses the t-axis.

 (b) In Figure 3, the peaks of the curve $t \to v(t)$ occur slightly before the curve $t \to y(t)$ crosses the t-axis. This is because the velocity has maximum magnitude where $v' = a = y'' = 0$. Since the damping constant $\mu > 0$, at these points the differential equation $my'' + \mu y' + ky = 0$ gives $y = -\mu v/k$. Thus $y \neq 0$ and has a sign opposite to that of v.

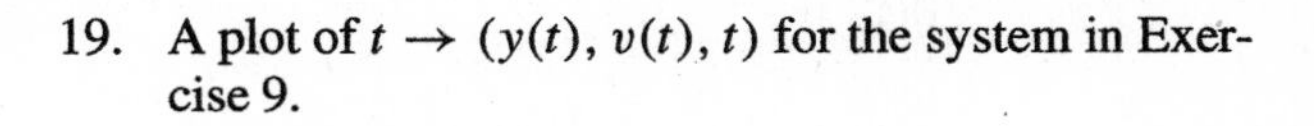

19. A plot of $t \to (y(t), v(t), t)$ for the system in Exercise 9.

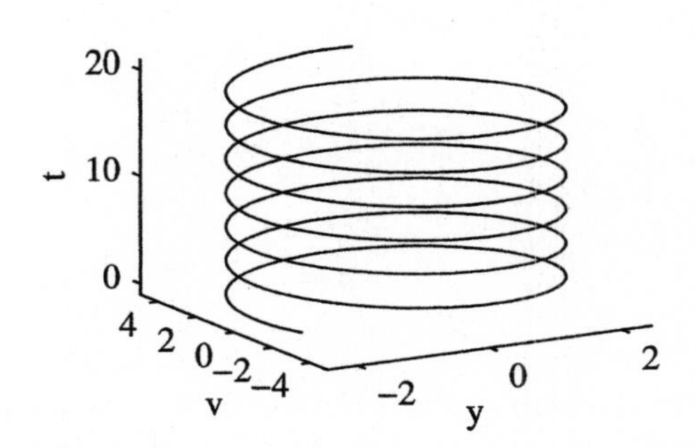

21. A plot of $t \to (y(t), v(t), t)$ for the system in Exercise 15.

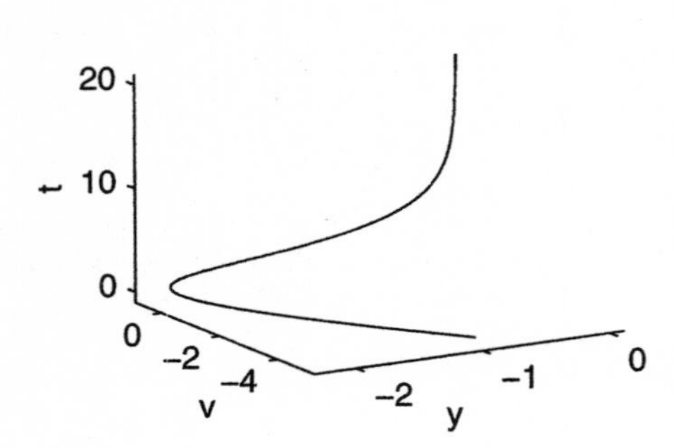

23. Our solver requires that we first change the second order equation $LQ'' + RQ' + (1/C)Q = 2\cos 2t$ into a system of two first order equations. If we let $Q' = I$, then

$$\begin{aligned} I' = Q'' &= -\frac{R}{L}Q' - \frac{1}{LC}Q + 2\cos 2t \\ &= -\frac{R}{L}I - \frac{1}{LC}Q + 2\cos 2t, \end{aligned}$$

which leads to the following system of first order equations.

$$\begin{aligned} Q' &= I \\ I' &= -\frac{R}{L}I - \frac{1}{LC}Q + 2\cos 2t \end{aligned}$$

Now, if $L = 1\,\text{H}$, $R = 0\,\Omega$, $C = 1\,\text{F}$, $Q(0) = -3\,\text{C}$, and $I(0) = -2\,\text{A}$, then

$$\begin{aligned} Q' &= I \\ I' &= -Q + 2\cos 2t, \end{aligned}$$

and

$$\begin{aligned} Q(0) &= -3 \\ I(0) &= -2. \end{aligned}$$

Entering this system in our solver, we generate plots of charge versus time and current versus time.

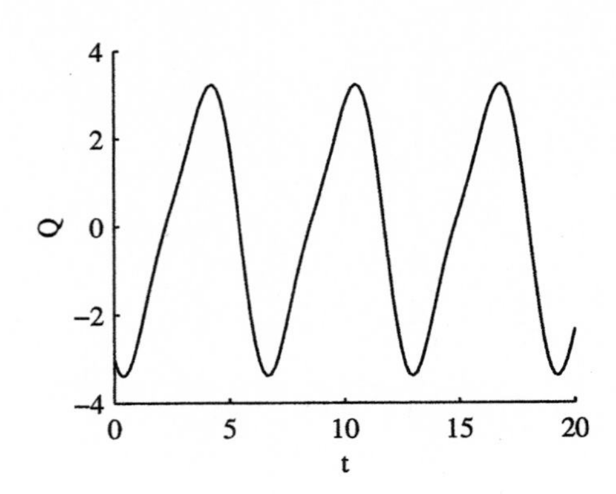

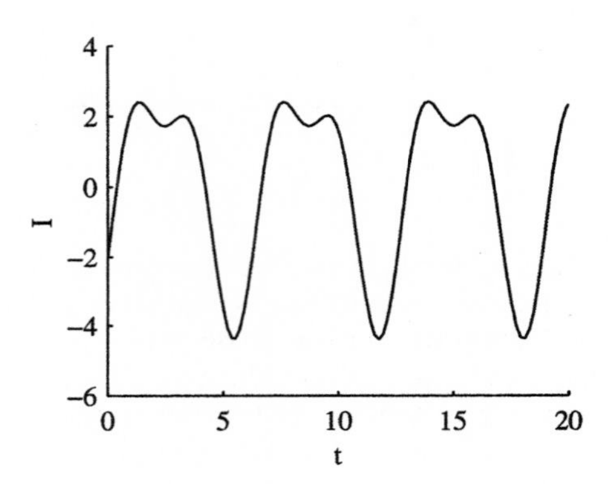

Our solver also generates a combined plot of both charge and current versus time, and a plot of current versus charge in the phase plane.

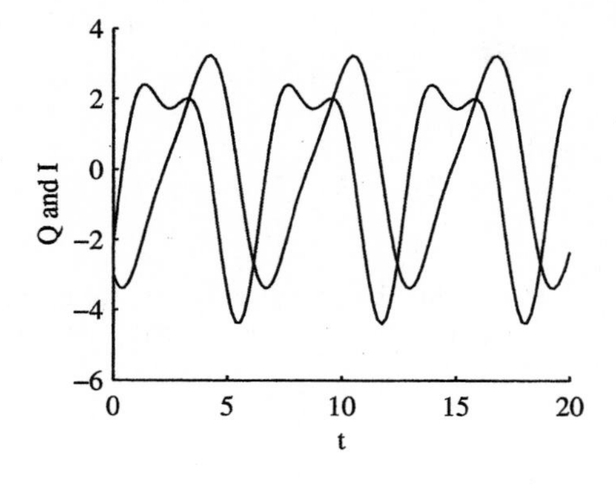

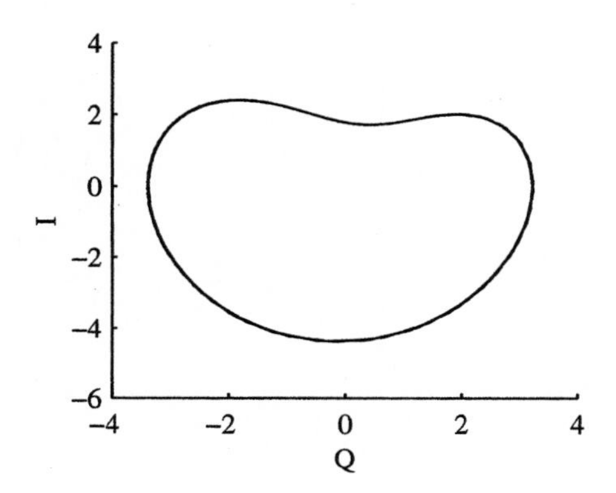

25. Our solver requires that we first change the second order equation $LQ'' + RQ' + (1/C)Q = 2\cos 2t$ into a system of two first order equations. If we let $Q' = I$, then

$$\begin{aligned} I' = Q'' &= -\frac{R}{L}Q' - \frac{1}{LC}Q + 2\cos 2t \\ &= -\frac{R}{L}I - \frac{1}{LC}Q + 2\cos 2t, \end{aligned}$$

which leads to the following system of first order equations.

$$\begin{aligned} Q' &= I \\ I' &= -\frac{R}{L}I - \frac{1}{LC}Q + 2\cos 2t \end{aligned}$$

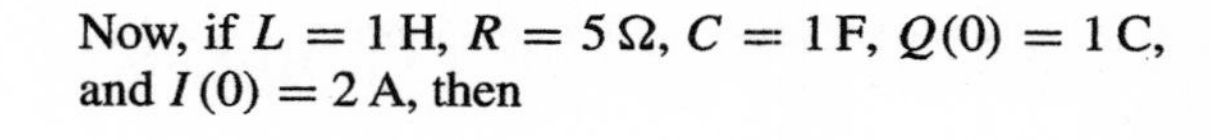

Now, if $L = 1\,\text{H}$, $R = 5\,\Omega$, $C = 1\,\text{F}$, $Q(0) = 1\,\text{C}$, and $I(0) = 2\,\text{A}$, then

$$\begin{aligned} Q' &= I \\ I' &= -5I - Q + 2\cos 2t, \end{aligned}$$

and

$$\begin{aligned} Q(0) &= 1 \\ I(0) &= 2. \end{aligned}$$

Entering this system in our solver, we generate plots of charge versus time and current versus time.

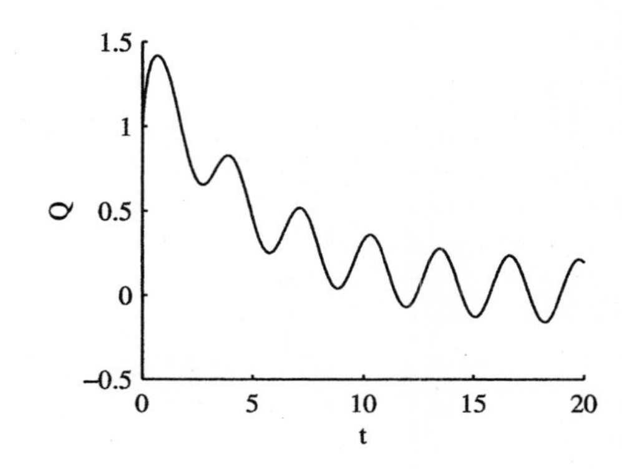

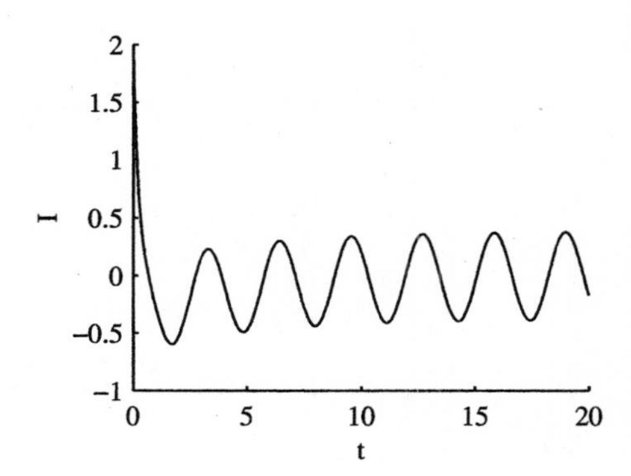

Our solver also generates a combined plot of both charge and current versus time, and a plot of current versus charge in the phase plane.

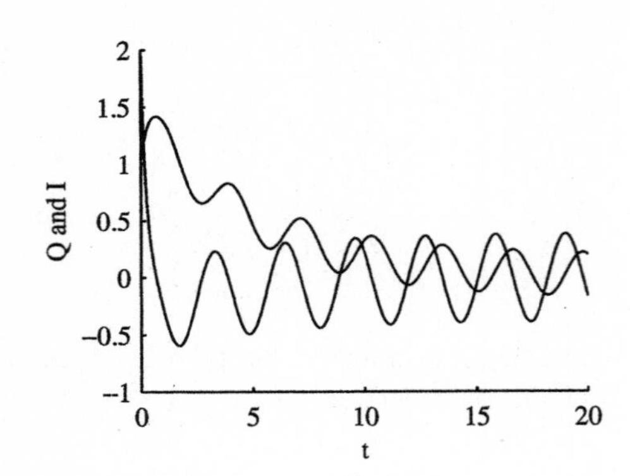

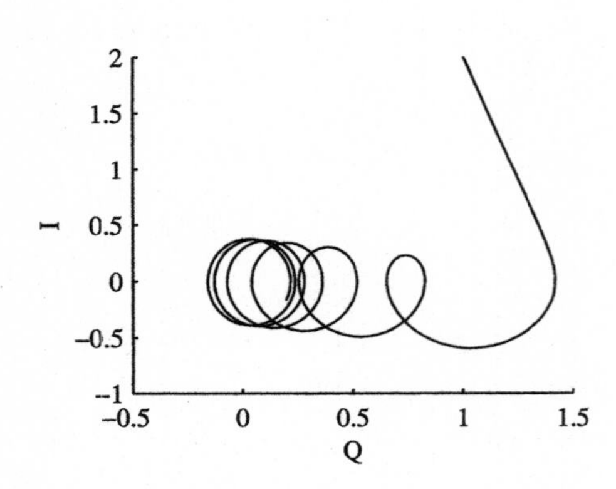

27. Our solver requires that we first change the second order equation $LQ'' + RQ' + (1/C)Q = 2\cos 2t$ into a system of two first order equations. If we let $Q' = I$, then

$$\begin{aligned} I' = Q'' &= -\frac{R}{L}Q' - \frac{1}{LC}Q + 2\cos 2t \\ &= -\frac{R}{L}I - \frac{1}{LC}Q + 2\cos 2t, \end{aligned}$$

which leads to the following system of first order equations.

$$\begin{aligned} Q' &= I \\ I' &= -\frac{R}{L}I - \frac{1}{LC}Q + 2\cos 2t \end{aligned}$$

Now, if $L = 1\,\text{H}$, $R = 0.5\,\Omega$, $C = 1\,\text{F}$, $Q(0) = 1\,\text{C}$, and $I(0) = 2\,\text{A}$, then

$$\begin{aligned} Q' &= I \\ I' &= -0.5I - Q + 2\cos 2t, \end{aligned}$$

and

$$\begin{aligned} Q(0) &= 1 \\ I(0) &= 2. \end{aligned}$$

Entering this system in our solver, we generate plots of charge versus time and current versus time.

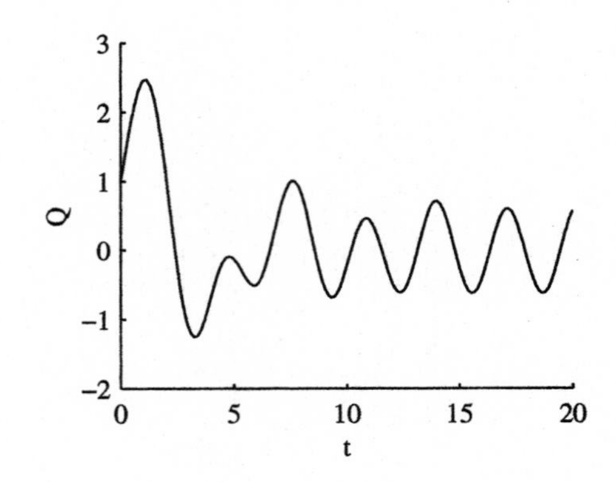

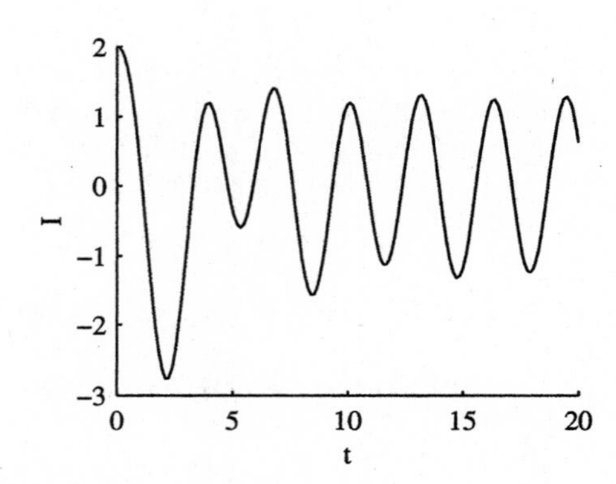

Our solver also generates a combined plot of both charge and current versus time, and a plot of current versus charge in the phase plane.

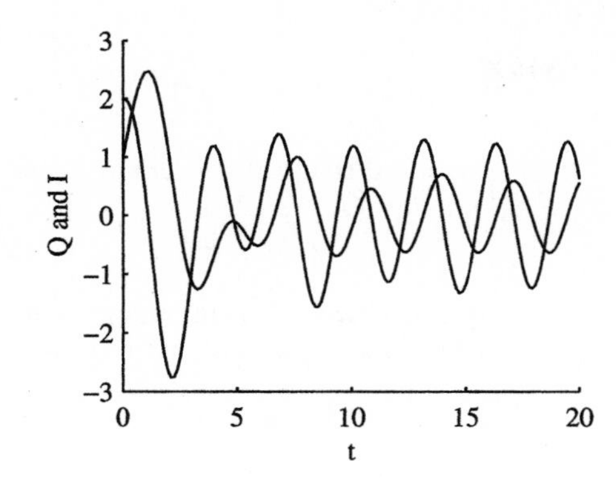

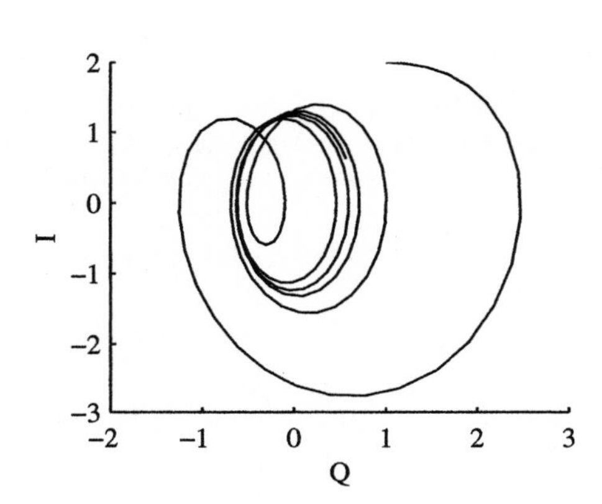

29. A plot of $t \to (Q(t), I(t), t)$ for the circuit of Exercise 23.

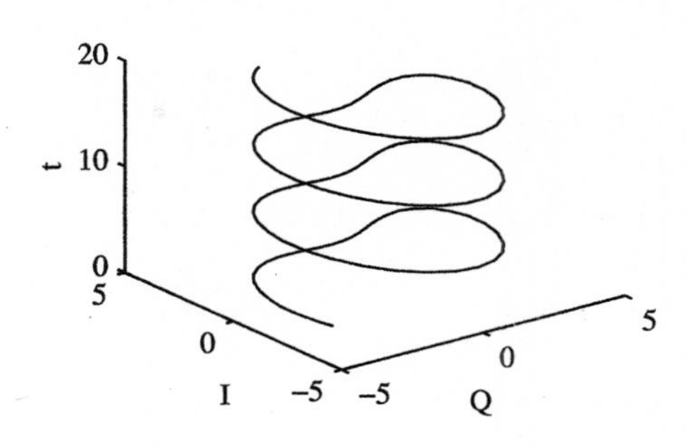

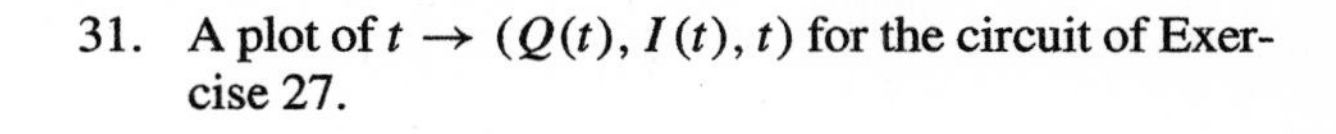

31. A plot of $t \to (Q(t), I(t), t)$ for the circuit of Exercise 27.

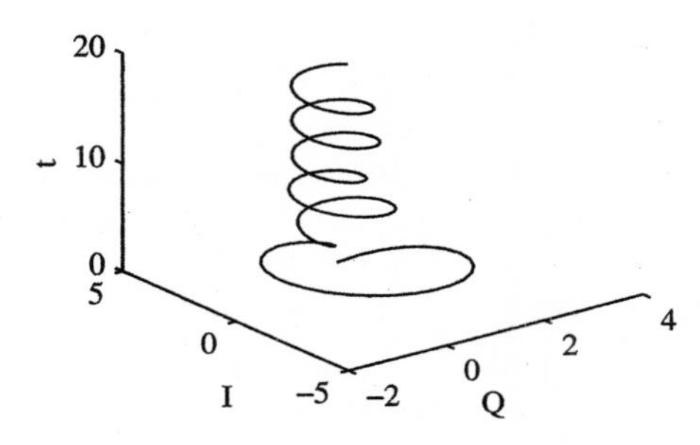

Section 4.3. Linear Homogeneous Equations with Constant Coefficients

1. Let $y = e^{\lambda t}$ in $y'' - y' - 2y = 0$ to obtain

$$\lambda^2 e^{\lambda t} - \lambda e^{\lambda t} - 2e^{\lambda t} = 0$$
$$e^{\lambda t}(\lambda^2 - \lambda - 2) = 0.$$

Because $e^{\lambda t} \neq 0$, we arrive at the characteristic equation

$$\lambda^2 - \lambda - 2 = 0$$
$$(\lambda - 2)(\lambda + 1) = 0,$$

and roots $\lambda = 2$ and $\lambda = -1$. Because the roots are distinct, the solutions $y_1(t) = e^{2t}$ and $y_2(t) = e^{-t}$ form a fundamental set of solutions and the general solution is

$$y(t) = C_1 e^{2t} + C_2 e^{-t}.$$

3. Let $y = e^{\lambda t}$ in $y'' + 5y' + 6y = 0$ to obtain

$$\lambda^2 e^{\lambda t} + 5\lambda e^{\lambda t} + 6e^{\lambda t} = 0,$$
$$e^{\lambda t}(\lambda^2 + 5\lambda + 6) = 0.$$

Because $e^{\lambda t} \neq 0$, we arrive at the characteristic equation

$$\lambda^2 + 5\lambda + 6 = 0,$$
$$(\lambda + 3)(\lambda + 2) = 0,$$

and roots $\lambda = -3$ and $\lambda = -2$. Because the roots are distinct, the solution $y_1(t) = e^{-3t}$ and $y_2(t) = e^{-2t}$ form a fundamental set of solutions and the general solution is

$$y(t) = C_1 e^{-3t} + C_2 e^{-2t}.$$

5. Let $y = e^{\lambda t}$ in $2y'' - y' - y = 0$ to obtain

$$2\lambda^2 e^{\lambda t} - \lambda e^{\lambda t} - e^{\lambda t} = 0,$$
$$e^{\lambda t}(2\lambda^2 - \lambda - 1) = 0.$$

Because $e^{\lambda t} \neq 0$, we arrive at the characteristic equation

$$2\lambda^2 - \lambda - 1 = 0,$$
$$(2\lambda + 1)(\lambda - 1) = 0,$$

and roots $\lambda = -1/2$ and $\lambda = 1$. Because the roots are distinct, the solutions $y_1(t) = e^{-(1/2)t}$ and $y_2(t) = e^t$ form a fundamental set of solutions and the general solution is

$$y(t) = C_1 e^{-(1/2)t} + C_2 e^t.$$

7. Let $y = e^{\lambda t}$ in $3y'' - 2y' - y = 0$ to obtain

$$3\lambda^2 e^{\lambda t} - 2\lambda e^{\lambda t} - e^{\lambda t} = 0$$
$$e^{\lambda t}(3\lambda^2 - 2\lambda - 1) = 0.$$

Because $e^{\lambda t} \neq 0$, we arrive at the characteristic equation

$$3\lambda^2 - 2\lambda - 1 = 0$$
$$(3\lambda + 1)(\lambda - 1) = 0,$$

and roots $\lambda = -1/3$ and $\lambda = 1$. Because the roots are distinct, the solutions $y_1(t) = e^{-t/3}$ and $y_2(t) = e^t$ form a fundamental set of solutions and the general solution is

$$y(t) = C_1 e^{-t/3} + C_2 e^t.$$

9. If $y'' + y = 0$, then the characteristic equation is

$$\lambda^2 + 1 = 0.$$

The roots of the characteristic equation are $\pm i$, leading to the complex solutions

$$z(t) = e^{it} \quad \text{and} \quad \overline{z}(t) = e^{-it}.$$

However, by Euler's identity,

$$z(t) = \cos t + i \sin t,$$

and the real and imaginary parts of z lead to a fundamental set of real solutions, $y_1(t) = \cos t$ and $y_2(t) = \sin t$. Hence, the general solution is

$$y(t) = C_1 \cos t + C_2 \sin t.$$

11. If $y'' + 4y' + 5y = 0$, then the characteristic equation is

$$\lambda^2 + 4\lambda + 5 = 0.$$

The roots of the characteristic equation are

$$z = \frac{-4 \pm \sqrt{16 - 20}}{2} = -2 \pm i,$$

leading to the complex solutions

$$z(t) = e^{(-2+i)t} \quad \text{and} \quad \overline{z}(t) = e^{(-2-i)t}.$$

However, by Euler's identity,

$$z(t) = e^{-2t} e^{it} = e^{-2t}(\cos t + i \sin t),$$

and the real and imaginary parts of z lead to a fundamental set of real solutions, $y_1(t) = e^{-2t} \cos t$ and $y_2(t) = e^{-2t} \sin t$. Hence, the general solution is

$$y(t) = C_1 e^{-2t} \cos t + C_2 e^{-2t} \sin t$$
$$= e^{-2t}(C_1 \cos t + C_2 \sin t).$$

13. If $y'' + 2y = 0$, then the characteristic equation is

$$\lambda^2 + 2 = 0.$$

The roots of the characteristic equation are $\pm\sqrt{2}i$, leading to the complex solutions

$$z(t) = e^{\sqrt{2}it} \quad \text{and} \quad \bar{z}(t) = e^{-\sqrt{2}it}.$$

However, by Euler's identity,

$$z(t) = \cos\sqrt{2}t + i\sin\sqrt{2}t,$$

and the real and imaginary parts of z lead to a fundamental set of real solutions, $y_1(t) = \cos\sqrt{2}t$ and $y_2(t) = \sin\sqrt{2}t$. Hence, the general solution is

$$y(t) = C_1\cos\sqrt{2}t + C_2\sin\sqrt{2}t.$$

15. If $y'' - 2y' + 4y = 0$, then the characteristic equation is

$$\lambda^2 - 2\lambda + 4 = 0.$$

The roots of the characteristic equation are $1 \pm \sqrt{3}i$, leading to the complex solutions

$$z(t) = e^{(1+\sqrt{3}i)t} \quad \text{and} \quad \bar{z}(t) = e^{(1-\sqrt{3}i)t}.$$

However, by Euler's identity,

$$z(t) = e^t e^{\sqrt{3}it} = e^t(\cos\sqrt{3}t + i\sin\sqrt{3}t),$$

and the real and imaginary parts of z lead to a fundamental set of real solutions $y_1(t) = e^t\cos\sqrt{3}t$ and $y_2(t) = e^t\sin\sqrt{3}t$. Hence the general solution is

$$y(t) = C_1e^t\cos\sqrt{3}t + C_2e^t\sin\sqrt{3}t.$$

17. If $y'' - 4y' + 4y = 0$, then the characteristic equation is

$$\lambda^2 - 4\lambda + 4 = (\lambda - 2)^2 = 0.$$

Hence, the characteristic equation has a single, double root, $\lambda = 2$. Therefore, $y_1(t) = e^{2t}$ and $y_2(t) = te^{2t}$ form a fundamental set of real solutions. Hence, the general solution is

$$y(t) = C_1e^{2t} + C_2te^{2t} = (C_1 + C_2t)e^{2t}.$$

19. If $4y'' + 4y' + y = 0$, then the characteristic equation is

$$4\lambda^2 + 4\lambda + 1 = (2\lambda + 1)^2 = 0.$$

Hence, the characteristic equation has a single, double root, $\lambda = -1/2$. Therefore, $y_1(t) = e^{-(1/2)t}$ and $y_2(t) = te^{-(1/2)t}$ form a fundamental set of real solutions. Hence, the general solution is

$$y(t) = C_1e^{-t/2} + C_2te^{-t/2} = (C_1 + C_2t)e^{-t/2}.$$

21. If $16y'' + 8y' + y = 0$, then the characteristic equation is

$$16\lambda^2 + 8\lambda + 1 = (4\lambda + 1)^2 = 0.$$

Hence, the characteristic equation has a single, double root, $\lambda = -1/4$. Therefore, $y_1(t) = e^{-(1/4)t}$ and $y_2(t) = te^{-(1/4)t}$ form a fundamental set of real solutions. Hence, the general solution is

$$y(t) = C_1e^{-t/4} + C_2te^{-t/4} = (C_1 + C_2t)e^{-t/4}.$$

23. If $16y'' + 24y' + 9y = 0$, then the characteristic equation is

$$16\lambda^2 + 24\lambda + 9 = (4\lambda + 3)^2 = 0.$$

Hence, the characteristic equation has a single, double root, $\lambda = -3/4$. Therefore, $y_1(t) = e^{-(3/4)t}$ and $y_2(t) = te^{-(3/4)t}$ form a fundamental set of real solutions. Hence, the general solution is

$$y(t) = C_1e^{-3t/4} + C_2te^{-3t/4} = (C_1 + C_2t)e^{-3t/4}.$$

25. If $y'' - y' - 2y = 0$, then the characteristic equation is

$$\lambda^2 - \lambda - 2 = (\lambda - 2)(\lambda + 1) = 0,$$

with roots $\lambda = 2$ and $\lambda = -1$. This leads to the general solution

$$y(t) = C_1e^{2t} + C_2e^{-t}.$$

Using the initial condition $y(0) = -1$ provides

$$-1 = C_1 + C_2.$$

Differentiating the general solution,

$$y'(t) = 2C_1e^{2t} - C_2e^{-t},$$

then using the initial condition $y'(0) = 2$ leads to

$$2 = 2C_1 - C_2.$$

These equations yield $C_1 = 1/3$ and $C_2 = -4/3$, giving the particular solution

$$y(t) = \frac{1}{3}e^{2t} - \frac{4}{3}e^{-t}.$$

27. If $y'' - 2y' + 17y = 0$, then the characteristic equation is

$$\lambda^2 - 2\lambda + 17 = 0,$$

with roots $1 \pm 4i$. The complex solution,

$$z(t) = e^{(1+4i)t} = e^t(\cos 4t + i \sin 4t)$$

leads to a fundamental set of real solutions and the general solution

$$y(t) = e^t(C_1 \cos 4t + C_2 \sin 4t).$$

The initial condition $y(0) = -2$ provides

$$-2 = C_1.$$

Differentiating the general solution,

$$y'(t) = e^t(C_1 \cos 4t + C_2 \sin 4t) + e^t(-4C_1 \sin 4t + 4C_2 \cos 4t),$$

then using the initial condition $y'(0) = 3$ leads to

$$3 = C_1 + 4C_2.$$

These equations yield $C_1 = -2$ and $C_2 = 5/4$, giving the solution

$$y(t) = e^t(-2 \cos 4t + (5/4) \sin 4t).$$

29. If $y'' + 10y' + 25y = 0$, then the characteristic equation is

$$\lambda^2 + 10\lambda + 25 = (\lambda + 5)^2 = 0,$$

with repeated root $\lambda = -5$. This leads to the fundamental set of solutions $y_1(t) = e^{-5t}$ and $y_2(t) = te^{-5t}$ and the general solution is

$$y(t) = C_1 e^{-5t} + C_2 t e^{-5t} = (C_1 + C_2 t)e^{-5t}.$$

Using the initial condition $y(0) = 2$ leads to

$$2 = C_1.$$

Differentiating the general solution,

$$y'(t) = C_2 e^{-5t} - 5(C_1 + C_2 t)e^{-5t},$$

then the initial condition $y'(0) = -1$ leads to

$$-1 = C_2 - 5C_1.$$

Thus, $C_1 = 2$ and $C_2 = 9$ and the final solutions is

$$y(t) = (2 + 9t)e^{-5t}.$$

31. If $y'' + 2y' + 3y = 0$, then the characteristic equation is $\lambda^2 + 2\lambda + 3 = 0$ with roots $z = -1 \pm \sqrt{2}i$. The complex solution

$$z(t) = e^{(-1+\sqrt{2}i)t} = e^{-t}(\cos \sqrt{2}t + i \sin \sqrt{2}t)$$

provides a fundamental set of real solutions, $y_1(t) = e^{-t} \cos \sqrt{2}t$ and $y_2(t) = e^{-t} \sin \sqrt{2}t$, and the general solution

$$y(t) = e^{-t}(C_1 \cos \sqrt{2}t + C_2 \sin \sqrt{2}t).$$

The initial condition $y(0) = 1$ provides

$$1 = C_1.$$

Differentiating the general solution,

$$y'(t) = e^{-t}(-C_1\sqrt{2} \cos \sqrt{2}t + C_2\sqrt{2} \sin \sqrt{2}t) - e^{-t}(C_1 \sin \sqrt{2}t + C_2 \cos \sqrt{2}t),$$

and the initial condition $y'(0) = 0$ provides

$$0 = -C_1 + \sqrt{2}C_2.$$

Thus, $C_1 = 1$ and $C_2 = \sqrt{2}/2$, giving the solution

$$y(t) = e^{-t}\left(\cos \sqrt{2}t + \frac{\sqrt{2}}{2} \sin \sqrt{2}t\right).$$

33. If $8y''+2y'-y=0$, then the characteristic equation is

$$8\lambda^2+2\lambda-1=(4\lambda-1)(2\lambda+1)=0,$$

with roots $\lambda=1/4$ and $\lambda=-1/2$. This leads to the general solution

$$y(t)=C_1e^{t/4}+C_2e^{-t/2}.$$

Using the initial condition $y(-1)=1$ provides

$$1=C_1e^{-1/4}+C_2e^{1/2}. \qquad (*)$$

Differentiating the general solution,

$$y'(t)=\frac{1}{4}C_1e^{t/4}-\frac{1}{2}C_2e^{-t/2},$$

then using the initial condition $y'(-1)=0$ leads to

$$0=\frac{1}{4}C_1e^{-1/4}-\frac{1}{2}C_2e^{1/2}. \qquad (**)$$

Subtracting 4 times equation $(**)$ from equation $(*)$, provides $1=3C_2e^{1/2}$, so $C_2=(1/3)e^{-1/2}$. Substituting this result in equation $(**)$ yields $C_1=(2/3)e^{1/4}$. Thus, the particular solution is

$$y(t)=\frac{2}{3}e^{1/4}e^{t/4}+\frac{1}{3}e^{-1/2}e^{-t/2}$$
$$y(t)=\frac{2}{3}e^{(t+1)/4}+\frac{1}{3}e^{-(t+1)/2}.$$

35. If $y''+12y'+36y=0$, then the characteristic equation is

$$\lambda^2+12\lambda+36=(\lambda+6)^2=0,$$

with double root $\lambda=-6$. This leads to a fundamental set of solutions, $y_1(t)=e^{-6t}$ and $y_2(t)=te^{-6t}$, and the general solution

$$y(t)=(C_1+C_2t)e^{-6t}.$$

Using the initial condition $y(1)=0$ provides

$$0=(C_1+C_2)e^{-6}.$$

Thus, $C_1=-C_2$. Differentiating the general solution,

$$\begin{aligned}y'(t)&=(C_1+C_2t)(-6e^{-6t})+C_2e^{-6t}\\&=e^{-6t}((C_2-6C_1)-6C_2t),\end{aligned}$$

then using the initial condition $y'(1)=-1$ leads to

$$\begin{aligned}-1&=e^{-6}((C_2-6C_1)-6C_2)\\-1&=e^{-6}(-6C_1-5C_2).\end{aligned}$$

Substituting $C_1=-C_2$ in this last equation,

$$\begin{aligned}-1&=e^{-6}(-6(-C_2)-5C_2)\\C_2&=-e^6.\end{aligned}$$

Then, $C_1=-C_2=e^6$. Thus, the particular solution is

$$\begin{aligned}y(t)&=(e^6-e^6t)e^{-6t}\\&=(1-t)e^{-6(t-1)}.\end{aligned}$$

37. (a) If λ_1 and λ_2 are are characteristic roots of the characteristic equation

$$\lambda^2+p\lambda+q=0,$$

then the characteristic equation factors as follows.

$$(\lambda-\lambda_1)(\lambda-\lambda_2)=0$$

Expanding this result, we obtain

$$\lambda_2-(\lambda_1+\lambda_2)\lambda+\lambda_1\lambda_2=0.$$

When this is compared with the characteristic equation,

$$\lambda_1\lambda_2=q.$$

(b) Similarly, when coefficients are compared,

$$\lambda_1+\lambda_2=-p.$$

Section 4.4. Harmonic Motion

1. The first of two images contains the plot of $y = \cos 2t + \sin 2t$ on the interval $[0, 3\pi]$.

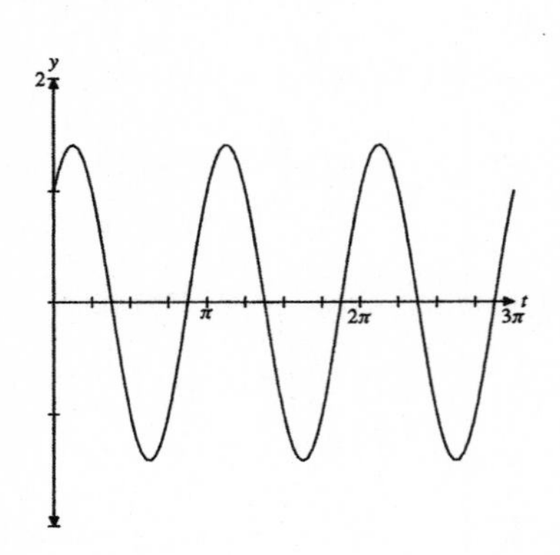

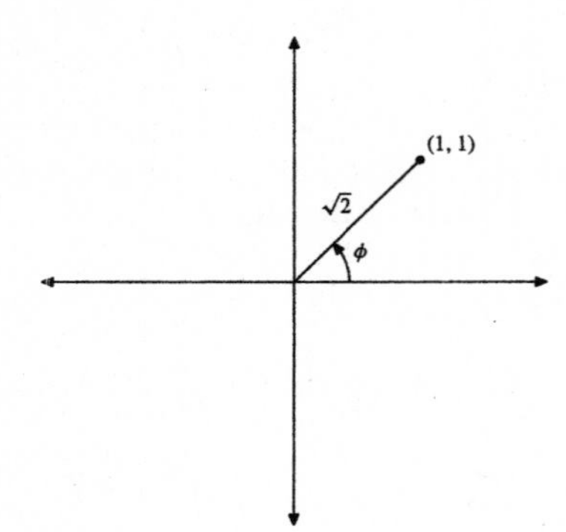

Plot the coefficients of

$$y = 1 \cdot \cos 2t + 1 \cdot \sin 2t$$

in the first quadrant, calculate the magnitude of the vector, and mark the angle. The angle ϕ is easily calculated.

$$\tan\phi = \frac{1}{1} = 1$$
$$\phi = \frac{\pi}{4}.$$

Factor out the magnitude $\sqrt{2}$ as follows.

$$y = \sqrt{2}\left(\frac{1}{\sqrt{2}}\cos 2t + \frac{1}{\sqrt{2}}\sin 2t\right)$$

But $\cos\phi = 1/\sqrt{2}$ and $\sin\phi = 1/\sqrt{2}$, so we can write

$$y = \sqrt{2}\,(\cos\phi\cos 2t + \sin\phi\sin 2t)$$
$$y = \sqrt{2}\cos(2t - \phi)$$
$$y = \sqrt{2}\cos\left(2t - \frac{\pi}{4}\right)$$
$$y = \sqrt{2}\cos 2\left(t - \frac{\pi}{8}\right).$$

Hence, the curve has amplitude $\sqrt{2}$, period $T = \pi$, and is shifted to the right $\pi/8$ units. This is clearly shown in the following image where the dashed curve is the unshifted $y = \sqrt{2}\cos 2t$.

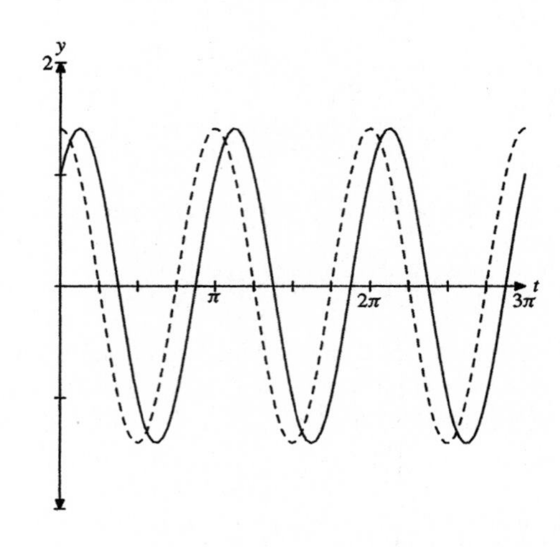

3. The first of two images contains the plot of $y = \cos 4t + \sqrt{3}\sin 4t$ on the interval $[0, \pi]$.

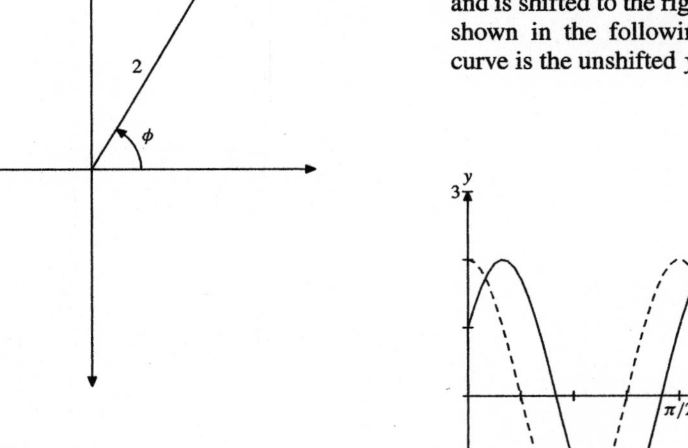

Plot the coefficients of

$$y = 1 \cdot \cos 4t + \sqrt{3} \cdot \sin 4t$$

in the first quadrant, calculate the magnitude of the vector, and mark the angle. The angle ϕ is easily calculated.

$$\tan \phi = \frac{\sqrt{3}}{1} = \sqrt{3}$$
$$\phi = \frac{\pi}{3}.$$

Factor out the magnitude 2 as follows.

$$y = 2\left(\frac{1}{2}\cos 4t + \frac{\sqrt{3}}{2}\sin 4t\right)$$

But $\cos \phi = 1/2$ and $\sin \phi = \sqrt{3}/2$, so we can write

$$y = 2\left(\cos \phi \cos 4t + \sin \phi \sin 4t\right)$$
$$y = 2\cos(4t - \phi)$$
$$y = 2\cos\left(4t - \frac{\pi}{3}\right)$$
$$y = 2\cos 4\left(t - \frac{\pi}{12}\right).$$

Hence, the curve has amplitude 2, period $T = \pi/2$, and is shifted to the right $\pi/12$ units. This is clearly shown in the following image where the dashed curve is the unshifted $y = 2\cos 4t$.

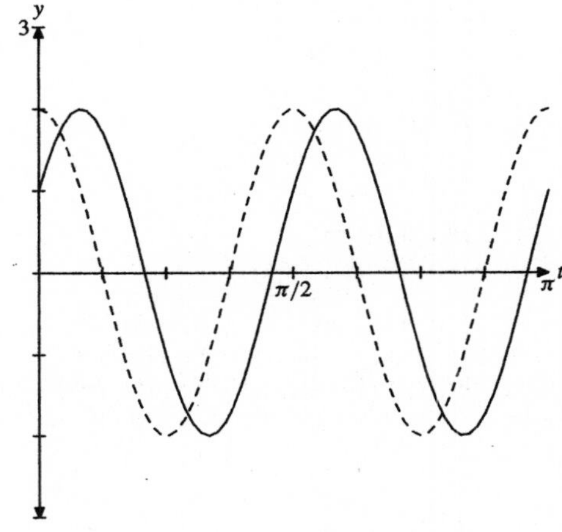

5. The first of the two images contains the plot of $y = 0.2\cos 2.5t - 0.1\sin 2.5t$ on the interval $[0, 8\pi/5]$.

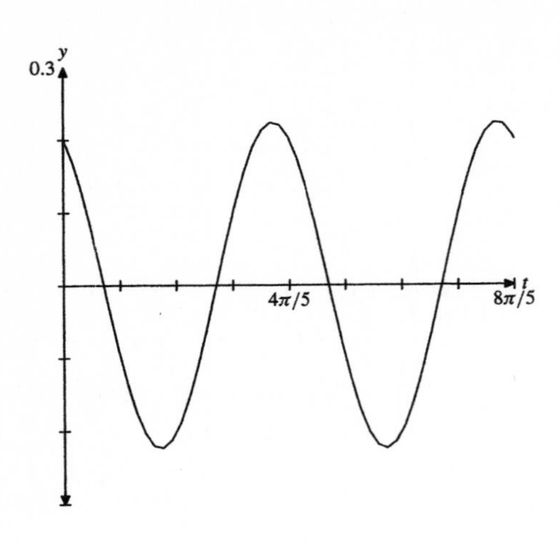

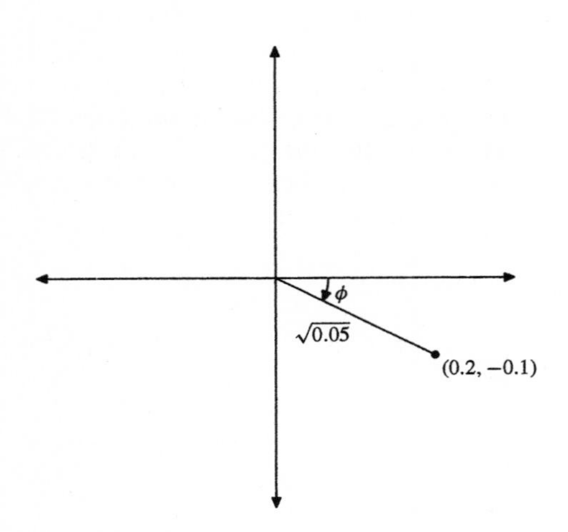

Plot the coefficients of

$$y = 0.2\cos 2.5t - 0.1\sin 2.5t$$

in the fourth quadrant, calculate the magnitude of the vector, and mark the angle. The angle ϕ is easily calculated.

$$\tan\phi = \frac{-1}{2} = -\frac{1}{2}$$
$$\phi \approx -0.4636.$$

Factor out the magnitude 2 as follows.

$$y = \sqrt{0.05}\left(\frac{0.2}{\sqrt{0.05}}\cos 2.5t - \frac{0.1}{\sqrt{0.05}}\sin 2.5t\right)$$

But $\cos\phi = 0.2/\sqrt{0.05}$ and $\sin\phi = -0.1/\sqrt{0.05}$, so we can write

$$y = \sqrt{0.05}\,(\cos\phi\cos 2.5t + \sin\phi\sin 2.5t)$$
$$y = 0.2236\cos(2.5t - \phi)$$
$$y = 0.2236\cos(2.5t + 0.4636)$$
$$y = .2236\cos 2.5\,(t + 0.1855)\,.$$

Hence, the curve has amplitude 0.2236, period $T = 2\pi/2.5 = 4\pi/5$, and is shifted to the left 0.1855 units. This is clearly shown in the image below, where the dashed curve is the unshifted $y = 0.2236\cos 2.5t$.

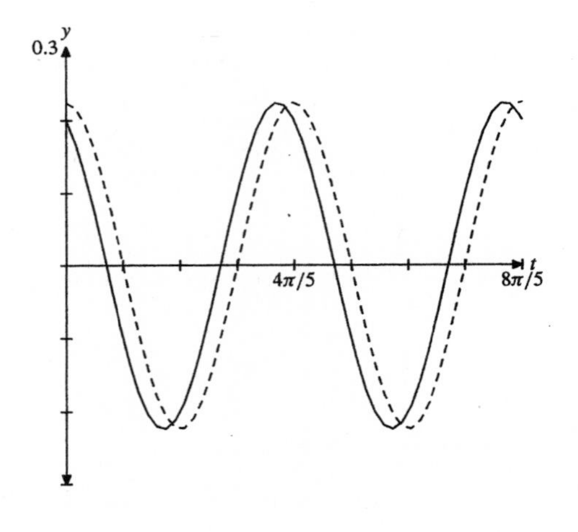

7. Plot the coefficients of $1 \cdot \cos 5t + 1 \cdot \sin 5t$ in the first quadrant, calculate the magnitude of the vector,

and mark the angle.

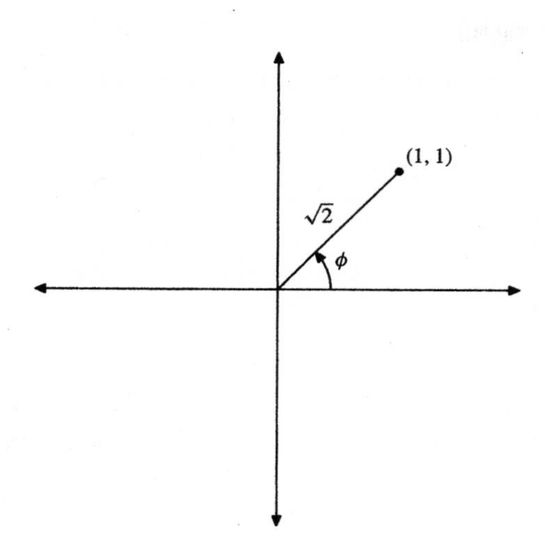

The angle ϕ is easily calculated.

$$\tan\phi = 1$$
$$\phi = \frac{\pi}{4}$$

Factor out the magnitude as follows:

$$y = \sqrt{2}e^{-t/2}\left(\frac{1}{\sqrt{2}}\cos 5t + \frac{1}{\sqrt{2}}\sin 5t\right)$$

But $\cos\phi = 1/\sqrt{2}$ and $\sin\phi = 1/\sqrt{2}$, so we can write

$$y = \sqrt{2}e^{-t/2}(\cos\phi\cos 5t + \sin\phi\sin 5t)$$
$$y = \sqrt{2}e^{-t/2}\cos(5t - \phi)$$
$$y = \sqrt{2}e^{-t/2}\cos(5t - \pi/4).$$

Thus, the amplitude is $\sqrt{2}e^{-t/2}$ which, along with $-\sqrt{2}e^{-t/2}$, clearly bound the graph of $y = \sqrt{2}e^{-t/2}\cos(5t - \pi/4)$.

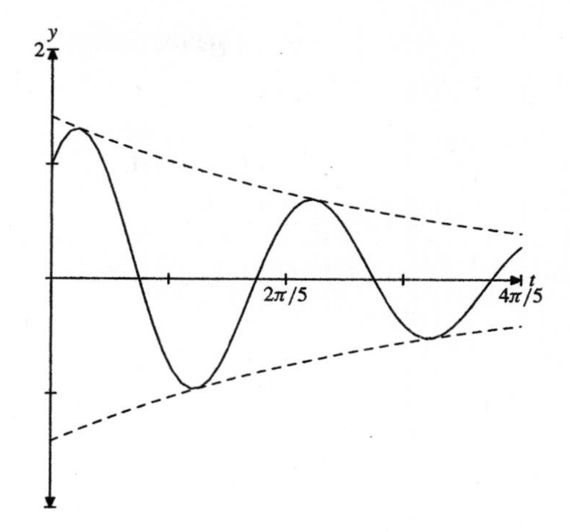

9. Plot the coefficients of $0.2 \cdot \cos 2t + 0.1 \cdot \sin 2t$ in the first quadrant, calculate the magnitude of the vector, and mark the angle.

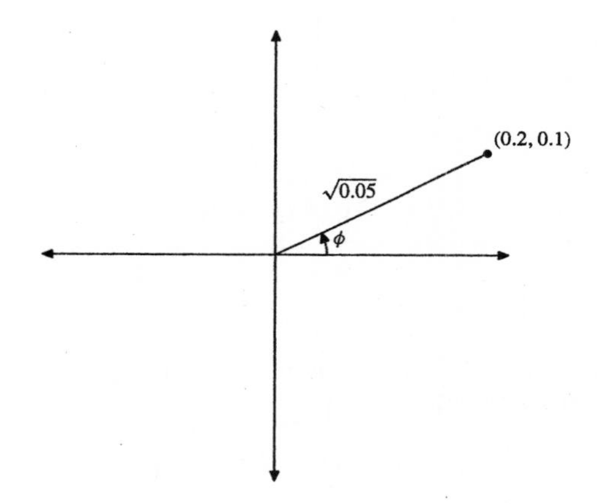

The angle ϕ is easily calculated.

$$\tan\phi = \frac{0.1}{0.2} = \frac{1}{2}$$
$$\phi = 0.4636$$

Factor out the magnitude as follows:

$$y = \sqrt{0.05}e^{-0.1t}\left(\frac{0.2}{\sqrt{0.05}}\cos 2t + \frac{0.1}{\sqrt{0.05}}\sin 2t\right)$$

But $\cos\phi = 0.2/\sqrt{0.05}$ and $\sin\phi = 0.1/\sqrt{0.05}$, so we can write

$$y = \sqrt{0.05}e^{-0.1t}(\cos\phi\cos 2t + \sin\phi\sin 2t)$$
$$y = 0.2236e^{-0.1t}\cos(2t - \phi)$$
$$y = 0.2236e^{-0.1t}\cos(2t - 0.4636).$$

Thus, the amplitude is $0.2236e^{-0.1t}$ which, along with $-0.2236e^{-0.1t}$, clearly bound the graph of $y = 0.2236e^{-0.1t}\cos(2t - 0.4636)$.

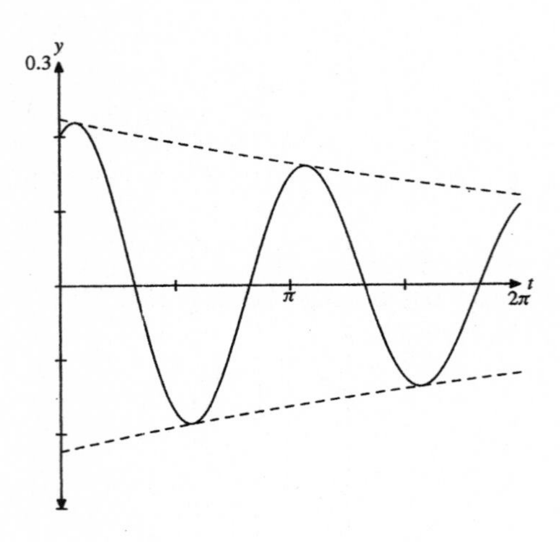

11. Substitute $m = 0.2$ kg and $k = 5$ kg/s^2 in $my'' + ky = 0$ to obtain $0.2y'' + 5y = 0$ or

$$y'' + 25y = 0.$$

The characteristic equation is $\lambda^2 + 25 = 0$, with zeros $\lambda = \pm 5i$, so

$$z(t) = e^{5it} = \cos 5t + i\sin 5t$$

is a complex solution. The real and imaginary parts of this solution form a fundamental set of real solutions, giving the general solution

$$y(t) = C_1\cos 5t + C_2\sin 5t.$$

The initial displacement is 0.5 m, so $y(0) = 0.5$ and $C_1 = 0.5$. Differentiating,

$$y'(t) = -5C_1\sin 5t + 5C_2\cos 5t.$$

The system is released from rest, so $y'(0) = 0$ and $C_2 = 0$. Thus, the solution is

$$y(t) = 0.5\cos 5t,$$

which has amplitude 0.5, frequency 5 rad/s, and zero phase.

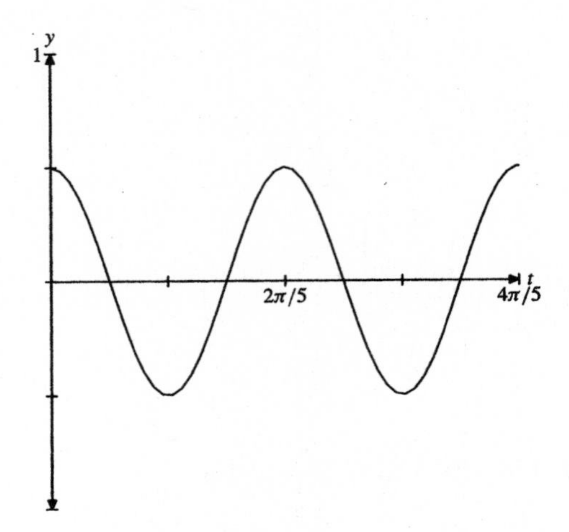

13. The system $(2/5)x'' + kx = 0$, or $x'' + (5k/2)x = 0$, is equivalent to

$$x'' + \omega_0^2 x = 0,$$

with $\omega_0^2 = 5k/2$. The characteristic equation is $\lambda^2 + \omega_0^2 = 0$ with roots $\lambda = \pm\omega_0 i$. Thus, we have complex solution

$$z(t) = e^{i\omega_0 t} = \cos\omega_0 t + i\sin\omega_0 t.$$

The real and imaginary parts of this solution form a fundamental set of solutions and provide the general solution

$$x(t) = C_1\cos\omega_0 t + C_2\sin\omega_0 t.$$

This solution is periodic with period $T = 2\pi/\omega_0$. Thus, if it is know that the period is $\pi/2$, then

$$T = \frac{2\pi}{\omega_0}$$
$$\frac{\pi}{2} = \frac{2\pi}{\omega_0}$$
$$\omega_0 = 4.$$

But $\omega_0^2 = 5k/2$,

$$4^2 = \frac{5k}{2}$$
$$k = \frac{32}{5}.$$

The initial condition $x(0) = 2$ gives $C_1 = 2$. Differentiating $x(t)$,

$$x'(t) = -C_1\omega_0 \sin \omega_0 t + C_2\omega_0 \cos \omega_0 t.$$

The initial condition $x'(0) = v_0$ leads to $C_2 = v_0/\omega_0$ and the solution

$$x(t) = 2\cos \omega_0 t + \frac{v_0}{\omega_0} \sin \omega_0 t = 2\cos 4t + \frac{v_0}{4} \sin 4t.$$

Plot the coefficients, calculate the magnitude of the vector, and mark the angle.

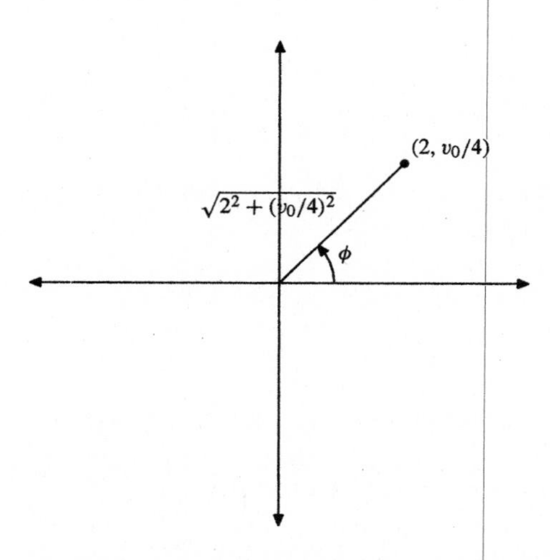

As in Exercises 1–6, factoring out the magnitude will place the solution in the form

$$x(t) = \sqrt{4 + v_0^2/16}\cos(4t - \phi),$$

where $\tan \phi = v_0/8$. If it is known that the amplitude is 2, then

$$\sqrt{4 + v_0^2/16} = 2$$
$$4 + v_0^2/16 = 4$$
$$v_0 = 0.$$

15. (a) In an LC circuit,

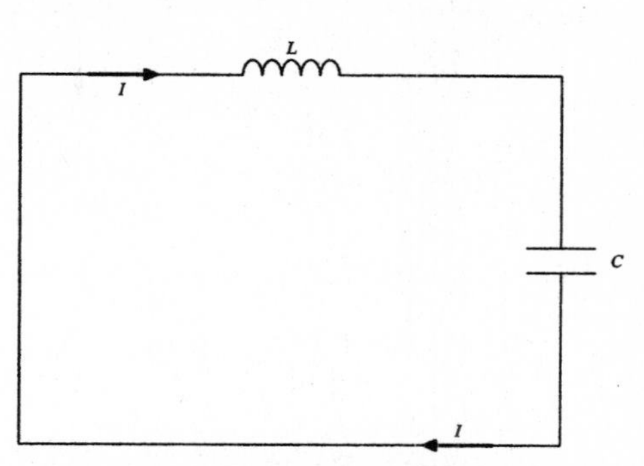

the sum of the voltage drops around the closed loop is zero. The voltage drop across the inductor is LI', where $L = 6\,\mu\text{H}$ and $I(t)$ is the current, and the voltage drop across the capacitor is $(1/C)Q$, where $C = 2\,\mu\text{F}$ and $Q(t)$ is the charge on the capacitor. Thus, we can write

$$LI' + \frac{1}{C}Q = 0,$$

or, since $I = dQ/dt$,

$$LQ'' + \frac{1}{C}Q = 0.$$

To find the initial charge on the capacitor, use

$$V_C = \frac{1}{C}Q$$
$$Q = CV_C$$
$$Q = (2 \times 10^{-6}\,\text{F})(20\,V)$$
$$Q = 4.0 \times 10^{-5}\text{coulombs}.$$

(b) If the initial current is zero, then $Q'(0) = I(0) = 0$. Thus, $LQ'' + (1/C)Q = 0$, or

$$Q'' + \frac{1}{LC}Q = 0,$$

with the intial conditions $Q(0) = 4.0 \times 10^{-5}$, and $Q'(0) = 0$. This last equation is equivalent to

$$Q'' + \omega_0^2 Q = 0,$$

where $\omega_0^2 = 1/(LC)$. This has characteristic equation $\lambda^2 + \omega_0^2 = 0$ and zeros $\lambda = \pm i\omega_0$, leading to the complex solution

$$z(t) = e^{i\omega_0 t} = \cos \omega_0 t + i \sin \omega_0 t.$$

The real and imaginary parts of this solution form a fundamental set of solutions, leading to the general solution

$$Q(t) = C_1 \cos \omega_0 t + C_2 \sin \omega_0 t.$$

The initial condition $Q(0) = 4.0 \times 10^{-5}$ gives $C_1 = 4.0 \times 10^{-5}$. Differentiating $Q(t)$,

$$Q'(t) = -C_1\omega_0 \sin \omega_0 t + C_2\omega_0 \cos \omega_0 t.$$

The initial condition $Q'(0) = 0$ gives $C_2 = 0$ and the solution

$$Q(t) = (4.0 \times 10^{-5}) \cos \omega_0 t.$$

Thus, the amplitude is 4.0×10^{-5}. Because $\omega_0^2 = 1/(LC)$,

$$\begin{aligned}\omega_0 &= \sqrt{\frac{1}{LC}} \\ &= \sqrt{\frac{1}{(6 \times 10^{-6})(2 \times 10^{-6})}} \\ &\approx 2.887 \times 10^5 \text{ rad/s}.\end{aligned}$$

The phase is zero. A plot of charge versus time follows.

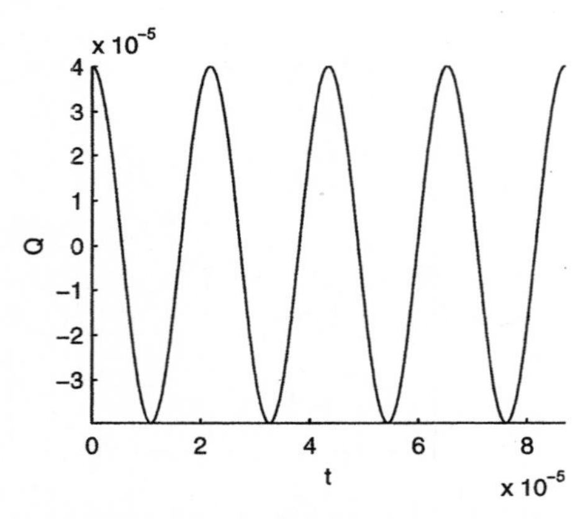

To find the current, differentiate the charge formula with respect to time.

$$I(t) = Q'(t) = -(4.0 \times 10^{-5})\omega_0 \sin \omega_0 t$$

Thus,

$$I(t) = -11.548 \sin \omega_0 t,$$

making the amplitude 11.548 and the frequency $\omega_0 \approx 2.887 \times 10^5$ rad/s. The phase is zero. A plot of current versus time follows.

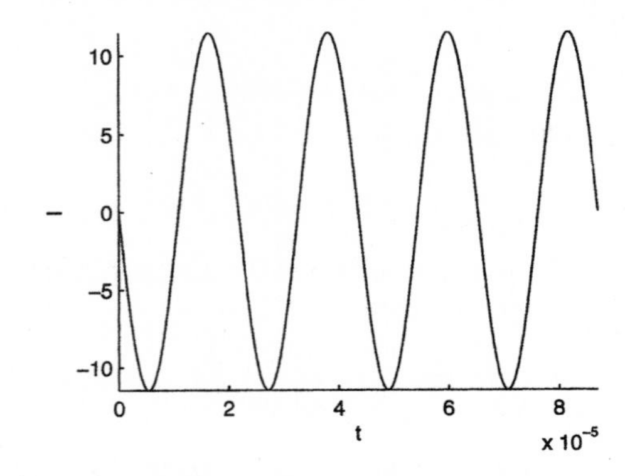

17. Suppose that the spring-mass system $my'' + \mu y' + ky = 0$ is overdamped. Then,

$$\begin{aligned}y'' + \frac{\mu}{m}y' + \frac{k}{m}y &= 0 \\ y'' + 2cy' + \omega_0^2 y &= 0,\end{aligned}$$

where $2c = \mu/m$ and $\omega_0^2 = k/m$. The characteristic equation is $\lambda^2 + 2c\lambda + \omega_0^2 = 0$ and the zeros are

$$\lambda_1 = -c - \sqrt{c^2 - \omega_0^2} \quad \text{and}$$
$$\lambda_2 = -c + \sqrt{c^2 - \omega_0^2}.$$

In order that the system be overdamped, we need $c^2 - \omega_0^2 > 0$. Of course, this condition results in $\lambda_1 < \lambda_2 < 0$ and the general solution is

$$y(t) = C_1 e^{\lambda_1 t} + C_2 e^{\lambda_2 t}.$$

To determine how many times this solution can cross the t-axis, set $y(t) = 0$ and solve for t.

$$\begin{aligned}0 &= C_1 e^{\lambda_1 t} + C_2 e^{\lambda_2 t} \\ 0 &= e^{\lambda_1 t}(C_1 + C_2 e^{(\lambda_2 - \lambda_1)t})\end{aligned}$$

Of course, $e^{\lambda_1 t}$ is never zero, so this leads to

$$\begin{aligned}C_1 + C_2 e^{(\lambda_2 - \lambda_1)t} &= 0 \\ e^{(\lambda_2 - \lambda_1)t} &= -\frac{C_1}{C_2}.\end{aligned}$$

If $C_1/C_2 > 0$, then there are no crossings, but if $C_1/C_2 < 0$, there will be exactly one crossing of the t-axis. The challenge is to find initial conditions y_0 and v_0 that lead to zero and one crossings. But $y(0) = y_0$ gives

$$y_0 = C_1 + C_2.$$

Differentiating $y(t)$,

$$y'(t) = C_1\lambda_1 e^{\lambda_1 t} + C_2\lambda_2 e^{\lambda_2 t},$$

the initial condition $y'(0) = v_0$ provides

$$v_0 = C_1\lambda_1 + C_2\lambda_2,$$

and this system is easily solved for C_1 and C_2,

$$C_1 = \frac{\lambda_2 y_0 - v_0}{\lambda_2 - \lambda_1} \quad \text{and} \quad C_2 = \frac{v_0 - \lambda_1 y_0}{\lambda_2 - \lambda_1},$$

which gives us

$$\frac{C_1}{C_2} = \frac{\lambda_2 y_0 - v_0}{v_0 - \lambda_1 y_0}.$$

Because there is exactly one zero crossing if $C_1/C_2 < 0$, we see two possible ways that this can happen:

- "Case 1:" We must have $\lambda_2 y_0 - v_0 < 0$ and $v_0 - \lambda_1 y_0 > 0$, which leads to the condition that $v_0 > \lambda_2 y_0$ and $v_0 > \lambda_1 y_0$, or
- "Case 2:" We must have $\lambda_2 y_0 - v_0 > 0$ and $v_0 - \lambda_1 y_0 < 0$, which leads to the condition that $v_0 < \lambda_2 y_0$ and $v_0 < \lambda_1 y_0$.

These conditions are easy to analyze if you think geometrically.

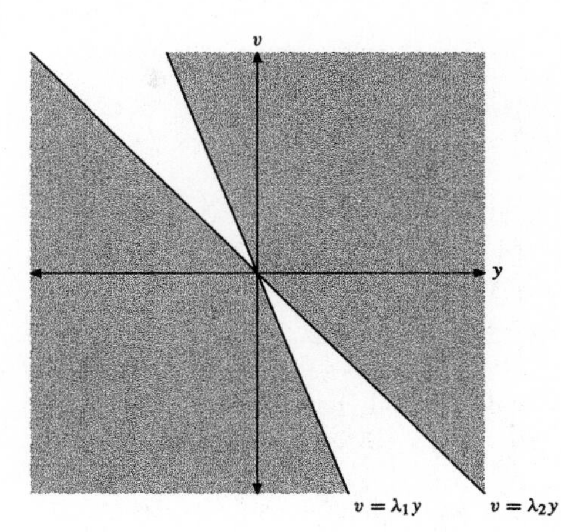

Thus, there will be one zero crossing if you select an initial condition from the shaded region. This is not especially intuitive until you give it a try in your solver. For example, the spring mass system $y'' + 6y'' + 5y = 0$, with characteristic equation $\lambda^2 + 6\lambda + 5 = 0$ and zeros $\lambda_1 = -5$ and $\lambda_2 = -1$ is overdamped. If we sketch the lines $v = -5y$ and $v = -y$ on our phase portrait, note, that as predicted, those initial conditions starting in the shaded region cross $y = 0$ exactly once.

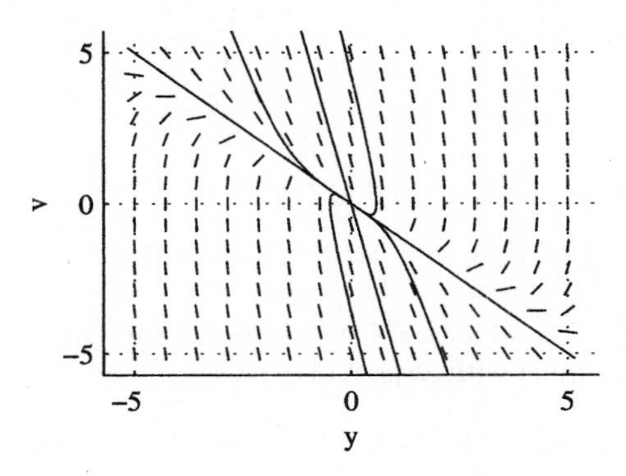

Of course, what we really want is a plot of y versus

t.

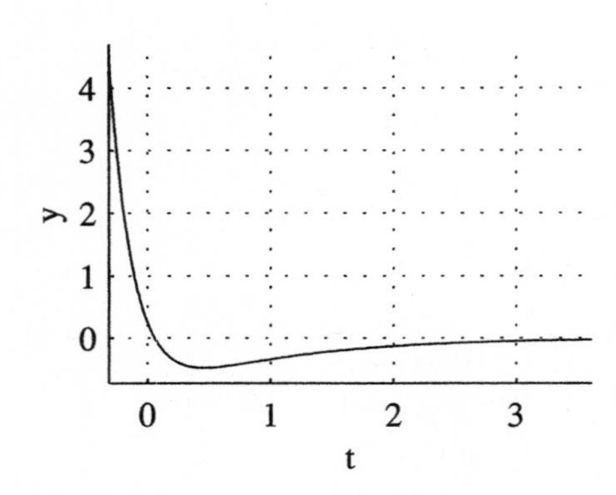

Note that the graph crosses the t-axis exactly once. Finally, by picking initial conditions from the unshaded region, you can find a solution that does not cross the time axis.

Further refinement of the above arguments is possible. For example, how would you limit the initial conditions so that the solution crosses at some time $t > 0$?

19. Suppose that the spring-mass system $my'' + \mu y' + ky = 0$ is critically damped. Then,

$$y'' + \frac{\mu}{m}y' + \frac{k}{m}y = 0$$
$$y'' + 2cy' + \omega_0^2 y = 0,$$

where $2c = \mu/m$ and $\omega_0^2 = k/m$. The characteristic equation is $\lambda^2 + 2c\lambda + \omega_0^2 = 0$ and the zeros are

$$\lambda_1 = -c - \sqrt{c^2 - \omega_0^2} \quad \text{and}$$
$$\lambda_2 = -c + \sqrt{c^2 - \omega_0^2}.$$

In order that the system be critically damped, we need $c^2 - \omega_0^2 = 0$. Of course, this condition results in $\lambda_1 = \lambda_2 = -c$ and the general solution is

$$y(t) = (C_1 + C_2 t)e^{-ct}.$$

To determine how many times this solution can cross the t-axis, set $y(t) = 0$ and solve for t.

$$0 = (C_1 + C_2 t)e^{-ct}$$

Of course, e^{-ct} is never zero, so this leads to

$$C_1 + C_2 t = 0$$
$$t = -\frac{C_1}{C_2}.$$

Thus, the solution will always cross the t-axis (except when $C_1 \neq 0$ and $C_2 = 0$), but let's find only those crossings when $t > 0$, which necessitates that $C_1/C_2 < 0$. The challenge is to find initial conditions y_0 and v_0 that lead to a crossing when $t > 0$. But $y(0) = y_0$ gives

$$y_0 = C_1.$$

Differentiating $y(t)$,

$$y'(t) = ((C_2 - cC_1) - cC_2 t)e^{-ct},$$

the initial condition $y'(0) = v_0$ provides

$$v_0 = C_2 - cC_1,$$

and this system is easily solved for C_1 and C_2,

$$C_1 = y_0 \quad \text{and} \quad C_2 = v_0 + cy_0,$$

which gives us

$$\frac{C_1}{C_2} = \frac{y_0}{v_0 + cy_0}.$$

Because there is a zero crossing at $t > 0$ if $C_1/C_2 < 0$, we see two possible ways that this can happen:

- "Case 1:" We must have $y_0 > 0$ and $v_0 + cy_0 < 0$, which leads to the condition that $y_0 > 0$ and $v_0 < -cy_0$, or
- "Case 2:" We must have $y_0 < 0$ and $v_0 + cy_0 > 0$, which leads to the condition that $y_0 < 0$ and $v_0 > -cy_0$.

These conditions are easy to analyze if you think geometrically.

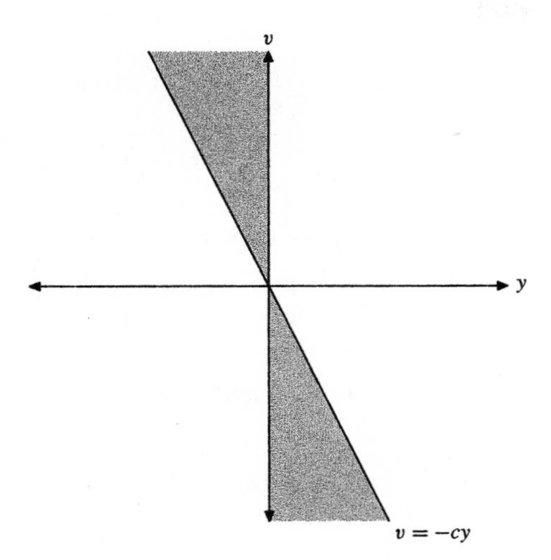

Thus, there will be one zero crossing if you select an initial condition from the shaded region. This is not especially intuitive until you give it a try in your solver. For example, the spring mass system $y'' + 4y'' + 4y = 0$, with characteristic equation $\lambda^2 + 4\lambda + 4 = 0$ and repeated zero $\lambda = -2$ is critically damped. If we sketch the lines $v = -2y$ on our phase portrait, note, that as predicted, those initial conditions starting in the shaded region cross $y = 0$ exactly once at $t > 0$.

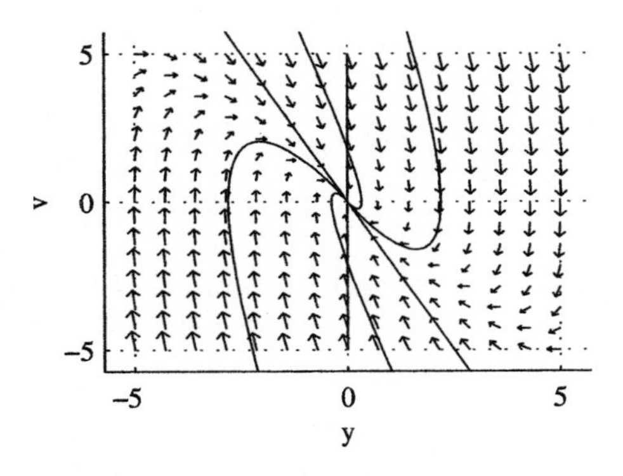

Of course, what we really want is a plot of y versus t.

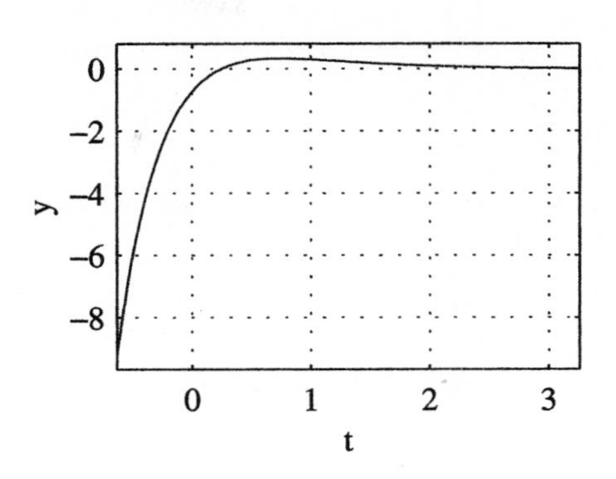

Note that the graph crosses the t-axis exactly once. Finally, by picking initial conditions from the unshaded region, you will note that this solution also crosses the y-axis exactly once, but at $t < 0$.

21. (a) Suppose that $mx'' + \mu x' + kx = 0$ is over-damped. We can write

$$x'' + \frac{\mu}{m}x' + \frac{k}{m}x = 0$$
$$x'' + 2cx' + \omega_0^2 = 0,$$

where $2c = \mu/m$ and $\omega_0^2 = k/m$. The system has characteristic equation $\lambda^2 + 2c\lambda + \omega_0^2 = 0$ and zeros

$$\lambda_1 = -c - \sqrt{c^2 - \omega_0^2} \quad \text{and}$$
$$\lambda_2 = -c + \sqrt{c^2 - \omega_0^2}.$$

If w4e set $\gamma = \sqrt{c^2 - \omega_0^2}$, then

$$\lambda_1 = -c - \gamma \quad \text{and} \quad \lambda_2 = -c + \gamma,$$

and $\lambda_2 - \lambda_1 = 2\gamma$. If the system is overdamped, note that

$$c^2 - \omega_0^2 > 0$$
$$\left(\frac{\mu}{2m}\right)^2 > \frac{k}{m}$$
$$\mu^2 > 4mk$$
$$\mu > 2\sqrt{mk}.$$

The general solution is

$$x(t) = C_1 e^{\lambda_1 t} + C_2 e^{\lambda_2 t}.$$

The initial condition $x(0) = 0$ gives $0 = C_1 + C_2$ and $C_1 = -C_2$. Differentiating $x(t)$,

$$x'(t) = C_1\lambda_1 e^{\lambda_1 t} + C_2\lambda_2 e^{\lambda_2 t},$$

and the initial condition $x'(0) = v_0$ provides $v_0 = C_1\lambda_1 + C_2\lambda_2$. This system is easily solved for

$$C_1 = \frac{v_0}{\lambda_1 - \lambda_2} = -\frac{v_0}{2\gamma} \quad \text{and}$$
$$C_2 = \frac{-v_0}{\lambda_1 - \lambda_2} = \frac{v_0}{2\gamma},$$

so

$$\begin{aligned} x(t) &= -\frac{v_0}{2\gamma}(e^{\lambda_1 t} - e^{\lambda_2 t}) \\ &= -\frac{v_0}{2\gamma}(e^{(-c-\gamma)t} - e^{(-c+\gamma)t}) \\ &= \frac{v_0}{\gamma} \cdot e^{-ct} \cdot \frac{e^{\gamma t} - e^{-\gamma t}}{2} \\ &= \frac{v_0}{\gamma} e^{-ct} \sinh \gamma t. \end{aligned}$$

But, $c = \mu/(2m)$, so

$$x(t) = \frac{v_0}{\gamma} e^{-\mu t/(2m)} \sinh \gamma t,$$

where $\gamma = \sqrt{c^2 - \omega_0^2} = \sqrt{\mu^2 - 4km}/(2m)$.

(b) The initial position is zero. The initial velocity is in the downward direction. Hence, due to the fact that the system is overdamped, we know that the solution has already crossed the t-axis once, and the motion is downward, after which the solution will approach the t-axis asymptotically. Therefore, the mass will reach its lowest point the *first* time the velocity is zero. Differentiating $x(t)$,

$$\begin{aligned} v(t) &= x'(t) \\ &= \frac{v_0}{\gamma} e^{-\mu t/(2m)} (\cosh \gamma t)(\gamma) \\ &\quad + \frac{v_0}{\gamma}\left(-\frac{\mu}{2m}\right) e^{-\mu t/(2m)} \sinh \gamma t \\ &= v_0 e^{\mu t/(2m)} \left(\cosh \gamma t - \frac{\mu}{2m\gamma} \sinh \gamma t\right). \end{aligned}$$

Setting the velocity equal to zero, it must be the case that

$$\begin{aligned} \cosh \gamma t - \frac{\mu}{2m\gamma} \sinh \gamma t &= 0 \\ \frac{\mu}{2m\gamma} \sinh \gamma t &= \cosh \gamma t \\ \tanh \gamma t &= \frac{2m\gamma}{\mu}. \end{aligned}$$

Taking the inverse hyperbolic tangent,

$$\begin{aligned} \gamma t &= \tanh^{-1}\left(\frac{2m\gamma}{\mu}\right) \\ t &= \frac{1}{\gamma} \tanh^{-1}\left(\frac{2m\gamma}{\mu}\right). \end{aligned}$$

(c) If the system is critically damped, then $c^2 - \omega_0^2 = 0$ and the characteristic equation produces a repeated zero, $\lambda = -c$. Thus, the general solution is

$$x(t) = (C_1 + C_2 t)e^{-ct}.$$

The initial condition $x(0) = 0$ produces $C_1 = 0$. The derivative of $x(t)$ is

$$\begin{aligned} v(t) &= x'(t) \\ &= (C_1 + C_2 t)(-ce^{-ct}) + C_2 e^{-ct} \\ &= ((C_2 - cC_1) - cC_2 t)e^{-ct}. \end{aligned}$$

The initial condition $x'(0) = v_0$ gives $v_0 = C_2 - cC_1$, so the fact that $C_1 = 0$ produces $C_2 = v_0$ and the solution

$$x(t) = v_0 t e^{-ct}$$

But $c = \mu/(2m)$, so

$$x(t) = v_0 t e^{-\mu t/(2m)}$$

and

$$\begin{aligned} v(t) &= x'(t) \\ &= v_0 t\left(-\frac{\mu}{2m}\right) e^{-\mu t/(2m)} + v_0 e^{-\mu t/(2m)} \\ &= v_0 e^{-\mu t/(2m)} \left(-\frac{\mu t}{2m} + 1\right). \end{aligned}$$

Because the system is critically damped, and the solution has already crossed the t-axis, the solution will travel downward, then approach the t-axis asymptotically. Hence, the solution reaches its lowest point when the velocity *first* equals zero. This occurs when

$$\frac{\mu t}{2m} = 1$$
$$t = \frac{2m}{\mu}.$$

23. The voltage across the capacitor is given by $V_C = (1/C)Q$. Because the initial voltage across the capacitor is $V_c(0) = 50\,\text{V}$,

$$Q(0) = CV_C(0) = (0.008\,\text{F})(50\,\text{V}) = 0.4\,\text{C},$$

where C is the abbreviation for coulombs. The sum of the voltage drops around the LRC circuit equals zero, so $LI' + RI + (1/C)Q = 0$. Because $I = Q'$, this equation becomes $LQ'' + RQ' + (1/C)Q = 0$. Substituting $L = 4\,\text{H}$, $R = 20\,\Omega$, and $C = 0.008\,\text{F}$, we get $4Q'' + 20Q' + 125Q = 0$. With no initial current this equation becomes

$$Q'' + 5Q' + \frac{125}{4}Q = 0,$$

with intitial conditions $Q(0) = 0.4$ and $Q'(0) = I(0) = 0$. The characteristic equation is $\lambda^2 + 5\lambda + (125/4) = 0$ with zeros $\lambda = -5/2 \pm 5i$. This leads to the complex solution

$$z(t) = e^{(-5/2+5i)t} = e^{-5t/2}(\cos 5t + i \sin 5t).$$

The real and imaginary parts of this solution provide a fundamental set of solutions and the general solution

$$Q(t) = e^{-5t/2}(C_1 \cos 5t + C_2 \sin 5t).$$

The initial condition $Q(0) = 0.4$ gives $C_1 = 2/5$. Differentiating $Q(t)$,

$$Q'(t) = e^{-5t/2}(-5C_1 \sin 5t + 5C_2 \cos 5t) - \frac{5}{2}e^{-5t/2}(C_1 \cos 5t + C_2 \sin 5t).$$

The initial condition $Q'(0) = 0$ leads to $0 = 5C_2 - (5/2)C_1$, and because $C_1 = 2/5$, we have $C_2 = 1/5$ and the solution

$$Q(t) = \frac{1}{5}e^{-5t/2}(2 \cos 5t + \sin 5t).$$

We find the current by differentiating Q to get

$$I(t) = -\frac{5}{2}e^{-5t/2} \sin 5t.$$

Since $-\sin 5t = \cos(5t + \pi/2)$, we see that the phase is $\phi = -\pi/2$. Clearly the amplitude is $5/2$, and the frequency is 5 rad/s.

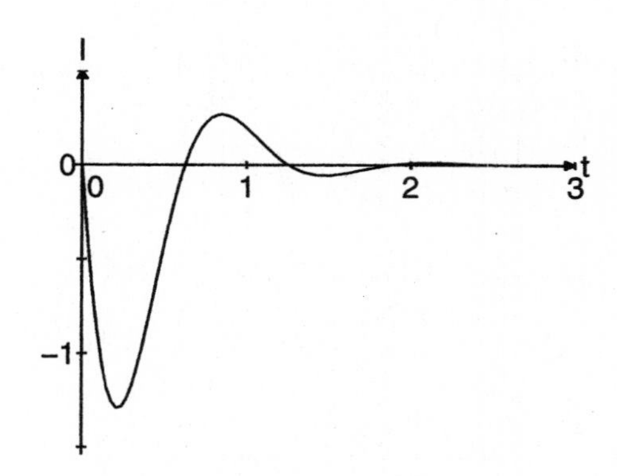

25. The voltage across the capacitor is given by $V_C = (1/C)Q$. Because the initial voltage across the capacitor is $V_c(0) = 1\,\text{V}$,

$$Q(0) = CV_C(0) = (0.02\,\text{F})(1\,\text{V}) = 0.02\,\text{C},$$

where C is the abbreviation for coulombs. The sum of the voltage drops around the LRC circuit equals zero, so $LI' + RI + (1/C)Q = 0$. Because $I = Q'$, this equation becomes $LQ'' + RQ' + (1/C)Q = 0$. Substituting $L = 10\,\text{H}$, $R = 40\,\Omega$, and $C = 0.02\,\text{F}$, we get $10Q'' + 40Q' + 50Q = 0$. With no initial current this equation becomes

$$Q'' + 4Q' + 5Q = 0,$$

with initial condtions $Q(0) = 0.02$ and $Q'(0) = I(0) = 0$. The characteristic equation is $\lambda^2 + 4\lambda + 5 = 0$ with zeros $\lambda = -2 \pm i$. This leads to the complex solution

$$z(t) = e^{(-2+i)t} = e^{-2t}(\cos t + i \sin t).$$

The real and imaginary parts of this solution provide a fundamental set of solutions and the general solution

$$Q(t) = e^{-2t}(C_1 \cos t + C_2 \sin t).$$

The initial condition $Q(0) = 0.02$ gives $C_1 = 1/50$. Differentiating $Q(t)$,

$$Q'(t) = e^{-2t}(-C_1 \sin t + C_2 \cos t) - 2e^{-2t}(C_1 \cos t + C_2 \sin t).$$

The initial condition $Q'(0) = 0$ leads to $0 = C_2 - 2C_1$, and because $C_1 = 1/50$, we have $C_2 = 1/25$ and the solution

$$Q(t) = \frac{1}{50}e^{-2t}(\cos t + 2 \sin t).$$

Plotting the coefficients of $\cos t$ and $\sin t$ and marking the angle,

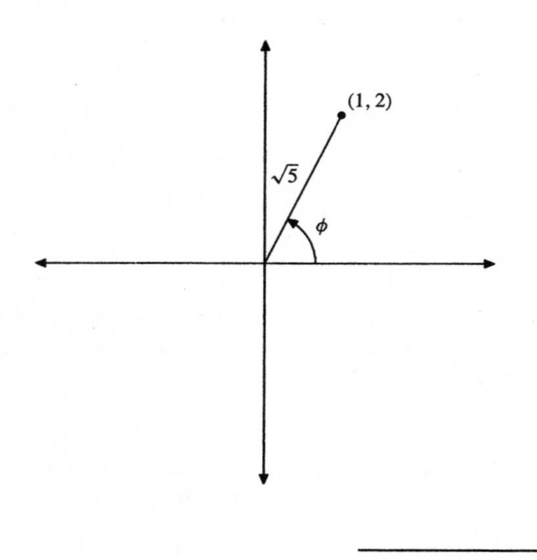

we see that $\tan \phi = 2$, so $\phi = \arctan 2 \approx 1.1071$. Factoring out the magnitude of the radial vector,

$$\begin{aligned} Q(t) &= \frac{\sqrt{5}}{50}e^{-2t}(\frac{1}{\sqrt{5}} \cos t + \frac{2}{\sqrt{5}} \sin t) \\ &= \frac{\sqrt{5}}{50}e^{-2t}(\cos \phi \cos t + \sin \phi \sin t) \\ &= \frac{\sqrt{5}}{50}e^{-2t} \cos(t - 1.1071). \end{aligned}$$

Thus, the amplitude is $\sqrt{5}/50$, the frequency is 1 rad/s, and the phase is 1.1071.

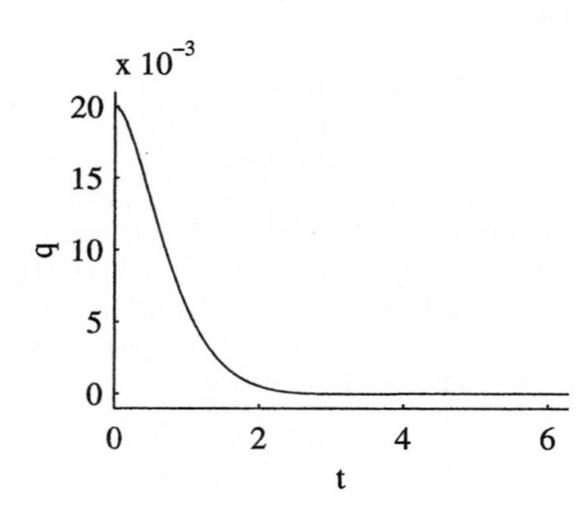

Section 4.5. Inhomogeneous Equations; the Method of Undetermined Coefficients

1. Let $y(t) = Ae^{-3t}$. Then

$$\begin{aligned} y'(t) &= -3Ae^{-3t} \\ y''(t) &= 9Ae^{-3t}, \end{aligned}$$

and $y'' + 3y' + 2y = 4e^{-3t}$ becomes

$$\begin{aligned} 9Ae^{-3t} + 3(-3Ae^{-3t}) + 2(Ae^{-3t}) &= 4e^{-3t} \\ 2A &= 4 \\ A &= 2. \end{aligned}$$

Thus, $y = 2e^{-3t}$ is a particular solution.

3. Let $y(t) = Ae^{-t}$. Then

$$\begin{aligned} y'(t) &= -Ae^{-t} \\ y''(t) &= Ae^{-t}, \end{aligned}$$

and $y'' + 2y' + 5y = 12e^{-t}$ becomes

$$\begin{aligned} Ae^{-t} + 2(-Ae^{-t}) + 5(Ae^{-t}) &= 12e^{-t} \\ 4A &= 12 \\ A &= 3. \end{aligned}$$

Thus, $y = 3e^{-t}$ is a particular solution.

5. Let $y_p = a\cos 3t + b\sin 3t$. Then

$$\begin{aligned} y_p' &= -3a\sin 3t + 3b\cos 3t \\ y_p'' &= -9a\cos 3t - 9b\sin 3t, \end{aligned}$$

and the equation $y'' + 4y = \cos 3t$ becomes

$$\begin{aligned} &-9a\cos 3t - 9b\sin 3t + 4(a\cos 3t + b\sin 3t) \\ &= \cos 3t, \end{aligned}$$

or, equivalently,

$$-5a\cos 3t - 5b\sin 3t = \cos 3t.$$

Thus, $-5a = 1$ and $-5b = 0$ lead to $a = -1/5$ and $b = 0$ and the particular solution $y_p = -1/5\cos 3t$.

7. Let $y_p = a\cos 2t + b\sin 2t$. Then

$$\begin{aligned} y_p' &= -2a\sin 2t + 2b\cos 2t \\ y_p'' &= -4a\cos 2t - 4b\sin 2t, \end{aligned}$$

and substituting these results into $y'' + 7y' + 6y = 3\sin 2t$ leads to

$$(2a + 14b)\cos 2t + (-14a + 2b)\sin 2t = 3\sin 2t.$$

Thus,

$$\begin{aligned} 2a + 14b &= 0 \\ -14a + 2b &= 3, \end{aligned}$$

leading to $a = -21/100$ and $b = 3/100$. Thus, the particular solution is $y_p = -(21/100)\cos 2t + (3/100)\sin 2t$.

9. If $z(t) = x(t) + iy(t)$, then $z'' + pz' + qz = Ae^{i\omega t}$ becomes

$$\begin{aligned} &(x'' + iy'') + p(x' + iy') + q(x + iy) \\ &\quad = A(\cos\omega t + i\sin\omega t) \\ &(x'' + px' + qx) + i(y'' + py' + qy) \\ &\quad = A\cos\omega t + iA\sin\omega t. \end{aligned}$$

Comparing real and imaginary parts,

$$\begin{aligned} x'' + px' + qx &= A\cos\omega t \\ y'' + py' + qy &= A\sin\omega t. \end{aligned}$$

Therefore, if $z(t) = x(t) + iy(t)$ is a solution of $z'' + pz' + qz = Ae^{i\omega t}$, then the real and imaginary parts of z are solutions of $x'' + px' + qx = A\cos\omega t$ and $y'' + py' + qy = A\sin\omega t$, respectively.

11. Let $z = Ae^{i2t}$. Then

$$\begin{aligned} z' &= 2iAe^{i2t} \\ z'' &= (2i)^2 Ae^{i2t}, \end{aligned}$$

and $z'' + 9z = e^{i2t}$ leads to

$$\begin{aligned} (2i)^2 Ae^{i2t} + 9Ae^{i2t} &= e^{i2t} \\ ((2i)^2 + 9)A &= 1 \\ A &= \frac{1}{5}. \end{aligned}$$

Thus, $z = (1/5)e^{i2t}$ is a particular solution of $z'' + 9z = e^{i2t}$ and the imaginary part of this solution, $y = (1/5)\sin 2t$, is the solution of $y'' + 9y = \sin 2t$.

13. Let $z = Ae^{i3t}$. Then

$$\begin{aligned} z' &= (3i)Ae^{i3t} \\ z'' &= (3i)^2 Ae^{i3t}, \end{aligned}$$

and $z'' + 7z' + 10z = -4e^{i3t}$ leads to

$$\begin{aligned} (3i)^2 Ae^{i3t} + 7(3i)Ae^{i3t} + 10Ae^{i3t} &= -4e^{i3t} \\ ((3i)^2 + 7(3i) + 10)A &= -4 \end{aligned}$$

Solving for A, we get

$$\begin{aligned} A &= -\frac{4}{1 + 21i} \\ &= -\frac{2}{221} + \frac{42}{221}i. \end{aligned}$$

Thus,

$$\begin{aligned} z &= \left(-\frac{2}{221}+\frac{42}{221}i\right)e^{i3t} \\ &= \left(-\frac{2}{221}+\frac{42}{221}i\right)(\cos 3t + i\sin 3t) \\ &= \left(-\frac{2}{221}\cos 3t - \frac{42}{221}\sin 3t\right) \\ &\quad + i\left(\frac{42}{221}\cos 3t - \frac{2}{221}\sin 3t\right) \end{aligned}$$

is a solution of $z''+7z'+10z=-4e^{i3t}$. The imaginary part,

$$y = \frac{42}{221}\cos 3t - \frac{2}{221}\sin 3t,$$

is a solution of $y''+7y'+10y=-4\sin 3t$.

15. Let $y = at+b$. Then, $y' = a$, $y''=0$, and $y''+6y'+8y=2t-3$ leads to $8at+(6a+8b) = 2t-3$. Thus, $8a=2$ and $6a+8b=-3$, which gives $a=1/4$ and $b=-9/16$. Therefore, the particular solution is $y=(1/4)t-9/16$.

17. Let $y=at^3+bt^2+ct+d$. Then, substituting

$$\begin{aligned} y' &= 3at^2+2bt+c \\ y'' &= 6at+2b \end{aligned}$$

into $y''+3y'+4y=t^3$, arranging terms,

$$\begin{aligned} &4at^3+(9a+4b)t^2+(6a+6b+4c)t \\ &\quad+(2b+3c+4d) \\ &= t^3, \end{aligned}$$

and comparing coefficients,

$$\begin{aligned} 4a &= 1 \\ 9a+4b &= 0 \\ 6a+6b+4c &= 0 \\ 2b+3c+4d &= 0, \end{aligned}$$

provides $a=1/4$, $b=-9/16$, $c=15/32$, and $d=-9/128$. Thus, a particular solution is

$$y = \frac{1}{4}t^3 - \frac{9}{16}t^2 + \frac{15}{32}t - \frac{9}{128}.$$

19. The homogeneous equation $y''-4y'-5y=0$ has characteristic equation $\lambda^2-4\lambda-5=(\lambda-5)(\lambda+1)=0$ with zeros $\lambda_1=5$ and $\lambda_2=-1$. This leads to the homogeneous solution

$$y_h = C_1e^{5t}+C_2e^{-t}.$$

The particular solution $y_p=Ae^{-2t}$ has derivatives $y_p' = -2Ae^{-2t}$ and $y_p''=4Ae^{-2t}$, which when substituted into the equation $y''-4y'-5y=4e^{-2t}$ provides

$$\begin{aligned} 4Ae^{-2t}+8Ae^{-2t}-5Ae^{-2t} &= 4e^{-2t} \\ 7A &= 4 \\ A &= \frac{4}{7}. \end{aligned}$$

Thus, a particular solution is $y_p=(4/7)e^{-2t}$. This leads to the general solution

$$y = C_1e^{5t}+C_2e^{-t}+\frac{4}{7}e^{-2t}.$$

The initial condition $y(0)=0$ gives

$$0 = C_1+C_2+\frac{4}{7}.$$

Differentiating,

$$y' = 5C_1e^{5t}-C_2e^{-t}-\frac{8}{7}e^{-2t}.$$

The initial condition $y'(0)=-1$ provides

$$-1 = 5C_1-C_2-\frac{8}{7}.$$

This system has solution $C_1=-1/14$ and $C_2=-1/2$, leading to the solution

$$y = -\frac{1}{14}e^{5t}-\frac{1}{2}e^{-t}+\frac{4}{7}e^{-2t}.$$

21. The homogeneous equation $y''-2y'+5y=0$ has characteristic equation $\lambda^2-2\lambda+5=0$ with zeros $\lambda_1=1+2i$ and $\lambda_2=1-2i$. This leads to the homogeneous solution

$$y_h = e^t(C_1\cos 2t + C_2\sin 2t).$$

The particular solution $z = Ae^{it}$ has derivatives

$$z' = iAe^{it}$$
$$z'' = i^2e^{it},$$

which, when inserted in the complex equation $z'' - 2z' + 5z = 3e^{it}$, gives

$$i^2Ae^{it} - 2iAe^{it} + 5Ae^{it} = 3e^{it}$$
$$(i^2 - 2i + 5)A = 3$$
$$A = \frac{3}{4-2i}$$
$$A = \frac{3}{5} + \frac{3}{10}i.$$

This gives the particular solution

$$z = \left(\frac{3}{5} + \frac{3}{10}i\right)e^{it}$$
$$z = \left(\frac{3}{5}\cos t + \frac{3}{10}i\right)(\cos t + i\sin t).$$

The real part of this solution is a particular solution of $y'' - 2y' + 5y = 3\cos t$.

$$y_p = \frac{3}{5}\cos t - \frac{3}{10}\sin t$$

Thus, the general solution is

$$y = e^t(C_1\cos 2t + C_2\sin 2t) + \frac{3}{5}\cos t - \frac{3}{10}\sin t.$$

The initial condition $y(0) = 0$ gives $0 = C_1 + 3/5$. Differentiating,

$$\begin{aligned} y' &= e^t(-2C_1\sin 2t + 2C_2\cos 2t) \\ &\quad + e^t(C_1\cos 2t + C_2\sin 2t) \\ &\quad - \frac{3}{5}\sin t - \frac{3}{10}\cos t. \end{aligned}$$

The initial condition $y'(0) = -2$ provides $-2 = 2C_2 + C_1 - 3/10$. The solution of this system is $C_1 = -3/5$ and $C_2 = -11/20$. Therefore, the solution is

$$\begin{aligned} y &= e^t\left(-\frac{3}{5}\cos 2t - \frac{11}{20}\sin 2t\right) \\ &\quad + \frac{3}{5}\cos t - \frac{3}{10}\sin t. \end{aligned}$$

23. The homogeneous equation $y'' - 2y' + y = 0$ has characteristic equation $\lambda^2 - 2\lambda + 1 = (\lambda - 1)^2$, with repeated zero $\lambda = 1$. Thus, the homogeneous solution is

$$y_h = (C_1 + C_2t)e^t.$$

The particular solution $y_p = at^3 + bt^2 + ct + d$ has derivatives

$$y_p' = 3at^2 + 2bt + c$$
$$y'' = 6at + 2b,$$

which when substituted in $y'' - 2y' + y = t^3$, rearranging, yields

$$at^3 + (-6a+b)t^2 + (6a-4b+c)t + (2b-2c+d) = t^3.$$

Thus,

$$\begin{aligned} a &= 1 \\ -6a + b &= 0 \\ 6a - 4b + c &= 0 \\ 2b - 2c + d &= 0, \end{aligned}$$

which has solution $a = 1$, $b = 6$, $c = 18$, and $d = 24$. Thus, the general solution is

$$y = (C_1 + C_2t)e^t + t^3 + 6t^2 + 18t + 24.$$

The initial condition $y(0) = 1$ gives $1 = C_1 + 24$. Differentiating,

$$y' = C_2e^t + (C_1 + C_2t)e^t + 3t^2 + 12t + 18.$$

The initial condition $y'(0) = 0$ gives $0 = C_2 + C_1 + 18$. The system has solution $C_1 = -23$ and $C_2 = 5$. Therefore, the solution is

$$y = (-23 + 5t)e^t + t^3 + 6t^2 + 18t + 24.$$

25. The homogeneous equation $y'' - y' - 2y = 0$ has characteristic equation $\lambda^2 - \lambda - 2 = (\lambda - 2)(\lambda + 1) = 0$ with zeros $\lambda_1 = 2$ and $\lambda_2 = -1$. Thus, the homogeneous solution is

$$y_h = C_1e^{2t} + C_2e^{-t}.$$

The forcing term of $y'' - y' - 2y = 2e^{-t}$ is also a solution of the homogeneous equation, so we multiply by a factor of t and try $y_p = Ate^{-t}$, which has derivatives

$$y' = Ae^{-t}(1 - t)$$
$$y'' = Ae^{-t}(t - 2).$$

Substituting these in $y'' - y' - 2y = 2e^{-t}$ gives

$$\begin{aligned} Ae^{-t}(t-2) - Ae^{-t}(1-t) - 2Ate^{-t} &= 2e^{-t} \\ A(t-2-1+t-2t) &= 2 \\ -3A &= 2 \\ A &= -\frac{2}{3}. \end{aligned}$$

Thus, a particular solution is $y_p = -(2/3)te^{-t}$.

27. The homogeneous equation $z'' + 9z = 0$ has characteristic equation $\lambda^2 + 9 = 0$, which has zeros $\lambda_1 = 3i$ and $\lambda_2 = -3i$. Thus, the homogeneous solution is

$$z_h = C_1 e^{3it} + C_2 e^{-3it}.$$

The forcing term of $z'' + 9z = e^{3it}$ is also a solution of the homogeneous equation, so multiply by a factor of t and try the particular solution $z_p = Ate^{3it}$. The particular solution has derivatives

$$\begin{aligned} z_p' &= Ae^{3it}(1+3it) \\ z_p'' &= 3iAe^{3it}(2+3it). \end{aligned}$$

If these are substituted into $z'' + 9z = e^{3it}$, then

$$\begin{aligned} 3iAe^{3it}(2+3it) + 9Ate^{3it} &= e^{3it} \\ 3iA(2+3it) + 9At &= 1 \\ 6iA &= 1 \\ A &= -\frac{1}{6}i. \end{aligned}$$

Thus,

$$\begin{aligned} z_p &= -\frac{1}{6}ite^{3it} \\ &= -\frac{1}{6}it(\cos 3t + i\sin 3t) \\ &= \frac{1}{6}t\sin 3t + i\left(-\frac{1}{6}t\cos 3t\right). \end{aligned}$$

The imaginary part of this solution is a particular solution of $y'' + 9y = \sin 3t$.

$$y_p = -\frac{1}{6}t\cos 3t$$

29. The homogeneous equation $y'' + 6y' + 9y = 0$ has characteristic equation $\lambda^2 + 6\lambda + 9 = (\lambda+3)^2 = 0$ and repeated zero $\lambda = -3$. Thus, the homogeneous solution is

$$y_h = (C_1 + C_2 t)e^{-3t}.$$

The forcing term of $y'' + 6y' + 9y = 5e^{-3t}$ is a solution of the homogeneous equation, as is te^{-3t}. Consequently, try $y_p = At^2e^{-3t}$, which has derivatives

$$\begin{aligned} y_p' &= -3At^2e^{-3t} + 2Ate^{-3t} \\ y_p'' &= 9At^2e^{-3t} - 12Ate^{-3t} + 2Ae^{-3t}. \end{aligned}$$

Substituting these into $y'' + 6y' + 9y = 5e^{-3t}$ yields

$$\begin{aligned} 2Ae^{-3t} &= 5e^{-3t} \\ 2A &= 5 \\ A &= \frac{5}{2}. \end{aligned}$$

Therefore, $y_p = (5/2)t^2e^{-3t}$ is a solution of $y'' + 6y' + 9y = 5e^{-3t}$.

31. It is easy to see that $y = 1$ is a solution of $y'' + 2y' + 2y = 2$. Next, $z_p = Ae^{i2t}$ has derivatives

$$\begin{aligned} z_p' &= 2iAe^{i2t} \\ z_p'' &= (2i)^2Ae^{i2t}. \end{aligned}$$

Substituting these into $z'' + 2z' + 2z = e^{i2t}$,

$$\begin{aligned} (2i)^2Ae^{i2t} + 2(2i)Ae^{i2t} + 2Ae^{i2t} &= e^{i2t} \\ ((2i)^2 + 2(2i) + 2)A &= 1 \\ A &= \frac{1}{-2+4i} \\ A &= -\frac{1}{10} - \frac{1}{5}i. \end{aligned}$$

Thus,

$$\begin{aligned} z_p &= \left(-\frac{1}{10} - \frac{1}{5}i\right)e^{i2t} \\ &= \left(-\frac{1}{10} - \frac{1}{5}i\right)(\cos 2t + i\sin 2t) \\ &= \left(-\frac{1}{10}\cos 2t + \frac{1}{5}\sin 2t\right) \\ &\quad + i\left(-\frac{1}{5}\cos 2t - \frac{1}{10}\sin 2t\right). \end{aligned}$$

The real part of this, $y = -(1/10)\cos 2t + (1/5)\sin 2t$, is a solution of $y'' + 2y' + 2y = \cos 2t$. Thus,

$$y = 1 - \frac{1}{10}\cos 2t + \frac{1}{5}\sin 2t$$

is a solution of $y'' + 2y' + 2y = 2 + \cos 2t$.

33. First, let $y = at + b$, with derivatives $y' = a$ and $y'' = 0$, be a solution of $y'' + 25y = 2 + 3t$. Substituting,

$$25at + 25b = 2 + 3t,$$

which, upon equating coefficients, leads to $a = 2/25$ and $b = 3/25$ and the particular solution $y = (2/25)t + 3/25$. Next, the homogeneous equation $z'' + 25z = 0$ has characteristic $\lambda^2 + 25 = 0$ and zeros $\lambda = \pm 5i$, giving the complex solution

$$z = C_1 e^{5it} + C_2 e^{-5it}.$$

Because $z = e^{5it}$ is a solution of the homogeneous equation, we try $z = Ate^{5it}$ as a solution of $z'' + 25z = e^{5it}$. The derivatives

$$z' = (5i)Ate^{5it} + Ae^{5it}$$
$$z'' = (5i)^2 Ate^{5it} + 2(5i)Ae^{5it},$$

when substituted in $z'' + 25z = e^{5it}$, provide

$$(5i)^2 Ate^{5it} + 2(5i)Ae^{5it} + 25Ate^{5it} = e^{5it}$$
$$10iA = 1$$
$$A = -\frac{1}{10}i.$$

Thus,

$$z = -\frac{1}{10}ite^{5it} = -\frac{1}{10}it(\cos 5t + i\sin 5t).$$

The real part of this solution, $y = (1/10)t\sin 5t$, is a solution of $y'' + 25y = \cos 5t$. Thus,

$$y = (2/25)t + 3/25 + (1/10)t\sin 5t$$

is a solution of $y'' + 25y = 2 + 3t + \cos 5t$.

35. Substitute $z = Ae^{i2t}$ and its derivatives

$$z' = (2i)Ae^{i2t}$$
$$z'' = (2i)^2 Ae^{i2t}$$

into $z'' + 4z' + 3z = e^{i2t}$.

$$(2i)^2 Ae^{i2t} + 4(2i)Ae^{i2t} + 3Ae^{i2t} = e^{i2t}$$
$$((2i)^2 + 4(2i) + 3)A = 1$$
$$A = -\frac{1}{65} - \frac{8}{65}i.$$

Thus,

$$\begin{aligned} z &= \left(-\frac{1}{65} - \frac{8}{65}i\right)e^{i2t} \\ &= \left(-\frac{1}{65} - \frac{8}{65}i\right)(\cos 2t + i\sin 2t) \\ &= \left(-\frac{1}{65}\cos 2t + \frac{8}{65}\sin 2t\right) \\ &\quad + i\left(-\frac{8}{65}\cos 2t - \frac{1}{65}\sin 2t\right). \end{aligned}$$

Thus, $y = -(1/65)\cos 2t + (8/65)\sin 2t$ is a solution of $y'' + 4y' + 3y = \cos 2t$ and $y = -(8/65)\cos 2t - (1/65)\sin 2t$ is a solution of $y'' + 4y' + 3y = \sin 2t$. This means that

$$\begin{aligned} y &= \left(-\frac{1}{65}\cos 2t + \frac{8}{65}\sin 2t\right) \\ &\quad + 3\left(-\frac{8}{65}\cos 2t - \frac{1}{65}\sin 2t\right) \\ &= -\frac{5}{13}\cos 2t + \frac{1}{13}\sin 2t \end{aligned}$$

is a solution of $y'' + 4y' + 3y = \cos 2t + 3\sin 2t$.

37. The homogeneous equation $y'' + 4y' + 4y = 0$ has characteristic equation $\lambda^2 + 4\lambda + 4 = (\lambda + 2)^2 = 0$ with repeated zero $\lambda = -2$ and solution $y = (C_1 + C_2 t)e^{-2t}$. Thus, both e^{-2t} and te^{-2t} are solutions of the homogeneous equation. Thus, try $y = At^2e^{-2t}$ as a particular solution of $y'' + 4y' + 4y = e^{-2t}$.

$$y' = -2At^2e^{-2t} + 2Ate^{-2t}$$
$$y'' = 4At^2e^{-2t} - 8Ate^{-2t} + 2Ae^{-2t}$$

Substitute y and its derivatives in $y'' + 4y' + 4y = e^{-2t}$ to get

$$2Ae^{-2t} = e^{-2t}$$
$$2A = 1$$
$$A = \frac{1}{2}.$$

Thus, $y = (1/2)t^2e^{-2t}$ is a solution of $y'' + 4y' + 4y = e^{-2t}$. Next, let $z = Ae^{i2t}$ and

$$z' = (2i)Ae^{i2t}$$
$$z'' = (2i)^2Ae^{i2t}.$$

Substituting in $z'' + 4z' + 4z = e^{i2t}$,

$$(2i)^2Ae^{i2t} + 4(2i)Ae^{i2t} + 4Ae^{i2t} = e^{i2t}$$
$$((2i)^2 + 4(2i) + 4)A = 1$$
$$A = -\frac{1}{8}i,$$

so

$$z = -\frac{1}{8}ie^{i2t} = -\frac{1}{8}i(\cos 2t + i\sin 2t).$$

The imaginary part of this solution, $y = -(1/8)\cos 2t$, is a solution of $y''+4y'+4y = \sin 2t$. Thus,

$$y = \frac{1}{2}t^2e^{-2t} - \frac{1}{8}\cos 2t$$

is a solution of $y'' + 4y' + 4y = e^{-2t} + \sin 2t$.

39. If $y = (at + b)e^{-4t}$, then

$$y' = e^{-4t}(a - 4at - 4b)$$
$$y'' = e^{-4t}(-8a + 16at + 16b),$$

and substituting in $y'' + 3y' + 2y = te^{-4t}$ gives

$$\begin{aligned}(-8a &+ 16at + 16b) + 3(a - 4at - 4b)\\ &+ 2(at + b)\\ &= t\end{aligned}$$

or $6at + (-5a + 6b) = t$. Comparing coefficients, $6a = 1$ and $-5a + 6b = 0$. This gives $a = 1/6$ and $b = 5/36$ and $y = ((1/6)t + 5/36)e^{-4t}$ as a solution of $y'' + 3y' + 2y = te^{-4t}$.

41. Let $y = (at^2 + bt + c)e^{-2t}$. Then,

$$y' = e^{-2t}(-2at^2 + (2a - 2b)t + (b - 2c))$$
$$y'' = e^{-2t}(4at^2 + (-8a + 4b)t + (2a - 4b + 4c)).$$

Substituting in $y'' + 2y' + y = t^2e^{-2t}$,

$$\begin{aligned}(4at^2 &+ (-8a + 4b)t + (2a - 4b + 4c))\\ &+ 2(-2at^2 + (2a - 2b)t + (b - 2c))\\ &+ (at^2 + bt + c)\\ &= t^2,\end{aligned}$$

or $at^2+(-4a+b)t+(2a-2b+c) = t^2$. Comparing coefficients,

$$a = 1$$
$$-4a + b = 0$$
$$2a - 2b + c = 0$$

and $a = 1$, $b = 4$, and $c = 6$. Thus, $y = (t^2+4t+6)e^{-2t}$ is a solution of $y''+2y'+y = t^2e^{-2t}$.

43. The homogeneous equation $y'' + 3y' + 2y = 0$ has characteristic equation $\lambda^2 + 3\lambda + 2 = (\lambda + 1)(\lambda + 2) = 0$ and solution

$$y = C_1e^{-t} + C_2e^{-2t}.$$

Thus, instead of $y = (at^2 + bt + c)e^{-2t}$, we will try $y = t(at^2 + bt + c)e^{-2t}$. After some work, the derivatives are found.

$$y' = e^{-2t}(-2at^3 + (3a - 2b)t^2 + (2b - 2c)t + c)$$
$$\begin{aligned}y'' = e^{-2t}(4at^3 &+ (-12a + 4b)t^2\\ &+ (6a - 8b + 4c)t + (2b - 4c))\end{aligned}$$

Substitute these into the differential equation $y'' + 3y' + 2y = t^2e^{-2t}$. After some computation,

$$-3at^2 + (6a - 2b)t + (2b - c) = t^2.$$

Comparing coefficients,

$$-3a = 1$$
$$6a - 2b = 0$$
$$2b - c = 0$$

leading to $a = -1/3$, $b = -1$, and $c = -2$. Thus,

$$y = t\left(-\frac{1}{3}t^2 - t - 2\right)e^{-2t}$$

is a solution of $y'' + 3y' + 2y = t^2e^{-2t}$.

45. If $z(t) = x(t)+iy(t)$ is a solution of $z''+pz'+qz = Ae^{(a+bi)t}$, then

$$(x'' + iy'') + p(x' + iy') + q(x + iy) = Ae^{at}e^{ibt}$$

or

$$(x'' + px' + qx) + i(y'' + py' + qy) = Ae^{at}\cos bt + iAe^{at}\sin bt.$$

Equating real and imaginary parts,

$$x'' + px' + qx = Ae^{at}\cos bt$$
$$y'' + py' + qy = Ae^{at}\sin bt.$$

47. If $z(t) = x(t) = iy(t)$ is a solution of $z'' + z' + z = te^{it}$, then

$$(x'' + iy'') + (x' + iy') + (x + iy) = t(\cos t + i\sin t)$$
$$(x'' + x' + x) + i(y'' + y' + y) = t\cos t + it\sin t.$$

Comparing imaginary parts,

$$y'' + y' + y = t\sin t.$$

Thus, if $z = (at + b)e^{it}$ and its derivatives,

$$z' = e^{it}(a + i(at + b))$$
$$z'' = e^{it}(-at - b + 2ai)$$

are substituted into $z'' + z' + z = te^{it}$, then

$$(-at - b + 2ai) + (a + i(at + b)) + (at + b) = t,$$

or $iat + ((1 + 2i)a + ib) = t$. Thus,

$$ia = 1$$
$$(1 + 2i)a + ib = 0,$$

and $a = -i$ and $b = 1 + 2i$. Thus, a particular solution of $z'' + z' + z = te^{it}$ is

$$\begin{aligned} z &= (-it + (1 + 2i))e^{it} \\ &= (1 + i(2 - t))(\cos t + i\sin t) \\ &= (\cos t - (2 - t)\sin t) + i((2 - t)\cos t + \sin t). \end{aligned}$$

Therefore, the imaginary part, $y = (2 - t)\cos t + \sin t$, is a solution of $y'' + y' + y = t\sin t$.

Section 4.6. Variation of Parameters

1. The homogeneous equation $y''+9y = 0$ has $y_1(t) = \cos 3t$ and $y_2(t) = \sin 3t$ as a fundamental set of solutions. The Wronskian is

$$\begin{aligned} W(\cos 3t, \sin 3t) &= \begin{vmatrix} \cos 3t & \sin 3t \\ -3\sin 3t & 3\cos 3t \end{vmatrix} \\ &= 3\cos^2 3t + 3\sin^2 3t \\ &= 3. \end{aligned}$$

Form the solution

$$y_p = v_1y_1 + v_2y_2$$

where v_1 and v_2 are to be determined. Indeed,

$$\begin{aligned} v_1' &= \frac{-y_2g(t)}{W(y_1, y_2)} \\ &= \frac{-\sin 3t\tan 3t}{3} \\ &= -\frac{1}{3}\cdot\frac{\sin^2 3t}{\cos 3t} \\ &= -\frac{1}{3}\cdot\frac{1-\cos^2 3t}{\cos 3t} \\ &= -\frac{1}{3}(\sec 3t - \cos 3t). \end{aligned}$$

Thus,

$$v_1 = -\frac{1}{9}\ln|\sec 3t + \tan 3t| + \frac{1}{9}\sin 3t.$$

Secondly,

$$\begin{aligned} v_2' &= \frac{y_1 g(t)}{W(y_1, y_2)} \\ &= \frac{\cos 3t \tan 3t}{3} \\ &= \frac{1}{3}\sin 3t. \end{aligned}$$

Thus,

$$v_2 = -\frac{1}{9}\cos 3t.$$

Inserting these results in $y_p = v_1 y_1 + v_2 y_2$,

$$\begin{aligned} y_p &= \left[-\frac{1}{9}\ln|\sec 3t + \tan 3t| + \frac{1}{9}\sin 3t\right]\cos 3t \\ &\quad + \left(-\frac{1}{9}\cos 3t\right)\sin 3t \\ &= -\frac{1}{9}\cos 3t \ln|\sec 3t + \tan 3t| \end{aligned}$$

is the particular solution we seek.

3. The homogeneous equation $y'' - y = 0$ has $y_1(t) = e^t$ and $y_2(t) = e^{-t}$ as a fundamental set of solutions. The Wronskian is

$$\begin{aligned} W(e^t, e^{-t}) &= \begin{vmatrix} e^t & e^{-t} \\ e^t & -e^{-t} \end{vmatrix} \\ &= -2. \end{aligned}$$

Form the solution

$$y_p = v_1 y_1 + v_2 y_2$$

where v_1 and v_2 are to be determined. Indeed,

$$\begin{aligned} v_1' &= \frac{-y_2 g(t)}{W(y_1, y_2)} \\ &= \frac{-e^{-t}(t+3)}{-2} \\ &= \frac{1}{2}e^{-t}(t+3). \end{aligned}$$

Integrating by parts,

$$v_1 = -\frac{1}{2}e^{-t}(t+3) - \frac{1}{2}e^{-t}.$$

Secondly,

$$\begin{aligned} v_2' &= \frac{y_1 g(t)}{W(y_1, y_2)} \\ &= \frac{e^t(t+3)}{-2} \\ &= -\frac{1}{2}e^t(t+3). \end{aligned}$$

Integrating by parts,

$$v_2 = -\frac{1}{2}e^t(t+3) + \frac{1}{2}e^t.$$

Inserting these results in $y_p = v_1 y_1 + v_2 y_2$,

$$\begin{aligned} y_p &= \left(-\frac{1}{2}e^{-t}(t+3) - \frac{1}{2}e^{-t}\right)e^t \\ &\quad + \left(-\frac{1}{2}e^t(t+3) + \frac{1}{2}e^t\right)e^{-t} \\ &= -\frac{1}{2}(t+3) - \frac{1}{2} - \frac{1}{2}(t+3) + \frac{1}{2} \\ &= -(t+3) \end{aligned}$$

is the particular solution we seek.

5. The homogeneous equation $y'' - 2y' + y = 0$ has $y_1(t) = e^t$ and $y_2(t) = te^t$ as a fundamental set of solutions. The Wronskian is

$$\begin{aligned} W(e^t, te^t) &= \begin{vmatrix} e^t & te^t \\ e^t & te^t + e^t \end{vmatrix} \\ &= e^{2t}. \end{aligned}$$

Form the solution

$$y_p = v_1 y_1 + v_2 y_2$$

where v_1 and v_2 are to be determined. Indeed,

$$\begin{aligned} v_1' &= \frac{-y_2 g(t)}{W(y_1, y_2)} \\ &= \frac{(-te^t)e^t}{e^{2t}} \\ &= -t. \end{aligned}$$

Thus,

$$v_1 = -\frac{1}{2}t^2.$$

Secondly,

$$\begin{aligned} v_2' &= \frac{y_1 g(t)}{W(y_1, y_2)} \\ &= \frac{e^t e^t}{e^{2t}} \\ &= 1 \end{aligned}$$

Thus,

$$v_2 = t.$$

Inserting these results in $y_p = v_1 y_1 + v_2 y_2$,

$$\begin{aligned} y_p &= -\frac{1}{2}t^2 e^t + t(te^t) \\ &= -\frac{1}{2}t^2 e^t + t^2 e^t \\ &= \frac{1}{2}t^2 e^t. \end{aligned}$$

is the particular solution we seek.

7. The homogeneous equation $x'' + x = 0$ has $x_1(t) = \cos t$ and $x_2(t) = \sin t$ as a fundamental set of solutions. The Wronskian is

$$\begin{aligned} W(\cos t, \sin t) &= \begin{vmatrix} \cos t & \sin t \\ -\sin t & \cos t \end{vmatrix} \\ &= 1. \end{aligned}$$

Form the solution

$$x_p = v_1 x_1 + v_2 x_2$$

where v_1 and v_2 are to be determined. Indeed,

$$\begin{aligned} v_1' &= \frac{-\sin t \tan^2 t}{1} \\ &= -\sin t(\sec^2 t - 1) \\ &= -\sec t \tan t + \sin t. \end{aligned}$$

Thus,

$$v_1 = -\sec t - \cos t.$$

Secondly,

$$\begin{aligned} v_2' &= \frac{\cos t \tan^2 t}{1} \\ &= \cos t(\sec^2 t - 1) \\ &= \sec t - \cos t. \end{aligned}$$

Thus,

$$v_2 = \ln|\sec t + \tan t| - \sin t.$$

Inserting these results in $x_p = v_1 x_1 + v_2 x_2$,

$$\begin{aligned} x_p &= (-\sec t - \cos t)\cos t \\ &\quad + (\ln|\sec t + \tan t| - \sin t)\sin t \\ &= -1 - \cos^2 t + \sin t \ln|\sec t + \tan t| - \sin^2 t \\ &= -2 + \sin t \ln|\sec t + \tan t| \end{aligned}$$

is the particular solution we seek.

9. A fundamental set of solutions to the homogeneous equation is $x_1(t) = \cos t$ and $x_2(t) = \sin t$. We look for a solution of the form

$$\begin{aligned} x(t) &= v_1(t)x_1(t) + v_2(t)x_2(t) \\ &= v_1(t)\cos t + v_2(t)\sin t. \end{aligned}$$

Differentiating we get

$$x' = v_1' \cos t + v_2' \sin t - v_1 \sin t + v_2 \cos t.$$

To simplify future calculations we set

$$v_1' \cos t + v_2' \sin t = 0.$$

Then $x' = -v_1 \sin t + v_2 \cos t$, and

$$x'' = -v_1' \sin t + v_2' \cos t - v_1 \cos t - v_2 \sin t.$$

Adding, we get

$$\begin{aligned} x'' + x &= -v_1' \sin t + v_2' \cos t \\ &= \tan^3 t. \end{aligned}$$

Thus we must solve the system

$$\begin{aligned} v_1' \cos t + v_2' \sin t &= 0, \\ -v_1' \sin t + v_2' \cos t &= \tan^3 t. \end{aligned}$$

The solutions are

$$v_1' = -\sin t \tan^3 t \quad \text{and} \quad v_2' = \cos t \tan^3 t.$$

Integrating we get

$$\begin{aligned} v_1(t) &= -\int \sin t \tan^3 t \, dt \\ &= \int \frac{\sin^4 t}{\cos^3 t} dt \\ &= -\int \frac{1 - 2\cos^2 t + \cos^4 t}{\cos^3 t} dt \\ &= -\int (\sec^3 t - 2\sec t + \cos t)\, dt. \end{aligned}$$

Using integration by parts,

$$\int \sec^3 t\,dt = \int \sec t \sec^2 t\,dt$$
$$= \sec t \tan t - \int \sec t \tan^2 t\,dt.$$

Using the identity $1 + \tan^2 t = \sec^2 t$ reveals

$$\int \sec^3 t\,dt = \frac{1}{2}\sec t \tan t + \frac{1}{2}\ln|\sec t \tan t|.$$

Thus,

$$\begin{aligned} v_1(t) &= -\frac{1}{2}\sec t \tan t - \frac{1}{2}\ln|\sec t + \tan t| \\ &\quad + 2\ln|\sec t + \tan t| - \sin t \\ &= -\frac{1}{2}\sec t \tan t + \frac{3}{2}\ln|\sec t + \tan t| - \sin t. \end{aligned}$$

Integrating again,

$$\begin{aligned} v_2(t) &= \int \cos t \tan^3 t\,dt \\ &= \int \frac{\sin^3 t}{\cos^2 t}\,dt \\ &= \int \frac{\sin t(1-\cos^2 t)}{\cos^2 t}\,dt \\ &= \int \left(\frac{\sin t}{\cos^2 t} - \sin t\right)dt \\ &= \int (\sec t \tan t - \sin t)\,dt \\ &= \sec t + \cos t. \end{aligned}$$

Thus, the solution is

$$\begin{aligned} x(t) &= v_1(t)\cos t + v_2(t)\sin t \\ &= \left(-\frac{1}{2}\sec t \tan t + \frac{3}{2}\ln|\sec t \right. \\ &\qquad \left. + \tan t| - \sin t\right)\cos t \\ &\quad + (\sec t + \cos t)\sin t \\ &= -\frac{1}{2}\tan t + \frac{3}{2}(\cos t)\ln|\sec t + \tan t| + \tan t \\ &= \frac{1}{2}\tan t + \frac{3}{2}(\cos t)(\ln|\sec t + \tan t|). \end{aligned}$$

11. The homogeneous equation $y'' + y = 0$ has $y_1(t) = \cos t$ and $y_2(t) = \sin t$ as a fundamental set of solutions. The Wronskian is

$$\begin{aligned} W(\cos t, \sin t) &= \begin{vmatrix} \cos t & \sin t \\ -\sin t & \cos t \end{vmatrix} \\ &= 1. \end{aligned}$$

Form the solution

$$y_p = v_1 y_1 + v_2 y_2$$

where v_1 and v_2 are to be determined. Indeed,

$$\begin{aligned} v_1' &= \frac{-\sin t(\tan t + \sin t + 1)}{1} \\ &= -\frac{\sin^2 t}{\cos t} - \sin^2 t - \sin t \\ &= \frac{\cos^2 t - 1}{\cos t} - \frac{1-\cos 2t}{2} - \sin t \\ &= \cos t - \sec t - \frac{1}{2} + \frac{1}{2}\cos 2t - \sin t. \end{aligned}$$

Thus,

$$v_1 = \sin t - \ln|\sec t + \tan t| - \frac{1}{2}t + \frac{1}{4}\sin 2t + \cos t.$$

Secondly,

$$\begin{aligned} v_2' &= \frac{\cos t(\tan t + \sin t + 1)}{1} \\ &= \sin t + \cos t \sin t + \cos t \\ &= \sin t + \frac{1}{2}\sin 2t + \cos t. \end{aligned}$$

Thus,

$$v_2 = -\cos t - \frac{1}{4}\cos 2t + \sin t.$$

Inserting these results in $y_p = v_1 y_1 + v_2 y_2$,

$$\begin{aligned} y_p &= \left(\sin t - \ln|\sec t + \tan t| - \frac{1}{2}t \right. \\ &\quad \left. + \frac{1}{4}\sin 2t + \cos t \right) \cos t \\ &\quad + \left(-\cos t - \frac{1}{4}\cos 2t + \sin t \right) \sin t \\ &= 1 - \frac{1}{2}t\cos t - \cos t \ln|\sec t + \tan t| \\ &\quad + \frac{1}{4}(\sin 2t \cos t - \sin t \cos 2t) \\ &= 1 - \frac{1}{2}t\cos t - \cos t \ln|\sec t + \tan t| \\ &\quad + \frac{1}{4}\sin(2t - t) \\ &= 1 - \frac{1}{2}t\cos t - \cos t \ln|\sec t + \tan t| \\ &\quad + \frac{1}{4}\sin t \end{aligned}$$

is the particular solution we seek.

13. The formulae given in the text depend upon the fact that the coefficient of y'' is 1. We start by dividing our equation by t^2.

$$y'' + \frac{3}{t}y' - \frac{3}{t^2}y = \frac{1}{t^3}.$$

First, check $y_1(t) = t$ is a solution.

$$y'' + \frac{3}{t}y' - \frac{3}{t^2}y = (0) + \frac{3}{t}(1) - \frac{3}{t^2}(t) = 0.$$

Check that $y_2(t) = t^{-3}$ is a solution.

$$\begin{aligned} y'' + \frac{3}{t}y' - \frac{3}{t^2}y &= (12t^{-5}) + \frac{3}{t}(-3t^{-4}) - \frac{3}{t^2}(t^{-3}) \\ &= 12t^{-5} - 9t^{-5} - 3t^{-5} \\ &= 0. \end{aligned}$$

Calculate the Wronskian.

$$W(t, t^{-3}) = \begin{vmatrix} t & t^{-3} \\ 1 & -3t^{-4} \end{vmatrix} = -4t^{-3}.$$

Next,

$$\begin{aligned} v_1' &= \frac{-y_2 g(t)}{W} \\ &= \frac{-t^{-3}t^{-3}}{-4t^{-3}} \\ &= \frac{1}{4}t^{-3}. \end{aligned}$$

Thus,

$$v_1 = -\frac{1}{8}t^{-2}.$$

Next,

$$\begin{aligned} v_2' &= \frac{y_1 g(t)}{W} \\ &= \frac{tt^{-3}}{-4t^{-3}} \\ &= -\frac{1}{4}t. \end{aligned}$$

Thus,

$$v_2 = -\frac{1}{8}t^2.$$

Form

$$\begin{aligned} y_p &= v_1 y_1 + v_2 y_2 \\ &= \left(-\frac{1}{8}t^{-2}\right)t + \left(-\frac{1}{8}t^2\right)t^{-3} \\ &= -\frac{1}{4t}. \end{aligned}$$

Thus, the general solution is

$$y(t) = C_1 t + \frac{C_2}{t^3} - \frac{1}{4t}.$$

Section 4.7. Forced Harmonic Motion

1. (a) If $x_p = a\cos\omega t$, then

$$x_p' = -a\omega\sin\omega t$$
$$x_p'' = -a\omega^2\cos\omega t.$$

Substituting these in $x'' + \omega_0^2 x = A\cos\omega t$,

$$-a\omega^2\cos\omega t + \omega_0^2 a\cos\omega t = A\cos\omega t$$
$$a(\omega_0^2 - \omega^2)\cos\omega t = A\cos\omega t.$$

Comparing coefficients, $a = A/(\omega_0^2 - \omega^2)$ and

$$x_p = \frac{A}{\omega_0^2 - \omega^2}\cos\omega t.$$

(b) Substitute $z = ae^{i\omega t}$ into $z'' + \omega_0^2 z = Ae^{i\omega t}$ to get

$$ae^{i\omega t} = Ae^{i\omega t}$$
$$(\omega_0^2 - \omega^2)z = Ae^{i\omega t}$$
$$z = \frac{A}{\omega_0^2 - \omega^2}e^{i\omega t}.$$

The real part of this is the solution we seek.

$$x_p = \frac{A}{\omega_0^2 - \omega^2}\cos\omega t$$

3. The mean frequency is $\overline{\omega} = 21/2$ and the half difference is $\delta = 1/2$. Thus,

$$\begin{aligned}\cos 10t - \cos 11t &= \cos\left(\frac{21}{2} - \frac{1}{2}\right)t \\ &\quad - \cos\left(\frac{21}{2} + \frac{1}{2}\right)t, \\ &= 2\sin\frac{1}{2}t\sin\frac{21}{2}t.\end{aligned}$$

The envelope is $y(t) = \pm 2\sin(1/2)t$. The graph of the envelope is dashed in the following figure.

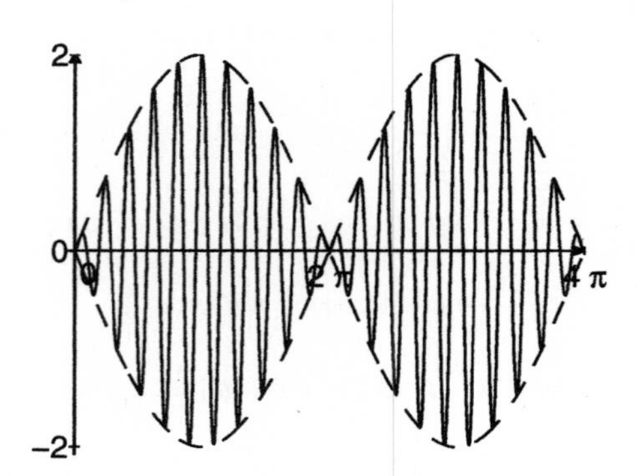

5. The mean frequency is $\overline{\omega} = 23/2$ and the half difference is $\delta = 1/2$. Thus,

$$\begin{aligned}\sin 12t - \sin 11t &= \sin\left(\frac{23}{2} + \frac{1}{2}\right)t \\ &\quad - \sin\left(\frac{23}{2} - \frac{1}{2}\right)t, \\ &= 2\sin\frac{1}{2}t\cos\frac{23}{2}t.\end{aligned}$$

The envelope is $y(t) = \pm 2\sin(1/2)t$. The graph of the envelope is dashed in the following figure.

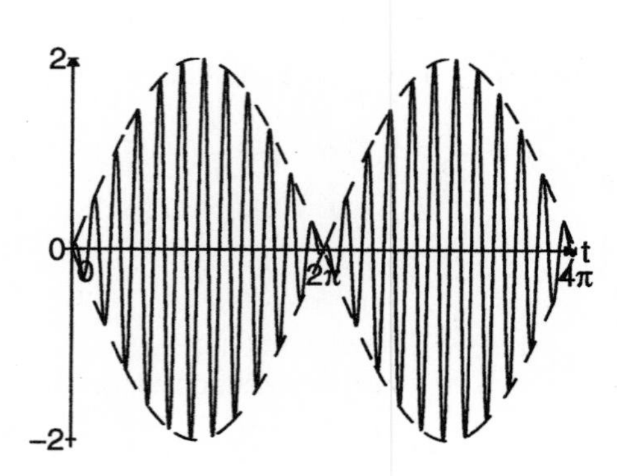

7. The graph of $x(t) = \cos 11t - \cos\omega t$, with $\omega =$

10.99, is shown below.

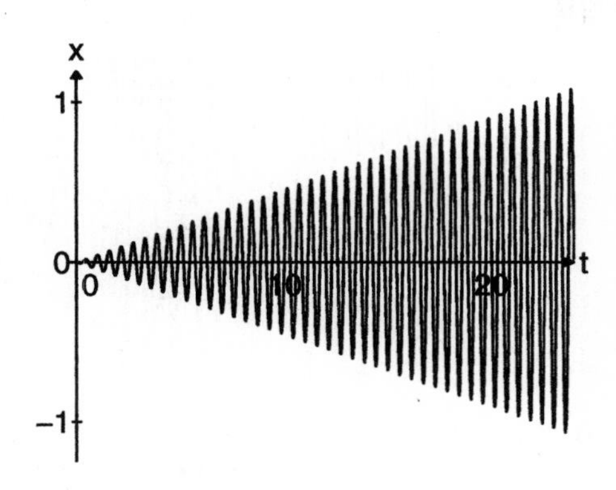

At first glance, this appears to be resonance. However, with

$$\overline{\omega} = \frac{11+\omega}{2} \quad \text{and} \quad \delta = \frac{11-\omega}{2},$$

$x(t) = \cos 11t - \cos \omega t$ becomes

$$\begin{aligned} x(t) &= \cos(\overline{\omega}+\delta) - \cos(\overline{\omega}-\delta)t \\ &= -2\sin\overline{\omega}t \sin\delta t. \end{aligned}$$

With $\omega = 10.99$,

$$\begin{aligned} \overline{\omega} &= \frac{11+\omega}{2} = 10.995 \\ \delta &= \frac{11-\omega}{2} = 0.005, \end{aligned}$$

and

$$x(t) = -2\sin 10.995t \sin 0.005t.$$

The period of the slow oscillation is

$$T = \frac{2\pi}{0.005} = 400\pi.$$

If we draw the graph of $x(t)$ on the interval $[0, 200\pi]$, we get the image below.

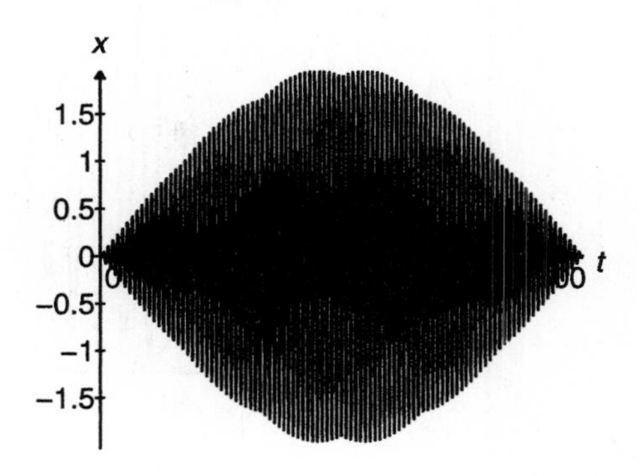

Note the aliasing in the image due to the frequency of the rapid oscillations. Still, note that "beats" are still present and this is non resonance. It's just that the beats occur over a long time period.

9. (a) The displacement satisfies the differential equation $x'' + 4x = 4\cos\omega t$, with initial conditions $x(0) = x'(0) = 0$. The solution is

$$x(t) = \frac{4}{4-\omega^2}(\cos\omega t - \cos 2t).$$

(b) If we set $\overline{\omega} = (2+\omega)/2$, and $\delta = (2-\omega)/2$, the solution becomes

$$x(t) = \frac{2}{\overline{\omega}\delta}\sin\delta t \sin\overline{\omega}t.$$

We will take $\omega = 1.8$, which is near to $\omega_0 = 2$. Then $\overline{\omega} = 1.9$, and $\delta = 0.1$, so

$$x(t) = \frac{2}{0.19}\sin 0.1t \sin 1.9t.$$

The graph of x and its envelope is presented in

the following figure.

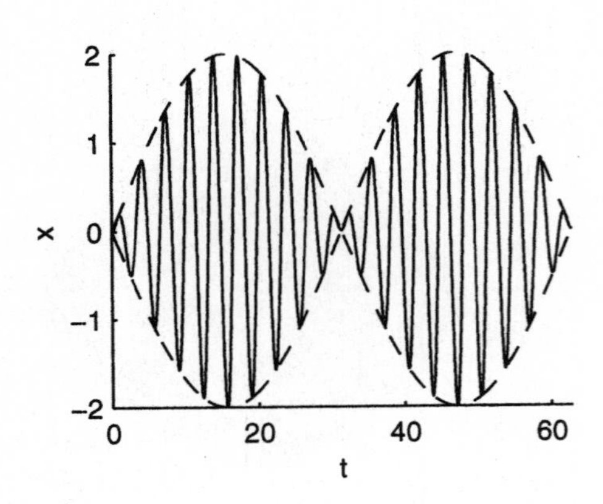

11. (a) The equation for the current is $I'' + 4I = -12\omega \sin \omega t$. We find a particular solution of the form $I_p(t) = A \cos \omega t + B \sin \omega t$. Solving for A and B we find that

$$I_p(t) = \frac{12\omega}{\omega^2 - 4} \sin \omega t.$$

Since the general solution of the homogeneous equation is

$$I_h(t) = C_1 \cos 2t + C_2 \sin 2t,$$

we know that the current has the form

$$I(t) = C_1 \cos 2t + C_2 \sin 2t + \frac{12\omega}{\omega^2 - 4} \sin \omega t.$$

The initial conditions $I(0) = 0$, and $I'(0) = 0$ imply that $C_1 = 0$, and $C_2 = -12\omega/(\omega^2 - 4)$. Hence the solution is

$$I(t) = \frac{6\omega}{\omega^2 - 4}(2 \sin \omega t - \omega \sin 2t).$$

With $\omega = 1.8$ we get the following graph of the solution.

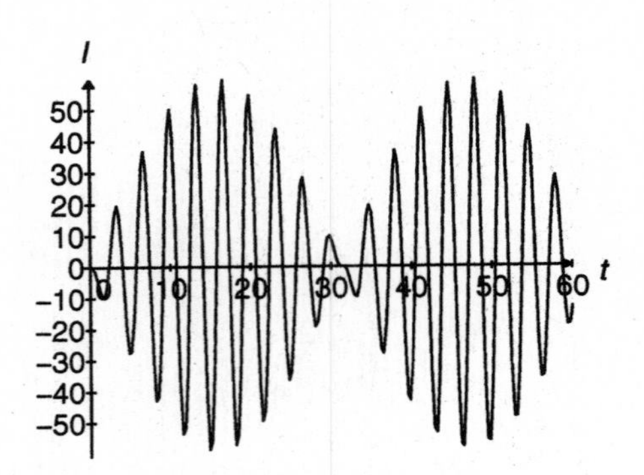

(b) The equation for the current is $I'' + 4I = -24 \sin 2t$. We find a particular solution of the form $I_p(t) = t[A \cos 2t + B \sin 2t]$. Solving for A and B we find that

$$I_p(t) = 6t \cos 2t.$$

The general solution is

$$I(t) = 6t \cos 2t + C_1 \cos 2t + C_2 \sin 2t.$$

The initial conditions $I(0) = 0$, and $I'(0) = 0$ imply that $C_1 = 0$, and $C_2 = -3$. Hence the solution is

$$I(t) = 6t \cos 2t - 3 \sin 2t.$$

It is plotted in the following figure.

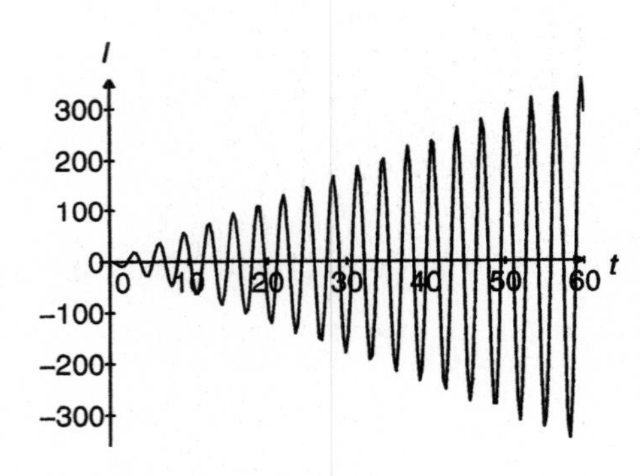

13. The characteristic polynomial is $P(\lambda) = \lambda^2 + 2c\lambda + \omega_0^2 = \lambda^2 + 2\lambda + 2$. Hence $2c = 2$ and $\omega_0^2 = 2$.

Therefore

$$R(\omega) = \sqrt{(\omega_0^2 - \omega^2)^2 + 4c^2\omega^2}$$
$$= \sqrt{(\omega^2 - 2)^2 + 4\omega^2}$$
$$\phi(\omega) = \text{arccot}\left(\frac{\omega_0^2 - \omega^2}{2c\omega}\right)$$
$$= \text{arccot}\left(\frac{2 - \omega^2}{2\omega}\right).$$

For the steady-state solution to $x'' + 2x' + 2x = 3\sin 4t$, we look first for a complex solution to $z'' + 2z' + 2z = 3e^{4it}$. The solution will be $x(t) = \text{Im}\, z(t)$. The complex solution is

$$z(t) = H(i\omega) \cdot 3e^{4it} = \frac{1}{R} \cdot 3e^{4it - i\phi}.$$

The real solution is then

$$x(t) = \text{Im}\, z(t) = \frac{3}{R}\sin(4t - \phi).$$

We use $\omega = 4$. Hence $R(4) = \sqrt{260} \approx 15.8114$ and $\phi(4) = \text{arccot}(-7/4) \approx 2.6224$. The solution is

$$x(t) = \frac{3}{\sqrt{260}}\sin(4t - \phi)$$
$$= \frac{3}{\sqrt{260}}\cos(4t - \phi - \pi/2).$$

15. We want to find the complex solution of $z'' + 4z' + 8z = 3e^{i2\pi t}$. Note that the frequency of the forcing term is $\omega = 2\pi$. The equation has characteristic polynomial $P(\lambda) = \lambda^2 + 4\lambda + 9$, so

$$P(i\omega) = (i\omega)^2 + 4(i\omega) + 8 = (8 - \omega^2) + 4i\omega,$$

which has magnitude and phase defined by

$$R(\omega) = \sqrt{(8 - \omega^2)^2 + 16\omega^2}$$
$$\cot\phi(\omega) = \frac{8 - \omega^2}{4\omega}.$$

With $\omega = 2\pi$, $R(2\pi) \approx 40.2808$ and $\phi(2\pi) \approx 2.4678$, leading to the complex solution

$$z(t) = H(i2\pi) \cdot 3e^{i2\pi t}$$
$$= \frac{1}{R(2\pi)}e^{-i\phi(2\pi)} \cdot 3e^{i2\pi t}$$
$$= \frac{1}{40.2808}e^{-i(2.4678)} \cdot 3e^{i2\pi t}$$
$$= 0.0745e^{i(2\pi t - 2.4678)}.$$

The steady-state solution of $x'' + 4x' + 8x = 3\cos 2\pi t$ is the real part of this solution, namely

$$x(t) = 0.0745\cos(2\pi t - 2.4678).$$

17. We will use the complex method, and look for a solution of the equation $z'' + 7z' + 10z = 3e^{3it}$ of the form $z(t) = ae^{3it}$. Then our particular solution will be $x_p = \text{Re}\, z$. Differentiating we get

$$z'' + 7z' + 10z = a((3i)^2 + 7(3i) + 10)e^{3it}$$
$$= a(1 + 21i)e^{3it} = 3e^{3it}.$$

Hence

$$a = \frac{3}{1 + 21i} = 3 \cdot \frac{1 - 21i}{442},$$

so

$$z(t) = 3 \cdot \frac{1 - 21i}{442}e^{3it}$$
$$= \frac{3}{442}[1 - 21i][\cos 3t + i\sin 3t]$$
$$= \frac{3}{442}[(\cos 3t + 21\sin 3t)$$
$$+ i(\sin 3t - 21\cos 3t)],$$

and $x_p(t) = \text{Re}\, z(t) = 3(\cos 3t + 21\sin 3t)/442$. The characteristic polynomial is $P(\lambda) = \lambda^2 + 7\lambda + 10 = (\lambda + 2)(\lambda + 5)$, which has roots -2 and -5. Hence the general solution to the homogenous equation is $x_h(t) = C_1e^{-2t} + C_2e^{-5t}$. The general solution to the inhomogeneous equation is

$$x(t) = \frac{3}{442}(\cos 3t + 21\sin 3t) + C_1e^{-2t} + C_2e^{-5t}.$$

The initial conditions imply that

$$-1 = x(0) = \frac{3}{442} + C_1 + C_2$$
$$0 = x'(0) = \frac{189}{442} - 2C_1 - 5C_2,$$

or

$$C_1 + C_2 = -\frac{445}{442}$$
$$2C_1 + 5C_2 = \frac{189}{442}.$$

The solutions are $C_1 = -71/39$ and $C_2 = 83/102$, so the solution to the initial value problem is

$$x(t) = \frac{3}{442}(\cos 3t + 21 \sin 3t) - \frac{71}{39}e^{-2t} + \frac{83}{102}e^{-5t}.$$

The steady-state solution is the particular solution

$$x_p(t) = \frac{3}{442}(\cos 3t + 21 \sin 3t),$$

and the transient response is

$$x_h(t) = -\frac{71}{39}e^{-2t} + \frac{83}{102}e^{-5t}.$$

In the following plot the graph of the solution to the initial value problem is the dashed curve, the transient response is dotted, and the steady-state solution is solid.

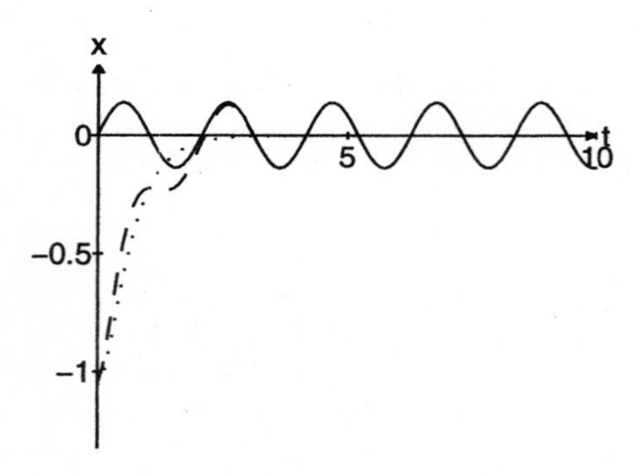

19. We will use the complex method, and look for a solution of the equation $z'' + 4z' + 5z = 3e^{it}$ of the form $z(t) = ae^{it}$. Then our particular solution will be $x_p = \text{Im}\, z$. Differentiating we get

$$z'' + 4z' + 5z = a[i^2 + 4i + 5]e^{it}$$
$$= a[4 + 4i]e^{it} = 3e^{it}.$$

Hence $a = 3/(4 + 4i) = 3(1 - i)/8$, so

$$z(t) = \frac{3}{8}[1 - i]e^{it}$$
$$= \frac{3}{8}[1 - i][\cos t + i \sin t]$$
$$= \frac{3}{8}[(\cos t + \sin t) + i(\sin t - \cos t)],$$

and $x_p(t) = \text{Im}\, z(t) = 3(\sin t - \cos t)/8$.

The characteristic polynomial is $P(\lambda) = \lambda^2 + 4\lambda + 5$, which has the complex roots $\lambda = -2 \pm i$. Hence the general solution to the homogeneous equation is $x_h(t) = e^{-2t}[C_1 \cos t + C_2 \sin t]$. The general solution of the inhomogeneous equation is

$$x(t) = \frac{3}{8}(\sin t - \cos t) + e^{-2t}[C_1 \cos t + C_2 \sin t].$$

The initial conditions imply

$$0 = x(0) = -\frac{3}{8} + C_1$$
$$-3 = x'(0) = \frac{3}{8} - 2C_1 + C_2.$$

We solve these equations, finding $C_1 = 3/8$ and $C_2 = -21/8$, so the solution to the initial value problem is

$$x(t) = \frac{3}{8}(\sin t - \cos t) + \frac{1}{8}e^{-2t}[3 \cos t - 21 \sin t].$$

The steady-state solution is the particular solution $x_p(t) = 3(\sin t - \cos t)/8$, and the transient response is $x_h = e^{-2t}[3 \cos t - 21 \sin t]/8$. In the following plot the graph of the solution to the initial value problem is the dashed curve, the transient response is dotted, and the steady-state solution is solid.

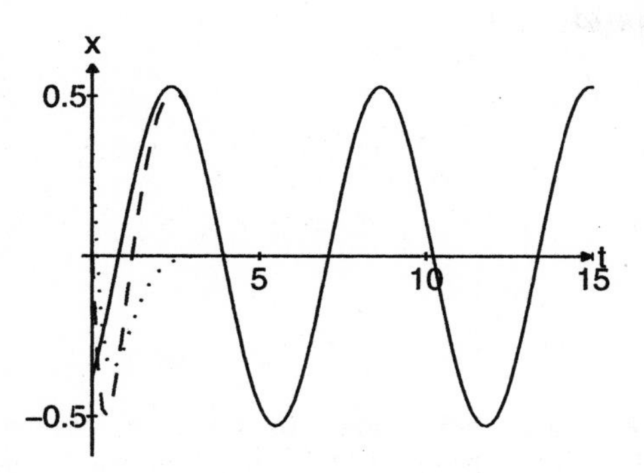

21. In Exercise 17, we found the transient solution

$$x_h(t) = -\frac{71}{39}e^{-2t} + \frac{83}{102}e^{-5t}.$$

Because there are two exponential terms, we will use the more slowing decaying term to determine a time constant. Thus, let $T_c = 1/2$. What follows is the plot of the transient solution on the interval $[0, 4T_c] = [0, 2]$.

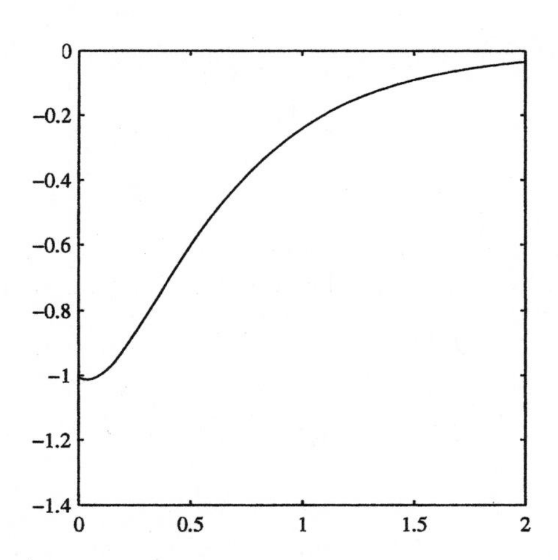

23. In Exercise 19, we found the transient solution

$$x_h = e^{-2t}[3\cos t - 21\sin t]/8.$$

Thus, the time constant is $T_c = 1/2$. What follows is the plot of the transient solution on the interval $[0, 4T_c] = [0, 2]$.

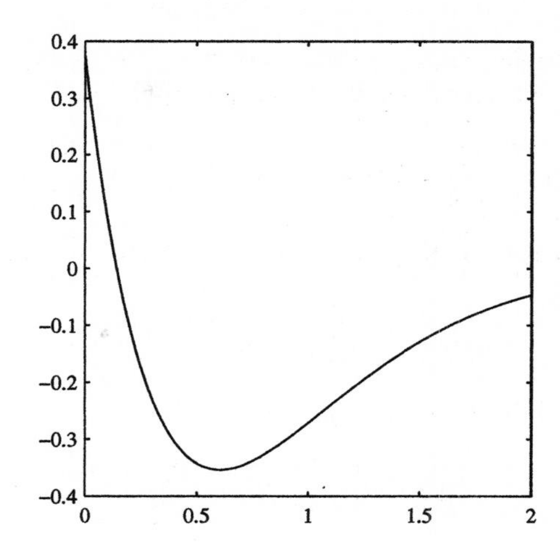

25. We want to find the complex solution of $z'' + 2z' + 5z = 2e^{i3t}$. Note that the frequency of the forcing term is $\omega = 3$. The equation has characteristic polynomial $P(\lambda) = \lambda^2 + 2\lambda + 5$, so

$$P(i\omega) = (i\omega)^2 + 2(i\omega) + 5 = (5 - \omega^2) + 2i\omega,$$

which has magnitude and phase defined by

$$R(\omega) = \sqrt{(5-\omega^2)^2 + 4\omega^2}$$

$$\cot\phi(\omega) = \frac{5-\omega^2}{2\omega}.$$

With $\omega = 3$, $R(3) = \sqrt{52} \approx 7.2111$ and $\phi(3) = \text{arccot}(-2/3) \approx 2.1588$, leading to the complex solution

$$z(t) = \frac{1}{\sqrt{52}}e^{-\phi i} \cdot 2e^{3it} = \frac{1}{\sqrt{13}}e^{i(3t-\phi)}.$$

The steady-state solution of $x'' + 2x' + 5x = 2\cos 3t$ is the real part of this solution, namely

$$x(t) = \frac{1}{\sqrt{13}}\cos(3t - \phi).$$

Note, in the following figure, that the solutions with the initial conditions $(-2, 0)$, $(-1, 0)$, $(0, 0)$, $(1, 0)$, and (2, 0), drawn as dashed lines, all converge to the steady-state solution (solid line).

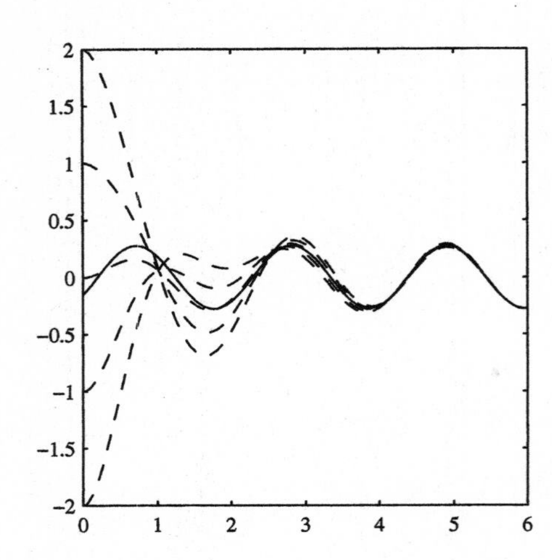

27. To find the complex solution of $z''+0.4z'+2z = e^{it}$, note that the frequency of the forcing term is $\omega = 1$. The equation has characteristic polynomial $P(\lambda) = \lambda^2 + 0.4\lambda + 2$, so

$$P(i\omega) = (i\omega)^2 + 0.4(i\omega) + 2 = (2 - \omega^2) + 0.4i\omega,$$

which has magnitude and phase defined by

$$R(\omega) = \sqrt{(2-\omega^2)^2 + 0.16\omega^2}$$
$$\cot\phi(\omega) = \frac{2-\omega^2}{0.4\omega}.$$

Thus, the transfer function is

$$\begin{aligned} H(i\omega) &= \frac{1}{P(i\omega)} = \frac{1}{R(\omega)e^{i\phi(\omega)}} = \frac{1}{R(\omega)}e^{-i\phi(\omega)} \\ &= G(\omega)e^{-i\phi(\omega)}, \end{aligned}$$

where $G(\omega) = 1/R(\omega)$ is the gain. Thus,

$$\begin{aligned} z(t) &= H(i\omega)e^{it} = G(\omega)e^{-i\phi(\omega)}e^{it} \\ &= G(\omega)e^{i(t-\phi(\omega))}, \end{aligned}$$

with $\omega = 1$, is the solution of the complex equation, and the real part, $x(t) = G(1)\cos(t - \phi(1))$, is the steady-state solution of $x''+0.4x'+2x = \cos t$. Using the formulae above, $G(1) = 1/\sqrt{1.16} \approx 0.9285$ and $\phi(1) = \operatorname{arccot}(2.5) \approx 0.3805$. Thus, the steady state solution is

$$x(t) = \frac{1}{\sqrt{1.16}}\cos(t - \phi).$$

Note that the amplitude of the steady-state is the gain (not always the case), due to the fact that the amplitude of the forcing function, $\cos t$, is 1. Thus, the gain is easily read in the following graph, where one can estimate the gain by recording the amplitude of the steady-state response (plotted as a dashed line).

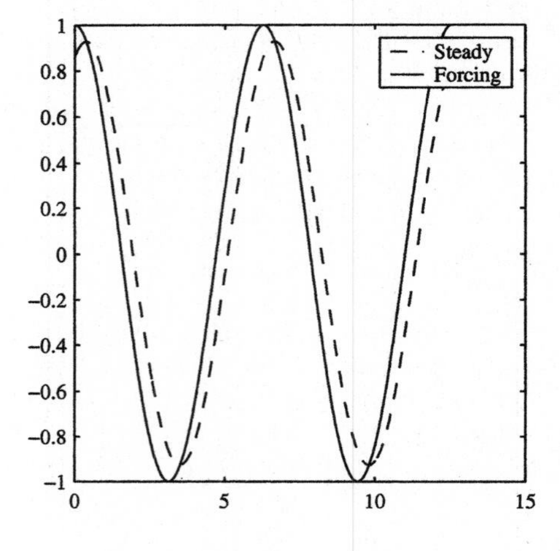

Further, note that the steady-state in the figure is shifted to the right of the forcing function. By zooming in on the upper left corner and adding a grid to the figure, one can estimate the phase shift.

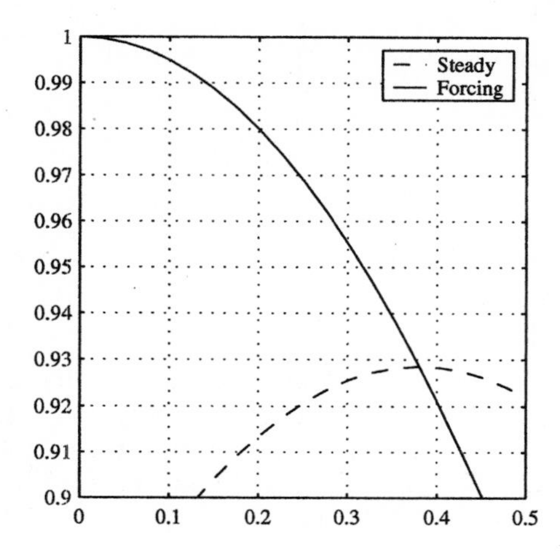

Note that the steady-state response is shifted about 0.38 units to the right of the forcing function. It is important to note that this reading of gain and phase is facilitated by the choice of forcing function. Because the forcing function $\cos t$ has amplitude 1, the gain is easily read as the amplitude of the steady-state response. Similarly, because $\omega = 1$, the forcing function is $\cos t$, the phase is a simple shift. Things would be more complicated (but still doable) if we used $2\cos 3t$ as the forcing function.

29. The equation $x'' + 0.4x' + 1.69x = \cos t$ has characteristic polynomial $P(\lambda) = \lambda^2 + 0.4\lambda + 1.69$, having roots $\lambda = -(1/5) \pm \sqrt{165}/10$. Consequently, the time constant is

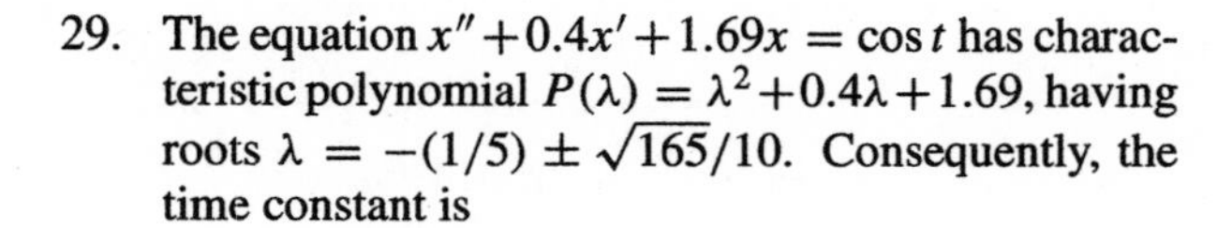

$$T_c = \frac{1}{c} = \frac{1}{1/5} = 5.$$

Thus, to avoid transients, we let at least $4T_c$ pass before examining the solution. Note that in the following figure that the solution settles down to its steady-state after about $[0, 4T_c] = [0, 20]$.

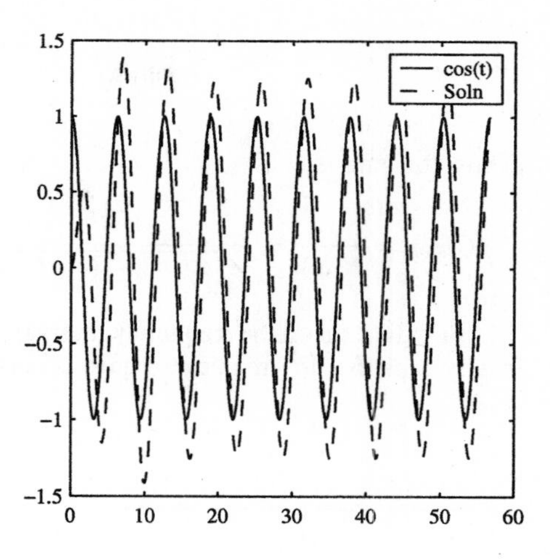

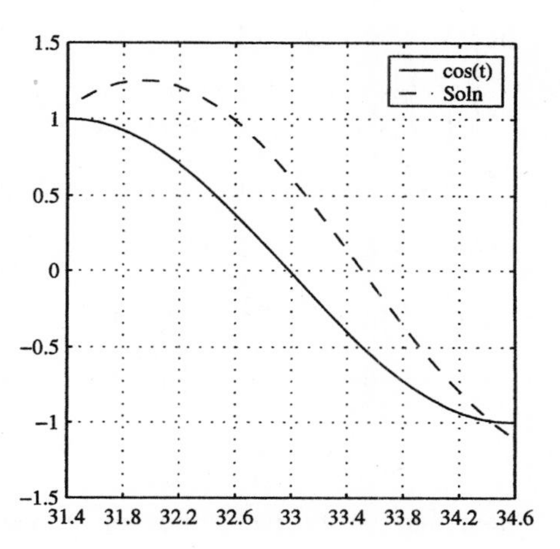

In the next image, we've zoomed in on the interval $[10\pi, 11\pi]$, which is well beyond $[0, 20]$. Note that the amplitude of the solution is about 1.25, which is the gain. Further, note that the cosine peaks at about 31.4, but the first peak of the solution occurs at approximately 32.0. Thus, the phase is $32.0 - 31.4$, or 0.6. The careful reader will want to compare these results with the actual gain and phase, 1.2538 and 0.5254.

31. Note that the characteristic polynomial of $z'' + 0.01z' + 49z = Ae^{i\omega t}$ is $P(\lambda) = \lambda^2 + 0.01\lambda + 49$.

Thus,

$$\begin{aligned} P(i\omega) &= (i\omega)^2 + 0.01(i\omega) + 49 \\ &= (49 - \omega^2) + 0.01i\omega. \end{aligned}$$

Thus, the gain is

$$G(\omega) = \frac{1}{R(\omega)} = \frac{1}{\sqrt{(49 - \omega^2)^2 + 0.0001\omega^2}}.$$

Plotting the gain on the frequency interval [6, 8], one easily sees that the maximum gain occurs at about $w \approx 7$.

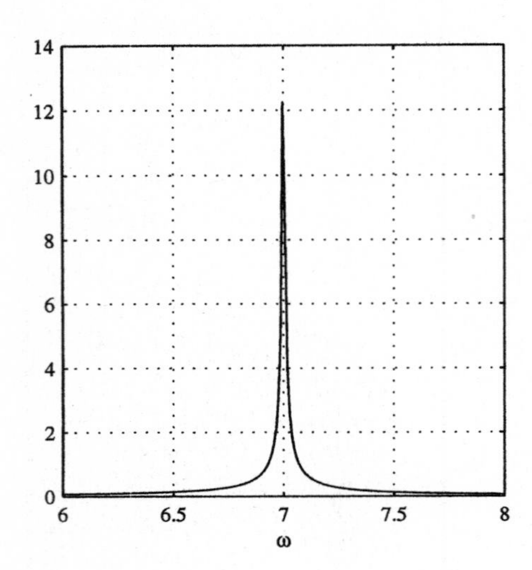

33. Note that the characteristic polynomial of $z'' + 0.05z' + 25z = Ae^{i\omega t}$ is $P(\lambda) = \lambda^2 + 0.05\lambda + 25$. Thus,

$$\begin{aligned} P(i\omega) &= (i\omega)^2 + 0.05(i\omega) + 25 \\ &= (25 - \omega^2) + 0.05i\omega. \end{aligned}$$

Thus, the gain is

$$G(\omega) = \frac{1}{R(\omega)} = \frac{1}{\sqrt{(25 - \omega^2)^2 + 0.0025\omega^2}}.$$

Plotting the gain on the frequency interval [4, 6], one easily sees that the maximum gain occurs at about $w \approx 5$.

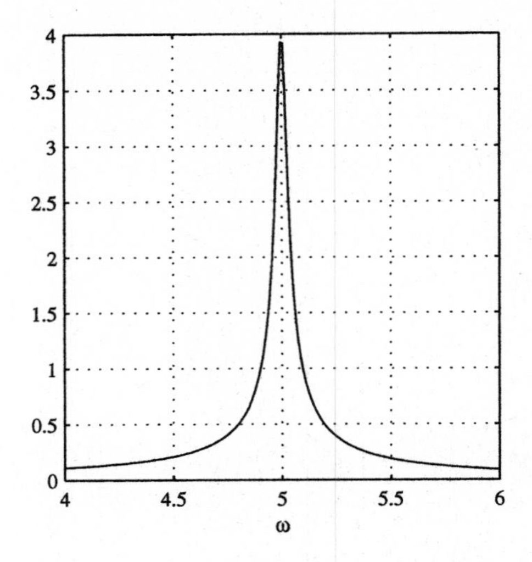

35. The characteristic polynomial of $z'' + 2cz' + \omega_0^2 z = Ae^{i\omega t}$ is $P(\lambda) = \lambda^2 + 2c\lambda + \omega_0^2$. Thus,

$$\begin{aligned} P(i\omega) &= (i\omega)^2 + 2c(i\omega) + \omega_0^2 \\ &= (\omega_0^2 - \omega^2) + 2ci\omega. \end{aligned}$$

Thus, the gain is

$$G(\omega) = \frac{1}{R(\omega)} = \frac{1}{\sqrt{(\omega_0^2 - \omega^2)^2 + 4c^2\omega^2}}.$$

Taking the derivative,

$$\begin{aligned} G'(\omega) &= -\frac{1}{2}((\omega_0^2 - \omega^2)^2 + 4c^2\omega^2)^{-3/2} \\ &\quad \times (-4\omega(\omega_0^2 - \omega^2) + 8c^2\omega) \\ &= ((\omega_0^2 - \omega^2)^2 + 4c^2\omega^2)^{-3/2} \\ &\quad \times (2\omega(\omega_0^2 - \omega^2) - 4c^2\omega). \end{aligned}$$

Note that the first factor of G is always positive, so critical values are determined by setting the remaining factors equal to zero. Thus,

$$2\omega(\omega_0^2 - \omega^2 - 2c^2) = 0$$

leads to the critical values

$$\omega = 0 \quad \text{and} \quad \omega = \sqrt{\omega_0^2 - 2c^2},$$

provided, of course, that $\omega_0^2 > 2c^2$. Practically, we are not interested in a forcing function with zero frequency, so we concentrate on $\omega = \sqrt{\omega_0^2 - 2c^2}$. It is not difficult to show that the derivative of G is positive to the left of this critical value and negative to the right. Thus, we have a maximum at $\omega_{res} = \sqrt{\omega_0^2 - 2c^2}$, which is the resonant frequency for the driven oscillator. Examining the equation $y'' + 0.01y' + 49y = A\cos\omega t$, the maximum gain occurs at

$$\omega_{res} = \sqrt{\omega_0^2 - 2c^2} = \sqrt{7^2 - 2(0.005)^2},$$

which, to four decimal places, equals 7.0000, agreeing nicely with the estimate found in Exercise 31.

37. See the derivation of the formula in Exercise 35. Because $y'' + 0.05y' + 25y = A\sin\omega t$, the maximum gain occurs at

$$\omega_{res} = \sqrt{\omega_0^2 - 2c^2} = \sqrt{5^2 - 2(0.025)^2} \approx 4.9999,$$

which agrees nicely with the result in Exercise 33.

39. The equation $y'' + 0.1y' + 25y = \cos\omega t$ has characteristic polynomial $P(\lambda) = \lambda^2 + 0.1\lambda + 25$. Thus,

$$\begin{aligned} P(i\omega) &= (i\omega)^2 + 0.1(i\omega) + 25 \\ &= (25 - \omega^2) + 0.1i\omega. \end{aligned}$$

The gain is

$$G(\omega) = \frac{1}{R(\omega)} = \frac{1}{\sqrt{(25-\omega^2)^2 + 0.01\omega^2}}.$$

Next, the gain is plotted versus the frequency, indicating a maximum gain near $\omega = 5$.

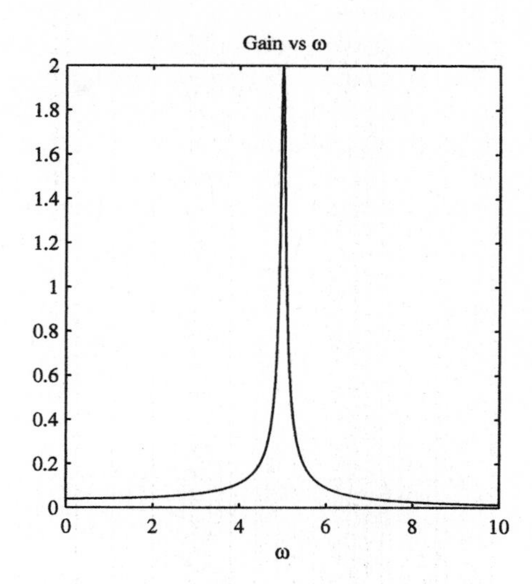

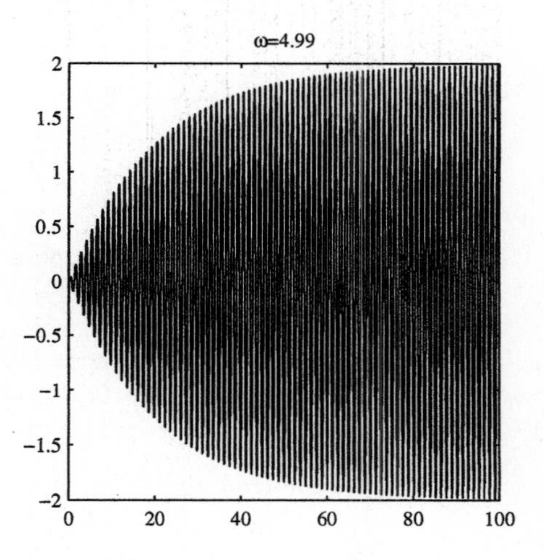

This estimate of the resonant frequency is verified with the calculation

$$\omega_{res} = \sqrt{\omega_0^2 - 2c^2} = \sqrt{5^2 - 2(0.05)^2} \approx 4.9995.$$

In the second figure above, careful attention was paid to eliminating transients. In this underdamped case, $T_c = 1/c = 1/0.05 = 20$. Thus, we must go beyond $4T_c = 80$ to eliminate most of the transient behavior.

Of course, this extended time interval allows truncation error to propagate, so we set the relative error tolerance of our Matlab solver to 1×10^{-8} to draw the solution in the second figure above, with $\omega = 4.99$ set near the resonant frequency. Note the gain in amplitude is about double that of the forcing function, $\cos 4.99t$. In the final figure, $\omega = 2$ was chosen at quite a distance from the resonant frequency.

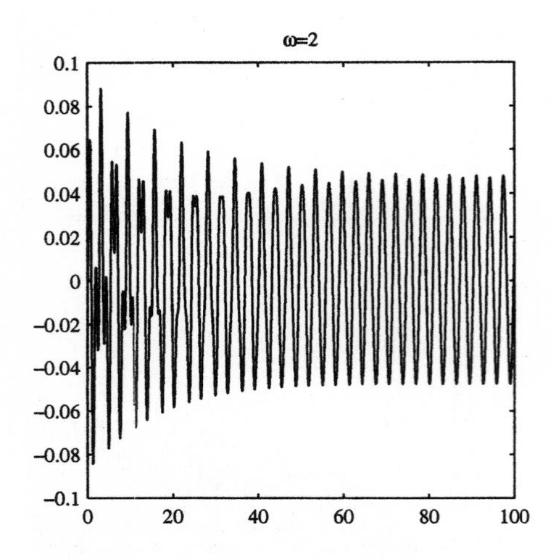

Note the severe attenuation of the amplitude.

41. The equation $y''+0.2y'+49y = \cos\omega t$ has characteristic polynomial $P(\lambda) = \lambda^2 + 0.2\lambda + 49$. Thus,

$$\begin{aligned} P(i\omega) &= (i\omega)^2 + 0.2(i\omega) + 49 \\ &= (49 - \omega^2) + 0.2i\omega. \end{aligned}$$

The gain is

$$G(\omega) = \frac{1}{R(\omega)} = \frac{1}{\sqrt{(49-\omega^2)^2 + 0.04\omega^2}}.$$

In the first figure, the gain is plotted versus the frequency, indicating a maximum gain near $\omega = 7$.

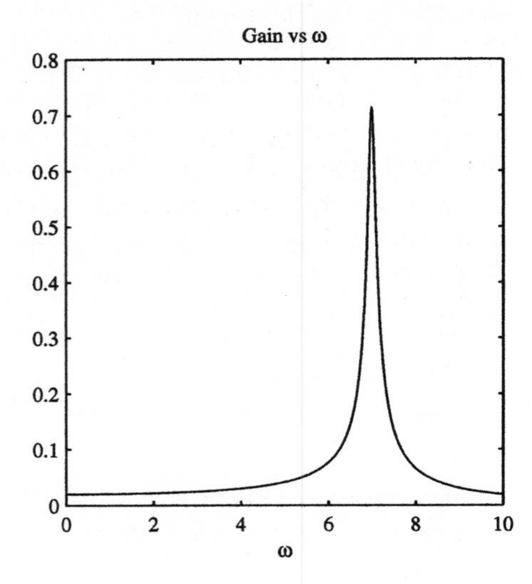

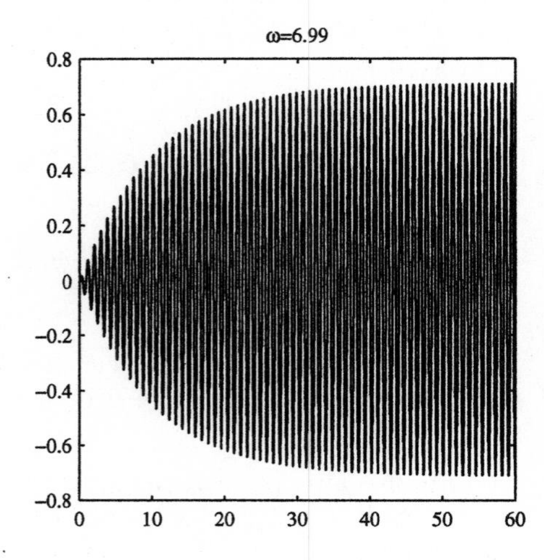

This estimate of the resonant frequency is verified with the calculation

$$\omega_{\text{res}} = \sqrt{\omega_0^2 - 2c^2} = \sqrt{7^2 - 2(0.1)^2} \approx 6.9986.$$

In the second figure above, careful attention was paid to eliminating transients. In this underdamped case, $T_c = 1/c = 1/0.1 = 10$. Thus, we must go beyond $4T_c = 40$ to eliminate most of the transient behavior.

Of course, this extended time interval allows truncation error to propagate, so we set the relative error tolerance of our Matlab solver to 1×10^{-8} to draw the solution in second figure above, with $\omega = 6.99$ set near the resonant frequency. Note the gain is a small attenuation of the amplitude of the forcing function, $\cos 6.99t$. This makes perfect sense in light of the graph of the gain in the first figure. Note that the maximum gain is about 0.72, so even near the resonant frequency, we can expect some small degradation of the amplitude of the forcing function. In the final figure, $\omega = 4$ was chosen at quite a distance from the resonant frequency.

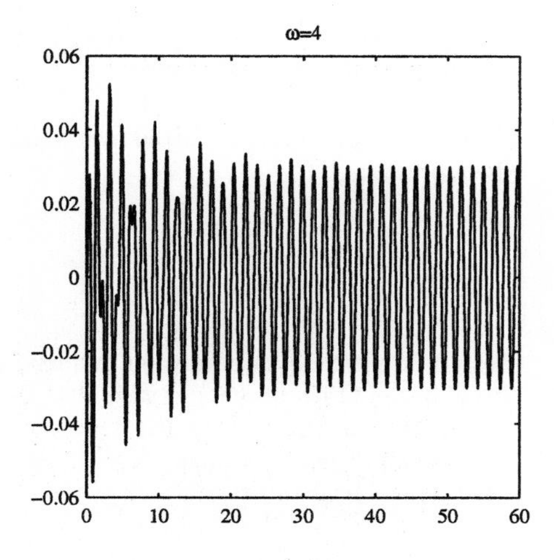

Note the severe attenuation of the amplitude.

43. (a) Because the current is the derivative of the charge ($I = Q'$), the equation $LI' + RI + (1/C)Q = A\cos\omega t$ becomes $LQ'' + RQ' + (1/C)Q = A\cos\omega t$, or upon dividing both sides by the inductance,

$$Q'' + \frac{R}{L}Q' + \frac{1}{LC}Q = \frac{A}{L}\cos\omega t.$$

The solution we seek is the real part of the complex solution of

$$z'' + \frac{R}{L}z' + \frac{1}{LC}z = \frac{A}{L}e^{i\omega t}.$$

Let $z = ae^{i\omega t}$ represent the steady-state solution. Substituting,

$$\left[(i\omega)^2 + \frac{R}{L}(i\omega) + \frac{1}{LC}\right]ae^{i\omega t} = \frac{A}{L}e^{i\omega t}$$

$$\left[\left(\frac{1}{LC} - \omega^2\right) + \frac{R}{L}i\omega\right]z = \frac{A}{L}e^{i\omega t}.$$

The complex coefficient of z can be written in polar form,

$$\sqrt{\left(\frac{1}{LC} - \omega^2\right)^2 + \frac{R^2}{L^2}\omega^2}\, e^{i\phi(\omega)}\, z = \frac{A}{L}e^{i\omega t},$$

where

$$\cot\phi(\omega) = \frac{1/(LC) - \omega^2}{R\omega/L}.$$

Thus,

$$z = \frac{(A/L)e^{i(\omega t - \phi(\omega))}}{\sqrt{\left(1/(LC) - \omega^2\right)^2 + R^2\omega^2/L^2}}.$$

The real part of z,

$$Q = \frac{(A/L)\cos(\omega t - \phi(\omega))}{\sqrt{\left(1/(LC) - \omega^2\right)^2 + R^2\omega^2/L^2}},$$

is the solution we seek. The charge will be maximized only if we can maximize

$$G(\omega) = \left[\left(\frac{1}{LC} - \omega^2\right)^2 + \frac{R^2}{L^2}\omega^2\right]^{-1/2}.$$

Taking the derivative,

$$\begin{aligned}G'(\omega) &= -\frac{1}{2}\left[\left(\frac{1}{LC} - \omega^2\right)^2 + \frac{R^2}{L^2}\omega^2\right]^{-3/2} \\ &\quad \times \left[2\left(\frac{1}{LC} - \omega^2\right)(-2\omega) + \frac{2R^2}{L^2}\omega\right] \\ &= \left[\left(\frac{1}{LC} - \omega^2\right)^2 + \frac{R^2}{L^2}\omega^2\right]^{-3/2} \\ &\quad \times \left[2\omega\left(\frac{1}{LC} - \omega^2\right) - \frac{R^2}{L^2}\omega\right].\end{aligned}$$

To find the critical value, set

$$2\omega\left(\frac{1}{LC}-\omega^2\right)-\frac{R^2}{L^2}\omega=0.$$

Assuming the forcing function has a nonzero frequency,

$$\frac{2}{LC}-2\omega^2-\frac{R^2}{L^2}=0$$
$$\omega^2=\frac{1}{LC}-\frac{R^2}{2L^2}$$
$$\omega=\sqrt{\frac{1}{LC}-\frac{R^2}{2L^2}}.$$

The careful reader will show that the oscillator is underdamped only if $1/(LC) > R^2/(4L^2)$. The resonant frequency occurs only if $1/(LC) > R^2/(2L^2)$, a bit more than underdamped.

(b) Start by differentiating $LI' + RI + (1/C)Q = A\cos\omega t$, remembering that $Q' = I$.

$$LI'' + RI' + \frac{1}{C}I = -A\omega\sin\omega t,$$

or, upon dividing by the inductance,

$$I'' + \frac{R}{L}I' + \frac{1}{LC}I = -\frac{A\omega}{L}\sin\omega t.$$

Substitute $z = ae^{i\omega t}$ in

$$z'' + \frac{R}{L}z' + \frac{1}{LC}z = -\frac{A\omega}{L}e^{i\omega t}$$

to get

$$\left[(i\omega)^2 + \frac{R}{L}(i\omega) + \frac{1}{LC}\right]ae^{i\omega t} = -\frac{A\omega}{L}e^{i\omega t}$$

$$\left[\left(\frac{1}{LC}-\omega^2\right)+\frac{R}{L}i\omega\right]z = -\frac{A\omega}{L}e^{i\omega t}$$

$$z = \frac{-A\omega/L}{(1/(LC)-\omega^2)+Ri\omega/L}e^{i\omega t} = \frac{-A}{(1/(\omega C)-L\omega)+Ri}e^{i\omega t}.$$

If we write the denominator of this last expression in polar form

$$\sqrt{\left(\frac{1}{\omega C}-L\omega\right)^2+R^2}\,e^{i\phi(\omega)},$$

then

$$z = \frac{-A}{\sqrt{(1/(\omega C)-L\omega)^2+R^2}}e^{i(\omega t-\phi(\omega))}.$$

The imaginary part of this is the steady-state solution for the current.

$$I = \frac{-A}{\sqrt{(1/(\omega C)-L\omega)^2+R^2}}\sin(\omega t-\phi(\omega))$$

Again, the current is maximized by maximizing the amplitude,

$$\frac{-A}{\sqrt{(1/(\omega C)-L\omega)^2+R^2}}.$$

We could proceed as before by taking derivatives, but in this case we can note that the amplitude is maximized when the denominator is minimized. The denominator's smallest possible value is R, which occurs when

$$\frac{1}{\omega C}-L\omega=0$$
$$LC\omega^2=1$$
$$\omega^2=\frac{1}{LC}$$
$$\omega=\sqrt{\frac{1}{LC}}.$$

45. First, $m = 50\,\text{g} = 0.05\,\text{kg}$ and $y = 10\,\text{cm} = 0.1\,\text{m}$. Use Hooke's law to compute the spring constant.

$$k=\frac{F}{y}=\frac{mg}{y}=\frac{(0.05)(9.8)}{0.1}=4.9\,\text{N/m}.$$

We are given that the damping force is opposite the velocity, with magnitude $0.1v$, and the driving force is $F(t) = 5\cos 4.4t$. Thus, $my'' + \mu y' + ky = F(t)$ becomes

$$0.05y'' + 0.01y' + 4.9y = 5\cos 4.4t,$$

or

$$y'' + 0.2y' + 98y = 100\cos 4.4t.$$

This has characteristic polynomial $P(\lambda) = \lambda^2 + 0.2\lambda + 98$ and zeros $\lambda = -0.1 \pm 9.8990$. Thus, the homogeneous solution is

$$y_h(t) = e^{-0.1t}(C_1\cos 9.8990t + C_2\sin 9.8990t).$$

To find the steady-state solution, substitute $z = ae^{i4.4t}$ in $z'' + 0.2z' + 98z = 100e^{i4.4t}$.

$$ae^{i4.4t} = 100e^{i4.4t}$$

$$(78.64 + 0.88i)z = 100e^{i4.4t}.$$

The coefficient of z has magnitude 78.6449 and phase 0.0112, so

$$78.6449e^{0.0112i}z = 100e^{i4.4t}$$

$$z = 1.2716e^{i(4.4t-0.0112)}.$$

The real part of this, $y_p(t) = 1.2716\cos(4.4t - 0.0112)$, is the steady-state solution. Thus, the solution is

$$\begin{aligned} y(t) &= y_h(t) + y_p(t) \\ &= e^{-0.1t}(C_1\cos 9.8990t + C_2\sin 9.98990t) \\ &\quad + 1.2716\cos(4.4t - 0.0112). \end{aligned}$$

The initial condition $y(0) = 0$ leads to

$$0 = C_1 + 1.2716\cos(-0.0112)$$

and $C_1 = -1.2715$. Differentiating,

$$\begin{aligned} y'(t) &= 9.8990e^{-0.1t}\big(-C_1\sin 9.8990t \\ &\quad + C_2\cos 9.8990t\big) \\ &\quad - 0.1e^{-0.1t}\big(C_1\cos 9.8990t \\ &\quad + C_2\sin 9.8990t\big) \\ &\quad - 5.5950\sin(4.4t - 0.0112). \end{aligned}$$

The initial condition $y'(0) = 0$ leads to

$$0 = 9.8990C_2 - 0.1C_1 - 5.5950\sin(-0.0112)$$

and $C_2 = -0.0192$. Thus, the solution is

$$\begin{aligned} y(t) &= e^{-0.1t}\big(-1.2715\cos 9.8990t \\ &\quad - 0.0192\sin 9.8990t\big) \\ &\quad + 1.2716\cos(4.4t - 0.0112). \end{aligned}$$

Chapter 5. The Laplace Transform

Section 5.1. The Definition of the Laplace Transform

1. Using Definition 1.1,

$$\begin{aligned}F(s) &= \int_0^\infty 3e^{-st}\,dt \\ &= \lim_{T\to\infty}\int_0^T 3e^{-st}\,dt \\ &= \lim_{T\to\infty}\left.\frac{-3e^{-st}}{s}\right|_0^T \\ &= \lim_{T\to\infty}\left(\frac{-3e^{-sT}}{s}+\frac{3}{s}\right) \\ &= \frac{3}{s},\end{aligned}$$

provided $s > 0$.

3. Using Definition 1.1,

$$\begin{aligned}F(s) &= \int_0^\infty e^{-2t}e^{-st}\,dt \\ &= \lim_{T\to\infty}\int_0^T e^{-(s+2)t}\,dt \\ &= \lim_{T\to\infty}\left.\frac{-e^{-(s+2)t}}{s+2}\right|_0^T \\ &= \lim_{T\to\infty}\left[\frac{-e^{-(s+2)T}}{s+2}+\frac{1}{s+2}\right] \\ &= \frac{1}{s+2},\end{aligned}$$

provided $s > -2$.

5. Using Definition 1.1,

$$\begin{aligned}F(s) &= \int_0^\infty (\cos 2t)e^{-st}\,dt \\ &= \lim_{T\to\infty}\int_0^T (\cos 2t)e^{-st}\,dt.\end{aligned}$$

Integrating by parts,

$$\begin{aligned}\int & e^{-st}\cos 2t\,dt \\ &= \frac{1}{2}\int e^{-st}d(\sin 2t) \\ &= \frac{1}{2}\left[e^{-st}\sin 2t - \int \sin 2t d(e^{-st})\right] \\ &= \frac{1}{2}\left[e^{-st}\sin 2t + s\int e^{-st}\sin 2t\,dt\right].\end{aligned}$$

Perform another integration by parts on the integral on the right.

$$\begin{aligned}\int & e^{-st}\cos 2t\,dt \\ &= \frac{1}{2}e^{-st}\sin 2t + \frac{s}{2}\left(-\frac{1}{2}\right)\int e^{-st}d(\cos 2t) \\ &= \frac{1}{2}e^{-st}\sin 2t \\ &\quad - \frac{s}{4}\left[e^{-st}\cos 2t - \int \cos 2t d(e^{-st})\right] \\ &= \frac{1}{2}e^{-st}\sin 2t \\ &\quad - \frac{s}{4}\left[e^{-st}\cos 2t + s\int e^{-st}\cos 2t\,dt\right] \\ &= \frac{1}{2}e^{-st}\sin 2t - \frac{s}{4}e^{-st}\cos 2t \\ &\quad - \frac{s^2}{4}\int e^{-st}\cos 2t\,dt.\end{aligned}$$

Transferring the last integral on the right to the left

side of the equation,

$$\left(1+\frac{s^2}{4}\right)\int e^{-st}\cos 2t$$
$$=\frac{1}{2}e^{-st}\sin 2t-\frac{s}{4}e^{-st}\cos 2t$$
$$\int e^{-st}\cos 2t\,dt$$
$$=\frac{2}{4+s^2}e^{-st}\sin 2t-\frac{s}{4+s^2}e^{-st}\cos 2t.$$

Inserting the limits 0 and T, we obtain

$$F(s)=\lim_{T\to\infty}\left(\frac{2}{4+s^2}e^{-st}\sin 2t - \frac{s}{4+s^2}e^{-st}\cos 2t\right)\Bigg|_{t=0}^{T}$$
$$=\lim_{T\to\infty}\left(\frac{2}{4+s^2}e^{-sT}\sin 2T - \frac{s}{4+s^2}e^{-sT}\cos 2T+\frac{s}{4+s^2}\right)$$
$$=\frac{s}{4+s^2},$$

provided $s>0$.

7. Using Definition 1.1,

$$F(s)=\int_0^\infty te^{2t}e^{-st}\,dt$$
$$=\int_0^\infty te^{-(s-2)t}\,dt$$
$$=\lim_{T\to\infty}\int_0^T te^{-(s-2)t}\,dt.$$

Integrating by parts,

$$\int te^{-(s-2)t}\,dt$$
$$=\frac{-1}{s-2}\int t\,d(e^{-(s-2)t})$$
$$=-\frac{1}{s-2}\left[te^{-(s-2)t}-\int e^{-(s-2)t}\,dt\right]$$
$$=-\frac{1}{s-2}\left[te^{-(s-2)t}+\frac{e^{-(s-2)t}}{s-2}\right]$$
$$=\frac{-te^{-(s-2)t}}{s-2}-\frac{e^{-(s-2)t}}{(s-2)^2}$$

Inserting the limits 0 and T,

$$F(s)=\lim_{T\to\infty}\left[\frac{-Te^{-(s-2)T}}{s-2}-\frac{e^{-(s-2)T}}{(s-2)^2}+\frac{1}{(s-2)^2}\right]$$
$$=\frac{1}{(s-2)^2},$$

provided $s>2$.

9. Using Definition 1.1,

$$F(s)=\int_0^\infty (e^{2t}\cos 3t)e^{-st}\,dt$$
$$=\int_0^\infty e^{-(s-2)t}\cos 3t\,dt.$$

Integrating by parts,

$$\int e^{-(s-2)t}\cos 3t\,dt$$
$$=\frac{1}{3}\int e^{-(s-2)t}\,d(\sin 3t)$$
$$=\frac{1}{3}\left[e^{-(s-2)t}\sin 3t-\int \sin 3t\,d(e^{-(s-2)t})\right]$$
$$=\frac{1}{3}e^{-(s-2)t}\sin 3t+\frac{s-2}{3}\int e^{-(s-2)t}\sin 3t\,dt.$$

We will compute the integral by parts.

$$\int e^{-(s-2)t}\cos 3t\,dt$$

$$= \frac{1}{3}e^{-(s-2)t}\sin 3t + \frac{s-2}{3}\cdot\frac{-1}{3}\int e^{-(s-2)t}\,d(\cos 3t)$$

$$= \frac{1}{3}e^{-(s-2)t}\sin 3t - \frac{s-2}{9}\int e^{-(s-2)t}\,d(\cos 3t)$$

$$= \frac{1}{3}e^{-(s-2)t}\sin 3t - \frac{s-2}{9}e^{-(s-2)t}\cos 3t + \frac{s-2}{9}\int \cos 3t\,d(e^{-(s-2)t})$$

$$= \frac{1}{3}e^{-(s-2)t}\sin 3t - \frac{s-2}{9}e^{-(s-2)t}\cos 3t - \frac{(s-2)^2}{9}\int e^{-(s-2)t}\cos 3t\,dt$$

Combining the integral on the right with the one on the left,

$$\left[1+\frac{(s-2)^2}{9}\right]\int e^{-(s-2)t}\cos 3t\,dt = \frac{1}{3}e^{-(s-2)t}\sin 3t - \frac{s-2}{9}e^{-(s-2)t}\cos 3t,$$

so

$$\int e^{-(s-t)t}\cos 3t\,dt = \frac{9}{9+(s-2)^2}\left[\frac{1}{3}e^{-(s-2)t}\sin 3t - \frac{s-2}{9}e^{-(s-2)}\cos 3t\right].$$

Inserting the limits 0 and T,

$$F(s) = \lim_{T\to\infty}\left[\frac{3}{9+(s-2)^2}e^{-(s-2)T}\cos 3T + \frac{s-2}{9+(s-2)^2}\right] = \frac{s-2}{9+(s-2)^2},$$

provided $s > 2$.

11. (a) Using Definition 1.1,

$$G(s) = \int_0^\infty t^2e^{-st}\,dt = \lim_{T\to\infty}\int_0^T t^2e^{-st}\,dt.$$

Integrating by parts,

$$\int t^2e^{-st}\,dt = -\frac{t^2e^{-st}}{s} + \frac{2}{3}\int te^{-st}\,dt.$$

Using integration by parts on the second integral,

$$\int t^2e^{-st}\,dt$$

$$= -\frac{t^2e^{-st}}{s} + \frac{2}{s}\left[-\frac{te^{-st}}{s} + \frac{1}{s}\int e^{-st}\,dt\right]$$

$$= -\frac{t^2e^{-st}}{s} - \frac{2te^{-st}}{s^2} + \frac{2}{s^2}\int e^{-st}\,dt$$

$$= -\frac{t^2e^{-st}}{s} - \frac{2te^{-st}}{s^2} - \frac{2e^{-st}}{s^3}.$$

Inserting the limits 0 and T,

$$G(s) = \lim_{T\to\infty}\left[-\frac{T^2e^{-sT}}{s} - \frac{2Te^{-sT}}{s^2} - \frac{2e^{-sT}}{s^3} + \frac{2}{s^3}\right] = \frac{2}{s^3} = \frac{2!}{s^3}.$$

(b) Using Definition 1.1,

$$H(s) = \int_0^\infty t^3e^{-st}\,dt = \lim_{T\to\infty}\int_0^T t^3e^{-st}\,dt.$$

Integrating by parts,

$$\int t^3e^{-st}\,dt = -\frac{1}{s}t^3e^{-st} + \frac{3}{s}\int t^2e^{-st}\,dt.$$

In part (a), we evaluated the integral on the right.

$$\int t^3 e^{-st}\,dt$$
$$= -\frac{t^3 e^{-st}}{s} + \frac{3}{s}\left[\frac{-t^2 e^{-st}}{s} - \frac{2te^{-st}}{s^2} - \frac{2e^{-st}}{s^3}\right]$$
$$= -e^{-st}\left[\frac{t^3}{s} + \frac{3t^2}{s^2} + \frac{6t}{s^3} + \frac{6}{s^4}\right].$$

Inserting the limits 0 and T,

$$H(s) = -\lim_{T\to\infty} e^{-sT}\left[\frac{T^3}{s} + \frac{3T^2}{s^2} + \frac{6T}{s^3} + \frac{6}{s^4}\right] + \frac{6}{s^4}$$
$$= \frac{6}{s^4} = \frac{3!}{s^4}.$$

(c) We've shown that $\mathcal{L}\{t\} = 1/s$. Assume that $\mathcal{L}\{t^k\} = k!/s^{k+1}$. Now,

$$\mathcal{L}\{t^{k+1}\}(s) = \int_0^\infty t^{k+1} e^{-st}\,dt$$
$$= \lim_{T\to\infty}\int_0^T t^{k+1} e^{-st}\,dt.$$

Integrating by parts,

$$\int_0^T t^{k+1} e^{-st}\,dt$$
$$= -\frac{t^{k+1}e^{-st}}{s}\Bigg|_0^T + \frac{k+1}{s}\int_0^T t^k e^{-st}\,dt$$
$$= \frac{-T^{k+1}e^{-sT}}{s} + \frac{k+1}{s}\int_0^T t^k e^{-st}\,dt.$$

Thus,

$$\mathcal{L}\{t^{k+1}\}(s)$$
$$= \lim_{T\to\infty}\left[\frac{-T^{k+1}e^{-sT}}{s} + \frac{k+1}{s}\int_0^T t^k e^{-st}\,dt\right]$$
$$= \frac{k+1}{s}\int_0^\infty t^k e^{-st}\,dt.$$

But this integral is the Laplace transform $\mathcal{L}\{t^k\}$, which by our induction hypothesis,

$$\mathcal{L}\{t^{k+1}\}(s) = \frac{k+1}{s}\cdot\frac{k!}{s^{k+1}}$$
$$= \frac{(k+1)!}{s^{k+2}}.$$

Thus, by induction,

$$\mathcal{L}\{t^k\}(s) = \frac{k!}{s^{k+1}}$$

for all $k = 1, 2, 3, \ldots$.

13. Using Definition 1.1,

$$F(s) = \int_0^\infty (e^{at}\cos\omega t)e^{-st}\,dt$$
$$= \lim_{T\to\infty}\int_0^T e^{-(s-a)t}\cos\omega t\,dt.$$

Integrating by parts,

$$\int e^{-(s-a)t}\cos\omega t\,dt$$
$$= \frac{e^{-(s-a)t}\sin\omega t}{\omega} + \frac{s-a}{\omega}\int e^{-(s-a)t}\sin\omega t\,dt.$$

Using integration by parts on the integral on the right,

$$\int e^{-(s-a)t}\sin\omega t\,dt$$
$$\frac{-e^{-(s-a)t}\cos\omega t}{\omega} - \frac{s-a}{\omega}\int e^{-(s-a)t}\cos\omega t\,dt$$

Substituting into the previous fomula,

$$\begin{aligned}\int e^{-(s-a)t} \cos \omega t \, dt &= \frac{e^{-(s-a)t}}{\omega^2}[\omega \sin \omega t - (s-a)\cos \omega t] \\ &\quad - \frac{(s-a)^2}{\omega^2}\int e^{-(s-a)t}\cos \omega t \, dt\end{aligned}$$

Transferring the last integral on the right to the left side of the equation,

$$\begin{aligned}\frac{\omega^2 + (s-a)^2}{\omega^2}\int e^{-(s-a)t}\cos \omega t \, dt &= \frac{e^{-(s-a)t}}{\omega^2}[\omega \sin \omega t - (s-a)\cos \omega t],\end{aligned}$$

Hence,

$$\begin{aligned}\int e^{-(s-a)t}\cos \omega t \, dt &= \frac{e^{-(s-a)t}}{\omega^2 + (s-a)^2}[\omega \sin \omega t - (s-a)\cos \omega t].\end{aligned}$$

Inserting the limits 0 and T,

$$\begin{aligned}F(s) &= \lim_{T\to\infty} \frac{e^{-(s-a)t}}{\omega^2 + (s-a)^2} \\ &\qquad \times [\omega \sin \omega t - (s-a)\cos \omega t]\Big|_0^T \\ &= \frac{s-a}{\omega^2 + (s-a)^2},\end{aligned}$$

provided $s > a$.

15. Using Table 1,

$$\begin{aligned}\mathcal{L}(3)(s) &= \int_0^\infty 3e^{-st}\, dt \\ &= 3\int_0^\infty 1e^{-st}\, dt \\ &= 3\,\mathcal{L}(1)(s) \\ &= 3 \cdot \frac{1}{s} \\ &= \frac{3}{s},\end{aligned}$$

provided $s > 0$.

17. Using Table 1,

$$\mathcal{L}(e^{-2t})(s) = \frac{1}{s-(-2)} = \frac{1}{s+2},$$

provided $s > -2$.

19. Using Table 1,

$$\mathcal{L}(\cos 2t)(s) = \frac{s}{s^2+2^2} = \frac{s}{s^2+4},$$

provided $s > 0$.

21. Using Table 1,

$$\mathcal{L}(te^{2t}) = \frac{1!}{(s-2)^2} = \frac{1}{(s-2)^2},$$

provided $s > 2$.

23. Using Table 1,

$$\begin{aligned}\mathcal{L}(e^{2t}\cos 3t)(s) &= \frac{s-2}{(s-2)^2+3^2} \\ &= \frac{s-2}{(s-2)^2+9},\end{aligned}$$

provided $s > 2$.

25. Using Definition 1.1,

$$\begin{aligned}F(s) &= \int_0^\infty f(t)e^{-st}\, dt \\ &= \int_0^2 0e^{-st}\, dt + \int_2^\infty 2e^{-st}\, dt \\ &= 2\lim_{T\to\infty} \frac{-e^{-st}}{s}\Big|_2^T \\ &= \frac{2e^{-2s}}{s},\end{aligned}$$

provided $s > 0$.

27. Using Definition 1.1,

$$F(s) = \int_0^2 te^{-st}\, dt + \int_2^\infty 2e^{-st}\, dt.$$

Using integration by parts, we get

$$\begin{aligned}\int_0^2 te^{-st}\,dt &= \frac{-te^{-st}}{s}\Big|_0^2 + \frac{1}{s}\int_0^2 e^{-st}\,dt\\ &= \frac{-2e^{-2s}}{s} - \frac{1}{s^2}e^{-st}\Big|_0^2\\ &= -\frac{2e^{-2s}}{s} - \frac{e^{-2s}-1}{s^2}\end{aligned}$$

For the second integral we have

$$\begin{aligned}\int_2^\infty 2e^{-st}\,dt &= \lim_{T\to\infty}\int_2^T 2e^{-st}\,dt\\ &= \lim_{T\to\infty}\frac{-2e^{-st}}{s}\Big|_2^T\\ &= \frac{2e^{-2s}}{s}\end{aligned}$$

provided $s > 0$. Hence

$$F(s) = -\frac{e^{-2s}-1}{s^2},$$

provided $s > 0$.

29. Because

$$y(t) = H(t-3)e^{0.2t} = \begin{cases}0, & t < 3\\ e^{0.2t}, & t \geq 3,\end{cases}$$

the graph is:

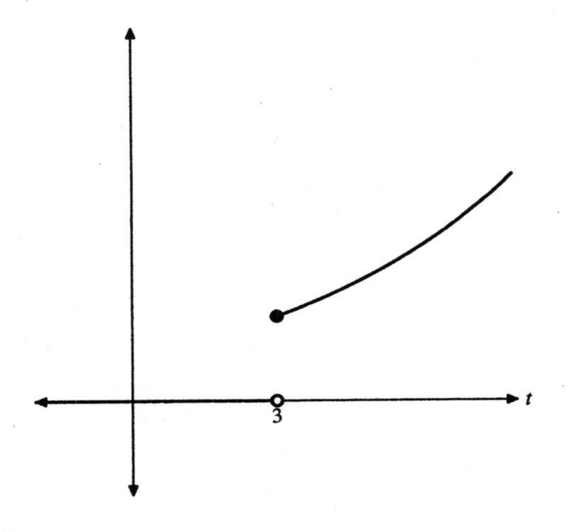

The Laplace transform is

$$\begin{aligned}Y(s) &= \int_0^\infty y(t)e^{-st}\,dt\\ &= \int_0^3 0e^{-st}\,dt + \int_3^\infty e^{0.2t}e^{-st}\,dt\\ &= \lim_{T\to\infty}\int_3^T e^{-(s-0.2)t}\,dt\\ &= \lim_{T\to\infty}\frac{-e^{-(s-0.2)t}}{s-0.2}\Big|_3^T\\ &= \lim_{T\to\infty}\left[\frac{-e^{-(s-0.2)T}}{s-0.2} + \frac{e^{-3(s-0.2)}}{s-0.2}\right]\\ &= \frac{e^{-3(s-0.2)}}{s-0.2},\end{aligned}$$

provided $s > 0.2$.

31. (a) By a property of integrals,

$$\begin{aligned}|F(s)| &= \left|\int_0^\infty f(t)e^{-st}\,dt\right|\\ &\leq \int_0^\infty |f(t)e^{-st}|\,dt\\ &= \int_0^\infty |f(t)|e^{-st}\,dt.\end{aligned}$$

But, $|f(t)| \leq Ce^{at}$ for all $t > 0$, so

$$\begin{aligned}|F(s)| &\leq \int_0^\infty Ce^{at}e^{-st}\,dt\\ &= C\int_0^\infty e^{at}e^{-st}\,dt\\ &= C\,\mathcal{L}(e^{at})(s).\end{aligned}$$

But, in Table 1, $\mathcal{L}(e^{at})(s) = 1/(s-a)$, provided $s > 0$. Thus,

$$F(s) \leq \frac{C}{s-a},$$

provided $s > a$.

(b) By part (a),

$$|F(s)| \leq \frac{C}{s-a}.$$

Thus,

$$\lim_{s\to\infty} |F(s)| \le \lim_{s\to\infty} \frac{C}{s-a} = 0.$$

Thus, $\lim_{s\to\infty} |F(s)| = 0$ and consequently,

$$\lim_{s\to\infty} F(s) = 0.$$

The function

$$F(s) = \frac{s+4}{s-2}$$

cannot be a Laplace transform of any function because

$$\begin{aligned}\lim_{s\to\infty} F(s) &= \lim_{s\to\infty} \frac{s+4}{s-2}\\ &= \lim_{s\to\infty} \frac{1+\frac{4}{s}}{1-\frac{2}{s}}\\ &= 1.\end{aligned}$$

———×———

Section 5.2. Basic Properties of the Laplace Transform

1. Using linearity and Table 1,

$$\begin{aligned}\mathcal{L}\{3t^2\}(s) &= 3\,\mathcal{L}\{t^2\}(s)\\ &= 3\cdot\frac{2!}{s^3}\\ &= \frac{6}{s^3},\end{aligned}$$

provided $s > 0$.

3. Using linearity and Table 1,

$$\begin{aligned}&\mathcal{L}\{t^2+4t+5\}(s)\\ &\quad= \mathcal{L}\{t^2\}(s) + 4\,\mathcal{L}\{t\}(s) + 5\,\mathcal{L}\{1\}(s)\\ &\quad= \frac{2!}{s^3} + 4\left(\frac{1}{s^2}\right) + 5\left(\frac{1}{s}\right)\\ &\quad= \frac{2}{s^3} + \frac{4}{s^2} + \frac{5}{s}\\ &\quad= \frac{2+4s+5s^2}{s^3},\end{aligned}$$

provided $s > 0$.

5. Using linearity and Table 1,

$$\begin{aligned}&\mathcal{L}\{-2\cos t + 4\sin 3t\}(s)\\ &\quad= -2\,\mathcal{L}\{\cos t\}(s) + 4\,\mathcal{L}\{\sin 3t\}(s)\\ &\quad= -2\left(\frac{s}{s^2+1}\right) + 4\left(\frac{3}{s^2+9}\right)\\ &\quad= \frac{-2s(s^2+9)+12(s^2+1)}{(s^2+1)(s^2+9)}\\ &\quad= \frac{-2s^3+12s^2-18s+12}{(s^2+1)(s^2+9)},\end{aligned}$$

provided $s > 0$.

7. Using linearity and Table 1,

$$\begin{aligned}&\mathcal{L}\{\cosh \omega t\}(s)\\ &\quad= \mathcal{L}\left\{\frac{e^{\omega t}+e^{-\omega t}}{2}\right\}(s)\\ &\quad= \frac{1}{2}\mathcal{L}\{e^{\omega t}\}(s) + \frac{1}{2}\mathcal{L}\{e^{-\omega t}\}(s)\\ &\quad= \frac{1}{2}\left(\frac{1}{s-\omega}\right) + \frac{1}{2}\left(\frac{1}{s+\omega}\right)\\ &\quad= \frac{1}{2}\left[\frac{s+\omega+s-\omega}{(s-\omega)(s+\omega)}\right]\\ &\quad= \frac{s}{s^2-\omega^2}.\end{aligned}$$

Next,

$$\begin{aligned}
\mathcal{L}\{\sinh \omega t\}(s) &= \mathcal{L}\left\{\frac{e^{\omega t}-e^{-\omega t}}{2}\right\}(s) \\
&= \frac{1}{2}\mathcal{L}\{e^{\omega t}\}(s) - \frac{1}{2}\mathcal{L}\{e^{-\omega t}\}(s) \\
&= \frac{1}{2}\left(\frac{1}{s-\omega}\right) - \frac{1}{2}\left(\frac{1}{s+\omega}\right) \\
&= \frac{1}{2}\left[\frac{(s+\omega)-(s-\omega)}{(s-\omega)(s+\omega)}\right] \\
&= \frac{\omega}{s^2-\omega^2}.
\end{aligned}$$

9. If $y(t) = t^3$, then

$$\begin{aligned}
\mathcal{L}\{y'(t)\}(s) &= \mathcal{L}\{3t^2\}(s) \\
&= 3\,\mathcal{L}\{t^2\}(s) \\
&= 3\left(\frac{2!}{s^3}\right) \\
&= \frac{6}{s^3}.
\end{aligned}$$

On the other hand,

$$\begin{aligned}
s\,\mathcal{L}\{y(t)\}(s) - y(0) &= s\,\mathcal{L}\{t^3\}(s) - 0 \\
&= s\left(\frac{3!}{s^4}\right) \\
&= \frac{6}{s^3}.
\end{aligned}$$

11. If $y(t) = e^{-3t}$, then

$$\begin{aligned}
\mathcal{L}\{y'(t)\}(s) &= \mathcal{L}\{-3e^{-3t}\}(s) \\
&= -3\,\mathcal{L}\{e^{-3t}\}(s) \\
&= -3\left(\frac{1}{s+3}\right) \\
&= \frac{-3}{s+3}.
\end{aligned}$$

On the other hand,

$$\begin{aligned}
s\,\mathcal{L}\{y(t)\}(s) - y(0) &= s\,\mathcal{L}\{e^{-3t}\}(s) - 1 \\
&= s\left(\frac{1}{s+3}\right) - 1 \\
&= \frac{s-(s+3)}{s+3} \\
&= \frac{-3}{s+3}.
\end{aligned}$$

13. If $y(t) = \sin 5t$, then

$$\begin{aligned}
\mathcal{L}\{y'(t)\}(s) &= \mathcal{L}\{5\cos 5t\}(s) \\
&= 5\,\mathcal{L}\{\cos 5t\}(s) \\
&= 5\left(\frac{s}{s^2+25}\right) \\
&= \frac{5s}{s^2+25}.
\end{aligned}$$

On the other hand,

$$\begin{aligned}
s\,\mathcal{L}\{y(t)\}(s) - y(0) &= s\,\mathcal{L}\{\sin 5t\}(s) - 0 \\
&= s\left(\frac{5}{s^2+25}\right) \\
&= \frac{5s}{s^2+25}.
\end{aligned}$$

15. If $y(t) = e^{-2t}$, then $y'(t) = -2e^{-2t}$ and $y''(t) = 4e^{-2t}$. On one hand,

$$\begin{aligned}
\mathcal{L}\{y''(t)\}(s) &= \mathcal{L}\{4e^{-2t}\}(s) \\
&= 4\,\mathcal{L}\{e^{-2t}\}(s) \\
&= 4\left(\frac{1}{s+2}\right) \\
&= \frac{4}{s+2}.
\end{aligned}$$

On the other hand,

$$\begin{aligned}
&s^2\,\mathcal{L}\{y(t)\}(s) - sy(0) - y'(0)\\
&= s^2\,\mathcal{L}\{e^{-2t}\}(s) - s(1) - (-2)\\
&= s^2\left(\frac{1}{s+2}\right) - s + 2\\
&= \frac{s^2 + (-s+2)(s+2)}{s+2}\\
&= \frac{s^2 - s^2 + 4}{s+2}\\
&= \frac{4}{s+2}.
\end{aligned}$$

17. If $y(t) = t^2 + 3t + 5$, then $y'(t) = 2t + 3$ and $y''(t) = 2$. On the one hand,

$$\begin{aligned}
\mathcal{L}\{y''(t)\}(s) &= \mathcal{L}\{2\}(s)\\
&= 2\,\mathcal{L}\{1\}(s)\\
&= 2\left(\frac{1}{s}\right)\\
&= \frac{2}{s}.
\end{aligned}$$

On the other hand,

$$\begin{aligned}
&s^2\,\mathcal{L}\{y(t)\}(s) - sy(0) - y'(0)\\
&= s^2\,\mathcal{L}\{t^2 + 3t + 5\}(s) - s(5) - 3\\
&= s^2\left(\mathcal{L}\{t^2\}(s) + 3\,\mathcal{L}\{t\}(s) + 5\,\mathcal{L}\{1\}(s)\right)\\
&\quad - 5s - 3\\
&= s^2\left(\frac{2!}{s^3} + 3\left(\frac{1}{s^2}\right) + 5\left(\frac{1}{s}\right)\right) - 5s - 3\\
&= s^2\left(\frac{2 + 3s + 5s^2}{s^3}\right) - 5s - 3\\
&= \frac{2 + 3s + 5s^2}{s} + \frac{-5s^2 - 3s}{s}\\
&= \frac{2}{s}.
\end{aligned}$$

19. If $y' - 5y = e^{-2t}$, with $y(0) = 1$, then

$$\begin{aligned}
\mathcal{L}\{y' - 5y\}(s) &= \mathcal{L}\{e^{-2t}\}(s)\\
\mathcal{L}\{y'\}(s) - 5\,\mathcal{L}\{y\}(s) &= \frac{1}{s+2}\\
s\,\mathcal{L}\{y\}(s) - y(0) - 5\,\mathcal{L}\{y\}(s) &= \frac{1}{s+2}.
\end{aligned}$$

If we let $Y(s) = \mathcal{L}\{y\}(s)$, then

$$\begin{aligned}
sY(s) - 1 - 5Y(s) &= \frac{1}{s+2}\\
(s-5)Y(s) &= 1 + \frac{1}{s+2}\\
Y(s) &= \frac{1}{s-5} + \frac{1}{(s-5)(s+2)}\\
Y(s) &= \frac{(s+2)+1}{(s-5)(s+2)}\\
Y(s) &= \frac{s+3}{(s-5)(s+2)}.
\end{aligned}$$

21. If $y' - 4y = \cos 2t$, with $y(0) = -2$, then

$$\begin{aligned}
\mathcal{L}\{y' - 4y\}(s) &= \mathcal{L}\{\cos 2t\}(s)\\
\mathcal{L}\{y'\}(s) - 4\,\mathcal{L}\{y\}(s) &= \frac{s}{s^2+4}\\
s\,\mathcal{L}\{y\}(s) - y(0) - 4\,\mathcal{L}\{y\}(s) &= \frac{s}{s^2+4}.
\end{aligned}$$

If we let $Y(s) = \mathcal{L}\{y\}(s)$, then

$$\begin{aligned}
sY(s) + 2 - 4Y(s) &= \frac{s}{s^2+4}\\
(s-4)Y(s) &= -2 + \frac{s}{s^2+4}\\
Y(s) &= -\frac{2}{s-4} + \frac{s}{(s-4)(s^2+4)}\\
Y(s) &= \frac{-2(s^2+4)+s}{(s-4)(s^2+4)}\\
Y(s) &= \frac{-2s^2 + s - 8}{(s-4)(s^2+4)}.
\end{aligned}$$

23. If $y'' + 2y' + 2y = \cos 2t$, with $y(0) = 1$ and $y'(0) = 0$, then

$$\begin{aligned}
\mathcal{L}\{y'' + 2y' + 2y\}(s) &= \mathcal{L}\{\cos 2t\}(s)\\
\mathcal{L}\{y''\}(s) + 2\,\mathcal{L}\{y'\}(s) + 2\,\mathcal{L}\{y\}(s) &= \frac{s}{s^2+4}
\end{aligned}$$

If we let $Y(s) = \mathcal{L}\{y\}(s)$, this becomes

$$\begin{aligned}
&[s^2Y(s) - s(1) - 0] + 2[sY(s) - 1] + 2Y(s)\\
&= \frac{s}{s^2+4}.
\end{aligned}$$

Collecting terms we get

$$(s^2 + 2s + 2)Y(s) - (s + 2) = \frac{s}{s^2 + 4},$$

which we solve for Y to get

$$Y(s) = \frac{s+2}{s^2 + 2s + 2} + \frac{s}{(s^2 + 4)(s^2 + 2s + 2)}$$

$$Y(s) = \frac{s^3 + 2s^2 + 5s + 8}{(s^2 + 4)(s^2 + 2s + 2)}.$$

25. If $y'' + 3y' + 5y = t + e^{-t}$,

$$\begin{aligned} &\mathcal{L}\{y'' + 3y' + 5y\}(s) \\ &= \mathcal{L}\{y''\}(s) + 3\,\mathcal{L}\{y'\}(s) + 5\,\mathcal{L}\{y\}(s) \\ &= [s^2\,\mathcal{L}\{y\}(s) - sy(0) - y'(0)] \\ &\quad + 3[s\,\mathcal{L}\{y\}(s) - y(0)] + s\,\mathcal{L}\{y\}(s) \end{aligned}$$

and

$$\mathcal{L}\{t + e^{-t}\}(s) = \frac{1}{s^2} + \frac{1}{s+1}.$$

If we equate these two, use the initial conditions $y(0) = -1$ and $y'(0) = 0$, and set $Y(s) = \mathcal{L}\{y\}(s)$, we get

$$s^2Y(s) + s + 3sY(s) + 3 + 5Y(s) = \frac{s^2 + s + 1}{s^2(s+1)}.$$

Solving for Y, we get

$$Y(s) = \frac{-s-3}{s^2 + 3s + 5} + \frac{s^2 + s + 1}{s^2(s+1)(s^2 + 3s + 5)}$$

$$Y(s) = \frac{(-s-3)s^2(s+1) + (s^2 + s + 1)}{s^2(s+1)(s^2 + 3s + 5)}$$

$$Y(s) = \frac{s^2(-s^2 - 4s - 3) + s^2 + s + 1}{s^2(s+1)(s^2 + 3s + 5)}$$

$$Y(s) = \frac{-s^4 - 4s^3 - 2s^2 + s + 1}{s^2(s+1)(s^2 + 3s + 5)}.$$

27. Because $f(t) = \cos 2t$ has transform $F(s) = s/(s^2 + 4)$, then $y(t) = e^{2t}\cos 2t$ will have transform

$$\begin{aligned} Y(s) &= F(s-2) \\ &= \frac{s-2}{(s-2)^2 + 4} \\ &= \frac{s-2}{s^2 - 4s + 8}. \end{aligned}$$

29. Because $f(t) = t^2 + 3t + 4$ has transform

$$\begin{aligned} F(s) &= \frac{2!}{s^3} + \frac{3}{s^2} + \frac{4}{s} \\ &= \frac{2 + 3s + 4s^2}{s^3}, \end{aligned}$$

then $y(t) = e^{-t}(t^2 + 3t + 4)$ will have transform

$$\begin{aligned} Y(s) &= F(s+1) \\ &= \frac{2 + 3(s+1) + 4(s+1)^2}{(s+1)^3} \\ &= \frac{2 + 3s + 3 + 4s^2 + 8s + 4}{(s+1)^3} \\ &= \frac{4s^2 + 11s + 9}{(s+1)^3}. \end{aligned}$$

31. Because $f(t) = e^{-t}$ has transform $F(s) = 1/(s+1)$, then $y(t) = te^{-t}$ has transform

$$\begin{aligned} Y(s) &= -F'(s) \\ &= -\left[\frac{-1}{(s+1)^2}\right] \\ &= \frac{1}{(s+1)^2}. \end{aligned}$$

33. Because $f(t) = e^{2t}$ has transform $F(s) = 1/(s-2)$, then $y(t) = t^2e^{2t}$ has transform

$$\begin{aligned} Y(s) &= (-1)^2F''(s) \\ &= \frac{d}{ds}\left(\frac{-1}{(s-2)^2}\right) \\ &= \frac{2}{(s-2)^3}. \end{aligned}$$

35. If $y' - y = t^2e^{-2t}$, with $y(0) = 0$, then with $Y(s) = \mathcal{L}\{y\}(s)$,

$$\begin{aligned} \mathcal{L}\{y' - y\}(s) &= s\,\mathcal{L}\{y\}(s) - y(0) - \mathcal{L}\{y\}(s) \\ &= sY(s) - Y(s) \\ &= (s-1)Y(s). \end{aligned}$$

Because the transform of $f(t) = e^{-2t}$ is $F(s) = 1/(s+2)$, the transform of t^2e^{-2t} is

$$\mathcal{L}\{t^2e^{-2t}\}(s) = (-1)^2F''(s) = \frac{2}{(s+2)^3}.$$

Equating,

$$\begin{aligned}(s-1)Y(s) &= \frac{2}{(s+2)^3}\\ Y(s) &= \frac{2}{(s-1)(s+2)^3}.\end{aligned}$$

37. If $y' - 2y = e^{2t}\cos t$, with $y(0) = -2$, then with $Y(s) = \mathcal{L}\{y\}(s)$,

$$\begin{aligned}\mathcal{L}\{y' - 2y\}(s) &= s\,\mathcal{L}\{y\}(s) - y(0) - 2\,\mathcal{L}\{y\}(s)\\ &= sY(s) + 2 - 2Y(s)\\ &= (s-2)Y(s) + 2.\end{aligned}$$

Because the transform of $f(t) = \cos t$ is $F(s) = s/(s^2+1)$, the transform of $e^{2t}\cos t$ is

$$F(s-2) = \frac{s-2}{(s-2)^2+1} = \frac{s-2}{s^2-4s+5}.$$

Equating,

$$(s-2)Y(s) + 2 = \frac{s-2}{s^2-4s+5}$$

and solving for Y, we get

$$\begin{aligned}Y(s) &= -\frac{2}{s-2} + \frac{s-2}{(s-2)(s^2-4s+5)}\\ &= \frac{-2(s^2-4s+5)+s-2}{(s-2)(s^2-4s+5)}\\ &= \frac{-2s^2+9s-12}{(s-2)(s^2-4s+5)}.\end{aligned}$$

39. If $y'' + y' + 2y = e^{-t}\cos 2t$, with $y(0) = 1$ and $y'(0) = -1$, then with $Y(s) = \mathcal{L}\{y\}(s)$,

$$\begin{aligned}&\mathcal{L}\{y'' + y' + 2y\}(s)\\ &\quad = s^2\,\mathcal{L}\{y\}(s) - sy(0) - y'(0)\\ &\qquad + s\,\mathcal{L}\{y\}(s) - y(0) + 2\,\mathcal{L}\{y\}(s)\\ &\quad = s^2Y(s) - s + 1 + sY(s) - 1 + 2Y(s)\\ &\quad = (s^2+s+2)Y(s) - s.\end{aligned}$$

Because the transform of $f(t) = \cos 2t$ is $F(s) = s/(s^2+4)$, the transform of $e^{-t}\cos 2t$ is

$$F(s+1) = \frac{s+1}{(s+1)^2+4} = \frac{s+1}{s^2+2s+5}.$$

Equating,

$$(s^2+s+2)Y(s) - s = \frac{s+1}{s^2+2s+5}.$$

Solving for Y

$$\begin{aligned}Y(s) &= \frac{s}{s^2+s+2} + \frac{s+1}{(s^2+s+2)(s^2+2s+5)}\\ &= \frac{s(s^2+2s+5)+s+1}{(s^2+s+2)(s^2+2s+5)}\\ &= \frac{s^3+2s^2+6s+1}{(s^2+s+2)(s^2+2s+5)}.\end{aligned}$$

41. If $y'' + 5y = 3e^{-t}\cos 4t$, with $y(0) = -1$ and $y'(0) = 2$, then with $Y(s) = \mathcal{L}\{y\}(s)$,

$$\begin{aligned}&\mathcal{L}\{y'' + 5y\}(s)\\ &\quad = s^2\,\mathcal{L}\{y\}(s) - sy(0) - y'(0) + 5\,\mathcal{L}\{y\}(s)\\ &\quad = s^2Y(s) + s - 2 + 5Y(s)\\ &\quad = (s^2+5)Y(s) + s - 2.\end{aligned}$$

Because the transform of $f(t) = 3\cos 4t$ is

$$F(s) = 3\left(\frac{s}{s^2+16}\right) = \frac{3s}{s^2+16},$$

the transform of $3e^{-t}\cos 4t$ is

$$F(s+1) = \frac{3(s+1)}{(s+1)^2+16} = \frac{3s+3}{s^2+2s+17}.$$

Equating,

$$(s^2+5)Y(s) + s - 2 = \frac{3s+3}{s^2+2s+17}.$$

Solving for Y

$$\begin{aligned}Y(s) &= -\frac{s-2}{s^2+5} + \frac{3s+3}{(s^2+5)(s^2+2s+17)}\\ &= \frac{(2-s)(s^2+2s+17)+3s+3}{(s^2+5)(s^2+2s+17)}\\ &= \frac{-s^3-10s+37}{(s^2+5)(s^2+2s+17)}.\end{aligned}$$

43. (a) If $\Gamma(\alpha) = \int_0^\infty e^{-t}t^{\alpha-1}\,dt$, then

$$
\begin{aligned}
\Gamma(1) &= \int_0^\infty e^{-t}t^{1-1}\,dt \\
&= \int_0^\infty e^{-t}\,dt \\
&= \lim_{T\to\infty}\int_0^T e^{-t}\,dt \\
&= \lim_{T\to\infty} -e^{-t}\Big|_0^T \\
&= -\lim_{T\to\infty}(e^{-T}-1) \\
&= 1.
\end{aligned}
$$

(b) Furthermore,

$$
\begin{aligned}
\Gamma(\alpha+1) &= \int_0^\infty e^{-t}t^{(\alpha+1)-1}\,dt \\
&= \int_0^\infty e^{-t}t^{\alpha}\,dt \\
&= \lim_{T\to\infty}\int_0^T e^{-t}t^{\alpha}\,dt.
\end{aligned}
$$

Integrating by parts,

$$\int e^{-t}t^{\alpha}\,dt = -e^{-t}t^{\alpha} + \alpha\int e^{-t}t^{\alpha-1}\,dt.$$

Inserting the limits 0 and T,

$$
\begin{aligned}
&\Gamma(\alpha+1) \\
&= \lim_{T\to\infty}\left[-e^{-t}t^{\alpha}\Big|_0^T + \alpha\int_0^T e^{-t}t^{\alpha-1}\,dt\right] \\
&= \lim_{T\to\infty}\left[-e^{-T}T^{\alpha} + \alpha\int_0^T e^{-t}t^{\alpha-1}\,dt\right] \\
&= 0 + \alpha\int_0^\infty e^{-t}t^{\alpha-1}\,dt \\
&= \alpha\Gamma(\alpha).
\end{aligned}
$$

Furthermore, if n is an integer, then

$$
\begin{aligned}
\Gamma(n+1) &= n\Gamma(n) \\
&= n(n-1)\Gamma(n-1) \\
&= n(n-1)(n-2)\Gamma(n-2) \\
&= n(n-1)(n-2)\cdots(2)\Gamma(1) \\
&= n(n-1)(n-2)\cdots(2)(1) \\
&= n!.
\end{aligned}
$$

(c) First, by definition,

$$\mathcal{L}\{t^{\alpha}\}(s) = \int_0^\infty t^{\alpha}e^{-st}\,dt.$$

Let $u = st$, then $du = s\,dt$ and

$$
\begin{aligned}
\int_0^\infty t^{\alpha}e^{-st}\,dt &= \int_0^\infty \left(\frac{u}{s}\right)^{\alpha} e^{-u}\left(\frac{du}{s}\right) \\
&= \frac{1}{s^{\alpha+1}}\int_0^\infty e^{-u}u^{(\alpha+1)-1}\,dy \\
&= \frac{1}{s^{\alpha+1}}\Gamma(\alpha+1) \\
&= \frac{\Gamma(\alpha+1)}{s^{\alpha+1}}.
\end{aligned}
$$

In the case where α is a positive integer n, then

$$\mathcal{L}\{t^{n}\}(s) = \frac{\Gamma(n+1)}{s^{n+1}} = \frac{n!}{s^{n+1}}.$$

Section 5.3. The Inverse Laplace Transform

1. Factor.

$$Y(s) = \frac{1}{3s+2} = \frac{1}{3} \cdot \frac{1}{s+2/3}.$$

Now,

$$\begin{aligned} y(t) &= \mathcal{L}^{-1}\left\{\frac{1}{3} \cdot \frac{1}{s+2/3}\right\} \\ &= \frac{1}{3}\mathcal{L}^{-1}\left\{\frac{1}{s+2/3}\right\} \\ &= \frac{1}{3}e^{-(2/3)t}. \end{aligned}$$

3. Adjust as follows:

$$Y(s) = \frac{1}{s^2+4} = \frac{1}{2} \cdot \frac{2}{s^2+4}.$$

Now,

$$\begin{aligned} y(t) &= \mathcal{L}^{-1}\left\{\frac{1}{2} \cdot \frac{2}{s^2+4}\right\} \\ &= \frac{1}{2}\mathcal{L}^{-1}\left\{\frac{2}{s^2+4}\right\} \\ &= \frac{1}{2}\sin 2t. \end{aligned}$$

5. Adjust as follows:

$$Y(s) = \frac{3}{s^2} = 3 \cdot \frac{1}{s^2}.$$

Now,

$$\begin{aligned} y(t) &= \mathcal{L}^{-1}\left\{3 \cdot \frac{1}{s^2}\right\} \\ &= 3\,\mathcal{L}^{-1}\left\{\frac{1}{s^2}\right\} \\ &= 3t. \end{aligned}$$

7. Adjust as follows:

$$\begin{aligned} Y(s) &= \frac{3s+2}{s^2+25} \\ &= \frac{3s}{s^2+25} + \frac{2}{s^2+25} \\ &= 3 \cdot \frac{s}{s^2+25} + \frac{2}{5} \cdot \frac{5}{s^2+25}. \end{aligned}$$

Thus,

$$\begin{aligned} y(t) &= \mathcal{L}^{-1}\left\{3 \cdot \frac{s}{s^2+25} + \frac{2}{5} \cdot \frac{5}{s^2+25}\right\} \\ &= 3\,\mathcal{L}^{-1}\left\{\frac{s}{s^2+25}\right\} + \frac{2}{5}\mathcal{L}^{-1}\left\{\frac{5}{s^2+25}\right\} \\ &= 3\cos 5t + \frac{2}{5}\sin 5t. \end{aligned}$$

9. Adjust as follows:

$$\begin{aligned} Y(s) &= \frac{1}{3-4s} + \frac{3-2s}{s^2+49} \\ &= \frac{1}{-4} \cdot \frac{1}{s-3/4} + \frac{3}{s^2+49} - \frac{2s}{s^2+49} \\ &= -\frac{1}{4} \cdot \frac{1}{s-3/4} + \frac{3}{7} \cdot \frac{7}{s^2+49} \\ &\quad - 2 \cdot \frac{s}{s^2+49}. \end{aligned}$$

Thus,

$$\begin{aligned} y(t) &= \mathcal{L}^{-1}\left\{-\frac{1}{4} \cdot \frac{1}{s-3/4} + \frac{3}{7} \cdot \frac{7}{s^2+49} \right. \\ &\quad \left. - 2 \cdot \frac{s}{s^2+49}\right\} \\ &= -\frac{1}{4}\mathcal{L}^{-1}\left\{\frac{1}{s-3/4}\right\} + \frac{3}{7}\mathcal{L}^{-1}\left\{\frac{7}{s^2+49}\right\} \\ &\quad - 2\,\mathcal{L}^{-1}\left\{\frac{s}{s^2+49}\right\} \\ &= -\frac{1}{4}e^{(3/4)t} + \frac{3}{7}\sin 7t - 2\cos 7t. \end{aligned}$$

11. Note the transform pair:

$$t^2 \Longleftrightarrow \frac{2}{s^3}$$

By Proposition 2.12,

$$e^{-2t}t^2 \Longleftrightarrow \frac{2}{(s+2)^3}.$$

Thus,

$$\begin{aligned} y(t) &= \mathcal{L}^{-1}\left\{\frac{5}{(s+2)^3}\right\} \\ &= \mathcal{L}^{-1}\left\{\frac{5}{2}\cdot\frac{2}{(s+2)^3}\right\} \\ &= \frac{5}{2}\mathcal{L}^{-1}\left\{\frac{2}{(s+2)^3}\right\} \\ &= \frac{5}{2}e^{-2t}t^2. \end{aligned}$$

13. Note the transform pair:

$$\sin 5t \Longleftrightarrow \frac{5}{s^2+25}$$

By Proposition 2.12,

$$e^{-2t}\sin 5t \Longleftrightarrow \frac{5}{(s+2)^2+25}.$$

Thus,

$$\begin{aligned} y(t) &= \mathcal{L}^{-1}\left\{\frac{3}{(s+2)^2+25}\right\} \\ &= \mathcal{L}^{-1}\left\{\frac{3}{5}\cdot\frac{5}{(s+2)^2+25}\right\} \\ &= \frac{3}{5}\mathcal{L}^{-1}\left\{\frac{5}{(s+2)^2+25}\right\} \\ &= \frac{3}{5}e^{-2t}\sin 5t. \end{aligned}$$

15. Note the transform pairs:

$$\begin{aligned} \cos\sqrt{5}t &\Longleftrightarrow \frac{s}{s^2+5} \\ \sin\sqrt{5}t &\Longleftrightarrow \frac{\sqrt{5}}{s^2+5} \end{aligned}$$

By Proposition 2.12,

$$\begin{aligned} e^t\cos\sqrt{5}t &\Longleftrightarrow \frac{s-1}{(s-1)^2+5} \\ e^t\sin\sqrt{5}t &\Longleftrightarrow \frac{\sqrt{5}}{(s-1)^2+5}. \end{aligned}$$

Thus,

$$\begin{aligned} y(t) &= \mathcal{L}^{-1}\left\{\frac{2s-3}{(s-1)^2+5}\right\} \\ &= \mathcal{L}^{-1}\left\{\frac{2s-2}{(s-1)^2+5} - \frac{1}{(s-1)^2+5}\right\} \\ &= \mathcal{L}^{-1}\left\{2\cdot\frac{s-1}{(s-1)^2+5} - \frac{1}{\sqrt{5}}\cdot\frac{\sqrt{5}}{(s-1)^2+5}\right\} \\ &= 2\mathcal{L}^{-1}\left\{\frac{s-1}{(s-1)^2+5}\right\} - \frac{1}{\sqrt{5}}\mathcal{L}^{-1}\left\{\frac{\sqrt{5}}{(s-1)^2+5}\right\} \\ &= 2e^t\cos\sqrt{5}t - \frac{1}{\sqrt{5}}e^t\sin\sqrt{5}t \\ &= e^t(2\cos\sqrt{5}t - \frac{\sqrt{5}}{5}\sin\sqrt{5}t). \end{aligned}$$

17. Complete the square.

$$Y(s) = \frac{3s+2}{s^2+4s+29} = \frac{3s+2}{(s+2)^2+25}$$

Note the transform pairs.

$$\begin{aligned} \cos 5t &\Longleftrightarrow \frac{s}{s^2+25} \\ \sin 5t &\Longleftrightarrow \frac{5}{s^2+25} \end{aligned}$$

By Proposition 2.12,

$$\begin{aligned} e^{-2t}\cos 5t &\Longleftrightarrow \frac{s+2}{(s+2)^2+25} \\ e^{-2t}\sin 5t &\Longleftrightarrow \frac{5}{(s+2)^2+25}. \end{aligned}$$

Thus,

$$\begin{aligned} y(t) &= \mathcal{L}^{-1}\left\{\frac{3s+2}{(s+2)^2+25}\right\} \\ &= \mathcal{L}^{-1}\left\{\frac{3s+6}{(s+2)^2+25} - \frac{4}{(s+2)^2+25}\right\} \\ &= \mathcal{L}^{-1}\left\{3\cdot\frac{s+2}{(s+2)^2+25} - \frac{4}{5}\cdot\frac{5}{(s+2)^2+25}\right\} \\ &= 3\,\mathcal{L}^{-1}\left\{\frac{s+2}{(s+2)^2+25}\right\} - \frac{4}{5}\mathcal{L}^{-1}\left\{\frac{5}{(s+2)^2+25}\right\} \\ &= 3e^{-2t}\cos 5t - \frac{4}{5}e^{-2t}\sin 5t \\ &= e^{-2t}(3\cos 5t - \frac{4}{5}\sin 5t). \end{aligned}$$

19. Find a partial fraction decomposition.

$$\frac{1}{(s+2)(s-1)} = \frac{A}{s+2} + \frac{B}{s-1}$$
$$1 = A(s-1) + B(s+2)$$

Now,

$$s = 1 \Rightarrow 1 = 3B \text{ or } B = \frac{1}{3}$$
$$s = -2 \Rightarrow 1 = -3A \text{ or } A = -\frac{1}{3}.$$

Thus,

$$\begin{aligned} y(t) &= \mathcal{L}^{-1}\left\{\frac{1}{(s+2)(s-1)}\right\} \\ &= \mathcal{L}^{-1}\left\{\frac{-1/3}{s+2} + \frac{1/3}{s-1}\right\} \\ &= -\frac{1}{3}\mathcal{L}^{-1}\left\{\frac{1}{s+2}\right\} + \frac{1}{3}\mathcal{L}^{-1}\left\{\frac{1}{s-1}\right\} \\ &= -\frac{1}{3}e^{-2t} + \frac{1}{3}e^{t}. \end{aligned}$$

21. Find a partial fraction decomposition.

$$\frac{2s-1}{(s+1)(s-2)} = \frac{A}{s+1} + \frac{B}{s-2}$$
$$2s-1 = A(s-2) + B(s+1)$$

Now,

$$s = 2 \Rightarrow 3 = 3B \text{ or } B = 1$$
$$s = -1 \Rightarrow -3 = -3A \text{ or } A = 1.$$

Thus,

$$\begin{aligned} y(t) &= \mathcal{L}^{-1}\left\{\frac{2s-1}{(s+1)(s-2)}\right\} \\ &= \mathcal{L}^{-1}\left\{\frac{1}{s+1} + \frac{1}{s-2}\right\} \\ &= \mathcal{L}^{-1}\left\{\frac{1}{s+1}\right\} + \mathcal{L}^{-1}\left\{\frac{1}{s-2}\right\} \\ &= e^{-t} + e^{2t}. \end{aligned}$$

23. Find a partial fraction decomposition.

$$\frac{7s+13}{s^2+2s-3} = \frac{A}{s+3} + \frac{B}{s-1}$$
$$7s+13 = A(s-1) + B(s+3)$$

Now,

$$s = 1 \Rightarrow 20 = 4B \text{ or } B = 5$$
$$s = -3 \Rightarrow -8 = -4A \text{ or } A = 2.$$

Thus,

$$\begin{aligned} y(t) &= \mathcal{L}^{-1}\left\{\frac{7s+13}{s^2+2s-3}\right\} \\ &= \mathcal{L}^{-1}\left\{\frac{2}{s+3} + \frac{5}{s-1}\right\} \\ &= 2\,\mathcal{L}^{-1}\left\{\frac{1}{s+3}\right\} + 5\,\mathcal{L}^{-1}\left\{\frac{1}{s-1}\right\} \\ &= 2e^{-3t} + 5e^{t}. \end{aligned}$$

25. Find a partial fraction decomposition.

$$\frac{13s-5}{2s^2-s} = \frac{A}{s} + \frac{B}{2s-1}$$
$$13s-5 = A(2s-1) + Bs$$

Now,

$$s = \frac{1}{2} \Rightarrow \frac{3}{2} = \frac{1}{2}B \text{ or } B = 3$$
$$s = 0 \Rightarrow -5 = -A \text{ or } A = 5.$$

Thus,

$$\begin{aligned} y(t) &= \mathcal{L}^{-1}\left\{\frac{13s-5}{2s^2-s}\right\} \\ &= \mathcal{L}^{-1}\left\{\frac{5}{s}+\frac{3}{2s-1}\right\} \\ &= \mathcal{L}^{-1}\left\{5\cdot\frac{1}{s}+\frac{3}{2}\cdot\frac{1}{s-1/2}\right\} \\ &= 5\,\mathcal{L}^{-1}\left\{\frac{1}{s}\right\}+\frac{3}{2}\mathcal{L}^{-1}\left\{\frac{1}{s-1/2}\right\} \\ &= 5+\frac{3}{2}e^{(1/2)t}. \end{aligned}$$

27. Find a partial fraction decomposition.

$$\frac{7s^2+3s+16}{(s+1)(s^2+4)} = \frac{A}{s+1}+\frac{Bs+C}{s^2+4}$$

Equating numerators, we get

$$\begin{aligned} &7s^2+3s+16 \\ &\quad = A(s^2+4)+(Bs+C)(s+1) \\ &\quad = (A+B)s^2+(B+C)s+(4A+C) \end{aligned}$$

Thus,

$$\begin{aligned} A+B&=7 \\ B+C&=3 \\ 4A+C&=16, \end{aligned}$$

leading to $A=4$, $B=3$ and $C=0$. Thus,

$$\begin{aligned} y(t) &= \mathcal{L}^{-1}\left\{\frac{7s^2+3s+16}{(s+1)(s^2+4)}\right\} \\ &= \mathcal{L}^{-1}\left\{\frac{4}{s+1}+\frac{3s}{s^2+4}\right\} \\ &= 4\,\mathcal{L}^{-1}\left\{\frac{1}{s+1}\right\}+3\,\mathcal{L}^{-1}\left\{\frac{s}{s^2+4}\right\} \\ &= 4e^{-t}+3\cos 2t. \end{aligned}$$

29. Find a partial fraction decomposition.

$$\frac{2s^2+9s+11}{(s+1)(s^2+4s+5)} = \frac{A}{s+1}+\frac{Bs+C}{s^2+4s+5}$$

$$\begin{aligned} &2s^2+9s+11 \\ &\quad = A(s^2+4s+5)+(Bs+C)(s+1) \\ &\quad = (A+B)s^2+(4A+B+C)s+(5A+C) \end{aligned}$$

Thus,

$$\begin{aligned} A+B&=2 \\ 4A+B+C&=9 \\ 5A+C&=11, \end{aligned}$$

leading to $A=2$, $B=0$, and $C=1$. Thus,

$$\begin{aligned} y(t) &= \mathcal{L}^{-1}\left\{\frac{2s^2+9s+11}{(s+1)(s^2+4s+5)}\right\} \\ &= \mathcal{L}^{-1}\left\{\frac{2}{s+1}+\frac{1}{s^2+4s+5}\right\} \\ &= \mathcal{L}^{-1}\left\{\frac{2}{s+1}+\frac{1}{(s+2)^2+1}\right\} \\ &= 2\,\mathcal{L}^{-1}\left\{\frac{1}{s+1}\right\}+\mathcal{L}^{-1}\left\{\frac{1}{(s+2)^2+1}\right\} \\ &= 2e^{-t}+e^{-2t}\sin t. \end{aligned}$$

31. Find a partial fraction decomposition.

$$\begin{aligned} &\frac{1}{(s-2)^2(s+1)^3} \\ &\quad = \frac{A}{s-2}+\frac{B}{(s-2)^2} \\ &\qquad + \frac{C}{s+1}+\frac{D}{(s+1)^2}+\frac{E}{(s+1)^3} \end{aligned}$$

Equating numerators we get

$$\begin{aligned} 1 &= A(s-2)(s+1)^3+B(s+1)^3 \\ &\quad + C(s-2)^2(s+1)^2 \\ &\quad + D(s-2)^2(s+1)+E(s-2)^2 \\ &= (A+C)s^4+(A+B-2C+D)s^3 \\ &\quad + (-3A+3B-3C-3D+E)s^2 \\ &\quad + (-5A+3B+4C-4E)s \\ &\quad + (-2A+B+4C+4D+4E) \end{aligned}$$

Thus,

$$\begin{aligned} A+C&=0 \\ A+B-2C+D&=0 \\ -3A+3B-3C-3D+E&=0 \\ -5A+3B+4C-4E&=0 \\ -2A+B+4C+4D+4E&=1, \end{aligned}$$

leading to $A = -1/27$, $B = 1/27$, $C = 1/27$, $D = 2/27$, and $E = 1/9$. Thus,

$$\begin{aligned}
y(t) &= \mathcal{L}^{-1}\left\{\frac{1}{(s-2)^2(s+1)^3}\right\} \\
&= \mathcal{L}^{-1}\left\{\frac{-1/27}{s-2} + \frac{1/27}{(s-2)^2} + \frac{1/27}{s+1} + \frac{2/27}{(s+1)^2} + \frac{1/9}{(s+1)^3}\right\} \\
&= -\frac{1}{27}\mathcal{L}^{-1}\left\{\frac{1}{s-2}\right\} + \frac{1}{27}\mathcal{L}^{-1}\left\{\frac{1}{(s-2)^2}\right\} + \frac{1}{27}\mathcal{L}^{-1}\left\{\frac{1}{s+1}\right\} + \frac{2}{27}\mathcal{L}^{-1}\left\{\frac{1}{(s+1)^2}\right\} \\
&\quad + \frac{1}{9}\cdot\frac{1}{2}\mathcal{L}^{-1}\left\{\frac{2}{(s+1)^3}\right\} \\
&= -\frac{1}{27}e^{2t} + \frac{1}{27}e^{2t}t + \frac{1}{27}e^{-t} + \frac{2}{27}e^{-t}t + \frac{1}{18}e^{-t}t^2 \\
&= \frac{1}{27}e^{2t}(t-1) + \frac{1}{27}e^{-t}(1 + 2t + \frac{3}{2}t^2).
\end{aligned}$$

33. Find a partial fraction decomposition.

$$\frac{1}{(s+2)^2(s^2+9)} = \frac{A}{s+2} + \frac{B}{(s+2)^2} + \frac{Cs+D}{s^2+9}$$

$$\begin{aligned}
1 &= A(s+2)(s^2+9) + B(s^2+9) + (Cs+D)(s+2)^2 \\
&= (A+C)s^3 + (2A+B+4C+D)s^2 + (9A+4C+4D)s + (18A+9B+4D)
\end{aligned}$$

Thus,

$$\begin{aligned}
A + C &= 0 \\
2A + B + 4C + D &= 0 \\
9A + 4C + 4D &= 0 \\
18A + 9B + 4D &= 1,
\end{aligned}$$

leading to $A = 4/169$, $B = 1/13$, $C = -4/169$, and $D = -5/169$. Thus,

$$\begin{aligned}
y(t) &= \mathcal{L}^{-1}\left\{\frac{1}{(s+2)^2(s^2+9)}\right\} \\
&= \mathcal{L}^{-1}\left\{\frac{4/169}{s+2} + \frac{1/13}{(s+2)^2} + \frac{-(4/169)s - 5/169}{s^2+9}\right\} \\
&= \mathcal{L}^{-1}\left\{\frac{4}{169}\cdot\frac{1}{s+2} + \frac{1}{13}\cdot\frac{1}{(s+2)^2} - \frac{4}{169}\cdot\frac{s}{s^2+9} - \frac{5}{169}\cdot\frac{1}{s^2+9}\right\} \\
&= \frac{4}{169}e^{-2t} + \frac{1}{13}e^{-2t}t - \frac{4}{169}\cos 3t - \frac{5}{169}\cdot\frac{1}{3}\mathcal{L}^{-1}\left\{\frac{3}{s^2+9}\right\} \\
&= \frac{4}{169}e^{-2t} + \frac{1}{13}te^{-2t} - \frac{4}{169}\cos 3t - \frac{5}{507}\sin 3t.
\end{aligned}$$

35. Find a partial fraction decomposition.

$$\frac{1}{(s+1)^2(s^2-4)} = \frac{A}{s+1} + \frac{B}{(s+1)^2} + \frac{C}{s+2} + \frac{D}{s-2}$$

Equating numerators

$$\begin{aligned}
1 &= A(s+1)(s^2-4) + B(s^2-4) \\
&\quad + C(s+1)^2(s-2) + D(s+1)^2(s+2)
\end{aligned}$$

Now,

$$\begin{aligned}
s = -1 &\Rightarrow 1 = -3B \text{ or } B = -1/3 \\
s = 2 &\Rightarrow 1 = 36D \text{ or } D = 1/36 \\
s = -2 &\Rightarrow 1 = -4C \text{ or } C = -1/4,
\end{aligned}$$

and

$$\begin{aligned}
s = 0 \Rightarrow 1 &= -4A - 4B - 2C + 2D \\
1 &= -4A + \frac{4}{3} + \frac{1}{2} + \frac{1}{18} \\
A &= \frac{2}{9}.
\end{aligned}$$

Thus,

$$\mathcal{L}^{-1}\left\{\frac{1}{(s+1)^2(s^2-4)}\right\}$$
$$=\mathcal{L}^{-1}\left\{\frac{2/9}{s+1}+\frac{-1/3}{(s+1)^2}+\frac{-1/4}{s+2}+\frac{1/36}{s-2}\right\}$$
$$=\frac{2}{9}\mathcal{L}^{-1}\left\{\frac{1}{s+1}\right\}-\frac{1}{3}\mathcal{L}^{-1}\left\{\frac{1}{(s+1)^2}\right\}$$
$$-\frac{1}{4}\mathcal{L}^{-1}\left\{\frac{1}{s+2}\right\}+\frac{1}{36}\mathcal{L}^{-1}\left\{\frac{1}{s-2}\right\}$$
$$=\frac{2}{9}e^{-t}-\frac{1}{3}e^{-t}t-\frac{1}{4}e^{-2t}+\frac{1}{36}e^{2t}.$$

Section 5.4. Using the Laplace Transform to Solve Differential Equations

1. Set $Y=\mathcal{L}(y)$. Then,

$$\begin{aligned}\mathcal{L}(y'+3y)&=\mathcal{L}(e^{2t})\\ \mathcal{L}(y')+3\,\mathcal{L}(y)&=\frac{1}{s-2}\\ s\cdot Y(s)-y(0)+3Y(s)&=\frac{1}{s-2}.\end{aligned}$$

But, $y(0)=-1$, so

$$\begin{aligned}(s+3)Y(s)+1&=\frac{1}{s-2}\\ Y(s)&=\frac{-1}{s+3}+\frac{1}{(s+3)(s-2)}.\end{aligned}$$

Find a partial fraction decomposition.

$$\begin{aligned}\frac{1}{(s+3)(s-2)}&=\frac{A}{s+3}+\frac{B}{s-2}\\ 1&=A(s-2)+B(s+3).\end{aligned}$$

Now,

$$\begin{aligned}s=2&\Rightarrow 1=5B \text{ or } B=1/5\\ s=-3&\Rightarrow 1=-5A \text{ or } A=-1/5.\end{aligned}$$

Thus,

$$\begin{aligned}Y(s)&=\frac{-1}{s+3}-\frac{1/5}{s+3}+\frac{1/5}{s-2}\\ &=\frac{-6/5}{s+3}+\frac{1/5}{s-2},\end{aligned}$$

and

$$\begin{aligned}y(t)&=-\frac{6}{5}\mathcal{L}^{-1}\left\{\frac{1}{s+3}\right\}+\frac{1}{5}\mathcal{L}^{-1}\left\{\frac{1}{s-2}\right\}\\ &=-\frac{6}{5}e^{-3t}+\frac{1}{5}e^{2t}.\end{aligned}$$

3. Set $Y=\mathcal{L}(y)$. Then,

$$\begin{aligned}\mathcal{L}(y'+4y)&=\mathcal{L}(\cos t)\\ \mathcal{L}(y')+4\,\mathcal{L}(y)&=\frac{s}{s^2+1}\\ s\cdot Y(s)-y(0)+4Y(s)&=\frac{s}{s^2+1}.\end{aligned}$$

But $y(0)=0$, so

$$\begin{aligned}(s+4)Y(s)&=\frac{s}{s^2+1}\\ Y(s)&=\frac{s}{(s+4)(s^2+1)}.\end{aligned}$$

Find a partial fraction decomposition.

$$\frac{s}{(s+4)(s^2+1)} = \frac{A}{s+4} + \frac{Bs+C}{s^2+1}$$

Equating numerators

$$\begin{aligned} s &= A(s^2+1) + (Bs+C)(s+4) \\ &= (A+B)s^2 + (4B+C)s + (A+4C). \end{aligned}$$

Thus,

$$\begin{aligned} A+B &= 0 \\ 4B+C &= 1 \\ A+4C &= 0, \end{aligned}$$

and $A = -4/17$, $B = 4/17$, and $C = 1/17$. Thus,

$$\begin{aligned} Y(s) &= \frac{-4/17}{s+4} + \frac{(4/17)s + 1/17}{s^2+1} \\ &= -\frac{4}{17} \cdot \frac{1}{s+4} + \frac{4}{17} \cdot \frac{s}{s^2+1} \\ &\quad + \frac{1}{17} \cdot \frac{1}{s^2+1}, \end{aligned}$$

and

$$y(t) = -\frac{4}{17}e^{-4t} + \frac{4}{17}\cos t + \frac{1}{17}\sin t.$$

5. Set $Y = \mathcal{L}(y)$. Then,

$$\begin{aligned} \mathcal{L}(y' + 6y) &= \mathcal{L}(2t+3) \\ \mathcal{L}(y') + 6\,\mathcal{L}(y) &= 2\,\mathcal{L}(t) + 3\,\mathcal{L}(1) \\ s \cdot Y(s) - y(0) + 6Y(s) &= \frac{2}{s^2} + \frac{3}{s}. \end{aligned}$$

But $y(0) = 1$, so

$$(s+6)Y(s) - 1 = \frac{2}{s^2} + \frac{3}{s}.$$

Solving for Y,

$$\begin{aligned} Y(s) &= \frac{1}{s+6} + \frac{2}{s^2(s+6)} + \frac{3}{s(s+6)} \\ &= \frac{1}{s+6} + \frac{3s+2}{s^2(s+6)}. \end{aligned}$$

Find a partial fraction decomposition.

$$\frac{3s+2}{s^2(s+6)} = \frac{A}{s} + \frac{B}{s^2} + \frac{Cs+D}{s+6}$$

Equating numerators,

$$\begin{aligned} 3s+2 &= As(s+6) + B(s+6) + (Cs+D)s^2 \\ 3s+2 &= Cs^3 + (A+D)s^2 + (6A+B)s + 6B. \end{aligned}$$

Thus,

$$\begin{aligned} C &= 0 \\ A+D &= 0 \\ 6A+B &= 3 \\ 6B &= 2, \end{aligned}$$

and $A = 4/9$, $B = 1/3$, $C = 0$, and $D = -4/9$. Thus,

$$\begin{aligned} Y(s) &= \frac{1}{s+6} + \frac{4/9}{s} + \frac{1/3}{s^2} - \frac{4/9}{s+6} \\ &= \frac{5/9}{s+6} + \frac{4/9}{s} + \frac{1/3}{s^2}. \end{aligned}$$

Therefore,

$$y(t) = \frac{5}{9}e^{-6t} + \frac{4}{9} + \frac{1}{3}t.$$

7. Let $\mathcal{L}(y) = Y$. Then,

$$\mathcal{L}(y' + 8y) = \mathcal{L}(e^{-2t}\sin t).$$

Note the transform pair:

$$\sin t \Longleftrightarrow \frac{1}{s^2+1}$$

Thus, by Proposition 2.12,

$$e^{-2t}\sin t \Longleftrightarrow \frac{1}{(s+2)^2+1}.$$

Thus,

$$\begin{aligned} \mathcal{L}(y') + 8\,\mathcal{L}(y) &= \frac{1}{(s+2)^2+1} \\ s \cdot Y(s) - y(0) + 8Y(s) &= \frac{1}{(s+2)^2+1}. \end{aligned}$$

But $y(0) = 0$, so

$$\begin{aligned} (s+8)Y(s) &= \frac{1}{(s+2)^2+1} \\ Y(s) &= \frac{1}{(s+8)((s+2)^2+1)}. \end{aligned}$$

Find a partial fraction decomposition.

$$\frac{1}{(s+8)((s+2)^2+1)} = \frac{A}{s+8} + \frac{Bs+C}{(s+2)^2+1}$$

$$1 = A((s+2)^2+1) + (Bs+C)(s+8)$$
$$1 = (A+B)s^2 + (4A+8B+C)s + (5A+8C).$$

Thus,

$$\begin{aligned} A+B &= 0 \\ 4A+8B+C &= 0 \\ 5A+8C &= 1, \end{aligned}$$

and $A = 1/37$, $B = -1/37$, and $C = 4/37$. Thus,

$$Y(s) = \frac{1/37}{s+8} + \frac{-(1/37)s+4/37}{(s+2)^2+1}.$$

Note the transform pairs.

$$\begin{aligned} \cos t &\Longleftrightarrow \frac{s}{s^2+1} \\ \sin t &\Longleftrightarrow \frac{1}{s^2+1}. \end{aligned}$$

Proposition 2.12 provides

$$\begin{aligned} e^{-2t}\cos t &= \frac{s+2}{(s+2)^2+1} \\ e^{-2t}\sin t &= \frac{1}{(s+2)^2+1}. \end{aligned}$$

Thus,

$$\begin{aligned} Y(s) &= \frac{1}{37}\cdot\frac{1}{s+8} - \frac{1}{37}\cdot\frac{s-4}{(s+2)^2+1} \\ &= \frac{1}{37}\cdot\frac{1}{s+8} \\ &\quad - \frac{1}{37}\left\{\frac{s+2}{(s+2)^2+1} - \frac{6}{(s+2)^2+1}\right\} \\ &= \frac{1}{37}\cdot\frac{1}{s+8} - \frac{1}{37}\cdot\frac{s+2}{(s+2)^2+1} \\ &\quad + \frac{6}{37}\cdot\frac{1}{(s+2)^2+1}. \end{aligned}$$

Therefore,

$$y(t) = \frac{1}{37}e^{-8t} - \frac{1}{37}e^{-2t}\cos t + \frac{6}{37}e^{-2t}\sin t.$$

9. Let $\mathcal{L}(y) = Y$. Then,

$$\begin{aligned} \mathcal{L}(y'+y) &= \mathcal{L}(te^t) \\ \mathcal{L}(y') + \mathcal{L}(y) &= \frac{1}{(s-1)^2} \\ s\cdot Y(s) - y(0) + Y(s) &= \frac{1}{(s-1)^2}. \end{aligned}$$

But $y(0) = -2$, so

$$\begin{aligned} (s+1)Y(s) + 2 &= \frac{1}{(s-1)^2} \\ Y(s) &= -\frac{2}{s+1} + \frac{1}{(s+1)(s-1)^2}. \end{aligned}$$

Find a partial fraction decomposition.

$$\frac{1}{(s+1)(s-1)^2} = \frac{A}{s+1} + \frac{B}{s-1} + \frac{C}{(s-1)^2}$$

Equating numerators,

$$\begin{aligned} 1 &= A(s-1)^2 + B(s+1)(s-1) + C(s+1) \\ &= (A+B)s^2 + (-2A-C)s + (A-B+C). \end{aligned}$$

Now,

$$\begin{aligned} s = 1 &\Rightarrow 1 = 2C \text{ or } C = 1/2 \\ s = -1 &\Rightarrow 1 = 4A \text{ or } A = 1/4. \end{aligned}$$

Finally, comparing coefficients of s^2 on each side,

$$0 = A + B = \frac{1}{4} + B \Rightarrow B = -\frac{1}{4}.$$

Thus,

$$\begin{aligned} Y(s) &= -\frac{2}{s+1} + \frac{1/4}{s+1} - \frac{1/4}{s-1} + \frac{1/2}{(s-1)^2} \\ &= -\frac{7/4}{s+1} - \frac{1/4}{s-1} + \frac{1/2}{(s-1)^2}. \end{aligned}$$

Therefore,

$$y(t) = -\frac{7}{4}e^{-t} - \frac{1}{4}e^t + \frac{1}{2}e^t t.$$

11. Let $\mathcal{L}(y) = Y$. Then,

$$\begin{aligned}\mathcal{L}(y'' - 4y) &= \mathcal{L}(e^{-t})\\ \mathcal{L}(y'') - 4\,\mathcal{L}(y) &= \frac{1}{s+1}\\ s^2Y(s) - s\cdot y(0) - y'(0) - 4Y(s) &= \frac{1}{s+1}.\end{aligned}$$

But $y(0) = -1$ and $y'(0) = 0$, so

$$\begin{aligned}(s^2-4)Y(s) + s &= \frac{1}{s+1}\\ Y(s) &= \frac{-s}{s^2-4} + \frac{1}{(s^2-4)(s+1)}.\end{aligned}$$

Both terms on the right have partial fraction decompositions.

$$\begin{aligned}\frac{-s}{s^2-4} &= \frac{-1/2}{s-2} - \frac{1/2}{s+2}\\ \frac{1}{(s^2-4)(s+1)} &= \frac{1/12}{s-2} + \frac{1/4}{s+2} - \frac{1/3}{s+1}.\end{aligned}$$

Thus,

$$Y(s) = \frac{-5/12}{s-2} - \frac{1/4}{s+2} - \frac{1/3}{s+1},$$

and

$$y(t) = -\frac{5}{12}e^{2t} - \frac{1}{4}e^{-2t} - \frac{1}{3}e^{-t}.$$

13. Let $\mathcal{L}(y) = Y$. Then,

$$\begin{aligned}\mathcal{L}(y'' + 4y) &= \mathcal{L}(\cos t)\\ \mathcal{L}(y'') + 4\,\mathcal{L}(y) &= \frac{s}{s^2+1}\\ s^2Y(s) - s\cdot y(0) - y'(0) + 4Y(s) &= \frac{s}{s^2+1}.\end{aligned}$$

But $y(0) = 1$ and $y'(0) = 0$, so

$$\begin{aligned}(s^2+4)Y(s) - s &= \frac{s}{s^2+1}\\ Y(s) &= \frac{s}{s^2+4} + \frac{s}{(s^2+4)(s^2+1)}.\end{aligned}$$

But the term on the right has a decomposition

$$\frac{s}{(s^2+4)(s^2+1)} = \frac{-(1/3)s}{s^2+4} + \frac{(1/3)s}{s^2+1}.$$

Thus,

$$Y(s) = \frac{(2/3)s}{s^2+4} + \frac{(1/3)s}{s^2+1},$$

and

$$\begin{aligned}y(t) &= \frac{2}{3}\mathcal{L}^{-1}\left\{\frac{s}{s^2+4}\right\} + \frac{1}{3}\mathcal{L}^{-1}\left\{\frac{s}{s^2+1}\right\}\\ &= \frac{2}{3}\cos 2t + \frac{1}{3}\cos t.\end{aligned}$$

15. Let $\mathcal{L}(y) = Y$. Then,

$$\begin{aligned}\mathcal{L}(y'' - y) &= \mathcal{L}(2t)\\ \mathcal{L}(y'') - \mathcal{L}(y) &= 2\,\mathcal{L}(t)\\ s^2Y(s) - s\cdot y(0) - y'(0) - Y(s) &= \frac{2}{s^2}.\end{aligned}$$

But $y(0) = 0$ and $y'(0) = -1$, so

$$\begin{aligned}(s^2-1)Y(s) + 1 &= \frac{2}{s^2}\\ Y(s) &= \frac{-1}{s^2-1} + \frac{2}{s^2(s^2-1)}.\end{aligned}$$

The terms on the right have the decompositions:

$$\begin{aligned}\frac{-1}{s^2-1} &= \frac{-1/2}{s-1} + \frac{1/2}{s+1}\\ \frac{2}{s^2(s^2-1)} &= \frac{2}{s^2} + \frac{1}{s-1} - \frac{1}{s+1}.\end{aligned}$$

Therefore,

$$Y(s) = -\frac{2}{s^2} + \frac{1/2}{s-1} - \frac{1/2}{s+1},$$

and

$$y(t) = -2t + \frac{1}{2}e^{t} - \frac{1}{2}e^{-t}.$$

17. Let $\mathcal{L}(y) = Y$. Since $y(0) = -2$ and $y'(0) = 0$,

$$\mathcal{L}(y'' + y') = s^2Y(s) + 2s + sY(s) + 2.$$

In addition, Note the transform pair:

$$f(t) = e^{-t} \Longleftrightarrow F(s) = \frac{1}{s+1}.$$

Thus, by Proposition 2.14,

$$te^{-t} \Longleftrightarrow -F'(s) = \frac{1}{(s+1)^2}.$$

Thus, Y must satisfy

$$s^2Y(s) + 2s + sY(s) + 2 = \frac{1}{(s+1)^2}.$$

Solving for Y, we get

$$Y(s) = -\frac{2}{s} + \frac{1}{s(s+1)^3}$$

Now,

$$\begin{aligned} s = 0 &\Rightarrow 1 = A \\ s = -1 &\Rightarrow 1 = -D \text{ or } D = -1. \end{aligned}$$

The coefficient of s^3 is $A + B$, so

$$A + B = 0 \quad \text{and} \quad B = -A = -1.$$

Finally,

$$\begin{aligned} s = 1 \Rightarrow 1 &= 8A + 4B + 2C + D \\ 1 &= 8 - 4 + 2C - 1 \\ -2C &= 2 \\ C &= -1. \end{aligned}$$

Thus,

$$\begin{aligned} Y(s) &= -\frac{2}{s} + \frac{1}{s} - \frac{1}{s+1} - \frac{1}{(s+1)^2} - \frac{1}{(s+1)^3} \\ &= -\frac{1}{s} - \frac{1}{s+1} - \frac{1}{(s+1)^2} - \frac{1}{(s+1)^3}, \end{aligned}$$

so

$$\begin{aligned} y(t) = -1 - e^{-t} - \mathcal{L}^{-1}&\left\{\frac{1}{(s+1)^2}\right\} \\ -\frac{1}{2}\mathcal{L}^{-1}&\left\{\frac{2}{(s+1)^3}\right\}. \end{aligned}$$

By Proposition 2.12,

$$y(t) = -1 - e^{-t} - e^{-t}t - \frac{1}{2}e^{-t}t^2.$$

19. Let $\mathcal{L}(y) = Y$. Since $y(0) = -1$ and $y'(0) = 0$,

$$\mathcal{L}(y''-4y'-5y) = s^2Y(s)+s-4[sY(s)+1]-5Y(s)$$

In addition,

$$\mathcal{L}(e^{2t}) = \frac{1}{s-2}.$$

Solving

$$(s^2 - 4s - 5)Y(s) + s - 4 = \frac{1}{s-2}$$

for Y, we get

$$Y(s) = \frac{-s+4}{s^2-4s-5} + \frac{1}{(s-2)(s^2-4s-5)}.$$

The terms on the right have decompositions:

$$\begin{aligned} \frac{-s+4}{s^2-4s-5} &= \frac{-5/6}{s+1} - \frac{1/6}{s-5} \\ \frac{1}{(s-2)(s^2-4s-5)} &= \frac{-1/9}{s-2} + \frac{1/18}{s+1} + \frac{1/18}{s-5}. \end{aligned}$$

Thus,

$$Y(s) = \frac{-7/9}{s+1} - \frac{1/9}{s-5} - \frac{1/9}{s-2},$$

and

$$y(t) = -\frac{7}{9}e^{-t} - \frac{1}{9}e^{5t} - \frac{1}{9}e^{2t}.$$

21. Let $\mathcal{L}(y) = Y$. Since $y(0) = -1$ and $y'(0) = 0$,

$$\begin{aligned} &\mathcal{L}(y'' - y' - 2y) \\ &= s^2Y(s) + s - [sY(s) + 1] - 2Y(s). \end{aligned}$$

In addition,

$$\mathcal{L}(e^{2t}) = \frac{1}{s-2}.$$

Solving

$$(s^2 - s - 2)Y(s) + s - 1 = \frac{1}{s-2}$$

for $Y(s)$, we get

$$Y(s) = \frac{-s+1}{s^2-s-2} + \frac{1}{(s-2)(s^2-s-2)}.$$

The terms on the right have decompositions.

$$\frac{-s+1}{s^2-s-2} = \frac{-2/3}{s+1} - \frac{1/3}{s-2}$$

and

$$\frac{1}{(s-2)(s^2-s-2)} = \frac{-1/9}{s-2} + \frac{1/3}{(s-2)^2} + \frac{1/9}{s+1}.$$

Thus,

$$Y(s) = \frac{-5/9}{s+1} - \frac{4/9}{s-2} + \frac{1/3}{(s-2)^2},$$

and

$$y(t) = -\frac{5}{9}e^{-t} - \frac{4}{9}e^{2t} + \frac{1}{3}\mathcal{L}^{-1}\left\{\frac{1}{(s-2)^2}\right\}.$$

Since

$$t \Longleftrightarrow \frac{1}{s^2},$$

Proposition 2.14 tells us that

$$e^{2t}t \Longleftrightarrow \frac{1}{(s-2)^2}.$$

Thus,

$$y(t) = -\frac{5}{9}e^{-t} - \frac{4}{9}e^{2t} + \frac{1}{3}te^{2t}.$$

23. Let $\mathcal{L}(y) = Y$. Since $y(0) = -1$ and $y'(0) = 0$,

$$\begin{aligned}\mathcal{L}(y'' + 2y' + 2y) \\ &= s^2Y(s) + s + 2[sY(s) + 1]) + 2Y(s).\end{aligned}$$

In addition,

$$\mathcal{L}(\cos t) = \frac{s}{s^2+1},$$

so

$$(s^2 + 2s + 2)Y(s) + s + 2 = \frac{s}{s^2+1}.$$

Solving for Y, we get

$$Y(s) = \frac{-s-2}{s^2+2s+2} + \frac{s}{(s^2+1)(s^2+2s+2)}.$$

The second term on the right has decomposition

$$\frac{s}{(s^2+1)(s^2+2s+2)} = \frac{(1/5)s + 2/5}{s^2+1} + \frac{-(1/5)s - 4/5}{s^2+2s+2},$$

so

$$\begin{aligned}Y(s) &= \frac{-(6/5)s - 14/5}{s^2+2s+2} + \frac{(1/5)s + 2/5}{s^2+1} \\ &= \frac{-(6/5)s - 14/5}{(s+1)^2+1} + \frac{1}{5}\cdot\frac{s}{s^2+1} \\ &\quad + \frac{2}{5}\cdot\frac{1}{s^2+1}.\end{aligned}$$

Note the pairs

$$\cos t \Longleftrightarrow \frac{s}{s^2+1}$$
$$\sin t \Longleftrightarrow \frac{1}{s^2+1}.$$

By Proposition 2.12,

$$e^{-t}\cos t \Longleftrightarrow \frac{s+1}{(s+1)^2+1}$$
$$e^{-t}\sin t \Longleftrightarrow \frac{1}{(s+1)^2+1}.$$

Thus, write

$$\begin{aligned}\frac{-(6/5)s - 14/5}{(s+1)^2+1} \\ &= -\frac{6}{5}\cdot\frac{s+7/3}{(s+1)^2+1} \\ &= -\frac{6}{5}\left[\frac{s+1}{(s+1)^2+1} + \frac{4}{3}\cdot\frac{1}{(s+1)^2+1}\right] \\ &= -\frac{6}{5}\cdot\frac{s+1}{(s+1)^2+1} - \frac{8}{5}\cdot\frac{1}{(s+1)^2+1}.\end{aligned}$$

Therefore,

$$\begin{aligned}Y(s) &= -\frac{6}{5}\cdot\frac{s+1}{(s+1)^2+1} - \frac{8}{5}\cdot\frac{1}{(s+1)^2+1} \\ &\quad + \frac{1}{5}\cdot\frac{s}{s^2+1} + \frac{2}{5}\cdot\frac{1}{s^2+1},\end{aligned}$$

and

$$y(t) = -\frac{6}{5}e^{-t}\cos t - \frac{8}{5}e^{-t}\sin t + \frac{1}{5}\cos t + \frac{2}{5}\sin t.$$

25. Let $\mathcal{L}(y) = Y$. Since $y(0) = 1$ and $y'(0) = 0$,

$$\mathcal{L}(y'' + 4y) = s^2Y(s) - s + 4Y(s).$$

On the other hand, we have the transform pair

$$\cos t \Longleftrightarrow \frac{s}{s^2+1},$$

so by Proposition 2.12,

$$e^{-2t}\cos 2t \Longleftrightarrow \frac{s+2}{(s+2)^2+1}.$$

Thus,

$$(s^2+4)Y(s) - s = \frac{s+2}{(s+2)^2+1}, \qquad (4.1)$$

so

$$Y(s) = \frac{s}{s^2+4} + \frac{s+2}{(s^2+4)((s+2)^2+1)}. \qquad (4.2)$$

The fraction on the right has decomposition

$$\frac{s+2}{(s^2+4)((s+2)^2+1)} = -\frac{1}{65}\cdot\frac{7s-18}{s^2+4} + \frac{1}{65}\cdot\frac{7s+10}{(s+2)^2+1}.$$

Thus,

$$\begin{aligned}Y(s) &= \frac{1}{65}\cdot\frac{58s+18}{s^2+4} + \frac{1}{65}\cdot\frac{7s+10}{(s+2)^2+1}\\ &= \frac{1}{65}\left[58\cdot\frac{s}{s^2+4} + 9\cdot\frac{2}{s^2+4}\right]\\ &\quad+\frac{1}{65}\left[7\cdot\frac{s+2}{(s+2)^2+1} - 4\cdot\frac{1}{(s+2)^2+1}\right]\end{aligned}$$

The transform pairs

$$\cos t \Longleftrightarrow \frac{s}{s^2+1}$$
$$\sin t \Longleftrightarrow \frac{1}{s^2+1}$$

and Proposition 2.12 yield

$$e^{-2t}\cos t \Longleftrightarrow \frac{s+2}{(s+2)^2+1}$$
$$e^{-2t}\sin t \Longleftrightarrow \frac{1}{(s+2)^2+1}.$$

Therefore,

$$\begin{aligned}y(t) =& \frac{58}{65}\cos 2t + \frac{9}{65}\sin 2t\\ &+\frac{7}{65}e^{-2t}\cos t - \frac{4}{65}e^{-2t}\sin t.\end{aligned}$$

27. (a) Let $y = e^{rt}$. Then

$$\begin{aligned}y'' - 4y' + 3y &= 0\\ (r^2 - 4r + 3)e^{rt} &= 0\\ r^2 - 4r + 3 &= 0\\ (r-3)(r-1) &= 0\end{aligned}$$

Thus, $r = 3$ and $r = 1$ lead to independent solutions $y = e^{3t}$ and $y = e^t$. The general solution is

$$y = C_1e^{3t} + C_2e^t.$$

The initial condition $y(0) = 1$ gives us

$$1 = C_1 + C_2.$$

Differentiating,

$$y' = 3C_1e^{3t} + C_2e^t.$$

The initial condition $y'(0) = -1$ gives

$$-1 = 3C_1 + C_2.$$

Therefore, $C_1 = -1$ and $C_2 = 2$ and the solution is $y = -e^{3t} + 2e^t$.

(b) Let $\mathcal{L}(y) = Y$. Since $y(0) = 1$ and $y'(0) = -1$,

$$\begin{aligned}0 &= \mathcal{L}(y'' - 4y' + 3y)\\ &= (s^2 - 4s + 3)Y(s) - s + 1 + 4 = 0.\end{aligned}$$

Hence

$$Y(s) = \frac{s-5}{s^2-4s+3} = \frac{2}{s-1} - \frac{1}{s-3}.$$

Thus, $y(t) = 2e^t - e^{3t}$.

29. Let $y = e^{rt}$. Then

$$\begin{aligned} y'' - 3y' + 2y &= 0 \\ (r^2 - 3r + 2)e^{rt} &= 0 \\ r^2 - 3r + 2 &= 0 \\ (r-2)(r-1) &= 0. \end{aligned}$$

Thus, $r = 2$ and $r = 1$ lead to independent solutions $y = e^{2t}$ and $y = e^{t}$. The general solution and its derivative are

$$\begin{aligned} y &= C_1 e^{2t} + C_2 e^{t} \\ y' &= 2C_1 e^{2t} + C_2 e^{t}. \end{aligned}$$

Thus,

$$\begin{aligned} y(0) = 0 &\Rightarrow 0 = C_1 + C_2 \\ y'(0) = -1 &\Rightarrow -1 = 2C_1 + C_2. \end{aligned}$$

Therefore, $C_1 = -1$, $C_2 = 1$, and

$$y = -e^{2t} + e^{t}.$$

Let $\mathcal{L}(y) = Y$. Since $y(0) = 0$ and $y'(0) = -1$,

$$\begin{aligned} 0 &= \mathcal{L}(y'' - 3y' + 2y) \\ &= (s^2 - 3s + 2)Y(s) + 1. \end{aligned}$$

Solving for Y, we get

$$Y(s) = \frac{-1}{s^2 - 3s + 2} = \frac{1}{s-1} - \frac{1}{s-2}.$$

Therefore,

$$y(t) = e^{t} - e^{-2t}.$$

31. Let $y = e^{rt}$. Then

$$\begin{aligned} y'' + 2y' + 2y &= 0 \\ (r^2 + 2r + 2)e^{rt} &= 0 \\ r^2 + 2r + 2 &= 0. \end{aligned}$$

Using the quadratic formula, $r = -1 \pm i$. Thus,

$$y = e^{(-1+i)t} = e^{-t}e^{it} = e^{-t}(\cos t + i \sin t).$$

Therefore, $e^{-t}\cos t$ and $e^{-t}\sin t$ are independent solutions. The general solution and its derivative are

$$\begin{aligned} y &= C_1 e^{-t}\cos t + C_2 e^{-t}\sin t \\ y' &= -e^{-t}\left((C_1 - C_2)\cos t + (C_1 + C_2)\sin t\right). \end{aligned}$$

But,

$$\begin{aligned} y(0) = -1 &\Rightarrow -1 = C_1 \\ y'(0) = 1 &\Rightarrow 1 = -(C_1 - C_2). \end{aligned}$$

Therefore, $C_1 = -1$ and $C_2 = 0$, and

$$y(t) = -e^{-t}\cos t.$$

Alternatively, let $\mathcal{L}(y) = Y$. Since $y(0) = -1$ and $y'(0) = 1$,

$$\begin{aligned} 0 &= \mathcal{L}(y'' + 2y' + 2y) \\ &= (s^2 + 2s + 2)Y(s) + s - 1 + 2. \end{aligned}$$

Solving for Y, we get

$$Y(s) = \frac{-s-1}{s^2 + 2s + 2} = -\frac{s+1}{(s+1)^2 + 1}.$$

But, the transform pair

$$\cos t \Longleftrightarrow \frac{s}{s^2 + 1},$$

and Proposition 2.12, give

$$e^{-t}\cos t \Longleftrightarrow \frac{s+1}{(s+1)^2 + 1}.$$

Thus,

$$y(t) = -e^{-t}\cos t.$$

33. Let $y = e^{rt}$. Then,

$$\begin{aligned} y'' + y &= 0 \\ (r^2 + 1)e^{rt} &= 0 \\ r^2 + 1 &= 0. \end{aligned}$$

Thus, $r = \pm i$ and

$$y = e^{it} = \cos t + i \sin t$$

produces a homogeneous solution

$$y_h = C_1 \cos t + C_2 \sin t.$$

We examine $z'' + z = -2e^{i2t}$. Let $z_p = Ae^{i2t}$. Then,

$$\begin{aligned} \left((2i)^2 + 1\right) Ae^{i2t} &= -2e^{i2t} \\ -3A &= -2 \\ A &= \frac{2}{3}. \end{aligned}$$

Thus, $z_p = (2/3)e^{i2t}$ and the real part of z_p is a solution of $y''+y = -2\cos 2t$. Thus, $y_p = (2/3)\cos 2t$. The general solution is

$$y = y_h + y_p = C_1 \cos t + C_2 \sin t + \frac{2}{3}\cos 2t,$$

and

$$y(0) = 1 \Rightarrow 1 = C_1 + \frac{2}{3}.$$

The derivative is

$$y' = -C_1 \sin t + C_2 \cos t - \frac{4}{3}\sin 2t,$$

and

$$y'(0) = -1 \Rightarrow -1 = C_2.$$

Thus, $C_1 = 1/3$ and $C_2 = -1$ and

$$y = \frac{1}{3}\cos t - \sin t + \frac{2}{3}\cos 2t.$$

Alternatively, let $\mathcal{L}(y) = Y$. Then,

$$\begin{aligned} \mathcal{L}(y'' + y) &= \mathcal{L}(-2\cos 2t) \\ s^2Y(s) - s \cdot y(0) - y'(0) + Y(s) &= -2 \cdot \frac{s}{s^2+4}. \end{aligned}$$

But $y(0) = 1$ and $y'(0) = -1$, so

$$(s^2+1)Y(s) - s + 1 = -\frac{2s}{s^2+4},$$

or

$$Y(s) = \frac{s-1}{s^2+1} - \frac{2s}{(s^2+1)(s^2+4)}.$$

The fraction on the right has decomposition

$$\frac{-2s}{(s^2+1)(s^2+4)} = -\frac{2}{3}\cdot\frac{s}{s^2+1} + \frac{2}{3}\cdot\frac{s}{s^2+4}.$$

Therefore,

$$\begin{aligned} Y(s) &= \frac{(1/3)s - 1}{s^2+1} + \frac{2}{3}\cdot\frac{s}{s^2+4} \\ &= \frac{1}{3}\cdot\frac{s-3}{s^2+1} + \frac{2}{3}\cdot\frac{s}{s^2+4} \\ &= \frac{1}{3}\left[\frac{s}{s^2+1} - 3\cdot\frac{1}{s^2+1}\right] + \frac{2}{3}\cdot\frac{s}{s^2+4}. \end{aligned}$$

Thus,

$$y(t) = \frac{1}{3}\cos t - \sin t + \frac{2}{3}\cos 2t.$$

35. Let $y = e^{rt}$. Then,

$$\begin{aligned} y'' + 4y &= 0 \\ (r^2+4)e^{rt} &= 0 \\ r^2 + 4 &= 0. \end{aligned}$$

Thus, $r = \pm 2i$ and $y = e^{2it} = \cos 2t + i\sin 2t$ produces independent real solutions $\cos 2t$ and $\sin 2t$. The homogeneous solution is

$$y_h = C_1\cos 2t + C_2 \sin 2t.$$

We examine $z'' + 4z = e^{i2t}$. Because $z = Ae^{i2t}$ is a solution of the homogeneous equation, we try $z_p = Ate^{i2t}$. Then,

$$\begin{aligned} z_p' &= 2iAte^{i2t} + Ae^{i2t} \\ z_p'' &= -4Ate^{i2t} + 4iAe^{i2t}. \end{aligned}$$

Thus,

$$\begin{aligned} -4Ate^{i2t} + 4iAe^{i2t} + 4Ate^{i2t} &= e^{i2t} \\ 4iA &= 1 \\ A &= -\frac{1}{4}i. \end{aligned}$$

Thus,

$$z_p = -\frac{1}{4}ite^{i2t} = -\frac{1}{4}it(\cos 2t + i\sin 2t),$$

and the real part of z_p,

$$y_p = \frac{1}{4}t\sin 2t,$$

is a particular solution of $y'' + 4y = \cos 2t$. The general solution is

$$y = y_h + y_p = C_1\cos 2t + C_2\sin 2t + \frac{1}{4}t\sin 2t,$$

and

$$y(0) = 1 \Rightarrow 1 = C_1.$$

Next, the derivative

$$\begin{aligned} y' &= -2C_1\sin 2t + 2C_2\cos 2t \\ &\quad + \frac{1}{4}\sin 2t + \frac{1}{2}t\cos 2t, \end{aligned}$$

and

$$y'(0) = -1 \Rightarrow -1 = 2C_2 \text{ or } C_2 = -1/2.$$

Thus, the solution is

$$y = \cos 2t - \frac{1}{2}\sin 2t + \frac{1}{4}t\sin 2t.$$

Alternatively, let $\mathcal{L}(y) = Y$. Since $y(0) = 1$ and $y'(0) = -1$,

$$\mathcal{L}(y'' + 4y) = (s^2+4)Y(s) - s + 1$$
$$\mathcal{L}(\cos 2t) = \frac{s}{s^2+4}.$$

Hence

$$(s^2+4)Y(s) - s + 1 = \frac{s}{s^2+4}.$$

Solving for Y we get

$$\begin{aligned} Y(s) &= \frac{s-1}{s^2+4} + \frac{s}{(s^2+4)^2} \\ &= \frac{s}{s^2+4} - \frac{1}{2}\cdot\frac{2}{s^2+4} + \frac{1}{4}\cdot\frac{4s}{(s^2+4)^2}. \end{aligned}$$

The last fraction is troublesome. However, the transform pair

$$\sin 2t \Longleftrightarrow \frac{2}{s^2+4},$$

and Proposition 2.14,

$$t\sin 2t \Longleftrightarrow -\frac{d}{ds}\frac{2}{s^2+4}$$
$$t\sin 2t \Longleftrightarrow \frac{4s}{(s^2+4)^2}.$$

Thus,

$$y(t) = \cos 2t - \frac{1}{2}\sin 2t + \frac{1}{4}t\sin 2t.$$

37. (a) Let $\mathcal{L}(y) = Y$. Since $y(0) = y_0$ and $y'(0) = v_0$,

$$\begin{aligned} 0 &= \mathcal{L}(y'' + \omega_0^2 y) \\ &= (s^2+\omega_0^2)Y(s) - sy_0 - v_0. \end{aligned}$$

Solving for Y, we get

$$\begin{aligned} Y(s) &= \frac{s\cdot y_0 + v_0}{s^2+\omega_0^2} \\ &= y_0\cdot\frac{s}{s^2+\omega_0^2} + \frac{v_0}{\omega_0}\cdot\frac{\omega_0}{s^2+\omega_0^2}. \end{aligned}$$

Thus,

$$y(t) = y_0\cos\omega_0 t + \frac{v_0}{\omega_0}\sin\omega_0 t.$$

(b) In the figure,

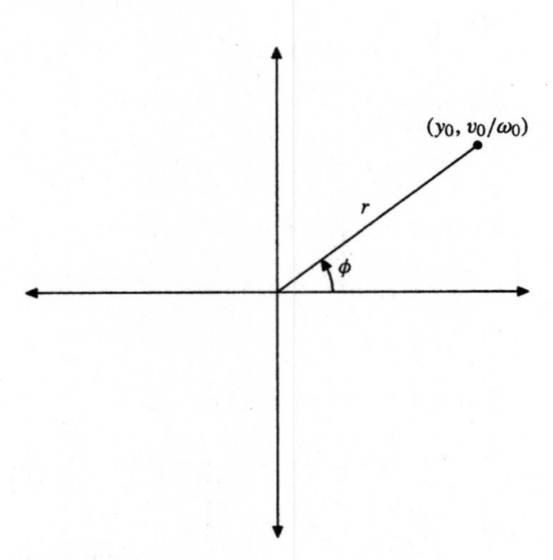

note that

$$r = \sqrt{y_0^2 + \frac{v_0^2}{\omega_0^2}} \quad \text{and} \quad \tan\phi = \frac{v_0}{y_0\omega_0}.$$

Factor out the magnitude r.

$$\begin{aligned} y(t) &= y_0\cos\omega_0 t + \frac{v_0}{\omega_0}\sin\omega_0 t \\ &= r\left[\frac{y_0}{r}\cos\omega_0 t + \frac{v_0/\omega_0}{r}\sin\omega_0 t\right] \\ &= r[\cos\phi\cos\omega_0 t + \sin\phi\sin\omega_0 t] \\ &= r\cos(\omega_0 t - \phi). \end{aligned}$$

39. Set $\omega = \sqrt{\omega_0^2 - c^2}$. This is possible because $\omega_0^2 > c^2$. In Exercise 38, we saw that $y'' + 2cy' + \omega_0^2 y = 0$,

$y(0) = y_0$, $y'(0) = v_0$, has transform

$$\begin{aligned} Y(s) &= \frac{y_0 s + v_0 + 2cy_0}{(s+c)^2 + (\omega_0^2 - c^2)} \\ &= \frac{y_0 s + v_0 + 2cy_0}{(s+c)^2 + \omega^2} \\ &= y_0 \cdot \frac{s+c}{(s+c)^2+\omega^2} \\ &+ \frac{y_0 c + v_0}{\omega} \cdot \frac{\omega}{(s+c)^2+\omega^2}. \end{aligned}$$

Thus,

$$y(t) = y_0 e^{-ct}\cos(\omega t) + \frac{y_0 c + v_0}{\omega} e^{-ct}\sin(\omega t).$$

41. Set $a = \sqrt{c^2 - \omega_0^2}$. This is ok, since $c^2 - \omega_0^2 > 0$. In Exercise 38, we saw that $y'' + 2cy' + \omega_0^2 y = 0$, $y(0) = y_0$, $y'(0) = v_0$, has transform

$$\begin{aligned} Y(s) &= \frac{y_0 s + v_0 + 2cy_0}{(s+c)^2 + (\omega_0^2 - c^2)} \\ &= \frac{y_0 s + v_0 + 2cy_0}{(s+c)^2 - a^2} \end{aligned}$$

With the additional condition $y(0) = y_0 = 0$, this becomes

$$\begin{aligned} Y(s) &= \frac{v_0}{(s+c)^2 - a^2} \\ &= \frac{v_0}{(s+c+a)(s+c-a)} \\ &= \frac{v_0}{2a}\left[\frac{1}{s+c-a} - \frac{1}{s+c+a}\right]. \end{aligned}$$

Hence

$$y(t) = \frac{v_0}{2a}\left[e^{-(c-a)t} - e^{-(c+a)t}\right],$$

where $a = \sqrt{c^2 - \omega_0^2}$.

Section 5.5. Discontinuous Forcing Terms

1. Since $f(t) = t$ has transform $F(s) = 1/s^2$, $g(t) = H(t-2)(t-2)$ has transform

$$\begin{aligned} G(s) &= e^{-2s}F(s) = e^{-2s}\cdot\frac{1}{s^2} \\ &= \frac{e^{-2s}}{s^2}. \end{aligned}$$

3. Since $f(t) = \sin 3t$ has transform $F(s) = 3/(s^2 + 9)$, $g(t) = H(t - \pi/4)\sin 3(t - \pi/4)$ has transform

$$\begin{aligned} G(S) &= e^{-\pi s/4}F(s) \\ &= e^{-\pi s/4}\cdot\frac{3}{s^2+9} \\ &= \frac{3e^{-\pi s/4}}{s^2+9}. \end{aligned}$$

5. Since $(t-1)^2 = t^2 - 2t + 1$, we can write

$$\begin{aligned} t^2 &= (t-1)^2 + 2t - 1 \\ &= (t-1)^2 + 2(t-1) + 1. \end{aligned}$$

Thus,

$$\begin{aligned} &H(t-1)t^2 \\ &= H(t-1)(t-1)^2 + 2H(t-1)(t-1) \\ &\quad + H(t-1), \end{aligned}$$

and

$$\begin{aligned}
&\mathcal{L}\{H(t-1)t^2\}(s) \\
&\quad = \mathcal{L}\{H(t-1)(t-1)^2\}(s) \\
&\qquad + 2\mathcal{L}\{H(t-1)(t-1)\}(s) \\
&\qquad + \mathcal{L}\{H(t-1)\}(s) \\
&\quad = e^{-s}\cdot\frac{2!}{s^3} + 2e^{-s}\cdot\frac{1}{s^2} + e^{-s}\cdot\frac{1}{s} \\
&\quad = \frac{e^{-s}(2+2s+s^2)}{s^3}.
\end{aligned}$$

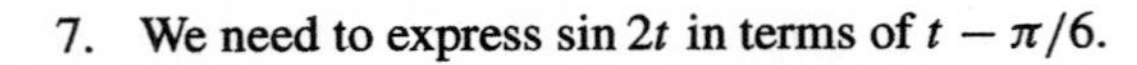

7. We need to express $\sin 2t$ in terms of $t - \pi/6$.

$$\begin{aligned}
\sin 2t &= \sin\left(2t - \frac{\pi}{3} + \frac{\pi}{3}\right) \\
&= \sin\left(2t - \frac{\pi}{3}\right)\cos\frac{\pi}{3} \\
&\qquad + \cos\left(2t - \frac{\pi}{3}\right)\sin\frac{\pi}{3} \\
&= \frac{1}{2}\sin 2\left(t - \frac{\pi}{6}\right) + \frac{\sqrt{3}}{2}\cos 2\left(t - \frac{\pi}{6}\right).
\end{aligned}$$

Now,

$$\begin{aligned}
&\mathcal{L}\left\{H\left(t - \frac{\pi}{6}\right)\sin 2t\right\}(s) \\
&\quad = \frac{1}{2}\mathcal{L}\left\{H\left(t - \frac{\pi}{6}\right)\sin 2\left(t - \frac{\pi}{6}\right)\right\}(s) \\
&\qquad + \frac{\sqrt{3}}{2}\mathcal{L}\left\{H\left(t - \frac{\pi}{6}\right)\cos 2\left(t - \frac{\pi}{6}\right)\right\}(s) \\
&\quad = \frac{1}{2}e^{-\pi s/6}\mathcal{L}\{\sin 2t\}(s) \\
&\qquad + \frac{\sqrt{3}}{2}e^{-\pi s/6}\mathcal{L}\{\cos 2t\}(s) \\
&\quad = \frac{1}{2}e^{-\pi s/6}\cdot\frac{4}{s^2+4} + \frac{\sqrt{3}}{2}e^{-\pi s/6}\cdot\frac{s^2}{s^2+4} \\
&\quad = \frac{1}{2}e^{-\pi s/6}\left[\frac{4}{s^2+4} + \frac{\sqrt{3}s^2}{s^2+4}\right] \\
&\quad = \frac{1}{2}e^{-\pi s/6}\frac{4+\sqrt{3}s^2}{s^2+4}.
\end{aligned}$$

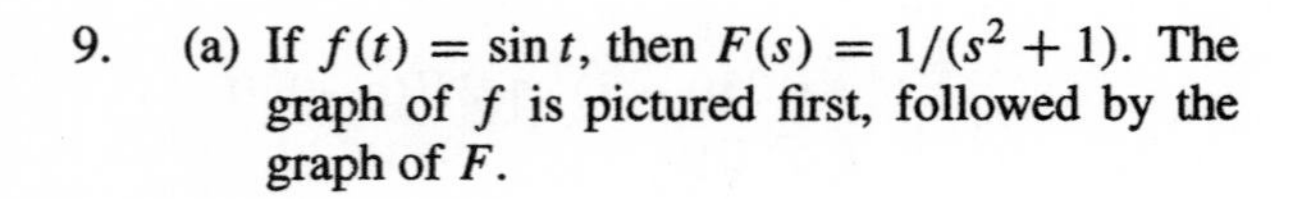

9. (a) If $f(t) = \sin t$, then $F(s) = 1/(s^2+1)$. The graph of f is pictured first, followed by the graph of F.

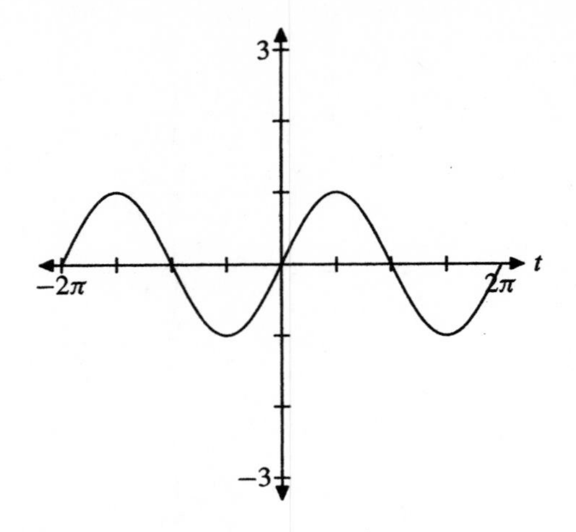

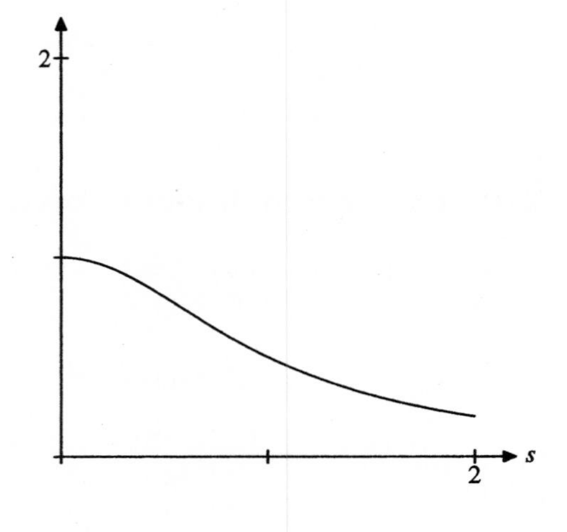

(b) If $g(t) = H(t-1)\sin(t-1)$, then $G(s) = e^{-s}/(s^2+1)$. The graph of g is pictured first, followed by the graph of G.

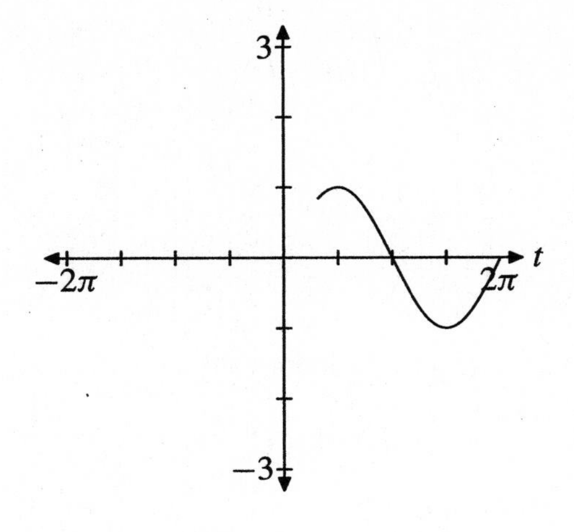

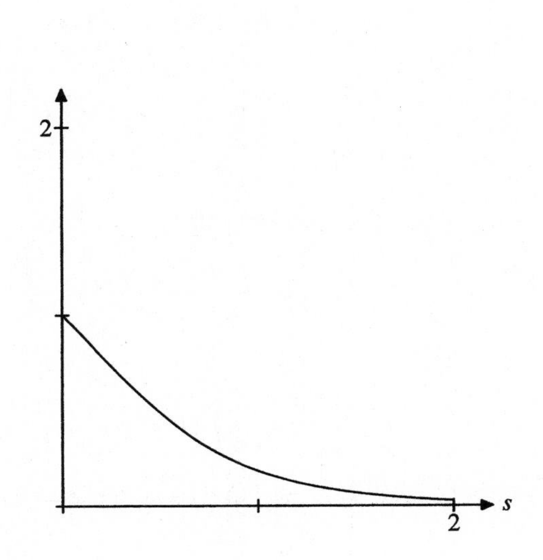

(c) If $g(t) = H(t-2)\sin(t-2)$, then $G(s) = e^{-2s}/(s^2+1)$. The graph of g is pictured first, followed by the graph of G.

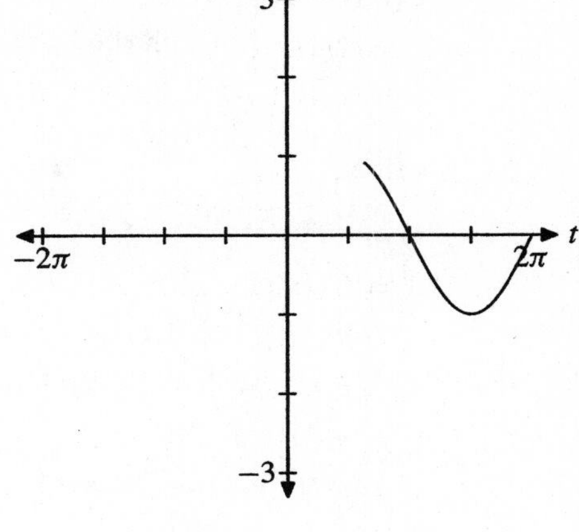

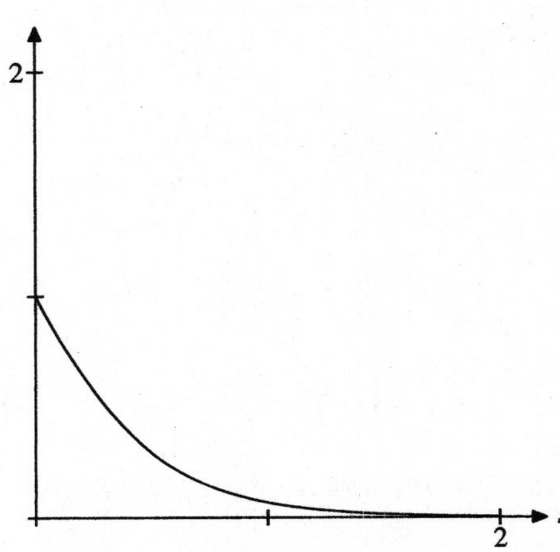

In all cases, note that a shift in the time domain causes a scaling in the s-domain.

11. The function

$$f(t) = \begin{cases} 5, & \text{if } 2 \le t < 4, \\ 0, & \text{otherwise,} \end{cases}$$

can be written $f(t) = 5H_{24}(t)$. Consequently,

$$\begin{aligned} \mathcal{L}\{f(t)\}(s) &= \mathcal{L}\{5H_{24}(t)\}(s) \\ &= 5(\mathcal{L}\{H_2(t)\}(s) - \mathcal{L}\{H_4(t)\}(s)) \\ &= 5\left(\frac{e^{-2s}}{s} - \frac{e^{-4s}}{s}\right) \\ &= \frac{5}{s}(e^{-2s} - e^{-4s}). \end{aligned}$$

13. The function

$$f(t)=\begin{cases}0, & \text{if } t<0,\\ t^2, & \text{if } 0\le t<2,\\ 4, & \text{if } t\ge 2,\end{cases}$$

can be written

$$\begin{aligned}f(t)&=t^2H_{02}(t)+4H_2(t)\\ &=t^2(H(t)-H(t-2))+4H(t-2)\\ &=t^2H(t)-t^2H(t-2)+4H(t-2)\\ &=t^2H(t)-(t-2+2)^2H(t-2)\\ &\quad -4H(t-2)\\ &=t^2H(t)-(t-2)^2H(t-2)\\ &\quad -4(t-2)H(t-2).\end{aligned}$$

Thus,

$$\begin{aligned}\mathcal{L}\{f(t)\}(s)&=\mathcal{L}\{t^2H(t)\}(s)\\ &\quad -\mathcal{L}\{(t-2)^2H(t-2)\}(s)\\ &\quad -4\,\mathcal{L}\{(t-2)H(t-2)\}(s)\\ &=\frac{2!}{s^3}-e^{-2s}\cdot\frac{2!}{s^3}-4e^{-2s}\cdot\frac{1}{s^2}\\ &=\frac{2}{s^3}-\frac{2e^{-2s}}{s^3}-\frac{4e^{-2s}}{s^2}.\end{aligned}$$

15. The function

$$f(t)=\begin{cases}t, & \text{if } 0\le t<1;\\ 2-t, & \text{if } 1\le t<3;\\ t-4, & \text{if } 3\le t<4;\\ 0, & \text{otherwise}\end{cases}$$

can be written

$$\begin{aligned}f(t)&=tH_{01}(t)+(2-t)H_{13}(t)+(t-4)H_{34}(t)\\ &=t(H(t)-H(t-1))\\ &\quad +(2-t)(H(t-1)-H(t-3))\\ &\quad +(t-4)(H(t-3)-H(t-4))\\ &=tH(t)+(2-2t)H(t-1)\\ &\quad +(2t-6)H(t-3)-(t-4)H(t-4)\\ &=tH(t)-2(t-1)H(t-1)\\ &\quad +2(t-3)H(t-3)-(t-4)H(t-4).\end{aligned}$$

Thus,

$$\begin{aligned}&\mathcal{L}\{f(t)\}(s)\\ &\quad=\mathcal{L}\{tH(t)\}(s)-2\,\mathcal{L}\{(t-1)H(t-1)\}(s)\\ &\qquad+2\,\mathcal{L}\{(t-3)H(t-3)\}(s)\\ &\qquad-\mathcal{L}\{(t-4)H(t-4)\}(s)\\ &\quad=\frac{1}{s^2}-2e^{-s}\cdot\frac{1}{s^2}+2e^{-3s}\cdot\frac{1}{s^2}-e^{-4s}\cdot\frac{1}{s^2}\\ &\quad=\frac{1}{s^2}(1-2e^{-s}+2e^{-3s}-e^{-4s}).\end{aligned}$$

17. Note the transform pair

$$f(t)=e^{2t}\Longleftrightarrow F(s)=\frac{1}{s-2}.$$

Thus,

$$\begin{aligned}&\mathcal{L}^{-1}\left\{e^{-s}\frac{1}{s-2}\right\}\\ &\quad=\mathcal{L}^{-1}\{e^{-s}F(s)\}\\ &\quad=H(t-1)f(t-1)\\ &\quad=H(t-1)e^{2(t-1)}\\ &\quad=\begin{cases}0, & \text{if } 0\le t<1,\\ e^{2(t-1)}, & \text{if } 1\le t<\infty.\end{cases}\end{aligned}$$

19. Note the transform pair

$$f(t)=t^2\Longleftrightarrow F(s)=\frac{2}{s^3}.$$

Thus,

$$\begin{aligned}&\mathcal{L}^{-1}\left\{\frac{2+e^{-2s}}{s^3}\right\}(t)\\ &\quad=\mathcal{L}^{-1}\left\{\frac{2}{s^3}\right\}(t)+\frac{1}{2}\mathcal{L}^{-1}\left\{e^{-2s}\cdot\frac{2}{s^3}\right\}(t)\\ &\quad=\mathcal{L}^{-1}\{F(s)\}(t)+\frac{1}{2}\mathcal{L}^{-1}\{e^{-2s}F(s)\}(t)\\ &\quad=t^2+\frac{1}{2}H(t-2)f(t-2)\\ &\quad=t^2+\frac{1}{2}H(t-2)(t-2)^2\\ &\quad=\begin{cases}t^2, & \text{if } 0\le t<2,\\ t^2+(1/2)(t-2)^2, & \text{if } 2\le t<\infty.\end{cases}\end{aligned}$$

21. Note the transform pair

$$f(t) = \cos 2t \Longleftrightarrow F(s) = \frac{s}{s^2+4}.$$

Thus,

$$\begin{aligned}
\mathcal{L}^{-1}&\left\{\frac{se^{-2s}}{s^2+4}\right\}(t)\\
&= \mathcal{L}^{-1}\{e^{-2s}F(s)\}(t)\\
&= H(t-2)f(t-2)\\
&= H(t-2)\cos 2(t-2)\\
&= \begin{cases} 0, & \text{if } 0 \le t < 2,\\ \cos 2(t-2), & \text{if } 2 \le t < \infty.\end{cases}
\end{aligned}$$

23. A partial fraction decomposition

$$\begin{aligned}
\frac{1}{s^2-2s-3} &= \frac{A}{s-3} + \frac{B}{s-1}\\
1 &= A(s-1) + B(s-3).
\end{aligned}$$

Then,

$$\begin{aligned}
s = 1 &\Rightarrow B = -1/2,\\
s = 3 &\Rightarrow A = 1/2,
\end{aligned}$$

and

$$\frac{1}{s^2-2s-3} = \frac{1/2}{s-3} - \frac{1/2}{s-1}.$$

Note the transform pairs

$$\begin{aligned}
f(t) = e^{3t} &\Longleftrightarrow F(s) = \frac{1}{s-3},\\
g(t) = e^{-t} &\Longleftrightarrow G(s) = \frac{1}{s+1}.
\end{aligned}$$

Thus,

$$\begin{aligned}
\mathcal{L}^{-1}&\left\{\frac{e^{-2s}}{s^2-2s-3}\right\}(t)\\
&= \mathcal{L}^{-1}\left\{e^{-2s}\left[\frac{1/2}{s-3} - \frac{1/2}{s+1}\right]\right\}(t)\\
&= \frac{1}{2}\mathcal{L}^{-1}\left\{e^{-2s}\cdot\frac{1}{s-3}\right\}(t)\\
&\quad - \frac{1}{2}\mathcal{L}^{-1}\left\{e^{-2s}\cdot\frac{1}{s+1}\right\}(t)\\
&= \frac{1}{2}\mathcal{L}^{-1}\{e^{-2s}F(s)\}(t)\\
&\quad - \frac{1}{2}\mathcal{L}^{-1}\{e^{-2s}G(s)\}(t)\\
&= \frac{1}{2}H(t-2)f(t-2)\\
&\quad - \frac{1}{2}H(t-2)g(t-2)\\
&= \frac{1}{2}H(t-2)\left[e^{3(t-2)} - e^{-(t-2)}\right]\\
&= \begin{cases} 0, & \text{if } 0 \le t < 2,\\ \left[e^{3(t-2)} - e^{-(t-2)}\right]/2, & \text{if } 2 \le t < \infty.\end{cases}
\end{aligned}$$

25. Complete the square.

$$F(s) = \frac{2-e^{-2s}}{s^2+2s+2} = \frac{2-e^{-2s}}{(s+1)^2+1}.$$

Note the transform pair:

$$\sin t \Longleftrightarrow \frac{1}{s^2+1}.$$

Thus, by Proposition 2.12, we have another transform pair:

$$e^{-t}\sin t \Longleftrightarrow \frac{1}{(s+1)^2+1}.$$

Thus,

$$\begin{aligned}
\mathcal{L}^{-1}&\left\{\frac{2-e^{-2s}}{(s+1)^2+1}\right\}(t)\\
&= 2\mathcal{L}^{-1}\left\{\frac{1}{(s+1)^2+1}\right\}(t)\\
&\quad - \mathcal{L}^{-1}\left\{e^{-2s}\cdot\frac{1}{(s+1)^2+1}\right\}(t)\\
&= 2e^{-t}\sin t - H(t-2)e^{-(t-2)}\sin(t-2),
\end{aligned}$$

or, equivalently,

$$\begin{cases} 2e^{-t}\sin t, & 0 \le t < 2, \\ 2e^{-t}\sin t - e^{-(t-2)}\sin(t-2), & 2 \le t < \infty. \end{cases}$$

27. The forcing function f is described by

$$f(t) = \begin{cases} 1, & \text{if } 0 \le t < 1, \\ 0, & \text{otherwise,} \end{cases}$$

or equivalently,

$$\begin{aligned} f(t) &= H_{01}(t) \\ &= H_0(t) - H_1(t) \\ &= H(t) - H(t-1). \end{aligned}$$

Hence,

$$F(s) = \mathcal{L}\{f(t)\}(s) = \frac{1}{s} - \frac{e^{-s}}{s}.$$

Take the Laplace transform of both sides of the given equation. Let $Y(s) = \mathcal{L}\{y(t)\}(s)$.

$$\begin{aligned} y'' + 4y &= f(t) \\ \mathcal{L}(y'')(s) + 4\mathcal{L}(y)(s) &= \mathcal{L}\{f(t)\}(s) \\ s^2Y(s) - sy(0) - y'(0) + 4Y(s) &= F(s). \end{aligned}$$

Use the initial conditions $y(0) = y'(0) = 0$ and the Laplace transform of f found earlier.

$$\begin{aligned} s^2Y(s) + 4Y(s) &= \frac{1}{s} - \frac{e^{-s}}{s} \\ Y(s) &= \frac{1}{s^2+4}\left[\frac{1}{s} - \frac{e^{-s}}{s}\right] \\ Y(s) &= \frac{1}{s(s^2+4)} - \frac{e^{-s}}{s(s^2+4)}. \end{aligned}$$

A partial fraction decomposition is needed.

$$\begin{aligned} \frac{1}{s(s^2+4)} &= \frac{A}{s} + \frac{Bs+C}{s^2+4} \\ 1 &= A(s^2+4) + (Bs+c)s \end{aligned}$$

Now,

$$s = 0 \Rightarrow A = 1/4.$$

Furthermore,

$$\begin{aligned} 1 &= As^2 + 4A + Bs^2 + Cs \\ 1 &= (A+B)s^2 + Cs + 4A. \end{aligned}$$

Thus,

$$A + B = 0 \Rightarrow \frac{1}{4} + B = 0 \Rightarrow B = -1/4.$$

Also, $C = 0$. Therefore,

$$\frac{1}{s(s^2+4)} = \frac{1/4}{s} - \frac{(1/4)s}{s^2+4},$$

which has transform

$$\frac{1}{4} - \frac{1}{4}\cos 2t.$$

Finally,

$$\begin{aligned} y(t) &= \mathcal{L}^{-1}\left\{\frac{1}{s(s^2+4)}\right\}(t) \\ &\quad - \mathcal{L}^{-1}\left\{e^{-s}\cdot\frac{1}{s(s^2+4)}\right\}(t) \\ &= \frac{1}{4} - \frac{1}{4}\cos 2t \\ &\quad - H(t-1)\left[\frac{1}{4} - \frac{1}{4}\cos 2(t-1)\right]. \end{aligned}$$

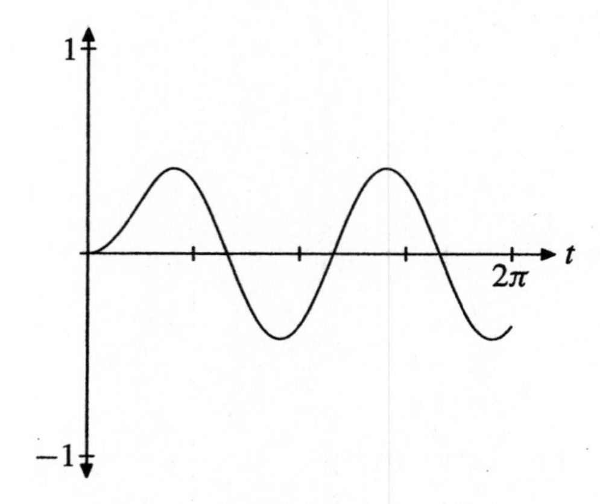

29. The forcing function f is described by

$$f(t) = \begin{cases} t, & \text{if } 0 \le t < 1, \\ 0, & \text{otherwise,} \end{cases}$$

or equivalently,

$$\begin{aligned} f(t) &= tH_{01}(t) \\ &= t(H(t) - H(t-1)) \\ &= tH(t) - (t-1)H(t-1) - H(t-1). \end{aligned}$$

Hence,

$$\begin{aligned} F(s) &= \frac{1}{s^2} - e^{-s} \cdot \frac{1}{s^2} - \frac{e^{-s}}{s} \\ &= \frac{1 - e^{-s}}{s^2} - \frac{e^{-s}}{s}. \end{aligned}$$

Take the Laplace transform of each side of the equation, letting $\mathcal{L}(y)(s) = Y(s)$.

$$\begin{aligned} y'' + 4y &= f(t) \\ \mathcal{L}(y'')(s) + 4\mathcal{L}(y)(s) &= \mathcal{L}\{f(t)\}(s) \\ s^2Y(s) - sy(0) - y'(0) + 4Y(s) &= F(s). \end{aligned}$$

Use the initial conditions $y(0) = y'(0) = 0$ and the Laplace transform of f found earlier.

$$\begin{aligned} s^2Y(s) + 4Y(s) &= \frac{1 - e^{-s}}{s^2} - \frac{e^{-s}}{s} \\ Y(s) &= \frac{1 - e^{-s}}{s^2(s^2+4)} - \frac{e^{-s}}{s(s^2+4)} \end{aligned}$$

Partial fraction decompositions are needed. We've seen (see Exercise 27)

$$\frac{1}{s(s^2+4)} = \frac{1/4}{s} - \frac{(1/4)s}{s^2+4},$$

which has inverse transform

$$\frac{1}{4} - \frac{1}{4}\cos 2t.$$

Next,

$$\frac{1}{s^2(s^2+4)} = \frac{A}{s} + \frac{B}{s^2} + \frac{Cs + D}{s^2+4},$$

and

$$\begin{aligned} 1 &= As(s^2+4) + B(s^2+4) + s^2(Cs + D) \\ 1 &= As^3 + 4As + Bs^2 + 4B + Cs^3 + Ds^2 \\ 1 &= (A + C)s^3 + (B + D)s^2 + 4As + 4B. \end{aligned}$$

Thus,

$$\begin{aligned} A + C &= 0 \\ B + D &= 0 \\ 4A &= 0 \\ 4B &= 1, \end{aligned}$$

and $A = 0$, $B = 1/4$, $C = 0$, and $D = -1/4$. Therefore,

$$\begin{aligned} \frac{1}{s^2(s^2+4)} &= \frac{1/4}{s^2} - \frac{1/4}{s^2+4} \\ &= \frac{1}{4} \cdot \frac{1}{s^2} - \frac{1}{8} \cdot \frac{2}{s^2+4}, \end{aligned}$$

which has inverse transform

$$\frac{1}{4}t - \frac{1}{8}\sin 2t.$$

Thus,

$$\begin{aligned} y(t) &= \mathcal{L}^{-1}\left\{\frac{1 - e^{-s}}{s^2(s^2+4)} - \frac{e^{-s}}{s(s^2+4)}\right\}(t) \\ &= \mathcal{L}^{-1}\left\{\frac{1}{s^2(s^2+4)}\right\}(t) \\ &\quad - \mathcal{L}^{-1}\left\{e^{-s} \cdot \frac{1}{s^2(s^2+4)}\right\}(t) \\ &\quad - \mathcal{L}^{-1}\left\{e^{-s} \cdot \frac{1}{s(s^2+4)}\right\}(t) \\ &= \frac{t}{4} - \frac{1}{8}\sin 2t \\ &\quad - H(t-1)\left[\frac{1}{4}(t-1) - \frac{1}{8}\sin 2(t-1)\right] \\ &\quad - H(t-1)\left[\frac{1}{4} - \frac{1}{4}\cos 2(t-1)\right] \\ &= \frac{t}{4} - \frac{1}{8}\sin 2t \\ &\quad - H(t-1)\left[\frac{1}{4}t - \frac{1}{8}\sin 2(t-1) \right. \\ &\quad \left. - \frac{1}{4}\cos 2(t-1)\right] \end{aligned}$$

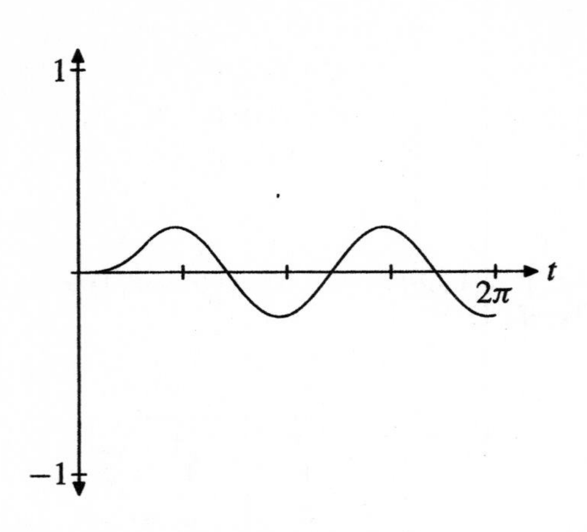

31. To begin with, the function f is periodic with period $T = 2c$. Define a window,

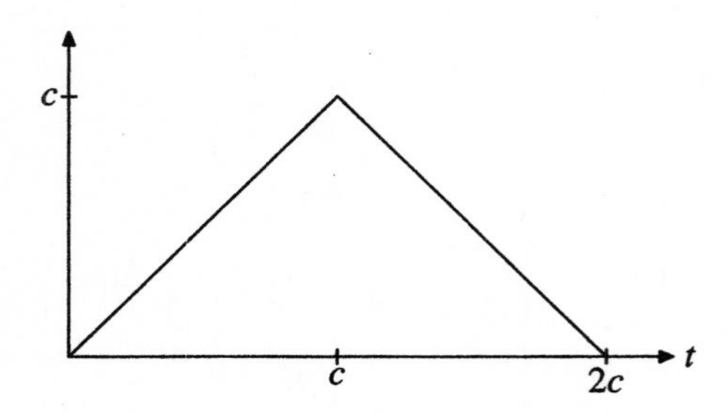

or equivalently,

$$f_T(t) = \begin{cases} t, & \text{if } 0 \le t < c, \\ 2c - t, & \text{if } c \le t < 2c, \\ 0, & \text{otherwise.} \end{cases}$$

Thus,

$$\begin{aligned} f_T(t) &= t(H(t) - H(t-c)) \\ &\quad + (2c - t)(H(t-c) - H(t-2c)) \\ &= tH(t) + (2c - 2t)H(t-c) \\ &\quad - (2c - t)H(t-2c) \\ &= tH(t) - 2(t-c)H(t-c) \\ &\quad + (t-2c)H(t-2c). \end{aligned}$$

Thus,

$$F_T(s) = \frac{1}{s^2} - 2e^{-cs} \cdot \frac{1}{s^2} + e^{-2cs} \cdot \frac{1}{s^2}.$$

Therefore, by Proposition 5.17,

$$\begin{aligned} F(s) &= \frac{F_T(s)}{1 - e^{-Ts}} \\ &= \frac{\dfrac{1}{s^2} - \dfrac{2e^{-cs}}{s^2} + \dfrac{e^{-2cs}}{s^2}}{1 - e^{-2cs}} \\ &\frac{1 - 2e^{-cs} + e^{-2cs}}{s^2(1 - e^{-2cs})} \\ &= \frac{(1 - e^{-cs})^2}{s^2(1 + e^{-cs})(1 - e^{-cs})} \\ &= \frac{1}{s^2} \cdot \frac{1 - e^{-cs}}{1 + e^{-cs}} \\ &= \frac{1}{s^2} \cdot \frac{e^{cs/2} - e^{-cs/2}}{e^{cs/2} + e^{-cs/2}} \\ &= \frac{1}{s^2} \tanh \frac{cs}{2}. \end{aligned}$$

33. (a) A little thought provides

$$f(t) = \begin{cases} 0, & \text{if } 0 \le t < 1, \\ 1, & \text{if } 1 \le t < 2, \\ 2, & \text{if } 2 \le t < 3, \\ 3, & \text{if } 3 \le t < 4, \\ \vdots, & \vdots \end{cases}$$

and the graph of f looks like a "staircase."

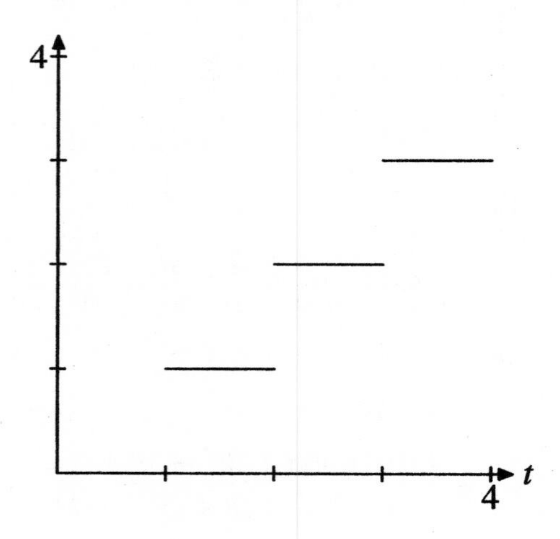

(b) If h has graph

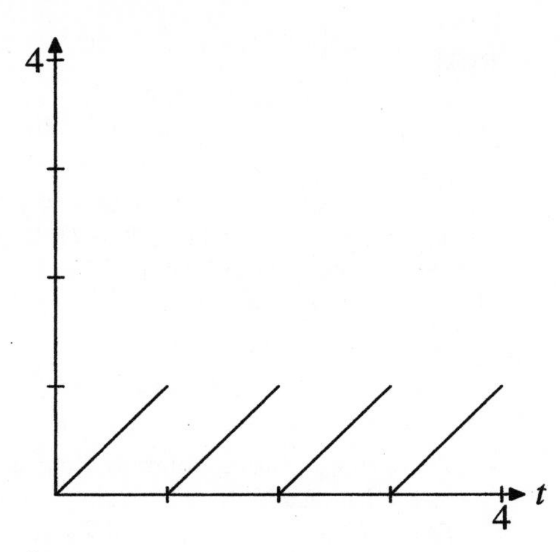

then

$$h(t) = \begin{cases} t, & \text{if } 0 \le t < 1, \\ t-1, & \text{if } 1 \le t < 2, \\ t-2, & \text{if } 2 \le t < 3, \\ t-3, & \text{if } 3 \le t < 4, \\ \vdots, & \vdots \end{cases}$$

Therefore,

$$t - h(t) = \begin{cases} 0, & \text{if } 0 \le t < 1, \\ 1, & \text{if } 0 \le t < 1, \\ 2, & \text{if } 0 \le t < 1, \\ 3, & \text{if } 0 \le t < 1, \\ \vdots, & \vdots \end{cases}$$

and $f(t) = t - h(t)$. By Exercise 32, with $A = 1$ and $c = 1$,

$$\mathcal{L}\{h(t)\}(s) = \frac{1}{s^2} - \frac{1}{s(e^s - 1)}.$$

Therefore,

$$\begin{aligned} \mathcal{L}\{f(t)\}(s) &= \mathcal{L}\{t\}(s) - \mathcal{L}\{h(t)\}(s) \\ &= \frac{1}{s^2} - \left[\frac{1}{s^2} - \frac{1}{s(e^s - 1)}\right] \\ &= \frac{1}{s(e^s - 1)}. \end{aligned}$$

35. (a) If

$$y(t) = \begin{cases} 0, & \text{if } t < 0, \\ t, & \text{if } 0 \le t < 1, \\ t-1, & \text{if } t \ge 1, \end{cases}$$

then y has graph

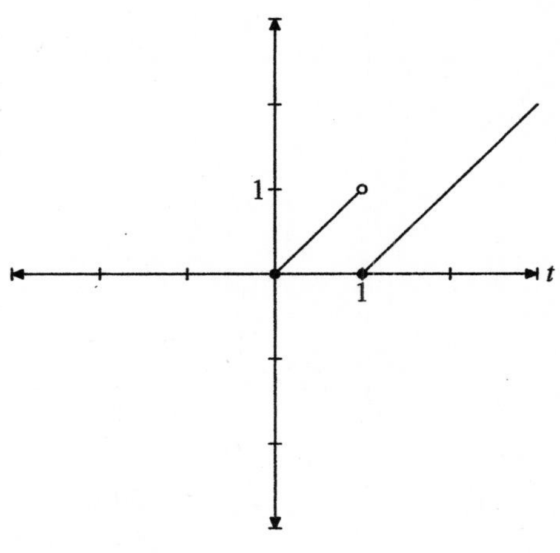

Moreover,

$$\begin{aligned} &y(t) \\ &= tH_{01}(t) + (t-1)H_1(t) \\ &= t(H(t) - H(t-1)) + (t-1)H(t-1) \\ &= tH(t) - H(t-1). \end{aligned}$$

Thus,

$$Y(s) = \frac{1}{s^2} - \frac{e^{-s}}{s}$$

and

$$\begin{aligned} s\,\mathcal{L}\{y(t)\}(s) - y(0) &= sY(s) - 0 \\ &= s\left[\frac{1}{s} - \frac{e^{-s}}{s}\right] \\ &= \frac{1}{s} - e^{-s}. \end{aligned}$$

(b) Because $y'(t)$ is the slope of the line tangent to the graph of $y(t)$ at t, the derivative is easily

seen to be

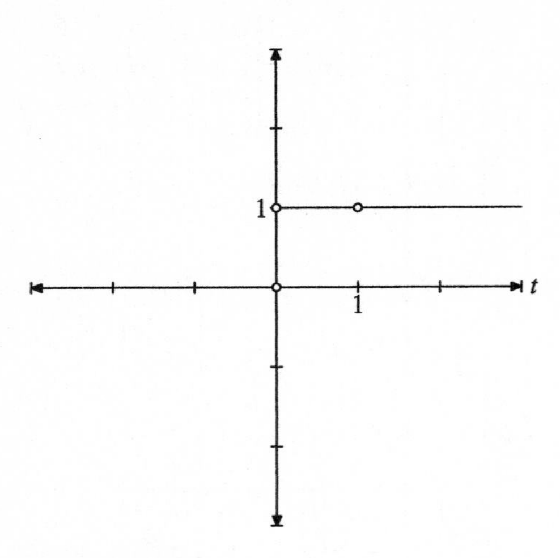

or equivalently,

$$y'(t) = \begin{cases} 0, & \text{if } t < 0, \\ 1, & \text{if } 0 < t < 1, \\ 1, & \text{if } t > 1. \end{cases}$$

Of course, the derivative of y does not exist at $t = 0$ because there are different right and left-hand derivatives at this point (note the sharp cusp in the graph in part (a) at $t = 0$). Nor is y differentiable at $t = 1$, because y fails to be continuous at $t = 1$. Thus,

$$\begin{aligned} \mathcal{L}\{y'(t)\}(s) &= \int_0^\infty y'(t)e^{-st}\,dt \\ &= \int_0^\infty 1e^{-st}\,dt, \end{aligned}$$

as the removable discontinuity at $t = 1$ will not affect the overall integral. Note that this is identical to the Laplace transform of 1, which is $1/s$. Therefore,

$$\mathcal{L}\{y'(t)\}(s) = \frac{1}{s}.$$

(c) There is a seeming contradiction, as parts (a) and (b) lead to different answers for $\mathcal{L}\{y'(t)\}(s)$. However, this does not contradict Proposition 2.1 as the hypotheses of Proposition 2.1 are not satisfied; namely, y is not piecewise differentiable (it isn't continuous).

(d) As we saw in part (b), $\mathcal{L}\{y'(t)\}(s) = 1/s$. Next,

$$\begin{aligned} & s\,\mathcal{L}\{y(t)\}(s) - \left[y(1^+) - y(1^-)\right]e^{-s} - y(0) \\ &= s\left[\frac{1}{s^2} - \frac{e^{-s}}{s}\right] - [0 - 1]e^{-s} - 0 \\ &= \frac{1}{s} - e^{-s} + e^{-s} \\ &= \frac{1}{s}. \end{aligned}$$

Section 5.6. The Delta Function

1. (a) If

$$\delta_p^\epsilon = \frac{1}{\epsilon}(H_p(t) - H_{p+\epsilon}(t)),$$

then

$$\begin{aligned} \mathcal{L}\{\delta_p^\epsilon(t)\}(s) &= \frac{1}{\epsilon}\mathcal{L}\{H_p(t)\}(s) \\ &\quad - \frac{1}{\epsilon}\mathcal{L}\{H_{p+\epsilon}(t)\}(s) \\ &= \frac{1}{\epsilon}\cdot\frac{e^{-ps}}{s} - \frac{1}{\epsilon}\cdot\frac{e^{-(p+\epsilon)s}}{s} \\ &= \frac{e^{-sp} - e^{-sp}e^{-s\epsilon}}{s\epsilon} \\ &= e^{-sp}\cdot\frac{1 - e^{-s\epsilon}}{s\epsilon}. \end{aligned}$$

(b) Taking the limit,

$$\begin{aligned}\delta_p(t) &= \lim_{\epsilon\to 0} \delta_p^\epsilon(t)\\ &= \lim_{\epsilon\to 0} e^{-sp} \cdot \frac{1-e^{s\epsilon}}{s\epsilon}\\ &= e^{-sp} \lim_{\epsilon\to 0} \frac{1-e^{-s\epsilon}}{s\epsilon}.\end{aligned}$$

Because both numerator and denominator approach zero as $\epsilon \to 0$, the indeterminate form $0/0$ allows the application of l'Hôpital's rule.

$$\begin{aligned}\delta_p(t) &= e^{-sp} \lim_{\epsilon\to 0} \frac{\dfrac{d}{d\epsilon}(1-e^{-s\epsilon})}{\dfrac{d}{d\epsilon}(s\epsilon)}\\ &= e^{-sp} \lim_{\epsilon\to 0} \frac{se^{-s\epsilon}}{s}\\ &= e^{-sp} \lim_{\epsilon\to 0} e^{-s\epsilon}\\ &= e^{-sp}.\end{aligned}$$

Thus, $\mathcal{L}\{\delta_p(t)\}(s) = e^{-sp}$.

3. By Theorem 6.10, the unit impulse response of

$$y'' - 4y' - 5y = \delta(t)$$

is $E(s) = 1/P(s)$, where $P(s) = s^2 - 4s - 5$ is the characteristic polynomial. Thus,

$$E(s) = \frac{1}{s^2 - 4s - 5}.$$

Using a partial fraction decomposition,

$$E(s) = \frac{1}{6} \cdot \frac{1}{s-5} - \frac{1}{6} \cdot \frac{1}{s+1},$$

so the unit impulse response is

$$e(t) = \frac{1}{6}e^{5t} - \frac{1}{6}e^{-t}.$$

5. By Theorem 6.10, the unit impulse response of

$$y'' - 9y = \delta(t)$$

is $E(s) = 1/P(s)$, where $P(s) = s^2 - 9$ is the characteristic polynomial. Thus,

$$E(s) = \frac{1}{s^2 - 9}.$$

Using a partial fraction decomposition,

$$E(s) = -\frac{1}{6} \cdot \frac{1}{s+3} + \frac{1}{6} \cdot \frac{1}{s-3},$$

so the unit impulse response is

$$e(t) = -\frac{1}{6}e^{-3t} + \frac{1}{6}e^{3t}.$$

7. By Theorem 6.10, the unit impulse response of

$$y'' + 2y' + 2y = \delta(t)$$

is $E(s) = 1/P(s)$, where $P(s) = s^2 + 2s + 2$ is the characteristic polynomial. Thus,

$$E(s) = \frac{1}{s^2 + 2s + 2}.$$

Completing the square,

$$E(s) = \frac{1}{(s+1)^2 + 1},$$

so the unit impulse response is

$$e(t) = e^{-t} \sin t.$$

9. (a) If

$$x'' + 2x' + 2x = \delta(t), \qquad x(0) = x'(0) = 0,$$

then, by Theorem 6.10, the unit impulse response is

$$X(s) = \frac{1}{P(s)} = \frac{1}{s^2 + 2s + 2}.$$

Completing the square,

$$X(s) = \frac{1}{(s+1)^2 + 1}.$$

Note the transform pair

$$\sin t \Longleftrightarrow \frac{1}{s^2 + 1}.$$

Thus, by Proposition 2.12,

$$x(t) = e^{-t} \sin t.$$

(b) The Laplace transform of the right-hand side, δ_0^ϵ, is

$$\begin{aligned}
&\mathcal{L}\{\delta_0^\epsilon(t)\}(s) \\
&= \mathcal{L}\left\{\frac{1}{\epsilon}(H(t) - H(t-\epsilon))\right\}(s) \\
&= \frac{1}{\epsilon}\left[\mathcal{L}\{H(t)\}(s) - \mathcal{L}\{H(t-\epsilon)\}(s)\right] \\
&= \frac{1}{\epsilon}\left[\frac{1}{s} - \frac{e^{-\epsilon s}}{s}\right] \\
&= \frac{1 - e^{\epsilon s}}{\epsilon s}.
\end{aligned}$$

Thus, if

$$x'' + 2x' + 2x = \delta_0^\epsilon(t), \quad \text{and} \quad x(0) = x'(0) = 0,$$

then

$$\begin{aligned}
\mathcal{L}(x'' + 2x' + 2x)(s) &= \mathcal{L}\{\delta_0^\epsilon(t)\}(s) \\
(s^2 + 2s + 2)X(s) &= \frac{1 - e^{-\epsilon s}}{\epsilon s} \\
X(s) &= \frac{1 - e^{-\epsilon s}}{\epsilon s(s^2 + 2s + 2)}.
\end{aligned}$$

A partial fraction decomposition gives

$$\begin{aligned}
&\frac{1}{s(s^2 + 2s + 2)} \\
&= \frac{1/2}{s} + \frac{-(1/2)s - 1}{s^2 + 2s + 2} \\
&= \frac{1}{2} \cdot \frac{1}{s} - \frac{1}{2} \cdot \frac{s + 2}{(s+1)^2 + 1} \\
&= \frac{1}{2} \cdot \frac{1}{s} - \frac{1}{2}\frac{s + 1}{(s+1)^2 + 1} \\
&\quad - \frac{1}{2}\frac{1}{(s+1)^2 + 1} \\
&= \frac{1}{2}\mathcal{L}\{f\}(s),
\end{aligned}$$

where

$$f(t) = 1 - e^{-t}(\cos t + \sin t).$$

Thus,

$$\begin{aligned}
x_\epsilon(t) &= \frac{1}{\epsilon}\mathcal{L}^{-1}\left\{\frac{1}{s(s^2 + 2s + 2)}\right\}(t) \\
&\quad - \frac{1}{\epsilon}\mathcal{L}^{-1}\left\{e^{-\epsilon s} \cdot \frac{1}{s(s^2 + 2s + 2)}\right\}(t) \\
&= \frac{1}{2\epsilon}\left[f(t) - H(t-\epsilon)f(t-\epsilon)\right] \\
&= \frac{1}{2\epsilon}\begin{cases} f(t)), & 0 \le t < \epsilon, \\ f(t) - f(t-\epsilon), & t \ge \epsilon. \end{cases}
\end{aligned}$$

(c) As $\epsilon \to 0$, the first interval $(0 \le t < \epsilon)$ in the piecewise definition in part (b) melts away. Let's concentrate on the second piece. By l'Hôpital's rule,

$$\begin{aligned}
\frac{1}{2}\lim_{\epsilon \to 0}\frac{f(t) - f(t-\epsilon)}{\epsilon} &= \frac{1}{2}\lim_{\epsilon \to 0} f'(t-\epsilon) \\
&= \frac{1}{2}f'(t) \\
&= e^{-t}\sin t.
\end{aligned}$$

11. If

$$x' = \delta_p(t), \qquad x(0) = 0,$$

then

$$\begin{aligned}
\mathcal{L}\{x'\}(s) &= \mathcal{L}\{\delta_p(t)\}(s) \\
sX(s) - x(0) &= e^{-ps} \\
X(s) &= \frac{1}{s}e^{-ps}.
\end{aligned}$$

Thus,

$$x(t) = H_p(t).$$

Section 5.7. Convolutions

1. By definition,

$$\begin{aligned}
&f * (g + h)(t) \\
&= \int_0^t f(u)(g + h)(t - u)\,du \\
&= \int_0^t f(u)(g(t - u) + h(t - u))\,du, \\
&= \int_0^t f(u)g(t - u)\,du + \int_0^t f(u)h(t - u)\,du \\
&= f * g(t) + f * h(t)
\end{aligned}$$

Thus,

$$f * (g + h) = f * g + f * h.$$

3. By definition,

$$F * 0(t) = \int_0^t f(u)z(t - u)\,du,$$

where z is the zero function. Of course, $z(t-u) = 0$, so

$$\begin{aligned}
f * 0(t) &= \int_0^t f(u) \cdot 0\,du \\
&= \int_0^t 0\,du \\
&= 0.
\end{aligned}$$

Since this is true for all t,

$$f * 0 = 0.$$

5. If $f(t) = e^t$ and $g(t) = e^{2t}$, then

$$\begin{aligned}
f * g(t) &= \int_0^t f(u)g(t - u)\,du \\
&= \int_0^t e^u e^{2(t-u)}\,du \\
&= \int_0^t e^{2t-u}\,du \\
&= -e^{2t-u}\Big|_0^t \\
&= -e^t - (-e^{2t}) \\
&= e^{2t} - e^t.
\end{aligned}$$

7. If $f(t) = t$ and $g(t) = 3 - t$, then

$$\begin{aligned}
f * g(t) &= \int_0^t f(u)g(t - u)\,du \\
&= \int_0^t u(3 - (t - u))\,du \\
&= \int_0^t (3u - ut + u^2)\,du \\
&= \frac{3}{2}u^2 - \frac{1}{2}u^2 t + \frac{1}{3}u^3\Big|_0^t \\
&= \frac{3}{2}t^2 - \frac{1}{2}t^3 + \frac{1}{3}t^3 \\
&= \frac{3}{2}t^2 - \frac{1}{6}t^3.
\end{aligned}$$

9. If $f(t) = t^2$ and $g(t) = e^{-t}$, then

$$\begin{aligned}
f * g(t) &= \int_0^t f(u)g(t - u)\,du \\
&= \int_0^t u^2 e^{-(t-u)}\,du \\
&= \int_0^t u^2 e^{u-t}\,du.
\end{aligned}$$

Integration by parts provides

$$f * g(t) = u^2 e^{u-t}\Big|_0^t - \int_0^t 2u e^{u-t}\,du.$$

A second integration by parts provides

$$\begin{aligned}
f * g(t) &= u^2 e^{u-t}\Big|_0^t - \left[2u e^{u-t}\Big|_0^t - \int_0^t 2e^{u-t}\,du\right] \\
&= u^2 e^{u-t} - 2u e^{u-t} + 2e^{u-t}\Big|_0^t \\
&= e^{u-t}(u^2 - 2u + 2)\Big|_0^t \\
&= (t^2 - 2t + 2) - e^{-t}(0 - 0 + 2) \\
&= t^2 - 2t + 2 - 2e^{-t}.
\end{aligned}$$

11. If $f(t) = \sin t$ and $g(t) = t$, then

$$\begin{aligned} f * g(t) &= \int_0^t f(u)g(t-u)\,du \\ &= \int_0^t (\sin u)(t-u)\,du. \end{aligned}$$

Integration by parts provides

$$\begin{aligned} f * g(t) &= \int_0^t (u-t)d(\cos u) \\ &= (u-t)\cos u\Big|_0^t - \int_0^t \cos u\,du \\ &= (u-t)\cos u - \sin u\Big|_0^t \\ &= (0-\sin t) - (-t-0) \\ &= t - \sin t. \end{aligned}$$

However,

$$\begin{aligned} \mathcal{L}\{f * g(t)\}(s) &= \mathcal{L}\{t - \sin t\}(s) \\ &= \mathcal{L}\{t\}(s) - \mathcal{L}\{\sin t\}(s) \\ &= \frac{1}{s^2} - \frac{1}{s^2+1} \\ &= \frac{1}{s^2(s^2+1)}. \end{aligned}$$

Alternatively, we have transform pairs

$$\begin{aligned} f(t) = \sin t &\Longleftrightarrow F(s) = \frac{1}{s^2+1} \\ g(t) = t &\Longleftrightarrow G(s) = \frac{1}{s^2}. \end{aligned}$$

Thus,

$$\begin{aligned} \mathcal{L}\{f * g(t)\}(s) &= F(s)G(s) \\ &= \frac{1}{s^2+1}\cdot\frac{1}{s^2} \\ &= \frac{1}{s^2(s^2+1)}. \end{aligned}$$

13. If $f(t) = e^{-2t}$ and $g(t) = t$, then

$$\begin{aligned} f * g(t) &= \int_0^t f(u)g(t-u)\,du \\ &= \int_0^t e^{-2u}(t-u)\,du \\ &= \int_0^t (t-u)d\left(-\frac{1}{2}e^{-2u}\right). \end{aligned}$$

Integrate by parts.

$$\begin{aligned} f * g(t) &= (t-u)\left(-\frac{1}{2}e^{-2u}\right)\Big|_0^t \\ &\quad - \int_0^t \left(-\frac{1}{2}e^{-2u}\right)(-du) \\ &= -\frac{1}{2}e^{-2u}(t-u)\Big|_0^t - \frac{1}{2}\int_0^t e^{-2u}\,du \\ &= -\frac{1}{2}e^{-2u}(t-u) + \frac{1}{4}e^{-2u}\Big|_0^t \\ &= \left(0 + \frac{1}{4}e^{-2t}\right) - \left(-\frac{1}{2}t + \frac{1}{4}\right) \\ &= \frac{1}{4}e^{-2t} + \frac{1}{2}t - \frac{1}{4}. \end{aligned}$$

However,

$$\begin{aligned} \mathcal{L}\{f * g(t)\}(s) &= \mathcal{L}\left\{\frac{1}{4}e^{-2t} + \frac{1}{2}t - \frac{1}{4}\right\}(s) \\ &= \frac{1}{4}\cdot\frac{1}{s+2} + \frac{1}{2}\cdot\frac{1}{s^2} - \frac{1}{4}\cdot\frac{1}{s} \\ &= \frac{s^2 + 2(s+2) - s(s+2)}{4s^2(s+2)} \\ &= \frac{4}{4s^2(s+2)} \\ &= \frac{1}{s^2(s+2)} \end{aligned}$$

Alternatively, we have the transform pairs

$$\begin{aligned} f(t) = e^{-2t} &\Longleftrightarrow F(s) = \frac{1}{s+2} \\ g(t) = t &\Longleftrightarrow G(s) = \frac{1}{s^2}. \end{aligned}$$

Thus,

$$\begin{aligned} \mathcal{L}\{f * g(t)\}(s) &= F(s)G(s) \\ &= \frac{1}{s+2}\cdot\frac{1}{s^2} \\ &= \frac{1}{s^2(s+2)}. \end{aligned}$$

15. If $f(t) = e^{-t}$ and $g(t) = \cos t$, then

$$f * g(t) = \int_0^t f(u)g(t-u)\,du = \int_0^t e^{-u}\cos(t-u)\,du.$$

One integration by parts provides

$$\begin{aligned}\int e^{-u}\cos(t-u)\,du &= \int \cos(t-u)d(-e^{-u}) \\ &= -e^{-u}\cos(t-u) - \int (-e^{-u})\sin(t-u)\,du \\ &= -e^{-u}\cos(t-u) + \int e^{-u}\sin(t-u)\,du \\ &= -e^{-u}\cos(t-u) + \int \sin(t-u)d(-e^{-u}).\end{aligned}$$

A second integration by parts leads to

$$\begin{aligned}\int e^{-u}\cos(t-u)\,du &= -e^{-u}\cos(t-u) + (-e^{-u})\sin(t-u) - \int (-e^{-u})(-\cos(t-u))\,du \\ &= -e^{-u}\cos(t-u) - e^{-u}\sin(t-u) - \int e^{-u}\cos(t-u)\,du.\end{aligned}$$

Thus,

$$2\int e^{-u}\cos(t-u)\,du = -e^{-u}\cos(t-u) - e^{-u}\sin(t-u).$$

Hence

$$\int e^{-u}\cos(t-u)\,du = -\frac{1}{2}e^{-u}\left[\cos(t-u) + \sin(t-u)\right],$$

and

$$\begin{aligned}f * g(t) &= -\frac{1}{2}e^{-u}\left[\cos(t-u) + \sin(t-u)\right]\Big|_0^t \\ &= -\frac{1}{2}e^{-t} + \frac{1}{2}[\cos t + \sin t]\end{aligned}$$

We have

$$\begin{aligned}&\mathcal{L}\{f * g(t)\}(s) \\ &= \frac{1}{2}\mathcal{L}\{-e^{-t} + \cos t + \sin t\} \\ &= \frac{1}{2}\left[\frac{-1}{s+1} + \frac{s}{s^2+1} + \frac{1}{s^2+1}\right] \\ &= \frac{2s}{2(s+1)(s^2+1)} \\ &= \frac{s}{(s+1)(s^2+1)}.\end{aligned}$$

Alternatively, we have transform pairs

$$f(t) = e^{-t} \Longleftrightarrow F(s) = \frac{1}{s+1}$$

$$g(t) = \cos t \Longleftrightarrow G(s) = \frac{s}{s^2+1}.$$

Thus,

$$\begin{aligned}\mathcal{L}\{f * g(t)\}(s) &= F(s)G(s) \\ &= \frac{1}{s+1}\cdot\frac{s}{s^2+1} \\ &= \frac{s}{(s+1)(s^2+1)}.\end{aligned}$$

17. The expression

$$\frac{1}{s(s+1)} = \frac{1}{s}\cdot\frac{1}{s+1} = F(s)G(s).$$

We have transform pairs

$$F(s) = \frac{1}{s} \Longleftrightarrow f(t) = 1$$

$$G(s) = \frac{1}{s+1} \Longleftrightarrow g(t) = e^{-t}.$$

Thus,

$$\mathcal{L}^{-1}\left\{\frac{1}{s(s+1)}\right\}(t) = \mathcal{L}^{-1}\{F(s)G(s)\}(t)$$
$$= f * g(t)$$
$$= \int_0^t f(u)g(t-u)\,du$$
$$= \int_0^t e^{-(t-u)}\,du$$
$$= e^{-(t-u)}\Big|_0^t$$
$$= 1 - e^{-t}.$$

19. The expression

$$\frac{1}{(s+1)(s-2)} = \frac{1}{s+1}\cdot\frac{1}{s-2} = F(s)G(s).$$

We have transform pairs

$$F(s) = \frac{1}{s+1} \iff f(t) = e^{-t}$$
$$G(s) = \frac{1}{s-2} \iff g(t) = e^{2t}.$$

Thus,

$$\mathcal{L}^{-1}\left\{\frac{1}{(s+1)(s-2)}\right\}(t)$$
$$= \mathcal{L}^{-1}\{F(s)G(s)\}(t)$$
$$= f * g(t)$$
$$= \int_0^t e^{-u}e^{2(t-u)}\,du$$
$$= -\frac{1}{3}e^{2t-3u}\Big|_0^t$$
$$= \frac{1}{3}(e^{2t} - e^{-t}).$$

21. The expression

$$\frac{s}{(s-1)(s^2+1)} = \frac{1}{s-1}\cdot\frac{s}{s^2+1} = F(s)G(s).$$

We have transform pairs

$$F(s) = \frac{1}{s-1} \iff f(t) = e^{t}$$
$$G(s) = \frac{s}{s^2+1} \iff g(t) = \cos t.$$

Thus,

$$\mathcal{L}^{-1}\left\{\frac{s}{(s-1)(s^2+1)}\right\}(t)$$
$$= \mathcal{L}^{-1}\{F(s)G(s)\}(t)$$
$$= f * g(t)$$
$$= \int_0^t e^{u}\cos(t-u)\,du.$$

Two applications of integration by parts can be used to show that

$$\int e^{u}\cos(t-u)\,du = \frac{1}{2}e^{u}\left[\cos(t-u) - \sin(t-u)\right].$$

Continuing,

$$\mathcal{L}^{-1}\left\{\frac{s}{(s-1)(s^2+1)}\right\}(t)$$
$$= \frac{1}{2}e^{u}\left[\cos(t-u) - \sin(t-u)\right]\Big|_0^t$$
$$= \frac{1}{2}e^{t} - \left(\frac{1}{2}\cos t - \frac{1}{2}\sin t\right)$$
$$= \frac{1}{2}e^{t} - \frac{1}{2}\cos t + \frac{1}{2}\sin t.$$

23. The expression

$$\frac{1}{(s^2+1)^2} = \frac{1}{s^2+1}\cdot\frac{1}{s^2+1} = F(s)G(s).$$

We have transform pairs

$$F(S) = \frac{1}{s^2+1} \iff f(t) = \sin t$$
$$G(s) = \frac{1}{s^2+1} \iff g(t) = \sin t.$$

Thus,

$$\mathcal{L}^{-1}\left\{\frac{1}{(s^2+1)^2}\right\}(t) = \mathcal{L}^{-1}\{F(s)G(s)\}(t)$$
$$= f * g(t)$$
$$= \int_0^t f(u)g(t-u)\,du$$
$$= \int_0^t \sin u\sin(t-u)\,du.$$

A trig product-to-sum identity is useful.

$$\sin A \sin B = \frac{1}{2}\left[\cos(A-B) - \cos(A+B)\right].$$

Continuing,

$$\begin{aligned}
&\mathcal{L}^{-1}\left\{\frac{1}{(s^2+1)^2}\right\}(t) \\
&= \frac{1}{2}\int_0^t (\cos(2u-t) - \cos t)\,du \\
&= \frac{1}{2}\left[\frac{1}{2}\sin(2u-t) - u\cos t\right]\Big|_0^t \\
&= \frac{1}{4}\sin(2u-t) - \frac{1}{2}u\cos t\Big|_0^t \\
&= \left[\frac{1}{4}\sin t - \frac{1}{2}t\cos t\right] - \left[\frac{1}{4}\sin(-t)\right] \\
&= \frac{1}{2}\sin t - \frac{1}{2}t\cos t.
\end{aligned}$$

25. (a) The expression

$$\frac{s\omega}{(s^2+\omega_0^2)(s^2+\omega^2)} = \frac{s}{s^2+\omega_0^2}\cdot\frac{\omega}{s^2+\omega^2} = F(s)G(s).$$

We have transform pairs

$$F(s) = \frac{s}{s^2+\omega_0^2} \Longleftrightarrow f(t) = \cos\omega_0 t$$

$$G(s) = \frac{\omega}{s^2+\omega^2} \Longleftrightarrow g(t) = \sin\omega t.$$

Thus,

$$\begin{aligned}
&\mathcal{L}^{-1}\left\{\frac{s\omega}{(s^2+\omega_0^2)(s^2+\omega^2)}\right\}(t) \\
&= \mathcal{L}^{-1}\{F(s)G(s)\}(t) \\
&= f * g(t) \\
&= \int_0^t f(u)g(t-u)\,du \\
&= \int_0^t \cos\omega_0 u \sin\omega(t-u)\,du \\
&= \int_0^t \sin\omega u \cos\omega_0(t-u)\,du.
\end{aligned}$$

The product-to-sum identity is useful.

$$\sin A \cos B = \frac{1}{2}\left[\sin(A+B) + \sin(A-B)\right].$$

Continuing,

$$\begin{aligned}
&\mathcal{L}^{-1}\left\{\frac{s\omega}{(s^2+\omega_0^2)(s^2+\omega^2)}\right\}(t) \\
&= \frac{1}{2}\int_0^t \sin((\omega-\omega_0)u + \omega_0 t)\,du \\
&\quad + \frac{1}{2}\int_0^t \sin((\omega+\omega_0)u - \omega_0 t)\,du
\end{aligned}$$

If $\omega = \omega_0$, then this is equal to

$$\begin{aligned}
&= \frac{1}{2}\int_0^t \sin\omega_0 t\,du \\
&\quad + \frac{1}{2}\int_0^t \sin(2\omega_0 u - \omega_0 t)\,du \\
&= \frac{1}{2}t\sin\omega_0 t - \frac{1}{4\omega_0}(\cos\omega_0 t - \cos(-\omega_0 t)) \\
&= \frac{1}{2}t\sin\omega_0 t.
\end{aligned}$$

(b) If $\omega \neq \omega_0$, then

$$\begin{aligned}
&f * g(t) \\
&= \frac{-1}{2(\omega-\omega_0)}\cos((\omega-\omega_0)u + \omega_0 t)\Big|_0^t \\
&\quad - \frac{1}{2(\omega+\omega_0)}\cos((\omega+\omega_0)u - \omega_0 t)\Big|_0^t \\
&= \frac{-1}{2(\omega-\omega_0)}[\cos\omega t - \cos\omega_0 t] \\
&\quad - \frac{1}{2(\omega+\omega_0)}[\cos\omega t - \cos(-\omega_0 t)] \\
&= -\frac{1}{2}\left[\frac{1}{\omega-\omega_0} + \frac{1}{\omega+\omega_0}\right] \\
&\quad \times (\cos\omega t - \cos\omega_o t) \\
&= -\frac{1}{2}\cdot\frac{2\omega}{\omega^2-\omega_0^2}(\cos\omega t - \cos\omega_0 t) \\
&= \frac{\omega}{\omega^2-\omega_0^2}(\cos\omega_0 t - \cos\omega t).
\end{aligned}$$

27. We start by computing the impulse response function $e(t)$.

$$e'' + 9e = \delta(t), \qquad e(0) = e'(0) = 0.$$

By Theorem 6.10, the Laplace transform of $e(t)$ is

$$E(s) = \frac{1}{P(s)} = \frac{1}{s^2+9}.$$

Thus,

$$\begin{aligned} e(t) &= \mathcal{L}^{-1}\left\{\frac{1}{s^2+9}\right\}(t) \\ &= \frac{1}{3}\mathcal{L}^{-1}\left\{\frac{3}{s^2+9}\right\}(t) \\ &= \frac{1}{3}\sin 3t. \end{aligned}$$

Compute the derivative.

$$e'(t) = \cos 3t$$

From Theorem 7.16, we know that the solution of

$$y'' + 9y = g(t), \qquad y(0) = -1, y'(0) = 2$$

is

$$\begin{aligned} y(t) &= e * g(t) + ay_0e'(t) \\ &\quad + (ay_1 + by_0)e(t) \\ &= e * g(t) + (1)(-1)e'(t) \\ &\quad + ((1)(2) + (0)(-1))e(t) \\ &= e * g(t) - e'(t) + 2e(t) \\ &= \frac{1}{3}\int_0^t (\sin 3u)g(t-u)\,du \\ &\quad - \cos 3t + \frac{2}{3}\sin 3t. \end{aligned}$$

29. We start by computing the impulse response function $e(t)$.

$$e'' + 4e' + 3e = \delta(t), \qquad e(0) = e'(0) = 0.$$

By Theorem 6.10, the Laplace transform of $e(t)$ is

$$\begin{aligned} E(s) &= \frac{1}{P(s)} \\ &= \frac{1}{s^2+4s+3} \\ &= \frac{1}{(s+1)(s+3)} \\ &= \frac{1/2}{s+1} - \frac{1/2}{s+3}. \end{aligned}$$

Thus,

$$e(t) = \frac{1}{2}e^{-t} - \frac{1}{2}e^{-3t}.$$

Compute the derivative.

$$e'(t) = -\frac{1}{2}e^{-t} + \frac{3}{2}e^{-3t}$$

From Theorem 7.16, we know that the solution of

$$y'' + 4y' + 3y = g(t), \qquad y(0) = -1, y'(0) = 1$$

is

$$\begin{aligned} y(t) &= e * g(t) + ay_0e'(t) + (ay_1 + by_0)e(t) \\ &= e * g(t) + (1)(-1)e'(t) \\ &\quad + ((1)(1) + (4)(-1))e(t) \\ &= e * g(t) - e'(t) - 3e(t) \\ &= \int_0^t \left(\frac{1}{2}e^{-u} - \frac{1}{2}e^{-3u}\right)g(t-u)\,du \\ &\quad + \frac{1}{2}e^{-t} - \frac{3}{2}e^{-3t} - \frac{3}{2}e^{-t} + \frac{3}{2}e^{-3t} \\ &= \int_0^t \left(\frac{1}{2}e^{-u} - \frac{1}{2}e^{-3u}\right)g(t-u)\,du \\ &\quad - e^{-t}. \end{aligned}$$

31. We start by computing the impulse response function $e(t)$.

$$e'' + 4e' + 29e = \delta(t), \qquad e(0) = e'(0) = 0$$

by Theorem 6.10, the Laplace transform of $e(t)$ is

$$\begin{aligned} E(s) &= \frac{1}{P(s)} \\ &= \frac{1}{s^2+4s+29} \\ &= \frac{1}{(s+2)^2+25} \\ &= \frac{1}{5}\cdot\frac{5}{(s+2)^2+25}. \end{aligned}$$

Thus,

$$e(t) = \frac{1}{5}e^{-2t}\sin 5t.$$

Differentiate.

$$e'(t) = \frac{1}{5}e^{-2t}(5\cos 5t) - \frac{2}{5}e^{-2t}\sin 5t$$
$$= e^{-2t}\cos 5t - \frac{2}{5}e^{-2t}\sin 5t.$$

From Theorem 7.16, we know that the solution of

$$y'' + 4y' + 29y = g(t), \qquad y(0) = -1, y'(0) = 2$$

is

$$y(t) = e * g(t) + ay_0e'(t) + (ay_1 + by_0)e(t)$$
$$= e * g(t) + (1)(-1)e'(t) + ((1)(2) + (4)(-1))e(t)$$
$$= e * g(t) - e'(t) - 2e(t)$$
$$= e * g(t) - e^{-2t}\cos 5t + \frac{2}{5}e^{-2t}\sin 5t - \frac{2}{5}e^{-2t}\sin 5t$$
$$= \frac{1}{5}\int_0^t e^{-2u}(\sin 5u)g(t-u)\,du - e^{-2t}\cos 5t.$$

Chapter 6. Numerical Methods

Section 6.1. Euler's Method

1. We have $t_0 = 0$, $y_0 = 1$ and $f(t, y) = t + y$. Thus, the first step of Euler's method is completed as follows.

$$y_1 = y_0 + (t_0 + y_0)\,h = 1 + (0 + 1) \times 0.1 = 1.1$$
$$t_1 = t_0 + h = 0 + 0.1 = 0.1.$$

The second step follows.

$$y_2 = y_1 + (t_1 + y_1)\,h = 1.1 + (0.1 + 1.1) \times 0.1 = 1.22$$
$$t_2 = t_1 + h = 0.1 + 0.1 = 0.2.$$

Continuing in this manner produces the results in the following table.

k	t_k	y_k	$f(t_k, y_k) = y_k$	h	$f(t_k, y_k)h$
0	0.0	1.0000	1.0000	0.1	0.1000
1	0.1	1.1000	1.2000	0.1	0.1200
2	0.2	1.2200	1.4200	0.1	0.1420
3	0.3	1.3620	1.6620	0.1	0.1662
4	0.4	1.5282	1.9282	0.1	0.1928
5	0.5	1.7210	2.2210	0.1	0.2221

3. We have $t_0 = 0$, $y_0 = 1$ and $f(t, y) = ty$. Thus, the first step of Euler's method is completed as follows.

$$y_1 = y_0 + t_0 y_0\,h = 1 + 0 \times 1 \times 0.1 = 1.0$$
$$t_1 = t_0 + h = 0 + 0.1 = 0.1.$$

The second step follows.

$$y_2 = y_1 + t_1 y_1\,h = 1.0 + 0.1 \times 1 \times 0.1 = 1.01$$
$$t_2 = t_1 + h = 0.1 + 0.1 = 0.2.$$

Continuing in this manner produces the results in the following table.

k	t_k	y_k	$f(t_k, y_k) = t_k y_k$	h	$f(t_k, y_k)h$
0	0.0	1.0000	0.0000	0.1	0.0000
1	0.1	1.0000	0.1000	0.1	0.0100
2	0.2	1.0100	0.2020	0.1	0.0202
3	0.3	1.0302	0.3091	0.1	0.0309
4	0.4	1.0611	0.4244	0.1	0.0424
5	0.5	1.1036	0.5518	0.1	0.0552

5. We have $x_0 = 0$, $z_0 = 1$ and $f(x, z) = x - 2z$. Thus, the first step of Euler's method is completed as follows.

$$z_1 = z_0 + (x_0 - 2z_0)\,h = 1 + (0 - 2 \times 1) \times 0.1 = 0.8$$
$$x_1 = x_0 + h = 0 + 0.1 = 0.1.$$

The second step follows.

$$z_2 = z_1 + (x_1 - 2z_1)\,h = 0.8 + (0.1 - 2 \times 0.8) \times 0.1 = 0.65$$
$$x_2 = x_1 + h = 0.1 + 0.1 = 0.2.$$

Continuing in this manner produces the results in the following table.

k	x_k	z_k	$f(x_k, z_k) = x_k - 2z_k$	h	$f(x_k, z_k)h$
0	0.0	1.0000	-2.0000	0.1	-0.2000
1	0.1	0.8000	-1.5000	0.1	-0.1500
2	0.2	0.6500	-1.1000	0.1	-0.1100
3	0.3	0.5400	-0.7800	0.1	-0.0780
4	0.4	0.4620	-0.5240	0.1	-0.0524
5	0.5	0.4096	-0.3192	0.1	-0.0319

7. An integrating factor for $y' + 2xy = x$ is e^{x^2}.

$$e^{x^2}y' + 2xe^{x^2}y = xe^{x^2}$$
$$(e^{x^2}y)' = xe^{x^2}$$

Integrate.

$$e^{x^2}y = \frac{1}{2}e^{x^2} + C$$
$$y = \frac{1}{2} + Ce^{-x^2}$$

The initial condition $y(0) = 8$ produces $C = 15/2$ and $y = 1/2 + (15/2)e^{-x^2}$. In the figure, three numerical solutions and the exact solution are pictured. The numerical solutions were calculated using Euler's method and step sizes $h = 0.2$, $h = 0.1$, and $h = 0.05$.

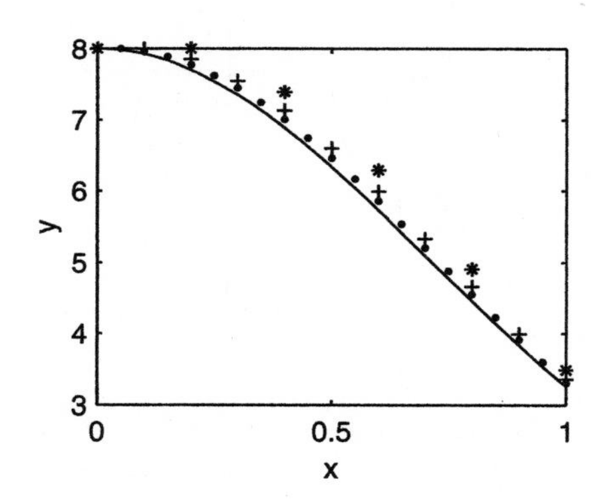

9. The equation $z' = (1 + t)z$ is separable.

$$\frac{dz}{z} = (1 + t)\,dt$$

$$\ln|z| = t + \frac{1}{2}t^2 + C$$

The initial condition $z(0) = -1$ produces $C = 0$ and $\ln|z| = t + (1/2)t^2$. Of course, we choose the negative branch.

$$|z| = e^{t+(1/2)t^2}$$

$$z = -e^{t+(1/2)t^2}$$

In the figure, three numerical solutions and the exact solution are pictured. The numerical solutions were calculated using Euler's method and step sizes $h = 0.2$, $h = 0.1$, and $h = 0.05$.

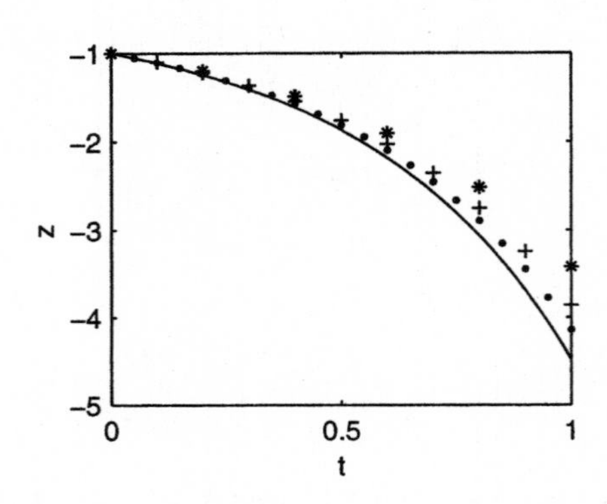

11. (a) Note that a plot of the error versus the step size signifies a linear relationship. Indeed, a line through the origin with the appropriate slope should pass through or close to each data point.

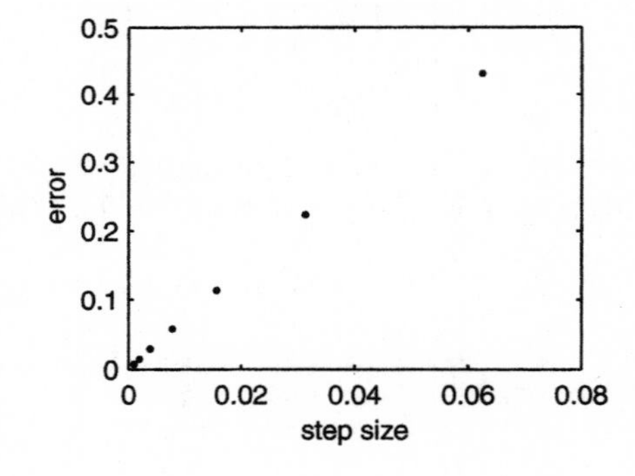

(b) We can estimate the proportionality constant by picking two points from the figure and calculating the slope of the line through the chosen points. Let's use the first and last points.

$$\begin{aligned}\lambda &= \frac{0.0072076631 - 0.4303893417}{0.0009765625 - 0.0625000000}\\ &\approx 6.8784\end{aligned}$$

We can use $E = \lambda h$ to calculate the step size.

$$\begin{aligned}E &= \lambda h\\ h &= \frac{E}{\lambda}\\ h &= \frac{0.001}{6.8784}\\ h &= 1.454 \times 10^{-4}\end{aligned}$$

Since $h = (b - a)/N$,

$$\begin{aligned}N &= \frac{b-a}{h}\\ N &\frac{2-0}{1.454 \times 10^{-4}}\\ N &= 13757.\end{aligned}$$

It will take about 13,757 iterations to achieve the required accuracy.

(c) We ran our Euler routine on a 300 MHz PC using the step size from part (b). The run took approximately 56 seconds and reported an error from the true value of 0.00107417614304.

13. A integrating factor for $y' = y + \sin t$ is e^{-t}. Thus,

$$\begin{aligned}e^{-t}y' - e^{-t}y &= e^{-t}\sin t\\ (e^{-t}y)' &= e^{-t}\sin t.\end{aligned}$$

Integrate, using integration by parts.

$$\begin{aligned}e^{-t}y &= -\frac{1}{2}e^{-t}\sin t - \frac{1}{2}e^{-t}\cos t + C\\ y &= -\frac{1}{2}\sin t - \frac{1}{2}\cos t + Ce^t\end{aligned}$$

The initial condition $y(0) = 1$ provides $C = 3/2$ and the exact solution $y = -(1/2)\sin t - (1/2)\cos t + (3/2)e^t$. Thus, $y(2) \approx 10.83701$. This value, plus our Euler routine, generated the data in the following table.

Step size h	Euler approx.	True value	Error E_h
0.06250	10.07789	10.83701	0.75912
0.03125	10.44404	10.83701	0.39297
0.01563	10.63700	10.83701	0.20001
0.00781	10.73610	10.83701	0.10091
0.00391	10.78632	10.83701	0.05069
0.00195	10.81161	10.83701	0.02540
0.00098	10.82429	10.83701	0.01271

A plot of the error versus the step size reveals a linear relationship.

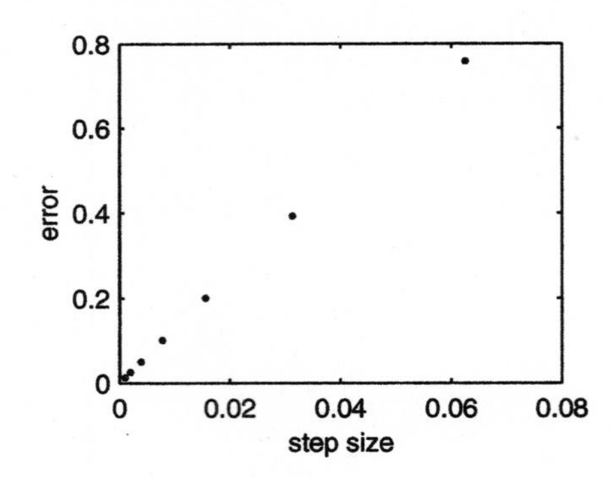

Choose two points and use the slope formula to estimate λ.

$$\lambda \approx \frac{0.01271 - 0.75912}{0.00098 - 0.06250}$$
$$\lambda \approx 12.1321$$

Calculate the step size required to maintain error less than 0.01 with the calculation

$$h = \frac{E}{\lambda}$$
$$h \approx \frac{0.01}{12.1321}$$
$$h \approx 8.2426 \times 10^{-4}.$$

Calculate the number of iterations with

$$N = \frac{b - a}{h}$$
$$N \approx \frac{2 - 0}{8.2426 \times 10^{-4}}$$
$$N \approx 2426.$$

Using a step size $h \approx 8.2426 \times 10^{-4}$, our Euler routine running on a 300 MHz PC produced an error of 0.01073284032773 in about 3.95 seconds.

15. The equation $y' = -(1/3)y^2$ is separable. Thus,

$$\begin{aligned}\frac{dy}{y^2} &= -\frac{1}{3}\,dt\\ -y^{-1} &= -\frac{1}{3}t + C\\ y &= \frac{1}{(1/3)t + C}\\ y &= \frac{3}{t + 3C}.\end{aligned}$$

The initial condition $y(0) = 1$ provides $C = -1$ and the exact solution $y = 3/(t-3)$. Thus, $y(2) = -3$. This value, plus our Euler routine, generated the data in the following table.

Step size h	Euler approx.	True value	Error E_h
0.06250	-2.82040	-3.00000	0.17960
0.03125	-2.90412	-3.00000	0.09588
0.01563	-2.95035	-3.00000	0.04965
0.00781	-2.97472	-3.00000	0.02528
0.00391	-2.98725	-3.00000	0.01275
0.00195	-2.99359	-3.00000	0.00641
0.00098	-2.99679	-3.00000	0.00321

A plot of the error versus the step size reveals a linear relationship.

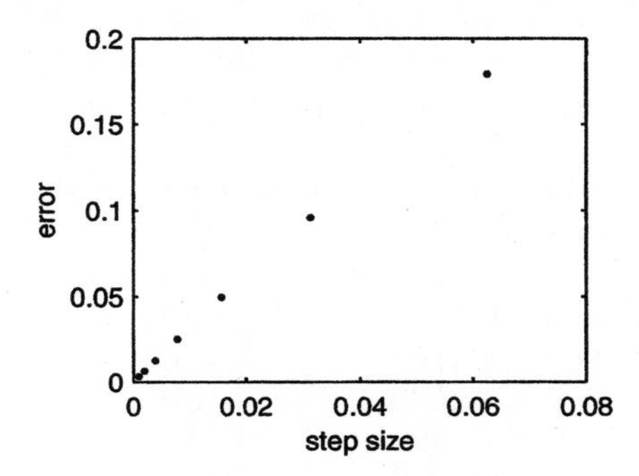

Choose two points and use the slope formula to estimate λ.

$$\begin{aligned}\lambda &\approx \frac{0.00321 - 0.17960}{0.00098 - 0.06250}\\ \lambda &\approx 2.8670\end{aligned}$$

Calculate the step size required to maintain error less than 0.01 with the calculation

$$h = \frac{E}{\lambda}$$
$$h \approx \frac{0.01}{2.8670}$$
$$h \approx 0.003487.$$

Calculate the number of iterations with

$$N = \frac{b-a}{h}$$
$$N \approx \frac{2-0}{0.003487}$$
$$N \approx 573.$$

Using a step size $h \approx 0.003487$, our Euler routine on a 300 MHz PC produced an error of 0.01139175100243 in about 0.66 seconds.

17. We have $t_0 = 0$, $x_0 = 1$, and $y_0 = 0$. We also have $f(t,x,y) = y$ and $g(t,x,y) = -x$. The first step of Euler's method is completed as follows.

$$x_1 = x_0 + y_0 h = 1 + 0 \times 0.1 = 1$$
$$y_1 = y_0 - x_0 h = 0 - 1 \times 0.1 = -0.1$$
$$t_1 = t_0 + h = 0 + 0.1 = 0.1$$

We iterate a second time as follows.

$$x_2 = x_1 + y_1 h = 1 - 1 \times 0.1 = 0.99$$
$$y_2 = y_1 - x_1 h = -0.1 - 1 \times 0.1 = -0.2$$
$$t_2 = t_1 + h = 0.1 + 0.1 = 0.2$$

Continuing in this manner produces the results in the following table.

t_k	x_k	y_k	$f(t_k, x_k, y_k)h = y_k h$	$g(t_k, x_k, y_k)h = -x_k h$
0.0	1.0000	0.0000	0.0000	-0.1000
0.1	1.0000	-0.1000	-0.0100	-0.1000
0.2	0.9900	-0.2000	-0.0200	-0.0990
0.3	0.9700	-0.2990	-0.0299	-0.0970
0.4	0.9401	-0.3960	-0.0396	-0.0940
0.5	0.9005	-0.4900	-0.0490	-0.0901

19. We have $t_0 = 0$, $x_0 = 0$, and $y_0 = -1$. We also have $f(t,x,y) = -2y$ and $g(t,x,y) = x$. The first step of Euler's method is completed as follows.

$$x_1 = x_0 - 2y_0 h = 0 - 2 \times -1 \times 0.1 = 0.2$$
$$y_1 = y_0 + x_0 h = -1 + 0 \times 0.1 = -1$$
$$t_1 = t_0 + h = 0 + 0.1 = 0.1$$

We iterate a second time as follows.

$$x_2 = x_1 - 2y_1h = 0.2 - 2 \times -1 \times 0.1 = 0.4$$
$$y_2 = y_1 + x_1h = -1 + 0.2 \times 0.1 = -0.98$$
$$t_2 = t_1 + h = 0.1 + 0.1 = 0.2$$

Continuing in this manner produces the results in the following table.

t_k	x_k	y_k	$f(t_k, x_k, y_k)h = -2y_kh$	$g(t_k, x_k, y_k)h = x_kh$
0.0	0.0000	-1.0000	-0.1000	0.0000
0.1	0.2000	-1.0000	-0.1000	-0.0200
0.2	0.4000	-0.9800	-0.0980	-0.0400
0.3	0.5960	-0.9400	-0.0940	-0.0596
0.4	0.7840	-0.8804	-0.0880	-0.0784
0.5	0.9601	-0.8020	-0.0802	-0.0960

21. We have $t_0 = 0$, $x_0 = 1$, and $y_0 = -1$. We also have $f(t, x, y) = -y$ and $g(t, x, y) = x + y$. The first step of Euler's method is completed as follows.

$$x_1 = x_0 - y_0h = 1 + 1 \times 0.1 = 1.1$$
$$y_1 = y_0 + (x_0 + y_0)h = -1 + (1 - 1) \times 0.1 = -1$$
$$t_1 = t_0 + h = 0 + 0.1 = 0.1$$

We iterate a second time as follows.

$$x_2 = x_1 - y_1h = 1.1 + 1 \times 0.1 = 1.2$$
$$y_2 = y_1 + (x_1 + y_1)h = -1 + (1.1 - 1) \times 0.1 = -0.99$$
$$t_2 = t_1 + h = 0.1 + 0.1 = 0.2$$

Continuing in this manner produces the results in the following table.

t_k	x_k	y_k	$f(t_k, x_k, y_k)h = -y_kh$	$g(t_k, x_k, y_k)h = (x_k + y_k)h$
0.0	1.0000	-1.0000	0.1000	0.0000
0.1	1.1000	-1.0000	0.1000	0.0100
0.2	1.2000	-0.9900	0.0990	0.0210
0.3	1.2990	-0.9690	0.0969	0.0330
0.4	1.3959	-0.9360	0.0936	0.0460
0.5	1.4895	-0.8900	0.0890	0.0599

23. Pictured below are two plots. The first shows the plots of x versus t (solid line) and y versus t (dashed line) on the time interval $[0, 2\pi]$. The second shows the plot of y versus x. The plots were generated with Euler's method, using a step size $h = 0.05$.

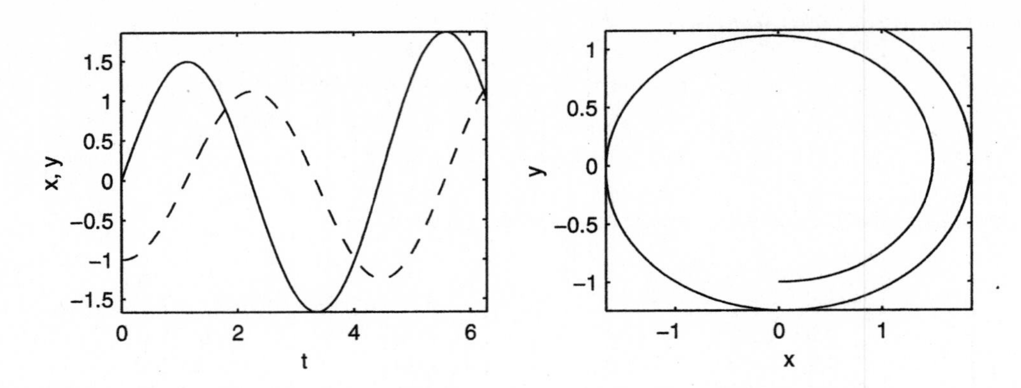

25. Pictured below are two plots. The first shows the plots of x versus t (solid line) and y versus t (dashed line) on the time interval $[0, 2\pi]$. The second shows the plot of y versus x. The plots were generated with Euler's method, using a step size $h = 0.05$.

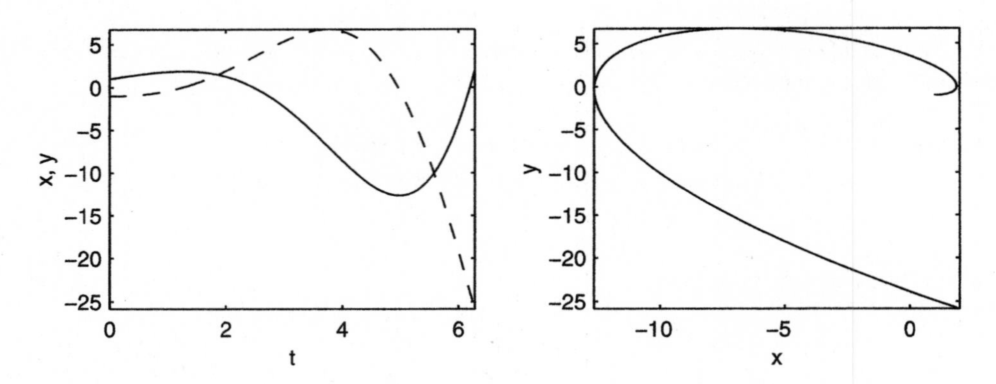

27. The first plot uses step size $h = 0.1$, the second uses $h = 0.01$

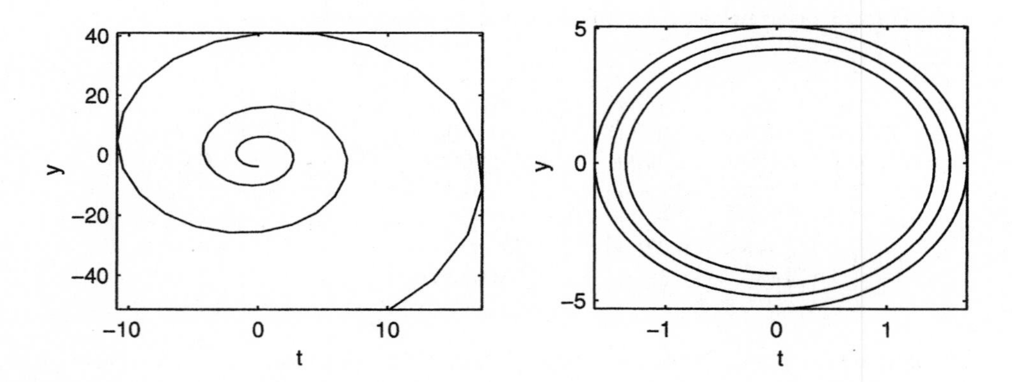

The third plot uses step size $h = 0.001$.

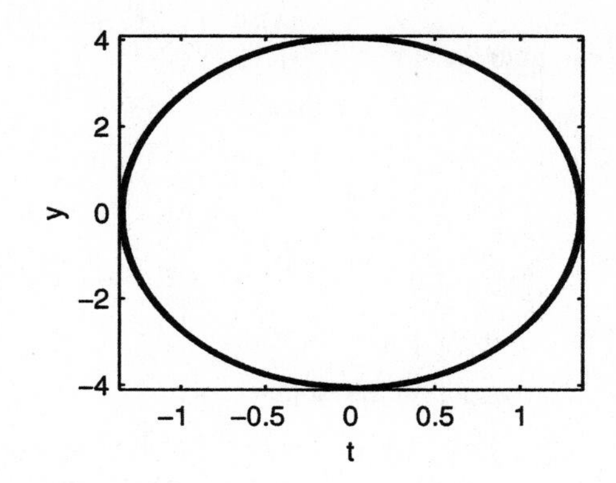

29. The first plot uses step size $h = 0.1$, the second uses $h = 0.01$

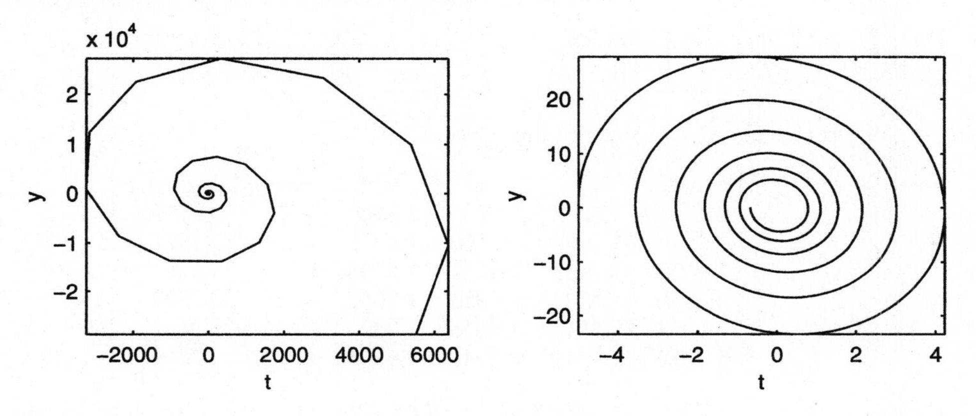

The third plot uses step size $h = 0.001$.

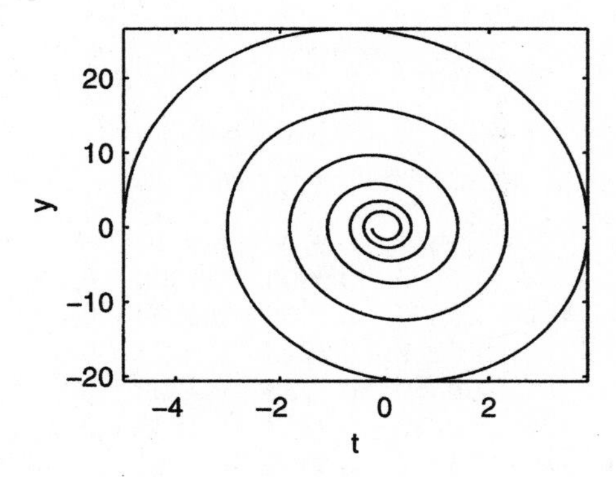

Section 6.2. Runge-Kutta Methods

1. We start with initial condition $t_0 = 0$ and $y_0 = 1$. Also, $f(t, y) = t + y$. The first step of the Runge-Kutta 2 algorithm follows. First we compute the slopes.

$$s_1 = f(t_0, y_0) = f(0, 1) = 0 + 1 = 1$$
$$s_2 = f(t_0 + h, y_0 + hs_1) = f(0.1, 1.1) = 0.1 + 1.1 = 1.2$$

You can now update y and t.

$$y_1 = y_0 + h\frac{s_1 + s_2}{2} = 1 + 0.1\frac{1 + 1.2}{2} = 1.1$$
$$t_1 = t_0 + h = 0 + 0.1 = 0.1$$

The second iteration begins with computing the slopes.

$$s_1 = f(t_1, y_1) = f(0.1, 1.11) = 0.1 + 1.11 = 1.21$$
$$s_2 = f(t_1 + h, y_1 + hs_1) = f(0.2, 1.231) = 0.2 + 1.231 = 1.431$$

You can now update y and t.

$$y_2 = y_1 + h\frac{s_1 + s_2}{2} = 1.11 + 0.1\frac{1.21 + 1.431}{2} = 1.24205$$
$$t_2 = t_1 + h = 0.1 + 0.1 = 0.2$$

Continuing in this manner, we can complete the table.

k	t_k	y_k	s_1	s_2	h	$h(s_1 + s_2)/2$
0	0.0	1.0000	1.0000	1.2000	0.1	0.1100
1	0.1	1.1100	1.2100	1.4310	0.1	0.1321
2	0.2	1.2421	1.4421	1.6863	0.1	0.1564
3	0.3	1.3985	1.6985	1.9683	0.1	0.1833
4	0.4	1.5818	1.9818	2.2800	0.1	0.2131
5	0.5	1.7949	2.2949	2.6244	0.1	0.2460

3. We begin with $t_0 = 0$, $y_0 = 1$ and $f(t, y) = ty$. First, the slopes.

$$s_1 = f(t_0, y_0) = f(0, 1) = 0 \times 1 = 0$$
$$s_2 = f(t_0 + h, y_0 + hs_1) = f(0.1, 1) = 0.1 \times 1 = 0.1$$

Update t and y.

$$y_1 = y_0 + h\frac{s_1 + s_2}{2} = 1 + 0.1\frac{0 + 0.1}{2} = 1.005$$
$$t_1 = t_0 + h = 0 + 0.1 = 0.1$$

Continuing in this manner, we arrive at the following table.

k	t_k	y_k	s_1	s_2	h	$h(s_1 + s_2)/2$
0	0.0	1.0000	0.0000	0.1000	0.1	0.0050
1	0.1	1.0050	0.1005	0.2030	0.1	0.0152
2	0.2	1.0202	0.2040	0.3122	0.1	0.0258
3	0.3	1.0460	0.3138	0.4309	0.1	0.0372
4	0.4	1.0832	0.4333	0.5633	0.1	0.0498
5	0.5	1.1331	0.5665	0.7138	0.1	0.0640

5. We begin with $x_0 = 0$, $z_0 = 1$ and $f(x, z) = x - 2z$. First, the slopes.

$$s_1 = f(x_0, z_0) = f(0, 1) = 0 - 2(1) = -2$$
$$s_2 = f(x_0 + h, z_0 + hs_1) = f(0.1, 0.8) = 0.1 - 2(0.8) = -1.5$$

Update x and z.

$$z_1 = z_0 + h\frac{s_1 + s_2}{2} = 1 + 0.1\frac{-2 - 1.5}{2} = 0.825$$
$$x_1 = x_0 + h = 0 + 0.1 = 0.1$$

Continuing in this manner, we arrive at the following table.

k	x_k	z_k	s_1	s_2	h	$h(s_1 + s_2)/2$
0	0.0	1.0000	-2.0000	-1.5000	0.1	-0.1750
1	0.1	0.8250	-1.5500	-1.1400	0.1	-0.1345
2	0.2	0.6905	-1.1810	-0.8448	0.1	-0.1013
3	0.3	0.5892	-0.8784	-0.6027	0.1	-0.0741
4	0.4	0.5152	-0.6303	-0.4042	0.1	-0.0517
5	0.5	0.4634	-0.4268	-0.2415	0.1	-0.0334

7. An integrating factor is e^x.

$$e^x z' + e^x z = e^x \cos x$$
$$(e^z)' = e^x \cos x$$

Integration is by parts.

$$e^x z = \frac{1}{2}e^x \sin x + \frac{1}{2}e^x \cos x + C$$
$$z = \frac{1}{2}\sin x + \frac{1}{2}\cos x + Ce^{-x}$$

The initial condition $z(0) = 1$ provides $C = 1/2$ and the exact solution $z = (1/2)\sin x + (1/2)\cos x + (1/2)e^{-x}$. In the image that follows, the exact solution is plotted on the interval $[0, 1]$. Further, Euler's method is used to superimpose three additional numeric solutions, using step sizes $h = 0.2$, $h = 0.1$, and $h = 0.05$, respectively.

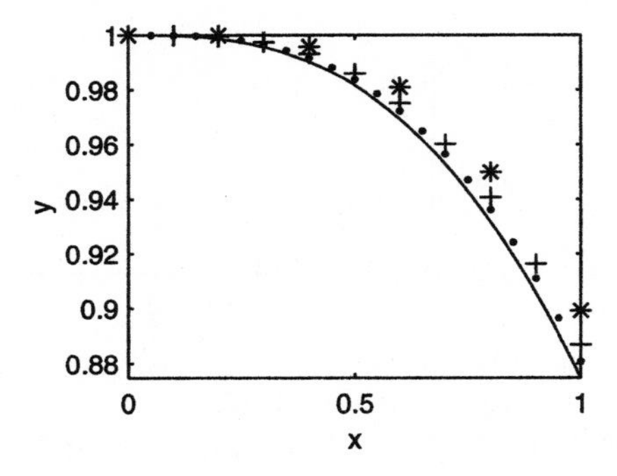

9. An integrating factor is e^x.

$$e^x w' + e^x w = x^2 e^x$$
$$(e^x w)' = x^2 e^x$$

Integrate by parts.

$$e^x w = x^2 e^x - 2xe^x + 2e^x + C$$
$$w = x^2 - 2x + 2 + Ce^{-x}$$

The initial condition $w(0) = 1/2$ provides $C = -3/2$ and the exact solution $w = x^2 - 2x + 2 - (3/2)e^{-x}$. In the image that follows, the exact solution is plotted on the interval [0, 1]. Further, Euler's method is used to superimpose three additional numeric solutions, using step sizes $h = 0.2$, $h = 0.1$, and $h = 0.05$, respectively.

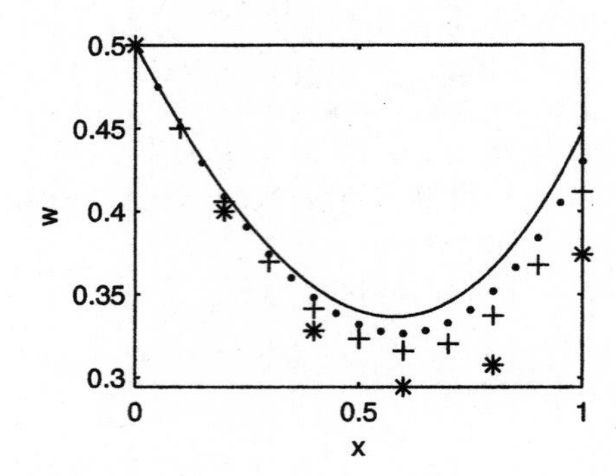

11. The plot of the power function $y = 100x^{-3}$ on [0, 1].

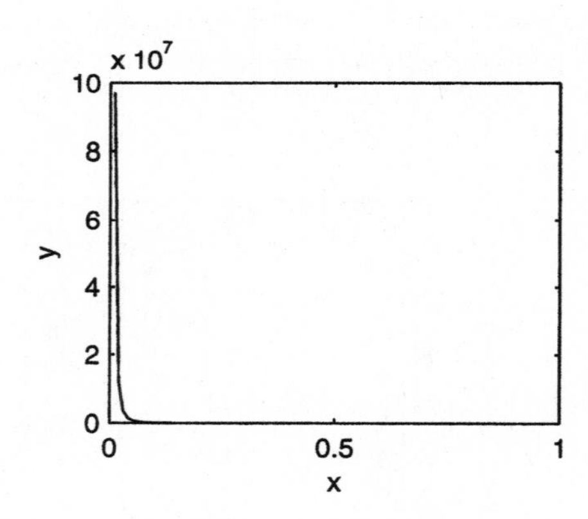

The difficulty most encounter is in circumventing the logarithm of zero. We actually made our plot on [0.01, 1] (close enough).

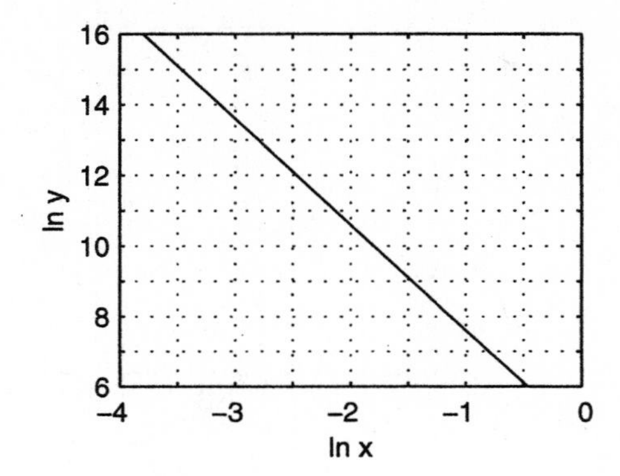

Pick any two points on the line and use the slope formula. We estimate $(-3, 13.6)$ and $(-1, 7.6)$ from the graph, giving a slope of

$$m \approx \frac{7.6 - 13.6}{-1 - (-3))} \approx -3.$$

Note that the slope of the line is identical to the exponent of the power function $y = 100x^{-3}$.

13. The plot of the power function $y = 5x^3$ on $[0, 1]$.

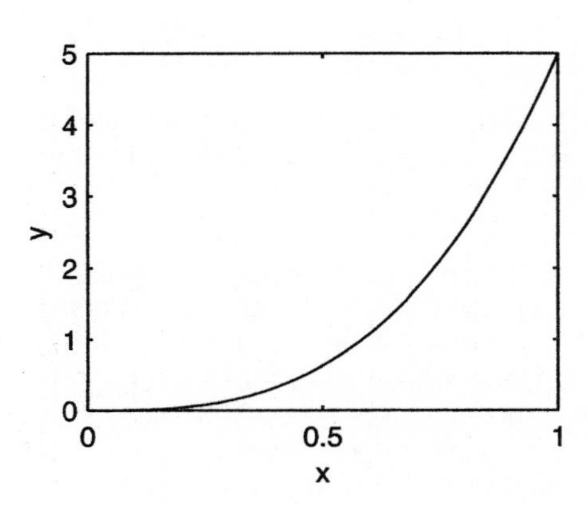

The difficulty most encounter is in circumventing the logarithm of zero. We actually made our plot on $[0.01, 1]$ (close enough).

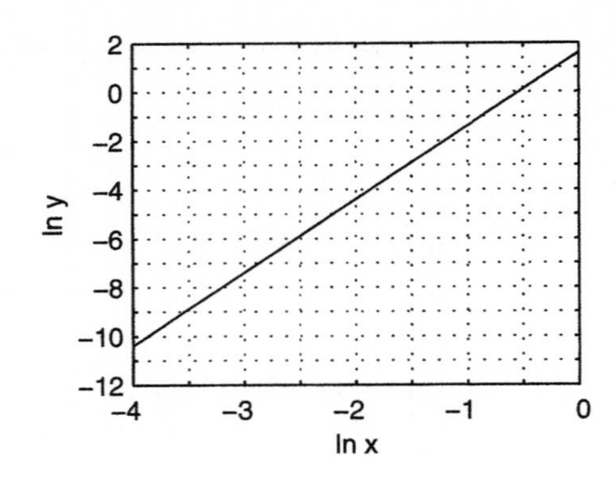

Pick any two points on the line and use the slope formula. We estimate (−3, −7.4) and (−1, −1.4) from the graph, giving a slope of

$$m \approx \frac{-1.4 - (-7.4)}{-1 - (-3))} \approx 3.$$

Note that the slope of the line is identical to the exponent of the power function $y = 5x^3$.

15. An integrating factor is e^t.

$$e^t y' + e^t y = e^t \sin t$$
$$(e^t y)' = e^t \sin t$$

Integrate by parts.

$$e^t y = \frac{1}{2} e^t \sin t - \frac{1}{2} e^t \cos t + C$$
$$y = \frac{1}{2} \sin t - \frac{1}{2} \cos t + Ce^{-t}$$

The initial condition $y(0) = 1$ gives us $C = 3/2$ and the exact solution $y = (1/2)\sin t - (1/2)\cos t + (3/2)e^{-t}$. Thus, the true solution at $t = 2$ is $y(2) \approx .8657250565$. This value allows us to use RK2 and a computer to construct the following table.

Step size h	RK2 approx.	True value	Error E_h
0.06250000	0.86557660	0.86572506	0.00014846
0.03125000	0.86568735	0.86572506	0.00003771
0.01562500	0.86571556	0.86572506	0.00000950
0.00781250	0.86572267	0.86572506	0.00000238
0.00390625	0.86572446	0.86572506	0.00000060
0.00195313	0.86572491	0.86572506	0.00000015
0.00097656	0.86572502	0.86572506	0.00000004

A logarithmic plot of the error in column 4 versus the step size in column 1 reveals the linear relationship between $\ln E_h$ and $\ln h$.

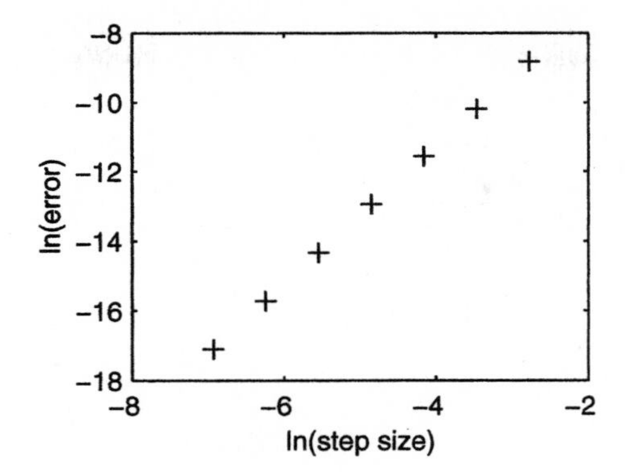

A computer was used to find a linear fit.

$$\ln E_h = 1.9938 \ln h - 3.278$$
$$E_h = e^{-3.278} e^{\ln h^{1.9938}}$$
$$E_h = 0.03770 h^{1.9938}$$

Note that the exponent is approximately 2, which is in keeping with the fact that RK2 is a second order method.

17. The equation is separable.

$$y^2\,dy = t\,dt$$
$$\frac{1}{3}y^2 = \frac{1}{2}t^2 + C$$

The initial condition $y(0) = 1$ gives $C = 1/3$. Thus,

$$\frac{1}{3}y^3 = \frac{1}{2}t^2 + \frac{1}{3}$$
$$y = \sqrt[3]{(3/2)t^2 + 1}$$

Thus, $y(2) \approx 1.91293118$. This value allows us to use RK2 and a computer to construct the following table.

Step size h	RK2 approx.	True value	Error E_h
0.06250000	1.91289601	1.91293118	0.00003517
0.03125000	1.91292139	1.91293118	0.00000979
0.01562500	1.91292861	1.91293118	0.00000257
0.00781250	1.91293053	1.91293118	0.00000066
0.00390625	1.91293102	1.91293118	0.00000017
0.00195313	1.91293114	1.91293118	0.00000004
0.00097656	1.91293117	1.91293118	0.00000001

A logarithmic plot of the error in column 4 versus the step size in column 1 reveals the linear relationship between $\ln E_h$ and $\ln h$.

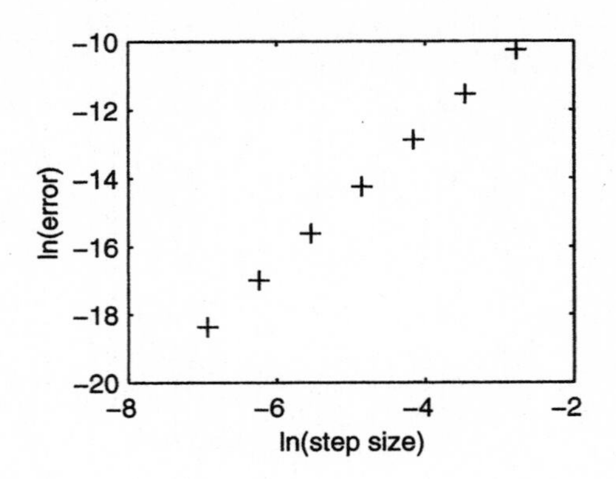

A computer was used to find a linear fit.

$$\ln E_h = 1.958 \ln h - 4.765$$
$$E_h = e^{-4.765} e^{\ln h^{1.958}}$$
$$E_h = 0.008522 h^{1.958}$$

Note that the exponent is approximately 2, which is in keeping with the fact that RK2 is a second order method.

19. We've seen that the error made by RK2 is related to the step size via $E_h = \lambda h^2$. So, if h leads to an error of E_h, then the error associated with step size $h/\sqrt{2}$ is

$$E = \lambda \left(\frac{h}{\sqrt{2}}\right)^2$$
$$E = \lambda \frac{h^2}{2}$$
$$E = \frac{1}{2} \lambda h^2$$
$$E = \frac{1}{2} E_h.$$

Thus, dividing the step size by the square root of 2 halves the error. Of course, if N_h is the number of iterations associated with step size h, then

$$N_h = \frac{b-a}{h}.$$

If we reduce the step size to $h/\sqrt{2}$, then the number of iterations is

$$N = \frac{b-a}{h/\sqrt{2}}$$
$$N = \sqrt{2} \frac{b-a}{h}$$
$$N = \sqrt{2} N_h.$$

Thus, the number of iterations is increased by a factor of $\sqrt{2}$.

This exercise may be better understood in reverse. That is, what happens when we halve the step size? If E_h is the error associated with step size h, then the error associated with step size $h/2$ is

$$E = \lambda\left(\frac{h}{2}\right)^2$$
$$E = \frac{1}{4}\lambda h^2$$
$$E = \frac{1}{4}E_h.$$

Thus, halving the step size quarters the error, which is considerably better than the performance given by Euler's method.

21. We have $t_0 = 0$, $y_0 = 1$ and $f(t, y) = t + y$. First compute four slopes.

$$s_1 = f(t_0, y_0) = f(0, 1) = 1$$
$$s_2 = f(t_0 + h/2, y_0 + hs_1/2) = f(0.05, 1.05) = 1.10$$
$$s_3 = f(t_0 + h/2, y_0 + hs_2/2) = f(0.05, 1.055) = 1.105$$
$$s_4 = f(t_0 + h, y_0 + hs_3) = f(0.1, 1.1105) = 1.2105$$

Update t and y.

$$y_1 = y_0 + h\frac{s_1 + 2s_2 + 2s_3 + s_4}{6}$$
$$= 1 + 0.1\frac{1 + 2(1.10) + 2(1.105) + 1.2105}{6} = 1.110341667$$
$$t_1 = t_0 + h = 0 + 0.1 = 0.1$$

Continuing in this manner produces the following table.

t_k	y_k	s_1	s_2	s_3	s_4	$h\frac{s_1+2s_2+2s_3+s_4}{6}$
0.0	1.0000	1.0000	1.1000	1.1050	1.2105	0.1103
0.1	1.1103	1.2103	1.3209	1.3264	1.4430	0.1325
0.2	1.2428	1.4428	1.5649	1.5711	1.6999	0.1569
0.3	1.3997					

23. We have $x_0 = 0$, $z_0 = 1$ and $f(x, z) = x - 2z$. First compute four slopes.

$$s_1 = f(x_0, z_0) = f(0, 1) = -2$$
$$s_2 = f(x_0 + h/2, z_0 + hs_1/2) = f(0.05, 0.9) = -1.75$$
$$s_3 = f(x_0 + h/2, z_0 + hs_2/2) = f(0.05, 0.9125) = -1.775$$
$$s_4 = f(x_0 + h, z_0 + hs_3) = f(0.1, 0.0.8225) = -1.545$$

Update x and z.

$$z_1 = z_0 + h\frac{s_1 + 2s_2 + 2s_3 + s_4}{6}$$
$$= 1 + 0.1\frac{-2 + 2(-1.75) + 2(-1.775) + (-1, 545)}{6} = 0.8234166667$$
$$x_1 = x_0 + h = 0 + 0.1 = 0.1$$

Continuing in this manner produces the following table.

s x_k	z_k	s_1	s_2	s_3	s_4	$h\frac{s_1+2s_2+2s_3+s_4}{6}$
0.0	1.0000	-2.0000	-1.7500	-1.7750	-1.5450	-0.1766
0.1	0.8234	-1.5468	-1.3422	-1.3626	-1.1743	-0.1355
0.2	0.6879	-1.1758	-1.0082	-1.0250	-0.8708	-0.1019
0.3	0.5860					

25. An integrating factor is e^t.

$$e^t y' + e^t y = e^t \sin t$$
$$(e^t y)' = e^t \sin t$$

Integrate by parts.

$$e^t y = \frac{1}{2}e^t \sin t - \frac{1}{2}e^t \cos t + C$$
$$y = \frac{1}{2}\sin t - \frac{1}{2}\cos t + Ce^{-t}$$

The initial condition $y(0) = 1$ gives us $C = 3/2$ and the exact solution $y = (1/2)\sin t - (1/2)\cos t + (3/2)e^{-t}$. Thus, the true solution at $t = 2$ is $y(2) \approx .8657250565$. This value allows us to use RK4 and a computer to construct the following table.

Step size h	RK4 approx.	True value	Error E_h
1.0000000	0.87036095021	0.86572505654	0.00463589367
0.5000000	0.86580167967	0.86572505654	0.00007662313
0.2500000	0.86572599485	0.86572505654	0.00000093831
0.1250000	0.86572502504	0.86572505654	0.00000003150
0.0625000	0.86572505215	0.86572505654	0.00000000439
0.0312500	0.86572505620	0.86572505654	0.00000000034
0.0156250	0.86572505652	0.86572505654	0.00000000002
0.0078125	0.86572505654	0.86572505654	0.00000000000

A logarithmic plot of the error in column 4 versus the step size in column 1 reveals the linear relationship between $\ln E_h$ and $\ln h$.

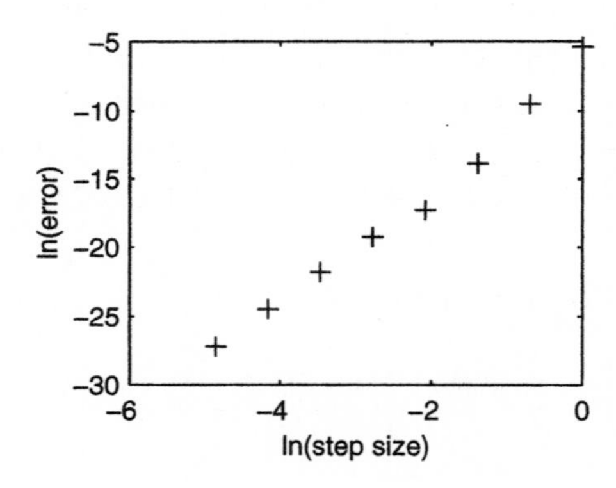

A computer was used to find a linear fit.

$$\begin{aligned} \ln E_h &= 4.352 \ln h - 6.778 \\ E_h &= e^{-6.778} e^{\ln h^{4.352}} \\ E_h &= 0.001138 h^{4.352} \end{aligned}$$

Note that the exponent is approximately 4, which is in keeping with the fact that RK4 is a fourth order method.

27. The equation is separable.

$$\begin{aligned} y^2\,dy &= t\,dt \\ \frac{1}{3}y^2 &= \frac{1}{2}t^2 + C \end{aligned}$$

The initial condition $y(0) = 1$ gives $C = 1/3$. Thus,

$$\begin{aligned} \frac{1}{3}y^3 &= \frac{1}{2}t^2 + \frac{1}{3} \\ y &= \sqrt[3]{(3/2)t^2 + 1} \end{aligned}$$

Thus, $y(2) \approx 1.91293118$. This value allows us to use RK4 and a computer to construct the following table.

Step size h	RK4 approx.	True value	Error E_h
0.25000000	1.91293958	1.91293118	0.00000840
0.12500000	1.91293164	1.91293118	0.00000046
0.06250000	1.91293121	1.91293118	0.00000003
0.03125000	1.91293118	1.91293118	0.00000000
0.01562500	1.91293118	1.91293118	0.00000000
0.00781250	1.91293118	1.91293118	0.00000000
0.00390625	1.91293118	1.91293118	0.00000000

A logarithmic plot of the error in column 4 versus the step size in column 1 reveals the linear relationship between $\ln E_h$ and $\ln h$.

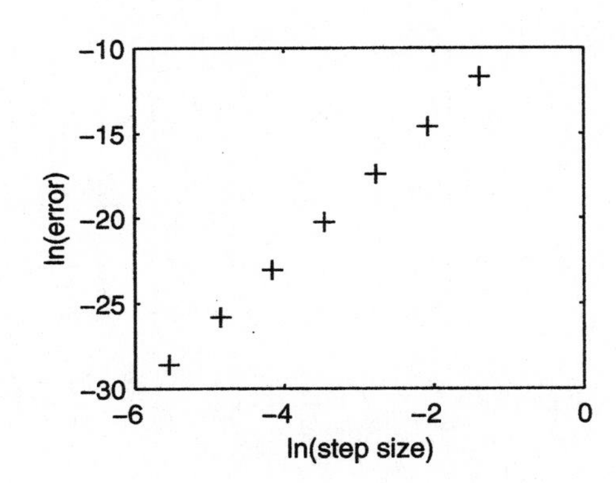

A computer was used to find a linear fit.

$$\ln E_h = 4.055 \ln h - 6.139$$
$$E_h = e^{-6.139} e^{\ln h^{4.055}}$$
$$E_h = 0.002157 h^{4.055}$$

Note that the exponent is approximately 4, which is in keeping with the fact that RK4 is a fourth order method.

29. Let $E_h = \lambda h^4$ be the error associated with the step size h. Now, what will be the error if we halve the step size?

$$E = \lambda \left(\frac{h}{2}\right)^4$$
$$E = \frac{1}{16}\lambda h^4$$
$$E = \frac{1}{16} E_h$$

Thus, having the step size reduces the error to 1/16 of its former size.

Section 6.3. Numerical Error Comparisons

1. The equation is separable.

$$\frac{dx}{x} = \sin t \, dt$$
$$\ln x = -\cos t + C$$

The initial condition $x(0) = 1$ gives us $C = 1$ and $\ln x = 1 - \cos t$. Solving for x, $x = e^{1-\cos t}$. Euler's method, with step sizes $h = 1$, 0.5, 0.1, and 0.05, produces the following result.

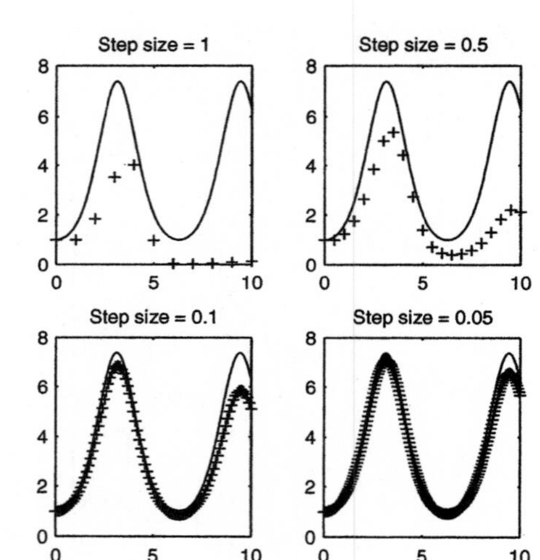

RK2 does a little better.

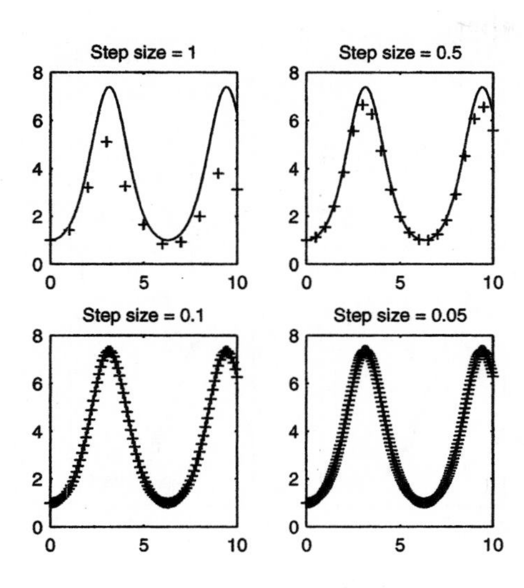

RK4 is the most accurate.

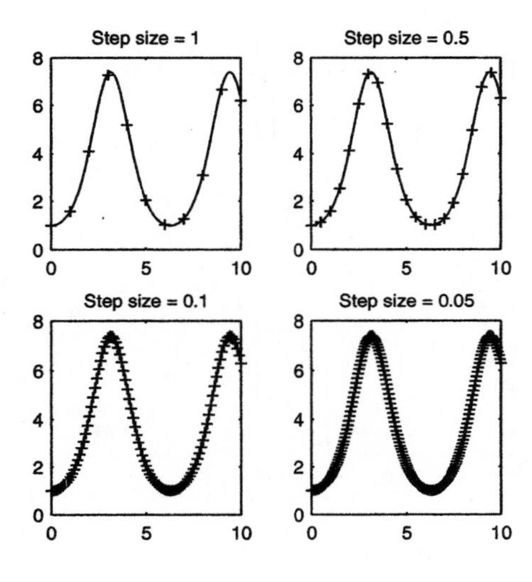

3. An integrating factor is e^t.

$$e^t x' + e^t x = e^t \sin 2t$$
$$(e^t x)' = e^t \sin 2t$$

Integration is by parts.

$$e^t x = \frac{1}{5} e^t \sin 2t - \frac{2}{5} e^t \cos 2t + C$$
$$x = \frac{1}{5} \sin 2t - \frac{2}{5} \cos 2t + ce^{-t}$$

The initial condition $x(0) = 1$ gives us $C = 7/5$ and $x = (1/5)\sin 2t - (2/5)\cos 2t + (7/5)e^{-t}$. Euler's method, with step sizes $h = 1$, 0.5, 0.1, and 0.05, produces the following result.

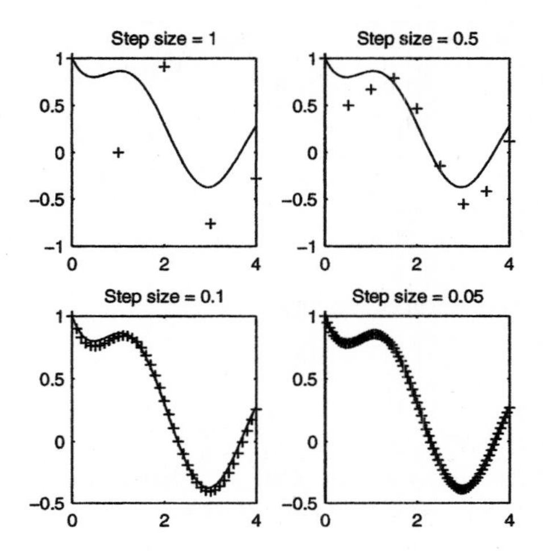

RK2 does a little better.

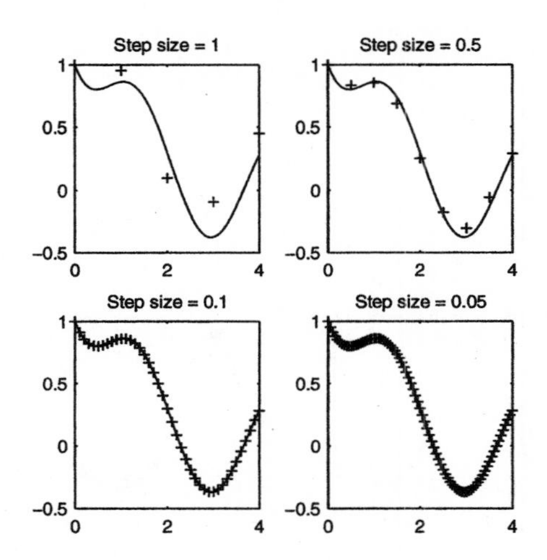

RK4 is the most accurate.

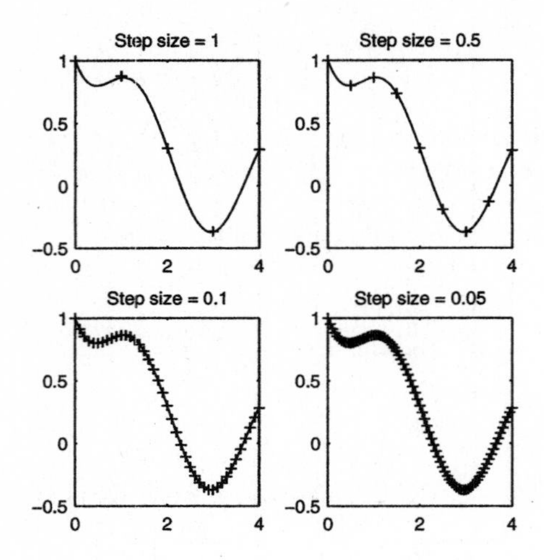

5. The equation is separable.

$$\frac{dx}{1+x} = \cos t \, dt$$
$$\ln(1+x) = \sin t + C$$

The initial condition $x(0) = 0$ gives us $C = 0$ and $\ln(1+x) = \sin t$. Solving for x, $x = -1 + e^{\sin t}$. Euler's method, with step sizes $h = 1$, 0.5, 0.1, and 0.05, produces the following result.

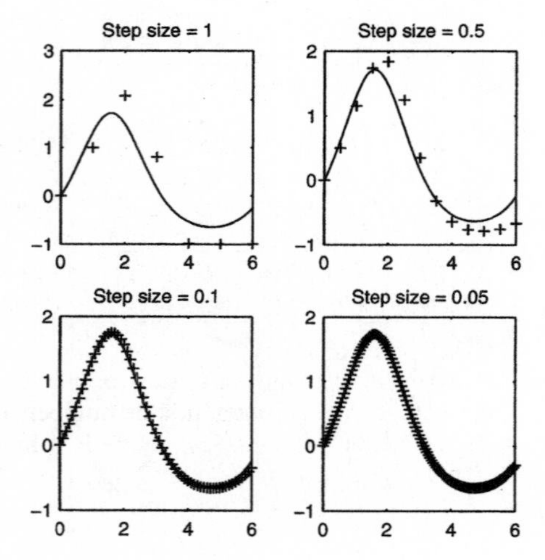

RK2 does a little better.

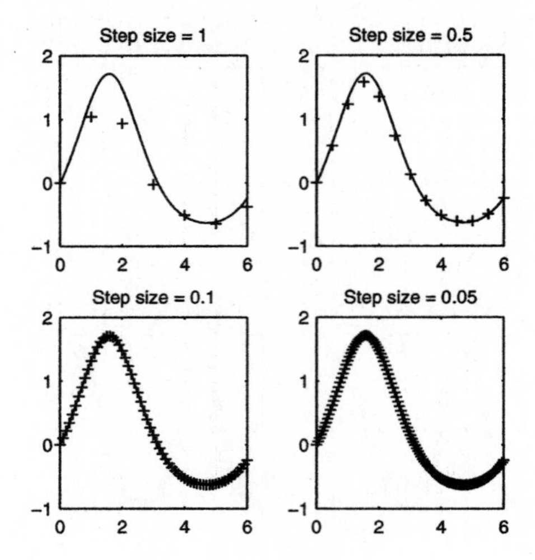

RK4 is the most accurate.

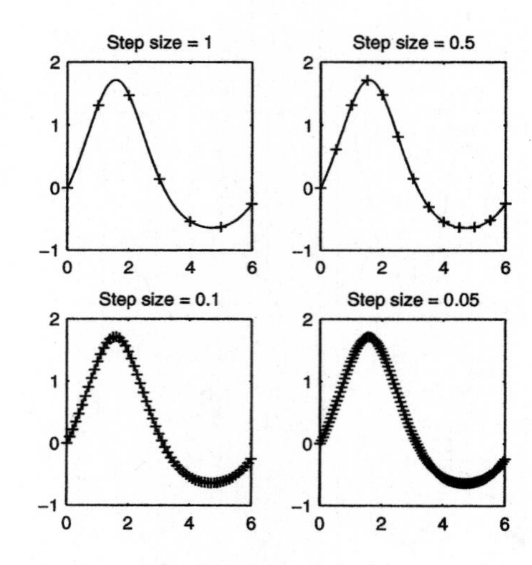

7. The equation is separable.

$$\frac{dx}{x} = dt$$
$$\ln x = t + C$$

The initial condition $x(0) = 1$ gives $C = 0$ and $x = e^t$. The figure that follows shows three numerical solutions on the interval $[0, 4]$ with step size $h = 0.5$. The exact solution is the solid line, the Euler solution is plotted with discrete dots, the RK2 solution is plotted with discrete plus signs, and the RK4 solution is plotted with discrete circles.

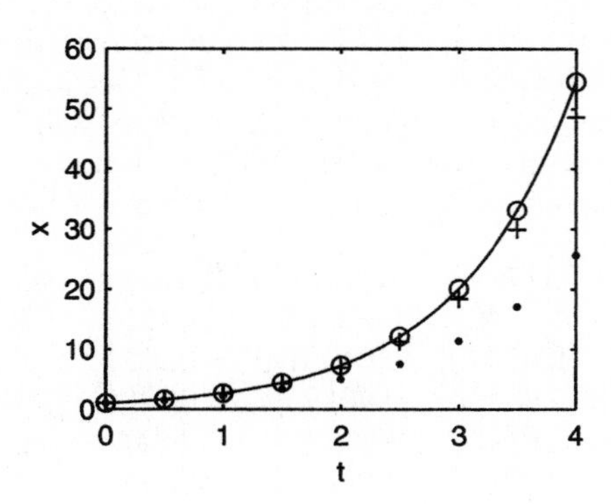

9. The equation is separable.

$$(x + 1)\,dx = \sin t\,dt$$
$$\frac{1}{2}x^2 + x = -\cos t + C$$

The initial condition $x(0) = 0$ gives $C = 1$ and

$$\frac{1}{2}x^2 + x = 1 - \cos t$$
$$x^2 + 2x = 2 - 2\cos t$$
$$(x + 1)^2 = 3 - 2\cos t$$
$$x = -1 \pm \sqrt{3 - 2\cos t}$$

Because $x(0) = 0$, choose $x = -1 + \sqrt{3 - 2\cos t}$. The figure that follows shows three numerical solutions on the interval $[0, 12]$ with step size $h = 1$. The exact solution is the solid line, the Euler solution is plotted with discrete dots, the RK2 solution is plotted with discrete plus signs, and the RK4 solution is plotted with discrete circles.

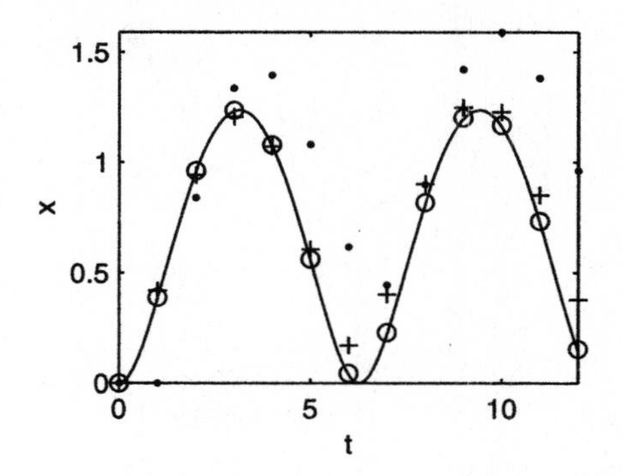

11. (a) First, we complete the entries in the table using a computer.

h	Euler	RK2	RK4
0.10000	2.59374	2.71408	2.71828
0.05000	2.65330	2.71719	2.71828
0.02500	2.68506	2.71800	2.71828
0.01250	2.70148	2.71821	2.71828
0.00625	2.70984	2.71826	2.71828

Next, since we will be plotting the magnitude of the error, we craft a new table with the errors at each step size. The error is calculated by taking the magnitude of the difference of $x(1) = e$ and the solution at $t = 1$ given by the numerical routine.

h	Eul error	RK2 error	RK4 error
0.10000	0.12454	0.00420	0.00000
0.05000	0.06498	0.00109	0.00000
0.02500	0.03322	0.00028	0.00000
0.01250	0.01680	0.00007	0.00000
0.00625	0.00845	0.00002	0.00000

We will work with the internal precision of the numbers in the computer, not the numbers displayed in the above tables. Here is a loglog graph of the Euler error versus the step size.

Note the logarithmic scale on each axis.

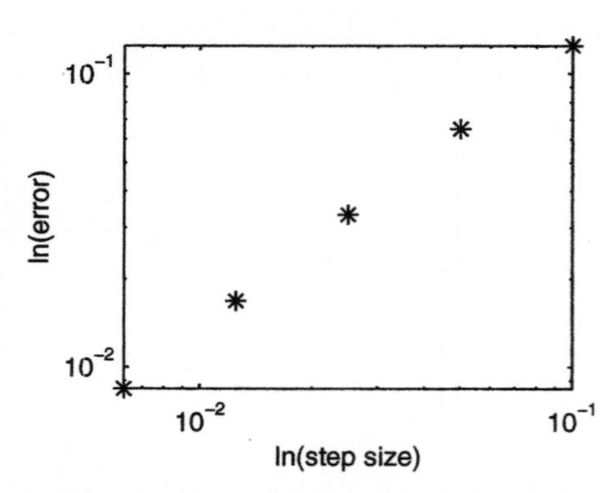

(b) What follows is a plot of the logarithm of the error versus the logarithm of the step size for each numerical method. Note the logarithmic scale on each axis. Euler (solid line), RK2 (dashed), RK4 (dotted).

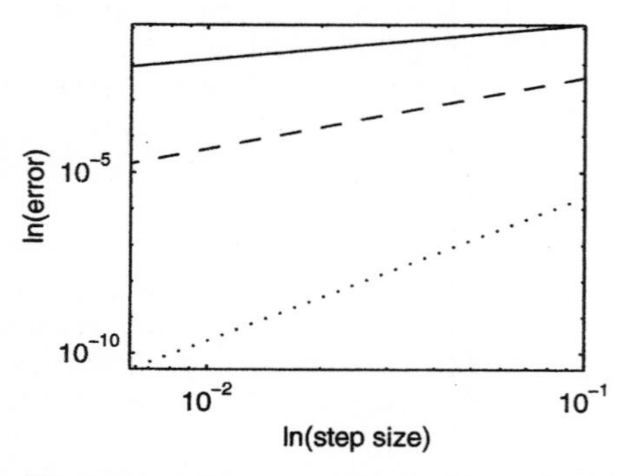

(c) We used a computer to find the slope of each line in part (b). For Euler's method, the slope was reported as 0.9716, which is close to 1, consistent with the fact that the Euler routine is a first order algorithm. The slope of the second line came in at 1.9755, which is close to 2, consistent with the fact that RK2 is a second order method. Finally, the slope of the third line was estimated as 3.9730, which is close to 4, consistent with the fact that RK4 is a fourth order method.

13. The equation is separable.

$$\frac{dx}{x^2} = -t\,dt$$
$$-\frac{1}{x} = -\frac{1}{2}t^2 + C$$
$$\frac{1}{x} = \frac{1}{2}t^2 - C$$

The initial condition $x(0) = 3$ gives $C = -(1/3)$. Thus,

$$\frac{1}{x} = \frac{1}{2}t^2 + \frac{1}{3}$$
$$x = \frac{6}{3t^2+2}.$$

A table is constructed to evaluate the error made at each step size. The error is the magnitude of the difference between $x(2) = 3/7$ and the approximation predicted by the particular method (Eul, RK2, RK4) at $t = 2$. The step size used in constructing the table is $h = 0.1$.

h	Eul error	RK2 error	RK4 error
0.10000	0.02045	0.00220	0.00000
0.05000	0.01001	0.00052	0.00000
0.02500	0.00495	0.00013	0.00000
0.01250	0.00246	0.00003	0.00000
0.00625	0.00123	0.00001	0.00000

Again, we use the internal precision of each number to continue with the analysis. A loglog plot of the error versus step size follows for each numerical method. Euler (solid line), RK2 (dashed), RK4 (dotted).

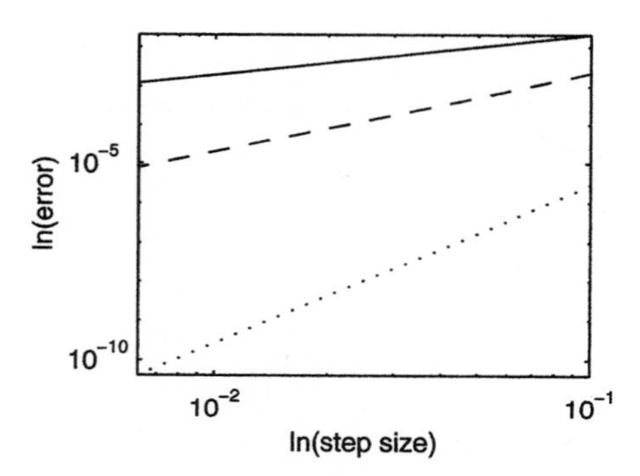

Finally, we used a computer routine to find the slope of each line. Euler: 1.0135, RK2: 2.0303, RK4: 4.0256. Note that each of these numbers is consistent with the order of the particular method.

Section 6.4. Practical Use of Solvers

1. The following figure shows the solution on $[0, 4]$, with step-size $h = 0.1$, and elapsed time $t \approx 0.11$ s. The next figure shows the solution on $[0, 6]$, with step-size $h = 0.01$, and elapsed time $t \approx 1.43$ s. The run was done on a P166 Pentium machine. Answers will vary on different systems.

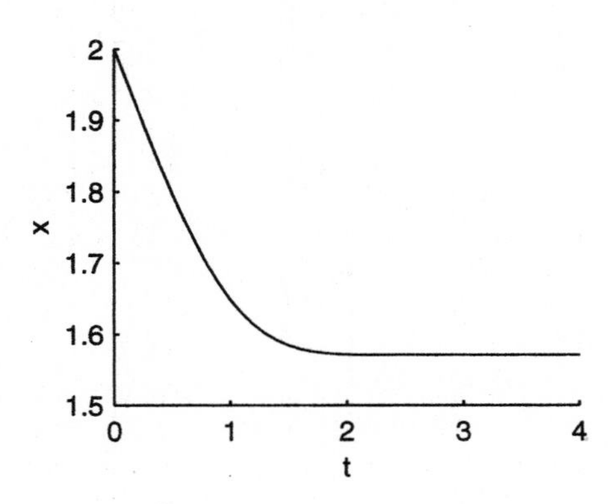

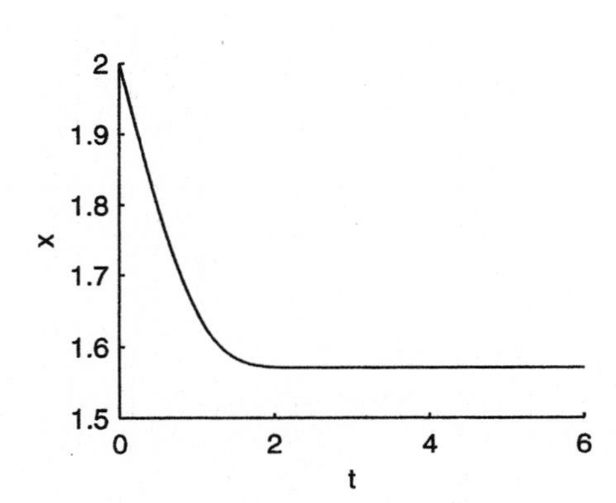

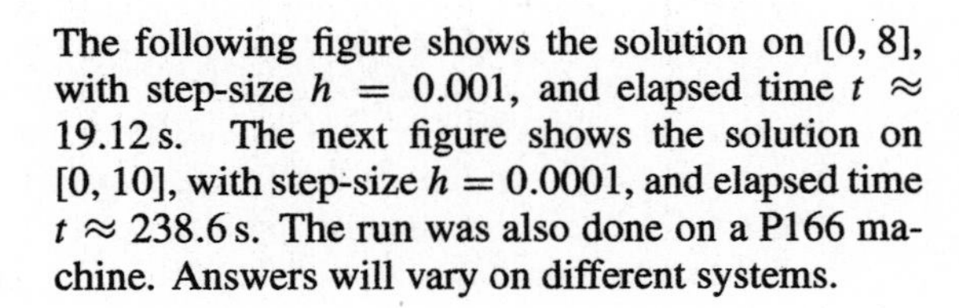
The following figure shows the solution on $[0, 8]$, with step-size $h = 0.001$, and elapsed time $t \approx 19.12$ s. The next figure shows the solution on $[0, 10]$, with step-size $h = 0.0001$, and elapsed time $t \approx 238.6$ s. The run was also done on a P166 machine. Answers will vary on different systems.

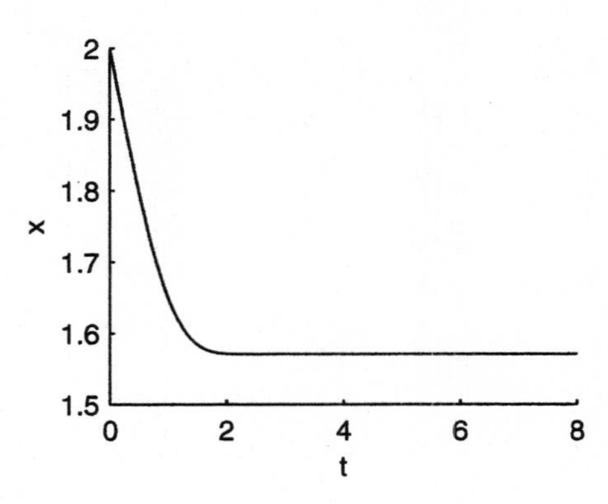

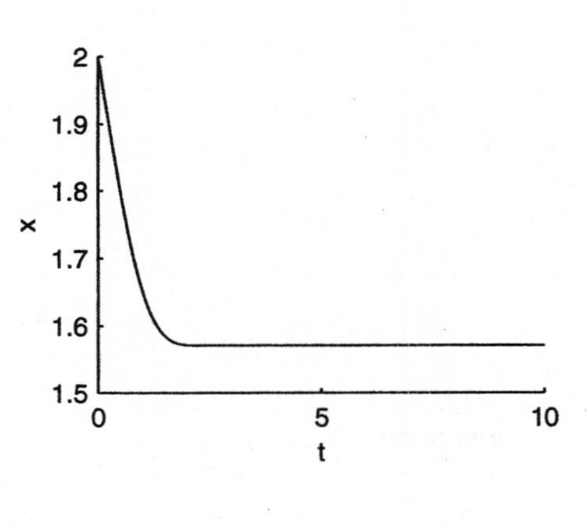

3. The following figure shows the solution on the interval $[0, 4]$. The elapsed time was $t \approx 0.72$ s, the minimum step-size was 0.2420, and the maximum step-size was 0.4232. The next figure shows the solution on $[0, 6]$. The elapsed times was $t \approx 0.22$ s, the minimum step-size was 0.2422, and the maximum step-size was 0.6304.

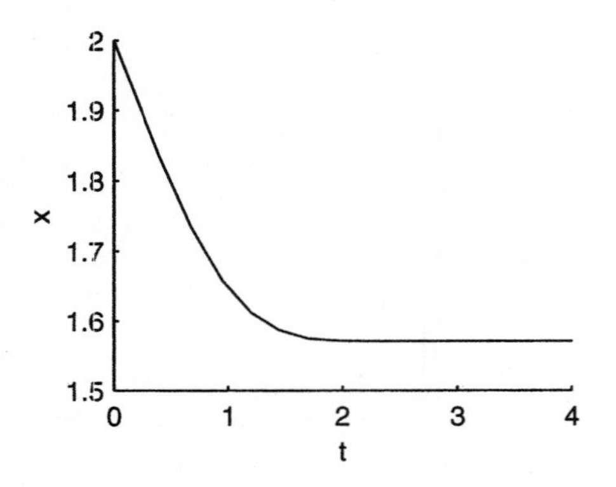

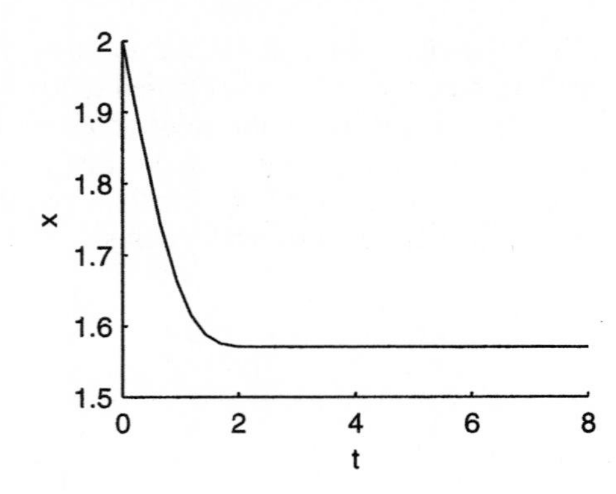

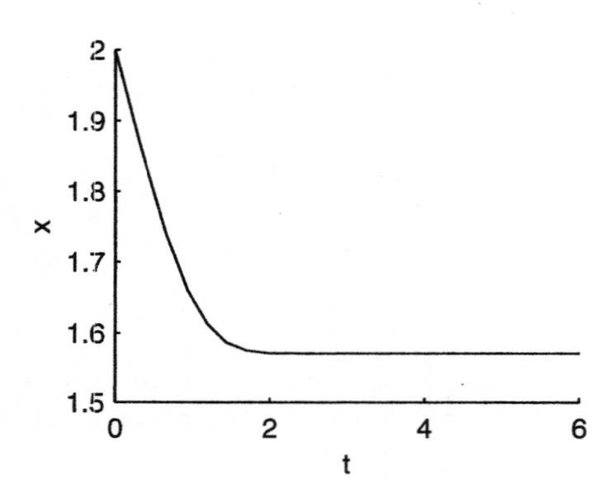

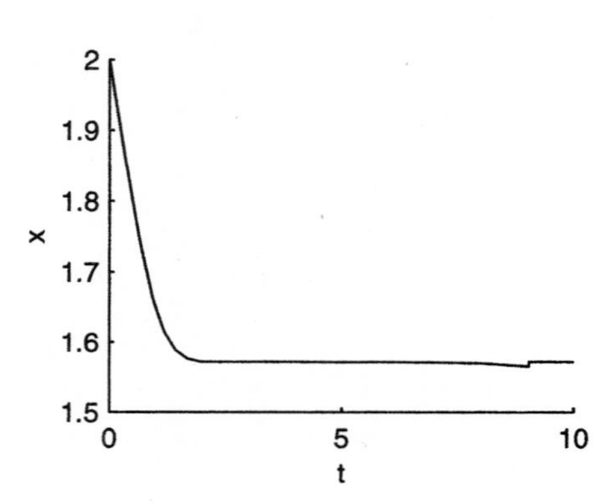

The following figure shows the solution on the interval $[0, 8]$. The elapsed times was $t \approx 0.28$ s. The minimum step-size was 0.1571, and the maximum step-size was 0.8. The next figure shows the solution on $[0, 10]$. The elapsed time was $t \approx 0.44$ s, the minimum step-size was 4.3×10^{-4}, and the maximum step-size was 1. Results were run on a 166 MHz system. Answers will vary on other systems.

5. Te following figure shows the solution for $\mu = 10$ on $I = [0, 20]$. The elapsed time was $t \approx 2.03$ s, the minimum step-size was 0.0051, and the maximum step-size was 0.0889. The next figure shows the solution for $\mu = 50$ on $I = [0, 100]$. The elapsed time was $t \approx 39.27$ s, the minimum step-size was 0.0010, and the maximum step-size was 0.0585.

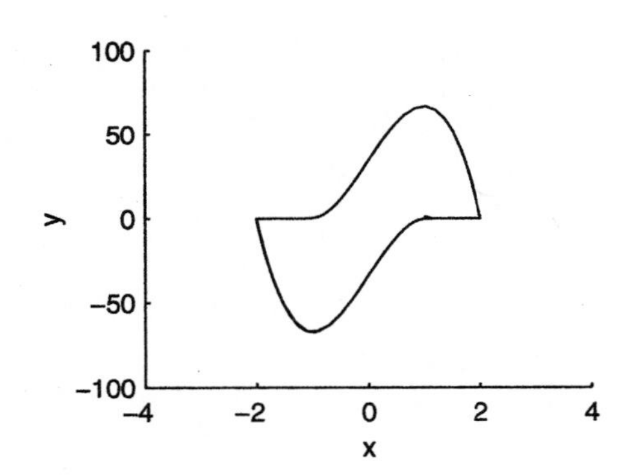

The following figure shows the solution for $\mu = 50$ on $I = [0, 200]$. The elapsed time was $t \approx 187.9$ s, the minimum step-size was 5.3×10^{-4}, and the maximum step-size was 0.0442. Answers will vary on different systems.

7. (a) If we substitute $x(t) = 2e^{-t} + \sin t$ into the left-hand side of the first equation,

$$x' = (2e^{-t} + \sin t)' = -2e^{-t} + \cos t.$$

If we substitute $x(t)$ and $y(t) = 2e^{-t} + \cos t$ into the right-hand side of the first equation,

$$\begin{aligned} &-2x + y + 2\sin t \\ &\quad = -2(2e^{-t} + \sin t) \\ &\qquad + (2e^{-t} + \cos t) + 2\sin t \\ &\quad = -2e^{-t} + \cos t. \end{aligned}$$

Thus, the solution satisfies the first equation. Substituting into the left-hand side of the second equation,

$$y' = (2e^{-t} + \cos t)' = -2e^{-t} - \sin t.$$

Substituting into the right-hand side of the second equation,

$$\begin{aligned} &x - 2y + 2(\cos t - \sin t) \\ &\quad = 2e^{-t} + \sin t - 2(2e^{-t} + \cos t) \\ &\qquad + 2\cos t - 2\sin t \\ &\quad = -2e^{-t} - \sin t. \end{aligned}$$

Thus, the solutions satisfy the second equation. Finally, $x(0) = 2e^0 + \sin 0 = 2$ and $y(0) = 2e^0 + \cos 0 = 3$.

Matlab's `ode45`, a variable-step solver, produced a numerical solution in about 0.22 seconds on a 300 MHz PC. The calculation required about 15,353 floating point operations, which was calculated with Matlab's `flops` command.

Two images follow. The first contains the exact and numerical solution of x versus t. The second contains that of y versus t.

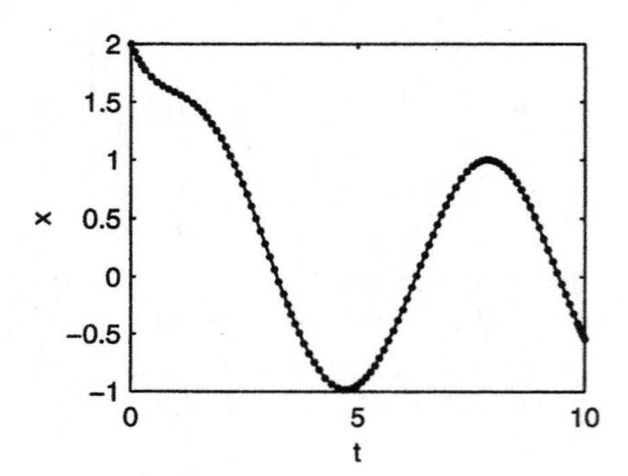

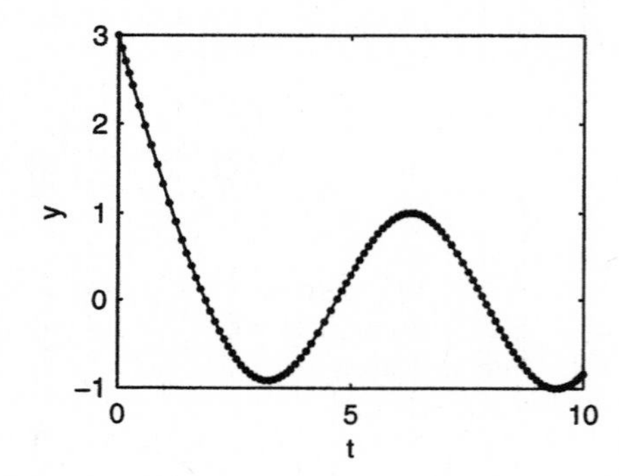

(b) First, we'll show that $x(t) = 2e^{-t} + \sin t$ and $y(t) = 2e^{-t} + \cos t$ are also solutions of the second system. Obviously, because the first equation of the second system is identical to the first equation of the first system, both $x(t)$ and $y(t)$ still satisfy this equation. Now for the second equation. Subbing $y(t)$ into the left-hand side of the second equation,

$$y' = (2e^{-t} + \cos t)' = -2e^{-t} - \sin t.$$

Substituting $x(t)$ and $y(t)$ into the right-hand side of the second equation,

$$\begin{aligned} &-2x + y + 2\sin t \\ &\quad = 998(2e^{-t} + \sin t) - 999(2e^{-t} + \cos t) \\ &\qquad + 999\cos t - 999\sin t \\ &\quad = -2e^{-t} - \sin t. \end{aligned}$$

Thus, the equations satisfy the second system. Finally, both $x(t)$ and $y(t)$ still satisfy the same initial conditions.

The same solver, `ode45`, was used to produce a numerical solution of the second system, same initial conditions, same interval, and, of course, same exact solutions.

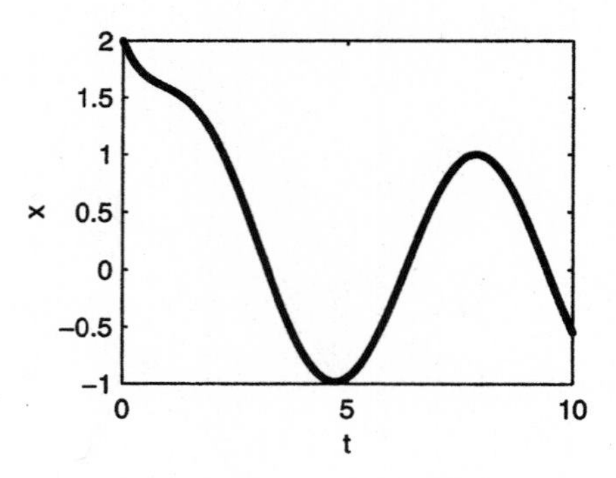

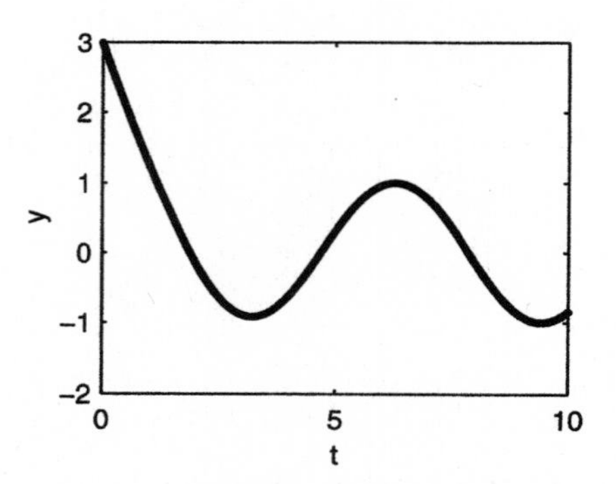

However, the time required elevated to about 12 seconds and the floating point operations required increased to 1,778,814! Two things are apparent. First, the second system is stiff. Secondly, stiffness is not a property of the solution, as both systems in this case have precisely the same solution. Rather, stiffness is some inherent part of the differential equations.

(c) This time we solved the second system with Matlab's `ode23s`, a stiff solver. The performance improved dramatically during the run, which lasted about 2.31 seconds, using approximately 84,703 floating point operations. This considerable savings in resources is evident is the less dense plots of the second system that follow.

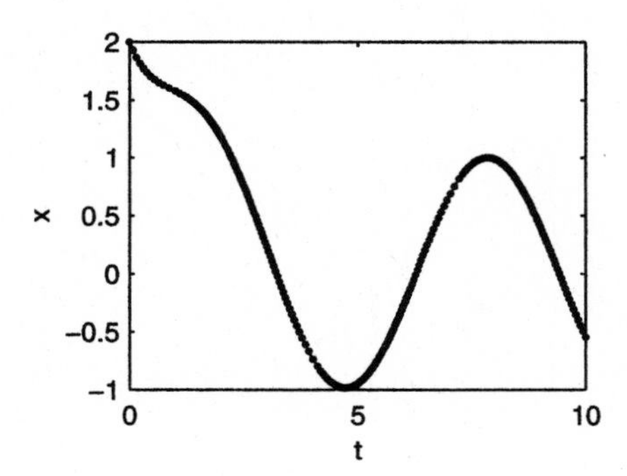

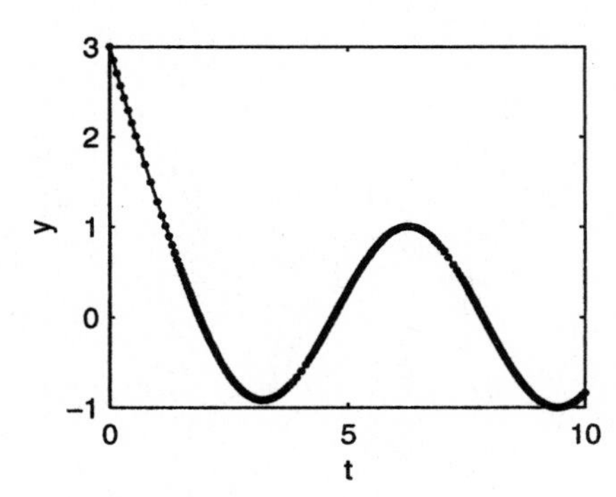

9. The initial value problem $x' = t(x-1), x(-10) = 0$, has solution $x(t) = 1 - e^{(t^2-100)/2}$. Note that $x(10) = 0$. We used Matlab and an RK4 routine to produce numerical solutions at step sizes $h = 0.1$, 0.01, and 0.001. Note that none of these solutions give the correct solution (namely, zero) at $t = 10$.

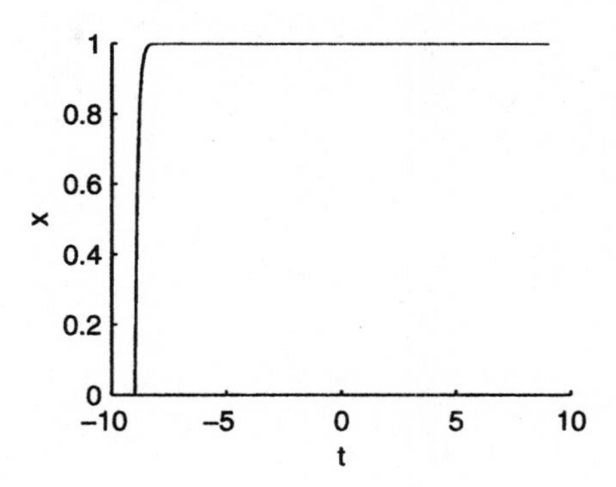

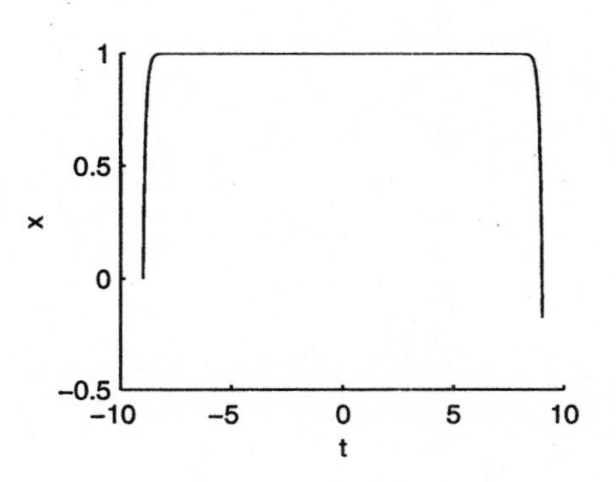

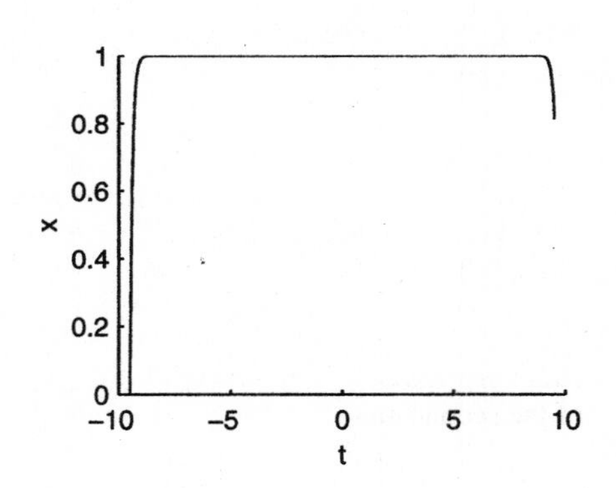

The run time and floating point operations at each step size are summarized in the following table.

Step size	Run Time	Flops
0.1	0.17	6,250
0.01	2.64	62,050
0.001	105.9	620,050

Chapter 7. Matrix Algebra

Section 7.1. Vectors and Matrices

1. The geometric sum

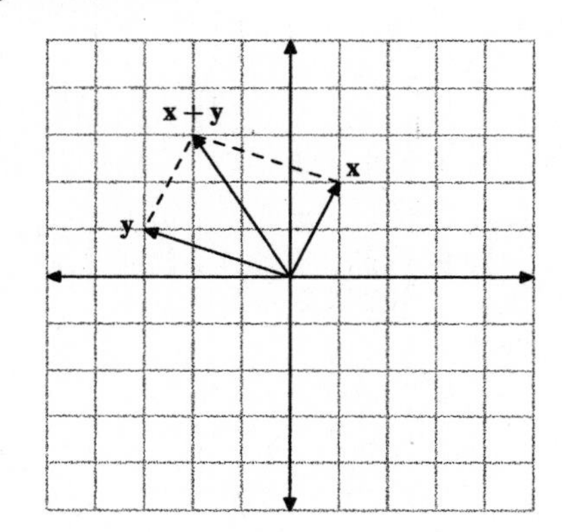

Analytically,

$$\mathbf{x}+\mathbf{y}=\begin{pmatrix}1\\2\end{pmatrix}+\begin{pmatrix}-3\\1\end{pmatrix}=\begin{pmatrix}-2\\3\end{pmatrix}$$

3. The geometric solution

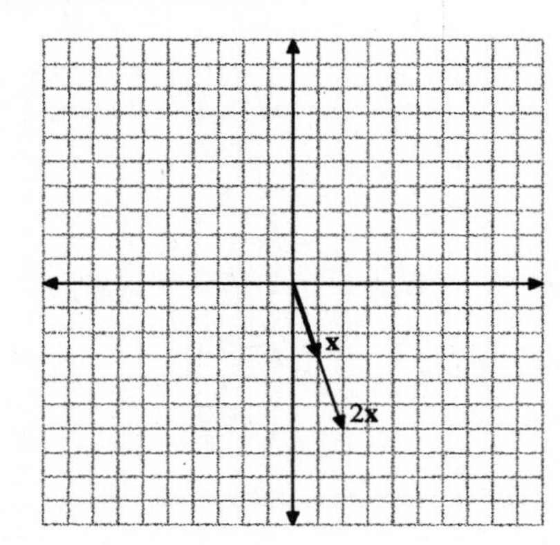

Analytically,

$$2\mathbf{x}=2\begin{pmatrix}1\\-3\end{pmatrix}=\begin{pmatrix}2\\-6\end{pmatrix}$$

5. The geometric solution

$$2\begin{pmatrix}1\\2\end{pmatrix}+1\begin{pmatrix}-3\\4\end{pmatrix}=\begin{pmatrix}-1\\8\end{pmatrix}$$

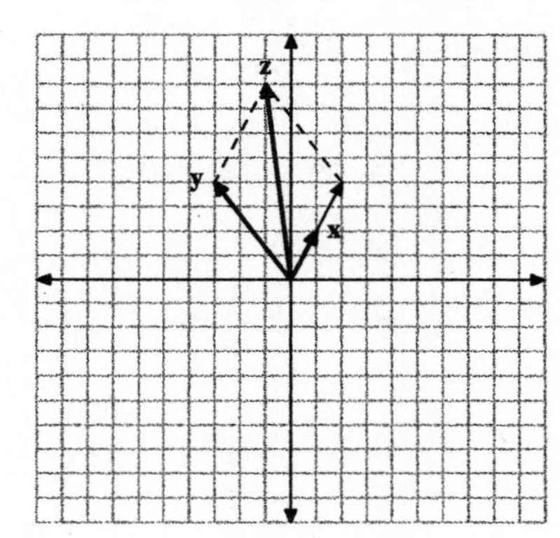

Checking,

$$2\mathbf{x}+1\mathbf{y}=2\begin{pmatrix}1\\2\end{pmatrix}+1\begin{pmatrix}-3\\4\end{pmatrix}=\begin{pmatrix}-1\\8\end{pmatrix}$$

7. The geometric solution

$$-2\begin{pmatrix}1\\3\end{pmatrix}+2\begin{pmatrix}4\\1\end{pmatrix}=\begin{pmatrix}6\\-4\end{pmatrix}$$

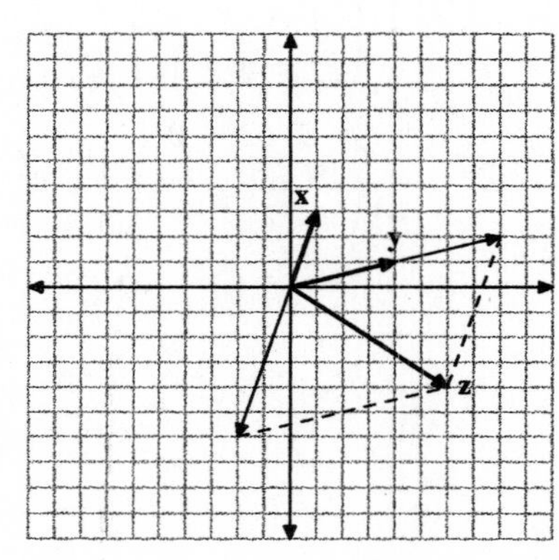

Checking,

$$-2\mathbf{x}+2\mathbf{y}=-2\begin{pmatrix}1\\3\end{pmatrix}+2\begin{pmatrix}4\\1\end{pmatrix}=1\begin{pmatrix}6\\-4\end{pmatrix}$$

9. If you multiply matrix A by $[2, 0, 0]^T$, then

$$A\begin{pmatrix}2\\0\\0\end{pmatrix}=[\mathbf{a}_1,\mathbf{a}_2,\mathbf{a}_3]\begin{pmatrix}2\\0\\0\end{pmatrix}=2\mathbf{a}_1.$$

You can use similar strategies to triple and quadruple the second and third columns, respectively. Thus,

$$\begin{pmatrix}-1&2&4\\0&5&2\\-1&-2&4\end{pmatrix}\begin{pmatrix}2&0&0\\0&3&0\\0&0&4\end{pmatrix}=\begin{pmatrix}-2&6&16\\0&15&8\\-2&-6&12\end{pmatrix}.$$

The columns of AB are

$$A\mathbf{b}_1=[\mathbf{a}_1,\mathbf{a}_2,\mathbf{a}_3]\begin{pmatrix}2\\0\\0\end{pmatrix}=2\mathbf{a}_1,$$

$$A\mathbf{b}_2=[\mathbf{a}_1,\mathbf{a}_2,\mathbf{a}_3]\begin{pmatrix}0\\3\\0\end{pmatrix}=3\mathbf{a}_2,$$

$$A\mathbf{b}_3=[\mathbf{a}_1,\mathbf{a}_2,\mathbf{a}_3]\begin{pmatrix}0\\0\\4\end{pmatrix}=4\mathbf{a}_3.$$

Thus,

$$B=\begin{pmatrix}2&0&0\\0&3&0\\0&0&4\end{pmatrix}.$$

11. First,

$$\begin{aligned}A(a\mathbf{x})&=\begin{pmatrix}-1&2&0\\3&0&-1\\2&2&1\end{pmatrix}\left(-3\begin{pmatrix}1\\-2\\0\end{pmatrix}\right)\\&=\begin{pmatrix}-1&2&0\\3&0&-1\\2&2&1\end{pmatrix}\begin{pmatrix}-3\\6\\0\end{pmatrix}=\begin{pmatrix}15\\-9\\6\end{pmatrix}.\end{aligned}$$

But,

$$\begin{aligned}a(A\mathbf{x})&=-3\left(\begin{pmatrix}-1&2&0\\3&0&-1\\2&2&1\end{pmatrix}\begin{pmatrix}1\\-2\\0\end{pmatrix}\right)\\&=-3\begin{pmatrix}-5\\3\\-2\end{pmatrix}=\begin{pmatrix}15\\-9\\6\end{pmatrix}.\end{aligned}$$

Secondly,

$$\begin{aligned}&A(\mathbf{x}+\mathbf{y})\\&=\begin{pmatrix}-1&2&0\\3&0&-1\\2&2&1\end{pmatrix}\left(\begin{pmatrix}1\\-2\\0\end{pmatrix}+\begin{pmatrix}0\\-1\\4\end{pmatrix}\right),\\&=\begin{pmatrix}-1&2&0\\3&0&-1\\2&2&1\end{pmatrix}\begin{pmatrix}1\\-3\\4\end{pmatrix},\\&=\begin{pmatrix}-7\\-1\\0\end{pmatrix}.\end{aligned}$$

But,

$$\begin{aligned}A\mathbf{x}+A\mathbf{y}&=\begin{pmatrix}-1&2&0\\3&0&-1\\2&2&1\end{pmatrix}\begin{pmatrix}1\\-2\\0\end{pmatrix}\\&\quad+\begin{pmatrix}-1&2&0\\3&0&-1\\2&2&1\end{pmatrix}\begin{pmatrix}0\\-1\\4\end{pmatrix},\\&=\begin{pmatrix}-5\\3\\-2\end{pmatrix}+\begin{pmatrix}-2\\-4\\-2\end{pmatrix}\\&=\begin{pmatrix}-7\\-1\\0\end{pmatrix}.\end{aligned}$$

13. If A is $m \times n$ and $\mathbf{x}, \mathbf{y} \in \mathbf{R}^n$, then

$$\begin{aligned}
A(\mathbf{x}+\mathbf{y}) &= [\mathbf{a}_1, \mathbf{a}_2, \ldots, \mathbf{a}_n] \begin{pmatrix} x_1 + y_1 \\ x_2 + y_2 \\ \vdots \\ x_n + y_n \end{pmatrix} \\
&= (x_1 + y_1)\mathbf{a}_1 + (x_2 + y_2)\mathbf{a}_2 + \ldots \\
&\quad + (x_n + y_n)\mathbf{a}_n \\
&= (x_1\mathbf{a}_1 + x_2\mathbf{a}_2 + \cdots + x_n\mathbf{a}_n) \\
&\quad + (y_1\mathbf{a}_1 + y_2\mathbf{a}_2 + \cdots + y_n\mathbf{a}_n) \\
&= [\mathbf{a}_1\mathbf{a}_2 \ldots \mathbf{a}_n] \begin{pmatrix} x_1 \\ x_2 \\ \vdots \\ x_n \end{pmatrix} \\
&\quad + [\mathbf{a}_1\mathbf{a}_2 \ldots \mathbf{a}_n] \begin{pmatrix} y_1 \\ y_2 \\ \vdots \\ y_n \end{pmatrix} \\
&= A\mathbf{x} + A\mathbf{y}.
\end{aligned}$$

If A is $m \times n$, $a \in \mathbf{R}$, and $\mathbf{x} \in \mathbf{R}^n$, then

$$\begin{aligned}
A(\alpha\mathbf{x}) &= [\mathbf{a}_1\mathbf{a}_2 \ldots \mathbf{a}_n] \begin{pmatrix} \alpha x_1 \\ \alpha x_2 \\ \vdots \\ \alpha x_n \end{pmatrix} \\
&= (\alpha x_1)\mathbf{a}_1 + (\alpha x_2)\mathbf{a}_2 + \cdots + (\alpha x_n)\mathbf{a}_n \\
&= \alpha(x_1\mathbf{a}_1) + \alpha(x_2\mathbf{a}_2) + \cdots + \alpha(x_n\mathbf{a}_n) \\
&= \alpha(x_1\mathbf{a}_1 + x_2\mathbf{a}_2 + \cdots + x_n\mathbf{a}_n) \\
&= \alpha\left([\mathbf{a}_1\mathbf{a}_2 \ldots \mathbf{a}_n] \begin{pmatrix} x_1 \\ x_2 \\ \vdots \\ x_n \end{pmatrix}\right) \\
&= \alpha(A\mathbf{x}).
\end{aligned}$$

15.

$$\begin{aligned}
(A+B)+C &= \begin{pmatrix} 0 & 3 \\ 0 & 1 \end{pmatrix} + \begin{pmatrix} 2 & -3 \\ 1 & 1 \end{pmatrix} \\
&= \begin{pmatrix} 2 & 0 \\ 1 & 2 \end{pmatrix} \\
A+(B+C) &= \begin{pmatrix} 1 & 2 \\ -1 & 0 \end{pmatrix} \\
&\quad + \begin{pmatrix} 1 & -2 \\ 2 & 2 \end{pmatrix} \begin{pmatrix} 2 & 0 \\ 1 & 2 \end{pmatrix}.
\end{aligned}$$

17.

$$\begin{aligned}
A(B+C) &= \begin{pmatrix} 1 & 2 \\ -1 & 0 \end{pmatrix} \begin{pmatrix} 1 & -2 \\ 2 & 2 \end{pmatrix} \\
&= \begin{pmatrix} 5 & 2 \\ -1 & 2 \end{pmatrix} \\
AB + AC &= \begin{pmatrix} 1 & 3 \\ 1 & -1 \end{pmatrix} + \begin{pmatrix} 4 & -1 \\ -2 & 3 \end{pmatrix} \\
&= \begin{pmatrix} 5 & 2 \\ -1 & 2 \end{pmatrix}.
\end{aligned}$$

19.

$$\begin{aligned}
(\alpha\beta)A &= -6\begin{pmatrix} 1 & 2 \\ -1 & 0 \end{pmatrix} = \begin{pmatrix} -6 & -12 \\ 6 & 0 \end{pmatrix} \\
\alpha(\beta A) &= 3\begin{pmatrix} -2 & -4 \\ 2 & 0 \end{pmatrix} = \begin{pmatrix} -6 & -12 \\ 6 & 0 \end{pmatrix}.
\end{aligned}$$

21.

$$\begin{aligned}
(\alpha+\beta)A &= 1\begin{pmatrix} 1 & 2 \\ -1 & 0 \end{pmatrix} = \begin{pmatrix} 1 & 2 \\ -1 & 0 \end{pmatrix} \\
\alpha A + \beta A &= \begin{pmatrix} 3 & 6 \\ -3 & 0 \end{pmatrix} + \begin{pmatrix} -2 & -4 \\ 2 & 0 \end{pmatrix} \\
&= \begin{pmatrix} 1 & 2 \\ -1 & 0 \end{pmatrix}.
\end{aligned}$$

23.

$$(A^T)^T = \begin{pmatrix} -2 & 4 \\ 4 & 0 \end{pmatrix}^T = \begin{pmatrix} -2 & 4 \\ 4 & 0 \end{pmatrix} = A.$$

25.

$$\begin{aligned}
(A+B)^T &= \begin{pmatrix} -2 & -1 \\ 2 & 3 \end{pmatrix}^T = \begin{pmatrix} -2 & 2 \\ -1 & 3 \end{pmatrix} \\
A^T + B^T &= \begin{pmatrix} -2 & 4 \\ 4 & 0 \end{pmatrix} + \begin{pmatrix} 0 & -2 \\ -5 & 3 \end{pmatrix} \\
&= \begin{pmatrix} -2 & 2 \\ -1 & 3 \end{pmatrix}.
\end{aligned}$$

27.

$$\begin{aligned}2\mathbf{x}_1 + 3\mathbf{x}_2 &= 2\begin{pmatrix}9\\5\end{pmatrix} + 3\begin{pmatrix}-6\\-1\end{pmatrix}\\ &= \begin{pmatrix}18\\10\end{pmatrix} + \begin{pmatrix}-18\\-3\end{pmatrix}\\ &= \begin{pmatrix}0\\7\end{pmatrix}\end{aligned}$$

$$\begin{pmatrix}9 & -6\\5 & -1\end{pmatrix}\begin{pmatrix}2\\3\end{pmatrix} = \begin{pmatrix}0\\7\end{pmatrix}$$

29.

$$\begin{aligned}&4\mathbf{x}_2 - 7\mathbf{x}_4 - 3\mathbf{x}_1\\ &= 4\begin{pmatrix}-6\\-1\end{pmatrix} - 7\begin{pmatrix}7\\-9\end{pmatrix} - 3\begin{pmatrix}9\\5\end{pmatrix}\\ &= \begin{pmatrix}-24\\-4\end{pmatrix} + \begin{pmatrix}-49\\63\end{pmatrix} + \begin{pmatrix}-27\\-15\end{pmatrix}\\ &= \begin{pmatrix}-100\\44\end{pmatrix}\end{aligned}$$

$$\begin{pmatrix}-6 & 7 & 9\\-1 & -9 & 5\end{pmatrix}\begin{pmatrix}4\\-7\\-3\end{pmatrix} = \begin{pmatrix}-100\\44\end{pmatrix}$$

31.

$$\begin{aligned}&4\mathbf{v}_1 - 3\mathbf{v}_4 + 3\mathbf{v}_2 + 4\mathbf{v}_3\\ &= 4\begin{pmatrix}10\\-5\\3\end{pmatrix} - 3\begin{pmatrix}-1\\3\\6\end{pmatrix} + 3\begin{pmatrix}0\\8\\6\end{pmatrix} + 4\begin{pmatrix}0\\-9\\7\end{pmatrix}\\ &= \begin{pmatrix}40\\-20\\12\end{pmatrix} + \begin{pmatrix}3\\-9\\-18\end{pmatrix} + \begin{pmatrix}0\\24\\18\end{pmatrix} + \begin{pmatrix}0\\-36\\28\end{pmatrix}\\ &= \begin{pmatrix}43\\-41\\40\end{pmatrix}\end{aligned}$$

33.

$$A\mathbf{y} = \begin{pmatrix}9 & -6\\5 & -1\end{pmatrix}\begin{pmatrix}3\\-2\end{pmatrix} = \begin{pmatrix}39\\17\end{pmatrix}.$$

35.

$$B\mathbf{z} = \begin{pmatrix}6 & -1 & 7\\-1 & 8 & -9\end{pmatrix}\begin{pmatrix}-1\\0\\2\end{pmatrix} = \begin{pmatrix}20\\-17\end{pmatrix}.$$

37.

$$C\mathbf{z} = \begin{pmatrix}10 & 0 & -1\\-5 & 8 & 3\\3 & 6 & 6\end{pmatrix}\begin{pmatrix}-1\\0\\2\end{pmatrix} = \begin{pmatrix}-12\\11\\9\end{pmatrix}.$$

39.

$$C\mathbf{v}_3 = \begin{pmatrix}10 & 0 & -1\\-5 & 8 & 3\\3 & 6 & 6\end{pmatrix}\begin{pmatrix}0\\-9\\7\end{pmatrix} = \begin{pmatrix}-7\\-51\\-12\end{pmatrix}.$$

41.

$$\mathbf{y}^T = \begin{pmatrix}3\\-2\end{pmatrix}^T = (3 \quad -2).$$

43.

$$A^T = \begin{pmatrix}9 & -6\\5 & -1\end{pmatrix}^T = \begin{pmatrix}9 & 5\\-6 & -1\end{pmatrix}.$$

45.

$$C^T = \begin{pmatrix}10 & 0 & -1\\-5 & 8 & 3\\3 & 6 & 6\end{pmatrix}^T = \begin{pmatrix}10 & -5 & 3\\0 & 8 & 6\\-1 & 3 & 6\end{pmatrix}.$$

47. $3x + 4y = 7$ is equivalent to $(3 \quad 4)\begin{pmatrix}x\\y\end{pmatrix} = (7)$.

49.

$$\begin{aligned}3x + 4y &= 7\\ -x + 3y &= 2\end{aligned}$$

is equivalent to

$$\begin{pmatrix}3 & 4\\-1 & 3\end{pmatrix}\begin{pmatrix}x\\y\end{pmatrix} = \begin{pmatrix}7\\2\end{pmatrix}.$$

51.

$$\begin{aligned}-x_1 + x_3 &= 0\\ 2x_1 + 3x_2 &= 3\end{aligned}$$

is equivalent to

$$\begin{pmatrix}-1 & 0 & 1\\2 & 3 & 0\end{pmatrix}\begin{pmatrix}x_1\\x_2\\x_3\end{pmatrix} = \begin{pmatrix}0\\3\end{pmatrix}.$$

53.

$$
\begin{aligned}
-x_1 + x_3 &= 0 \\
2x_1 + 3x_2 &= 3 \\
x_2 - x_3 &= 4
\end{aligned}
$$

is equivalent to

$$
\begin{pmatrix} -1 & 0 & 1 \\ 2 & 3 & 0 \\ 0 & 1 & -1 \end{pmatrix} \begin{pmatrix} x_1 \\ x_2 \\ x_3 \end{pmatrix} = \begin{pmatrix} 0 \\ 3 \\ 4 \end{pmatrix}.
$$

Section 7.2. Systems of Linear Equations with Two or Three Variables

1. Let $(x, y)^T$ represent a solution, the solve $3x - 4y = 12$ for x, obtaining $x = (4y + 12)/3$. Hence,

$$
\begin{pmatrix} x \\ y \end{pmatrix} = \begin{pmatrix} (4y + 12)/3 \\ y \end{pmatrix} = \begin{pmatrix} 4 \\ 0 \end{pmatrix} + y \begin{pmatrix} 4/3 \\ 1 \end{pmatrix}.
$$

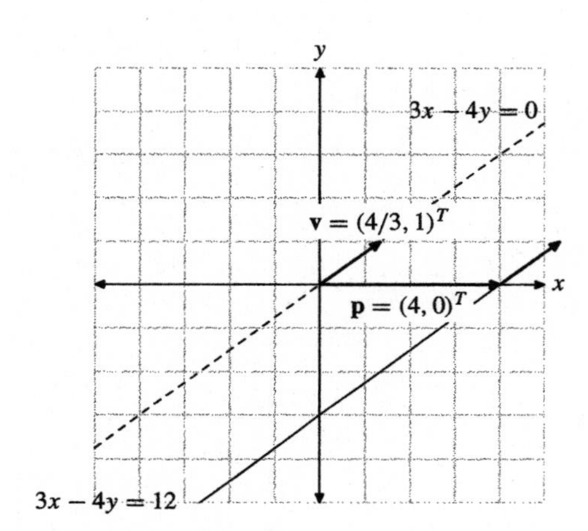

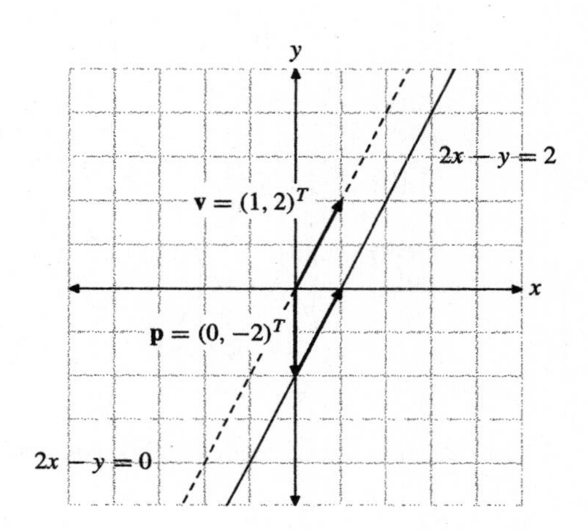

3. Let $(x, y)^T$ represent a solution, the solve $2x - y = 2$ for y, obtaining $y = 2x - 2$. Hence,

$$
\begin{pmatrix} x \\ y \end{pmatrix} = \begin{pmatrix} x \\ 2x - 2 \end{pmatrix} = \begin{pmatrix} 0 \\ -2 \end{pmatrix} + x \begin{pmatrix} 1 \\ 2 \end{pmatrix}.
$$

5. The plot of the system shows a single point of intersection, approximately $(4.3, 0.8)$.

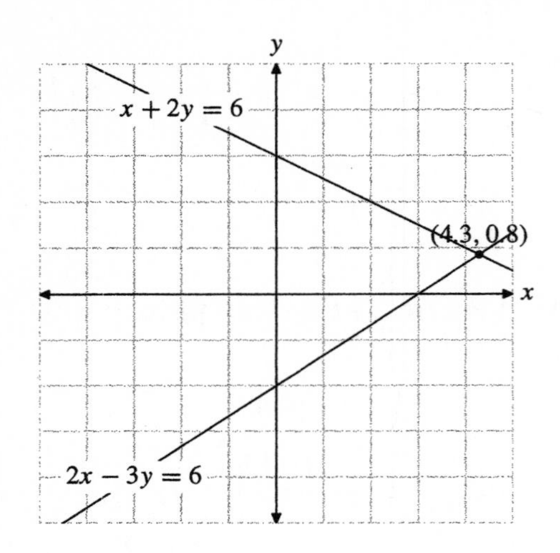

The augmented matrix for the system $x + 2y = 6$, $2x - 3y = 6$ is

$$\begin{pmatrix} 1 & 2 & 6 \\ 2 & -3 & 6 \end{pmatrix}.$$

Add -2 times the first row to the second row to obtain

$$\begin{pmatrix} 1 & 2 & 6 \\ 0 & -7 & -6 \end{pmatrix}.$$

Solve the second equation for y.

$$\begin{aligned} -7y &= -6 \\ y &= \frac{6}{7} \end{aligned}$$

Substitute $y = 6/7$ in the first equation, $x + 2y = 6$, then solve for x.

$$\begin{aligned} x + 2\left(\frac{6}{7}\right) &= 6 \\ x &= \frac{30}{7} \end{aligned}$$

Hence, the solution is $(30/7, 6/7) \approx (4.28, 0.85)$.

7. The plot of the system shows a single point of intersection, approximately $(3, 0)$.

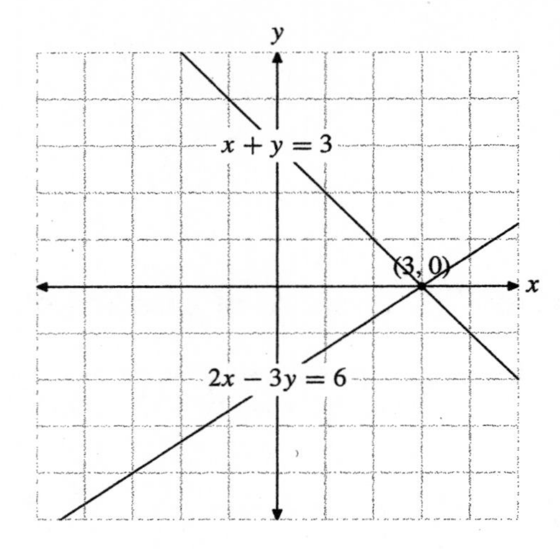

The augmented matrix for the system $x + y = 3$, $2x - 3y = 6$ is

$$\begin{pmatrix} 1 & 1 & 3 \\ 2 & -3 & 6 \end{pmatrix}.$$

Add -2 times the first row to the second row to obtain

$$\begin{pmatrix} 1 & 1 & 3 \\ 0 & -5 & 0 \end{pmatrix}.$$

Solve the second equation for y.

$$\begin{aligned} -5y &= 0 \\ y &= 0 \end{aligned}$$

Substitute $y = 0$ in the first equation, $x + y = 3$, then solve for x.

$$\begin{aligned} x + 0 &= 3 \\ x &= 3 \end{aligned}$$

Hence, the solution is $(3, 0)$.

9. The lines having equations $x + y = 3$ and $2x + 2y = -4$ are parallel, as seen in the following figure.

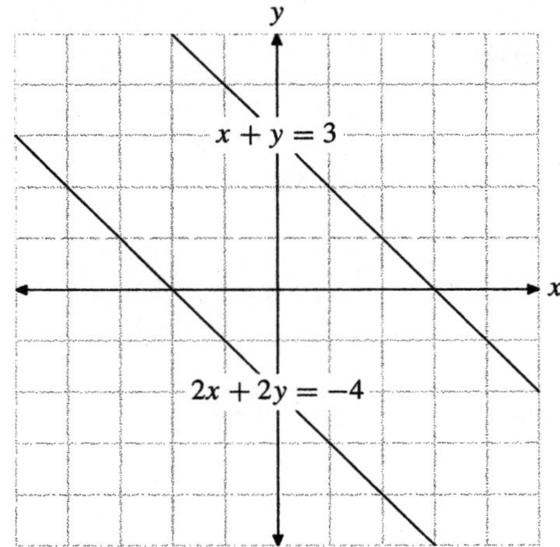

The system has no solutions. The system is inconsistent.

11. It is easy to see that the second equation is a multiple of the first. Thus, the lines having equations $x + 2y = 4$ and $3x + 6y = 12$ are the same line, as seen in the following figure.

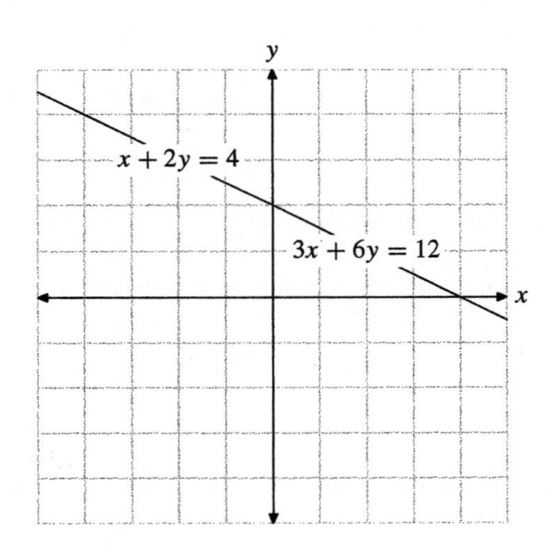

Any point on the line is a solution. The system is consistent.

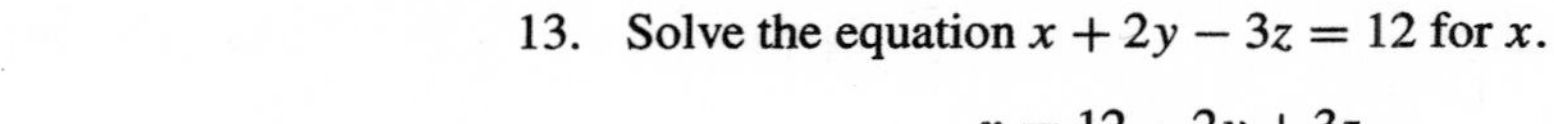

13. Solve the equation $x + 2y - 3z = 12$ for x.

$$x = 12 - 2y + 3z$$

Thus,

$$\begin{aligned} \begin{pmatrix} x \\ y \\ z \end{pmatrix} &= \begin{pmatrix} 12 - 2y + 3z \\ y \\ z \end{pmatrix} \\ &= \begin{pmatrix} 12 \\ 0 \\ 0 \end{pmatrix} + y \begin{pmatrix} -2 \\ 1 \\ 0 \end{pmatrix} + z \begin{pmatrix} 3 \\ 0 \\ 1 \end{pmatrix}. \end{aligned}$$

15. In this case, it is easier to solve the equation $3x - 4y + z = 12$ for z.

$$z = 12 - 3x + 4y$$

Thus,

$$\begin{aligned} \begin{pmatrix} x \\ y \\ z \end{pmatrix} &= \begin{pmatrix} x \\ y \\ 12 - 3x + 4y \end{pmatrix} \\ &= \begin{pmatrix} 0 \\ 0 \\ 12 \end{pmatrix} + x \begin{pmatrix} 1 \\ 0 \\ -3 \end{pmatrix} + y \begin{pmatrix} 0 \\ 1 \\ 4 \end{pmatrix}. \end{aligned}$$

17. The augmented matrix for the system is

$$\begin{pmatrix} 1 & 2 & -3 & 6 \\ 2 & 5 & 8 & 40 \end{pmatrix}$$

Subtract 2 times the first row from the second row.

$$\begin{aligned} x + 2y - \ 3z &= 6 \\ y + 14z &= 28 \end{aligned} \qquad \begin{pmatrix} 1 & 2 & -3 & 6 \\ 0 & 1 & 14 & 28 \end{pmatrix}$$

Note that z is free. Let $z = t$ in the second equation and solve for y.

$$\begin{aligned} y + 14t &= 28 \\ y &= 28 - 4t \end{aligned}$$

Sub $z = t$ and $y = 28 - 14t$ in the first equation and solve for x.

$$\begin{aligned} x + 2y - 3z &= 6 \\ x + 2(28 - 14t) - 3t &= 6 \\ x &= 31t - 50 \end{aligned}$$

Hence,

$$\begin{pmatrix} x \\ y \\ z \end{pmatrix} = \begin{pmatrix} 31t - 50 \\ 28 - 14t \\ t \end{pmatrix} = \begin{pmatrix} -50 \\ 28 \\ 0 \end{pmatrix} + t\begin{pmatrix} 31 \\ -14 \\ 1 \end{pmatrix}.$$

19. The augmented matrix for the system is

$$\begin{pmatrix} 2 & -4 & 5 & 40 \\ -4 & 10 & -4 & 20 \end{pmatrix}$$

Add 2 times the first row to the second row.

$$\begin{aligned} 2x - 4y + 5z &= 40 \\ 2y + 6z &= 100 \end{aligned} \qquad \begin{pmatrix} 2 & -4 & 5 & 40 \\ 0 & 2 & 6 & 100 \end{pmatrix}$$

Note that z is free. Let $z = t$ in the second equation and solve for y.

$$\begin{aligned} 2y + 6t &= 100 \\ y &= 50 - 3t \end{aligned}$$

Sub $z = t$ and $y = 50 - 3t$ in the first equation and solve for x.

$$\begin{aligned} 2x - 4y + 5z &= 40 \\ 2x - 4(50 - 3t) + 5t &= 40 \\ x &= 120 - \frac{17}{2}t \end{aligned}$$

Hence,

$$\begin{pmatrix} x \\ y \\ z \end{pmatrix} = \begin{pmatrix} 120 - (17/2)t \\ 50 - 3t \\ t \end{pmatrix} = \begin{pmatrix} 120 \\ 50 \\ 0 \end{pmatrix} + t\begin{pmatrix} -17/2 \\ -3 \\ 1 \end{pmatrix}.$$

21. The augmented matrix for the system is

$$\begin{pmatrix} 1 & 1 & 1 & 3 \\ 2 & -1 & 1 & 4 \\ -1 & 2 & 2 & 6 \end{pmatrix}.$$

Subtract 2 times row 1 from row 2. Add row 1 to row 3.

$$\begin{pmatrix} 1 & 1 & 1 & 3 \\ 0 & -3 & -1 & -2 \\ 0 & 3 & 3 & 9 \end{pmatrix}$$

Add row 2 to row 3.

$$\begin{aligned} x + y + z &= 3 \\ -3y - z &= -2 \\ 2z &= 7 \end{aligned} \qquad \begin{pmatrix} 1 & 1 & 1 & 3 \\ 0 & -3 & -1 & -2 \\ 0 & 0 & 2 & 7 \end{pmatrix}$$

Solve the last equation for z.

$$\begin{aligned} 2z &= 7 \\ z &= \frac{7}{2} \end{aligned}$$

Substitute $z = 7/2$ in the second equation and solve for y.

$$\begin{aligned} -3y - \frac{7}{2} &= -2 \\ y &= -\frac{1}{2} \end{aligned}$$

Substitute $y = -1/2$ and $z = 7/2$ in the first equation and solve for x.

$$\begin{aligned} x - \frac{1}{2} + \frac{7}{2} &= 3 \\ x &= 0 \end{aligned}$$

Hence, the solution is $(x, y, z) = (0, -1/2, 7/2)$.

23. The augmented matrix for the system is

$$\begin{pmatrix} 1 & -2 & 5 & 10 \\ -2 & 1 & 1 & 12 \\ 2 & -1 & -2 & 6 \end{pmatrix}$$

Add 2 times row 1 to row 2. Subtract 2 times row 1 from row 3.

$$\begin{pmatrix} 1 & -2 & 5 & 10 \\ 0 & -3 & 11 & 32 \\ 0 & 3 & -12 & -14 \end{pmatrix}$$

Add row 2 to row 3.

$$\begin{aligned} x - 2y + 5z &= 10 \\ -3y + 11z &= 32 \\ -z &= 18 \end{aligned} \qquad \begin{pmatrix} 1 & -2 & 5 & 10 \\ 0 & -3 & 11 & 32 \\ 0 & 0 & -1 & 18 \end{pmatrix}$$

Solve the last equation for z.

$$\begin{aligned} -z &= 18 \\ z &= -18 \end{aligned}$$

Substitute $z = -18$ in the second equation and solve for y.

$$\begin{aligned} -3y + 11(-18) &= 32 \\ y &= -\frac{230}{3} \end{aligned}$$

Substitute $y = -230/3$ and $z = -18$ in the first equation and solve for x.

$$\begin{aligned} x - 2\left(-\frac{230}{3}\right) + 5(-18) &= 10 \\ x &= -\frac{160}{3} \end{aligned}$$

Hence, the solution is $(x, y, z) = (-160/3, -230/3, -18)$.

25. No. There are only three possible solution sets: a point, a line, or no solutions. A circle is none of these.

27. The set $S = \left\{ \begin{pmatrix} t \\ 0 \end{pmatrix} : t > 0 \right\}$ in the right half plane in $\mathbf{R}^2$, not including the vertical axis. This set is not a point or a line, so it cannot be the solution set of a system of linear equations in two unknowns.

29. As $y = t\begin{pmatrix} 2 \\ -3 \end{pmatrix}$ in a line in $\mathbf{R}^2$, one would expect to find a system whose solution set is this line. Indeed, if

$$\begin{pmatrix} x \\ y \end{pmatrix} = t\begin{pmatrix} 2 \\ -3 \end{pmatrix},$$

then $x = 2t$ and $y = -3t$. Eliminating the parameter t by solving one equation for t and substituting in the other equation. Then, $t = x/2$ and

$$\begin{aligned} y &= -3\left(\frac{x}{2}\right) \\ 2y &= -3x \\ 3x + 2y &= 0. \end{aligned}$$

Of course, there are other systems having this line as their solution set. For example, consider the system

$$\begin{aligned} 3x + 2y &= 0 \\ 6x + 4y &= 0. \end{aligned}$$

31. The line passes through $\begin{pmatrix} 0 \\ 1 \end{pmatrix}$ in the direction $\begin{pmatrix} 2 \\ -3 \end{pmatrix}$. Write

$$\begin{pmatrix} x \\ y \end{pmatrix} = \begin{pmatrix} 0 \\ 1 \end{pmatrix} + t\begin{pmatrix} 2 \\ -3 \end{pmatrix},$$

so that

$$\begin{aligned} x &= 2t, \\ y &= 1 - 3t. \end{aligned}$$

Multiply the first equation by 3 and the second equation by 2.

$$\begin{aligned} 3x &= 6t \\ 2y &= 2 - 6t. \end{aligned}$$

Add these to get $3x + 2y = 2$. Of course, other systems are possible. For example, consider

$$\begin{aligned} 3x + 2y &= 2 \\ 6x + 4y &= 4. \end{aligned}$$

33. A point has zero dimension. A line is one-dimensional.

35. In $\mathbf{R}^4$ we can expect dimensions 0, 1, 2, and 3.

37. Solve the second equation for x_2

$$x_2 = 3 + 3x_3 + x_4.$$

Substitute in the first equation

$$\begin{aligned} x_1 + 2(3 + 3x_3 + x_4) - 2x_3 + x_4 &= 2 \\ x_1 + 6 + 4x_3 + 3x_4 &= 2, \end{aligned}$$

and solve for

$$x_1 = -4 - 4x_3 - 3x_4.$$

Thus, solutions have the form

$$\begin{pmatrix} x_1 \\ x_2 \\ x_3 \\ x_4 \end{pmatrix} = \begin{pmatrix} -4 - 4x_3 - 3x_4 \\ 3 + 3x_3 + x_4 \\ x_3 \\ x_4 \end{pmatrix}$$

$$= \begin{pmatrix} -4 \\ 3 \\ 0 \\ 0 \end{pmatrix} + x_3 \begin{pmatrix} -4 \\ 3 \\ 1 \\ 0 \end{pmatrix} + x_4 \begin{pmatrix} -3 \\ 1 \\ 0 \\ 1 \end{pmatrix}.$$

This set has dimension 2. Think of the set as a translated plane in $\mathbf{R}^4$.

Section 7.3. Solving Systems of Equations

1. The augmented matrix for the system is

$$\begin{pmatrix} 1 & 1 & 1 & 4 \\ 2 & -1 & -1 & 6 \\ 4 & 1 & 1 & 14 \end{pmatrix}$$

Use the pivot in the first row to zero the entries below it in the first column. Add -2 times row 1 to row 2. Add -4 times row 1 to row 3.

$$\begin{pmatrix} 1 & 1 & 1 & 4 \\ 0 & -3 & -3 & -2 \\ 0 & -3 & -3 & -3 \end{pmatrix}$$

Use the pivot in the second row to zero the entries below it in the second column. Add -1 times row 2 to row 3.

$$\begin{pmatrix} 1 & 1 & 1 & 4 \\ 0 & -3 & -3 & -2 \\ 0 & 0 & 0 & 0 \end{pmatrix} \qquad \begin{aligned} x_1 + \quad x_2 + \ x_3 &= \ 4 \\ -3x_2 - 3x_3 &= -2 \end{aligned}$$

Note that x_1 and x_2 are pivot variables and x_3 is a free variable. Set $x_3 = t$ in the second equation and solve for x_2.

$$\begin{aligned} -3x_2 - 3t &= -2 \\ x_2 &= -t + 2/3 \end{aligned}$$

Set $x_3 = t$ and $x_2 = -t + 2/3$ in the first equation and solve for x_1.

$$\begin{aligned} x_1 + (-t + 2/3) + t &= 4 \\ x_1 &= 10/3 \end{aligned}$$

Thus,

$$\begin{pmatrix} x_1 \\ x_2 \\ x_3 \end{pmatrix} = \begin{pmatrix} 10/3 \\ -t + 2/3 \\ t \end{pmatrix} = \begin{pmatrix} 10/3 \\ 2/3 \\ 0 \end{pmatrix} + t \begin{pmatrix} 0 \\ -1 \\ 1 \end{pmatrix}.$$

3. The augmented matrix for the system is

$$\begin{pmatrix} 1 & 2 & 2 & 2 \\ -1 & -1 & 2 & 4 \\ 1 & 3 & 6 & 7 \end{pmatrix}$$

Use the pivot in the first row to zero the entries below it in the first column. Add 1 times row 1 to row 2. Add -1 times row 1 to row 3.

$$\begin{pmatrix} 1 & 2 & 2 & 2 \\ 0 & 1 & 4 & 6 \\ 0 & 1 & 4 & 5 \end{pmatrix}$$

Use the pivot in the second row to zero the entries below it in the second column. Add -1 times row 2 to row 3.

$$\begin{pmatrix} 1 & 2 & 2 & 2 \\ 0 & 1 & 4 & 6 \\ 0 & 0 & 0 & -1 \end{pmatrix}$$

Note that there is a pivot in the last column of the third row. Therefore, the system is inconsistent.

5. The augmented matrix for the system is

$$\begin{pmatrix} 1 & 2 & -2 & 6 \\ 2 & 4 & -4 & 12 \\ 3 & 6 & -6 & 18 \end{pmatrix}$$

Use the pivot in the first row to zero the entries below it in the first column. Add -2 times row 1 to row 2. Add -3 times row 1 to row 3.

$$\begin{pmatrix} 1 & 2 & -2 & 6 \\ 0 & 0 & 0 & 0 \\ 0 & 0 & 0 & 0 \end{pmatrix} \qquad x_1 + 2x_2 - 2x_3 = 6$$

Note that x_1 is a pivot variable and x_2 and x_3 are free variables. Set $x_2 = s$ and $x_3 = t$ in the first equation and solve for x_1.

$$x_1 + 2s - 2t = 6$$
$$x_1 = 6 - 2s + 2t$$

Thus,

$$\begin{pmatrix} x_1 \\ x_2 \\ x_3 \end{pmatrix} = \begin{pmatrix} 6 - 2s + 2t \\ s \\ t \end{pmatrix} = \begin{pmatrix} 6 \\ 0 \\ 0 \end{pmatrix} + s\begin{pmatrix} -2 \\ 1 \\ 0 \end{pmatrix} + t\begin{pmatrix} 2 \\ 0 \\ 1 \end{pmatrix}.$$

7. The augmented matrix for the system is

$$\begin{pmatrix} 0 & 1 & -2 & 4 \\ 1 & 2 & -2 & 6 \\ 1 & 4 & -6 & 14 \end{pmatrix}$$

Exchange rows 1 and 2.

$$\begin{pmatrix} 1 & 2 & -2 & 6 \\ 0 & 1 & -2 & 4 \\ 1 & 4 & -6 & 14 \end{pmatrix}$$

Use the pivot in the first row to zero the entries below it in the first column. Add -1 times row 1 to row 3.

$$\begin{pmatrix} 1 & 2 & -2 & 6 \\ 0 & 1 & -2 & 4 \\ 0 & 2 & -4 & 8 \end{pmatrix}$$

Use the pivot in the second row to zero the entries below it in the second column. Add -2 times row 2 to row 3.

$$\begin{pmatrix} 1 & 2 & -2 & 6 \\ 0 & 1 & -2 & 4 \\ 0 & 0 & 0 & 0 \end{pmatrix} \qquad \begin{aligned} x_1 + 2x_2 - 2x_3 &= 6 \\ x_2 - 2x_3 &= 4 \end{aligned}$$

Note that x_1 and x_2 are pivot variables and x_3 is a free variable. Set $x_3 = t$ in the second equation and solve for x_2.

$$x_2 - 2t = 4$$
$$x_2 = 4 + 2t$$

Set $x_2 = 4 + 2t$ and $x_3 = t$ in the first equation and solve for x_1.

$$x_1 + 2(4 + 2t) - 2t = 6$$
$$x_1 = -2 - 2t$$

Thus,

$$\begin{pmatrix} x_1 \\ x_2 \\ x_3 \end{pmatrix} = \begin{pmatrix} -2 - 2t \\ 4 + 2t \\ t \end{pmatrix} = \begin{pmatrix} -2 \\ 4 \\ 0 \end{pmatrix} + t\begin{pmatrix} -2 \\ 2 \\ 1 \end{pmatrix}.$$

9. The augmented matrix for the system is

$$\begin{pmatrix} 1 & 2 & -3 & 1 & 6 \\ 2 & 1 & -2 & -1 & 4 \\ 0 & 6 & 4 & -1 & 4 \end{pmatrix}$$

Use the pivot in the first row to zero the entries below it in the first column. Add -2 times row 1 to row 2.

$$\begin{pmatrix} 1 & 2 & -3 & 1 & 6 \\ 0 & -3 & 4 & -3 & -8 \\ 0 & 6 & 4 & -1 & 4 \end{pmatrix}$$

Use the pivot in the second row to zero the entries below it in the second column. Add 2 times row 2 to row 3.

$$\begin{pmatrix} 1 & 2 & -3 & 1 & 6 \\ 0 & -3 & 4 & -3 & -8 \\ 0 & 0 & 12 & -7 & -12 \end{pmatrix}.$$

$$\begin{aligned} x_1 + 2x_2 - 3x_3 + x_4 &= 6 \\ -3x_2 + 4x_3 - 3x_4 &= -8 \\ +12x_3 - 7x_4 &= -12 \end{aligned}$$

Note that x_1, x_2, and x_3 are pivot variables and x_4 is a free variable. Set $x_4 = t$ in the third equation and solve for x_3.

$$\begin{aligned} 12x_3 - 7t &= -12 \\ x_3 &= -1 + 7t/12 \end{aligned}$$

Set $x_3 = -1 + 7t/12$ and $x_4 = t$ in the second equation and solve for x_2.

$$\begin{aligned} -3x_2 + 4(-1 + 7t/12) - 3t &= -8 \\ x_2 &= 4/3 - 2t/9 \end{aligned}$$

Finally, set $x_2 = 4/3 - 2t/9$, $x_3 = -1 + 7t/12$, and $x_4 = t$ in the first equation and solve

$$x_1 + 2(4/3 - 2t/9) - 3(-1 + 7t/12) + t = 6$$

for x_1, getting

$$x_1 = 1/3 + 43t/36.$$

Thus,

$$\begin{aligned} \begin{pmatrix} x_1 \\ x_2 \\ x_3 \end{pmatrix} &= \begin{pmatrix} 1/3 + 43t/36 \\ 4/3 - 2t/9 \\ -1 + 7t/12 \\ t \end{pmatrix} \\ &= \begin{pmatrix} 1/3 \\ 4/3 \\ -1 \\ 0 \end{pmatrix} + t \begin{pmatrix} 43/36 \\ -2/9 \\ 7/12 \\ 1 \end{pmatrix}. \end{aligned}$$

11. We prefer to make a "forward pass" through the matrix to row echelon form, then make a "backward pass" through the matrix to reduced row echelon form. The augmented matrix is

$$\begin{pmatrix} 1 & 2 & -3 & 1 \\ 2 & 5 & -1 & 0 \end{pmatrix}.$$

Use the pivot in the first row to zero each entry in column 1 that falls below the pivot. Add -2 times row 1 to row 2.

$$\begin{pmatrix} 1 & 2 & -3 & 1 \\ 0 & 1 & 5 & -2 \end{pmatrix}.$$

This matrix is in row echelon form. We will now use the pivot in the second row to zero every entry in column 2 that lies above the pivot. Add -2 times row 2 to row 1.

$$\begin{pmatrix} 1 & 0 & -13 & 5 \\ 0 & 1 & 5 & -2 \end{pmatrix}$$

This last matrix is in reduced row echelon form.

13. We prefer to make a "forward pass" through the matrix to row echelon form, then make a "backward pass" through the matrix to reduced row echelon form. The augmented matrix is

$$\begin{pmatrix} 1 & -2 & 4 & 0 \\ -2 & 1 & 1 & 4 \end{pmatrix}$$

Use the pivot in the first row to zero each entry in column 1 that falls below the pivot. Add 2 times row 1 to row 2.

$$\begin{pmatrix} 1 & -2 & 4 & 0 \\ 0 & -3 & 9 & 4 \end{pmatrix}$$

This matrix is in row echelon form. We will now use the pivot in the second row to zero every entry in column 2 that lies above the pivot. Add $-2/3$ times row 2 to row 1.

$$\begin{pmatrix} 1 & 0 & -2 & -8/3 \\ 0 & -3 & 9 & 4 \end{pmatrix}$$

Finally, we need to scale each pivot to 1. Multiply the second row by $-1/3$.

$$\begin{pmatrix} 1 & 0 & -2 & -8/3 \\ 0 & 1 & -3 & -4/3 \end{pmatrix}$$

This last matrix is in reduced row echelon form.

15. We prefer to make a "forward pass" through the matrix to row echelon form, then make a "backward pass" through the matrix to reduced row echelon form. The augmented matrix is

$$\begin{pmatrix} 1 & 1 & -1 & 2 & 3 \\ 2 & 2 & -3 & 4 & 5 \end{pmatrix}$$

Use the pivot in the first row to zero each entry in column 1 that falls below the pivot. Add -2 times row 1 to row 2.

$$\begin{pmatrix} 1 & 1 & -1 & 2 & 3 \\ 0 & 0 & -1 & 0 & -1 \end{pmatrix}$$

This matrix is in row echelon form. We will now use the pivot in the second row to zero every entry that lies above the pivot in the third column. Add -1 times row 2 to row 1.

$$\begin{pmatrix} 1 & 1 & 0 & 2 & 4 \\ 0 & 0 & -1 & 0 & -1 \end{pmatrix}$$

We need to scale each pivot to 1. Multiply the second row by -1.

$$\begin{pmatrix} 1 & 1 & 0 & 2 & 4 \\ 0 & 0 & 1 & 0 & 1 \end{pmatrix}$$

This last matrix is in reduced row echelon form.

17. We prefer to make a "forward pass" through the matrix to row echelon form, then make a "backward pass" through the matrix to reduced row echelon form. The augmented matrix is

$$\begin{pmatrix} 1 & 2 & 2 & -2 & 3 \\ 1 & 2 & -2 & 4 & 2 \\ -1 & -3 & 0 & 0 & 2 \end{pmatrix}$$

Use the pivot in the first row to zero each entry in column 1 that falls below the pivot. Add -1 times row 1 to row 2. Add 1 times row 1 to row 3.

$$\begin{pmatrix} 1 & 2 & 2 & -2 & 3 \\ 0 & 0 & -4 & 6 & -1 \\ 0 & -1 & 2 & -2 & 5 \end{pmatrix}$$

Exchange rows 2 and 3.

$$\begin{pmatrix} 1 & 2 & 2 & -2 & 3 \\ 0 & -1 & 2 & -2 & 5 \\ 0 & 0 & -4 & 6 & -1 \end{pmatrix}$$

This matrix is in row echelon form. We will now use the pivot in the third row to zero every entry that lies above the pivot in the third column. Add $1/2$ times row 3 to row 2. Add $1/2$ times row 3 to row 1.

$$\begin{pmatrix} 1 & 2 & 0 & 1 & 5/2 \\ 0 & -1 & 0 & 1 & 9/2 \\ 0 & 0 & -4 & 6 & -1 \end{pmatrix}$$

We will now use the pivot in the second row to zero each entry in column 2 that lies above the pivot. Add 2 times row 2 to row 1.

$$\begin{pmatrix} 1 & 0 & 0 & 3 & 23/2 \\ 0 & -1 & 0 & 1 & 9/2 \\ 0 & 0 & -4 & 6 & -1 \end{pmatrix}$$

We need to scale each pivot to 1. Multiply the second row by -1, the third row by $-1/4$.

$$\begin{pmatrix} 1 & 0 & 0 & 3 & 23/2 \\ 0 & 1 & 0 & -1 & -9/2 \\ 0 & 0 & 1 & -3/2 & 1/4 \end{pmatrix}$$

This last matrix is in reduced row echelon form.

19. The augmented matrix and corresponding system.

$$\begin{pmatrix} 1 & 0 & -2 & 2 & -2 \\ 0 & 1 & 1 & -1 & 1 \\ 0 & 0 & 0 & 0 & 0 \end{pmatrix}$$

$$\begin{aligned} x_1 \quad - 2x_3 + 2x_4 &= -2 \\ x_2 + x_3 - x_4 &= 1 \end{aligned}$$

Thus, x_1 and x_2 are pivot variables, while x_3 and x_4 are free. Set $x_3 = s$ and $x_4 = t$ and solve each equation for its pivot variable.

$$\begin{aligned} x_1 &= -2 + 2s - 2t \\ x_2 &= 1 - s + t \end{aligned}$$

Thus, the solution is

$$\begin{pmatrix} x_1 \\ x_2 \\ x_3 \\ x_4 \end{pmatrix} = \begin{pmatrix} -2 + 2s - 2t \\ 1 - s + t \\ s \\ t \end{pmatrix} = \begin{pmatrix} -2 \\ 1 \\ 0 \\ 0 \end{pmatrix} + s \begin{pmatrix} 2 \\ -1 \\ 1 \\ 0 \end{pmatrix} + t \begin{pmatrix} -2 \\ 1 \\ 0 \\ 1 \end{pmatrix}.$$

21. The augmented matrix and corresponding system.

$$\begin{pmatrix} 1 & -1 & 1 & 0 & -1 & 1 \\ 0 & 0 & 0 & 1 & 1 & 0 \\ 0 & 0 & 0 & 0 & 0 & 0 \\ 0 & 0 & 0 & 0 & 0 & 0 \end{pmatrix}$$

$$\begin{aligned} x_1 - x_2 + x_3 \quad - x_5 &= 1 \\ x_4 + x_5 &= 0 \end{aligned}$$

Thus, x_1 and x_4 are pivot variables, while x_2, x_3, and x_5 are free. Set $x_2 = u$, $x_3 = v$, and $x_5 = w$ and solve each equation for its pivot variable.

$$x_1 = 1 + u - v + w$$
$$x_4 = -w$$

Thus, the solution is

$$\begin{pmatrix} x_1 \\ x_2 \\ x_3 \\ x_4 \\ x_5 \end{pmatrix} = \begin{pmatrix} 1+u-v+w \\ u \\ v \\ -w \\ w \end{pmatrix} = \begin{pmatrix} 1 \\ 0 \\ 0 \\ 0 \\ 0 \end{pmatrix} + u \begin{pmatrix} 1 \\ 1 \\ 0 \\ 0 \\ 0 \end{pmatrix} + v \begin{pmatrix} -1 \\ 0 \\ 1 \\ 0 \\ 0 \end{pmatrix} + w \begin{pmatrix} 1 \\ 0 \\ 0 \\ -1 \\ 1 \end{pmatrix}.$$

23. Set up the augmented matrix and reduce.

$$\begin{pmatrix} -6 & 8 & 0 & 2 \\ 4 & 8 & 8 & 20 \\ -2 & 2 & 7 & 7 \end{pmatrix} \sim \begin{pmatrix} 1 & 0 & 0 & 1 \\ 0 & 1 & 0 & 1 \\ 0 & 0 & 1 & 1 \end{pmatrix}$$

This gives $\mathbf{x} = (1, 1, 1)^T$.

25. The augmented matrix is

$$M = \begin{pmatrix} 4 & 2 & -5 & -5 \\ -14 & -8 & 18 & 16 \\ -3 & -2 & 4 & 3 \end{pmatrix}$$

Using a computer we reduce this to row echelon form

$$\begin{pmatrix} 4 & 2 & -5 & -5 \\ 0 & -1 & 1/2 & -3/2 \\ 0 & 0 & 0 & 0 \end{pmatrix}.$$

The simplified system is

$$4y_1 + 2y_2 - 5y_3 = -5$$
$$-y_2 + y_3/2 = -3/2.$$

To backsolve, we first set the free variable $y_3 = t$. Next we solve for $y_2 = 3/2 + y_3/2 = (3+t)/2$. Finally, we solve for $y_1 = (-5 - 2y_2 + 5y_3)/4 = [-5-(3+t)+5t]/4 = -2+t$. Hence our solutions are all vectors of the form

$$\mathbf{y} = \begin{pmatrix} y_1 \\ y_2 \\ y_3 \end{pmatrix} = \begin{pmatrix} -2+t \\ (3+t)/2 \\ t \end{pmatrix} = \begin{pmatrix} -2 \\ 3/2 \\ 0 \end{pmatrix} + t \begin{pmatrix} 1 \\ 1/2 \\ 1 \end{pmatrix}.$$

27. The augmented matrix is

$$M = \begin{pmatrix} -3 & -3 & 1 & 4 \\ 8 & 7 & -2 & -8 \\ 8 & 6 & -1 & -5 \end{pmatrix}$$

Using a computer we reduce this to the row echelon form

$$\begin{pmatrix} -3 & -3 & 1 & 4 \\ 0 & -1 & 2/3 & 8/3 \\ 0 & 0 & 1/3 & 1/3 \end{pmatrix}$$

the simplified system is

$$-3y_1 - 3y_2 + y_3 = 4$$
$$-y_2 + 2y_3/3 = 8/3$$
$$y_3/3 = 1/3.$$

Backsolving we get $y_3 = 1$. Next $y_2 = -8/3 + 2/3 = -2$. Finally, $y_1 = (-4 - 3y_2 + y_3)/3 = 1$. Hence the only solution is $\mathbf{y} = (1, -2, 1)^T$

29. Set up the augmented matrix and reduce.

$$\begin{pmatrix} -12 & 12 & -8 & -8 \\ -16 & 16 & -10 & -10 \\ -3 & 3 & -1 & -1 \end{pmatrix} \sim \begin{pmatrix} 1 & -1 & 0 & 0 \\ 0 & 0 & 1 & 1 \\ 0 & 0 & 0 & 0 \end{pmatrix}$$

This gives

$$x - y = 0,$$
$$z = 1.$$

Thus, x and z are pivot variables and y is free. Solve each equation for its pivot variable.

$$x = y$$
$$z = 1$$

Thus

$$\mathbf{x} = \begin{pmatrix} x \\ y \\ z \end{pmatrix} = \begin{pmatrix} y \\ y \\ 1 \end{pmatrix} = \begin{pmatrix} 0 \\ 0 \\ 1 \end{pmatrix} + y \begin{pmatrix} 1 \\ 1 \\ 0 \end{pmatrix},$$

where y is free.

31. Set up the augmented matrix and reduce.

$$\begin{pmatrix} -5 & -4 & 4 & 3 & 17 \\ 7 & 6 & -5 & 3 & 12 \end{pmatrix}$$
$$\sim \begin{pmatrix} 1 & 0 & -2 & -15 & -75 \\ 0 & 1 & 3/2 & 18 & 179/2 \end{pmatrix}$$

This gives

$$x_1 - 2x_3 - 15x_4 = -75,$$
$$x_2 + \frac{3}{2}x_3 + 18x_4 = \frac{179}{2}.$$

Thus, x_1 and x_2 are pivot variables, while x_3 and x_4 are free. Solve each equation for its pivot variable.

$$x_1 = -75 + 2x_3 + 15x_4$$
$$x_2 = \frac{179}{2} - \frac{3}{2}x_3 - 18x_4$$

Thus,

$$\mathbf{x} = \begin{pmatrix} x_1 \\ x_2 \\ x_3 \\ x_4 \end{pmatrix} = \begin{pmatrix} -75 + 2x_3 + 15x_4 \\ 179/2 - (3/2)x_3 - 18x_4 \\ x_3 \\ x_4 \end{pmatrix},$$
$$= \begin{pmatrix} -75 \\ 179/2 \\ 0 \\ 0 \end{pmatrix} + x_3 \begin{pmatrix} 2 \\ -3/2 \\ 1 \\ 0 \end{pmatrix} + x_4 \begin{pmatrix} 15 \\ -18 \\ 0 \\ 1 \end{pmatrix},$$

where x_3 and x_4 are free.

33. The augmented matrix is

$$M = \begin{pmatrix} -7 & 7 & -8 & -3 & 37 \\ 9 & -5 & 8 & -2 & -35 \\ 5 & 0 & 2 & 8 & -9 \end{pmatrix}.$$

Using a computer to perform row operations this is reduced to the row echelon form

$$\begin{pmatrix} -7 & 7 & -8 & -3 & 37 \\ 0 & 4 & -16/7 & -41/7 & 88/7 \\ 0 & 0 & -6/7 & 369/28 & 12/7 \end{pmatrix}$$

The simplified system of equations is

$$-7y_1 + 7y_2 - 8y_3 - 3y_4 = 37$$
$$4y_2 - 16y_3/7 - 41y_4/7 = 88/7$$
$$-6y_3/7 + 369y_4/28 = 12/7$$

To backsolve, we first set the free variable $y_4 = t$. Then we solve for $y_3 = (-12 + 369t/4)/6 = -2 + 123t/8$. Next,

$$\begin{aligned} y_2 &= (88/7 + 16y_3/7 + 41y_4/7)/4 \\ &= [88 + 16(-2 + 123t/8) + 41t]/28 \\ &= [56 + 287t]/28 \\ &= 2 + 41t/4. \end{aligned}$$

Finally

$$\begin{aligned} y_1 &= [-37 + 7y_2 - 8y_3 - 3y_4]/7 \\ &= [-37 + 7(2 + 41t/4) \\ &\quad - 8(-2 + 123t/8) - 3t]/7 \\ &= -1 - 31t/4. \end{aligned}$$

Hence the solutions are the vectors

$$\mathbf{y} = \begin{pmatrix} -1 - 31t/4 \\ 2 + 41t/4 \\ -2 + 123t/8 \\ t \end{pmatrix}$$
$$= \begin{pmatrix} -1 \\ 2 \\ -2 \\ 0 \end{pmatrix} + t \begin{pmatrix} -31/4 \\ 41/4 \\ 123/8 \\ 1 \end{pmatrix}.$$

35. The augmented matrix is

$$M = \begin{pmatrix} 8 & -6 & 9 & 8 & -1 & 15 \\ -9 & 5 & -7 & 9 & 0 & -30 \\ 1 & -4 & 1 & -3 & -7 & 9 \end{pmatrix}$$

With a computer we reduce this to row echelon form

$$\begin{pmatrix} 8 & -6 & 9 & 8 & -1 & 15 \\ 0 & -14 & 25 & 144 & -9 & -105 \\ 0 & 0 & -83 & -524 & -67 & 441 \end{pmatrix}$$

The simplified system of equation is

$$8y_1 - 6y_2 + 9y_3 + 8y_4 - y_5 = 15$$
$$-14y_2 + 25y_3 + 144y_4 - 9y_5 = -105$$
$$-83y_3 - 524y_4 - 67y_5 = 441$$

To backsolve, we first set the free variables $y_4 = s$ and $y_5 = t$. The we solve for

$$y_3 = [-441 - 524s - 67t]/83.$$

Next,

$$\begin{aligned} y_2 &= [105 + 25y_3 + 144y_4 - 9y_5]/14 \\ &= [105 + 25(-441 - 524s - 67t)/83 \\ &\quad + 144s - 9t]/14 \\ &= [-165 - 82s - 173t]/83. \end{aligned}$$

and finally,

$$\begin{aligned} y_1 &= [15 + 6y_2 - 9y_3 - 8y_4 + y_5]/8 \\ &= [15 + 6(-165 + 82s + 173t)/83 \\ &\quad - 9(-441 - 524s - 67t)/83 - 8s + t]/8 \\ &= [528 + 445s - 44t]/83. \end{aligned}$$

Hence the solutions are the vectors

$$\mathbf{y} = \begin{pmatrix} (528 + 445s - 44t)/83 \\ (-165 - 82s - 173t)/83 \\ (-441 - 524s - 67t)/83 \\ s \\ t \end{pmatrix} = \begin{pmatrix} 528/83 \\ -165/83 \\ -441/83 \\ 0 \\ 0 \end{pmatrix} + s \begin{pmatrix} 445/83 \\ -82/83 \\ -524/83 \\ 1 \\ 0 \end{pmatrix} + t \begin{pmatrix} -44/83 \\ -173/83 \\ -67/83 \\ 0 \\ 1 \end{pmatrix}.$$

Section 7.4. Homogeneous and Inhomogeneous Systems

1. We know that $A(c\mathbf{b}) = c(A\mathbf{b})$, where A is a matrix, c is a scalar, and $\mathbf{b}$ is a vector. Thus,

$$\begin{aligned}
&\begin{pmatrix} -3 & -2 & 4 \\ 14 & 8 & -18 \\ 4 & 2 & -5 \end{pmatrix}\left(t\begin{pmatrix} 1 \\ 1/2 \\ 1 \end{pmatrix}\right) \\
&= t\begin{pmatrix} -3 & -2 & 4 \\ 14 & 8 & -18 \\ 4 & 2 & -5 \end{pmatrix}\begin{pmatrix} 1 \\ 1/2 \\ 1 \end{pmatrix} \\
&== t\left(1\begin{pmatrix} -3 \\ 14 \\ 4 \end{pmatrix} + \frac{1}{2}\begin{pmatrix} -2 \\ 8 \\ 2 \end{pmatrix} + \begin{pmatrix} 4 \\ -18 \\ -4 \end{pmatrix}\right) \\
&== t\begin{pmatrix} -3-1+4 \\ 14+4-18 \\ 4+1-5 \end{pmatrix} \\
&== t\begin{pmatrix} 0 \\ 0 \\ 0 \end{pmatrix} \\
&= \mathbf{0}.
\end{aligned}$$

Hence the solution in Example 4.1 is in the null space of matrix A.

3. The matrix

$$A = \begin{pmatrix} 1 & 0 & -2 \\ 0 & 1 & 3 \\ 0 & 0 & 0 \end{pmatrix}$$

$$\begin{aligned} x_1 \quad - 2x_3 &= 0 \\ x_2 + 3x_3 &= 0 \end{aligned}$$

is in reduced row echelon form. The variables x_1 and x_2 are pivot variables, while x_3 is free. Set $x_3 = t$ and solve each equation for its pivot variable.

$$\begin{aligned} x_1 &= 2t \\ x_2 &= -3t \end{aligned}$$

Thus, all solutions of $A\mathbf{x} = \mathbf{0}$ are given parametrically by

$$\mathbf{x} = \begin{pmatrix} x_1 \\ x_2 \\ x_3 \end{pmatrix} = \begin{pmatrix} 2t \\ -3t \\ t \end{pmatrix} = t\begin{pmatrix} 2 \\ -3 \\ 1 \end{pmatrix}.$$

5. The matrix

$$A = \begin{pmatrix} 1 & 0 & 2 & -2 \\ 0 & 1 & 3 & -1 \end{pmatrix}$$

$$\begin{aligned} x_1 \quad + 2x_3 - 2x_4 &= 0 \\ x_2 + 3x_3 - x_4 &= 0 \end{aligned}$$

is in reduced row echelon form. The variables x_1 and x_2 are pivot variables, while x_3 and x_4 are free. Set $x_3 = s$ and $x_4 = t$ and solve each equation for its pivot variable.

$$\begin{aligned} x_1 &= -2s + 2t \\ x_2 &= -3s + t \end{aligned}$$

Thus, all solutions of $A\mathbf{x} = \mathbf{0}$ are given parametrically by

$$\mathbf{x} = \begin{pmatrix} x_1 \\ x_2 \\ x_3 \\ x_4 \end{pmatrix} = \begin{pmatrix} -2s + 2t \\ -3s + t \\ s \\ t \end{pmatrix} = s\begin{pmatrix} -2 \\ -3 \\ 1 \\ 0 \end{pmatrix} + t\begin{pmatrix} 2 \\ 1 \\ 0 \\ 1 \end{pmatrix}.$$

7. The matrix

$$A = \begin{pmatrix} 1 & 0 & -1 & 0 & 3 \\ 0 & 1 & 2 & 0 & -5 \\ 0 & 0 & 0 & 1 & 2 \end{pmatrix}$$

$$\begin{aligned} x_1 \quad - x_3 \quad + 3x_5 &= 0 \\ x_2 + 2x_3 \quad - 5x_5 &= 0 \\ x_4 + 2x_5 &= 0 \end{aligned}$$

is in reduced row echelon form. The variables x_1, x_2, and x_4 are pivot variables, while x_3 and x_5 are free. Set $x_3 = s$ and $x_5 = t$ and solve each equation for its pivot variable.

$$\begin{aligned} x_1 &= s - 3t \\ x_2 &= -2s + 5t \\ x_4 &= -2t \end{aligned}$$

Thus, all solutions of $A\mathbf{x} = \mathbf{0}$ are given parametrically by

$$\mathbf{x} = \begin{pmatrix} x_1 \\ x_2 \\ x_3 \\ x_4 \\ x_5 \end{pmatrix} = \begin{pmatrix} s - 3t \\ -2s + 5t \\ s \\ -2t \\ t \end{pmatrix} = s \begin{pmatrix} 1 \\ -2 \\ 1 \\ 0 \\ 0 \end{pmatrix} + t \begin{pmatrix} -3 \\ 5 \\ 0 \\ -2 \\ 1 \end{pmatrix}.$$

9. The matrix

$$A = \begin{pmatrix} 1 & 0 & -2 & 4 \end{pmatrix} \qquad x_1 \quad - 2x_3 + 4x_4 = 0$$

is in reduced row echelon form. The variable x_1 is a pivot variable, while x_2, x_3, and x_4 are free. Set $x_2 = u$, $x_3 = v$, and $x_4 = w$ and solve the equation for its pivot variable.

$$x_1 = 2v - 4w$$

Thus, all solutions of $A\mathbf{x} = \mathbf{0}$ are given parametrically by

$$vecx = \begin{pmatrix} x_1 \\ x_2 \\ x_3 \\ x_4 \end{pmatrix} = \begin{pmatrix} 2v - 4w \\ u \\ v \\ w \end{pmatrix} = u \begin{pmatrix} 0 \\ 1 \\ 0 \\ 0 \end{pmatrix} + v \begin{pmatrix} 2 \\ 0 \\ 1 \\ 0 \end{pmatrix} + w \begin{pmatrix} -4 \\ 0 \\ 0 \\ 1 \end{pmatrix}.$$

11. The reduced row echelon form of A is

$$A = \begin{pmatrix} 2 & -1 & 1 & -1 \\ 2 & 0 & 0 & -2 \\ 3 & -4 & 4 & 1 \end{pmatrix} \xrightarrow{\text{rref}} \begin{pmatrix} 1 & 0 & 0 & -1 \\ 0 & 1 & -1 & -1 \\ 0 & 0 & 0 & 0 \end{pmatrix}$$

Note that x_1 and x_2 are pivot variables, while x_3 and x_4 are free. Solve the first two equations (represented by the first two rows of the reduced row echelon form) for the pivot variables in terms of the free variables.

$$x_1 = x_4$$
$$x_2 = x_3 + x_4$$

Thus,

$$\mathbf{x} = \begin{pmatrix} x_1 \\ x_2 \\ x_3 \\ x_4 \end{pmatrix} = \begin{pmatrix} x_4 \\ x_3 + x_4 \\ x_3 \\ x_4 \end{pmatrix} = x_3 \begin{pmatrix} 0 \\ 1 \\ 1 \\ 0 \end{pmatrix} + x_4 \begin{pmatrix} 1 \\ 1 \\ 0 \\ 1 \end{pmatrix}.$$

Thus, the vectors $(0, 1, 1, 0)^T$ and $(1, 1, 0, 1)^T$ are "special." Note that there are two of these special vectors. It is also important to note that there are two free variables in the reduced row echelon form of the matrix A. This is not a coincidence.

13. The reduced row echelon form of A is

$$A = \begin{pmatrix} 0 & -2 & -2 & 0 & 2 \\ -3 & 1 & 4 & 0 & -1 \\ 1 & 0 & -1 & 0 & 0 \end{pmatrix} \xrightarrow{\text{rref}} \begin{pmatrix} 1 & 0 & -1 & 0 & 0 \\ 0 & 1 & 1 & 0 & -1 \\ 0 & 0 & 0 & 0 & 0 \end{pmatrix}$$

Note that x_1 and x_2 are pivot variables, while x_3, x_4, and x_5 are free. Solve the first two equations (represented by the first two rows of the reduced row echelon form) for the pivot variables in terms of the free variables.

$$x_1 = x_3$$
$$x_2 = -x_3 + x_5$$

Thus,

$$\mathbf{x} = \begin{pmatrix} x_1 \\ x_2 \\ x_3 \\ x_4 \\ x_5 \end{pmatrix} = \begin{pmatrix} x_3 \\ -x_3 + x_5 \\ x_3 \\ x_4 \\ x_5 \end{pmatrix} = x_3 \begin{pmatrix} 1 \\ -1 \\ 1 \\ 0 \\ 0 \end{pmatrix} + x_4 \begin{pmatrix} 0 \\ 0 \\ 0 \\ 1 \\ 0 \end{pmatrix} + x_5 \begin{pmatrix} 0 \\ 1 \\ 0 \\ 0 \\ 1 \end{pmatrix}.$$

Thus, the vectors $(1,-1,1,0,0)^T$, $(0,0,0,1,0)^T$, and $(0,1,0,0,1)^T$ are "special." Note that there are three of these special vectors. It is also important to note that there are three free variables in the reduced row echelon form of the matrix A. This is not a coincidence.

15. (a) There are six variables, but only three equations. There must be at least three free variables. The reduced form of the augmented matrix will contain only three rows. Therefore, there can be at most three pivot variables, leaving room for three or more free variables.

    (b) The augmented matrix reduces.

$$\begin{pmatrix} -1 & 1 & 1 & -1 & 0 & -1 \\ 1 & -4 & 2 & 4 & -3 & -2 \\ 3 & -3 & -3 & 3 & 0 & 3 \end{pmatrix} \xrightarrow{\text{rref}} \begin{pmatrix} 1 & 0 & -2 & 0 & 1 & 2 \\ 0 & 1 & -1 & -1 & 1 & 1 \\ 0 & 0 & 0 & 0 & 0 & 0 \end{pmatrix}$$

As you can see, x_1 and x_2 are pivot variables. So there are actually four free variables, x_3, x_4, x_5, and x_6.

17. Let $\mathbf{x}$ be in null(A) and α be a scalar (number). Then, since $\mathbf{x} \in$ null(A), we know that $A\mathbf{x} = \mathbf{0}$. Consequently,

$$\begin{aligned} A(\alpha\mathbf{x}) &= \alpha(A\mathbf{x}) \\ &= \alpha\mathbf{0} \\ &= \mathbf{0}. \end{aligned}$$

Thus, $\alpha\mathbf{x}$ is in null(A).

19. Reduce the augmented matrix.

$$\begin{pmatrix} 5 & -2 & -5 & 2 \\ -3 & 0 & 3 & 0 \\ 0 & -3 & 0 & 3 \end{pmatrix} \xrightarrow{\text{rref}} \begin{pmatrix} 1 & 0 & -1 & 0 \\ 0 & 1 & 0 & -1 \\ 0 & 0 & 0 & 0 \end{pmatrix}$$

Thus x_1 and x_2 are pivot variables and x_3 is free. Set $x_3 = t$, then solve the first two equations (rows) for their pivot variables.

$$\begin{aligned} x_1 &= t \\ x_2 &= -1 \end{aligned}$$

Thus, the solution is

$$\mathbf{x} = \begin{pmatrix} x_1 \\ x_2 \\ x_3 \end{pmatrix} = \begin{pmatrix} t \\ -1 \\ t \end{pmatrix} = \begin{pmatrix} 0 \\ -1 \\ 0 \end{pmatrix} + t\begin{pmatrix} 1 \\ 0 \\ 1 \end{pmatrix}.$$

Note that $\mathbf{p} = (0,-1,0)^T$ and $\mathbf{v} = t(1,0,1)^T$ and

$$\begin{aligned} &\begin{pmatrix} 5 & -2 & -5 \\ -3 & 0 & 3 \\ 0 & -3 & 0 \end{pmatrix} \left[t\begin{pmatrix} 1 \\ 0 \\ 1 \end{pmatrix} \right] \\ &= t\begin{pmatrix} 5 & -2 & -5 \\ -3 & 0 & 3 \\ 0 & -3 & 0 \end{pmatrix}\begin{pmatrix} 1 \\ 0 \\ 1 \end{pmatrix} \\ &= t\begin{pmatrix} 0 \\ 0 \\ 0 \end{pmatrix} \\ &= \mathbf{0}. \end{aligned}$$

21. Reduce the augmented matrix.

$$\begin{pmatrix} 2 & 0 & 0 & -4 & 0 \\ -1 & 1 & -1 & 1 & 1 \\ -1 & 1 & -1 & 1 & 1 \end{pmatrix} \xrightarrow{\text{rref}} \begin{pmatrix} 1 & 0 & 0 & -2 & 0 \\ 0 & 1 & -1 & -1 & 1 \\ 0 & 0 & 0 & 0 & 0 \end{pmatrix}$$

Thus x_1 and x_2 are pivot variables and x_3 and x_4 are free. Set $x_3 = s$ and $x_4 = t$, then solve the first two equations (rows) for their pivot variables.

$$\begin{aligned} x_1 &= 2t \\ x_2 &= 1 + s + t \end{aligned}$$

Thus, the solution is

$$\begin{aligned} \mathbf{x} = \begin{pmatrix} x_1 \\ x_2 \\ x_3 \\ x_4 \end{pmatrix} &= \begin{pmatrix} 2t \\ 1+s+t \\ s \\ t \end{pmatrix} \\ &= \begin{pmatrix} 0 \\ 1 \\ 0 \\ 0 \end{pmatrix} + s\begin{pmatrix} 0 \\ 1 \\ 1 \\ 0 \end{pmatrix} + t\begin{pmatrix} 2 \\ 1 \\ 0 \\ 1 \end{pmatrix}. \end{aligned}$$

Note that $\mathbf{p} = (0, 1, 0, 0)^T$ and $\mathbf{v} = s(0, 1, 1, 0)^T + t(2, 1, 0, 1)^T$ and

$$\begin{pmatrix} 2 & 0 & 0 & -4 \\ -1 & 1 & -1 & 1 \\ -1 & 1 & -1 & 1 \end{pmatrix} \left[s \begin{pmatrix} 0 \\ 1 \\ 1 \\ 0 \end{pmatrix} + t \begin{pmatrix} 2 \\ 1 \\ 0 \\ 1 \end{pmatrix} \right]$$

$$= s \begin{pmatrix} 2 & 0 & 0 & -4 \\ -1 & 1 & -1 & 1 \\ -1 & 1 & -1 & 1 \end{pmatrix} \begin{pmatrix} 0 \\ 1 \\ 1 \\ 0 \end{pmatrix}$$

$$+ t \begin{pmatrix} 2 & 0 & 0 & -4 \\ -1 & 1 & -1 & 1 \\ -1 & 1 & -1 & 1 \end{pmatrix} \begin{pmatrix} 2 \\ 1 \\ 0 \\ 1 \end{pmatrix}$$

$$= s \begin{pmatrix} 0 \\ 0 \\ 0 \end{pmatrix} + t \begin{pmatrix} 0 \\ 0 \\ 0 \end{pmatrix}$$

$$= \mathbf{0}.$$

23. Reduce the given matrix.

$$A = \begin{pmatrix} 1 & 2 \\ -2 & -4 \end{pmatrix} \xrightarrow{\text{rref}} \begin{pmatrix} 1 & 2 \\ 0 & 0 \end{pmatrix}$$

Thus, x_1 is a pivot variable and x_2 is free. Setting $x_2 = t$ and solving the first equation for x_1,

$$\mathbf{x} = \begin{pmatrix} x_1 \\ x_2 \end{pmatrix} = \begin{pmatrix} -2t \\ t \end{pmatrix} = t \begin{pmatrix} -2 \\ 1 \end{pmatrix}.$$

This is a line through the origin in $\mathbf{R}^2$, as shown in the following figure.

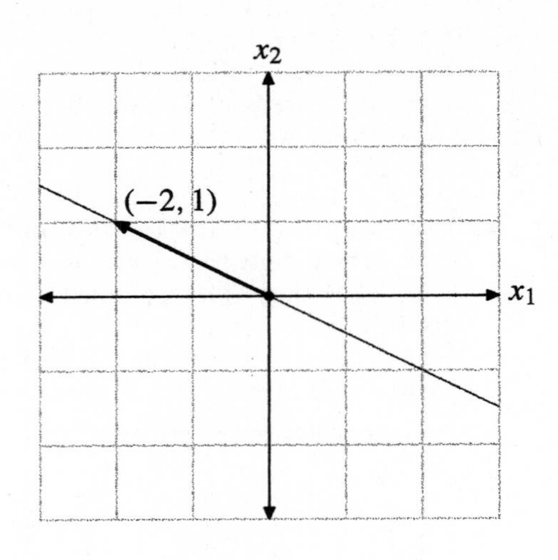

25. Reduce the given matrix.

$$A = \begin{pmatrix} 3 & -2 \\ 6 & -4 \end{pmatrix} \xrightarrow{\text{rref}} \begin{pmatrix} 1 & -2/3 \\ 0 & 0 \end{pmatrix}$$

Thus, x_1 is a pivot variable and x_2 is free. Setting $x_2 = t$ and solving the first equation for x_1,

$$\mathbf{x} = \begin{pmatrix} x_1 \\ x_2 \end{pmatrix} = \begin{pmatrix} 2t/3 \\ t \end{pmatrix} = t \begin{pmatrix} 2/3 \\ 1 \end{pmatrix}.$$

However, any multiple of $(2/3, 1)^T$ will do, so let's use $(2, 3)^T$ and write

$$\mathbf{x} = t \begin{pmatrix} 2 \\ 3 \end{pmatrix}.$$

This is a line through the origin in $\mathbf{R}^2$, as shown in the following figure.

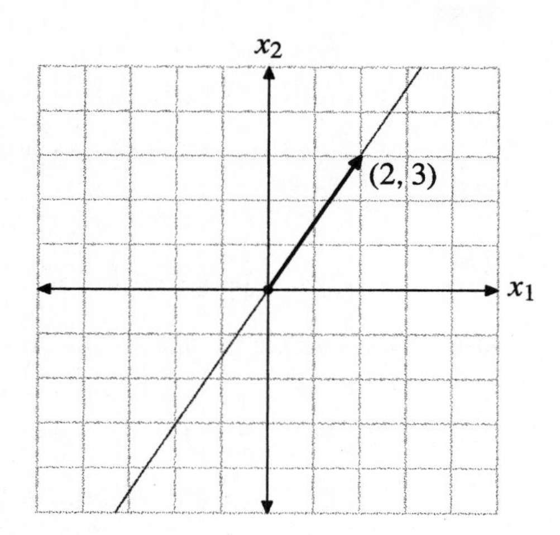

27. Reduce the given matrix.

$$\begin{pmatrix} 2 & 1 & -1 \\ 1 & -1 & -2 \\ 2 & 0 & -2 \end{pmatrix} \xrightarrow{\text{rref}} \begin{pmatrix} 1 & 0 & -1 \\ 0 & 1 & 1 \\ 0 & 0 & 0 \end{pmatrix}$$

Thus, x_1 and x_2 are pivot variables and x_3 is free. Set $x_3 = t$ and solve the first two equations (rows) for their pivot variables.

$$x_1 = t$$
$$x_2 = -t$$

Thus, the null space solution is

$$\mathbf{x} = \begin{pmatrix} x_1 \\ x_2 \\ x_3 \end{pmatrix} = \begin{pmatrix} t \\ -t \\ t \end{pmatrix} = t \begin{pmatrix} 1 \\ -1 \\ 1 \end{pmatrix}$$

This is a line through the origin in $\mathbf{R}^3$.

29. Reduce the given matrix.

$$\begin{pmatrix} 1 & -1 & -1 \\ 2 & -2 & -2 \\ -4 & 4 & 4 \end{pmatrix} \xrightarrow{\text{rref}} \begin{pmatrix} 1 & -1 & -1 \\ 0 & 0 & 0 \\ 0 & 0 & 0 \end{pmatrix}$$

Thus, x_1 is a pivot variable, but x_2 and x_3 are free. Set $x_2 = s$ and $x_3 = t$ and solve the first equation (row) for its pivot variable.

$$x_1 = s + t$$

Thus, the null space solution is

$$\mathbf{x} = \begin{pmatrix} x_1 \\ x_2 \\ x_3 \end{pmatrix} = \begin{pmatrix} s+t \\ s \\ t \end{pmatrix} = s \begin{pmatrix} 1 \\ 1 \\ 0 \end{pmatrix} + t \begin{pmatrix} 1 \\ 0 \\ 1 \end{pmatrix}$$

This is a plane passing through the origin in $\mathbf{R}^3$.

Section 7.5. Bases of a Subspace

1. Arrange the appropriate augmented matrix and reduce.

$$[\mathbf{u}_1, \mathbf{u}_2, \mathbf{w}] = \begin{pmatrix} 1 & 3 & 5 \\ -2 & 0 & -2 \end{pmatrix}$$
$$\xrightarrow{\text{rref}} \begin{pmatrix} 1 & 0 & 1 \\ 0 & 1 & 4/3 \end{pmatrix}$$

This system has a unique solution $(1, 4/3)^T$, so

$$\begin{pmatrix} 5 \\ -2 \end{pmatrix} = \mathbf{w} = 1\mathbf{u}_1 + \frac{4}{3}\mathbf{u}_2 = 1\begin{pmatrix} 1 \\ -2 \end{pmatrix} + \frac{4}{3}\begin{pmatrix} 3 \\ 0 \end{pmatrix},$$

which is easily checked.

3. Arrange the appropriate augmented matrix and reduce.

$$\mathbf{u}_1, \mathbf{u}_3, \mathbf{w}t = \begin{pmatrix} 1 & 2 & 3 \\ -2 & -4 & -3 \end{pmatrix}$$
$$\xrightarrow{\text{rref}} \begin{pmatrix} 1 & 2 & 0 \\ 0 & 0 & 1 \end{pmatrix}$$

This system is inconsistent. Hence, $\mathbf{w}$ is not in span$\{\mathbf{u}_1, \mathbf{u}_3\}$.

5. Arrange the appropriate augmented matrix and reduce.

$$[\mathbf{v}_1, \mathbf{v}_3, \mathbf{w}] = \begin{pmatrix} 1 & 1 & 1 \\ -4 & -2 & 4 \\ 4 & 3 & 1 \end{pmatrix} \xrightarrow{\text{rref}} \begin{pmatrix} 1 & 0 & 0 \\ 0 & 1 & 0 \\ 0 & 0 & 1 \end{pmatrix}$$

This system is inconsistent. Hence, $\mathbf{w}$ is not in $\text{span}\{\mathbf{v}_1, \mathbf{v}_3\}$.

7. Arrange the appropriate augmented matrix and reduce.

$$[\mathbf{v}_1, \mathbf{v}_2, \mathbf{v}_3, \mathbf{w}] = \begin{pmatrix} 1 & 0 & 1 & 1 \\ -4 & -2 & -2 & 0 \\ 4 & 1 & 3 & 2 \end{pmatrix} \xrightarrow{\text{rref}} \begin{pmatrix} 1 & 0 & 1 & 1 \\ 0 & 1 & -1 & -2 \\ 0 & 0 & 0 & 0 \end{pmatrix}$$

This system has solution $(1-t, -2+t, t)^T$. For example, choose $t = 3$ and $(1-t, -2+t, t)^T = (-2, 1, 3)^T$. Hence,

$$\begin{pmatrix} 1 \\ 0 \\ 2 \end{pmatrix} = \mathbf{w} = -2\mathbf{v}_1 + 1\mathbf{v}_2 + 3\mathbf{v}_3 = -2\begin{pmatrix} 1 \\ -4 \\ 4 \end{pmatrix} + 1\begin{pmatrix} 0 \\ -2 \\ 1 \end{pmatrix} + 3\begin{pmatrix} 1 \\ -2 \\ 3 \end{pmatrix},$$

as is easily checked. There are other solutions. For example, choose $t = 0$ and $(1-t, -2+t, t)^T = (1, -2, 0)^T$. Hence,

$$\begin{pmatrix} 1 \\ 0 \\ 2 \end{pmatrix} = \mathbf{w} = 1\mathbf{v}_1 - 2\mathbf{v}_2 + 0\mathbf{v}_3 = 1\begin{pmatrix} 1 \\ -4 \\ 4 \end{pmatrix} - 2\begin{pmatrix} 0 \\ -2 \\ 1 \end{pmatrix} + 0\begin{pmatrix} 1 \\ -2 \\ 3 \end{pmatrix},$$

as is easily checked.

9. Set up the augmented matrix

$$[\mathbf{v}_1, \mathbf{v}_2, \mathbf{w}] = \begin{pmatrix} 1 & 2 & w_1 \\ -2 & 3 & w_2 \end{pmatrix}.$$

Add 2 times row 1 to row 2.

$$\begin{pmatrix} 1 & 2 & w_1 \\ 0 & 7 & 2w_1 + w_2 \end{pmatrix}$$

Backsolving, the second equation (row) gives

$$7x_2 = 2w_1 + w_2$$
$$x_2 = \frac{2}{7}w_1 + \frac{1}{7}w_2.$$

Substitute this result in the first equation (row).

$$x_1 + 2\left(\frac{2}{7}w_1 + \frac{1}{7}w_2\right) = w_1$$
$$x_1 = \frac{3}{7}w_1 - \frac{2}{7}w_2.$$

Hence,

$$\begin{pmatrix} w_1 \\ w_2 \end{pmatrix} = \mathbf{w} = \left(\frac{3}{7}w_1 - \frac{2}{7}w_2\right)\begin{pmatrix} 1 \\ -2 \end{pmatrix} + \left(\frac{2}{7}w_1 + \frac{1}{7}w_2\right)\begin{pmatrix} 2 \\ 3 \end{pmatrix}.$$

11. Set up the augmented matrix and reduce.

$$[\mathbf{v}_1, \mathbf{v}_2, \mathbf{v}_3] = \begin{pmatrix} 1 & 1 & 3 \\ 1 & 2 & 4 \\ -2 & 2 & -2 \end{pmatrix} \xrightarrow{\text{rref}} \begin{pmatrix} 1 & 0 & 2 \\ 0 & 1 & 1 \\ 0 & 0 & 0 \end{pmatrix}$$

This system has solutions $(-2t, -t, t)^T$, t a number. Thus, the nullspace is spanned by the single vector$(-2, -1, 1)^T$. Thus,

$$-2\mathbf{v}_1 - \mathbf{v}_2 + \mathbf{v}_3 = -2\begin{pmatrix} 1 \\ 1 \\ -2 \end{pmatrix} - \begin{pmatrix} 1 \\ 2 \\ 2 \end{pmatrix} + \begin{pmatrix} 3 \\ 4 \\ -2 \end{pmatrix} = \begin{pmatrix} 0 \\ 0 \\ 0 \end{pmatrix} = \mathbf{0}.$$

13. Set up the augmented matrix and reduce.

$$[\mathbf{v}_1, \mathbf{v}_2, \mathbf{v}_3] = \begin{pmatrix} 1 & 0 & 2 \\ -2 & 1 & -1 \\ -2 & 5 & 11 \end{pmatrix} \xrightarrow{\text{rref}} \begin{pmatrix} 1 & 0 & 2 \\ 0 & 1 & 3 \\ 0 & 0 & 0 \end{pmatrix}$$

This system has solutions $(-2t, -3t, t)^T$, t a number. Thus, the nullspace is spanned by the single vector$(-2, -3, 1)^T$. Thus,

$$-2\mathbf{v}_1 - 3\mathbf{v}_2 + \mathbf{v}_3 = -2\begin{pmatrix} 1 \\ -2 \\ -2 \end{pmatrix} - 3\begin{pmatrix} 0 \\ 1 \\ 5 \end{pmatrix} + \begin{pmatrix} 2 \\ -1 \\ 11 \end{pmatrix} = \begin{pmatrix} 0 \\ 0 \\ 0 \end{pmatrix} = \mathbf{0}.$$

15. Set up the augmented matrix and reduce.

$$[\mathbf{v}_1, \mathbf{v}_2, \mathbf{v}_3, \mathbf{v}_4] = \begin{pmatrix} 1 & 2 & -2 & 6 \\ 2 & 0 & 4 & -4 \\ -2 & 2 & -8 & 12 \\ 0 & 3 & -6 & 12 \end{pmatrix} \xrightarrow{\text{rref}} \begin{pmatrix} 1 & 0 & 2 & -2 \\ 0 & 1 & -2 & 4 \\ 0 & 0 & 0 & 0 \\ 0 & 0 & 0 & 0 \end{pmatrix}$$

This system has solutions $(-2s+2t, 2s-4t, s, t)^T$, s, t numbers. This can be written $s(-2, 2, 1, 0)^T + t(2, -4, 0, 1)^T$, so the null space is spanned by the vectors $(-2, 2, 1, 0)^T$ and $(2, -4, 0, 1)^T$. If we choose $s = 2$ and $t = 1$, then $s(-2, 2, 1, 0)^T + t(2, -4, 0, 1)^T = (-2, 0, 2, 1)^T$. Thus,

$$-2\mathbf{v}_1 + 0\mathbf{v}_2 + 2\mathbf{v}_3 + \mathbf{v}_4 = -2\begin{pmatrix} 1 \\ 2 \\ -2 \\ 0 \end{pmatrix} + 0\begin{pmatrix} 2 \\ 0 \\ 2 \\ 3 \end{pmatrix} + 2\begin{pmatrix} -2 \\ 4 \\ -8 \\ -6 \end{pmatrix} + \begin{pmatrix} 6 \\ -4 \\ 12 \\ 12 \end{pmatrix} = \begin{pmatrix} 0 \\ 0 \\ 0 \\ 0 \end{pmatrix} = \mathbf{0}.$$

17. It is easy to see that $(1, 2)^T$ is not a scalar multiple of $(-1, 3)^T$. Thus, the vectors are independent. More formally, if

$$c_1\begin{pmatrix} 1 \\ 2 \end{pmatrix} + c_2\begin{pmatrix} -1 \\ 3 \end{pmatrix} = \begin{pmatrix} 0 \\ 0 \end{pmatrix},$$

then the augmented matrix reduces as follows.

$$\begin{pmatrix} 1 & -1 & 0 \\ 2 & 3 & 0 \end{pmatrix} \sim \begin{pmatrix} 1 & 0 & 0 \\ 0 & 1 & 0 \end{pmatrix}$$

Thus, $c_1 = c_2 = 0$ and the vectors are independent.

19. The vector $(-1, 7, 7)^T$ is not a scalar multiple of the vector $(-3, 7, -4)^T$. They are independent.

21. Set up the equation

$$c_1\begin{pmatrix} -1 \\ 7 \\ 7 \end{pmatrix} + c_2\begin{pmatrix} -3 \\ 7 \\ -4 \end{pmatrix} + c_3\begin{pmatrix} -4 \\ -14 \\ 23 \end{pmatrix} = \begin{pmatrix} 0 \\ 0 \\ 0 \end{pmatrix}.$$

The augmented matrix reduces as follows.

$$\begin{pmatrix} -1 & -3 & -4 & 0 \\ 7 & 7 & -14 & 0 \\ 7 & -4 & 23 & 0 \end{pmatrix} \sim \begin{pmatrix} 1 & 0 & 0 & 0 \\ 0 & 1 & 0 & 0 \\ 0 & 0 & 1 & 0 \end{pmatrix}$$

Therefore, $c_1 = c_2 = c_3 = 0$ and the vectors are independent.

23. Set up the equation

$$c_1\begin{pmatrix} -1 \\ 7 \\ 7 \end{pmatrix} + c_2\begin{pmatrix} -3 \\ 8 \\ -4 \end{pmatrix} + c_3\begin{pmatrix} -4 \\ -14 \\ 23 \end{pmatrix} = \begin{pmatrix} 0 \\ 0 \\ 0 \end{pmatrix}.$$

Reduce the coefficient matrix.

$$\begin{pmatrix} -1 & -3 & -4 \\ 7 & 8 & -14 \\ 7 & -4 & 23 \end{pmatrix} \sim \begin{pmatrix} 1 & 0 & 0 \\ 0 & 1 & 0 \\ 0 & 0 & 1 \end{pmatrix}$$

Thus, $c_1 = c_2 = c_3 = 0$ and the vectors are independent.

25. Reduce.

$$\begin{pmatrix} 2 & -1 \end{pmatrix} \xrightarrow{\text{rref}} \begin{pmatrix} 1 & -1/2 \end{pmatrix}$$

This system has solutions $(t/2, t)^T$, t a number. This can be written $t(1/2, 1)^T$, so the nullspace is spanned by the single vector $(1/2, 1)^T$. Since a single vector is linearly independent, a basis for the nullspace is

$$B = \left\{ \begin{pmatrix} 1/2 \\ 1 \end{pmatrix} \right\}.$$

27. Reduce.

$$\begin{pmatrix} 4 & 4 \\ -2 & -2 \end{pmatrix} \xrightarrow{\text{rref}} \begin{pmatrix} 1 & 1 \\ 0 & 0 \end{pmatrix}$$

This system has solutions $(-t, t)^T$, t a number. This can be written $t(-1, 1)^T$, so the nullspace is spanned by the single vector $(-1, 1)^T$. Since a single vector is linearly independent, a basis for the nullspace is

$$B = \left\{ \begin{pmatrix} -1 \\ 1 \end{pmatrix} \right\}.$$

29. Reduce.

$$\begin{pmatrix} 1 & 1 & 1 \\ -5 & -2 & -5 \\ 1 & 0 & 1 \end{pmatrix} \xrightarrow{\text{rref}} \begin{pmatrix} 1 & 0 & 1 \\ 0 & 1 & 0 \\ 0 & 0 & 0 \end{pmatrix}$$

This system has solutions $(-t, 0, t)^T$, t a number. This can be written $t(-1, 0, 1)^T$, so the nullspace is spanned by the single vector $(-1, 0, 1)^T$. Since a single vector is linearly independent, a basis for the nullspace is

$$B = \left\{ \begin{pmatrix} -1 \\ 0 \\ 1 \end{pmatrix} \right\}$$

31. Reduce.

$$\begin{pmatrix} 2 & -1 & 0 & 1 \\ -1 & 1 & 1 & 0 \\ 1 & 1 & 3 & 2 \\ -3 & 3 & 3 & 0 \end{pmatrix} \xrightarrow{\text{rref}} \begin{pmatrix} 1 & 0 & 1 & 1 \\ 0 & 1 & 2 & 1 \\ 0 & 0 & 0 & 0 \\ 0 & 0 & 0 & 0 \end{pmatrix}$$

This system has solutions $(-s-t, -2s-t, s, t)^T$, s, t numbers. This can be written $s(-1, -2, 1, 0)^T + t(-1, -1, 0, 1)^T$, so the nullspace is spanned by the vectors $(-1, -2, 1, 0)^T$ and $(-1, -1, 0, 1)^T$. Because these vectors are not multiples of one another, they are independent. Thus, a basis for this nullspace is

$$B = \left\{ \begin{pmatrix} -1 \\ -2 \\ 1 \\ 0 \end{pmatrix}, \begin{pmatrix} -1 \\ -1 \\ 0 \\ 1 \end{pmatrix} \right\}$$

.

33. Since the vectors $\mathbf{v}_1 = (1, 2)^T$ and $\mathbf{v}_2 = (-1, 3)^T$ are not multiples of one another, they are linearly independent and they form a basis for their span. Since there are two vectors in the basis, the dimension is 2.

$$B = \left\{ \begin{pmatrix} 1 \\ 2 \end{pmatrix}, \begin{pmatrix} -1 \\ 3 \end{pmatrix} \right\}.$$

35. Since the vectors $\mathbf{v}_1 = (-1, 7, 7)^T$ and $\mathbf{v}_2 = (-3, 7, -4)^T$ are not multiples of one another, they are linearly independent and they form a basis for their span. Since there are two vectors in the basis, the dimension is 2.

$$B = \left\{ \begin{pmatrix} -1 \\ 7 \\ 7 \end{pmatrix}, \begin{pmatrix} -3 \\ 7 \\ -4 \end{pmatrix} \right\}.$$

37. Place the vectors in a matrix and reduce.

$$\begin{pmatrix} -1 & -3 & -4 \\ 7 & 7 & -14 \\ 7 & -4 & 23 \end{pmatrix} \xrightarrow{\text{rref}} \begin{pmatrix} 1 & 0 & 0 \\ 0 & 1 & 0 \\ 0 & 0 & 1 \end{pmatrix}$$

Thus, the nullspace contains only the zero vector and the vectors are independent. Thus, a basis for $\text{span}\{\mathbf{v}_1, \mathbf{v}_2, \mathbf{v}_3\}$ is

$$B = \left\{ \begin{pmatrix} -1 \\ 7 \\ 7 \end{pmatrix}, \begin{pmatrix} -3 \\ 7 \\ -4 \end{pmatrix}, \begin{pmatrix} -4 \\ -14 \\ 23 \end{pmatrix} \right\}.$$

The dimension is 3.

39. Place the vectors in a matrix and reduce.

$$\begin{pmatrix} -1 & -3 & -4 \\ 7 & 8 & -14 \\ 7 & -4 & 23 \end{pmatrix} \xrightarrow{\text{rref}} \begin{pmatrix} 1 & 0 & 0 \\ 0 & 1 & 0 \\ 0 & 0 & 1 \end{pmatrix}$$

Thus, the nullspace contains only the zero vector and the vectors are independent and form a basis for the span$\{\mathbf{v}_1, \mathbf{v}_2, \mathbf{v}_3\}$.

$$B = \left\{ \begin{pmatrix} -1 \\ 7 \\ 7 \end{pmatrix}, \begin{pmatrix} -3 \\ 8 \\ -4 \end{pmatrix}, \begin{pmatrix} -4 \\ -14 \\ 23 \end{pmatrix} \right\}.$$

The dimension is 3.

41. Note that $\mathbf{v}_1 = (1, -2)^T$ is a multiple of $\mathbf{v}_2 = (2, -4)^T$. Therefore, we can throw away the vector $\mathbf{v}_2$ without losing any information and span$\{\mathbf{v}_1, \mathbf{v}_2\}$ = span$\{\mathbf{v}_1\}$. Geometrically, all multiples of $\mathbf{v}_1 = (1, 2)^T$ form a line through the origin in $\mathbf{R}^2$.

43. Place the vectors in a matrix and reduce.

$$\begin{pmatrix} -1 & -3 & 4 \\ 3 & -2 & -1 \\ 3 & 2 & -5 \end{pmatrix} \xrightarrow{\text{rref}} \begin{pmatrix} 1 & 0 & -1 \\ 0 & 1 & -1 \\ 0 & 0 & 0 \end{pmatrix}$$

This system has solutions $(t, t, t)^T$, t a number. This can be written $t(1, 1, 1)^T$, so the nullspace is spanned by the single vector $(1, 1, 1)^T$. Hence, $\mathbf{v}_1 + \mathbf{v}_2 + \mathbf{v}_3 = \mathbf{0}$, or equivalently, $\mathbf{v}_3 = -\mathbf{v}_1 - \mathbf{v}_2$. Hence, $\mathbf{v}_3$ is redundant information and span$\{\mathbf{v}_1, \mathbf{v}_2 . \mathbf{v}_3\}$ = span$\{\mathbf{v}_1, \mathbf{v}_2\}$. Because the remaining vectors $\mathbf{v}_1$ and $\mathbf{v}_2$ are not multiples of one another, they are independent and form a basis of dimension two for the span. Hence, this describes a plane in $\mathbf{R}^3$ that passes through the origin.

Section 7.6. Square Matrices

1. Reduce the coefficient matrix.

$$A = \begin{pmatrix} 1 & -2 \\ 2 & -4 \end{pmatrix} \xrightarrow{\text{rref}} \begin{pmatrix} 1 & -2 \\ 0 & 0 \end{pmatrix}$$

It is easy to see that $(2, 1)^T$ is in the nullspace. Hence, $2\mathbf{a}_1 + \mathbf{a}_2 = \mathbf{0}$, or equivalently, $\mathbf{a}_2 = -2\mathbf{a}_1$, where $\mathbf{a}_1$ and $\mathbf{a}_2$ are the columns of the coefficient matrix A. Hence, span$\{\mathbf{a}_1, \mathbf{a}_2\}$ = span$\{\mathbf{a}_1\}$, so the span is a line through the origin in the direction of $(1, 2)^T$. Thus, the system $A\mathbf{x} = \mathbf{b}$ has solutions only if $\mathbf{b}$ lies on this line. Thus, $(b_1, b_2)^T$ must be a multiple of $(1, 2)^T$. However, $(b_1, b_2)^T = \lambda(1, 2)^T$ implies that $b_1 = \lambda$ and $b_2 = 2\lambda$. Dividing, $b_1/b_2 = 1/2$, which leads to $-2b_1 + b_2 = 0$. Because the system only has solutions for $\mathbf{b} = (b_1, b_2)^T$ on this line, and not for all values of $\mathbf{b}$, the coefficient matrix A is singular.

Alternatively, start with the augmented matrix

$$\begin{pmatrix} 1 & -2 & b_1 \\ 2 & -4 & b_2 \end{pmatrix}.$$

Add -2 times row 1 to row 2 to obtain

$$\begin{pmatrix} 1 & -2 & b_1 \\ 0 & 0 & -2b_1 + b_2 \end{pmatrix}.$$

This system is inconsistent unless $-2b_1 + b_2 = 0$.

3. Reduce the coefficient matrix.

$$\begin{pmatrix} 3 & 3 & -3 \\ -1 & -1 & 1 \\ 3 & 5 & -1 \end{pmatrix} \xrightarrow{\text{rref}} \begin{pmatrix} 1 & 0 & -2 \\ 0 & 1 & 1 \\ 0 & 0 & 0 \end{pmatrix}$$

It is easy to see that $(2, -1, 1)$ is in the nullspace of the coefficient matrix A. Thus, $2\mathbf{a}_1 - \mathbf{a}_2 + \mathbf{a}_3 = \mathbf{0}$, or $\mathbf{a}_3 = -2\mathbf{a}_1 + \mathbf{a}_2$, where $\mathbf{a}_1$, $\mathbf{a}_2$, and $\mathbf{a}_3$ are the columns of the matrix A. Thus, $\mathbf{a}_3$ is redundant, and span$\{\mathbf{a}_1, \mathbf{a}_2, \mathbf{a}_3\}$ = span$\{\mathbf{a}_1, \mathbf{a}_2\}$. Thus, the span is the set of all linear combinations of $\mathbf{a}_1$ and $\mathbf{a}_2$, which is a plane in $\mathbf{R}^3$ passing through the origin. Thus, unless $\mathbf{b} = (b_1, b_2, b_3)^T$ lies on this plane, the system is inconsistent.

Now, set up the augmented matrix

$$\begin{pmatrix} 3 & 3 & -3 & b_1 \\ -1 & -1 & 1 & b_2 \\ 3 & 5 & -1 & b_3 \end{pmatrix}$$

Add 1/3 times row 1 to row 2.

$$\begin{pmatrix} 3 & 3 & -3 & b_1 \\ 0 & 0 & 0 & (1/3)b_1 + b_2 \\ 3 & 5 & -1 & b_3 \end{pmatrix}$$

Note the second row. The system will be inconsistent unless $(1/3)b_1 + b_2 = 0$, which is a plane in $\mathbf{R}^3$. Therefore, the coefficient matrix is singular.

Can you show that the plane $(1/3)b_1 + b_2 = 0$ is spanned by $\mathbf{a}_1$ and $\mathbf{a}_2$? *Hint: Consider the nullspace of the matrix* $\begin{pmatrix} 1/3 & 1 & 0 \end{pmatrix}$.

5. Subtract 3 times row 1 from row 2.

$$\begin{pmatrix} 1 & 2 \\ 3 & -4 \end{pmatrix} \rightarrow \begin{pmatrix} 1 & 2 \\ 0 & -10 \end{pmatrix}$$

Each diagonal element is nonzero. Thus, the matrix in nonsingular.

7. Add 2 times row 1 to row 2. Add 2 times row 1 to row 3.

$$\begin{pmatrix} 1 & 0 & -1 \\ -2 & 3 & 3 \\ -2 & 3 & 1 \end{pmatrix} \rightarrow \begin{pmatrix} 1 & 0 & -1 \\ 0 & 3 & 1 \\ 0 & 3 & -1 \end{pmatrix}$$

Subtract row 2 from row 3.

$$\begin{pmatrix} 1 & 0 & -1 \\ 0 & 3 & 1 \\ 0 & 3 & -1 \end{pmatrix} \rightarrow \begin{pmatrix} 1 & 0 & -1 \\ 0 & 3 & 1 \\ 0 & 0 & -2 \end{pmatrix}$$

Each diagonal element is nonzero. Thus, the matrix in nonsingular.

9. Add 1/2 times row 1 to row 2. Add 3/2 times row 1 to row 3.

$$\begin{pmatrix} 2 & 1 & -1 \\ -1 & -3 & -2 \\ -3 & -2 & 1 \end{pmatrix} \rightarrow \begin{pmatrix} 2 & 1 & -1 \\ 0 & -5/2 & -5/2 \\ 0 & -1/2 & -1/2 \end{pmatrix}$$

Add $-1/5$ times row 2 to row 3.

$$\begin{pmatrix} 2 & 1 & -1 \\ 0 & -5/2 & -5/2 \\ 0 & -1/2 & -1/2 \end{pmatrix} \rightarrow \begin{pmatrix} 2 & 1 & -1 \\ 0 & -5/2 & -5/2 \\ 0 & 0 & 0 \end{pmatrix}$$

Note the presence of a zero element on the diagonal. Thus, the matrix in singular.

11. Add 3 times row 1 to row 3.

$$\begin{pmatrix} -1 & -1 & 1 \\ 0 & -2 & 4 \\ 3 & 0 & 3 \end{pmatrix} \rightarrow \begin{pmatrix} -1 & -1 & 1 \\ 0 & -2 & 4 \\ 0 & -3 & 6 \end{pmatrix}$$

Add $-3/2$ times row 2 to row 3.

$$\begin{pmatrix} -1 & -1 & 1 \\ 0 & -2 & 4 \\ 0 & -3 & 6 \end{pmatrix} \rightarrow \begin{pmatrix} -1 & -1 & 1 \\ 0 & -2 & 4 \\ 0 & 0 & 0 \end{pmatrix}$$

Note the presence of a zero element on the diagonal. Thus, the matrix in singular.

13. Reduce A.

$$A = \begin{pmatrix} 1 & -2 \\ 2 & -4 \end{pmatrix} \xrightarrow{\text{rref}} \begin{pmatrix} 1 & -2 \\ 0 & 0 \end{pmatrix}.$$

The solutions are

$$\mathbf{x} = \begin{pmatrix} x_1 \\ x_2 \end{pmatrix} = \begin{pmatrix} 2x_2 \\ x_2 \end{pmatrix} = x_2 \begin{pmatrix} 2 \\ 1 \end{pmatrix},$$

where x_2 is free. The coefficient matrix is singular.

15. Reduce A.

$$A = \begin{pmatrix} 1 & 1 & 2 & 0 \\ 1 & 0 & 1 & 0 \\ 1 & 1 & 2 & 0 \end{pmatrix} \xrightarrow{\text{rref}} \begin{pmatrix} 1 & 0 & 1 & 0 \\ 0 & 1 & 1 & 0 \\ 0 & 0 & 0 & 0 \end{pmatrix}$$

The solutions are

$$\mathbf{x} = \begin{pmatrix} x_1 \\ x_2 \\ x_3 \end{pmatrix} = \begin{pmatrix} -x_3 \\ -x_3 \\ x_3 \end{pmatrix} = x_3 \begin{pmatrix} -1 \\ -1 \\ 0 \end{pmatrix},$$

where x_3 is free. The coefficient matrix is singular.

17. Reduce A.

$$A = \begin{pmatrix} 0 & -1 & -2 \\ -5 & 2 & -1 \\ -4 & 2 & 0 \end{pmatrix} \xrightarrow{\text{rref}} \begin{pmatrix} 1 & 0 & 1 \\ 0 & 1 & 2 \\ 0 & 0 & 0 \end{pmatrix}$$

The solutions are

$$\mathbf{x} = \begin{pmatrix} x_1 \\ x_2 \\ x_3 \end{pmatrix} = \begin{pmatrix} -x_3 \\ -2x_3 \\ x_3 \end{pmatrix} = x_3 \begin{pmatrix} -1 \\ -2 \\ 1 \end{pmatrix},$$

where x_3 is free. The coefficient matrix is singular.

19. Reduce A.

$$A = \begin{pmatrix} 0 & 1 & 3 & 0 \\ -2 & 1 & -3 & 1 \\ -1 & -2 & 3 & 0 \\ 2 & -1 & -1 & 2 \end{pmatrix} \xrightarrow{\text{rref}} \begin{pmatrix} 1 & 0 & 0 & 0 \\ 0 & 1 & 0 & 0 \\ 0 & 0 & 1 & 0 \\ 0 & 0 & 0 & 1 \end{pmatrix}$$

The only solution is $\mathbf{x} = (0, 0, 0, 0)^T$. The matrix is nonsingular.

21. The augmented matrix $[A\ I]$ reduces to

$$\begin{pmatrix} 0 & -4 & 1 & 0 \\ -1 & 2 & 0 & 1 \end{pmatrix} \xrightarrow{\text{rref}} \begin{pmatrix} 1 & 0 & -1/2 & -1 \\ 0 & 1 & -1/4 & 0 \end{pmatrix}$$

Hence A is nonsingular and

$$A^{-1} = \begin{pmatrix} -1/2 & -1 \\ -1/4 & 0 \end{pmatrix}.$$

23. The augmented matrix $[A\ I]$ reduces to

$$\begin{pmatrix} 1 & 1 & 1 & 1 & 0 & 0 \\ 0 & 1 & 1 & 0 & 1 & 0 \\ 0 & 0 & 1 & 0 & 0 & 1 \end{pmatrix} \xrightarrow{\text{rref}} \begin{pmatrix} 1 & 0 & 0 & 1 & -1 & 0 \\ 0 & 1 & 0 & 0 & 1 & -1 \\ 0 & 0 & 1 & 0 & 0 & 1 \end{pmatrix}.$$

Thus, matrix A is nonsingular and

$$A^{-1} = \begin{pmatrix} 1 & -1 & 0 \\ 0 & 1 & -1 \\ 0 & 0 & 1 \end{pmatrix}.$$

25. The augmented matrix $[A\ I]$ reduces to

$$\begin{pmatrix} 1 & 2 & -3 & 1 & 0 & 0 \\ 0 & 0 & 0 & 0 & 1 & 0 \\ 0 & 1 & 1 & 0 & 0 & 1 \end{pmatrix} \xrightarrow{\text{rref}} \begin{pmatrix} 1 & 0 & -5 & 1 & 0 & -2 \\ 0 & 1 & 1 & 0 & 0 & 1 \\ 0 & 0 & 0 & 0 & 1 & 0 \end{pmatrix}.$$

Matrix A is singular and has no inverse.

27. The augmented matrix $[A\ I]$ reduces to

$$\begin{pmatrix} 3 & -1 & -3 & -1 & 1 & 0 & 0 & 0 \\ 0 & -3 & -4 & 1 & 0 & 1 & 0 & 0 \\ -2 & 1 & 2 & 1 & 0 & 0 & 1 & 0 \\ 1 & 1 & 3 & -3 & 0 & 0 & 0 & 1 \end{pmatrix} \xrightarrow{\text{rref}} \begin{pmatrix} 1 & 0 & 0 & -1 & 0 & -1/12 & -5/12 & 1/6 \\ 0 & 1 & 0 & 1 & 0 & -2/3 & -1/3 & -2/3 \\ 0 & 0 & 1 & -1 & 0 & 1/4 & 1/4 & 1/2 \\ 0 & 0 & 0 & 0 & 1 & 1/3 & 5/3 & 1/3 \end{pmatrix}$$

Hence A is singular.

29. The coefficient matrix, in row echelon form, is

$$\begin{pmatrix} 1 & 2 \\ 2 & 4 \end{pmatrix} \sim \begin{pmatrix} 1 & 2 \\ 0 & 0 \end{pmatrix}.$$

The presence of a zero diagonal element in the row echelon form indicates that the coefficient matrix is singular. The system does not have a unique solution.

31. In matrix form,

$$\begin{pmatrix} 1 & 2 \\ 2 & 1 \end{pmatrix} \begin{pmatrix} x_1 \\ x_2 \end{pmatrix} = \begin{pmatrix} 1 \\ 0 \end{pmatrix},$$

and the coefficient matrix has row echelon form

$$\begin{pmatrix} 1 & 2 \\ 2 & 1 \end{pmatrix} \sim \begin{pmatrix} 1 & 2 \\ 0 & -3 \end{pmatrix}.$$

All diagonal elements in the row echelon form are nonzero, so the coefficient matrix is nonsingular and the system has a unique solution.

33. The coefficient matrix has row echelon form

$$\begin{pmatrix} 1 & 0 & 3 \\ -1 & 1 & -1 \\ 0 & 2 & 4 \end{pmatrix} \sim \begin{pmatrix} 1 & 0 & 3 \\ 0 & 1 & 2 \\ 0 & 0 & 0 \end{pmatrix}.$$

The presence of a zero on the diagonal of the row echelon form indicates that the coefficient matrix in singular. The system does not have a unique solution.

35. Suppose that A is invertible. Then:

- A is nonsingular.
- The only solution of the homogeneous system $A\mathbf{y} = \mathbf{0}$ is the zero vector $\mathbf{0}$.
- The equation $A\mathbf{x} = \mathbf{b}$ has a unique solution for any right hand side $\mathbf{b}$.
- If A is put into row echelon form then the diagonal entries of he result are nonzero.
- If A is put into reduced row echelon form then the result is the identity matrix.

Section 7.7. Determinants

1. The area of the triangle spanned by the vectors $\mathbf{x}_1 = (x_1, y_1)^T$ and $\mathbf{x}_2 = (x_2, y_2)^T$ is calculated by subtracting the areas of the triangular regions I, II, and III from the area of the bounding rectangle.

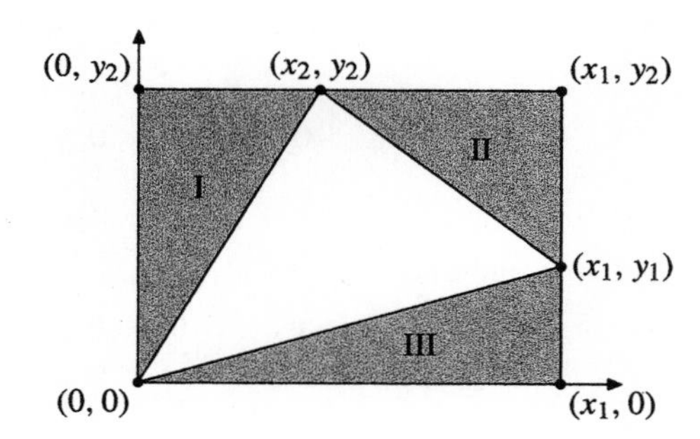

Thus, the area of the triangle is

$$\begin{aligned} A &= x_1y_2 - \frac{1}{2}x_2y_2 - \frac{1}{2}(x_1 - x_2)(y_2 - y_1) \\ &\quad - \frac{1}{2}x_1y_1, \\ &= \frac{1}{2}x_1y_2 - \frac{1}{2}x_2y_1, \\ &= \frac{1}{2}\begin{vmatrix} x_1 & x_2 \\ y_1 & y_2 \end{vmatrix}. \end{aligned}$$

3. Estimate the area by counting square units inside the parallelogram in

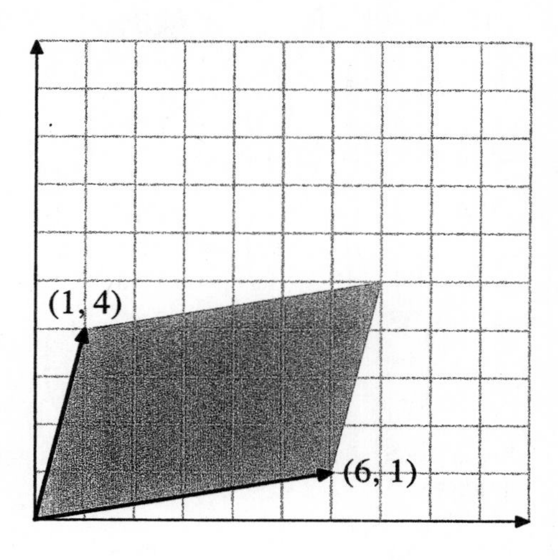

The determinant is

$$\begin{aligned} |\mathbf{v}_1, \mathbf{v}_2| &= \begin{vmatrix} 1 & 6 \\ 4 & 1 \end{vmatrix} = (1)(1) - (4)(6) \\ &= 1 - 24 = -23. \end{aligned}$$

Note that the determinant is the negative of the area.

5. Estimate the area by counting square units inside the parallelogram in

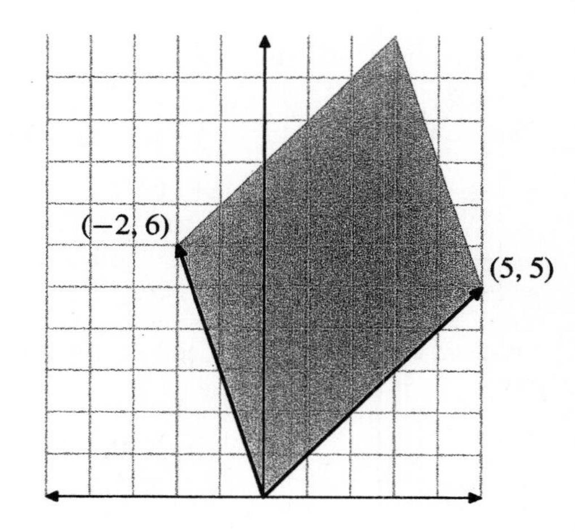

The determinant is

$$|\mathbf{v}_1, \mathbf{v}_2| = \begin{vmatrix} 5 & -2 \\ 5 & 6 \end{vmatrix} = (5)(6) - (5)(-2)$$
$$= 30 + 10 = 40.$$

7. Using row operations we get

$$\begin{pmatrix} 3 & 0 & 0 \\ 3 & 6 & 3 \\ -18 & -18 & -9 \end{pmatrix} \rightarrow \begin{pmatrix} 3 & 0 & 0 \\ 0 & 6 & 3 \\ -18 & -18 & -9 \end{pmatrix}$$
$$\rightarrow \begin{pmatrix} 3 & 0 & 0 \\ 0 & 6 & 3 \\ 0 & -18 & -9 \end{pmatrix} \rightarrow \begin{pmatrix} 3 & 0 & 0 \\ 0 & 6 & 3 \\ 0 & 0 & 0 \end{pmatrix}$$

Since the final matrix has a row of zeros, the determinant is 0.

9. Add 3 times row 1 to row 2; add -4 times row 1 to row 3.

$$\begin{vmatrix} 1 & 0 & 4 \\ -3 & 3 & -2 \\ 4 & -1 & -2 \end{vmatrix} = \begin{vmatrix} 1 & 0 & 4 \\ 0 & 3 & 10 \\ 0 & -1 & -18 \end{vmatrix}$$

Add $1/3$ times row 2 to row 3.

$$= \begin{vmatrix} 1 & 0 & 4 \\ 0 & 3 & 10 \\ 0 & 0 & -44/3 \end{vmatrix}$$
$$= (1)(3)\left(-\frac{44}{3}\right)$$
$$= -44$$

11. Swap rows 1 and 3.

$$\begin{vmatrix} 2 & -1 & 3 & 4 \\ 0 & 2 & -2 & 0 \\ -1 & 2 & 0 & 0 \\ -1 & 3 & 1 & 2 \end{vmatrix} = -\begin{vmatrix} -1 & 2 & 0 & 0 \\ 0 & 2 & -2 & 0 \\ 2 & -1 & 3 & 4 \\ -1 & 3 & 1 & 2 \end{vmatrix}$$

Add 2 times row 1 to row 3; add -1 times row 1 to row 4.

$$= -\begin{vmatrix} -1 & 2 & 0 & 0 \\ 0 & 2 & -2 & 0 \\ 0 & 3 & 3 & 4 \\ 0 & 1 & 1 & 2 \end{vmatrix}$$

Factor a 2 out of row 2.

$$= -2\begin{vmatrix} -1 & 2 & 0 & 0 \\ 0 & 1 & -1 & 0 \\ 0 & 3 & 3 & 4 \\ 0 & 1 & 1 & 2 \end{vmatrix}$$

Add -3 times row 2 to row 3; add -1 times row 2 to row 4.

$$= -2\begin{vmatrix} -1 & 2 & 0 & 0 \\ 0 & 1 & -1 & 0 \\ 0 & 0 & 6 & 4 \\ 0 & 0 & 2 & 2 \end{vmatrix}$$

Swap rows 3 and 4; then factor out a 2 from the new row 3.

$$= (2)(2)\begin{vmatrix} -1 & 2 & 0 & 0 \\ 0 & 1 & -1 & 0 \\ 0 & 0 & 1 & 1 \\ 0 & 0 & 6 & 4 \end{vmatrix}$$

Add -6 times row 3 to row 4.

$$= (2)(2)\begin{vmatrix} -1 & 2 & 0 & 0 \\ 0 & 1 & -1 & 0 \\ 0 & 0 & 1 & 1 \\ 0 & 0 & 0 & -2 \end{vmatrix}$$
$$= (2)(2)(-1)(1)(1)(-2)$$
$$= 8$$

13. (a) Suppose that row i is a scalar multiple of row j, say

$$|A| = \begin{vmatrix} \vdots & \vdots & & \vdots \\ ra_{j1} & ra_{j2} & \dots & ra_{jn} \\ \vdots & \vdots & & \vdots \\ a_{j1} & a_{j2} & \dots & a_{jn} \\ \vdots & \vdots & & \vdots \end{vmatrix}$$

Then, adding $-r$ times for j to row i does not change the value of the determinant. Thus,

$$= \begin{vmatrix} \vdots & \vdots & & \vdots \\ 0 & 0 & \dots & 0 \\ \vdots & \vdots & & \vdots \\ a_{j1} & a_{j2} & \dots & a_{jn} \\ \vdots & \vdots & & \vdots \end{vmatrix}$$
$$= 0,$$

because it has a zero row. The same result applies to the columns. Suppose column i is a scalar multiple of column j, say

$$|A| = \begin{vmatrix} \dots & ra_{1j} & \dots & a_{1j} & \dots \\ \dots & ra_{2j} & \dots & a_{2j} & \dots \\ & \vdots & & \vdots & \\ \dots & ra_{nj} & \dots & a_{nj} & \dots \end{vmatrix}$$

But adding $-r$ times column j to column i does not change the value of the determinant. Thus,

$$= \begin{vmatrix} \dots & 0 & \dots & a_{1j} & \dots \\ \dots & 0 & \dots & a_{2j} & \dots \\ & \vdots & & \vdots & \\ \dots & 0 & \dots & a_{nj} & \dots \end{vmatrix}$$
$$= 0,$$

because A has a zero column.

(b) The first matrix has zero determinant because the third row has all zeros. Similarly, the second matrix has zero determinant because the third column has all zeros. The third matrix has zero determinant because its second row is a scalar multiple of its first row. The fourth matrix has zero determinant because the second column is a scalar multiple of its first column.

15. The transpose of a lower triangular matrix is upper triangular. The determinant of the transpose equals the determinant of the original. Note that transposing a lower triangular matrix does not dislodge the diagonal elements. Thus, the determinant of a lower triangular matrix is the product of its diagonal elements.

$$|A| = \begin{vmatrix} 1 & 0 & 0 & 0 \\ 1 & 2 & 0 & 0 \\ 1 & 1 & 3 & 0 \\ 1 & 1 & 1 & 4 \end{vmatrix} = (1)(2)(3)(4) = 24$$

17. Expand across the first row.

$$\begin{vmatrix} 5 & 6 & 4 \\ -4 & -9 & -8 \\ 4 & 6 & 5 \end{vmatrix}$$
$$= 5\begin{vmatrix} -9 & -8 \\ 6 & 5 \end{vmatrix} - 6\begin{vmatrix} -4 & -8 \\ 4 & 5 \end{vmatrix} + 4\begin{vmatrix} -4 & -9 \\ 4 & 6 \end{vmatrix}$$
$$= 5(-45+48) - 6(-20+32) + 4(-24+36)$$
$$= 15 - 72 + 48$$
$$= -9$$

19. There is a zero in the first column, so we will expand by that column.

$$\det A = 1 \cdot \det\begin{pmatrix} 6 & -2 \\ 3 & 2 \end{pmatrix} + (-2) \cdot \det\begin{pmatrix} 2 & -3 \\ 6 & -2 \end{pmatrix}$$
$$= 18 - 28 = -10$$

21. Expand across the third row.

$$\begin{vmatrix} 3 & -3 & -2 & 1 \\ 2 & 0 & -2 & -1 \\ 1 & -2 & 0 & 0 \\ 4 & -1 & -4 & -2 \end{vmatrix}$$
$$= 1\begin{vmatrix} -3 & -2 & -1 \\ 0 & -2 & -1 \\ -1 & -4 & -2 \end{vmatrix} + 2\begin{vmatrix} 3 & -2 & -1 \\ 2 & -2 & -1 \\ 4 & -4 & -2 \end{vmatrix}$$

Expand down the first column of the first matrix and the first row of the second matrix.

$$= 1\left\{-3\begin{vmatrix} -2 & -1 \\ -4 & -2 \end{vmatrix} - 1\begin{vmatrix} -2 & -1 \\ -2 & -1 \end{vmatrix}\right\}$$
$$+ 2\left\{3\begin{vmatrix} -2 & -1 \\ -4 & -2 \end{vmatrix} + 2\begin{vmatrix} 2 & -1 \\ 4 & -2 \end{vmatrix} - 1\begin{vmatrix} 2 & -2 \\ 4 & -4 \end{vmatrix}\right\}$$

Note that in each 2×2 determinant, the second row is a multiple of the first. Therefore, they all equal zero and the determinant of the original 4×4 is also zero.

23. The determinant of A is

$$|A| = \begin{vmatrix} -1 & 2 \\ 2 & -4 \end{vmatrix} = 0.$$

To find the nullspace, reduce the augmented matrix $[A\ \vec{0}]$.

$$\begin{pmatrix} -1 & 2 & 0 \\ 2 & -4 & 0 \end{pmatrix} \sim \begin{pmatrix} 1 & -2 & 0 \\ 0 & 0 & 0 \end{pmatrix}$$

The solutions are

$$\mathbf{x} = \begin{pmatrix} x_1 \\ x_2 \end{pmatrix} = \begin{pmatrix} 2x_2 \\ x_2 \end{pmatrix} = x_2 \begin{pmatrix} 2 \\ 1 \end{pmatrix},$$

where x_2 is free. A basis for the nullspace is

$$B = \left\{ \begin{pmatrix} 2 \\ 1 \end{pmatrix} \right\}.$$

Because the nullspace is nontrivial,

$$x_1 \begin{pmatrix} -1 \\ 2 \end{pmatrix} + x_4 \begin{pmatrix} 2 \\ -4 \end{pmatrix} = \begin{pmatrix} 0 \\ 0 \end{pmatrix}$$

has nontrivial solutions, so the columns of A are dependent.

25. The determinant of A is

$$|A| = \begin{vmatrix} 2 & -1 \\ 1 & 0 \end{vmatrix} = 1$$

To find the nullspace, reduce the augmented matrix $[A\ \vec{0}]$.

$$\begin{pmatrix} 2 & -1 & 0 \\ 1 & 0 & 0 \end{pmatrix} \sim \begin{pmatrix} 1 & 0 & 0 \\ 0 & 1 & 0 \end{pmatrix}$$

The nullspace contains only the zero vector. Thus, the only solution of

$$x_1 \begin{pmatrix} 2 \\ 1 \end{pmatrix} + x_2 \begin{pmatrix} -1 \\ 0 \end{pmatrix} = \begin{pmatrix} 0 \\ 0 \end{pmatrix}$$

is $x_1 = x_2 = 0$ and the columns are independent.

27. The determinant is 0. The reduced row echelon form is

$$\begin{pmatrix} 1 & -2 & -4 \\ 2 & 1 & 2 \\ 3 & 0 & 0 \end{pmatrix} \sim \begin{pmatrix} 1 & 0 & 0 \\ 0 & 1 & 2 \\ 0 & 0 & 0 \end{pmatrix}.$$

A basis for the nullspace is

$$B = \left\{ \begin{pmatrix} 0 \\ -2 \\ 1 \end{pmatrix} \right\}.$$

Thus, the equation

$$x_1 \begin{pmatrix} 1 \\ 2 \\ 3 \end{pmatrix} + x_2 \begin{pmatrix} -2 \\ 1 \\ 0 \end{pmatrix} + x_3 \begin{pmatrix} -4 \\ 2 \\ 0 \end{pmatrix} = \begin{pmatrix} 0 \\ 0 \\ 0 \end{pmatrix}$$

has nontrivial solutions and the columns are dependent.

29. The determinant is 1. The reduced row echelon form is

$$\begin{pmatrix} 1 & 1 & 2 \\ -1 & 1 & 5 \\ 1 & 0 & -1 \end{pmatrix} \sim \begin{pmatrix} 1 & 0 & 0 \\ 0 & 1 & 0 \\ 0 & 0 & 1 \end{pmatrix}.$$

The nullspace contains only the zero vector. Thus, the equation

$$x_1 \begin{pmatrix} 1 \\ -1 \\ 1 \end{pmatrix} + x_2 \begin{pmatrix} 1 \\ 1 \\ 0 \end{pmatrix} + x_3 \begin{pmatrix} 2 \\ 5 \\ -1 \end{pmatrix} = \begin{pmatrix} 0 \\ 0 \\ 0 \end{pmatrix}$$

has only the trivial solution and the columns are independent.

31. To have a nontrivial nullspace, the determinant must equal zero.

$$0 = \begin{vmatrix} 2 & x \\ x & 3 \end{vmatrix}$$
$$0 = 6 - x^2$$
$$x = \pm\sqrt{6}$$

33. To have a nontrivial nullspace, the determinant must equal zero.

$$0 = \begin{vmatrix} -1 & x \\ -x & 4 \end{vmatrix}$$
$$0 = -4 + x^2$$
$$x^2 = 4$$
$$x = \pm 2$$

35. To have a nontrivial nullspace, the determinant must equal zero.

$$0 = \begin{vmatrix} -1 - x & 0 \\ 3 & 2 - x \end{vmatrix}$$
$$0 = (-1 - x)(2 - x)$$
$$x = -1, 2$$

37. To have a nontrivial nullspace, the determinant must equal zero.

$$0 = \begin{vmatrix} 2-x & 0 & 0 \\ -1 & -x & 2 \\ 0 & -2 & 5-x \end{vmatrix}$$

Expand across the first row.

$$0 = (2-x)\begin{vmatrix} -x & 2 \\ -2 & 5-x \end{vmatrix}$$
$$0 = (2-x)\{-x(5-x)+4\}$$
$$0 = -(x-2)(x^2-5x+4)$$
$$0 = -(x-2)(x-1)(x-4)$$

Thus, $x = 2, 1, 4$.

39. To have a nontrivial nullspace, the determinant must equal zero.

$$0 = \begin{vmatrix} -1-x & 2 & 2 \\ 0 & -2-x & 0 \\ -1 & 4 & 2-x \end{vmatrix}$$

Expand across the second row.

$$0 = (-2-x)\begin{vmatrix} -1-x & 2 \\ -1 & 2-x \end{vmatrix}$$
$$0 = -(x+2)\{(-1-x)(2-x)+2\}$$
$$0 = -(x+2)(-x+x^2)$$
$$0 = -(x+2)x(x-1)$$

Thus, $x = -2, 0, 1$.

41. The determinant is

$$\begin{vmatrix} -2 & 3 \\ -2 & 4 \end{vmatrix} = -2.$$

Therefore, the nullspace is trivial, containing only the zero vector.

43. Add 3 times the first row to the second; add -4 times the first row to the third.

$$\begin{vmatrix} 1 & 0 & 4 \\ -3 & 3 & -2 \\ 4 & 0 & -2 \end{vmatrix} = \begin{vmatrix} 1 & 0 & 4 \\ 0 & 3 & 10 \\ 0 & 0 & -18 \end{vmatrix}$$
$$= -54$$

The nullspace is trivial, containing only the zero vector.

45. Swap rows 1 and 3.

$$\begin{vmatrix} 0 & 1 & 2 \\ 2 & 0 & -2 \\ -1 & 0 & 3 \end{vmatrix} = -\begin{vmatrix} -1 & 0 & 3 \\ 2 & 0 & -2 \\ 0 & 1 & 2 \end{vmatrix}$$

Add 2 times row 1 to row 2.

$$= -\begin{vmatrix} -1 & 0 & 3 \\ 0 & 0 & 4 \\ 0 & 1 & 2 \end{vmatrix}$$

Swap rows 2 and 3.

$$= \begin{vmatrix} -1 & 0 & 3 \\ 0 & 1 & 2 \\ 0 & 0 & 4 \end{vmatrix}$$
$$= (-1)(1)(4)$$
$$= -4$$

The nullspace is trivial, containing only the zero vector.

47. Swap rows 1 and 2.

$$\begin{vmatrix} 3 & 0 & 20 & -8 \\ 2 & 3 & -2 & 0 \\ 6 & 4 & 17 & -8 \\ 16 & 10 & 50 & -23 \end{vmatrix} = -\begin{vmatrix} 2 & 3 & -2 & 0 \\ 3 & 0 & 20 & -8 \\ 6 & 4 & 17 & -8 \\ 16 & 10 & 50 & -23 \end{vmatrix}$$

Add $-3/2$ times row 1 to row 2; add -3 times row 1 to row 3; add -8 times row 1 to row 4.

$$= -\begin{vmatrix} 2 & 3 & -2 & 0 \\ 0 & -9/2 & 23 & -8 \\ 0 & -5 & 23 & -8 \\ 0 & -14 & 66 & -23 \end{vmatrix}$$

Too hard to continue with reduction. Expand down first column.

$$= -2\begin{vmatrix} -9/2 & 23 & -8 \\ -5 & 23 & -8 \\ -14 & 66 & -23 \end{vmatrix}$$
$$= \begin{vmatrix} 9 & -46 & 16 \\ -5 & 23 & -8 \\ -14 & 66 & -23 \end{vmatrix}$$

Expand across first row.

$$= 9\begin{vmatrix} 23 & -8 \\ 66 & -23 \end{vmatrix} + 46\begin{vmatrix} -5 & -8 \\ -14 & -23 \end{vmatrix} + 16\begin{vmatrix} -5 & 23 \\ -14 & 66 \end{vmatrix}$$

$$= 9(-1) + 46(3) + 16(-8)$$

$$= 1$$

The nullspace is trivial, containing only the zero vector.

49. Rows 2 and 5 are identical, so the determinant is 0. Therefore there is a nonzero vector in the nullspace.

51. False. For a counterexample, let

$$A = \begin{bmatrix} 1 & 1 \\ 1 & 0 \end{bmatrix} \quad \text{and} \quad B = \begin{bmatrix} 1 & 1 \\ 1 & 1 \end{bmatrix}.$$

Then

$$\det(A + B) = \det\left(\begin{bmatrix} 2 & 2 \\ 2 & 1 \end{bmatrix}\right) = -2.$$

But,

$$\det(A) + \det(B)$$
$$= \det\left(\begin{bmatrix} 1 & 1 \\ 1 & 0 \end{bmatrix}\right) + \det\left(\begin{bmatrix} 1 & 1 \\ 1 & 1 \end{bmatrix}\right)$$
$$= -1 + 0 = -1.$$

53. $\det A \neq 0$ is equivalent to each of the following:

- A is nonsingular.
- A is invertible.
- null(A) is trivial.
- The system $A\mathbf{x} = \vec{b}$ has a unique solution for every right hand side $\vec{b}$.
- If A is $n \times n$, the column vectors in A are a basis for $\mathbf{R}^n$.
- When A is reduced to row echelon form the diagonal entries are all nonzero.
- When A is reduced to reduced row echelon form the result is the identity matrix.

Chapter 8. An Introduction to Systems

Section 8.1. Definitions and Examples

1. The system

$$x' = v$$
$$v' = -x - 0.02v + 2\cos t,$$

has dependent variables (unknowns) x and v. Therefore, the dimension is 2. The right side depends explicitly on the independent variable t, so the system is nonautonomous.

3. The system

$$x' = -ax + ay$$
$$y' = rx - y - xz$$
$$z' = -bz + xy,$$

has dependent variables (unknowns) x, y, and z. Therefore, the dimension is 3. The right side does not depend explicitly on t, so the system is autonomous.

5. The system

$$u_1' = u_2$$
$$u_2' = -\frac{1}{2}u_1 + \frac{1}{2}u_3$$
$$u_3' = u_4$$
$$u_4' = \frac{3}{2}u_1 + \frac{1}{2}u_3,$$

has dependent variables (unknowns) u_1, u_2, u_3 and u_4. Therefore, the dimension is 4. The right side does not depend explicitly on t, so the system is autonomous.

7. If $x(t) = 2e^{2t} - 2e^{-t}$ and $y(t) = -e^{-t} + 2e^{2t}$, then

$$x' = (2e^{2t} - 2e^{-t})' = 4e^{2t} + 2e^{-t},$$

and

$$-4x + 6y = -4(2e^{2t} - 2e^{-t}) + 6(-e^{-t} + 2e^{2t})$$
$$= 4e^{2t} + 2e^{-t},$$

so the first equation is satisfied. Further,

$$y' = (-e^{-t} + 2e^{2t})' = e^{-t} + 4e^{2t},$$

and

$$-3x + 5y = -3(2e^{2t} - 2e^{-t}) + 5(-e^{-t} + 2e^{2t})$$
$$= 4e^{2t} + e^{-t},$$

so the second equation is satisfied. Finally,

$$x(0) = 2e^{2(0)} - 2e^{-0} = 0$$
$$y(0) = -e^{-0} + 2e^{2(0)} = 1,$$

so the initial conditions are also satisfied.

9. If $x(t) = e^{-t}(-\cos t - \sin t)$ and $v(t) = 2e^{-t}\sin t$, then

$$x' = (e^{-t}(-\cos t - \sin t))'$$
$$= e^{-t}(\sin t - \cos t) - e^{-t}(-\cos t - \sin t)$$
$$= 2e^{-t}\sin t$$
$$= v,$$

and the first equation is satisfied. Further,

$$v' = (2e^{-t}\sin t)' = 2e^{-t}\cos t - 2e^{-t}\sin t,$$

and

$$-2x - 2v = -2(e^{-t}(-\cos t - \sin t))$$
$$- 2(2e^{-t}\sin t)$$
$$= 2e^{-t}\cos t - 2e^{-t}\sin t,$$

so the second equation is satisfied. Finally,

$$x(0) = e^{-0}(-\cos 0 - \sin 0) = -1$$
$$v(0) = 2e^{-0}\sin 0 = 0,$$

so the initial conditions are satisfied.

11. We know that $y'' = -2y' - 4y + 3\cos 2t$, $y(0) = 1$, $y'(0) = 0$. If we let $u_1 = y$ and $u_2 = y'$, then

$$u_1' = u_2$$
$$u_2' = -2u_2 - 4u_1 + 3\cos 2t.$$

Hence with $\mathbf{u} = (u_1, u_2)^T$,

$$\mathbf{u}' = \begin{pmatrix} u_2 \\ -2u_2 - 4u_1 + 3\cos 2t \end{pmatrix}.$$

Furthermore $\mathbf{u}(0) = (u_1(0), u_2(0))^T = (y(0), y'(0))^T = (1, 0)^T$.

13. We know that $x'' = -\delta x' + x - x^3 + \gamma \cos \omega t$, $x(0) = x_0, x'(0) = v_0$. If we let $u_1 = x$ and $u_2 = x'$, then

$$u_1' = u_2$$
$$u_2' = -\delta u_2 + u_1 - u_1^3 + \gamma \cos \omega t.$$

Furthermore, $u_1(0) = x(0) = x_0$ and $u_2(0) = x'(0) = v_0$.

15. We know that $\omega''' = \omega$, $\omega(0) = \omega_0$, $\omega'(0) = \alpha_0$, and $\omega''(0) = \gamma_0$. If we let $u_1 = \omega$, $u_2 = \omega'$, and $u_3 = \omega''$, then

$$u_1' = u_2$$
$$u_2' = u_3$$
$$u_3' = u_1.$$

Further, $u_1(0) = \omega(0) = \omega_0$, $u_2(0) = \omega'(0) = \alpha_0$, and $u_3(0) = \omega''(0) = \gamma_0$.

17. The right side of the system

$$u' = v$$
$$v' = -3u - 2v + 5\cos t$$

depends explicitly on the independent variable t. Therefore, the system is nonautonomous.

19. The right side of the system

$$u' = v\cos u$$
$$v' = tv$$

depends explicitly on the independent variable t. Therefore, the system is nonautonomous.

21. In the system

$$u' = v + \cos u$$
$$v' = v - t\omega$$
$$\omega' = 5u - 9v + 8\omega,$$

the right side depends explicitly on the independent variable t. Therefore, the system is nonautonomous.

23. Let $x_1 = u$, $x_2 = v$. If we let $\mathbf{x} = (x_1, x_2)'$, then the system

$$x_1' = x_2$$
$$x_2' = -3x_1 - 2x_2 + 5\cos t$$

can be written

$$\mathbf{x}' = \mathbf{f}(t, \mathbf{x}),$$

where

$$\mathbf{x}' = \begin{pmatrix} x_1' \\ x_2' \end{pmatrix} \quad \text{and}$$

$$\mathbf{f}(t, \mathbf{x}) = \begin{pmatrix} x_2 \\ -3x_1 - 2x_2 + 5\cos t \end{pmatrix}.$$

25. Let $x_1 = u$, $x_2 = v$. If we let $\mathbf{x} = (x_1, x_2)'$, then the system

$$x_1' = x_2 \cos x_1$$
$$x_2' = tx_2$$

can be written

$$\mathbf{x}' = \mathbf{f}(t, \mathbf{x}),$$

where

$$\mathbf{x}' = \begin{pmatrix} x_1' \\ x_2' \end{pmatrix} \quad \text{and} \quad \mathbf{f}(t, \mathbf{x}) = \begin{pmatrix} x_2 \cos x_1 \\ tx_2 \end{pmatrix}.$$

27. Let $x_1 = u$, $x_2 = v$, and $x_3 = \omega$. If we let $\mathbf{x} = (x_1, x_2, x_3)'$, then the system

$$x_1' = x_2 + \cos x_1$$
$$x_2' = x_2 - tx_3$$
$$x_3' = 5x_1 - 9x_2 + 8x_3$$

can be written

$$\mathbf{x}' = \mathbf{f}(t, \mathbf{x}),$$

where

$$\mathbf{x}' = \begin{pmatrix} x_1' \\ x_2' \\ x_3' \end{pmatrix} \quad \text{and} \quad \mathbf{f}(t, \mathbf{x}) = \begin{pmatrix} x_2 + \cos x_1 \\ x_2 - tx_3 \\ 5x_1 - 9x_2 + 8x_3 \end{pmatrix}.$$

29. Let $x_1 = S$, $x_2 = I$, and $x_3 = R$. If we let $\mathbf{x} = (x_1, x_2, x_3)^T$, then the system

$$x_1' = -x_1x_2$$
$$x_2' = x_1x_2 - x_2$$
$$x_3' = x_1$$

can be written

$$\mathbf{x} = f(\mathbf{x}),$$

where

$$\mathbf{x}' = \begin{pmatrix} x_1' \\ x_2' \\ x_3' \end{pmatrix} \quad \text{and} \quad f(\mathbf{x}) = \begin{pmatrix} -x_1x_2 \\ x_1x_2 - x_2 \\ x_1 \end{pmatrix}.$$

Furthermore, $(x_1(0), x_2(0), x_3(0))^T = (S(0), I(0), R(0))^T = (4, 0, 1, 0)^T$. These equations can be used to produce the following component solutions, $x_1(t) = S(t)$, $x_2(t) = I(t)$, and $x_3(t) = R(t)$.

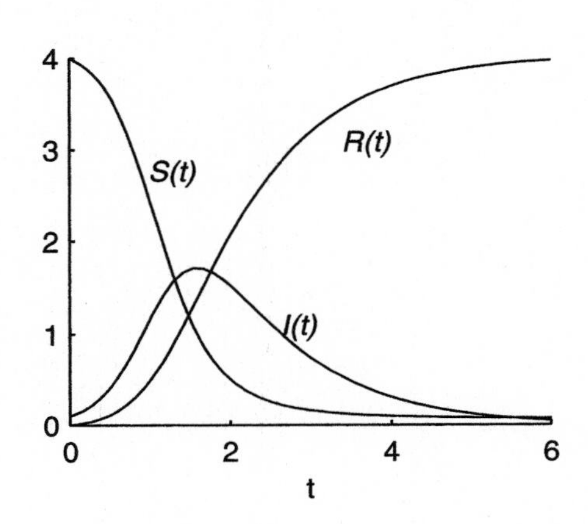

31. First, break each velocity vector into vertical and horizontal components, letting b represent the magnitude of the vector $\mathbf{b}$.

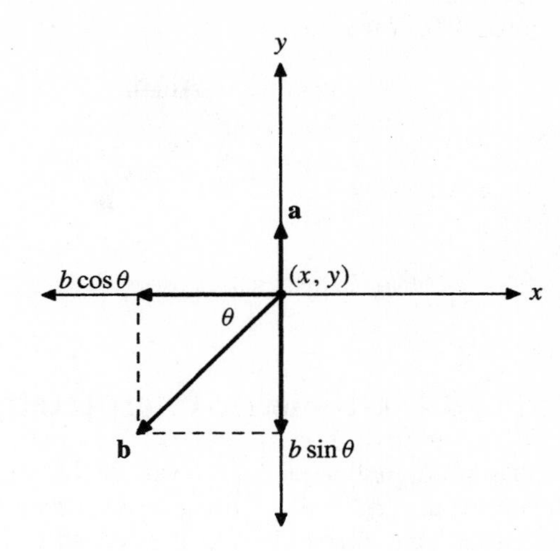

As you can see in the figure, the velocity in the x-direction is $-b\cos\theta$. Thus,

$$\frac{dx}{dt} = -b\cos\theta.$$

In the y-direction, we resolve two components.

$$\frac{dy}{dt} = a - b\sin\theta$$

A second image,

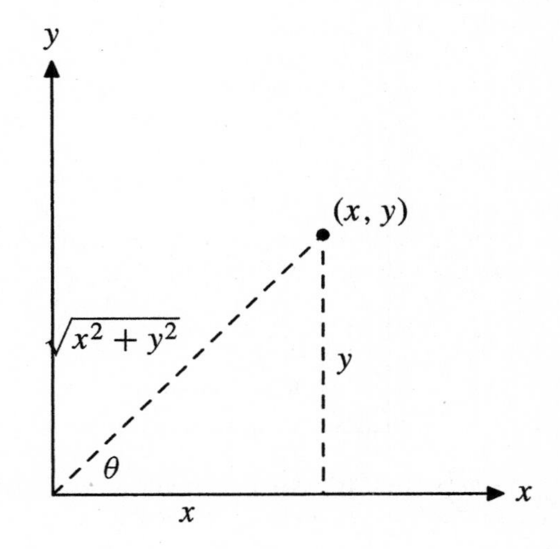

gives us

$$\cos\theta = \frac{x}{\sqrt{x^2+y^2}}$$
$$\sin\theta = \frac{y}{\sqrt{x^2+y^2}}.$$

Substituting,

$$\frac{dx}{dt} = \frac{-bx}{\sqrt{x^2+y^2}}$$
$$\frac{dy}{dt} = a - \frac{by}{\sqrt{x^2+y^2}}.$$

Section 8.2. Geometric Interpretation of Solutions

1. The first figure is a plot of $x_1(t) = 2e^t - e^{-t}$ (the solid curve) and $x_2(t) = e^{-t}$ (the dashed curve) versus t on the time interval $[0, 2]$. The second figure uses the same time interval, but it is a plot of x_2 versus x_1 in the phase plane.

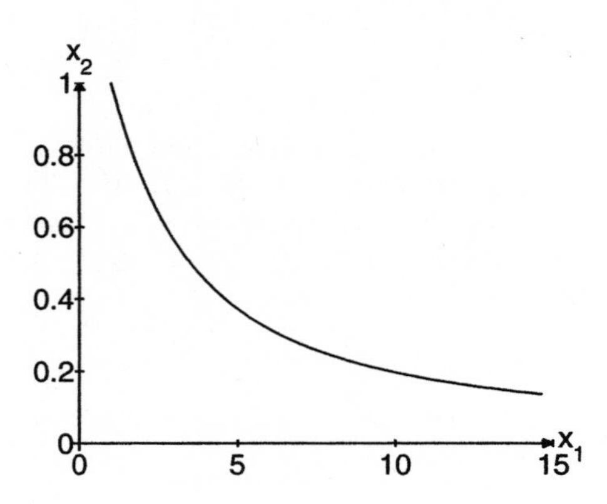

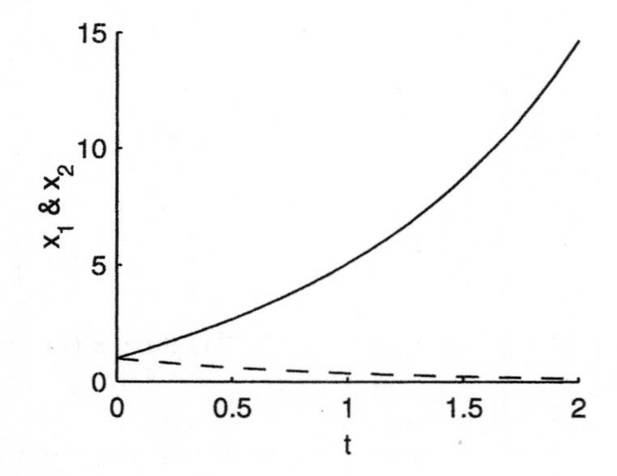

3. The first figure is a plot of $x_1(t) = \cos t$ (the solid curve) and $x_2(t) = \sin t$ (the dashed curve) versus t on the time interval $[0, 2\pi]$. The second figure uses the same time interval, but it is a plot of x_2 versus x_1 in the phase plane.

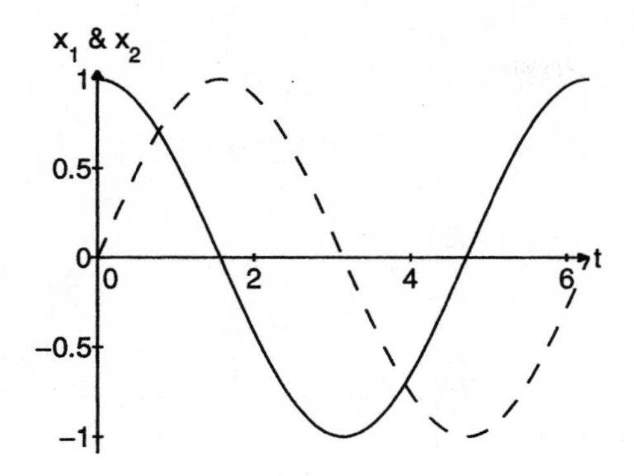

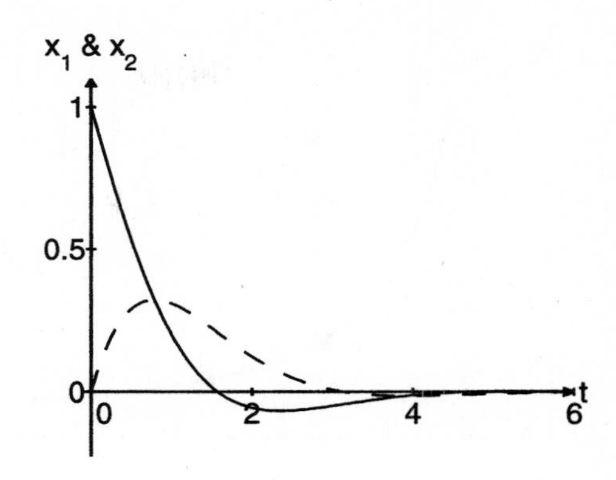

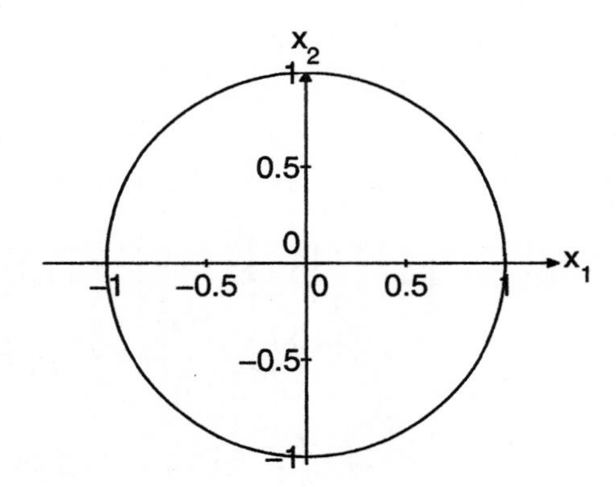

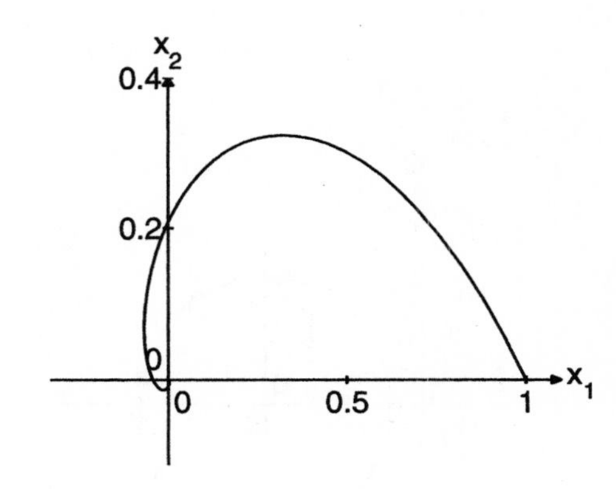

5. The first figure is a plot of $x_1(t) = e^{-t} \cos t$ (the solid curve) and $x_2(t) = e^{-t} \sin t$ (the dashed curve) versus t on the time interval $[0, 2\pi]$. The second figure uses the same time interval, but it is a plot of x_2 versus x_1 in the phase plane.

7. If $\mathbf{x}(t) = (2e^t - e^{-t}, e^{-t})$, then $\mathbf{x}'(t) = (2e^t + e^{-t}, -e^{-t})$. In the figure that follows, the derivative was used to plot vectors tangent to the curve at selected points. Tangent vectors are plotted at 25% of their actual length.

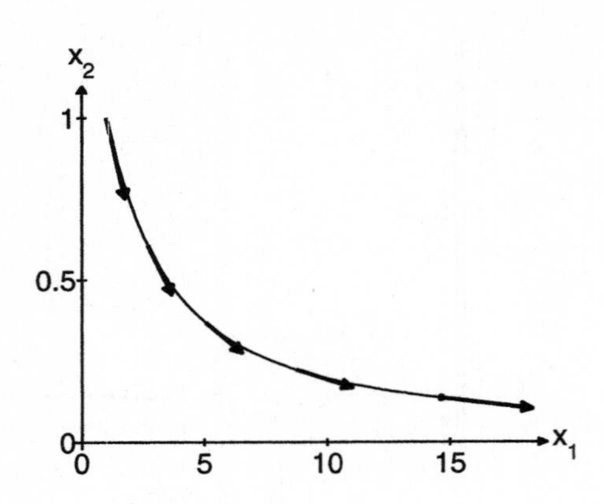

9. If $\mathbf{x}(t) = (\cos t, \sin t)$, then $\mathbf{x}'(t) = (-\sin t, \cos t)$. In the figure that follows, the derivative was used to plot vectors tangent to the curve at selected points. Tangent vectors are plotted at 25% of their actual length.

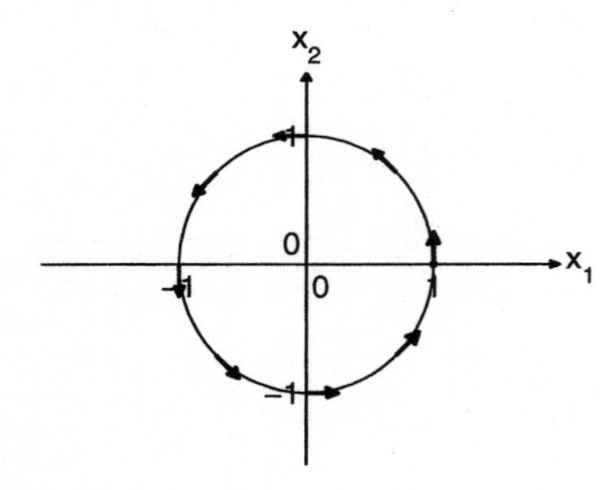

11. If $\mathbf{x}(t) = (e^{-t}\cos t, e^{-t}\sin t)$, then $\mathbf{x}'(t) = (e^{-t}(-\sin t - \cos t), e^{-t}(\cos t - \sin t))$. In the figure that follows, the derivative was used to plot vectors tangent to the curve at selected points. Tangent vectors are plotted at 25% of their actual length.

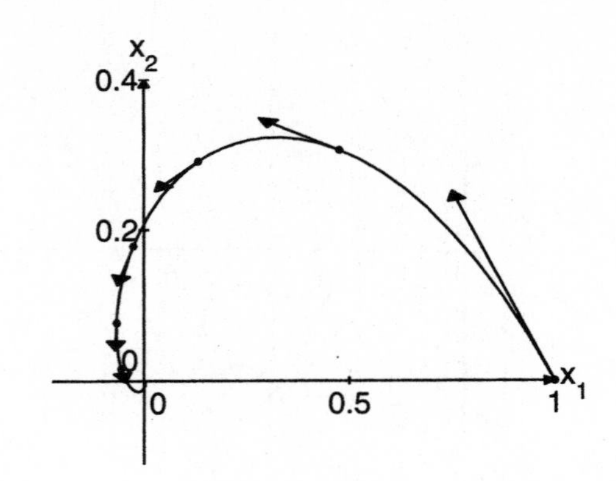

13. The directions and several solution trajectories for $\theta' = \omega$ and $\omega' = -\sin\theta$ follow.

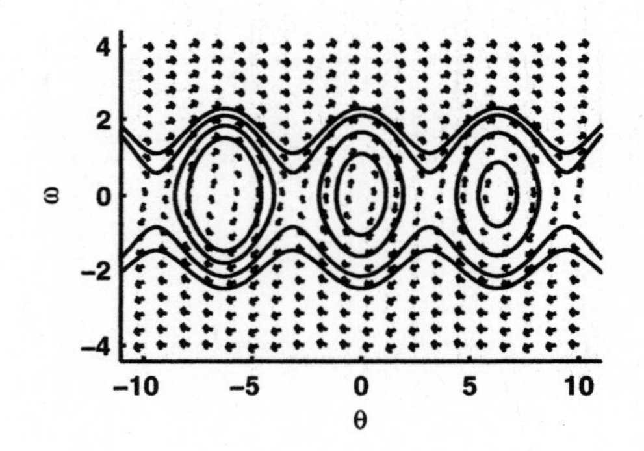

15. The directions and several solution trajectories for $x' = (0.4 - 0.01y)x$ and $y' = (0.005x - 0.3)y$ follow.

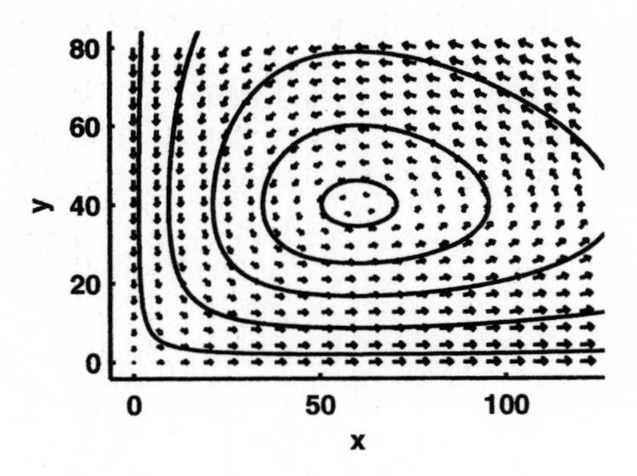

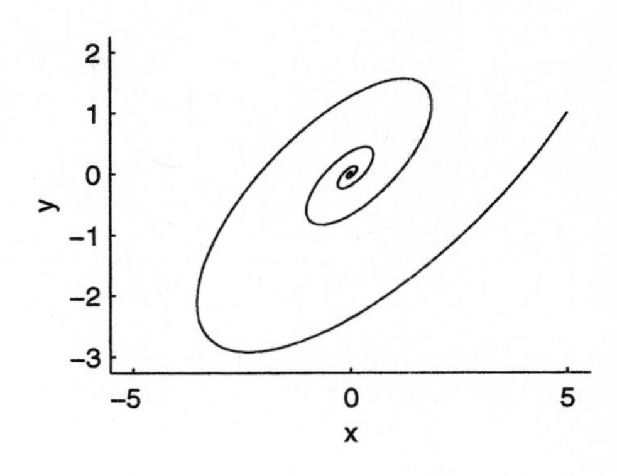

Finally, a composite plot, with the 3D plot solid and the others dashed.

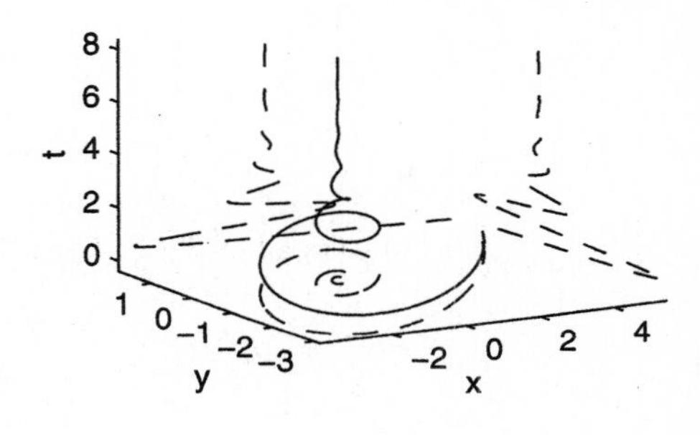

17. In the first figure, component plots, with x the solid curve and y the dashed curve. In the second figure, the solution in the phase plane.

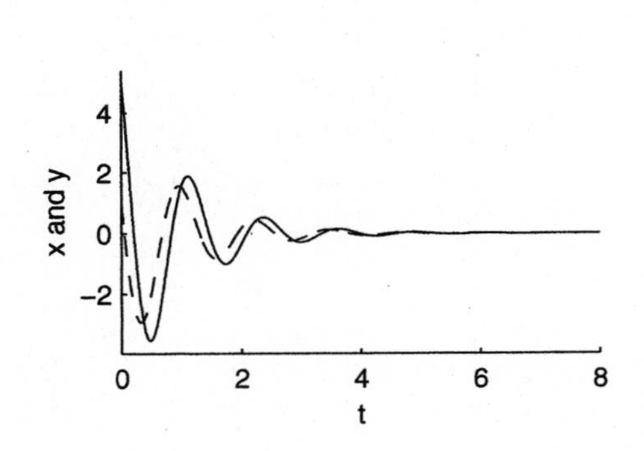

19. In the first figure, component plots, with x the solid curve and y the dashed curve. In the second figure, the solution in the phase plane.

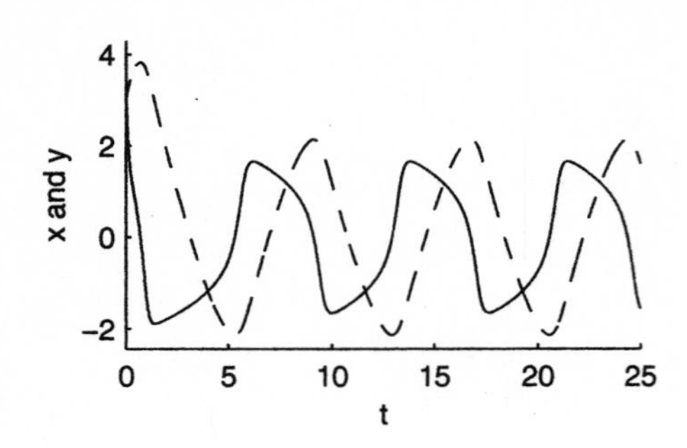

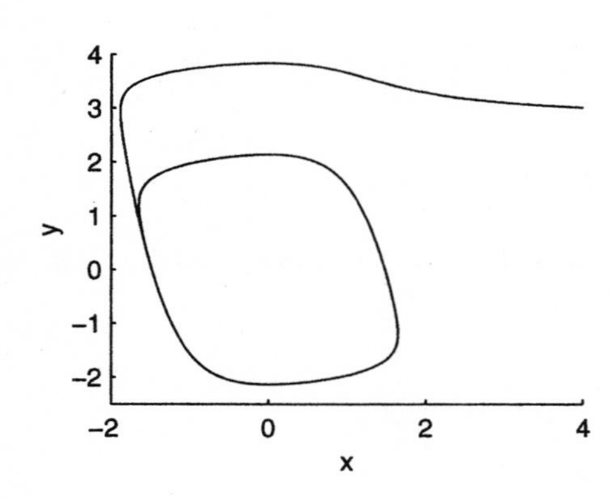

Finally, a composite plot, with the 3D plot solid and the others dashed.

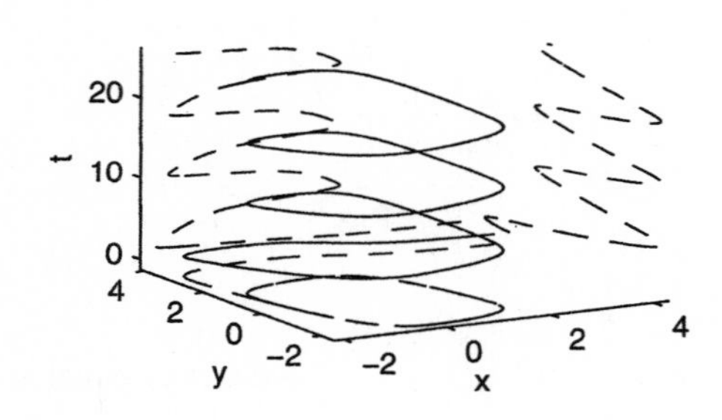

21. In I and IV, x and y exhibit oscillatory behavior. In I, the oscillatory behavior decays, while in IV, the oscillatory behavior eventually becomes periodic. Thus, I $\mapsto$ D and IV $\mapsto$ C. In II and III, one component decays to zero while the other component levels asymptotically to some value. In II, y decays to zero, so II $\mapsto$ A. But, in III, x decays to zero, so III $\mapsto$ B.

23. Initially, $x(0) = 0$ and $y(0) = 2$. Shortly thereafter, y decays as x increases. Soon, both x and y begin a seemingly periodic motion.

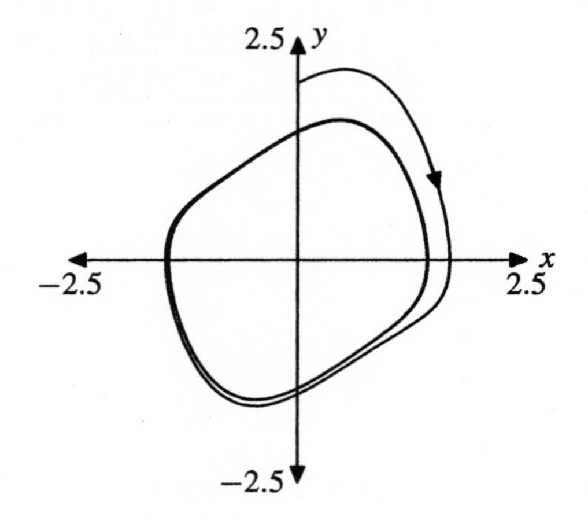

25. Initially, $x(0) = 0$ and $y(0) = 2$. Thereafter, x increases rapidly, then decays asymptotically in an oscillatory manner to about 5 or 6. Meanwhile, y decays, eventually oscillating about zero. One possible solution follows.

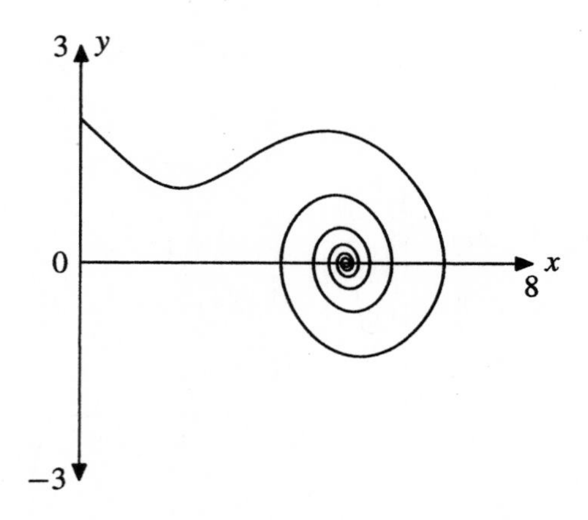

27. If the prey population grows according to the logistic model, then

$$F' = r\left(1 - \frac{F}{K}\right)F.$$

Expanding, $F = rF - rF^2/K$, and replacing r with a and r/K with e,

$$F' = aF - eF^2.$$

Now, because interaction between species is harmful to the prey,

$$F' = aF - eF^2 - bFS.$$

If predators still follow the Malthusian model, and interactions between species is still considered beneficial to the predators, then

$$S' = -cS + dFS.$$

29. (a) Given that $dF/dt = aF - bFS$ and $dS/dt = -cS + dFS$,

$$\begin{aligned}\frac{dS}{dF} &= \frac{dS/dt}{dF/dt}\\ &= \frac{-cS + dFS}{aF - bFS}\\ &= \frac{(-c + dF)S}{(a - bS)F}.\end{aligned}$$

(b) Using the result in part (a), we separate the variables.

$$\begin{aligned}\frac{a - bS}{S}\,dS &= \frac{-c + dF}{F}\,dF\\ \left(\frac{a}{S} - b\right)dS &= \left(-\frac{c}{F} + d\right)dF\end{aligned}$$

Integrate.

$$a\ln S - bS = -c\ln F + dF + C,$$

or, equivalently,

$$a\ln S - bS + c\ln F - dF = C.$$

(c) An implicit function plotter was used to sketch

$$0.2\ln S - 0.1S + 0.3\ln F - 0.1F = C,$$

for $C = -0.2, -0.4, -0.6, -0.8$, and -1.

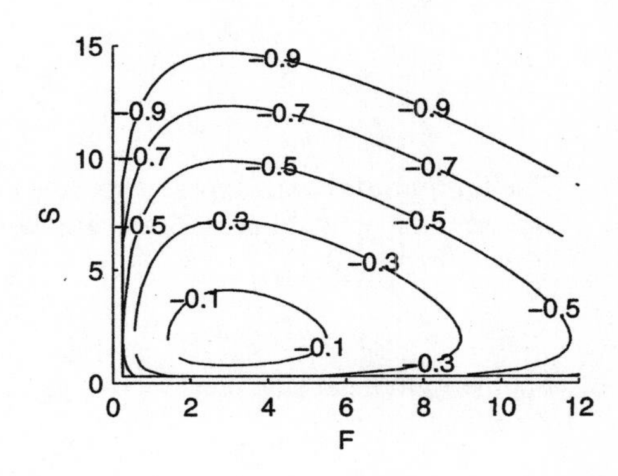

Section 8.3. Qualitative Analysis

1. Set the right hand side of $x' = 0.2x - 0.04xy$ equal to zero.

$$\begin{aligned}0.2x - 0.04xy &= 0\\ 20x - 4xy &= 0\\ 4x(5 - y) &= 0\end{aligned}$$

Thus, $x = 0$ and $y = 5$ are the x-nullclines. They appear in a solid line style in the figure. Set the right hand side of $y' = -0.1y + 0.005xy$ equal to zero.

$$\begin{aligned}-0.1y + 0.005xy &= 0\\ -100y + 5xy &= 0\\ 5y(-20 + x) &= 0\end{aligned}$$

Thus, $y = 0$ and $x = 20$ are the y-nullclines. They appear in a dashed line style in the figure.

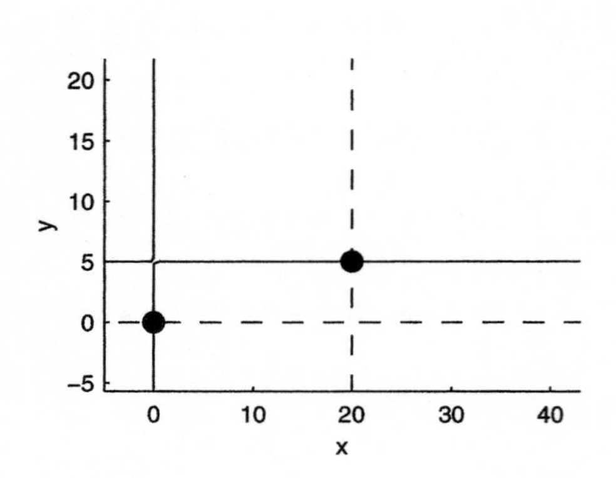

The equilibrium points appear where the x-nullclines intersect the y-nullclines. These are easily seen to be (0, 0) and (20, 5).

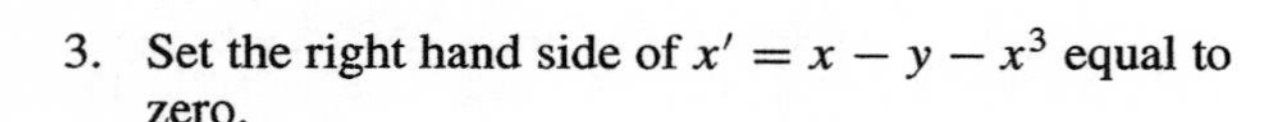

3. Set the right hand side of $x' = x - y - x^3$ equal to zero.

$$x - y - x^3 = 0$$
$$y = x - x^3$$

Thus, $y = x - x^3$ is the x-nullcline. It appears in a solid line style in the figure. Set the right hand side of $y' = x$ equal to zero.

$$x = 0$$

Thus, $x = 0$ is the y-nullcline. It appears in a dashed line style in the figure.

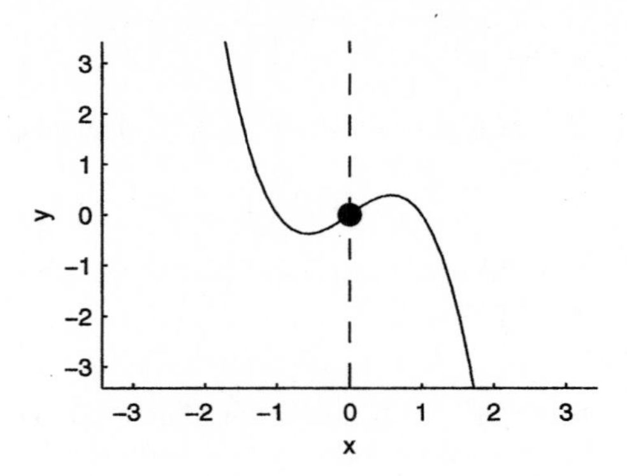

The equilibrium point appears where the x-nullcline intersects the y-nullcline. This is easily seen to be (0, 0).

5. Set the right hand side of $x' = y$ equal to zero.

$$y = 0$$

Thus, $y = 0$ is the x-nullcline. It appears in a solid line style in the figure. Set the right hand side of $y' = -\sin x - y$ equal to zero.

$$-\sin x - y = 0$$
$$y = -\sin x$$

Thus, $y = -\sin x$ is the y-nullcline. It appears in a dashed line style in the figure.

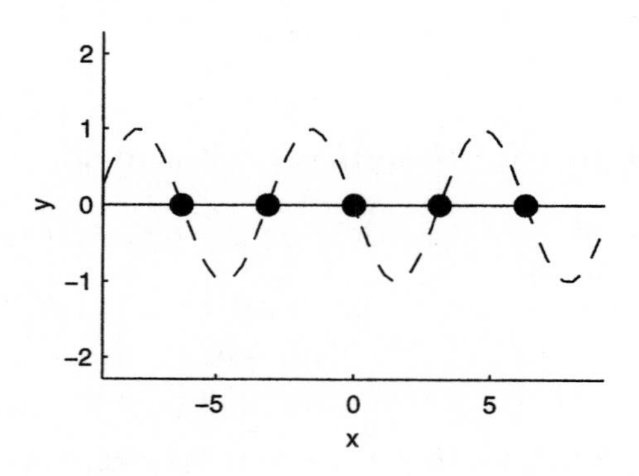

The equilibrium points appear where the x-nullcline intersects the y-nullcline. These occur at the points $(k\pi, 0)$, where k is an integer. A few are shown in the figure.

7. (a) If $x(t) = t$ and $y(t) = \sin t$, then

$$x' = (t)' = 1,$$

and

$$1-(y-\sin x)\cos x = 1-(\sin t-\sin t)\cos t = 1,$$

so the first equation is satisfied. Further,

$$y' = (\sin t)' = \cos t,$$

and

$$\cos x - y + \sin x = \cos t - \sin t + \sin t = \cos t,$$

so the second equation is satisfied.

(b) See the figure in part (c).

(c) Because of uniqueness, the solution with initial condition $x(0) = \pi/2$, $y(0) = 0$, cannot cross the solution $x = t$, $y = \sin t$ found in part (a). Thus, it must remain below the solution in part (a) for all time. Therefore, if $(x(t), y(t))$ denotes the second solution, we must have $y(t) < \sin x(t)$ for all time, as shown in the figure.

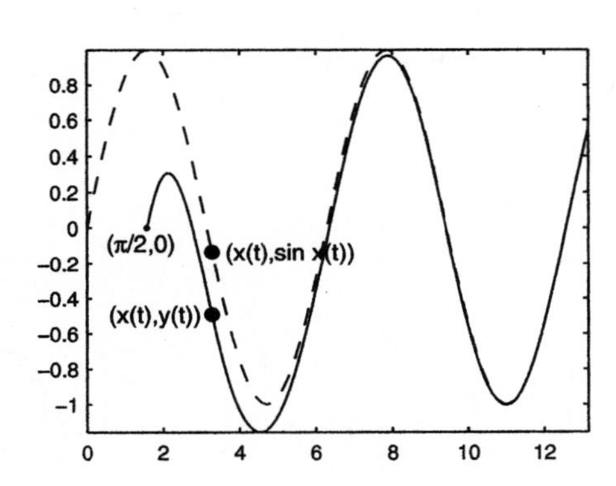

9. (a) If $x = e^t$ and $y = -e^t$, then

$$x' = (e^t)' = e^t,$$

and

$$\begin{aligned} &-y - x(x^2 - y^2) \\ &= -(-e^t) - e^t((e^t)^2 - (-e^t)^2) \\ &= e^t - e^t(e^{2t} - e^{2t}) \\ &= e^t, \end{aligned}$$

so the first equation is satisfied. Further,

$$y' = (-e^t)' = -e^t,$$

and

$$\begin{aligned} &-x - y(x^2 - y^2) \\ &= -e^t - (-e^t)((e^t)^2 - (-e^t)^2) \\ &= -e^t + e^t(e^{2t} - e^{2t}) \\ &= -e^t, \end{aligned}$$

so the second equation is satisfied.

(b) If $x(t) = e^{-t}$ and $y(t) = e^{-t}$, then

$$x' = (e^{-t})' = -e^{-t},$$

and

$$\begin{aligned} &-y - x(x^2 - y^2) \\ &= -e^{-t} - e^{-t}((e^{-t})^2 - (e^{-t})^2) \\ &= -e^{-t} - e^{-t}(e^{-2t} - e^{-2t}) \\ &= -e^{-t}, \end{aligned}$$

so the first equation is satisfied. Further,

$$y' = (e^{-t})' = -e^{-t},$$

and

$$\begin{aligned} &-x - y(x^2 - y^2) \\ &= -e^{-t} - e^{-t}((e^{-t})^2 - (e^{-t})^2) \\ &= -e^{-t} - e^{-t}(e^{-2t} - e^{-2t}) \\ &= -e^{-t}, \end{aligned}$$

and the second equation is satisfied.

(c) See the figure in part (d).

(d) It is important to recognize that solution (a), $x(t) = e^t$, $y(t) = -e^t$, is a parametric description of the line $y = -x$. Also, solution (b) lies on the line $y = -x$. Uniqueness tells us that this third solution must remain trapped between the solutions of parts (a) and (b) (solutions may not cross). Thus, if $(x(t), y(t))$ is a point on this third solution, the figure shows why $-x(t) < y(t) < x(t)$ for all t.

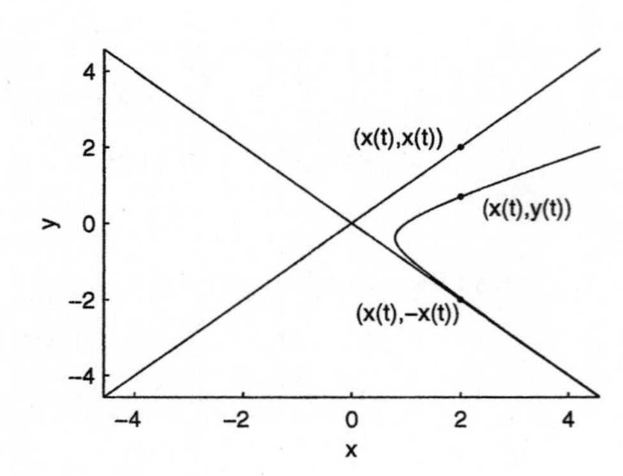

11. (a) Set the right side of $x' = ax - bxy - \epsilon x$ equal to zero.

$$ax - bxy - \epsilon x = 0$$
$$x((a - \epsilon) - by) = 0$$

Thus, $x = 0$ and $y = (a - \epsilon)/b$ are the x-nullclines. Set the right side of $y' = -cy + dxy - \epsilon y$ equal to zero.

$$-cy + dxy - \epsilon y = 0$$
$$y((-c - \epsilon) + dx) = 0$$

Thus, $y = 0$ and $x = (c + \epsilon)/d$ are the y-nullclines. The equilibrium points are intersections of the x- and y-nullclines, easily seen to be $(0, 0)$ and $((c + \epsilon)/d, (a - \epsilon)/b)$.

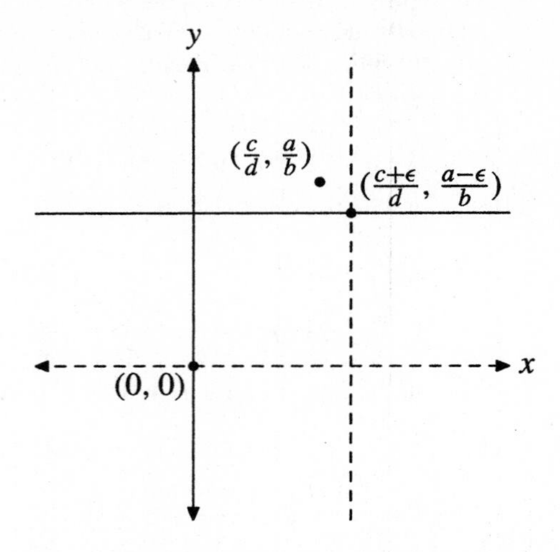

(b) As you can see in the figure in part (a), the harvesting strategy has shifted the equilibrium point. It has moved from

$$\left(\frac{c}{d}, \frac{a}{b}\right) \quad \text{to} \quad \left(\frac{c+\epsilon}{d}, \frac{a-\epsilon}{b}\right).$$

Therefore, the predator population decreases somewhat, but the prey population actually increases. So, this harvesting strategy is a poor one if the intent was to decrease the level of prey (such as spraying with insecticides).

(c) First, the original system, $x' = (0.4 - 0.01y)x$, $y' = (-0.3 + 0.005x)y$, without harvesting, provides this image.

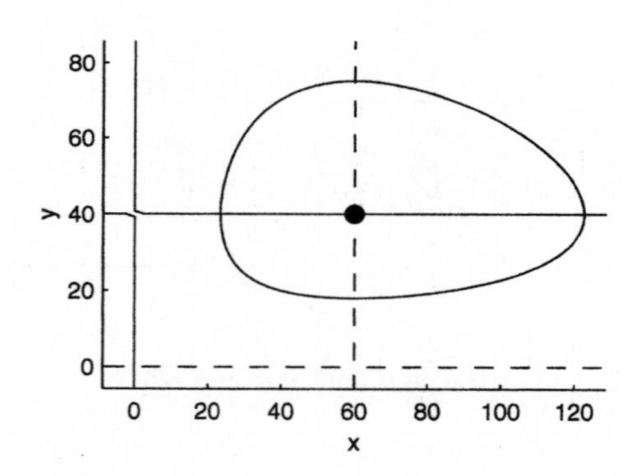

Note that the equilibrium point is (60, 40) for this system. Now, if we add harvesting, as in system 11c, then we get this result. Note that both pictures were crafted with the same initial condition; namely, $x(0) = 40$ and $y(0) = 20$.

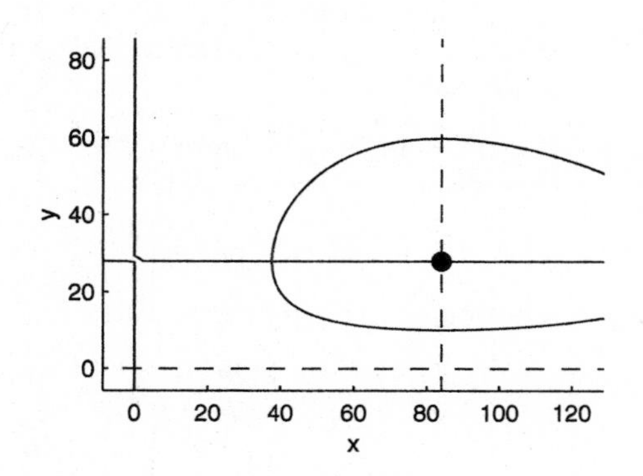

Clearly, the predator population has decreased on average, while the prey population has increased on average. The equilibrium point is now at (84, 28).

13. If x grows according to the logistic model, then

$$x' = r\left(1 - \frac{x}{K}\right)x = rx - \frac{rx^2}{K}.$$

Of course, this becomes

$$x' = ax - bx^2,$$

if we replace r with a and r/K with b. Similarly, if y grows according to the logistic model, we can eventually write

$$y' = Ay - By^2.$$

Now, in light of the competition for resources, interaction between the two species is harmful to each. Thus, a certain percentage of contacts between species will decrease the rates dx/dt and dy/dt (perhaps differently for each rate), so we write

$$x' = ax - bx^2 - cxy$$
$$y' = Ay - By^2 - Cxy.$$

15. Setting the right side of $x' = \sigma(y - x)$ equal to zero,

$$\sigma(y - x) = 0$$
$$y = x.$$

Setting the right side of $z' = -\beta z + xy$ equal to zero,

$$-\beta z + xy = 0$$
$$xy = \beta z.$$

However, $y = x$, so we can write

$$x^2 = \beta z.$$

Setting the right side of $y' = \rho x - y - xz$ equal to zero,

$$\rho x - y - xz = 0.$$

However, $y = x$, so

$$\rho x - x - xz = 0$$
$$x(\rho - 1 - z) = 0.$$

Thus, either $x = 0$ or $z = \rho - 1$. If $x = 0$, then y and z are easily shown to also equal zero, giving $(0, 0, 0)$ as one equilibrium point. If $z = \rho - 1$, then

$$x^2 = \beta z$$
$$x^2 = \beta(\rho - 1)$$
$$x = \pm\sqrt{\beta(\rho - 1)}.$$

Of course, this last expression is real only if $\rho > 1$, which it is. Thus, the second equilibrium point is $(\sqrt{\beta(\rho - 1)}, \sqrt{\beta(\rho - 1)}, \rho - 1)$ and the third equilibrium point is $(-\sqrt{\beta(\rho - 1)}, -\sqrt{\beta(\rho - 1)}, \rho - 1)$.

Section 8.4. Linear Systems

1. Note that the system

$$x_1' = -x_2$$
$$x_2' = -x_1 - 2x_2 + 5\sin t$$

has the form

$$x_1' = a_{11}(t)x_1 + a_{12}(t)x_2 + f_1(t)$$
$$x_2' = a_{21}(t)x_2 + a_{22}(t)x_2 + f_2(t).$$

Hence, the system is linear. Although $f_1(t) = 0$, because $f_2(t) = 5\sin t$ is not zero, the system is inhomogeneous.

3. Note that the system

$$x_1' = -x_2$$
$$x_2' = \sin x_1,$$

because of the nonlinear term $\sin x_1$, does not have the form

$$x_1' = a_{11}(t)x_1 + a_{12}(t)x_2 + f_1(t)$$
$$x_2' = a_{21}(t)x_2 + a_{22}(t)x_2 + f_2(t).$$

Hence, the system is nonlinear.

5. Note that the system

$$x_1' = x_2$$
$$x_2' = x_3$$
$$x_3' = -x_1 - x_2 + \sin t$$

has the form

$$x_1' = a_{11}(t)x_1 + a_{12}(t)x_2 + a_{13}(t)x_3 + f_1(t)$$
$$x_2' = a_{21}(t)x_2 + a_{22}(t)x_2 + a_{23}(t)x_3 + f_2(t)$$
$$x_3' = a_{31}(t)x_1 + a_{32}(t)x_2 + a_{33}(t)x_3 + f_3(t).$$

Hence, the system is linear. Although $f_1(t) = 0$ and $f_2(t) = 0$, because $f_3(t) = \sin t$ is nonzero, the system is inhomogeneous.

7. If $x_1(t) = e^t$ and $x_2(t) = -e^t$, then the left side of $x_1' = 3x_1 + 2x_2$ is

$$x_1' = e^t.$$

Substituting in the right hand side,

$$3x_1 + 2x_2 = 3(e^t) + 2(-e^t) = e^t.$$

Hence, the first equation is satisfied. In a similar fashion, the left side of $x_2' = -4x_1 - 3x_2$ is $x_2' = -e^t$ and the right side is $-4x_1 - 3x_2 = -4(e^t) - 3(-e^t) = -e^t$. Hence, the second equation is satisfied.

9. If $x_1(t) = 2e^{-t} - e^{-2t}$ and $x_2(t) = -2e^{-t}$, then the left side of $x_1' = -2x_1 - x_2$ is

$$x_1' = -2e^{-t} + 2e^{-2t}.$$

Substituting in the right hand side,

$$\begin{aligned} -2x_1 - x_2 &= -2(2e^{-t} - e^{-2t}) - (-2e^{-t}) \\ &= -2e^{-t} + 2e^{-2t}. \end{aligned}$$

Hence, the first equation is satisfied. In a similar fashion, the left side of $x_2' = -x_2$ is $x_2' = 2e^{-t}$ and the right side is $-x_2 = 2e^{-t}$. Hence, the second equation is satisfied.

11. We can write

$$x_1' = -2x_1 + 3x_2$$
$$x_2' = x_1 - 4x_2$$

as

$$\begin{pmatrix} x_1 \\ x_2 \end{pmatrix}' = \begin{pmatrix} -2 & 3 \\ 1 & -4 \end{pmatrix} \begin{pmatrix} x_1 \\ x_2 \end{pmatrix}.$$

Therefore, the system is linear.

13. The system

$$x_1' = -2x_1 + x_2^2$$
$$x_2' = 3x_1 - x_2$$

is nonlinear. Note the term x_2^2. It cannot be written in the form $\mathbf{x}' = A(t)\mathbf{x} + \mathbf{f}(t)$.

15. The system

$$t^2 x_1' = -2tx_1x_2 + 3x_2$$
$$(1/t)x_2' = tx_1 - (4/t)x_2$$

is nonlinear. Note the nonlinear term $-2tx_1x_2$. Therefore, the system cannot be written in the form $\mathbf{x}' = A(t)\mathbf{x} + \mathbf{f}(t)$.

17. Exercises 11, 12, 14, and 16 are linear; 13 and 15 are nonlinear. Of the linear systems, only 11 and 12 are homogeneous.

19. Write the system $x_1' = 8x_1 - 10x_2$, $x_2' = 5x_1 - 7x_2$ in matrix-vector form.

$$\begin{pmatrix} x_1 \\ x_2 \end{pmatrix}' = \begin{pmatrix} 8 & -10 \\ 5 & -7 \end{pmatrix} \begin{pmatrix} x_1 \\ x_2 \end{pmatrix}$$

Note that this system is in the form $\mathbf{x}' = A\mathbf{x}$. Next, if $\mathbf{x} = (2e^{3t}, e^{3t})^T$, then $\mathbf{x}' = (6e^{3t}, 3e^{3t})^T$ and

$$A\mathbf{x} = \begin{pmatrix} 8 & -10 \\ 5 & -7 \end{pmatrix} \begin{pmatrix} 2e^{3t} \\ e^{3t} \end{pmatrix} = \begin{pmatrix} 6e^{3t} \\ 3e^{3t} \end{pmatrix}.$$

Thus, $\mathbf{x}' = A\mathbf{x}$ and $\mathbf{x} = (2e^{3t}, e^{3t})^T$ is a solution.

21. Write the system$x_1' = -x_1 + 4x_2$, $x_2' = 3x_2$ in matrix-vector form.

$$\begin{pmatrix} x_1 \\ x_2 \end{pmatrix}' = \begin{pmatrix} -1 & 4 \\ 0 & 3 \end{pmatrix} \begin{pmatrix} x_1 \\ x_2 \end{pmatrix}$$

Note that this system is in the form $\mathbf{x}' = A\mathbf{x}$. Next, if $\mathbf{x} = (e^{3t} - e^{-t}, e^{3t})^T$, then $\mathbf{x}' = (3e^{3t} + e^{-t}, 3e^{3t})^T$ and

$$A\mathbf{x} = \begin{pmatrix} & 4 \\ 0 & 3 \end{pmatrix} \begin{pmatrix} e^{3t} - e^{-t} \\ e^{3t} \end{pmatrix} = \begin{pmatrix} 3e^{3t} + e^{-t} \\ 3e^{3t} \end{pmatrix}.$$

Thus, $\mathbf{x}' = A\mathbf{x}$ and $\mathbf{x} = (e^{3t} - e^{-t}, e^{3t})^T$ is a solution.

23. Label the currents coming out of node "a", as shown in Figure 8. The sum of the current flowing out of node "a" must equal zero.

$$I + I_2 + I_3 = 0 \tag{4.1}$$

Proceeding in a clockwise direction around the top loop, the sum of the voltage drops across each element must equal zero. Thus,

$$R_1 I + LI' - R_2 I_2 = 0. \tag{4.2}$$

Because we are moving in the direction of the current I, the voltage drops across the resistor R_1 and the inductor L are positive. However, as we continue in a clockwise direction, we move **against** the current I_2 and thus experience a voltage **gain** across the resistor R_2. Hence, the negative sign in equation (4.2).

Secondly, we proceed in a clockwise direction around the bottom loop. Again, the sum of the voltage drops must equal zero.

$$R_2 I_2 - V = 0 \tag{4.3}$$

Again, we move opposite the current when moving clockwise through the capacitor. This, coupled with the fact that V represents the voltage drop across the capacitor, explains $-V$ in equation (4.3). Solve equation (4.2) for LI'.

$$LI' = -R_1 I + R_2 I_2 \tag{4.4}$$

>From equation (4.3), $R_2 I_2 = V$, which, when substituted in (4.4), produces the following result.

$$LI' = -R_1 I + V \tag{4.5}$$

Note that this is our first equation (4.20).

Next, solve equation (4.3) for I_2.

$$I_2 = \frac{V}{R_2} \tag{4.6}$$

>From equation (4.1), $I_2 = -I - I_3$. With this substitution, equation (4.6) becomes

$$\frac{V}{R_2} = -I - I_3. \tag{4.7}$$

However, the voltage drop across the capacitor is given by

$$V = \frac{q}{C},$$

where q is the charge on the capacitor. Differentiating,

$$CV' = q',$$

and using the fact that $I_3 = q'$,

$$CV' = I_3.$$

Substituting this result in equation (4.7), we obtain

$$\frac{V}{R_2} = -I - CV',$$

or equivalently,

$$CV' = -I - \frac{V}{R_2}, \tag{4.8}$$

which is identical to equation (4.21).

25. Let $x_1(t)$ and $x_2(t)$ represent the salt content (in pounds) of the upper and lower tanks, respectively. Because pure water flows into the upper tank, no salt is coming into the tank. However, the rate at which salt is leaving the tank is

$$\begin{aligned}\text{Rate Out} &= 4\text{ gal/min} \times \frac{x_1(t)}{100}\text{ lb/gal}\\ &= \frac{x_1(t)}{25}\text{ lb/min.}\end{aligned}$$

Because $dx_1/dt = \text{Rate In} - \text{Rate Out}$, we write

$$\frac{dx_1}{dt} = -\frac{x_1}{25}. \tag{4.9}$$

Salt enters the lower tank at the same rate as it is leaving the upper tank. Further, the rate at which salt leaves the lower tank is

$$\begin{aligned}\text{Rate Out} &= 4\text{ gal/min} \times \frac{x_2(t)}{100}\text{ lb/gal}\\ &= \frac{x_2(t)}{25}\text{ lb/min.}\end{aligned}$$

Again, because $dx_2/dt = \text{Rate In} - \text{Rate Out}$, we write

$$\frac{dx_2}{dt} = \frac{x_1}{25} - \frac{x_2}{25}. \tag{4.10}$$

In matrix form, the system defined by equations (4.9) and (4.10) is

$$\begin{pmatrix} x_1 \\ x_2 \end{pmatrix}' = \begin{pmatrix} -1/25 & 0 \\ 1/25 & -1/25 \end{pmatrix}\begin{pmatrix} x_1 \\ x_2 \end{pmatrix}. \tag{4.11}$$

Recall that the initial salt content in the upper and lower tanks is 10 and 20 pounds, respectively. Thus, the initial condition is $\mathbf{x}(0) = (x_1(0), x_2(0)) = (10, 20)$.

Finally, when the system (4.11) is compared with $\mathbf{x}' = A\mathbf{x} + \mathbf{f}$, we see that $\mathbf{f} = \mathbf{0}$ and the system is homogeneous.

27. Let $x_1(t)$, $x_2(t)$, and $x_3(t)$ represent the salt content (in pounds) of the first, second, and third tanks, respectively. Salt enters the first tank at a rate

$$5\text{ gal/min} \times 2\text{ lb/gal} = 10\text{ lb/min.}$$

Salt leaves the first tank at a rate

$$5\text{ gal/min} \times \frac{x_1}{100}\text{ lb/gal} = \frac{x_1}{20}\text{ lb/gal.}$$

Because $dx_1/dt = \text{Rate In} - \text{Rate Out}$, the differential equation governing the salt content in the first tank is

$$\frac{dx_1}{dt} = 10 - \frac{x_1}{20}. \tag{4.12}$$

Similarly, the equations governing the salt content in the second and third tanks are

$$\frac{dx_2}{dt} = \frac{x_1}{20} - \frac{x_2}{16}, \tag{4.13}$$

and

$$\frac{dx_3}{dt} = \frac{x_2}{16} - \frac{x_3}{12}, \tag{4.14}$$

respectively. In matrix-vector form, the system created by equations (4.12), (4.13), and (4.14), becomes

$$\begin{pmatrix} x_1 \\ x_2 \\ x_3 \end{pmatrix}' = \begin{bmatrix} -1/20 & 0 & 0 \\ 1/20 & -1/16 & 0 \\ 0 & 1/16 & -1/12 \end{bmatrix}\begin{pmatrix} x_1 \\ x_2 \\ x_3 \end{pmatrix} + \begin{bmatrix} 10 \\ 0 \\ 0 \end{bmatrix}.$$

Comparing this system with $\mathbf{x}' = A\mathbf{x}+\mathbf{f}$, we see that $\mathbf{f}(t) = (10, 0, 0)^T$, so the system is inhomogeneous. Because each tank contains pure water at time $t = 0$, the initial condition is $\mathbf{x}(0) = (0, 0, 0)^T$.

29. As in Example 4.8, the forces on mass the leftmost mass are $-kx$ and $k(y-x)$. Thus, by Newton's law,

$$mx'' = -kx + k(y - x) = -2kx + ky,$$

or equivalently,

$$x'' = -\frac{2k}{m}x + \frac{k}{m}y. \tag{4.15}$$

On the middle mass m, the spring connecting it to the left most mass must exert an equal but opposite force to that which it exerts on the leftmost mass; i.e.,

$-k(y-x)$. On the other side, the spring connecting the middle mass to the rightmost mass exerts a force $k(z-y)$. Thus, by Newton's law,

$$my'' = -k(y-x) + k(z-y) = kx - 2ky + kz,$$

or equivalently,

$$y'' = \frac{k}{m}x - \frac{2k}{m}y + \frac{k}{m}z. \quad (4.16)$$

Finally, the spring connecting the rightmost mass to the middle mass must exert an equal but opposite force to that which it exerts on the middle mass; i.e., $-k(z-y)$. On the other side, the spring connecting the right most mass to the vertical support on the right exerts a force of $-kz$. Hence, by Newton's law,

$$mz'' = -k(z-y) - kz = ky - 2kz,$$

or equivalently,

$$z'' = \frac{k}{m}y - \frac{2k}{m}z. \quad (4.17)$$

To change to a system of first order equations, let $x_1 = x$, $x_2 = x'$, $x_3 = y$, $x_4 = y'$, $x_5 = z$, and $x_6 = z'$. Then,

$$\begin{aligned} x_1' &= x_2 \\ x_2' &= -\frac{2k}{m}x_1 + \frac{k}{m}x_3 \\ x_3' &= x_4 \\ x_4' &= \frac{k}{m}x_1 - \frac{2k}{m}x_3 + \frac{k}{m}x_5 \\ x_5' &= x_6 \\ x_6' &= \frac{k}{m}x_3 - \frac{2k}{m}x_5. \end{aligned} \quad (4.18)$$

In matrix form, this can be written $\mathbf{x}' = A\mathbf{x}$, where $\mathbf{x} = (x_1, x_2, x_3, x_4, x_5, x_6)^T$ and

$$A = \begin{pmatrix} 0 & 1 & 0 & 0 & 0 & 0 \\ -2k/m & 0 & k/m & 0 & 0 & 0 \\ 0 & 0 & 0 & 1 & 0 & 0 \\ k/m & 0 & -2k/m & 0 & k/m & 0 \\ 0 & 0 & 0 & 0 & 0 & 1 \\ 0 & 0 & k/m & 0 & -2k/m & 0 \end{pmatrix}.$$

Note that this has the form $\mathbf{x}' = A\mathbf{x} + \mathbf{f}$, with $\mathbf{f}(t) = \mathbf{0}$. Hence, the system is homogeneous. Because the masses are released from rest, each displaced 10 cm to the right of their equilibrium positions,

$$\begin{aligned} \mathbf{x}(0) &= (x_1(0), x_2(0), x_3(0), x_4(0), x_5(0), x_6(0))^T \\ &= (x(0), x'(0), y(0), y'(0), z(0), z'(0))^T \\ &= (10, 0, 10, 0, 10, 0)^T. \end{aligned}$$

———×———

Section 8.5. Properties of Linear Systems

1. The system

$$\begin{aligned} x_1' &= -x_1 + 3x_2 \\ x_2' &= 2x_2 \end{aligned}$$

can be written

$$\begin{pmatrix} x_1 \\ x_2 \end{pmatrix}' = \begin{pmatrix} -1 & 3 \\ 0 & 2 \end{pmatrix}\begin{pmatrix} x_1 \\ x_2 \end{pmatrix}.$$

3. The system

$$\begin{aligned} x_1' &= x_1 + x_2 \\ x_2' &= -x_1 + x_2 \end{aligned}$$

can be written

$$\begin{pmatrix} x_1 \\ x_2 \end{pmatrix}' = \begin{pmatrix} 1 & 1 \\ -1 & 1 \end{pmatrix}\begin{pmatrix} x_1 \\ x_2 \end{pmatrix}.$$

5. The system

$$x_1' = x_1 + x_2$$
$$x_2' = -x_1 + x_2 + e^t$$

can be written

$$\begin{pmatrix} x_1 \\ x_2 \end{pmatrix}' = \begin{pmatrix} 1 & 1 \\ -1 & 1 \end{pmatrix}\begin{pmatrix} x_1 \\ x_2 \end{pmatrix} + \begin{pmatrix} 0 \\ e^t \end{pmatrix}.$$

7. We saw that the system in Exercise 1 can be written as $\mathbf{x}' = A\mathbf{x}$ where

$$A = \begin{pmatrix} -1 & 3 \\ 0 & 2 \end{pmatrix}.$$

If $\mathbf{x}(t) = (e^{-t}, 0)^T$, then

$$\mathbf{x}'(t) = \begin{pmatrix} e^{-t} \\ 0 \end{pmatrix}' = \begin{pmatrix} -e^{-t} \\ 0 \end{pmatrix},$$

and

$$A\mathbf{x}(t) = \begin{pmatrix} -1 & 3 \\ 0 & 2 \end{pmatrix}\begin{pmatrix} e^{-t} \\ 0 \end{pmatrix} = \begin{pmatrix} -e^{-t} \\ 0 \end{pmatrix}.$$

Thus, $\mathbf{x}$ satisfies the system. If $\vec{y}(t) = (e^{2t}, e^{2t})^T$, then

$$\mathbf{y}(t) = \begin{pmatrix} e^{2t} \\ e^{2t} \end{pmatrix}' = \begin{pmatrix} 2e^{2t} \\ 2e^{2t} \end{pmatrix},$$

and

$$A\mathbf{y}(t) = \begin{pmatrix} -1 & 3 \\ 0 & 2 \end{pmatrix}\begin{pmatrix} e^{2t} \\ e^{2t} \end{pmatrix} = \begin{pmatrix} 2e^{2t} \\ 2e^{2t} \end{pmatrix}.$$

Thus, $\mathbf{y}$ satisfies the system. Now,

$$C_1\mathbf{x}(t) + C_2\mathbf{y}(t) = \begin{pmatrix} C_1e^{-t} + C_2e^{2t} \\ C_2e^{2t} \end{pmatrix}.$$

Thus,

$$(C_1\mathbf{x}(t) + C_2\mathbf{y}(t))' = \begin{pmatrix} -C_1e^{-t} + 2C_2e^{2t} \\ 2C_2e^{2t} \end{pmatrix}.$$

But,

$$\begin{aligned} A(C_1\mathbf{x}(t) + C_2\mathbf{y}(t)) &= \begin{pmatrix} -1 & 3 \\ 0 & 2 \end{pmatrix}\begin{pmatrix} C_1e^{-t} + C_2e^{2t} \\ C_2e^{2t} \end{pmatrix} \\ &= \begin{pmatrix} -C_1e^{-t} + 2C_2e^{2t} \\ 2C_2e^{2t} \end{pmatrix}, \end{aligned}$$

So $C_1\mathbf{x}(t) + C_2\mathbf{y}(t)$ is a solution of the system.

9. We saw that the system in Exercise 3 can be written as $\mathbf{x}' = A\mathbf{x}$, where

$$A = \begin{pmatrix} 1 & 1 \\ -1 & 1 \end{pmatrix}.$$

If $\mathbf{x}(t) = (e^t\cos t, -e^t\sin t)^T$, then

$$\mathbf{x}(t) = \begin{pmatrix} e^t\cos t \\ -e^t\sin t \end{pmatrix}' = \begin{pmatrix} -e^t\sin t + e^t\cos t \\ -e^t\cos t - e^t\sin t \end{pmatrix}.$$

But,

$$\begin{aligned} A\mathbf{x}(t) &= \begin{pmatrix} 1 & 1 \\ -1 & 1 \end{pmatrix}\begin{pmatrix} e^t\cos t \\ -e^t\sin t \end{pmatrix} \\ &= \begin{pmatrix} e^t\cos t - e^t\sin t \\ -e^t\cos t - e^t\sin t \end{pmatrix}, \end{aligned}$$

so $\mathbf{x}(t)$ is a solutions of the system. Next, if $\mathbf{y}(t) = (e^t\sin t, e^t\cos t)^T$, then

$$\mathbf{y}'(t) = \begin{pmatrix} e^t\sin t \\ e^t\cos t \end{pmatrix}' = \begin{pmatrix} e^t\cos t + e^t\sin t \\ -e^t\sin t + e^t\cos t \end{pmatrix}.$$

But,

$$\begin{aligned} A\mathbf{y}(t) &= \begin{pmatrix} 1 & 1 \\ -1 & 1 \end{pmatrix}\begin{pmatrix} e^t\sin t \\ e^t\cos t \end{pmatrix} \\ &= \begin{pmatrix} e^t\sin t + e^t\cos t \\ -e^t\sin t + e^t\cos t \end{pmatrix}, \end{aligned}$$

so $\mathbf{y}(t)$ is a solution of the system. Now,

$$C_1\mathbf{x}(t) + C_2\mathbf{y}(t) = \begin{pmatrix} C_1e^t\cos t + C_2e^t\sin t \\ -C_1e^t\sin t + C_2e^t\cos t \end{pmatrix}.$$

Thus,

$$\begin{aligned} &(C_1\mathbf{x}(t) + C_2\mathbf{y}(t))' \\ &= \begin{pmatrix} -C_1e^t\sin t + C_1e^t\cos t \\ +C_2e^t\cos t + C_2e^t\sin t \\ -C_1e^t\cos t - C_1e^t\sin t \\ -C_2e^t\sin t + C_2e^t\cos t \end{pmatrix} \\ &= \begin{pmatrix} (C_1 + C_2)e^t\cos t \\ +(-C_1 + C_2)e^t\sin t \\ (-C_1 + C_2)e^t\cos t \\ +(-C_1 - C_2)e^t\sin t \end{pmatrix}. \end{aligned}$$

But,

$$\begin{aligned}
&A(C_1\mathbf{x}(t) + C_2\mathbf{y}(t)) \\
&= \begin{pmatrix} 1 & 1 \\ -1 & 1 \end{pmatrix} \begin{pmatrix} C_1e^t \cos t + C_2e^t \sin t \\ -C_1e^t \sin t + C_2e^t \cos t \end{pmatrix} \\
&= \begin{pmatrix} (C_1 + C_2)e^t \cos t \\ +(-C_1 + C_2)e^t \sin t \\ (-C_1 + C_2)e^t \cos t \\ +(-C_1 - C_2)e^t \sin t \end{pmatrix},
\end{aligned}$$

so $C_1\mathbf{x}(t) + C_2\mathbf{y}(t)$ is a solution of the system.

11. We saw in Exercise 5 that the system can be written as $\mathbf{x}' = A\mathbf{x} + \mathbf{w}$, where

$$A' = \begin{pmatrix} 1 & 1 \\ -1 & 1 \end{pmatrix} \quad \text{and} \quad \mathbf{w} = \begin{pmatrix} 0 \\ e^t \end{pmatrix}.$$

If $\mathbf{x}_p(t) = (e^t, 0)^T$, then

$$\mathbf{x}_p'(t) = \begin{pmatrix} e^t \\ 0 \end{pmatrix}' = \begin{pmatrix} e^t \\ 0 \end{pmatrix}.$$

But,

$$\begin{aligned}
A\mathbf{x}_p + \mathbf{w} &= \begin{pmatrix} 1 & 1 \\ -1 & 1 \end{pmatrix} \begin{pmatrix} e^t \\ 0 \end{pmatrix} + \begin{pmatrix} 0 \\ e^t \end{pmatrix} \\
&= \begin{pmatrix} e^t \\ -e^t \end{pmatrix} + \begin{pmatrix} 0 \\ e^t \end{pmatrix} \\
&= \begin{pmatrix} e^t \\ 0 \end{pmatrix}.
\end{aligned}$$

Therefore, $\mathbf{x}_p$ is a solution of the system. Next, if $\mathbf{z}(t) = \mathbf{x}_p + C_1\mathbf{x}(t) + C_2\mathbf{y}(t)$, then

$$\begin{aligned}
\mathbf{z}' &= \mathbf{x}_p' + C_1\mathbf{x}'(t) + C_2\mathbf{y}'(t) \\
&= A\mathbf{x}_p + \mathbf{w} + C_1A\mathbf{x} + C_2A\mathbf{y} \\
&= A[\mathbf{x}_p + C_1\mathbf{x} + C_2\mathbf{y}] + \mathbf{w} \\
&= A\mathbf{z} + \mathbf{w}.
\end{aligned}$$

Thus, $\mathbf{z}$ is a solution of the system.

13. In Exercise 7 we saw that $\mathbf{x}(t) = (e^{-t}, 0)^T$ and $\mathbf{y}(t) = (e^{2t}, e^{2t})^T$ were solutions of

$$\mathbf{x}' = \begin{pmatrix} -1 & 3 \\ 0 & 2 \end{pmatrix} \mathbf{x}.$$

If we evaluate these solutions at $t = 0$, then

$$\mathbf{x}(0) = \begin{pmatrix} 1 \\ 0 \end{pmatrix} \quad \text{and} \quad \mathbf{y}(0) = \begin{pmatrix} 1 \\ 1 \end{pmatrix}.$$

Because,

$$\det([\mathbf{x}(0), \mathbf{y}(0)]) = \det \begin{pmatrix} 1 & 1 \\ 0 & 1 \end{pmatrix} = 1 \neq 0,$$

these vectors are independent and form a fundamental solution set for the system. Hence, the general solution of the system is

$$\mathbf{z}(t) = C_1 \begin{pmatrix} e^{-t} \\ 0 \end{pmatrix} + C_2 \begin{pmatrix} e^{2t} \\ e^{2t} \end{pmatrix}.$$

To find the solution with initial condition $\mathbf{z}(0) = (0, 1)^T$, substitute the initial condition in the general solution to get

$$\begin{aligned}
\begin{pmatrix} 0 \\ 1 \end{pmatrix} &= C_1 \begin{pmatrix} 1 \\ 0 \end{pmatrix} + C_2 \begin{pmatrix} 1 \\ 1 \end{pmatrix} \\
\begin{pmatrix} 0 \\ 1 \end{pmatrix} &= \begin{pmatrix} 1 & 1 \\ 0 & 1 \end{pmatrix} \begin{pmatrix} C_1 \\ C_2 \end{pmatrix}.
\end{aligned}$$

This system has solutions $C_1 = -1$ and $C_2 = 1$. Thus, the solution is

$$\mathbf{z}(t) = -1 \begin{pmatrix} e^{-t} \\ 0 \end{pmatrix} + 1 \begin{pmatrix} e^{2t} \\ e^{2t} \end{pmatrix} = \begin{pmatrix} -e^{-t} + e^{2t} \\ e^{2t} \end{pmatrix}.$$

15. In Exercise 9, we saw that $\mathbf{x}(t) = (e^t \cos t, -e^t \sin t)^T$ and $\mathbf{y}(t) = (e^t \sin t, e^t \cos t)^T$ were solutions of

$$\mathbf{x}' = \begin{pmatrix} 1 & 1 \\ -1 & 1 \end{pmatrix} \mathbf{x}.$$

If we evaluate the solutions at $t = 0$, then

$$\mathbf{x}(0) = \begin{pmatrix} 1 \\ 0 \end{pmatrix} \quad \text{and} \quad \mathbf{y}(0) = \begin{pmatrix} 0 \\ 1 \end{pmatrix}.$$

Because $\det([\mathbf{x}(0), \mathbf{y}(0)]) = \det \begin{pmatrix} 1 & 0 \\ 0 & 1 \end{pmatrix} = 1 \neq 0,$ the vectors are independent and form a fundamental set of solutions. Hence, then general solution of the system is

$$\mathbf{z}(t) = C_1 \begin{pmatrix} e^t \cos t \\ -e^t \sin t \end{pmatrix} + C_2 \begin{pmatrix} e^t \sin t \\ e^t \cos t \end{pmatrix}.$$

To find the particular solution, substitute the initial condition $\mathbf{z}(0) = (-2, 3)^T$.

$$\begin{pmatrix} -2 \\ 3 \end{pmatrix} = C_1 \begin{pmatrix} 1 \\ 0 \end{pmatrix} + C_2 \begin{pmatrix} 0 \\ 1 \end{pmatrix}.$$

Thus, $C_1 = -2$ and $C_2 = 3$ and the solution is

$$\begin{aligned} \mathbf{z}(t) &= -2 \begin{pmatrix} e^t \cos t \\ -e^t \sin t \end{pmatrix} + 3 \begin{pmatrix} e^t \sin t \\ e^t \cos t \end{pmatrix} \\ &= \begin{pmatrix} -2e^t \cos t + 3e^t \sin t \\ 2e^t \sin t + 3e^t \cos t \end{pmatrix}. \end{aligned}$$

17. (a) If $\mathbf{x} = (x_1, x_2)^T = (0, e^t)^T$, then

$$\begin{aligned} x_1' &= 0 = 0 \cdot e^t = x_1 x_2, \text{ and} \\ x_2' &= e^t = x_2. \end{aligned}$$

Hence, $\mathbf{x}$ is a solution of the system

$$\begin{aligned} x_1' &= x_1 x_2 \\ x_2' &= x_2. \end{aligned}$$

If $\mathbf{y} = (y_1, y_2)^T = (1, 0)^T$, then

$$\begin{aligned} x_1' &= 0 = 1 \cdot 0 = x_1 x_2, \text{ and} \\ x_2' &= 0 = x_2, \end{aligned}$$

so $\mathbf{y}$ is also a solution of the system.

(b) If $\mathbf{z}(t) = \mathbf{x}(t) + \mathbf{y}(t)$, then $\vec{z} = (z_1, z_2)^T = (1, e^t)^T$. But,

$$\mathbf{z}_1' = 0 \neq 1 \cdot e^t = z_1 z_2,$$

so $\mathbf{z}$ is *not* a solution of the system. There is no contradiction of Theorem 5.1 because the system in *not* linear.

19. Evaluate each solution $\mathbf{y}_1(t) = (-e^{-t}, -e^{-t}, e^{-t})^T$, $\mathbf{y}_2(t) = (0, -e^t, 2e^t)^T$, and $\mathbf{y}_3(t) = (e^{2t}, 0, 2e^{2t})^T$ at $t = 0$.

$$\mathbf{y}_1(0) = \begin{pmatrix} -1 \\ -1 \\ 1 \end{pmatrix}, \quad \mathbf{y}_2(0) = \begin{pmatrix} 0 \\ -1 \\ 2 \end{pmatrix}, \quad \text{and}$$

$$\mathbf{y}_3(0) = \begin{pmatrix} 1 \\ 0 \\ 2 \end{pmatrix}$$

Because

$$\det \begin{pmatrix} -1 & 0 & 1 \\ -1 & -1 & 0 \\ 1 & 2 & 2 \end{pmatrix} = 1 \neq 0,$$

these vectors are independent. Therefore, by Proposition 5.12, the solutions $\mathbf{y}_1$, $\mathbf{y}_2$, and $\mathbf{y}_3$ are independent for all t.

21. Evaluate each solution $\mathbf{y}_1(t) = (-e^{-t} + e^{2t}, -e^{-t} + e^{2t}, e^{2t})^T$, $\mathbf{y}_2(t) = (-e^t + e^{-t}, e^{-t}, 0)^T$, and $\mathbf{y}_3(t) = (e^t, 0, 0)^T$ at $t = 0$.

$$\mathbf{y}_1(0) = \begin{pmatrix} 0 \\ 0 \\ 1 \end{pmatrix}, \quad \mathbf{y}_2(0) = \begin{pmatrix} 0 \\ 1 \\ 0 \end{pmatrix}, \quad \text{and}$$

$$\mathbf{y}_3(0) = \begin{pmatrix} 1 \\ 0 \\ 0 \end{pmatrix}$$

Because

$$\det \begin{pmatrix} 0 & 0 & 1 \\ 0 & 1 & 0 \\ 1 & 0 & 0 \end{pmatrix} = -1 \neq 0,$$

then these vectors are independent. Therefore, by Proposition 5.12, the solutions $\mathbf{y}_1$, $\mathbf{y}_2$, and $\mathbf{y}_3$ are independent for all t.

23. If $\mathbf{y}_1(t) = (-e^{2t}, 2e^{2t})^T$, then

$$\mathbf{y}_1' = \begin{pmatrix} -e^{2t} \\ 2e^{2t} \end{pmatrix}' = \begin{pmatrix} -2e^{2t} \\ 4e^{2t} \end{pmatrix},$$

and

$$\begin{aligned} \begin{pmatrix} -6 & -4 \\ 8 & 6 \end{pmatrix} \mathbf{y}_1 &= \begin{pmatrix} -6 & -4 \\ 8 & 6 \end{pmatrix} \begin{pmatrix} -e^{2t} \\ 2e^{2t} \end{pmatrix} \\ &= \begin{pmatrix} -2e^{2t} \\ 4e^{2t} \end{pmatrix}. \end{aligned}$$

Therefore, $\mathbf{y}_1$ is a solution. In a similar manner, you can also show that $\mathbf{y}_2(t) = (-e^{-2t}, e^{-2t})^T$ is a solution. Evaluating each solution at $t = 0$,

$$\mathbf{y}_1(0) = \begin{pmatrix} -1 \\ 2 \end{pmatrix} \quad \text{and} \quad \mathbf{y}_2(0) = \begin{pmatrix} -1 \\ 1 \end{pmatrix}.$$

These vectors are independent ($\mathbf{y}_1$ is not a multiple of $\mathbf{y}_2$), so the solutions are independent for all t and form a fundamental set of solutions. Thus, the general solution is

$$\mathbf{y}(t) = C_1 \begin{pmatrix} -e^{2t} \\ 2e^{2t} \end{pmatrix} + C_2 \begin{pmatrix} -e^{-2t} \\ e^{-2t} \end{pmatrix}.$$

Substitute the initial condition $\mathbf{y}(0) = (-5, 8)^T$.

$$\begin{pmatrix} -5 \\ 8 \end{pmatrix} = C_1 \begin{pmatrix} -1 \\ 2 \end{pmatrix} + C_2 \begin{pmatrix} -1 \\ 1 \end{pmatrix}$$

$$\begin{pmatrix} -5 \\ 8 \end{pmatrix} = \begin{pmatrix} -1 & -1 \\ 2 & 1 \end{pmatrix} \begin{pmatrix} C_1 \\ C_2 \end{pmatrix}$$

This system has solution $C_1 = 3$ and $C_2 = 2$, so the final solution is

$$\begin{aligned} \mathbf{y}(t) &= 3 \begin{pmatrix} -e^{2t} \\ 2e^{2t} \end{pmatrix} + 2 \begin{pmatrix} -e^{-2t} \\ e^{-2t} \end{pmatrix} \\ &= \begin{pmatrix} -3e^{2t} - 2e^{-2t} \\ 6e^{2t} + 2e^{-2t} \end{pmatrix}. \end{aligned}$$

25. If $\mathbf{y}_1(t) = (e^{2t}, e^{2t})^T$, then

$$\mathbf{y}_1' = \begin{pmatrix} e^{2t} \\ e^{2t} \end{pmatrix}' = \begin{pmatrix} 2e^{2t} \\ 2e^{2t} \end{pmatrix},$$

and

$$\begin{pmatrix} 3 & -1 \\ 1 & 1 \end{pmatrix} \mathbf{y}_1 = \begin{pmatrix} 3 & -1 \\ 1 & 1 \end{pmatrix} \begin{pmatrix} e^{2t} \\ e^{2t} \end{pmatrix} = \begin{pmatrix} 2e^{2t} \\ 2e^{2t} \end{pmatrix}.$$

In a similar manner, you can also show that $\vec{y}_2(t) = (e^{2t}(t+2), e^{2t}(t+1))^T$ is a solution of the system. Further, evaluating each solution at $t = 0$,

$$\mathbf{y}_1(0) = \begin{pmatrix} 1 \\ 1 \end{pmatrix} \quad \text{and} \quad \mathbf{y}_2(0) = \begin{pmatrix} 2 \\ 1 \end{pmatrix}.$$

The vectors are independent ($\mathbf{y}_1$ is not a multiple of $\mathbf{y}_2$), so the solutions $\mathbf{y}_1$ and $\mathbf{y}_2$ are independent for all t and form a fundamental set of solutions. Thus, the general solution is

$$\mathbf{y}(t) = C_1 \begin{pmatrix} e^{2t} \\ e^{2t} \end{pmatrix} + C_2 \begin{pmatrix} e^{2t}(t+2) \\ e^{2t}(t+1) \end{pmatrix}.$$

Substitute the initial condition $\mathbf{y}(0) = (0, 1)^T$.

$$\begin{pmatrix} 0 \\ 1 \end{pmatrix} = C_1 \begin{pmatrix} 1 \\ 1 \end{pmatrix} + C_2 \begin{pmatrix} 2 \\ 1 \end{pmatrix}$$

$$\begin{pmatrix} 0 \\ 1 \end{pmatrix} = \begin{pmatrix} 1 & 2 \\ 1 & 1 \end{pmatrix} \begin{pmatrix} C_1 \\ C_2 \end{pmatrix}$$

This system has solution $C_1 = 2$ and $C_2 = -1$. Thus, the final solution is

$$\mathbf{y}(t) = 2 \begin{pmatrix} e^{2t} \\ e^{2t} \end{pmatrix} - \begin{pmatrix} e^{2t}(t+2) \\ e^{2t}(t+1) \end{pmatrix} = \begin{pmatrix} -te^{2t} \\ (1-t)e^{2t} \end{pmatrix}.$$

27. A careful examination of the flow rates in and out of each tank will indicate that the volume of solution in each tank remains unchanged (3L) over time. Let $x_A(t)$ and $x_B(t)$ represent the salt content (kg) of tanks A and B, respectively. Let's begin with an analysis of tank A. We use the so-called "balance law,"

$$\frac{dx_A}{dt} = \text{Rate in} - \text{Rate out}.$$

Pure water enters tank A at 5 L/min, so there is no salt entering there. However, solution from tank B enters tank A at a rate of 4 L/min. The salt concentration in tank B is $x_B(t)/3$ kg/L, so the rate at which salt in entering tank A from tank B is

$$4\,\text{L/min} \times x_B(t)/3\,\text{kg/L} = 4x_B(t)/3\,\text{kg/min}.$$

Solution is leaving tank A for tank B at a rate of 9 L/min. The salt concentration of tank A is $x_A(t)/3$ kg/L. Thus, salt is leaving tank A at a rate of

$$9\,\text{L/min} \times x_A(t)/3\,\text{kg/L} = 3x_A(t)\,\text{kg/min}.$$

Thus, using the balance law above,

$$\frac{dx_A}{dt} = \frac{4x_B}{3} - 3x_A.$$

On tank B,

$$\frac{dx_B}{dt} = \text{Rate in} - \text{Rate out}.$$

We already know that salt enters tank B from tank A at a rate of $3x_A(t)$ kg/min. We also know that salt is leaving tank B for tank A at a rate of $4x_B(t)/3$ kg/min. Let's examine the drain, where solution leaves tank B at a rate of 5 L/min. Because the concentration of salt in tank B is $x_B(t)/3$ kg/L, salt leaves tank B through the drain at

$$5\,\text{L/min} \times x_B/3 = 5x_B/3\,\text{kg/min}.$$

Using the balance law,

$$\frac{dx_B}{dt} = 3x_A - \frac{4x_B}{3} - \frac{5x_B}{3} = 3x_A - 3x_B.$$

29. In Exercise 28, the salt content in tanks A, B, and C was modeled by the system

$$\begin{aligned}\frac{dx_A}{dt} &= \frac{x_B}{50} - \frac{9x_A}{200}\\ \frac{dx_B}{dt} &= \frac{9x_A}{200} - \frac{13x_B}{200} + \frac{x_C}{50}\\ \frac{dx_C}{dt} &= \frac{9x_B}{200} - \frac{9x_C}{200}.\end{aligned}$$

Setting $x_A(0) = x_B(0) = x_C(0) = 40$, the salt content in each tank is pictured below.

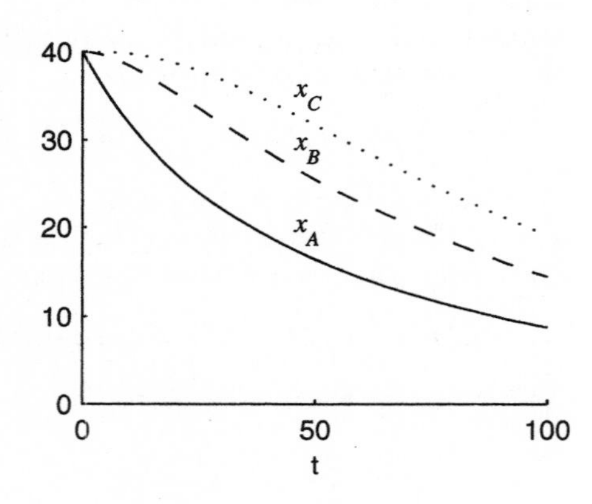

31. In Exercise 30, we saw that

$$\begin{aligned}\frac{dx}{dt} &= -\frac{x}{20}\\ \frac{dy}{dt} &= \frac{x}{20} - \frac{y}{20}\\ \frac{dz}{dt} &= \frac{y}{20} - \frac{z}{20}\end{aligned}$$

modeled the salt content in three cascading tanks. If $x(0) = y(0) = z(0) = 40$, then the following figure shows the salt content in each tank over time.

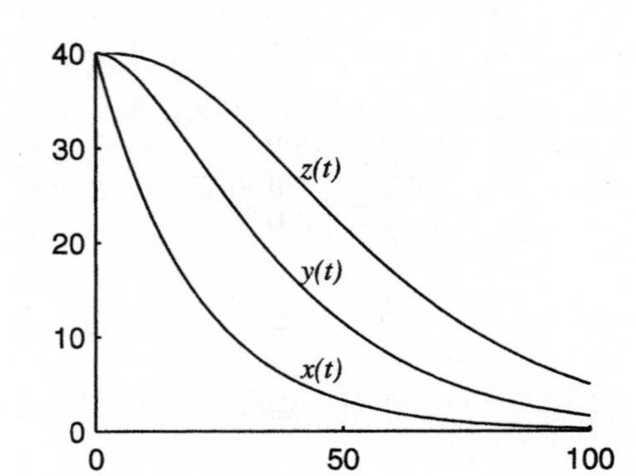

33. In the figure,

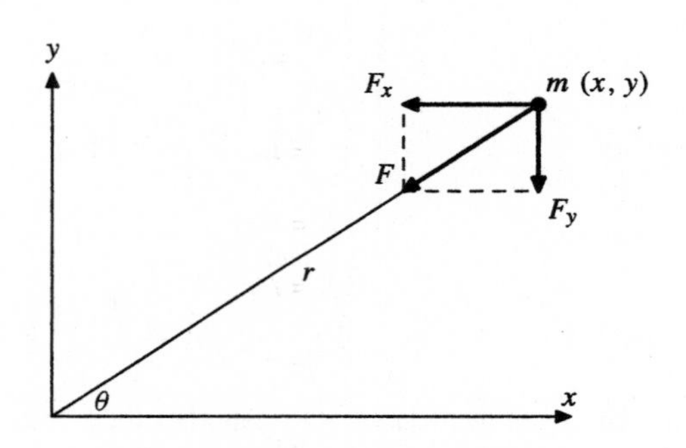

F represents the magnitude of the radially inward force. It is easy to see that

$$\begin{aligned}F_x &= F\cos\theta\\ F_y &= F\sin\theta.\end{aligned}$$

Furthermore,

$$\begin{aligned}\cos\theta &= \frac{x}{r}\\ \sin\theta &= \frac{y}{r},\end{aligned}$$

and with these substitutions and the fact that $F = km/r^2$, the former equations become

$$F_x = -\frac{km}{r^2} \cdot \frac{x}{r} = -\frac{kmx}{r^3},$$
$$F_y = -\frac{km}{r^2} \cdot \frac{y}{r} = -\frac{kmy}{r^3},$$

the minus signs being present because the force is directed opposite the displacement. Next, Newton gives us $F_x = ma_x = mx''$ and $F_y = ma_y = my''$, so

$$mx'' = -\frac{kmx}{r^3}$$
$$my'' = -\frac{kmy}{r^3},$$

or,

$$x'' = -\frac{kx}{r^3}$$
$$y'' = -\frac{ky}{r^3}.$$

If we now let $u_1 = x$, $u_2 = x'$, $u_3 = y$, and $u_4 = y'$, then

$$u_1' = u_2$$
$$u_2' = -\frac{ku_1}{r^3}$$
$$u_3' = u_4$$
$$u_4' = -\frac{ku_3}{r^3}.$$

Note that $r = (x^2 + y^2)^{1/2} = (u_1^2 + u_3^2)^{1/2}$, so we actually have

$$u_1' = u_2$$
$$u_2' = -\frac{ku_1}{(u_1^2 + u_3^2)^{3/2}}$$
$$u_3' = u_4$$
$$u_4' = -\frac{ku_3}{(u_1^2 + u_3^2)^{3/2}}.$$

35. In the figure,

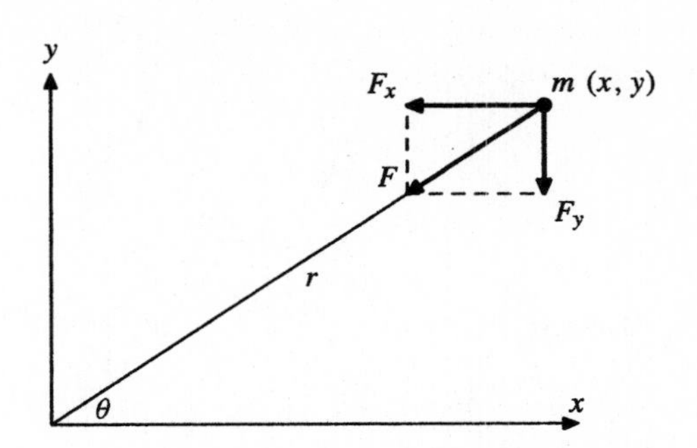

F represents the magnitude of the radially inward force. It is easy to see that

$$F_x = F\cos\theta$$
$$F_y = F\sin\theta.$$

Furthermore,

$$\cos\theta = \frac{x}{r}$$
$$\sin\theta = \frac{y}{r},$$

and with these substitutions and the fact that $F = km/r^3$, the former equations become

$$F_x = -\frac{km}{r^3} \cdot \frac{x}{r} = -\frac{kmx}{r^4},$$
$$F_y = -\frac{km}{r^3} \cdot \frac{y}{r} = -\frac{kmy}{r^4},$$

the minus signs being present because the force is directed opposite the displacement. Next, Newton gives us $F_x = ma_x = mx''$ and $F_y = ma_y = my''$, so

$$mx'' = -\frac{kmx}{r^4}$$
$$my'' = -\frac{kmy}{r^4},$$

or,

$$x'' = -\frac{kx}{r^4}$$
$$y'' = -\frac{ky}{r^4}.$$

If we now let $u_1 = x$, $u_2 = x'$, $u_3 = y$, and $u_4 = y'$,

then

$$\begin{aligned} u_1' &= u_2 \\ u_2' &= -\frac{ku_1}{r^4} \\ u_3' &= u_4 \\ u_4' &= -\frac{ku_3}{r^4}. \end{aligned}$$

Note that $r = (x^2 + y^2)^{1/2} = (u_1^2 + u_3^2)^{1/2}$, so we actually have

$$\begin{aligned} u_1' &= u_2 \\ u_2' &= -\frac{ku_1}{(u_1^2 + u_3^2)^2} \\ u_3' &= u_4 \\ u_4' &= -\frac{ku_3}{(u_1^2 + u_3^2)^2}. \end{aligned}$$

Chapter 9. Linear Systems with Constant Coefficients

Section 9.1. Overview of the Technique

1. If
$$A = \begin{pmatrix} 12 & 14 \\ -7 & -9 \end{pmatrix},$$
then the characteristic polynomial is
$$\begin{aligned} p(\lambda) &= \det(A - \lambda I) \\ &= \det \begin{pmatrix} 12-\lambda & 14 \\ -7 & -9-\lambda \end{pmatrix} \\ &= (12-\lambda)(-9-\lambda) + 98 \\ &= \lambda^2 - 3\lambda - 10 \\ &= (\lambda - 5)(\lambda + 2). \end{aligned}$$
Thus, the eigenvalues are $\lambda_1 = 5$ and $\lambda_2 = -2$.

3. If
$$A = \begin{pmatrix} -2 & 3 \\ 0 & -5 \end{pmatrix},$$
then the characteristic polynomial is
$$\begin{aligned} p(\lambda) &= \det(A - \lambda I) \\ &= \det \begin{pmatrix} -2-\lambda & 3 \\ 0 & -5-\lambda \end{pmatrix} \\ &= (-2-\lambda)(-5-\lambda) \\ &= (\lambda + 2)(\lambda + 5). \end{aligned}$$
Thus, the eigenvalues are $\lambda_1 = -2$ and $\lambda_2 = -5$.

5. If
$$A = \begin{pmatrix} 5 & 3 \\ -6 & -4 \end{pmatrix},$$
then the characteristic polynomial is
$$\begin{aligned} p(\lambda) &= \det(A - \lambda I) \\ &= \det \begin{pmatrix} 5-\lambda & 3 \\ -6 & -4-\lambda \end{pmatrix} \\ &= (5-\lambda)(-4-\lambda) + 18 \\ &= \lambda^2 - \lambda - 2 \\ &= (\lambda - 2)(\lambda + 1) \end{aligned}$$
Thus, the eigenvalues are $\lambda_1 = 2$ and $\lambda_2 = -1$.

7. If
$$A = \begin{pmatrix} -3 & 0 \\ 0 & -3 \end{pmatrix},$$
then the characteristic polynomial is
$$\begin{aligned} p(\lambda) &= \det(A - \lambda I) \\ &= \det \begin{pmatrix} -3-\lambda & 0 \\ 0 & -3-\lambda \end{pmatrix} \\ &= (-3-\lambda)(-3-\lambda) \\ &= (\lambda + 3)^2. \end{aligned}$$
Thus, $\lambda = -3$ is a repeated eigenvalue of algebraic multiplicity 2.

9. If
$$A = \begin{pmatrix} 1 & 2 & 3 \\ 0 & 0 & 2 \\ 0 & 3 & 1 \end{pmatrix},$$
then
$$p(\lambda) = -\begin{vmatrix} 1-\lambda & 2 & 3 \\ 0 & 0-\lambda & 2 \\ 0 & 3 & 1-\lambda \end{vmatrix}.$$
Expanding down the first column,
$$\begin{aligned} p(\lambda) &= -(1-\lambda)\begin{vmatrix} 0-\lambda & 2 \\ 3 & 1-\lambda \end{vmatrix} \\ &= (\lambda - 1)(-\lambda(1-\lambda) - 6) \\ &= (\lambda - 1)(\lambda^2 - \lambda - 6) \\ &= (\lambda - 1)(\lambda - 3)(\lambda + 2). \end{aligned}$$
Thus, the eigenvalues are $\lambda_1 = 1$, $\lambda_2 = 3$, and $\lambda_3 = -2$.

11. If

$$A = \begin{pmatrix} -1 & -4 & -2 \\ 0 & 1 & 1 \\ -6 & -12 & 2 \end{pmatrix},$$

then

$$p(\lambda) = -\begin{vmatrix} -1-\lambda & -4 & -2 \\ 0 & 1-\lambda & 1 \\ -6 & -12 & 2-\lambda \end{vmatrix}.$$

Expanding down the first column,

$$\begin{aligned} p(\lambda) & \\ &= -\left\{(-1-\lambda)\begin{vmatrix} 1-\lambda & 1 \\ -12 & 2-\lambda \end{vmatrix}\right. \\ &\qquad \left. 6\begin{vmatrix} -4 & -2 \\ 1-\lambda & 1 \end{vmatrix}\right\} \\ &= (\lambda+1)((1-\lambda)(2-\lambda)+12) \\ &\qquad +6(-4+2(1-\lambda)) \\ &= (\lambda+1)((1-\lambda)(2-\lambda)+12)+6(-2-2\lambda) \\ &= (\lambda+1)(\lambda^2-3\lambda+14)-12(\lambda+1) \\ &= (\lambda+1)(\lambda^2-3\lambda+14-12) \\ &= (\lambda+1)(\lambda^2-3\lambda+2) \\ &= (\lambda+1)(\lambda-1)(\lambda-2). \end{aligned}$$

Thus, the eigenvalues are $\lambda_1 = -1$, $\lambda_2 = 1$, and $\lambda_3 = 2$.

13. We used a computer to calculate the characteristic polynomial of matrix A.

$$p_A(\lambda) = \lambda^3 - 3\lambda^2 - 13\lambda + 15$$

A computer was used to calculate the eigenvalues: $\lambda_1 = -3$, $\lambda_2 = 1$, and $\lambda_3 = 5$. Next, a computer was used to draw the plot of p_A.

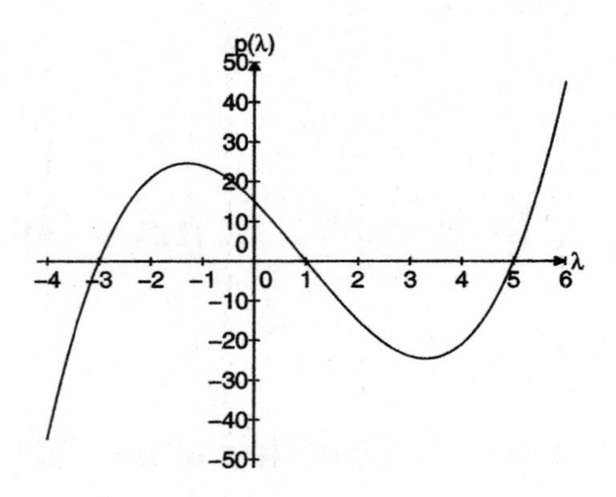

The graph of the characteristic polynomial appears to cross the horizontal axis at -3, 1, and 5. Thus, the zeros of the characteristic polynomial p_A are the eigenvalues of the matrix A. In a similar manner, the characteristic polynomial of matrix B is

$$p_B(\lambda) = \lambda^3 + 3\lambda^2 - 13\lambda - 15.$$

A computer was used to calculate the eigenvalues: $\lambda_1 = -5$, $\lambda_2 = -1$, and $\lambda_3 = 3$. A computer drawn graph of p_B follows.

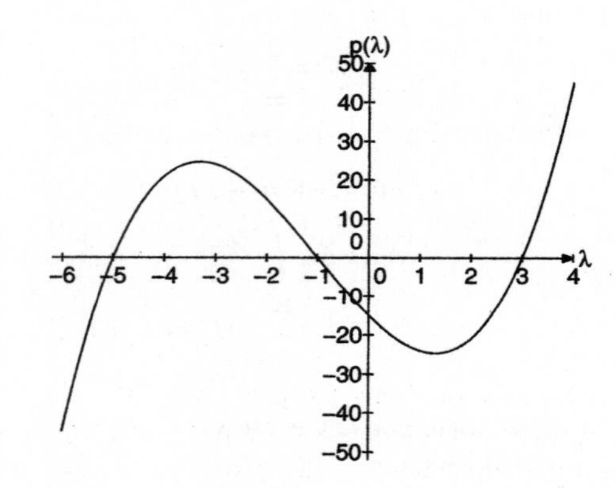

The graph of the characteristic polynomial p_B crosses the horizontal axis at -5, -1, and 3. Again, the zeros of the polynomial are the eigenvalues.

15. Using Matlab, for example, you would execute the commands

```
>> A=[12,14;-7,-9]; p=poly(A);
polyvalm(p,A)
```

for the matrix in Exercise 1. This will result in the zero matrix. A similar command works for the matrices in the other problems.

17. If

$$A = \begin{pmatrix} 6 & -8 \\ 0 & -2 \end{pmatrix},$$

then

$$\begin{aligned} p(\lambda) &= \det \begin{pmatrix} 6-\lambda & -8 \\ 0 & -2-\lambda \end{pmatrix} \\ &= -12 - 6\lambda + 2\lambda + \lambda^2 \\ &= \lambda^2 - 4\lambda - 12 \\ &= (\lambda - 6)(\lambda + 2). \end{aligned}$$

Thus, the eigenvalues are $\lambda_1 = 6$ and $\lambda_2 = -2$. For $\lambda_1 = 6$,

$$A - 6I = \begin{pmatrix} 0 & -8 \\ 0 & -8 \end{pmatrix}.$$

It is easily seen that the nullspace of $A - 6I$ is generated by the vector $(1, 0)^T$. Thus,

$$\mathbf{y}_1(t) = e^{6t}\begin{pmatrix} 1 \\ 0 \end{pmatrix}$$

is a solution. For $\lambda_2 = -2$,

$$A + 2I = \begin{pmatrix} 8 & -8 \\ 0 & 0 \end{pmatrix}.$$

It is easily seen that the nullspace of $A + 2I$ is generated by the vector $(1, 1)^T$. Thus,

$$\mathbf{y}_2(t) = e^{-2t}\begin{pmatrix} 1 \\ 1 \end{pmatrix}$$

is a solution. Because $\mathbf{y}_1(0) = (1, 0)^T$ and $\vec{y}_2(0) = (1, 1)^T$ are independent, the solutions $\mathbf{y}_1(t)$ and $\mathbf{y}_2(t)$ are independent for all t and form a fundamental set of solutions.

19. If

$$A = \begin{pmatrix} -1 & 0 \\ 0 & -1 \end{pmatrix},$$

then

$$\begin{aligned} p(\lambda) &= \det \begin{pmatrix} -1-\lambda & 0 \\ 0 & -1-\lambda \end{pmatrix} \\ &= (-1-\lambda)^2 \\ &= (1+\lambda)^2. \end{aligned}$$

Thus, $\lambda = -1$ is an eigenvalue. For $\lambda = -1$,

$$A + I = \begin{pmatrix} 0 & 0 \\ 0 & 0 \end{pmatrix}.$$

It is easily seen that both $(1, 0)^T$ and $(0, 1)^T$ are elements of the nullspace of $A + I$. Thus,

$$\mathbf{y}_1(t) = e^t\begin{pmatrix} 1 \\ 0 \end{pmatrix} \quad \text{and} \quad \mathbf{y}_2(t) = e^t\begin{pmatrix} 0 \\ 1 \end{pmatrix}$$

are solutions. Because $\mathbf{y}_1(0) = (1, 0)^T$ and $\mathbf{y}_2(0) = (0, 1)^T$ are independent, $\mathbf{y}_1(t)$ and $\mathbf{y}_2(t)$ are independent for all t and form a fundamental set of solutions.

21. If

$$A = \begin{pmatrix} 7 & 10 \\ -5 & -8 \end{pmatrix},$$

then

$$\begin{aligned} p(\lambda) &= \det \begin{pmatrix} 7-\lambda & 10 \\ -5 & -8-\lambda \end{pmatrix} \\ &= -56 - 7\lambda + 8\lambda + \lambda^2 + 50 \\ &= \lambda^2 + \lambda - 6 \\ &= (\lambda + 3)(\lambda - 2). \end{aligned}$$

Thus, $\lambda_1 = -3$ and $\lambda_2 = 2$ are eigenvalues. For $\lambda_1 = -3$,

$$A + 3I = \begin{pmatrix} 10 & 10 \\ -5 & -5 \end{pmatrix}.$$

It is easily seen that the nullspace of $A + 3I$ is generated $(1, -1)^T$. Thus,

$$\mathbf{y}_1(t) = e^{-3t}\begin{pmatrix} 1 \\ -1 \end{pmatrix}$$

is a solution. For $\lambda = 2$,

$$A - 2I = \begin{pmatrix} 5 & 10 \\ -5 & -10 \end{pmatrix}.$$

It is easily seen that the nullspace of $A - 2I$ is generated by $(2, -1)^T$. Thus,

$$\mathbf{y}_2(t) = e^{2t}\begin{pmatrix} 2 \\ -1 \end{pmatrix}$$

is a solution. Because $\mathbf{y}_1(0) = (1, -1)^T$ and $\mathbf{y}_2(0) = (2, -1)^T$ are independent, the solutions $\mathbf{y}_1(t)$ and $\mathbf{y}_2(t)$ are independent for all t and form a fundamental set of solutions.

23. If

$$A = \begin{pmatrix} 5 & -4 \\ 8 & -7 \end{pmatrix},$$

then

$$\begin{aligned} p(\lambda) &= \det\begin{pmatrix} 5-\lambda & -4 \\ 8 & -7-\lambda \end{pmatrix} \\ &= -35 - 5\lambda + 7\lambda + \lambda^2 + 32 \\ &= \lambda^2 + 2\lambda - 3 \\ &= (\lambda+3)(\lambda-1). \end{aligned}$$

Thus, $\lambda_1 = -3$ and $\lambda_2 = 1$ are eigenvalues. For $\lambda_1 = -3$,

$$A + 3I = \begin{pmatrix} 8 & -4 \\ 8 & -4 \end{pmatrix}.$$

It is easily seen that the nullspace of $A + 3I$ is generated $(1, 2)^T$. Thus,

$$\mathbf{y}_1(t) = e^{-3t}\begin{pmatrix} 1 \\ 2 \end{pmatrix}$$

is a solution. For $\lambda = 1$,

$$A - I = \begin{pmatrix} 4 & -4 \\ 8 & -8 \end{pmatrix}.$$

It is easily seen that the nullspace of $A - I$ is generated by $(1, 1)^T$. Thus,

$$\mathbf{y}_2(t) = e^{t}\begin{pmatrix} 1 \\ 1 \end{pmatrix}$$

is a solution. Because $\mathbf{y}_1(0) = (1, 2)^T$ and $\mathbf{y}_2(0) = (1, 1)^T$ are independent, the solutions $\mathbf{y}_1(t)$ and $\mathbf{y}_2(t)$ are independent for all t and form a fundamental set of solutions.

25. If

$$A = \begin{pmatrix} -1 & 0 & 0 \\ 2 & -5 & -6 \\ -2 & 3 & 4 \end{pmatrix},$$

then

$$\begin{aligned} p(\lambda) &= -\det\begin{pmatrix} -1-\lambda & 0 & 0 \\ 2 & -5-\lambda & -6 \\ -2 & 3 & 4-\lambda \end{pmatrix} \\ &= -(-1-\lambda)\begin{vmatrix} -5-\lambda & -6 \\ 3 & 4-\lambda \end{vmatrix} \\ &= (\lambda+1)(-20 + 5\lambda - 4\lambda + \lambda^2 + 18) \\ &= (\lambda+1)(\lambda^2 + \lambda - 2) \\ &= (\lambda+1)(\lambda+2)(\lambda-1) \end{aligned}$$

Thus, $\lambda_1 = -1$, $\lambda_2 = -2$, and $\lambda_3 = 1$ are eigenvalues. For $\lambda_1 = -1$,

$$A + I = \begin{pmatrix} 0 & 0 & 0 \\ 2 & -4 & -6 \\ -2 & 3 & 5 \end{pmatrix},$$

which has reduced row echelon form

$$\begin{pmatrix} 1 & 0 & -1 \\ 0 & 1 & 1 \\ 0 & 0 & 0 \end{pmatrix}.$$

It is easily seen that the nullspace of $A + I$ is generated $(1, -1, 1)^T$. Thus,

$$\mathbf{y}_1(t) = e^{-t}\begin{pmatrix} 1 \\ -1 \\ 1 \end{pmatrix}$$

is a solution. For $\lambda = -2$,

$$A + 2I = \begin{pmatrix} 1 & 0 & 0 \\ 2 & -3 & -6 \\ -2 & 3 & 6 \end{pmatrix},$$

which has reduced row echelon form

$$\begin{pmatrix} 1 & 0 & 0 \\ 0 & 1 & 2 \\ 0 & 0 & 0 \end{pmatrix}.$$

It is easily seen that the nullspace of $A+2I$ is generated by $(0,-2,1)^T$. Thus,

$$\mathbf{y}_2(t)=e^{-2t}\begin{pmatrix}0\\-2\\1\end{pmatrix}$$

is a solution. For $\lambda_3=1$,

$$A-I=\begin{pmatrix}-2&0&0\\2&-6&-6\\-2&3&3\end{pmatrix},$$

which has reduced row echelon form

$$\begin{pmatrix}1&0&0\\0&1&1\\0&0&0\end{pmatrix}.$$

It is easily seen that the nullspace of $A-I$ is generated by $(0,-1,1)$. Thus,

$$\mathbf{y}_3(t)=e^{t}\begin{pmatrix}0\\-1\\1\end{pmatrix}$$

is a solution. Because

$$\begin{aligned}\det[\mathbf{y}_1(0),\mathbf{y}_2(0),\mathbf{y}_3(0)]&=\det\begin{pmatrix}1&0&0\\-1&-2&-1\\1&1&1\end{pmatrix}\\&=-1,\end{aligned}$$

the solutions $\mathbf{y}_1(t)$, $\mathbf{y}_2(t)$, and $\mathbf{y}_3(t)$ are independent for all t and form a fundamental set of solutions.

27. If

$$A=\begin{pmatrix}-3&0&2\\6&3&-12\\2&2&-6\end{pmatrix},$$

then

$$\begin{aligned}p(\lambda)&=-\det\begin{pmatrix}-3-\lambda&0&2\\6&3-\lambda&-12\\2&2&-6-\lambda\end{pmatrix}\\&=-\left\{(-3-\lambda)\begin{vmatrix}3-\lambda&-12\\2&-6-\lambda\end{vmatrix}\right.\\&\qquad\left.+2\begin{vmatrix}6&3-\lambda\\2&2\end{vmatrix}\right\}\\&=(\lambda+3)(\lambda^2+3\lambda+6)-2(6+2\lambda)\\&=(\lambda+3)(\lambda^2+3\lambda+6)-4(\lambda+3)\\&=(\lambda+3)(\lambda^2+3\lambda+6-4)\\&=(\lambda+3)(\lambda^2+3\lambda+2)\\&=(\lambda+3)(\lambda+2)(\lambda+1).\end{aligned}$$

Thus, $\lambda_1=-3$, $\lambda_2=-2$, and $\lambda_3=-1$ are eigenvalues. For $\lambda_1=-3$,

$$A+3I=\begin{pmatrix}0&0&2\\6&6&-12\\2&2&-3\end{pmatrix},$$

which has reduced row echelon form

$$\begin{pmatrix}1&1&0\\0&0&1\\0&0&0\end{pmatrix}.$$

It is easily seen that the nullspace of $A+3I$ is generated $(-1,1,0)^T$. Thus,

$$\mathbf{y}_1(t)=e^{-3t}\begin{pmatrix}-1\\1\\0\end{pmatrix}$$

is a solution. For $\lambda=-2$,

$$A+2I=\begin{pmatrix}-1&0&2\\6&5&-12\\2&2&-4\end{pmatrix},$$

which has reduced row echelon form

$$\begin{pmatrix}1&0&-2\\0&1&0\\0&0&0\end{pmatrix}.$$

It is easily seen that the nullspace of $A + 2I$ is generated by $(2, 0, 1)^T$. Thus,

$$\mathbf{y}_2(t) = e^{-2t}\begin{pmatrix}2\\0\\1\end{pmatrix}$$

is a solution. For $\lambda_3 = -1$,

$$A + I = \begin{pmatrix}-2 & 0 & 2\\6 & 4 & -12\\2 & 2 & -5\end{pmatrix},$$

which has reduced row echelon form

$$\begin{pmatrix}1 & 0 & -1\\0 & 1 & -3/2\\0 & 0 & 0\end{pmatrix}.$$

It is easily seen that the nullspace of $A + I$ is generated by $(1, 3/2, 1)$. Thus,

$$\mathbf{y}_3(t) = e^{-t}\begin{pmatrix}1\\3/2\\1\end{pmatrix}$$

is a solution. Because

$$\det[\mathbf{y}_1(0), \mathbf{y}_2(0), \mathbf{y}_3(0) = \det\begin{pmatrix}-1 & 2 & 1\\1 & 0 & 3/2\\0 & 1 & 1\end{pmatrix} = \frac{1}{2},$$

the solutions $\mathbf{y}_1(t)$, $\mathbf{y}_2(t)$, and $\mathbf{y}_3(t)$ are independent for all t and form a fundamental set of solutions.

29. Using a computer, we find the following eigenvalue-eigenvector pairs.

$$3 \to \begin{pmatrix}1\\0\\1\end{pmatrix}, \quad 1 \to \begin{pmatrix}1\\-1\\1\end{pmatrix}, \quad -2 \to \begin{pmatrix}2\\0\\1\end{pmatrix}$$

31. Using a computer, we find the following eigenvalue-eigenvector pairs.

$$3 \to \begin{pmatrix}1\\-2\\0\end{pmatrix}, \quad 2 \to \begin{pmatrix}1\\-2\\-1\end{pmatrix}, \quad -1 \to \begin{pmatrix}1\\-1\\3\end{pmatrix}$$

33. Using a computer, we find the following eigenvalue-eigenvector pairs.

$$1 \to \begin{pmatrix}-1\\-1\\1\end{pmatrix}, \quad -3 \to \begin{pmatrix}-1\\1\\2\end{pmatrix}, \quad -1 \to \begin{pmatrix}-2\\1\\3\end{pmatrix}$$

35. Using a computer, we find the following eigenvalue-eigenvector pairs.

$$3 \to \begin{pmatrix}1\\1\\2\\2\end{pmatrix}, \quad -1 \to \begin{pmatrix}0\\1\\3/2\\1\end{pmatrix},$$

$$-2 \to \begin{pmatrix}0\\1\\1\\1\end{pmatrix}, \quad -4 \to \begin{pmatrix}-1\\2\\1\\2\end{pmatrix}$$

37. Using a computer, we find the following eigenvalue-eigenvector pairs.

$$4 \to \begin{pmatrix}1\\1\\1\\0\end{pmatrix}, \quad -2 \to \begin{pmatrix}0\\0\\-1\\1\end{pmatrix},$$

$$2 \to \begin{pmatrix}-1\\-1\\-1\\1\end{pmatrix}, \quad -1 \to \begin{pmatrix}-1\\0\\1\\0\end{pmatrix}$$

39. Using a computer, a fundamental set of solutions is found.

$$\mathbf{y}_1(t) = e^{3t}\begin{pmatrix}2\\1\\0\end{pmatrix}, \quad \mathbf{y}_2(t) = e^{-4t}\begin{pmatrix}-1\\-1\\1\end{pmatrix},$$

$$\mathbf{y}_3(t) = e^{-2t}\begin{pmatrix}2\\1\\1\end{pmatrix}$$

41. Using a computer, a fundamental set of solutions is found.

$$\mathbf{y}_1(t) = e^{-t}\begin{pmatrix}0\\0\\1\end{pmatrix}, \quad \mathbf{y}_2(t) = e^{-3t}\begin{pmatrix}-1\\0\\1\end{pmatrix},$$

$$\mathbf{y}_3(t) = e^{2t}\begin{pmatrix}-2\\1\\0\end{pmatrix}$$

43. Using a computer, a fundamental set of solutions is found.

$$\mathbf{y}_1(t) = e^{t}\begin{pmatrix}1\\-2\\2\end{pmatrix},\quad \mathbf{y}_2(t) = e^{-3t}\begin{pmatrix}1\\-2\\1\end{pmatrix},$$

$$\mathbf{y}_3(t) = e^{4t}\begin{pmatrix}-1\\1\\-1\end{pmatrix}$$

45. Using a computer, a fundamental set of solutions is found.

$$\mathbf{y}_1(t) = e^{-4t}\begin{pmatrix}1\\-1\\1\\0\end{pmatrix},\quad \mathbf{y}_2(t) = e^{-2t}\begin{pmatrix}-1\\1\\0\\0\end{pmatrix},$$

$$\mathbf{y}_3(t) = e^{2t}\begin{pmatrix}-1\\-1\\-1\\1\end{pmatrix},\quad \mathbf{y}_4(t) = e^{-t}\begin{pmatrix}-3/2\\1\\-1\\0\end{pmatrix}$$

47. Using a computer, a fundamental set of solutions is found.

$$\mathbf{y}_1(t) = e^{-5t}\begin{pmatrix}0\\2\\1\\2\end{pmatrix},\quad \mathbf{y}_2(t) = e^{-2t}\begin{pmatrix}1\\0\\1\\2\end{pmatrix},$$

$$\mathbf{y}_3(t) = e^{4t}\begin{pmatrix}0\\2\\1\\1\end{pmatrix},\quad \mathbf{y}_4(t) = e^{2t}\begin{pmatrix}1\\1\\1\\2\end{pmatrix}$$

49. If

$$A = \begin{pmatrix}6 & -8\\4 & -6\end{pmatrix},$$

then A has eigenvalues 2 and -2, and determinant $D = -4$. Note that the product of the eigenvalues equals the determinant. If

$$B = \begin{pmatrix}-11 & -16\\8 & 13\end{pmatrix},$$

then B has eigenvalues -3 and 5, and determinant $D = -15$. Note that the product of the eigenvalues equals the determinant. If

$$C = \begin{pmatrix}7 & -21 & -11\\5 & -13 & -5\\-5 & 9 & 1\end{pmatrix},$$

then C has eigenvalues 2, -3, and -4, and determinant $D = 24$. Note that the product of the eigenvalues equals the determinant.

51. If

$$A = \begin{pmatrix}2 & 3\\0 & -4\end{pmatrix},$$

then the eigenvalues of A are 2 and -4. Note that the eigenvalues lie on the main diagonal. If

$$B = \begin{pmatrix}1 & 2 & 3\\0 & -1 & 4\\0 & 0 & 5\end{pmatrix},$$

then the eigenvalues of B are 1, -1, and 5. Note that the eigenvalues lie on the main diagonal. If

$$C = \begin{pmatrix}2 & -1 & 1 & 1\\0 & 3 & -1 & 0\\0 & 0 & -4 & 1\\0 & 0 & 0 & 2\end{pmatrix},$$

then the eigenvalues of C are 2, 3, -4 and 2. Note that the eigenvalues lie on the main diagonal. Here is an example of a lower triangular matrix.

$$\begin{pmatrix}1 & 0 & 0\\2 & -2 & 0\\3 & 1 & 4\end{pmatrix}$$

A computer shows that the eigenvalues are 1, -2, and 4. Again, note that the main diagonal contains the eigenvalues.

53. If

$$V = \begin{pmatrix}-2 & 1\\1 & 0\end{pmatrix} \quad\text{and}\quad D = \begin{pmatrix}-2 & 0\\0 & 3\end{pmatrix},$$

then

$$\begin{aligned} VDV^{-1} &= \begin{pmatrix} -2 & 1 \\ 1 & 0 \end{pmatrix} \begin{pmatrix} -2 & 0 \\ 0 & 3 \end{pmatrix} \begin{pmatrix} -2 & 1 \\ 1 & 0 \end{pmatrix}^{-1} \\ &= \begin{pmatrix} 4 & 3 \\ -2 & 0 \end{pmatrix} \begin{pmatrix} -2 & 1 \\ 1 & 0 \end{pmatrix}^{-1} \\ &= \begin{pmatrix} 4 & 3 \\ -2 & 0 \end{pmatrix} \begin{pmatrix} 0 & 1 \\ 1 & 2 \end{pmatrix} \\ &= \begin{pmatrix} 3 & 10 \\ 0 & -2 \end{pmatrix} \\ &= A. \end{aligned}$$

55. If

$$A = \begin{pmatrix} -1 & -2 \\ 4 & -7 \end{pmatrix},$$

then a computer reveals the following eigenvalue–eigenvector pairs.

$$-5 \to \begin{pmatrix} 1 \\ 2 \end{pmatrix} \quad \text{and} \quad -3 \to \begin{pmatrix} 1 \\ 1 \end{pmatrix}.$$

Thus, the matrices

$$V = \begin{pmatrix} 1 & 1 \\ 2 & 1 \end{pmatrix} \quad \text{and} \quad D = \begin{pmatrix} -5 & 0 \\ 0 & -3 \end{pmatrix}$$

diagonalize matrix A. That is, $A = VDV^{-1}$.

Section 9.2. Planar Systems

1. The matrix

$$A = \begin{pmatrix} 2 & -6 \\ 0 & -1 \end{pmatrix},$$

has the following eigenvalue-eigenvector pairs.

$$\lambda_1 = 2 \to \begin{pmatrix} 1 \\ 0 \end{pmatrix} \quad \text{and} \quad \lambda_2 = -1 \to \begin{pmatrix} 2 \\ 1 \end{pmatrix}.$$

Thus, the general solution is

$$\mathbf{y}(t) = C_1 e^{2t} \begin{pmatrix} 1 \\ 0 \end{pmatrix} + C_2 e^{-t} \begin{pmatrix} 2 \\ 1 \end{pmatrix}.$$

3. The matrix

$$A = \begin{pmatrix} -5 & 1 \\ -2 & -2 \end{pmatrix}$$

has the following eigenvalue-eigenvector pairs.

$$\lambda_1 = -4 \to \begin{pmatrix} 1 \\ 1 \end{pmatrix} \quad \text{and} \quad \lambda_2 = -3 \to \begin{pmatrix} 1 \\ 2 \end{pmatrix}.$$

Thus, the general solution is

$$\mathbf{y}(t) = C_1 e^{-4t} \begin{pmatrix} 1 \\ 1 \end{pmatrix} + C_2 e^{-3t} \begin{pmatrix} 1 \\ 2 \end{pmatrix}.$$

5. The matrix

$$A = \begin{pmatrix} 1 & 2 \\ -1 & 4 \end{pmatrix}$$

has the following eigenvalue-eigenvector pairs.

$$\lambda_1 = 2 \to \begin{pmatrix} 2 \\ 1 \end{pmatrix} \quad \text{and} \quad \lambda_2 = 3 \to \begin{pmatrix} 1 \\ 1 \end{pmatrix}.$$

Thus, the general solution is

$$\mathbf{y}(t) = C_1 e^{2t} \begin{pmatrix} 2 \\ 1 \end{pmatrix} + C_2 e^{3t} \begin{pmatrix} 1 \\ 1 \end{pmatrix}.$$

7. The system in Exercise 1 had general solution

$$\mathbf{y}(t) = C_1 e^{2t} \begin{pmatrix} 1 \\ 0 \end{pmatrix} + C_2 e^{-t} \begin{pmatrix} 2 \\ 1 \end{pmatrix}.$$

Thus, if $\mathbf{y}(0) = (0, 1)^T$, then

$$\begin{pmatrix} 0 \\ 1 \end{pmatrix} = C_1 \begin{pmatrix} 1 \\ 0 \end{pmatrix} + C_2 \begin{pmatrix} 2 \\ 1 \end{pmatrix} = \begin{pmatrix} 1 & 2 \\ 0 & 1 \end{pmatrix} \begin{pmatrix} C_1 \\ C_2 \end{pmatrix}.$$

The augmented matrix reduces.

$$\begin{pmatrix} 1 & 2 & 0 \\ 0 & 1 & 1 \end{pmatrix} \to \begin{pmatrix} 1 & 0 & -2 \\ 0 & 1 & 1 \end{pmatrix}$$

Therefore, $C_1 = -2$ and $C_2 = 1$, giving particular solution

$$\mathbf{y}(t) = -2e^{2t} \begin{pmatrix} 1 \\ 0 \end{pmatrix} + e^{-t} \begin{pmatrix} 2 \\ 1 \end{pmatrix}.$$

9. The system in Exercise 3 had general solution

$$\mathbf{y}(t) = C_1 e^{-4t} \begin{pmatrix} 1 \\ 1 \end{pmatrix} + C_2 e^{-3t} \begin{pmatrix} 1 \\ 2 \end{pmatrix}.$$

Thus, if $\mathbf{y}(0) = (0, -1)^T$, then

$$\begin{pmatrix} 0 \\ -1 \end{pmatrix} = C_1 \begin{pmatrix} 1 \\ 1 \end{pmatrix} + C_2 \begin{pmatrix} 1 \\ 2 \end{pmatrix} = \begin{pmatrix} 1 & 1 \\ 1 & 2 \end{pmatrix} \begin{pmatrix} C_1 \\ C_2 \end{pmatrix}.$$

The augmented matrix reduces.

$$\begin{pmatrix} 1 & 1 & 0 \\ 1 & 2 & -1 \end{pmatrix} \to \begin{pmatrix} 1 & 0 & 1 \\ 0 & 1 & -1 \end{pmatrix}$$

Therefore, $C_1 = 1$ and $C_2 = -1$, giving particular solution

$$\mathbf{y}(t) = e^{-4t} \begin{pmatrix} 1 \\ 1 \end{pmatrix} - e^{-3t} \begin{pmatrix} 1 \\ 2 \end{pmatrix}.$$

11. The system in Exercise 5 had general solution

$$\mathbf{y}(t) = C_1 e^{2t} \begin{pmatrix} 2 \\ 1 \end{pmatrix} + C_2 e^{3t} \begin{pmatrix} 1 \\ 1 \end{pmatrix}.$$

Thus, if $\mathbf{y}(0) = (3, 2)^T$, then

$$\begin{pmatrix} 3 \\ 2 \end{pmatrix} = C_1 \begin{pmatrix} 2 \\ 1 \end{pmatrix} + C_2 \begin{pmatrix} 1 \\ 1 \end{pmatrix} = \begin{pmatrix} 2 & 1 \\ 1 & 1 \end{pmatrix} \begin{pmatrix} C_1 \\ C_2 \end{pmatrix}.$$

The augmented matrix reduces.

$$\begin{pmatrix} 2 & 1 & 3 \\ 1 & 1 & 2 \end{pmatrix} \to \begin{pmatrix} 1 & 0 & 1 \\ 0 & 1 & 1 \end{pmatrix}$$

Therefore, $C_1 = 1$ and $C_2 = 1$, giving particular solution

$$\mathbf{y}(t) = e^{2t} \begin{pmatrix} 2 \\ 1 \end{pmatrix} + e^{3t} \begin{pmatrix} 1 \\ 1 \end{pmatrix}.$$

13. If $z(t) = e^{2it}(1, 1+i)^T$, then

$$\begin{aligned} z(t) &= (\cos 2t + i\sin 2t)\begin{bmatrix} 1 \\ 1+i \end{bmatrix} \\ &= (\cos 2t + i\sin 2t)\left[\begin{pmatrix} 1 \\ 1 \end{pmatrix} + i\begin{pmatrix} 0 \\ 1 \end{pmatrix}\right] \\ &= \left[\cos 2t\begin{pmatrix} 1 \\ 1 \end{pmatrix} - \sin 2t\begin{pmatrix} 0 \\ 1 \end{pmatrix}\right] \\ &\quad + i\left[\cos 2t\begin{pmatrix} 0 \\ 1 \end{pmatrix} + \sin 2t\begin{pmatrix} 1 \\ 1 \end{pmatrix}\right]. \end{aligned}$$

Therefore, $\mathrm{Re}(\mathbf{z}(t)) = (\cos 2t, \cos 2t - \sin 2t)^T$ and $\mathrm{Im}(\mathbf{z}(t)) = (\sin 2t, \cos 2t + \sin 2t)^T$.

15. If $z(t) = e^{3it}(-1-i, 2)^T$, then

$$\begin{aligned} z(t) &= (\cos 3t + i\sin 3t)\left[\begin{pmatrix} -1 \\ 2 \end{pmatrix} + i\begin{pmatrix} -1 \\ 0 \end{pmatrix}\right] \\ &= \left[\cos 3t\begin{pmatrix} -1 \\ 2 \end{pmatrix} - \sin 3t\begin{pmatrix} -1 \\ 0 \end{pmatrix}\right] \\ &\quad + i\left[\sin 3t\begin{pmatrix} -1 \\ 2 \end{pmatrix} + \cos 3t\begin{pmatrix} -1 \\ 0 \end{pmatrix}\right] \end{aligned}$$

The real part of $z(t)$ is

$$\mathbf{y}_1(t) = \begin{pmatrix} -\cos 3t + \sin 3t \\ 2\cos 3t \end{pmatrix}$$

and

$$\mathbf{y}_1'(t) = \begin{pmatrix} 3\sin 3t + 3\cos 3t \\ -6\sin 3t \end{pmatrix}.$$

However,

$$\begin{pmatrix} 3 & 3 \\ -6 & -3 \end{pmatrix}\begin{pmatrix} -\cos 3t + \sin 3t \\ 2\cos 3t \end{pmatrix} = \begin{pmatrix} 3\sin 3t + 3\cos 3t \\ -6\sin 3t \end{pmatrix}$$

as well, so $\mathbf{y}_1$ is a solution of $\mathbf{y}' = A\mathbf{y}$. The imaginary part of $z(t)$ is

$$\mathbf{y}_2(t) = \begin{pmatrix} -\sin 3t - \cos 3t \\ 2\sin 3t \end{pmatrix}$$

and

$$\mathbf{y}_2'(t) = \begin{pmatrix} -3\cos 3t + 3\sin 3t \\ 6\cos 3t \end{pmatrix}.$$

However,

$$\begin{pmatrix} 3 & 3 \\ -6 & -3 \end{pmatrix}\begin{pmatrix} -\sin 3t - \cos 3t \\ 2\sin 3t \end{pmatrix} = \begin{pmatrix} -3\cos 3t + 3\sin 3t \\ 6\cos 3t \end{pmatrix}$$

as well, so $\mathbf{y}_2$ is a solution of $\mathbf{y}' = A\mathbf{y}$. Finally, because

$$\mathbf{y}_1(0) = \begin{pmatrix} -1 \\ 2 \end{pmatrix} \quad \text{and} \quad \mathbf{y}_2(0) = \begin{pmatrix} -1 \\ 0 \end{pmatrix}$$

are independent, $\mathbf{y}_1(t)$ and $\mathbf{y}_2(t)$ are independent for all values of t and form a fundamental set of solutions.

17. If

$$A = \begin{pmatrix} -1 & -2 \\ 4 & 3 \end{pmatrix},$$

then the characteristic polynomial is $p(\lambda) = \lambda^2 - 2\lambda + 5$ and the eigenvalues are $1 \pm 2i$. Trusting that

$$A - (1+2i)I = \begin{pmatrix} -2-2i & -2 \\ 4 & 2-2i \end{pmatrix}$$

is singular, examination of the first row reveals the eigenvector $\mathbf{v} = (1, -1-i)^T$, Thus,

$$\begin{aligned} z(t) &= e^{(1+2i)t}\begin{pmatrix} 1 \\ -1-i \end{pmatrix} \\ &= e^t(\cos 2t + i\sin 2t)\left[\begin{pmatrix} 1 \\ -1 \end{pmatrix} + i\begin{pmatrix} 0 \\ -1 \end{pmatrix}\right] \\ &= e^t\left[\cos 2t\begin{pmatrix} 1 \\ -1 \end{pmatrix} - \sin 2t\begin{pmatrix} 0 \\ -1 \end{pmatrix}\right] \\ &\quad + ie^t\left[\cos 2t\begin{pmatrix} 0 \\ -1 \end{pmatrix} + \sin 2t\begin{pmatrix} 1 \\ -1 \end{pmatrix}\right]. \end{aligned}$$

Therefore,

$$\mathbf{y}_1(t) = e^t\begin{pmatrix} \cos 2t \\ -\cos 2t + \sin 2t \end{pmatrix} \quad \text{and}$$

$$\mathbf{y}_2(t) = e^t\begin{pmatrix} \sin 2t \\ -\cos 2t - \sin 2t \end{pmatrix}$$

form a fundamental set of solutions.

19. The characteristic polynomial of

$$A = \begin{pmatrix} 0 & 4 \\ -2 & -4 \end{pmatrix}$$

is $p(\lambda) = \lambda^2 - 4\lambda + 8$, which has complex roots $\lambda = -2 \pm 2i$. For the eigenvalue $\lambda = 2 + 2i$, we have the eigenvector $\mathbf{w} = (-1 - i, 1)^T$. The corresponding exponential solution is

$$\begin{aligned}
\mathbf{z}(t) &= e^{(-2+2i)t}\begin{pmatrix} -1-i \\ 1 \end{pmatrix} \\
&= e^{-2t}[\cos 2t + i \sin 2t]\left[\begin{pmatrix} -1 \\ 1 \end{pmatrix} + i\begin{pmatrix} -1 \\ 0 \end{pmatrix}\right] \\
&= e^{-2t}\left[\cos 2t \cdot \begin{pmatrix} -1 \\ 1 \end{pmatrix} - \sin 2t \cdot \begin{pmatrix} -1 \\ 0 \end{pmatrix}\right] \\
&\quad + ie^{-2t}\left[\cos 2t \cdot \begin{pmatrix} -1 \\ 0 \end{pmatrix} + \sin 2t \cdot \begin{pmatrix} -1 \\ 1 \end{pmatrix}\right].
\end{aligned}$$

The real and imaginary parts of $\mathbf{z}$,

$$\mathbf{y}_1(t) = e^{-2t}\begin{pmatrix} -\cos 2t + \sin 2t \\ \cos 2t \end{pmatrix}$$

$$\mathbf{y}_2(t) = e^{-2t}\begin{pmatrix} -\cos 2t - \sin 2t \\ \sin 2t \end{pmatrix}$$

are a fundamental set of solutions.

21. If

$$A = \begin{pmatrix} 3 & -6 \\ 3 & 5 \end{pmatrix},$$

then the characteristic polynomial is $p(\lambda) = \lambda^2 - 8\lambda + 33$ and the eigenvalues are $4 \pm \sqrt{17}i$. Trusting that

$$A - \lambda I = \begin{pmatrix} 3-\lambda & -6 \\ 3 & 5-\lambda \end{pmatrix}$$

is singular, examination of the first row reveals the eigenvector $\mathbf{v} = (6, 3 - \lambda)^T$. Substituting $\lambda = 4 + \sqrt{17}i$ give $\mathbf{v} = (6, -1 - \sqrt{17}i)^T$. Thus,

$$\begin{aligned}
&z(t) \\
&= e^{(4+\sqrt{17}i)t}\begin{pmatrix} 6 \\ -1-\sqrt{17}i \end{pmatrix} \\
&= e^{4t}\left[\cos\sqrt{17}t + i\sin\sqrt{17}t\right] \\
&\quad \times\left[\begin{pmatrix} 6 \\ -1 \end{pmatrix} + i\begin{pmatrix} 0 \\ -\sqrt{17} \end{pmatrix}\right] \\
&= e^{4t}\left[\cos\sqrt{17}t\begin{pmatrix} 6 \\ -1 \end{pmatrix} - \sin\sqrt{17}t\begin{pmatrix} 0 \\ -\sqrt{17} \end{pmatrix}\right] \\
&\quad + ie^{4t}\left[\cos\sqrt{17}t\begin{pmatrix} 0 \\ -\sqrt{17} \end{pmatrix}\right. \\
&\quad \left. + \sin\sqrt{17}t\begin{pmatrix} 6 \\ -1 \end{pmatrix}\right]
\end{aligned}$$

Therefore,

$$\mathbf{y}_1(t) = e^{4t}\begin{pmatrix} 6\cos\sqrt{17}t \\ -\cos\sqrt{17}t + \sqrt{17}\sin\sqrt{17}t \end{pmatrix}$$

and

$$\mathbf{y}_2(t) = e^{4t}\begin{pmatrix} 6\sin\sqrt{17}t \\ -\sqrt{17}\cos\sqrt{17}t - \sin\sqrt{17}t \end{pmatrix}$$

form a fundamental set of solutions.

23. The fundamental solutions found in Exercise 17 allows the formation of the general solution

$$\begin{aligned}
\mathbf{y}(t) &= C_1e^t\begin{pmatrix} \cos 2t \\ -\cos 2t + \sin 2t \end{pmatrix} \\
&\quad + C_2e^t\begin{pmatrix} \sin 2t \\ -\cos 2t - \sin 2t \end{pmatrix}.
\end{aligned}$$

If $\mathbf{y}(0) = (0, 1)^T$, then

$$\begin{pmatrix} 0 \\ 1 \end{pmatrix} = C_1\begin{pmatrix} 1 \\ -1 \end{pmatrix} + C_2\begin{pmatrix} 0 \\ -1 \end{pmatrix}.$$

The augmented matrix reduces.

$$\begin{pmatrix} 1 & 0 & 0 \\ -1 & -1 & 1 \end{pmatrix} \rightarrow \begin{pmatrix} 1 & 0 & 0 \\ 0 & 1 & -1 \end{pmatrix}.$$

Thus, $C_1 = 0$ and $C_2 = -1$ and

$$\mathbf{y}(t) = -e^t\begin{pmatrix} \sin 2t \\ -\cos 2t - \sin 2t \end{pmatrix}.$$

25. A fundamental set of solutions was found in Exercise 19, so the solution has the form $\mathbf{y}(t) = C_1\mathbf{y}_1(t) + C_2\mathbf{y}_2(t)$, where

$$\mathbf{y}_1(t) = e^{-2t}\begin{pmatrix} -\cos 2t + \sin 2t \\ \cos 2t \end{pmatrix}$$
$$\mathbf{y}_2(t) = e^{-2t}\begin{pmatrix} -\cos 2t - \sin 2t \\ \sin 2t \end{pmatrix}.$$

At $t = 0$ we have

$$\begin{pmatrix} -1 \\ 2 \end{pmatrix} = \mathbf{y}(0) = C_1\begin{pmatrix} -1 \\ 1 \end{pmatrix} + C_2\begin{pmatrix} -1 \\ 0 \end{pmatrix}$$
$$= \begin{pmatrix} -1 & -1 \\ 1 & 0 \end{pmatrix}\begin{pmatrix} C_1 \\ C_2 \end{pmatrix}.$$

This system can be readily solved, getting $C_1 = 2$ and $C_2 = -1$. Hence the solution is

$$\mathbf{y}(t) = 2\mathbf{y}_1(t) - \mathbf{y}_2(t) = e^{-2t}\begin{pmatrix} -\cos 2t + 3\sin 2t \\ 2\cos 2t - \sin 2t \end{pmatrix}.$$

27. The fundamental solutions found in Exercise 21 allows the formation of the general solution.

$$\mathbf{y}(t) = C_1 e^{4t}\begin{pmatrix} 6\cos\sqrt{17}t \\ -\cos\sqrt{17}t + \sqrt{17}\sin\sqrt{17}t \end{pmatrix}$$
$$+ C_2 e^{4t}\begin{pmatrix} 6\sin\sqrt{17}t \\ -\sqrt{17}\cos\sqrt{17}t - \sin\sqrt{17}t \end{pmatrix}.$$

If $\mathbf{y}(0) = (1, 3)^T$, then

$$\begin{pmatrix} 1 \\ 3 \end{pmatrix} = C_1\begin{pmatrix} 6 \\ -1 \end{pmatrix} + C_2\begin{pmatrix} 0 \\ -\sqrt{17} \end{pmatrix}.$$

The augmented matrix reduces .

$$\begin{pmatrix} 6 & 0 & 1 \\ -1 & -\sqrt{17} & 3 \end{pmatrix} \rightarrow \begin{pmatrix} 1 & 0 & 1/6 \\ 0 & 1 & -19\sqrt{17}/102 \end{pmatrix}$$

Thus, $C_1 = 1/6$ and $C_2 = -19\sqrt{17}/102$ and

$$\mathbf{y}(t) = \frac{1}{6}e^{4t}\begin{pmatrix} 6\cos\sqrt{17}t \\ -\cos\sqrt{17}t + \sqrt{17}\sin\sqrt{17}t \end{pmatrix}$$
$$- \frac{19\sqrt{17}}{102}\begin{pmatrix} 6\sin\sqrt{17}t \\ -\sqrt{17}\cos\sqrt{17}t - \sin\sqrt{17}t \end{pmatrix}.$$

29. The matrix

$$A = \begin{pmatrix} -2 & 0 \\ 0 & -2 \end{pmatrix}$$

has a single eigenvalue $\lambda = -2$. However,

$$A - (-2)I = \begin{pmatrix} 0 & 0 \\ 0 & 0 \end{pmatrix},$$

so all nonzero vectors are eigenvectors. Choose $\mathbf{e}_1 = (1, 0)^T$ and $\mathbf{e}_2 = (0, 1)^T$ as eigenvectors. Then,

$$\mathbf{y}(t) = C_1 e^{-2t}\begin{pmatrix} 1 \\ 0 \end{pmatrix} + C_2 e^{-2t}\begin{pmatrix} 0 \\ 1 \end{pmatrix}$$

is the general solution.

31. The matrix

$$A = \begin{pmatrix} 3 & -1 \\ 1 & 1 \end{pmatrix}$$

has one eigenvalue, $\lambda = 2$. However, the nullspace of

$$A - 2I = \begin{pmatrix} 1 & -1 \\ 1 & -1 \end{pmatrix}$$

is generated by the single eigenvector, $\mathbf{v}_1 = (1, 1)^T$, with corresponding solution

$$\mathbf{y}_1(t) = e^{2t}\begin{pmatrix} 1 \\ 1 \end{pmatrix}.$$

To find another solution, we need to find a vector $\mathbf{v}_2$ which satisfies $(A - 2I)\mathbf{v}_2 = \mathbf{v}_1$. Choose $\mathbf{w} = (1, 0)^T$, which is independent of $\mathbf{v}_1$, and note that

$$(A - 2I)\mathbf{w} = \begin{pmatrix} 1 & -1 \\ 1 & -1 \end{pmatrix}\begin{pmatrix} 1 \\ 0 \end{pmatrix} = \begin{pmatrix} 1 \\ 0 \end{pmatrix} = \mathbf{v}_1.$$

Thus, choose $\mathbf{v}_2 = \mathbf{w} = (1, 0)^T$. Our second solution is

$$\mathbf{y}_2(t) = e^{2t}(\mathbf{v}_2 + t\mathbf{v}_1)$$
$$= e^{2t}\left[\begin{pmatrix} 1 \\ 0 \end{pmatrix} + t\begin{pmatrix} 1 \\ 1 \end{pmatrix}\right].$$

Thus, the general solution can be written

$$\mathbf{y}(t) = C_1 e^{2t}\begin{pmatrix} 1 \\ 1 \end{pmatrix} + C_2 e^{2t}\left[\begin{pmatrix} 1 \\ 0 \end{pmatrix} + t\begin{pmatrix} 1 \\ 1 \end{pmatrix}\right]$$
$$= e^{2t}\left[(C_1 + C_2 t)\begin{pmatrix} 1 \\ 1 \end{pmatrix} + C_2\begin{pmatrix} 1 \\ 0 \end{pmatrix}\right]$$

33. The matrix

$$A = \begin{pmatrix} -2 & 1 \\ -9 & 4 \end{pmatrix}$$

has one eigenvalue, $\lambda = 1$. However, the nullspace of

$$A - I = \begin{pmatrix} -3 & 1 \\ -9 & 3 \end{pmatrix}$$

is generated by the single eigenvector, $\mathbf{v}_1 = (1, 3)^T$, with corresponding solution

$$\mathbf{y}_1(t) = e^t \begin{pmatrix} 1 \\ 3 \end{pmatrix}.$$

To find another solution, we need to find a vector $\mathbf{v}_2$ which satisfies $(A - I)\mathbf{v}_2 = \mathbf{v}_1$. Choose $\mathbf{w} = (1, 0)^T$, which is independent of $\mathbf{v}_1$, and note that

$$\begin{aligned}(A - I)\mathbf{w} &= \begin{pmatrix} -3 & 1 \\ -9 & 3 \end{pmatrix}\begin{pmatrix} 1 \\ 0 \end{pmatrix} \\ &= \begin{pmatrix} -3 \\ -9 \end{pmatrix} = -3\begin{pmatrix} 1 \\ 3 \end{pmatrix} = -3\mathbf{v}_1.\end{aligned}$$

Thus, choose $\mathbf{v}_2 = -(1/3)\mathbf{w} = (-1/3, 0)^T$. Our second solution is

$$\begin{aligned}\mathbf{y}_2(t) &= e^t(\mathbf{v}_2 + t\mathbf{v}_1) \\ &= e^t\left[\begin{pmatrix} -1/3 \\ 0 \end{pmatrix} + t\begin{pmatrix} 1 \\ 3 \end{pmatrix}\right].\end{aligned}$$

Thus, the general solution can be written

$$\begin{aligned}\mathbf{y}(t) &= C_1e^t\begin{pmatrix} 1 \\ 3 \end{pmatrix} + C_2e^t\left[\begin{pmatrix} -1/3 \\ 0 \end{pmatrix} + t\begin{pmatrix} 1 \\ 3 \end{pmatrix}\right] \\ &= e^t\left[(C_1 + C_2t)\begin{pmatrix} 1 \\ 3 \end{pmatrix} + C_2\begin{pmatrix} -1/3 \\ 0 \end{pmatrix}\right]\end{aligned}$$

35. From Exercise 29,

$$\mathbf{y}(t) = C_1e^{-2t}\begin{pmatrix} 1 \\ 0 \end{pmatrix} + C_2e^{-2t}\begin{pmatrix} 0 \\ 1 \end{pmatrix}.$$

If $\mathbf{y}(0) = (3, -2)^T$, then

$$\begin{pmatrix} 3 \\ -2 \end{pmatrix} = C_1\begin{pmatrix} 1 \\ 0 \end{pmatrix} + C_2\begin{pmatrix} 0 \\ 1 \end{pmatrix},$$

and $C_1 = 3$ and $C_2 = -2$. Thus, the particular solution is

$$\begin{aligned}\mathbf{y}(t) &= 3e^{-2t}\begin{pmatrix} 1 \\ 0 \end{pmatrix} - 2e^{-2t}\begin{pmatrix} 0 \\ 1 \end{pmatrix} \\ &= e^{-2t}\begin{pmatrix} 3 \\ -2 \end{pmatrix}.\end{aligned}$$

37. From Exercise 31,

$$\mathbf{y}(t) = e^{2t}\left[(C_1 + C_2t)\begin{pmatrix} 1 \\ 1 \end{pmatrix} + C_2\begin{pmatrix} 1 \\ 0 \end{pmatrix}\right].$$

If $\mathbf{y}(0) = (2, -1)^T$, then

$$\begin{pmatrix} 2 \\ -1 \end{pmatrix} = C_1\begin{pmatrix} 1 \\ 1 \end{pmatrix} + C_2\begin{pmatrix} 1 \\ 0 \end{pmatrix}.$$

The augmented matrix reduces,

$$\begin{pmatrix} 1 & 1 & 2 \\ 1 & 0 & -1 \end{pmatrix} \rightarrow \begin{pmatrix} 1 & 0 & -1 \\ 0 & 1 & 3 \end{pmatrix},$$

and $C_1 = -1$ and $C_2 = 3$. Thus, the particular solution is

$$\begin{aligned}\mathbf{y}(t) &= e^{2t}\left[(-1 + 3t)\begin{pmatrix} 1 \\ 1 \end{pmatrix} + 3\begin{pmatrix} 1 \\ 0 \end{pmatrix}\right] \\ &= e^{2t}\begin{pmatrix} 2 + 3t \\ -1 + 3t \end{pmatrix}.\end{aligned}$$

39. From Exercise 33,

$$\mathbf{y}(t) = e^t\left[(C_1 + C_2t)\begin{pmatrix} 1 \\ 3 \end{pmatrix} + C_2\begin{pmatrix} -1/3 \\ 0 \end{pmatrix}\right].$$

If $\mathbf{y}(0) = (5, 3)^T$, then

$$\begin{pmatrix} 5 \\ 3 \end{pmatrix} = C_1\begin{pmatrix} 1 \\ 3 \end{pmatrix} + C_2\begin{pmatrix} -1/3 \\ 0 \end{pmatrix}.$$

The augmented matrix reduces,

$$\begin{pmatrix} 1 & -1/3 & 5 \\ 3 & 0 & 3 \end{pmatrix} \rightarrow \begin{pmatrix} 1 & 0 & 1 \\ 0 & 1 & -12 \end{pmatrix},$$

and $C_1 = 1$ and $C_2 = -12$. Thus, the particular solution is

$$\begin{aligned}\mathbf{y}(t) &= e^t\left[(1 - 12t)\begin{pmatrix} 1 \\ 3 \end{pmatrix} - 12\begin{pmatrix} -1/3 \\ 0 \end{pmatrix}\right] \\ &= e^t\begin{pmatrix} 5 - 12t \\ 3 - 36t \end{pmatrix}.\end{aligned}$$

41. The matrix

$$A = \begin{pmatrix} 2 & 4 \\ -1 & 6 \end{pmatrix}$$

has characteristic polynomial $p(\lambda) = \lambda^2 - 8\lambda + 16$ and one eigenvalue, $\lambda = 4$. Moreover, the nullspace of

$$A - 4I = \begin{pmatrix} -2 & 4 \\ -1 & 2 \end{pmatrix}$$

is generated by the single eigenvector, $\mathbf{v}_1 = (2, 1)^T$, with corresponding solution

$$\mathbf{y}_1(t) = e^{4t}\begin{pmatrix} 2 \\ 1 \end{pmatrix}.$$

To find another solution, we need to find a vector $\mathbf{v}_2$ which satisfies $(A - 4I)\mathbf{v}_2 = \mathbf{v}_1$. Choose $\vec{w} = (1, 0)^T$, which is independent of $\mathbf{v}_1$, and note that

$$(A - 4I)\mathbf{w} = \begin{pmatrix} -2 & 4 \\ -1 & 2 \end{pmatrix}\begin{pmatrix} 1 \\ 0 \end{pmatrix} = -\begin{pmatrix} 2 \\ 1 \end{pmatrix} = -\mathbf{v}_1.$$

Thus, choose $\mathbf{v}_2 = -\mathbf{w} = (-1, 0)^T$. Our second solution is

$$\begin{aligned} \mathbf{y}_2(t) &= e^{4t}(\mathbf{v}_2 + t\mathbf{v}_1) \\ &= e^{4t}\left[\begin{pmatrix} -1 \\ 0 \end{pmatrix} + t\begin{pmatrix} 2 \\ 1 \end{pmatrix}\right]. \end{aligned}$$

Thus, the general solution can be written

$$\begin{aligned} \mathbf{y}(t) &= C_1e^{4t}\begin{pmatrix} 2 \\ 1 \end{pmatrix} + C_2e^{4t}\left[\begin{pmatrix} -1 \\ 0 \end{pmatrix} + t\begin{pmatrix} 2 \\ 1 \end{pmatrix}\right] \\ &= e^{4t}\left[(C_1 + C_2t)\begin{pmatrix} 2 \\ 1 \end{pmatrix} + C_2\begin{pmatrix} -1 \\ 0 \end{pmatrix}\right] \end{aligned}$$

43. The matrix

$$A = \begin{pmatrix} 5 & 12 \\ -4 & -9 \end{pmatrix}$$

has characteristic polynomial $p(\lambda) = \lambda^2 + 4\lambda + 3$ and eigenvalues $\lambda_1 = -1$ and $\lambda_2 = -3$. The nullspace of

$$A - (-1)I = \begin{pmatrix} 6 & 12 \\ -4 & -8 \end{pmatrix}$$

is generated by the single eigenvector, $\mathbf{v}_1 = (-2, 1)^T$, with corresponding solution

$$\mathbf{y}_1(t) = e^{-t}\begin{pmatrix} -2 \\ 1 \end{pmatrix}.$$

The nullspace of

$$A - (-3)I = \begin{pmatrix} 8 & 12 \\ -4 & -6 \end{pmatrix}$$

is generated by the single eigenvector, $\mathbf{v}_2 = (-3/2, 1)^T$, with corresponding solution

$$\mathbf{y}_2(t) = e^{-3t}\begin{pmatrix} -3/2 \\ 1 \end{pmatrix}.$$

Thus, the general solution can be written

$$\mathbf{y}(t) = C_1e^{-t}\begin{pmatrix} -2 \\ 1 \end{pmatrix} + C_2e^{-3t}\begin{pmatrix} -3/2 \\ 1 \end{pmatrix}.$$

45. The matrix

$$A = \begin{pmatrix} -4 & -5 \\ 2 & 2 \end{pmatrix}$$

has characteristic polynomial $p(\lambda) = \lambda^2 + 2\lambda + 2$ and eigenvalues $\lambda_1 = -1 + i$ and $\lambda_2 = -1 - i$. The nullspace of

$$A - (-1 + i)I = \begin{pmatrix} -3 - i & -5 \\ 2 & 3 - i \end{pmatrix}$$

is generated by the single eigenvector, $\mathbf{v}_1 = (5, -3-i)^T$, with corresponding solution

$$\mathbf{z}(t) = e^{(-1+i)t}\begin{pmatrix} 5 \\ -3 - i \end{pmatrix}.$$

Breaking this solution into real and imaginary parts,

$$\begin{aligned} \mathbf{z}(t) &= e^{(-1+i)t}\begin{pmatrix} 5 \\ -3 - i \end{pmatrix} \\ &= e^{-t}(\cos t + i \sin t)\left[\begin{pmatrix} 5 \\ -3 \end{pmatrix} + i\begin{pmatrix} 0 \\ -1 \end{pmatrix}\right] \\ &= e^{-t}\left[\cos t\begin{pmatrix} 5 \\ -3 \end{pmatrix} - \sin t\begin{pmatrix} 0 \\ -1 \end{pmatrix}\right. \\ &\quad \left. + i\cos t\begin{pmatrix} 0 \\ -1 \end{pmatrix} + i\sin t\begin{pmatrix} 5 \\ -3 \end{pmatrix}\right] \\ &= e^{-t}\begin{pmatrix} 5\cos t \\ -3\cos t + \sin t \end{pmatrix} \\ &\quad + ie^{-t}\begin{pmatrix} 5\sin t \\ -\cos t - 3\sin t \end{pmatrix}. \end{aligned}$$

Thus, the general solution is

$$\mathbf{y}(t) = C_1 e^{-t}\begin{pmatrix} 5\cos t \\ -3\cos t + \sin t \end{pmatrix} + C_2 e^{-t}\begin{pmatrix} 5\sin t \\ -\cos t - 3\sin t \end{pmatrix}.$$

47. The matrix

$$A = \begin{pmatrix} -10 & 4 \\ -12 & 4 \end{pmatrix}$$

has characteristic polynomial $p(\lambda) = \lambda^2+6\lambda+8$ and eigenvalues $\lambda_1 = -4$ and $\lambda_2 = -2$. The nullspace of

$$A - (-4)I = \begin{pmatrix} -6 & 4 \\ -12 & 8 \end{pmatrix}$$

is generated by the single eigenvector, $\mathbf{v}_1 = (2, 3)^T$, with corresponding solution

$$\mathbf{y}_1(t) = e^{-4t}\begin{pmatrix} 2 \\ 3 \end{pmatrix}.$$

The nullspace of

$$A - (-2)I = \begin{pmatrix} -8 & 4 \\ -12 & 6 \end{pmatrix}$$

is generated by a single eigenvector, $\mathbf{v}_2 = (1, 2)^T$, with corresponding solution

$$\mathbf{y}_2(t) = e^{-2t}\begin{pmatrix} 1 \\ 2 \end{pmatrix}.$$

Thus, the general solution is

$$\mathbf{y}(t) = C_1 e^{-4t}\begin{pmatrix} 2 \\ 3 \end{pmatrix} + C_2 e^{-2t}\begin{pmatrix} 1 \\ 2 \end{pmatrix}.$$

49. From Exercise 41, the general solution is

$$\mathbf{y}(t) = e^{4t}\left[(C_1 + C_2 t)\begin{pmatrix} 2 \\ 1 \end{pmatrix} + C_2\begin{pmatrix} -1 \\ 0 \end{pmatrix}\right].$$

Because $\mathbf{y}(0) = (3, 1)^T$,

$$\begin{pmatrix} 3 \\ 1 \end{pmatrix} = C_1\begin{pmatrix} 2 \\ 1 \end{pmatrix} + C_2\begin{pmatrix} -1 \\ 0 \end{pmatrix}.$$

Reduce the augmented matrix.

$$\begin{pmatrix} 2 & -1 & 3 \\ 1 & 1 & 0 \end{pmatrix} \rightarrow \begin{pmatrix} 1 & 0 & 1 \\ 0 & 1 & -1 \end{pmatrix}$$

Thus, $C_1 = 1$ and $C_2 = -1$ and the particular solution is

$$\mathbf{y}(t) = e^{4t}\left[(1-t)\begin{pmatrix} 2 \\ 1 \end{pmatrix} - \begin{pmatrix} -1 \\ 0 \end{pmatrix}\right] = e^{4t}\begin{pmatrix} 3-2t \\ 1-t \end{pmatrix}.$$

51. From Exercise 43, the general solution is

$$\mathbf{y}(t) = C_1 e^{-t}\begin{pmatrix} -2 \\ 1 \end{pmatrix} + C_2 e^{-3t}\begin{pmatrix} -3/2 \\ 1 \end{pmatrix}.$$

Because $\mathbf{y}(0) = (1, 0)^T$,

$$\begin{pmatrix} 1 \\ 0 \end{pmatrix} = C_1\begin{pmatrix} -2 \\ 1 \end{pmatrix} + C_2\begin{pmatrix} -3/2 \\ 1 \end{pmatrix}.$$

Reduce the augmented matrix.

$$\begin{pmatrix} -2 & -3/2 & 1 \\ 1 & 1 & 0 \end{pmatrix} \rightarrow \begin{pmatrix} 1 & 0 & -2 \\ 0 & 1 & 2 \end{pmatrix}$$

Thus, $C_1 = -2$ and $C_2 = 2$ and the particular solution is

$$\mathbf{y}(t) = -2e^{-t}\begin{pmatrix} -2 \\ 1 \end{pmatrix} + 2e^{-3t}\begin{pmatrix} -3/2 \\ 1 \end{pmatrix} = \begin{pmatrix} 4e^{-t} - 3e^{-3t} \\ -2e^{-t} + 2e^{-3t} \end{pmatrix}.$$

53. From Exercise 45, the general solution is

$$\mathbf{y}(t) = C_1 e^{-t}\begin{pmatrix} 5\cos t \\ -3\cos t + \sin t \end{pmatrix} + C_2 e^{-t}\begin{pmatrix} 5\sin t \\ -\cos t - 3\sin t \end{pmatrix}.$$

Because $\mathbf{y}(0) = (-3, 2)^T$,

$$\begin{pmatrix} -3 \\ 2 \end{pmatrix} = C_1\begin{pmatrix} 5 \\ -3 \end{pmatrix} + C_2\begin{pmatrix} 0 \\ -1 \end{pmatrix}.$$

Reduce the augmented matrix.

$$\begin{pmatrix} 5 & 0 & -3 \\ -3 & -1 & 2 \end{pmatrix} \rightarrow \begin{pmatrix} 1 & 0 & -3/5 \\ 0 & 1 & -1/5 \end{pmatrix}$$

Thus, $C_1 = -3/5$ and $C_2 = -1/5$ and the particular solution is

$$\begin{aligned} \mathbf{y}(t) &= -\frac{3}{5}e^{-t}\begin{pmatrix} 5\cos t \\ -3\cos t + \sin t \end{pmatrix} \\ &\quad -\frac{1}{5}e^{-t}\begin{pmatrix} 5\sin t \\ -\cos t - 3\sin t \end{pmatrix} \\ &= e^{-t}\begin{pmatrix} -3\cos t - \sin t \\ 2\cos t \end{pmatrix}. \end{aligned}$$

55. From Exercise 47, the general solution is

$$\mathbf{y}(t) = C_1 e^{-4t}\begin{pmatrix} 2 \\ 3 \end{pmatrix} + C_2 e^{-2t}\begin{pmatrix} 1 \\ 2 \end{pmatrix}.$$

Because $\mathbf{y}(0) = (2, 1)^T$,

$$\begin{pmatrix} 2 \\ 1 \end{pmatrix} = C_1\begin{pmatrix} 2 \\ 3 \end{pmatrix} + C_2\begin{pmatrix} 1 \\ 2 \end{pmatrix}.$$

Reduce the augmented matrix.

$$\begin{pmatrix} 2 & 1 & 2 \\ 3 & 2 & 1 \end{pmatrix} \rightarrow \begin{pmatrix} 1 & 0 & 3 \\ 0 & 1 & -4 \end{pmatrix}$$

Thus, $C_1 = 3$ and $C_2 = -4$ and the particular solution is

$$\begin{aligned} \mathbf{y}(t) &= 3e^{-4t}\begin{pmatrix} 2 \\ 3 \end{pmatrix} - 4e^{-2t}\begin{pmatrix} 1 \\ 2 \end{pmatrix} \\ &= \begin{pmatrix} 6e^{-4t} - 4e^{-2t} \\ 9e^{-4t} - 8e^{-2t} \end{pmatrix} \end{aligned}$$

57. (a) Let

$$(A - \lambda I)^2 = \begin{pmatrix} a & b \\ c & d \end{pmatrix}$$

and assume that $(A - \lambda I)^2\mathbf{v} = \mathbf{0}$ for all $\mathbf{v}$ in $\mathbf{R}$. Then

$$\begin{pmatrix} a & b \\ c & d \end{pmatrix}\begin{pmatrix} 1 \\ 0 \end{pmatrix} = \begin{pmatrix} 0 \\ 0 \end{pmatrix} \Rightarrow \begin{pmatrix} a \\ c \end{pmatrix} = \begin{pmatrix} 0 \\ 0 \end{pmatrix}$$

and

$$\begin{pmatrix} a & b \\ c & d \end{pmatrix}\begin{pmatrix} 0 \\ 1 \end{pmatrix} = \begin{pmatrix} 0 \\ 0 \end{pmatrix} \Rightarrow \begin{pmatrix} b \\ d \end{pmatrix} = \begin{pmatrix} 0 \\ 0 \end{pmatrix}.$$

Thus,

$$\begin{pmatrix} a & b \\ c & d \end{pmatrix} = \begin{pmatrix} 0 & 0 \\ 0 & 0 \end{pmatrix},$$

so $(A - \lambda I)^2 = 0I$.

(b) Let $\mathbf{v}_1$ be an eigenvector of A associated with the eigenvalue λ. Note that this means that $A\mathbf{v}_1 - \lambda\mathbf{v}_1 = \mathbf{0}$. Let $\mathbf{v} = \alpha\mathbf{v}_1$ be a multiple of the eigenvector $\mathbf{v}_1$. Then

$$\begin{aligned} (A - \lambda I)^2\mathbf{v} &= (A - \lambda I)^2(\alpha\mathbf{v}_1) \\ &= \alpha(A - \lambda I)^2\mathbf{v}_1 \\ &= \alpha(A - \lambda I)(A\mathbf{v}_1 - \lambda\mathbf{v}_1) \\ &= \alpha(A - \lambda I)\mathbf{0} \\ &= \alpha\mathbf{0} \\ &= \mathbf{0}. \end{aligned}$$

(c) Now choose $\mathbf{v}$ in $\mathbf{R}^2$ such that $\mathbf{v}$ is **not** a multiple of the eigenvector $\mathbf{v}_1$. Note that this means that $\mathbf{v}$ is not an eigenvector associated with the eigenvalue λ. The set $B = \{\mathbf{v}, \mathbf{v}_1\}$ is independent with dimension two. Therefore, it must span all of $\mathbf{R}^2$ and is a basis for $\mathbf{R}^2$.

(d) Set $\mathbf{w} = (A - \lambda I)\mathbf{v}$. Note that this means that $\mathbf{w}$ is nonzero, for otherwise $\mathbf{v}$ would be an eigenvector associated with the eigenvalue λ. In part (c), we saw that $B = \{\mathbf{v}, \mathbf{v}_1\}$ was a basis for $\mathbf{R}^2$. Thus, B spans $\mathbf{R}^2$ and we can find a and b such that

$$\mathbf{w} = a\mathbf{v}_1 + b\mathbf{v}.$$

(e) From (d), $\mathbf{w} = (A - \lambda I)\mathbf{v}$ and $\mathbf{w} = a\mathbf{v}_1 + b\mathbf{v}$. Thus,

$$\begin{aligned} (A - \lambda I)\mathbf{w} &= (A - \lambda I)(a\mathbf{v}_1 + b\mathbf{v}) \\ &= a(A - \lambda I)\mathbf{v}_1 + b(A - \lambda I)\mathbf{v} \\ &= \mathbf{0} + b\mathbf{w} \\ &= b\mathbf{w}, \end{aligned}$$

Hence,

$$\begin{aligned} (A - \lambda I)\mathbf{w} &= b\mathbf{w} \\ A\mathbf{w} - \lambda\mathbf{w} &= b\mathbf{w} \\ A\mathbf{w} &= (\lambda + b)\mathbf{w}. \end{aligned}$$

Thus, $\mathbf{w}$, being nonzero, is an eigenvector of A with eigenvalue $\lambda + b$. But, λ is the only eigenvalue, so b must equal zero and $\mathbf{w}$ must be a multiple of $\mathbf{v}_1$.

(f) Finally, because $b = 0$ and $(A - \lambda I)\mathbf{v} = \mathbf{w}$,

$$\begin{aligned}(A - \lambda I)^2\mathbf{v} &= (A - \lambda I)(A - \lambda I)\mathbf{v}\\ &= (A - \lambda I)\mathbf{w}\\ &= b\mathbf{w}\\ &= \mathbf{0}.\end{aligned}$$

Consequently, whether $\mathbf{v}$ is a multiple of $\mathbf{v}_1$ or not, $(A - \lambda I)^2\mathbf{v} = \mathbf{0}$. Since this is true for any arbitrary $\mathbf{v}$ in $\mathbf{R}^2$, by part (a), $(A - \lambda I)^2 = 0I$.

59. (a) Let $x_A(t)$ and $x_B(t)$ represent the number of pounds of salt as a function of time in the left and right tanks, respectively. The rate at which salt is changing the left tank is

$$\frac{dx_A}{dt} = \text{Rate In} - \text{Rate out} = -\frac{x_A}{40} + \frac{x_B}{90}.$$

The rate at which salt content is changing in the right tank is

$$\frac{dx_B}{dt} = \text{Rate In} - \text{Rate out} = \frac{x_A}{40} - \frac{x_B}{40}.$$

In matrix-vector form, this becomes

$$\begin{pmatrix} x_A \\ x_B \end{pmatrix}' = \begin{pmatrix} -1/40 & 1/90 \\ 1/40 & -1/40 \end{pmatrix} \begin{pmatrix} x_A \\ x_B \end{pmatrix}.$$

The initial condition is $\mathbf{x}(0) = (x_A(0), x_B(0))^T = (60, 0)^T$, as there is initially 60 pounds of salt in the left tank, but none in the right tank.

(b) The trace of the coefficient matrix is $T = -2/40 = -1/20$ and the determinant is $D = 5/14400$. Thus, the characteristic polynomial is

$$\begin{aligned}p(\lambda) &= \lambda^2 + \frac{1}{20}\lambda + \frac{5}{14400}\\ &= \left(\lambda + \frac{1}{120}\right)\left(\lambda + \frac{1}{24}\right),\end{aligned}$$

so the eigenvalues are $\lambda = -1/120$ and $\lambda = -1/24$. With the first eigenvalue $\lambda = -1/120$,

$$\begin{aligned}A + \frac{1}{120}I &= \begin{pmatrix} -1/60 & 1/90 \\ 1/40 & -1/60 \end{pmatrix}\\ &\xrightarrow{\text{rref}} \begin{pmatrix} 1 & -2/3 \\ 0 & 0 \end{pmatrix}.\end{aligned}$$

Thus, $(2/3, 1)^T$ will serve as an eigenvalue. As any multiple of an eigenvalue is also an eigenvalue, let's choose $\mathbf{v}_1 = (2, 3)^T$. This leads to the solution $\mathbf{x}_1 = e^{-t/120}(2, 3)^T$. Next, with the second eigenvalue $\lambda = -1/24$,

$$\begin{aligned}A + \frac{1}{24}I &= \begin{pmatrix} 1/60 & 1/90 \\ 1/40 & 1/60 \end{pmatrix}\\ &\xrightarrow{\text{rref}} \begin{pmatrix} 1 & 2/3 \\ 0 & 0 \end{pmatrix}.\end{aligned}$$

Thus, $(-2/3, 1)^T$ is an eigenvector, but we again choose a multiple, $\mathbf{v}_2 = (-2, 3)^T$. Thus, a second indpendent solution (distinct eigenvalues) is $\mathbf{x}_2 = e^{-t/24}(-2, 3)^T$. Thus, the general solution is

$$\mathbf{x} = C_1 e^{-t/120}\begin{pmatrix} 2 \\ 3 \end{pmatrix} + C_2 e^{-t/24}\begin{pmatrix} -2 \\ 3 \end{pmatrix}.$$

The initial condition provides

$$\begin{pmatrix} 60 \\ 0 \end{pmatrix} = \mathbf{x}(0) = C_1\begin{pmatrix} 2 \\ 3 \end{pmatrix} + C_2\begin{pmatrix} -2 \\ 3 \end{pmatrix}.$$

Solving, $C_1 = 15$ and $C_2 = -15$, so the solution of the initial value problem is

$$\mathbf{x} = 15e^{-t/120}\begin{pmatrix} 2 \\ 3 \end{pmatrix} - 15e^{-t/24}\begin{pmatrix} -2 \\ 3 \end{pmatrix}.$$

(c) The component solutions are

$$\begin{aligned}x_A &= 30e^{-t/120} + 30e^{-t/24}\\ x_B &= 45e^{-t/120} - 45e^{-t/24}.\end{aligned}$$

The plot of the salt content in each tank is shown below.

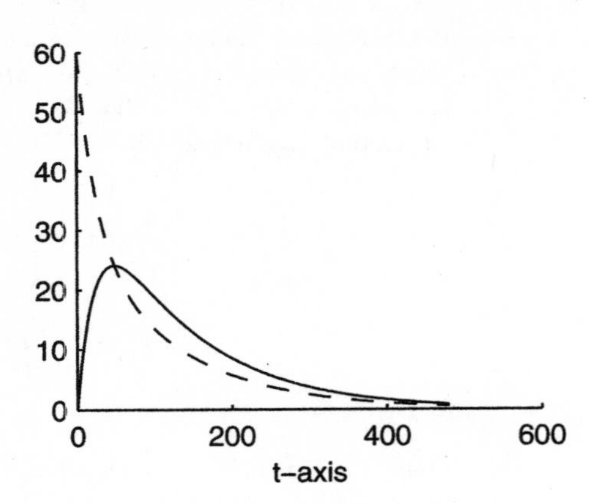

Note that the time constant on the exponential $e^{-t/24}$ is $T_c = 24$, but on the exponential $e^{-t/120}$, the time constant is $T_c = 120$. We chose the larger of these two time constants and made our plot over four time constants $[0, T_c] = [0, 480]$. This allows enough time to show both components decaying to zero. Physically, our intuition tells us that if we keep pouring pure water into the leftmost tank, eventually the system will purge itself of all salt content. Mathematically, both

$$30e^{-t/120} + 30e^{-t/24} \to 0$$
$$45e^{-t/120} - 45e^{-t/24} \to 0,$$

as $t \to \infty$.

61. We've labeled current and added a note at a in the figure below.

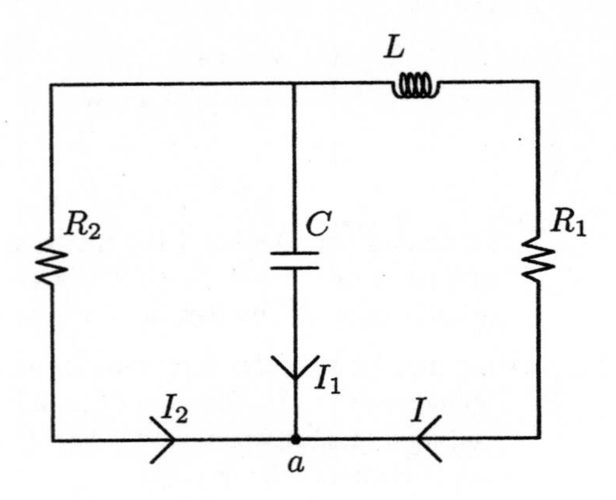

The current coming into the node at a must equal the current coming out, so

$$I + I_1 + I_2 = 0.$$

Moving counterclockwise around the leftmost loop containing the resistor R_2 and capacitor,

$$-R_2 I_2 + V = 0.$$

However, $I_2 = -I - I_1$, so

$$-R_2(-I - I_1) + V = 0.$$

Now, the voltage across the capacitor follows the law $V = q_1/C$, where q_1 is the charge on the capacitor. Differentiating, $CV' = q_1' = I_1$. Substituting this in the last equation,

$$R_2 I + R_2 C V' + V = 0$$
$$V' = -\frac{1}{R_2 C} V - \frac{1}{C} I.$$

Passing around the loop (counterclockwise) at the far right, containing capacitor, inductor, and resistor R_1,

$$-V + LI' + R_1 I = 0$$
$$I' = \frac{1}{L} V - \frac{R_1}{L} I.$$

In matrix form, and using $C = 1$, $L = 1$, $R_1 = 5$,

and $R_2 = 1$,

$$\begin{pmatrix} V \\ I \end{pmatrix}' = \begin{pmatrix} -1/(R_2C) & -1/C \\ 1/L & -R_1/L \end{pmatrix}\begin{pmatrix} V \\ I \end{pmatrix} = \begin{pmatrix} -1 & -1 \\ 1 & -5 \end{pmatrix}\begin{pmatrix} V \\ I \end{pmatrix}.$$

Now, the trace is $T = -6$, the determinant is $D = 6$, and the characteristic polynomial is

$$p(\lambda) = \lambda^2 + 6\lambda + 6.$$

The quadratic formula provides eigenvalues $\lambda = -3 \pm \sqrt{3}$, and

$$A - \lambda I = \begin{pmatrix} -1-\lambda & -1 \\ 1 & -5-\lambda \end{pmatrix}.$$

Trusting that $A - \lambda I$ is singular, we look for a vector that will zero the second row, so $\mathbf{v} = (5 + \lambda, 1)^T$. With $\lambda = -3 + \sqrt{3}$, this gives the solution $\mathbf{x}_1 = e^{(-3+\sqrt{3})t}(2 + \sqrt{3}, 1)^T$. With $\lambda = -3 - \sqrt{3}$, this gives the solution $\mathbf{x}_2 = e^{(-3-\sqrt{3})t}(2-\sqrt{3}, 1)^T$. Thus, the general solution is

$$\mathbf{x} = C_1 e^{(-3+\sqrt{3})t}\begin{pmatrix} 2+\sqrt{3} \\ 1 \end{pmatrix} + C_2 e^{(-3-\sqrt{3})t}\begin{pmatrix} 2-\sqrt{3} \\ 1 \end{pmatrix}.$$

The initial voltage is 12 volts, the initial current is 0 amps, so

$$\begin{pmatrix} 12 \\ 0 \end{pmatrix} = \begin{pmatrix} V(0) \\ I(0) \end{pmatrix} = \mathbf{x}(0) = C_1\begin{pmatrix} 2+\sqrt{3} \\ 1 \end{pmatrix} + C_2\begin{pmatrix} 2-\sqrt{3} \\ 1 \end{pmatrix}.$$

Solving, $C_1 = 2\sqrt{3}$ and $C_2 = -2\sqrt{3}$, giving the final solution $\mathbf{x}(t) = 2\sqrt{3}e^{(-3+\sqrt{3})t}(2 + \sqrt{3}, 1)^T - 2\sqrt{3}e^{(-3-\sqrt{3})t}(2 - \sqrt{3}, 1)^T$, which leads to the component solutions

$$V = (4\sqrt{3} + 6)e^{(-3+\sqrt{3})t} - (4\sqrt{3} - 6)e^{(-3-\sqrt{3})t}$$
$$I = 2\sqrt{3}e^{(-3+\sqrt{3})t} - 2\sqrt{3}e^{(-3-\sqrt{3})t}.$$

The solution is plotted below.

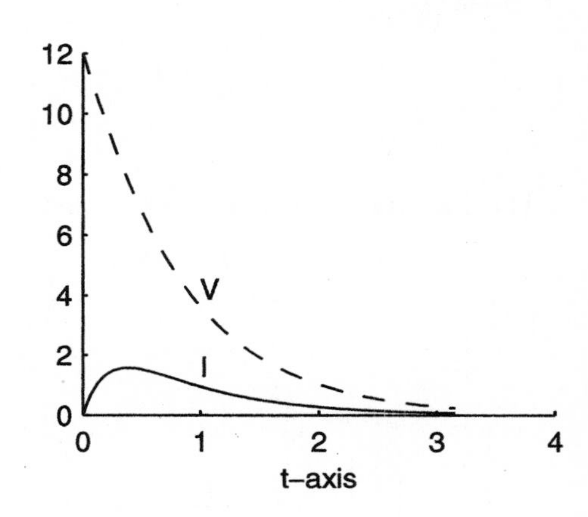

———×———

Section 9.3. Phase Plane Portraits

1. If

$$A = \begin{pmatrix} -10 & -25 \\ 5 & 10 \end{pmatrix},$$

then $T = 0$ and $D = 25$, leading to

$$p(\lambda) = \lambda^2 - T\lambda + D = \lambda^2 + 25.$$

On the other hand,

$$\begin{aligned} p(\lambda) &= \det(A - \lambda I) \\ &= \begin{vmatrix} -10-\lambda & -25 \\ 5 & 10-\lambda \end{vmatrix} \\ &= (-10-\lambda)(10-\lambda) + 125 \\ &= \lambda^2 + 25. \end{aligned}$$

3. Position of

$$\mathbf{y}(t) = 1.2e^{0.6t}\begin{pmatrix} 1 \\ 1 \end{pmatrix}$$

on halfline solution at times $t = 0$, 1, and 2.

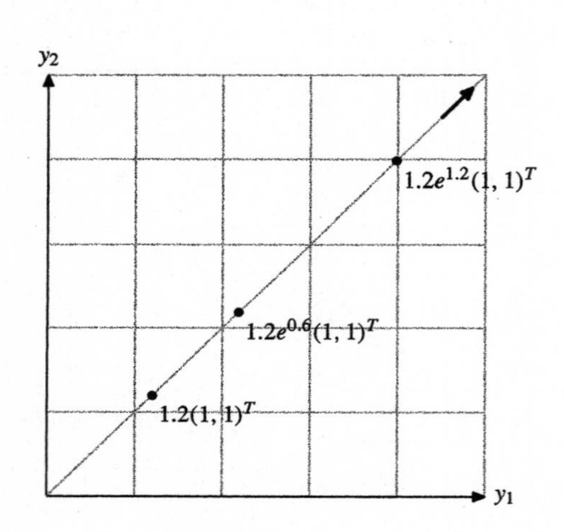

5. Position of

$$\mathbf{y}(t) = 0.8e^{-0.6t}\begin{pmatrix} 4 \\ 4 \end{pmatrix}$$

on halfline solution at times $t = 0$, 1, and 2.

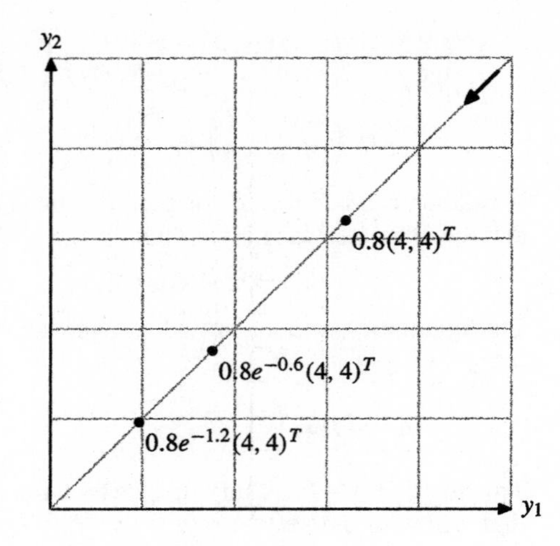

7. Position of

$$\mathbf{y}(t) = 0.8e^{0.4t}\begin{pmatrix} -2 \\ 1 \end{pmatrix} + 1.2e^{-0.2t}\begin{pmatrix} 4 \\ 4 \end{pmatrix}$$

at times $t = -5, -4, \ldots, 5$.

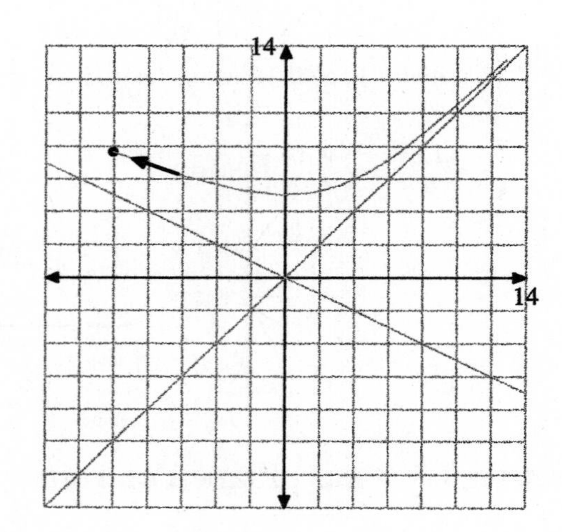

9. Position of

$$\mathbf{y}(t) = -0.4e^{-0.4t}\begin{pmatrix} -2 \\ 1 \end{pmatrix} + 0.4e^{-0.2t}\begin{pmatrix} 4 \\ 4 \end{pmatrix}$$

at times $t = -5, -4, \ldots, 5$.

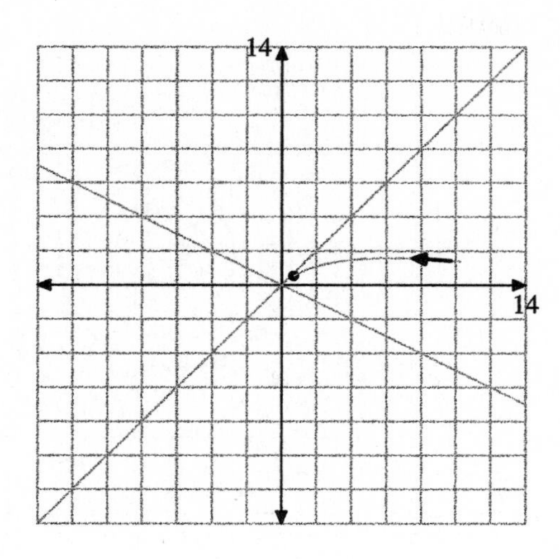

11. Both eigenvalues are positive, so the equilibrium point at the origin is a source. Solutions emanate from the origin tangent to the slow exponential solution, $e^t(-1, -2)^T$, eventually paralleling the fast exponential solution, $e^{2t}(3, -1)^T$.

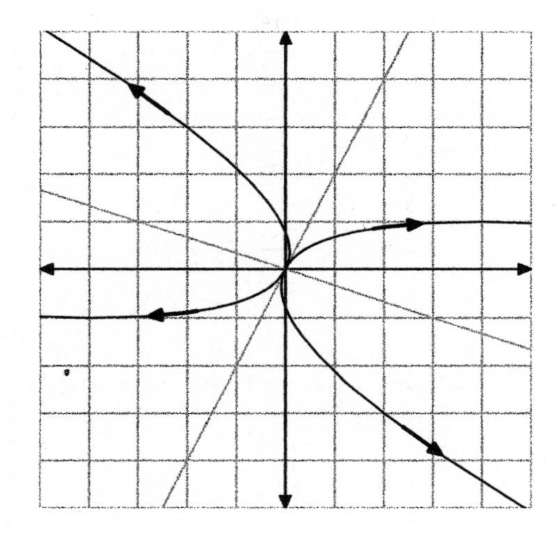

13. Both eigenvalues are negative, so the equilibrium point at the origin is a sink. Solutions dive toward the origin tangent to the slow exponential solution, $e^{-t}(1, 2)^T$. As solutions move backward in time, they eventually parallel the fast exponential solution, $e^{-3t}(-4, 1)^T$.

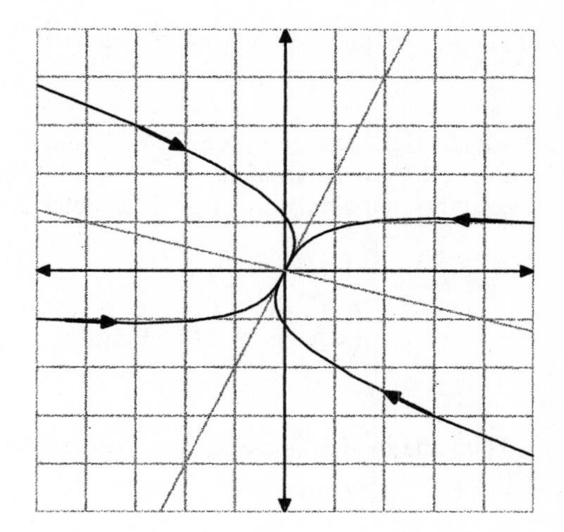

15. Both eigenvalues are positive, so the equilibrium point at the origin is a source. Solutions emanate from the origin tangent to the slow exponential solution, $e^t(1, 5)^T$, eventually paralleling the fast exponential solution, $e^{3t}(4, 1)^T$.

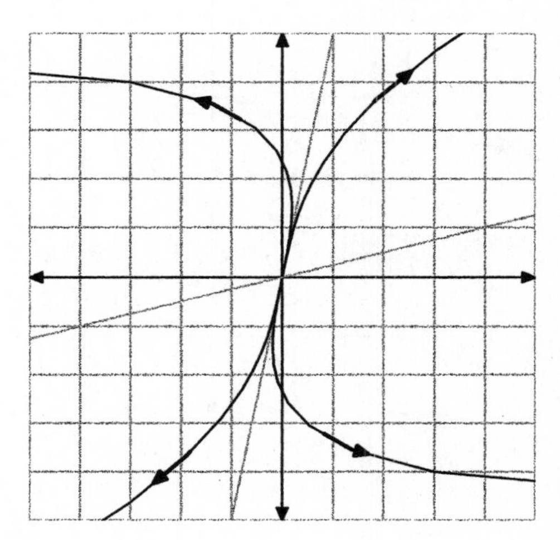

17. Matrix

$$A = \begin{pmatrix} 0 & 3 \\ -3 & 0 \end{pmatrix}$$

has trace $T = 0$ and determinant $D = 9$. Thus, the characteristic polynomial is

$$p(\lambda) = \lambda^2 - T\lambda + D = \lambda^2 + 9,$$

which produces eigenvalues $\lambda_1 = 3i$ and $\lambda_2 = -3i$. Because the real part of these eigenvalues is zero, the equilibrium point at the origin is a center. At $(1, 0)$,

$$\begin{pmatrix} 0 & 3 \\ -3 & 0 \end{pmatrix} \begin{pmatrix} 1 \\ 0 \end{pmatrix} = \begin{pmatrix} 0 \\ -3 \end{pmatrix}.$$

Thus, the motion is clockwise. A hand sketch follows.

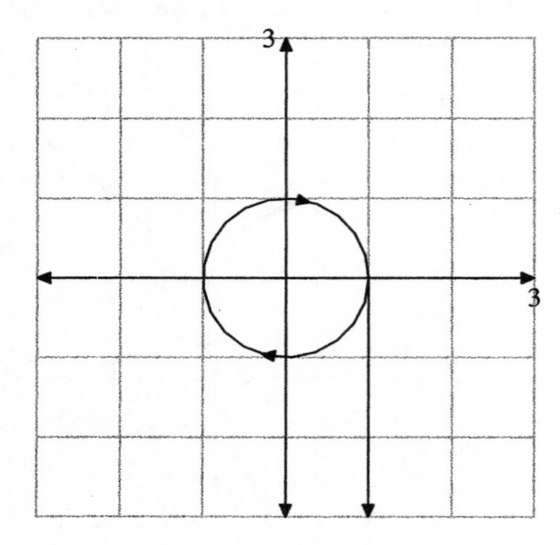

The phase portrait, drawn using a numerical solver, follows.

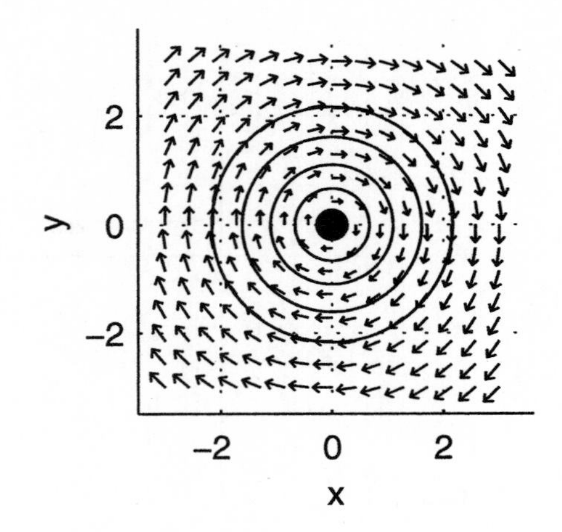

19. Matrix

$$A = \begin{pmatrix} 0 & 1 \\ -4 & 0 \end{pmatrix}$$

has trace $T = 0$ and determinant $D = 4$. Thus, the characteristic polynomial is

$$p(\lambda) = \lambda^2 - T\lambda + D = \lambda^2 + 4,$$

which produces eigenvalues $\lambda_1 = 2i$ and $\lambda_2 = -2i$. Because the real part of these eigenvalues is zero, the equilibrium point at the origin is a center. At $(1, 0)$,

$$\begin{pmatrix} 0 & 1 \\ -4 & 0 \end{pmatrix} \begin{pmatrix} 1 \\ 0 \end{pmatrix} = \begin{pmatrix} 0 \\ -4 \end{pmatrix}.$$

Thus, the motion is clockwise. A hand sketch follows.

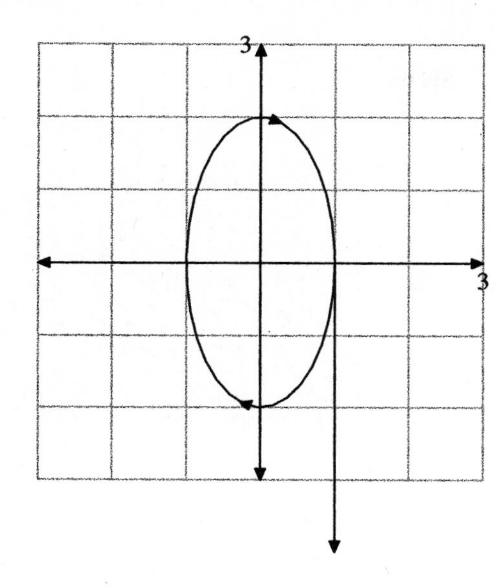

The phase portrait, drawn using a numerical solver, follows.

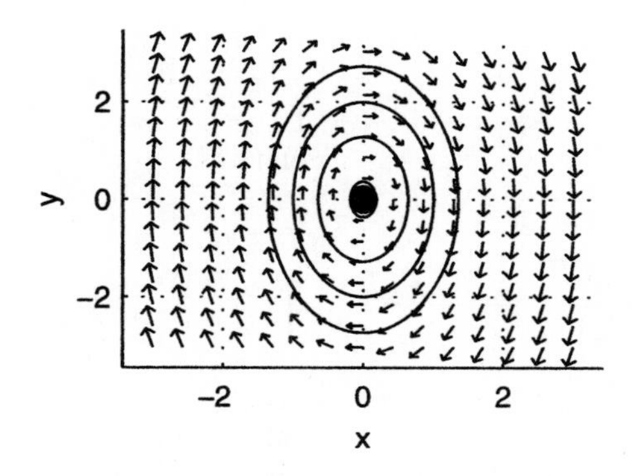

21. Matrix

$$A = \begin{pmatrix} -1 & 1 \\ -5 & 3 \end{pmatrix}$$

has trace $T = 2$ and determinant $D = 2$. Thus, the characteristic polynomial is

$$p(\lambda) = \lambda^2 - T\lambda + D = \lambda^2 - 2\lambda + 2,$$

which produces eigenvalues $\lambda_1 = 1 + i$ and $\lambda_2 = 1 - i$. Because the real part of the eigenvalues is positive, the equilibrium point at the origin is a spiral source. At $(1, 0)$,

$$\begin{pmatrix} -1 & 1 \\ -5 & 3 \end{pmatrix}\begin{pmatrix} 1 \\ 0 \end{pmatrix} = \begin{pmatrix} -1 \\ -5 \end{pmatrix},$$

so the motion is clockwise. A hand sketch follows.

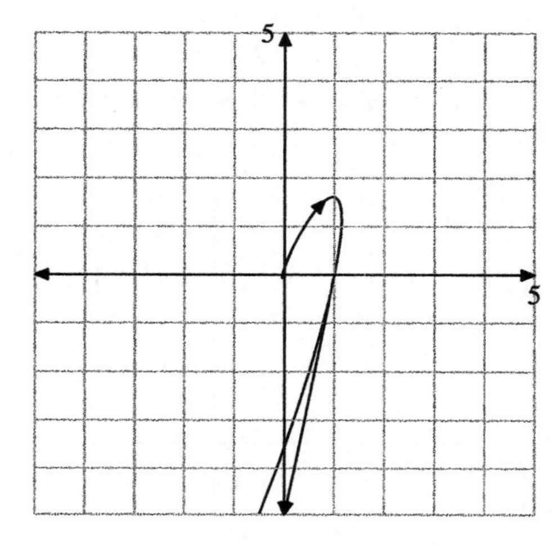

The phase portrait, drawn in a numerical solver, follows.

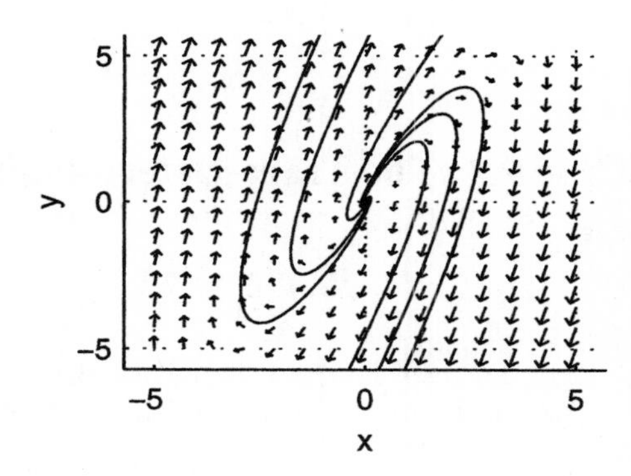

23. Matrix

$$A = \begin{pmatrix} -3 & 2 \\ -4 & 1 \end{pmatrix}$$

has trace $T = -2$ and determinant $D = 5$. Thus, the characteristic polynomial is

$$p(\lambda) = \lambda^2 - T\lambda + D = \lambda^2 + 2\lambda + 5,$$

which produces eigenvalues $\lambda_1 = -1 + 2i$ and $\lambda_2 = -1 - 2i$. Because the real part of the eigenvalues is negative, the equilibrium point at the origin is a spiral sink. At $(1, 0)$,

$$\begin{pmatrix} -3 & 2 \\ -4 & 1 \end{pmatrix}\begin{pmatrix} 1 \\ 0 \end{pmatrix} = \begin{pmatrix} -3 \\ -4 \end{pmatrix},$$

so the motion is clockwise. A hand sketch follows.

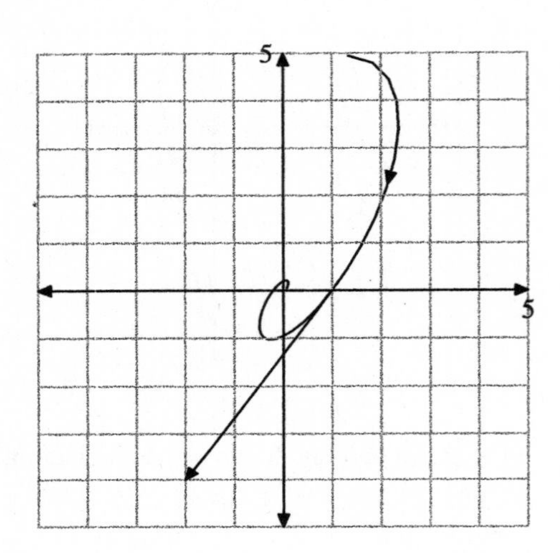

The phase portrait, drawn in a numerical solver, follows.

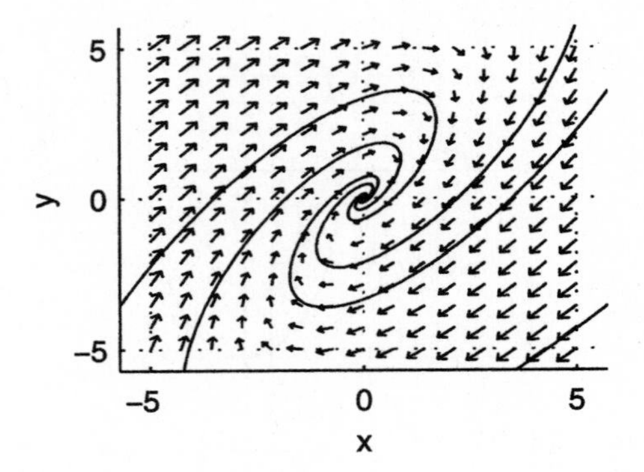

Section 9.4. The Trace-Determinant Plane

1. If

$$A = \begin{pmatrix} 8 & 20 \\ -4 & -8 \end{pmatrix},$$

then the trace is $T = 0$ and the determinant is $D = 16$. Further, the characteristic polynomial is

$$p(\lambda) = \lambda^2 - T\lambda + D = \lambda^2 + 16,$$

which produces eigenvalues $\lambda_1 = 4i$ and $\lambda_2 = -4i$. Therefore, the equilibrium point at the origin is a center. At $(1, 0)$,

$$\begin{pmatrix} 8 & 20 \\ -4 & -8 \end{pmatrix}\begin{pmatrix} 1 \\ 0 \end{pmatrix} = \begin{pmatrix} 8 \\ -4 \end{pmatrix},$$

so the motion is clockwise. A hand sketch follows.

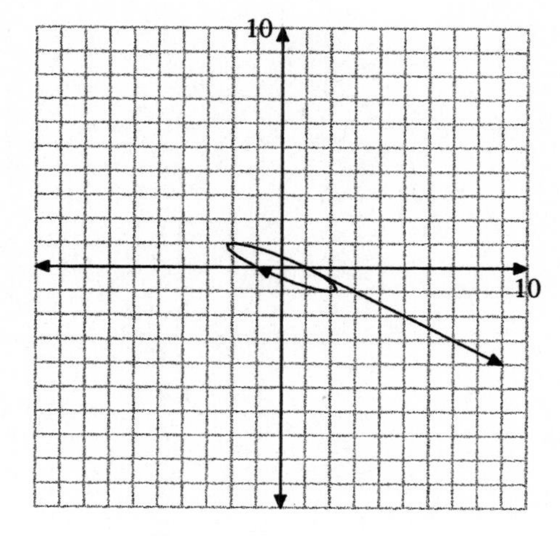

The phase portrait, draw with a numerical solver, follows.

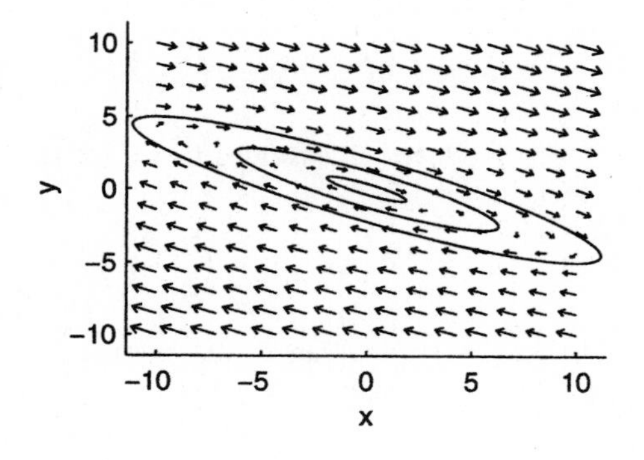

3. If

$$A = \begin{pmatrix} 2 & -4 \\ 8 & -6 \end{pmatrix}$$

then the trace is $T = -4$ and the determinant is $D = 20$. Further, $T^2 - 4D = (-4)^2 - 4(20) = -64 < 0$, so the equilibrium point at the origin is a spiral sink. At $(1, 0)$,

$$\begin{pmatrix} 2 & -4 \\ 8 & -6 \end{pmatrix} \begin{pmatrix} 1 \\ 0 \end{pmatrix} = \begin{pmatrix} 2 \\ 8 \end{pmatrix},$$

so the motion is counterclockwise. A hand sketch follows.

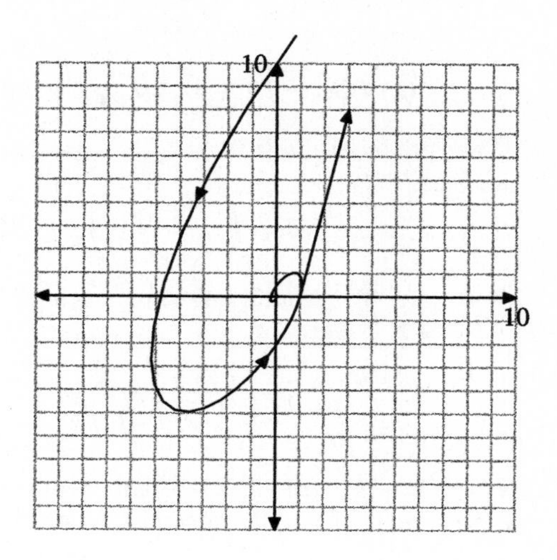

The phase portrait, draw with a numerical solver, follows.

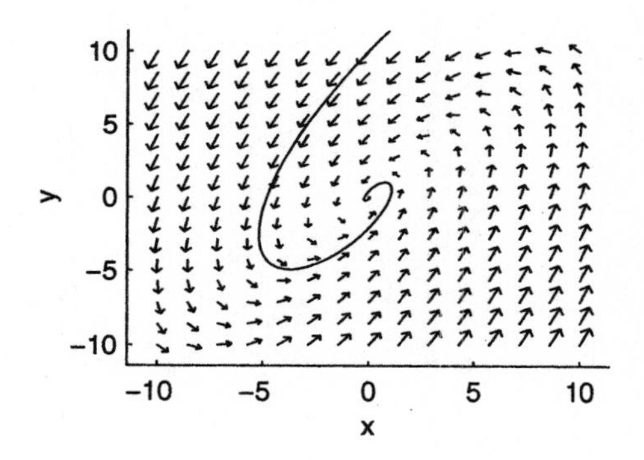

5. If

$$A = \begin{pmatrix} -11 & -5 \\ 10 & 4 \end{pmatrix},$$

then the trace is $T = -7$ and the determinant is $D = 6$. Further, $T^2 - 4D = (-7)^2 - 4(6) = 25 > 0$, so the equilibrium point at the origin is a nodal sink. Further, the characteristic polynomial is

$$p(\lambda) = \lambda^2 - T\lambda + D = \lambda^2 + 7\lambda + 6,$$

which produces eigenvalues $\lambda_1 = -1$ and $\lambda_2 = -6$. Because

$$A + I = \begin{pmatrix} -10 & -5 \\ 10 & 5 \end{pmatrix},$$

$\mathbf{v}_1 = (1, -2)^T$, leading to the exponential solution $e^{-t}(1, -2)^T$. Because

$$A + 6I = \begin{pmatrix} -5 & -5 \\ 10 & 10 \end{pmatrix},$$

$\mathbf{v}_2 = (1, -1)^T$, leading to the exponential solution $e^{-6t}(1, -1)^T$. Thus, the general solution is

$$\mathbf{y}(t) = C_1 e^{-t} \begin{pmatrix} 1 \\ -2 \end{pmatrix} + C_2 e^{-6t} \begin{pmatrix} 1 \\ -1 \end{pmatrix}.$$

Solutions approach the origin tangent to the "slow" halfline solution generated by $C_1(1, -2)^T$. As time moves backwards, solutions eventually parallel the halfline generated by $C_2(1, -1)^T$, the "fast" solution. A hand sketch follows.

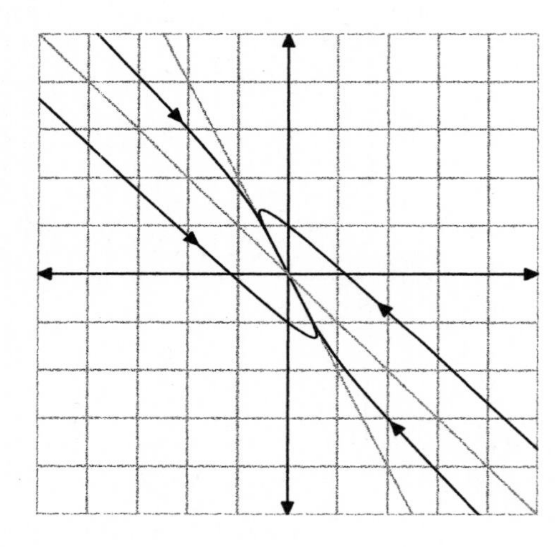

The phase portrait, draw in a numerical solver, follows.

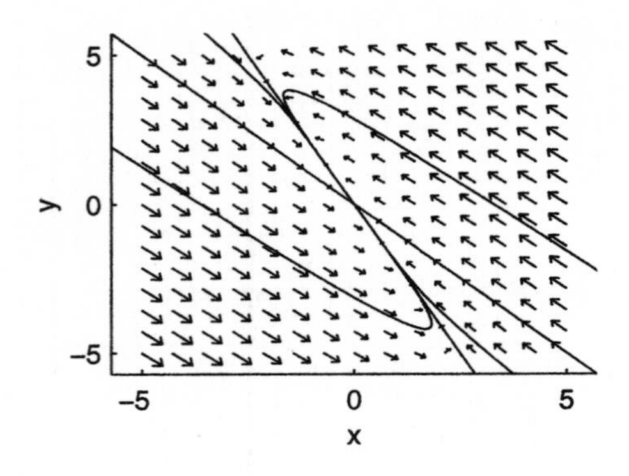

7. If

$$A = \begin{pmatrix} -7 & 10 \\ -5 & 8 \end{pmatrix},$$

then the trace is $T = 1$ and the determinant is $D = -6$, so the origin is a saddle point. Further, the characteristic polynomial is

$$p(\lambda) = \lambda^2 - T\lambda + D = \lambda^2 - \lambda - 6.$$

which produces eigenvalues $\lambda_1 = -2$ and $\lambda_2 = 3$. Because

$$A + 2I = \begin{pmatrix} -5 & 10 \\ -5 & 10 \end{pmatrix}$$

$\mathbf{v}_1 = (2, 1)^T$, leading to the exponential solution $e^{-2t}(2, 1)^T$. Because

$$A - 3I = \begin{pmatrix} -10 & 10 \\ -5 & 5 \end{pmatrix},$$

$\mathbf{v}_2 = (1, 1)^T$, leading to the exponential solution $e^{3t}(1, 1)^T$. Thus, the general solution is

$$\mathbf{y}(t) = C_1 e^{-2t} \begin{pmatrix} 2 \\ 1 \end{pmatrix} + C_2 e^{3t} \begin{pmatrix} 1 \\ 1 \end{pmatrix}.$$

Solutions approach the halfline $C_2(1, 1)$ as they move forward in time, but they approach the halfline generated by $C_1(2, 1)$ as they move backward in time. A hand sketch follows.

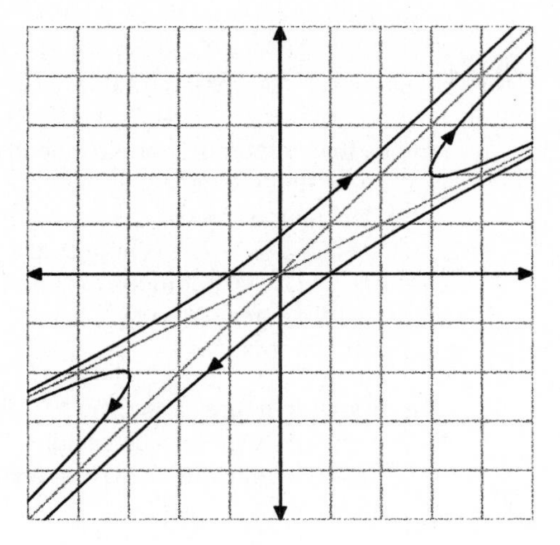

The phase portrait, drawn in a numerical solver, follows.

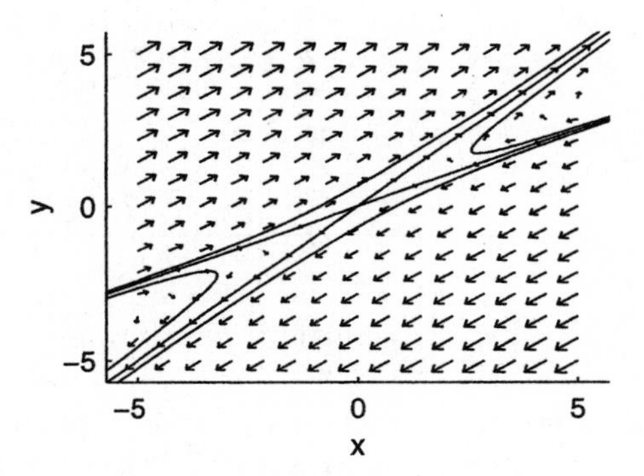

9. If

$$A = \begin{pmatrix} 3 & 2 \\ -4 & -1 \end{pmatrix},$$

then the trace $T = 2$ and the determinant is $D = 5$, and the discriminant is $T^2 - 4D = (2)^2 - 4(5) = -16 < 0$. Thus, the origin is a spiral source. At $(1, 0)$,

$$\begin{pmatrix} 3 & 2 \\ -4 & -1 \end{pmatrix} \begin{pmatrix} 1 \\ 0 \end{pmatrix} = \begin{pmatrix} 3 \\ -4 \end{pmatrix},$$

so the motion is clockwise. A hand sketch follows.

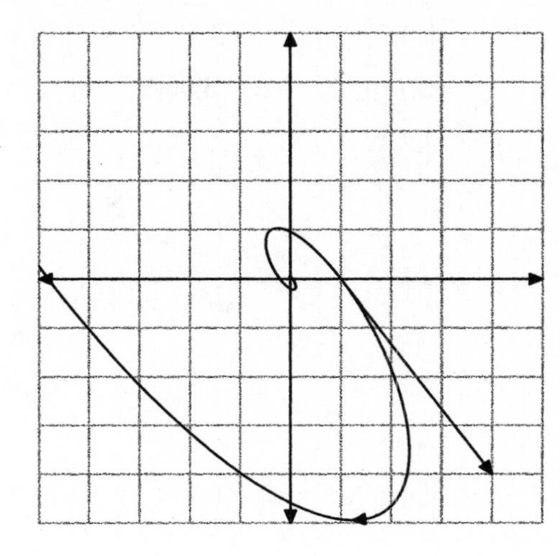

The phase portrait, drawn in a numerical solver, follows.

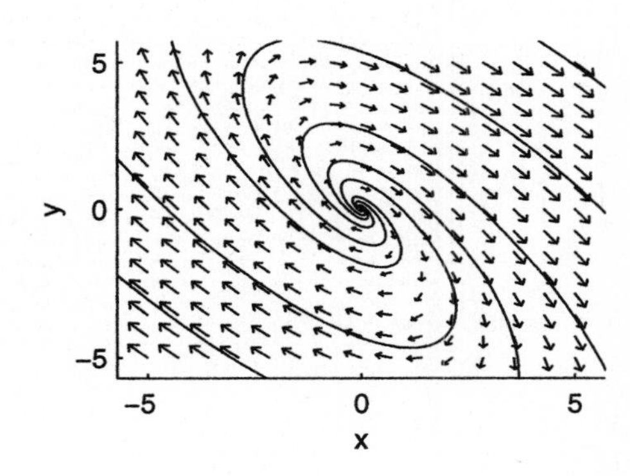

11. If

$$A = \begin{pmatrix} -4 & 10 \\ -2 & 4 \end{pmatrix}$$

then the trace is $T = 0$ and the determinant is $D = 4$, so the origin is a center. At $(1, 0)$,

$$\begin{pmatrix} -4 & 10 \\ -2 & 4 \end{pmatrix} \begin{pmatrix} 1 \\ 0 \end{pmatrix} = \begin{pmatrix} -4 \\ -2 \end{pmatrix},$$

so the rotation is clockwise. A hand sketch follows.

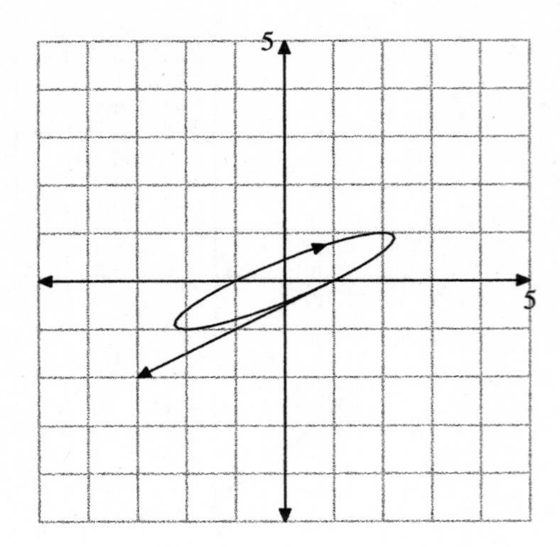

The phase portrait, drawn in a numerical solver, follows.

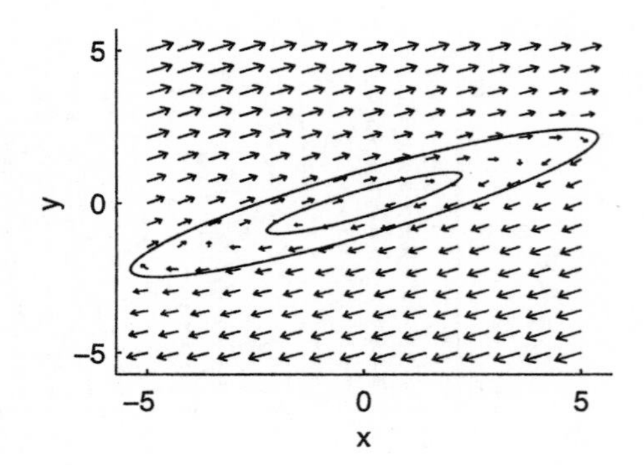

13. (a) For

$$A = \begin{pmatrix} 1 & 4 \\ -1 & -3 \end{pmatrix}$$

we have $T = \text{tr}(A) = -2$ and $D = \det(A) = 1$. Since the discriminant $T^2 - 4D = 0$ the point (T, D) lies on the parabola that divides nodal sinks from spiral sinks in the trace-determinant plane.

(b) The general solution can be written

$$\mathbf{y}(t) = e^{-t}\left((C_1 + C_2 t)\begin{pmatrix} 2 \\ -1 \end{pmatrix} + C_2 \begin{pmatrix} 0 \\ 1/2 \end{pmatrix}\right)$$

Because $te^{-t} \to 0$ as $t \to \infty$ (use l'Hôpital's rule), both $e^{-t}(C_1 + C_2 t)(2, -1)^T \to 0$ and $C_2 e^{-t}(0, 1/2)^T \to 0$ as $t \to \infty$. However, the first term is larger for large values of t. Thus, as $t \to \infty$, $\mathbf{y}(t) \approx e^{-t}(C_1 + C_2 t)(2, -1)^T$, which implies that solutions approach the origin tangent to the halfline generated by $(2, -1)^T$. In a similar manner, as $t \to -\infty$, the term $e^{-t}(C_1 + C_2 t)(2, -1)^T$ is larger than the term $C_2 e^{-t}(0, 1/2)^T$, so solutions eventually parallel the halfline generated by $(2, -1)^T$ as time moves backwards.

(c) The following figure shows the half-line solutions and one other in each sector. The solutions clearly exhibit the behavior predicted in part (a).

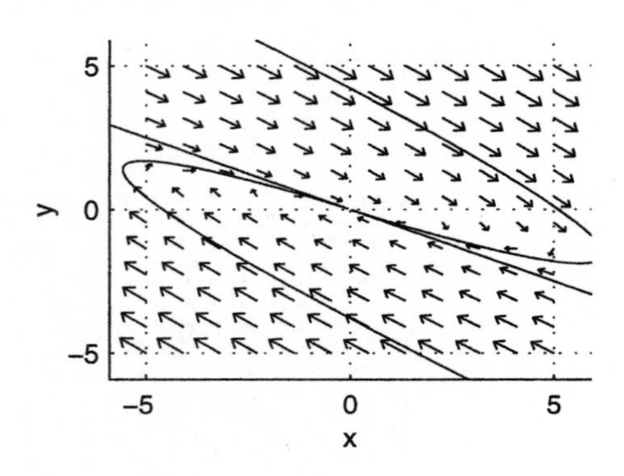

15. In general everything moves in the opposite direction in comparison to the situation in Exercise 14.

(a) As $t \to \infty$ the exponential solutions tends to ∞ along the half-line generated by $C_1 \mathbf{v}_1$.

(b) As $t \to \infty$ the general solution tends to ∞ and becomes parallel to the half-line generated by $C_2 \mathbf{v}_1$.

(c) "As $t \to -\infty$ the general solution tends to $\mathbf{0}$ tangent to the half-line generated by $-C_2 \mathbf{v}_1$.

(d) The origin is a degenerate nodal source.

17. If $\mathbf{y}' = A\mathbf{y}$, where

$$A = \begin{pmatrix} 6 & 4 \\ -1 & 2 \end{pmatrix},$$

then the trace is $T = 8$ and the determinant is $D = 16$. Further, $T^2 - 4D = 8^2 - 4(16) = 0$, so this system lies on the parabola $T^2 - 4D = 0$ that separates spiral sources and sinks from nodal sources and sinks in the trace determinant plane. Thus, the equilibrium point at the origin is a degenerate nodal source ($T = 8$).

The characteristic equation is

$$p(\lambda) = \lambda^2 - T\lambda + D = \lambda^2 - 8\lambda + 16,$$

which produces a single eigenvalue $\lambda = 4$. Because

$$A - 4I = \begin{pmatrix} 2 & 4 \\ -1 & -2 \end{pmatrix},$$

$\mathbf{v}_1 = (2, -1)^T$ and we have the exponential solution $e^{4t}(2, -1)^T$. To find another solution, we must solve $(A - \lambda I)\mathbf{v}_2 = \mathbf{v}_1$. Start with any vector that is not a multiple of $\mathbf{v}_1$, say $\mathbf{w} = (1, 0)^T$. Then

$$(A - 4I)\mathbf{w} = \begin{pmatrix} 2 & 4 \\ -1 & -2 \end{pmatrix}\begin{pmatrix} 1 \\ 0 \end{pmatrix} = \begin{pmatrix} 2 \\ -1 \end{pmatrix} = \mathbf{v}_1.$$

Thus, let $\mathbf{v}_2 = \mathbf{w} = (1, 0)^T$. Thus, a second, independent solution is

$$e^{4t}(\mathbf{v}_2 + t\mathbf{v}_1) = e^{4t}\left[\begin{pmatrix} 1 \\ 0 \end{pmatrix} + t\begin{pmatrix} 2 \\ -1 \end{pmatrix}\right],$$

and the general solution is

$$\begin{aligned} \mathbf{y}(t) &= C_1 e^{4t}\begin{pmatrix} 2 \\ -1 \end{pmatrix} + C_2 e^{4t}\left[\begin{pmatrix} 1 \\ 0 \end{pmatrix} + t\begin{pmatrix} 2 \\ -1 \end{pmatrix}\right] \\ &= e^{4t}\left[(C_1 + C_2 t)\begin{pmatrix} 2 \\ -1 \end{pmatrix} + C_2\begin{pmatrix} 1 \\ 0 \end{pmatrix}\right]. \end{aligned}$$

We know that solutions must emanate from the origin parallel to the halflines generated by $C_1(2, -1)$. Not only that, the solutions must also turn parallel to the halflines as time marches forward. At $(1, 0)$,

$$\begin{pmatrix} 6 & 4 \\ -1 & 2 \end{pmatrix}\begin{pmatrix} 1 \\ 0 \end{pmatrix} = \begin{pmatrix} 6 \\ -1 \end{pmatrix},$$

so the rotation is clockwise. A hand sketch follows.

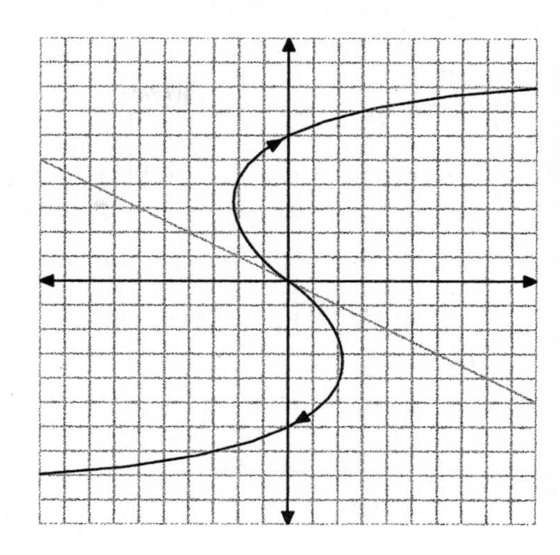

The phase portrait, drawn in a numerical solver, follows.

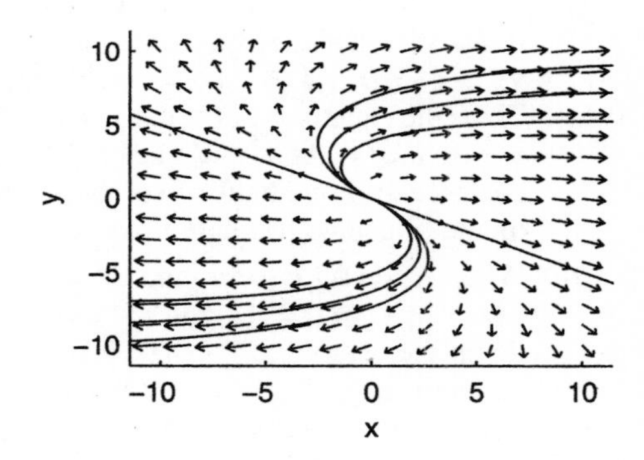

19. (a) In matrix form, the system

$$\begin{aligned} x' &= x + ay \\ y' &= x + y \end{aligned}$$

is written

$$\begin{pmatrix} x \\ y \end{pmatrix}' = \begin{pmatrix} 1 & a \\ 1 & 1 \end{pmatrix}\begin{pmatrix} x \\ y \end{pmatrix}.$$

The trace of the coefficient matrix is $T = 2$ and the determinant is $D = 1-a$. The discriminant is

$$T^2 - 4D = (2)^2 - 4(1-a) = 4a.$$

If the origin is a nodal source, then we must have $D > 0$ and $T^2 - 4D > 0$. Thus,

$$1 - a > 0 \quad \text{and} \quad 4a > 0.$$

This leads to the requirement $0 < a < 1$.

(b) Let

$$A = \begin{pmatrix} 1 & a \\ 1 & 1 \end{pmatrix}.$$

In the case that $0 < a < 1$,

$$p(\lambda) = \lambda^2 - T\lambda + D = \lambda^2 - 2\lambda + (1-a).$$

The quadratic formula reveals the eigenvalues, $\lambda_1 = 1 + \sqrt{a}$ and $\lambda_2 = 1 - \sqrt{a}$. Because

$$A - \lambda I = \begin{pmatrix} 1-\lambda & a \\ 1 & 1-\lambda \end{pmatrix},$$

$\mathbf{v} = (\lambda - 1, 1)^T$ is the eigenvector associated with λ. If $\lambda_1 = 1 + \sqrt{a}$, then $\mathbf{v}_1 = (\sqrt{a}, 1)^T$ is its associated eigenvector. If $\lambda_2 = 1 - \sqrt{a}$, then $\mathbf{v}_2 = (-\sqrt{a}, 1)^T$ is its associated eigenvector. Thus, the equations of the halfline solutions are $y = \pm x/\sqrt{a}$. As $a \to 0$, the halfline solutions coalesce into one halfline solution, which lies on the y-axis with equation $x = 0$.

(c) When $a = 0$, $T = 2$ and $D = 1$. Moreover, $T^2 - 4D = (2)^2 - 4(1) = 0$, and we lie on the parabola $T^2 - 4D = 0$ in the trace-determinant plane. By part (b), the eigenvalues and eigenvectors coalesce, and we have a degenerate nodal source. If $a < 0$, then $T^2 - 4D = 4a < 0$, and we move above the parabola $T^2 - 4D = 0$ into the land of spiral sources.

21. Let A be a 2×2 matrix with real entries. If $D = \det(A) = 0$, then the characteristic polynomial becomes

$$\begin{aligned} p(\lambda) &= \lambda^2 - T\lambda + D \\ &= \lambda^2 - T\lambda \\ &= \lambda(\lambda - T). \end{aligned}$$

Thus, $\lambda = 0$ is an eigenvalue. On the other hand, if one eigenvalue is $\lambda = 0$, then λ must be a factor of the characteristic equation $\lambda^2 - T\lambda + D$. This can only happen if $D = 0$.

23. (i) If

$$A = \begin{pmatrix} 8 & 4 \\ -10 & -5 \end{pmatrix}$$

then the trace is $T = 3$ and the determinant is $D = 0$. Thus, this degenerate case lies on the horizontal axis, separating saddles from the nodal sources.

(ii) To find the equilibrium points, we set the right-hand side of $\mathbf{y}' = A\mathbf{y}$ equal to zero, as in $A\mathbf{y} = \mathbf{0}$. Consequently, the equilibrium points are simply the nullspace of A, which is generated by a single vector, $\mathbf{v}_1 = (1, -2)^T$. Thus, everything on the line $y = -2x$ is an equilibrium point.

(iii) The characteristic polynomial is

$$p(\lambda) = \lambda^2 - T\lambda + D = \lambda^2 - 3\lambda,$$

which produces eigenvalues $\lambda_1 = 0$ and $\lambda_2 = 3$. Because

$$A + 0I = \begin{pmatrix} 8 & 4 \\ -10 & -5 \end{pmatrix},$$

the eigenvector is $\mathbf{v}_1 = (1, -2)^T$, the same vector that produces a line of equilibrium points. Because

$$A - 3I = \begin{pmatrix} 5 & 4 \\ -10 & -8 \end{pmatrix},$$

$\mathbf{v}_2 = (4, -5)^T$. Thus, the general solution is

$$\mathbf{y}(t) = C_1 e^{0t} \begin{pmatrix} 1 \\ -2 \end{pmatrix} + C_2 e^{3t} \begin{pmatrix} 4 \\ -5 \end{pmatrix},$$

or

$$\mathbf{y}(t) = C_1 \begin{pmatrix} 1 \\ -2 \end{pmatrix} + C_2 e^{3t} \begin{pmatrix} 4 \\ -5 \end{pmatrix}.$$

Note that each solutions in this family is the sum of a fixed multiple of $(1, -2)^T$ and an increasing multiple of $(4, -5)^T$. Thus, as $t \to \infty$, solutions move away from the line of equilibrium points along lines parallel to $(4, -5)^T$, as shown in the following figure.

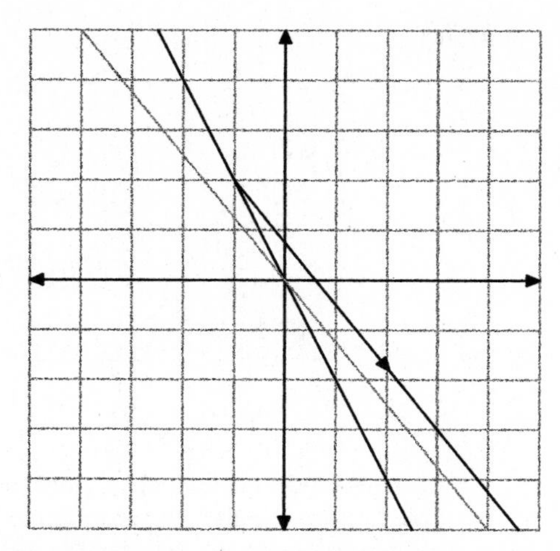

Our numerical solver provides further evidence of this behavior.

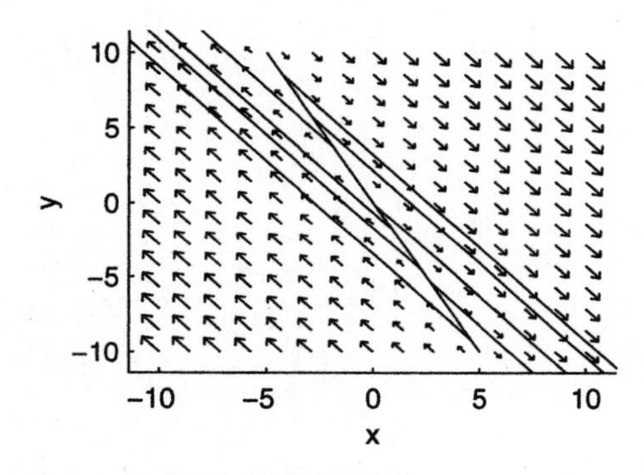

25. Kirchoff's Current Law (KCL) tells us that the current entering the node at "a" equals the current leaving the node at "a". Thus,

$$i_R = i_L + i_C. \tag{4.1}$$

Traversing (clockwise) rightmost loop containing capacitor and inductor, Kirchoff's Voltage Law (KVL) tells us that

$$v_C - Li_L' = 0. \tag{4.2}$$

Traversing (clockwise) the outermost loop containing resistor and capacitor, KVL provides us with

$$Ri_R + v_C = 0. \tag{4.3}$$

First, substitute equation (4.1) in (4.3) to obtain

$$Ri_L + Ri_C + v_C = 0. \tag{4.4}$$

The charge on the capacitor is given by $q_C = Cv_C$, which when differentiated, provides us with the current across the capacitor, $i_C = Cv_C'$. With this substitution, equation (4.4) becomes

$$Ri_L + RCv_C' + v_C = 0. \tag{4.5}$$

Solving equations (4.2) and (4.5) for i_L' and v_C' gives us the requested system.

$$\begin{aligned} v_C' &= -\frac{1}{RC}v_C - \frac{1}{C}i_L \\ i_L' &= \frac{1}{L}v_C. \end{aligned} \tag{4.6}$$

In matrix form,

$$\begin{bmatrix} v_C \\ i_L \end{bmatrix}' = \begin{bmatrix} -1/RC & -1/C \\ 1/L & 0 \end{bmatrix} \begin{bmatrix} v_C \\ i_L \end{bmatrix}. \tag{4.7}$$

Note that the trace and determinant of the coefficient matrix are

$$T = -\frac{1}{RC} < 0 \quad \text{and} \quad D = \frac{1}{LC} > 0,$$

respectively, indicating that the equilibrium point at (0, 0) is either a nodal sink, a degenerate sink, or a spiral sink. Moreover, the characteristic polynomial is

$$\lambda^2 - T\lambda + D = \lambda^2 + \frac{1}{RC}\lambda + \frac{1}{LC}. \tag{4.8}$$

The discriminant

$$T^2 - 4D = \frac{1}{R^2C^2} - \frac{4}{LC} = \frac{L - 4R^2C}{LR^2C^2}$$

has sign completely determined by the sign of its numerator (the denominator LR^2C^2 is always a positive number in this application).

- If $L - 4R^2C > 0$, then the system has two real, negative roots, and the equilibrium point at (0, 0) is a nodal sink. As an example, consider the case where $L = 5\,\text{H}$, $R = 2/5\,\Omega$, and $C = 5\,\text{F}$. Then,

$$L - 4R^2C = (5) - 4\left(\frac{2}{5}\right)^2(5) = \frac{9}{5} > 0,$$

and the system is *overdamped*, as shown below.

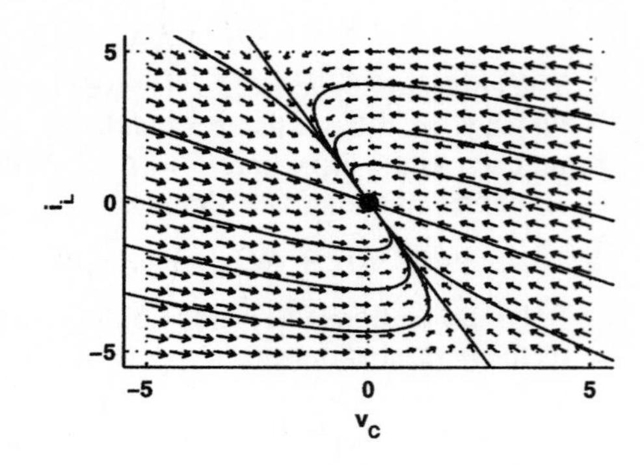

- If $L - 4R^2C = 0$, then the system has one real, negative root, and the equilibirum point lies in the trace-determinant plane precisely on the parabola that divides the region of spiral sinks from nodal sinks. We say that the node is a degenerate sink. As an example, consider $L = 1$ H, $R = 1/4\,\Omega$, and $C = 4$ F, then

$$L - 4R^2C = 1 - 4\left(\frac{1}{4}\right)^2(4) = 0,$$

and the system is *critically damped*, as shown below.

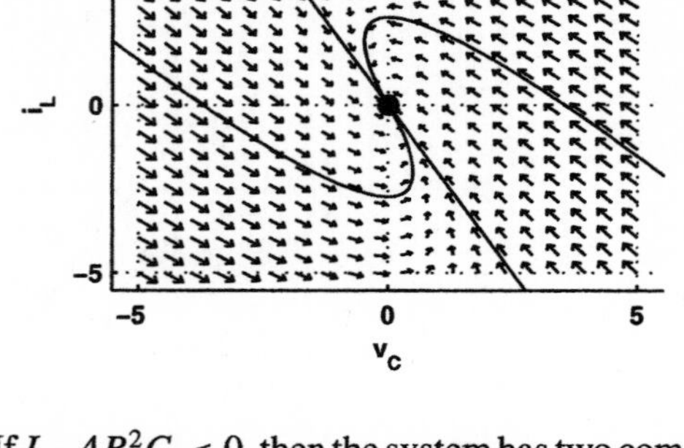

- If $L - 4R^2C < 0$, then the system has two complex roots, and the equilibrium point at $(0, 0)$ is a spiral sink. As an example, let $L = 2$ H, $R = 5\,\Omega$, and $C = 1$ F, then

$$L - 4R^2C = 2 - 4(5)^2(1) = -98 < 0,$$

and the system is *underdamped*, as shown below.

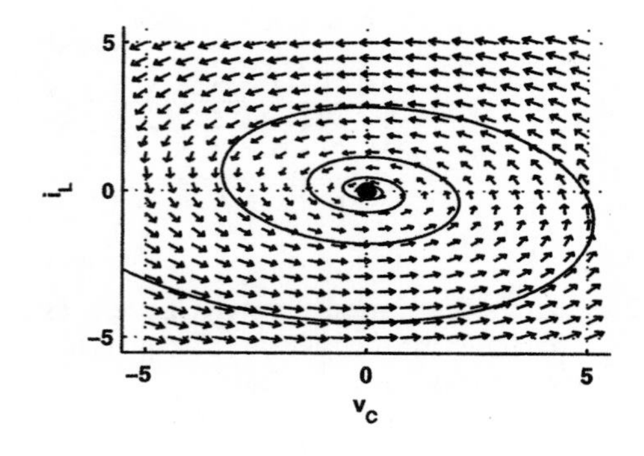

———————×———————

Section 9.5. Higher Dimensional Systems

1. If

$$A = \begin{pmatrix} 2 & 1 & 0 \\ 0 & 1 & 0 \\ 6 & 10 & -1 \end{pmatrix},$$

then

$$\begin{aligned} p(\lambda) &= -\det(A - \lambda I) \\ &= -\begin{vmatrix} 2-\lambda & 1 & 0 \\ 0 & 1-\lambda & 0 \\ 6 & 10 & -1-\lambda \end{vmatrix} \end{aligned}$$

Expanding down the third column,

$$\begin{aligned} p(\lambda) &= -(-1-\lambda)\begin{vmatrix} 2-\lambda & 1 \\ 0 & 1-\lambda \end{vmatrix} \\ &= (\lambda+1)(2-\lambda)(1-\lambda) \\ &= (\lambda+1)(\lambda-2)(\lambda-1). \end{aligned}$$

Thus, the eigenvalues are -1, 2, and 1, respectively. The graph of the characteristic polynomial follows. Note that the graph crosses the horizontal axis at the eigenvalues -1, 2, and 1.

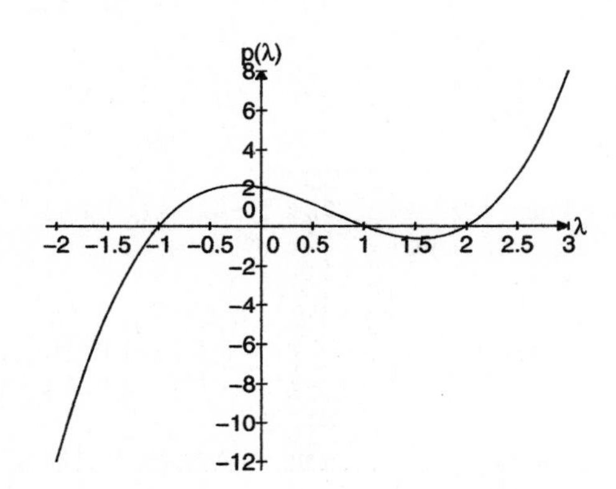

3. If

$$A = \begin{pmatrix} 2 & 0 & 0 \\ -6 & 1 & -4 \\ -3 & 0 & -1 \end{pmatrix},$$

then

$$\begin{aligned} p(\lambda) &= -\det(A - \lambda I) \\ &= -\begin{vmatrix} 2-\lambda & 0 & 0 \\ -6 & 1-\lambda & -4 \\ -3 & 0 & -1-\lambda \end{vmatrix} \end{aligned}$$

Expanding across the first row,

$$\begin{aligned} p(\lambda) &= -(2-\lambda)\begin{vmatrix} 1-\lambda & -4 \\ 0 & -1-\lambda \end{vmatrix} \\ &= (\lambda-2)(1-\lambda)(-1-\lambda) \\ &= (\lambda-2)(\lambda-1)(\lambda+1). \end{aligned}$$

Thus, the eigenvalues are 2, 1, and -1, respectively. Because $A - 2I$ reduces,

$$A - 2I = \begin{pmatrix} 0 & 0 & 0 \\ -6 & -1 & -4 \\ -3 & 0 & -3 \end{pmatrix} \rightarrow \begin{pmatrix} 1 & 0 & 1 \\ 0 & 1 & -2 \\ 0 & 0 & 0 \end{pmatrix},$$

it is easily seen that the nullspace of $A - 2I$ is generated by the eigenvector $\mathbf{v}_1 = (-1, 2, 1)^T$. In a similar manner, we arrive at the following eigenvalue-eigenvector pairs.

$$1 \rightarrow \begin{pmatrix} 0 \\ 1 \\ 0 \end{pmatrix} \quad \text{and} \quad -1 \rightarrow \begin{pmatrix} 0 \\ 2 \\ 1 \end{pmatrix}$$

Because

$$\det \begin{pmatrix} -1 & 0 & 0 \\ 2 & 1 & 2 \\ 1 & 0 & 1 \end{pmatrix} = -1,$$

the eigenvectors are independent.

5. If

$$A = \begin{pmatrix} -4 & 0 & 2 \\ 12 & 2 & -6 \\ -6 & 0 & 3 \end{pmatrix},$$

then

$$\begin{aligned} p(\lambda) &= -\det(A - \lambda I) \\ &= -\begin{vmatrix} -4-\lambda & 0 & 2 \\ 12 & 2-\lambda & -6 \\ -6 & 0 & 3-\lambda \end{vmatrix} \end{aligned}$$

Expanding down the second column,

$$\begin{aligned} p(\lambda) &= -(2-\lambda)\begin{vmatrix} -4-\lambda & 2 \\ -6 & 3-\lambda \end{vmatrix} \\ &= (\lambda-2)(\lambda^2+\lambda) \\ &= \lambda(\lambda-2)(\lambda+1). \end{aligned}$$

Thus, the eigenvalues are 0, 2, and -1, respectively. Because $A - 0I$ reduces,

$$A - 0I = \begin{pmatrix} -4 & 0 & 2 \\ 12 & 2 & -6 \\ -6 & 0 & 3 \end{pmatrix} \rightarrow \begin{pmatrix} 1 & 0 & -1/2 \\ 0 & 1 & 0 \\ 0 & 0 & 0 \end{pmatrix},$$

it is easily seen that the nullspace of $A - 0I$ is generated by the eigenvector $\mathbf{v}_1 = (1, 0, 2)^T$. In a similar manner, we arrive at the following eigenvalue-eigenvector pairs.

$$2 \rightarrow \begin{pmatrix} 0 \\ 1 \\ 0 \end{pmatrix} \quad \text{and} \quad -1 \rightarrow \begin{pmatrix} -2 \\ 2 \\ -3 \end{pmatrix}$$

Because

$$\det \begin{pmatrix} 1 & 0 & -2 \\ 0 & 1 & 2 \\ 2 & 0 & -3 \end{pmatrix} = 1,$$

the eigenvectors are independent.

7. The system in matrix form,

$$\begin{pmatrix} x \\ y \\ z \end{pmatrix}' = \begin{pmatrix} 4 & -5 & 4 \\ 0 & -1 & 4 \\ 0 & 0 & 1 \end{pmatrix} \begin{pmatrix} x \\ y \\ z \end{pmatrix},$$

reveals that the matrix

$$A = \begin{pmatrix} 4 & -5 & 4 \\ 0 & -1 & 4 \\ 0 & 0 & 1 \end{pmatrix}$$

is upper triangular. Thus, the eigenvalues are located on the main diagonal and are $-1, 4$, and 1. Because

$$A + I = \begin{pmatrix} 5 & -5 & 4 \\ 0 & 0 & 4 \\ 0 & 0 & 2 \end{pmatrix} \rightarrow \begin{pmatrix} 1 & -1 & 0 \\ 0 & 0 & 1 \\ 0 & 0 & 0 \end{pmatrix},$$

it is easily seen that $-1 \rightarrow (1, 1, 0)^T$ is an eigenvalue-eigenvector pair. Similarly,

$$4 \rightarrow \begin{pmatrix} 1 \\ 0 \\ 0 \end{pmatrix} \quad \text{and} \quad 1 \rightarrow \begin{pmatrix} 2 \\ 2 \\ 1 \end{pmatrix}$$

are the remaining eigenvalue-eigenvector pairs. These lead to the general solution

$$\begin{pmatrix} x \\ y \\ z \end{pmatrix} = C_1 e^{-t} \begin{pmatrix} 1 \\ 1 \\ 0 \end{pmatrix} + C_2 e^{4t} \begin{pmatrix} 1 \\ 0 \\ 0 \end{pmatrix} + C_3 e^{t} \begin{pmatrix} 2 \\ 2 \\ 1 \end{pmatrix}.$$

9. In matrix form,

$$\begin{pmatrix} x \\ y \\ z \end{pmatrix}' = \begin{pmatrix} -3 & 0 & 0 \\ -5 & 6 & -4 \\ -5 & 2 & 0 \end{pmatrix} \begin{pmatrix} x \\ y \\ z \end{pmatrix},$$

the characteristic polynomial of matrix

$$A = \begin{pmatrix} -3 & 0 & 0 \\ -5 & 6 & -4 \\ -5 & 2 & 0 \end{pmatrix},$$

is found by calculating

$$\begin{aligned} p(\lambda) &= -\det(A - \lambda I) \\ &= -\begin{vmatrix} -3-\lambda & 0 & 0 \\ -5 & 6-\lambda & -4 \\ -5 & 2 & -\lambda \end{vmatrix}. \end{aligned}$$

Expanding across the first row,

$$\begin{aligned} p(\lambda) &= -(-3-\lambda) \begin{vmatrix} 6-\lambda & -4 \\ 2 & -\lambda \end{vmatrix} \\ &= (\lambda + 3)(\lambda^2 - 6\lambda + 8) \\ &= (\lambda + 3)(\lambda - 4)(\lambda - 2). \end{aligned}$$

Thus, the eigenvalues are 4, -3, and 2. Because $A - 4I$ reduces,

$$A - 4I = \begin{pmatrix} -7 & 0 & 0 \\ -5 & 2 & -4 \\ -5 & 2 & -4 \end{pmatrix} \rightarrow \begin{pmatrix} 1 & 0 & 0 \\ 0 & 1 & -2 \\ 0 & 0 & 0 \end{pmatrix},$$

it is easily seen that the nullspace of $A - 4I$ is generated by the eigenvector $\mathbf{v}_1 = (0, 2, 1)^T$. In a similar manner, we arrive at the following eigenvalue-eigenvector pairs.

$$-3 \rightarrow \begin{pmatrix} 1 \\ 1 \\ 1 \end{pmatrix} \quad \text{and} \quad 2 \rightarrow \begin{pmatrix} 0 \\ 1 \\ 1 \end{pmatrix}$$

Thus, the general solution is

$$\begin{pmatrix} x \\ y \\ z \end{pmatrix} = C_1 e^{4t} \begin{pmatrix} 0 \\ 2 \\ 1 \end{pmatrix} + C_2 e^{-3t} \begin{pmatrix} 1 \\ 1 \\ 1 \end{pmatrix} + C_3 e^{2t} \begin{pmatrix} 0 \\ 1 \\ 1 \end{pmatrix}.$$

11. The characteristic polynomial of matrix

$$A = \begin{pmatrix} -3 & 4 & 8 \\ -2 & 3 & 2 \\ 0 & 0 & 2 \end{pmatrix}$$

is found by calculating

$$\begin{aligned} p(\lambda) &= -\det(A - \lambda I) \\ &= -\begin{vmatrix} -3-\lambda & 4 & 8 \\ -2 & 3-\lambda & 2 \\ 0 & 0 & 2-\lambda \end{vmatrix}. \end{aligned}$$

Expanding across the third row,

$$\begin{aligned} p(\lambda) &= -(2-\lambda)\begin{vmatrix} -3-\lambda & 4 \\ -2 & 3-\lambda \end{vmatrix} \\ &= (\lambda - 2)(\lambda^2 - 1) \\ &= (\lambda - 2)(\lambda + 1)(\lambda - 1). \end{aligned}$$

Thus, the eigenvalues are -1, 1, and 2. Because $A + I$ reduces,

$$A + I = \begin{pmatrix} -2 & 4 & 8 \\ -2 & 4 & 2 \\ 0 & 0 & 3 \end{pmatrix} \to \begin{pmatrix} 1 & -2 & 0 \\ 0 & 0 & 1 \\ 0 & 0 & 0 \end{pmatrix},$$

it is easily seen that the nullspace of $A + I$ is generated by the eigenvector $\mathbf{v}_1 = (2, 1, 0)^T$. In a similar manner, we arrive at the following eigenvalue-eigenvector pairs.

$$1 \to \begin{pmatrix} 1 \\ 1 \\ 0 \end{pmatrix} \quad \text{and} \quad 2 \to \begin{pmatrix} 0 \\ -2 \\ 1 \end{pmatrix}$$

Thus, the general solution is

$$\mathbf{y}(t) = C_1 e^{-t}\begin{pmatrix} 2 \\ 1 \\ 0 \end{pmatrix} + C_2 e^{t}\begin{pmatrix} 1 \\ 1 \\ 0 \end{pmatrix} + C_3 e^{2t}\begin{pmatrix} 0 \\ -2 \\ 1 \end{pmatrix}.$$

13. The general solution in Exercise 7 was

$$\begin{pmatrix} x \\ y \\ z \end{pmatrix} = C_1 e^{-t}\begin{pmatrix} 1 \\ 1 \\ 0 \end{pmatrix} + C_2 e^{4t}\begin{pmatrix} 1 \\ 0 \\ 0 \end{pmatrix} + C_3 e^{t}\begin{pmatrix} 2 \\ 2 \\ 1 \end{pmatrix}.$$

If $x(0) = 1$, $y(0) = -1$, and $z(0) = 2$, then

$$\begin{pmatrix} 1 \\ -1 \\ 2 \end{pmatrix} = C_1\begin{pmatrix} 1 \\ 1 \\ 0 \end{pmatrix} + C_2\begin{pmatrix} 1 \\ 0 \\ 0 \end{pmatrix} + C_3\begin{pmatrix} 2 \\ 2 \\ 1 \end{pmatrix}.$$

The augmented matrix reduces.

$$\begin{pmatrix} 1 & 1 & 2 & 1 \\ 1 & 0 & 2 & -1 \\ 0 & 0 & 1 & 2 \end{pmatrix} \to \begin{pmatrix} 1 & 0 & 0 & -5 \\ 0 & 1 & 0 & 2 \\ 0 & 0 & 1 & 2 \end{pmatrix}$$

Thus, $C_1 = -5$, $C_2 = 2$, and $C_3 = 2$, and the particular solution is

$$\begin{aligned} \begin{pmatrix} x(t) \\ y(t) \\ z(t) \end{pmatrix} &= -5e^{-t}\begin{pmatrix} 1 \\ 1 \\ 0 \end{pmatrix} + 2e^{4t}\begin{pmatrix} 1 \\ 0 \\ 0 \end{pmatrix} + 2e^{t}\begin{pmatrix} 2 \\ 2 \\ 1 \end{pmatrix} \\ &= \begin{pmatrix} -5e^{-t} + 2e^{4t} + 4e^{t} \\ -5e^{-t} + 4e^{t} \\ 2e^{t} \end{pmatrix} \end{aligned}$$

15. The general solution in Exercise 9 was

$$\begin{pmatrix} x \\ y \\ z \end{pmatrix} = C_1 e^{4t}\begin{pmatrix} 0 \\ 2 \\ 1 \end{pmatrix} + C_2 e^{-3t}\begin{pmatrix} 1 \\ 1 \\ 1 \end{pmatrix} + C_3 e^{2t}\begin{pmatrix} 0 \\ 1 \\ 1 \end{pmatrix}.$$

If $x(0) = -2$, $y(0) = 0$, and $z(0) = 2$, then

$$\begin{pmatrix} -2 \\ 0 \\ 2 \end{pmatrix} = C_1\begin{pmatrix} 0 \\ 2 \\ 1 \end{pmatrix} + C_2\begin{pmatrix} 1 \\ 1 \\ 1 \end{pmatrix} + C_3\begin{pmatrix} 0 \\ 1 \\ 1 \end{pmatrix}.$$

The augmented matrix reduces.

$$\begin{pmatrix} 0 & 1 & 0 & -2 \\ 2 & 1 & 1 & 0 \\ 1 & 1 & 1 & 2 \end{pmatrix} \to \begin{pmatrix} 1 & 0 & 0 & -2 \\ 0 & 1 & 0 & -2 \\ 0 & 0 & 1 & 6 \end{pmatrix}$$

Thus, $C_1 = -2$, $C_2 = -2$, and $C_3 = 6$, and the particular solution is

$$\begin{aligned} \begin{pmatrix} x(t) \\ y(t) \\ z(t) \end{pmatrix} &= -2e^{4t}\begin{pmatrix} 0 \\ 2 \\ 1 \end{pmatrix} - 2e^{-3t}\begin{pmatrix} 1 \\ 1 \\ 1 \end{pmatrix} + 6e^{2t}\begin{pmatrix} 0 \\ 1 \\ 1 \end{pmatrix} \\ &= \begin{pmatrix} -2e^{-3t} \\ -4e^{4t} - 2e^{-3t} + 6e^{2t} \\ -2e^{4t} - 2e^{-3t} + 6e^{2t} \end{pmatrix} \end{aligned}$$

17. The general solution in Exercise 11 was

$$\mathbf{y}(t) = C_1 e^{-t}\begin{pmatrix}2\\1\\0\end{pmatrix} + C_2 e^{t}\begin{pmatrix}1\\1\\0\end{pmatrix} + C_3 e^{2t}\begin{pmatrix}0\\-2\\1\end{pmatrix}.$$

If $\mathbf{y}(0) = (1, -2, 1)^T$, then

$$\begin{pmatrix}1\\-2\\1\end{pmatrix} = C_1\begin{pmatrix}2\\1\\0\end{pmatrix} + C_2\begin{pmatrix}1\\1\\0\end{pmatrix} + C_3\begin{pmatrix}0\\-2\\1\end{pmatrix}.$$

The augmented matrix reduces.

$$\begin{pmatrix}2 & 1 & 0 & 1\\ 1 & 1 & -2 & -2\\ 0 & 0 & 1 & 1\end{pmatrix} \to \begin{pmatrix}1 & 0 & 0 & 1\\ 0 & 1 & 0 & -1\\ 0 & 0 & 1 & 1\end{pmatrix}$$

Thus, $C_1 = 1$, $C_2 = -1$, and $C_3 = 1$, and the particular solution is

$$\mathbf{y}(t) = e^{-t}\begin{pmatrix}2\\1\\0\end{pmatrix} - e^{t}\begin{pmatrix}1\\1\\0\end{pmatrix} + e^{2t}\begin{pmatrix}0\\-2\\1\end{pmatrix}$$

$$= \begin{pmatrix}2e^{-t} - e^{t}\\ e^{-t} - e^{t} - 2e^{2t}\\ e^{2t}\end{pmatrix}$$

19. Using Euler's formula

$$\mathbf{y}(t) = e^{2it}\begin{pmatrix}1\\1+2i\\-3i\end{pmatrix}$$

$$= (\cos 2t + i\sin 2t)\left[\begin{pmatrix}1\\1\\-3\end{pmatrix} + i\begin{pmatrix}0\\2\\-3\end{pmatrix}\right]$$

$$= \cos 2t\begin{pmatrix}1\\1\\-3\end{pmatrix} - \sin 2t\begin{pmatrix}0\\2\\-3\end{pmatrix}$$

$$+ i\cos 2t\begin{pmatrix}0\\2\\-3\end{pmatrix} + i\sin 2t\begin{pmatrix}1\\1\\-3\end{pmatrix}$$

$$= \begin{pmatrix}\cos 2t\\ \cos 2t - 2\sin 2t\\ -3\cos 2t + 3\sin 2t\end{pmatrix}$$

$$+ i\begin{pmatrix}\sin 2t\\ 2\cos 2t + \sin 2t\\ -3\cos 2t - 3\sin 2t\end{pmatrix}.$$

Thus, the real and imaginary parts of the complex solution $\mathbf{y}(t) = \mathbf{y}_1(t) + i\mathbf{y}_2(t)$ are

$$\mathbf{y}_1(t) = \begin{pmatrix}\cos 2t\\ \cos 2t - 2\sin 2t\\ -3\cos 2t + 3\sin 2t\end{pmatrix} \quad \text{and}$$

$$\mathbf{y}_2(t) = \begin{pmatrix}\sin 2t\\ 2\cos 2t + \sin 2t\\ -3\cos 2t - 3\sin 2t\end{pmatrix}.$$

21. In matrix form,

$$\begin{pmatrix}x\\y\\z\end{pmatrix}' = \begin{pmatrix}-4 & 8 & 8\\ -4 & 4 & 2\\ 0 & 0 & 2\end{pmatrix}\begin{pmatrix}x\\y\\z\end{pmatrix}.$$

Using a computer, matrix

$$A = \begin{pmatrix}-4 & 8 & 8\\ -4 & 4 & 2\\ 0 & 0 & 2\end{pmatrix}$$

has eigenvalues 2, $4i$, and $-4i$. For the eigenvalue 2, we look for an vector in the nullspace (eigenspace) of

$$A - 2I = \begin{pmatrix}-6 & 8 & 8\\ -4 & 2 & 2\\ 0 & 0 & 0\end{pmatrix}.$$

The computer tells us that $(0, -1, 1)^T$ is in the nullspace of $A - 2I$. Thus, one solution is $\mathbf{y}_1(t) = e^{2t}(0, -1, 1)^T$. In a similar vein, our computer tells us that $(1 - i, 1, 0)^T$ is in the nullspace of $A - (4i)I$. Thus, we have conjugate solutions

$$\mathbf{z}(t) = e^{4it}\begin{pmatrix}1-i\\1\\0\end{pmatrix} \quad \text{and} \quad \overline{\mathbf{z}}(t) = e^{-4it}\begin{pmatrix}1+i\\1\\0\end{pmatrix}.$$

Using Euler's formula, we find the real and imaginary parts of the solution $\mathbf{z}(t)$.

$$\mathbf{z}(t) = e^{4it}\begin{pmatrix}1-i\\1\\0\end{pmatrix}$$

$$= (\cos 4t + i\sin 4t)\left[\begin{pmatrix}1\\1\\0\end{pmatrix} + i\begin{pmatrix}-1\\0\\0\end{pmatrix}\right]$$

$$= \begin{pmatrix}\cos 4t + \sin 4t\\ \cos 4t\\ 0\end{pmatrix} + i\begin{pmatrix}-\cos 4t + \sin 4t\\ \sin 4t\\ 0\end{pmatrix}.$$

The real and imaginary parts of $\mathbf{z}$ are solutions and we can write the general solution

$$\begin{pmatrix}x(t)\\y(t)\\z(t)\end{pmatrix} = C_1e^{2t}\begin{pmatrix}0\\-1\\1\end{pmatrix} + C_2\begin{pmatrix}\cos 4t + \sin 4t\\ \cos 4t\\ 0\end{pmatrix} + C_3\begin{pmatrix}-\cos 4t + \sin 4t\\ \sin 4t\\ 0\end{pmatrix}.$$

23. In matrix form,

$$\begin{pmatrix}x\\y\\z\end{pmatrix}' = \begin{pmatrix}6 & 0 & -4\\ 8 & -2 & 0\\ 8 & 0 & -2\end{pmatrix}\begin{pmatrix}x\\y\\z\end{pmatrix}.$$

Using a computer, matrix

$$A = \begin{pmatrix}6 & 0 & -4\\ 8 & -2 & 0\\ 8 & 0 & -2\end{pmatrix}$$

has eigenvalues -2, $2+4i$, and $2-4i$. For the eigenvalue -2, we look for a vector in the nullspace (eigenspace) of

$$A + 2I = \begin{pmatrix}8 & 0 & -4\\ 8 & 0 & 0\\ 8 & 0 & 0\end{pmatrix}.$$

The computer tells us that $(0,1,0)^T$ is in the nullspace of $A+2I$. Thus, one solution is $\mathbf{y}_1(t) = e^{-2t}(0,1,0)^T$. In a similar vein, our computer tells us that $(1+i,2,2)^T$ is in the nullspace of $A-(2+4i)I$. Thus, we have conjugate solutions

$$\mathbf{z}(t) = e^{(2+4i)t}\begin{pmatrix}1+i\\2\\2\end{pmatrix} \quad \text{and}$$

$$\overline{\mathbf{z}}(t) = e^{(2-4i)t}\begin{pmatrix}1-i\\2\\2\end{pmatrix}.$$

Using Euler's formula, we find the real and imaginary parts of the solution $\mathbf{z}(t)$.

$$\mathbf{z}(t) = e^{2t}e^{4it}\begin{pmatrix}1+i\\2\\2\end{pmatrix}$$

$$= e^{2t}(\cos 4t + i\sin 4t)\left[\begin{pmatrix}1\\2\\2\end{pmatrix} + i\begin{pmatrix}1\\0\\0\end{pmatrix}\right]$$

$$= e^{2t}\begin{pmatrix}\cos 4t - \sin 4t\\ 2\cos 4t\\ 2\cos 4t\end{pmatrix} + ie^{2t}\begin{pmatrix}\cos 4t + \sin 4t\\ 2\sin 4t\\ 2\sin 4t\end{pmatrix}$$

The real and imaginary parts of $\mathbf{z}$ are solutions and we can write the general solution

$$\begin{pmatrix}x(t)\\y(t)\\z(t)\end{pmatrix} = C_1e^{-2t}\begin{pmatrix}0\\1\\0\end{pmatrix} + C_2e^{2t}\begin{pmatrix}\cos 4t - \sin 4t\\ 2\cos 4t\\ 2\cos 4t\end{pmatrix} + C_3e^{2t}\begin{pmatrix}\cos 4t + \sin 4t\\ 2\sin 4t\\ 2\sin 4t\end{pmatrix}.$$

25. In system $\mathbf{y}' = A\mathbf{y}$, where

$$A = \begin{pmatrix}-7 & -13 & 0\\ 2 & 3 & 0\\ 3 & 8 & -2\end{pmatrix},$$

we have

$$A - \lambda I = \begin{pmatrix}-7-\lambda & -13 & 0\\ 2 & 3-\lambda & 0\\ 3 & 8 & -2-\lambda\end{pmatrix}.$$

We can compute the characteristic polynomial by expanding along the third column. We get

$$\begin{aligned} p(\lambda) &= -\det(A - \lambda I) \\ &= -(-2-\lambda)\det\begin{pmatrix} -7-\lambda & -13 \\ 2 & 3-\lambda \end{pmatrix} \\ &= (\lambda+2)(\lambda^2+4\lambda+5). \end{aligned}$$

Hence we have one real eigenvalue $\lambda_1 = -2$, and the quadratic $\lambda^2+4\lambda+5$ has complex roots $\lambda_2 = -2+i$, and $\overline{\lambda_2} = -2-i$. For the eigenvalue $\lambda_1 = -2$, we look for a vector in the nullspace (eigenspace) of

$$A - \lambda_1 I = A + 2I = \begin{pmatrix} -5 & -13 & 0 \\ 2 & 5 & 0 \\ 3 & 8 & 0 \end{pmatrix}.$$

The eigenspace is generated by $\mathbf{v}_1 = (0, 0, 1)^T$. Thus, one solution is

$$\mathbf{y}_1(t) = e^{-2t}\begin{pmatrix} 0 \\ 0 \\ 1 \end{pmatrix}.$$

For the eigenvalue $\lambda_2 = -2+i$, we look for an vector in the nullspace (eigenspace) of

$$A - \lambda_1 I = A + (2-i)I = \begin{pmatrix} -5-i & -13 & 0 \\ 2 & 5-i & 0 \\ 3 & 8 & -i \end{pmatrix}.$$

The eigenspace is generated by $(-5+i, 2, 3-i)^T$. Thus, we have the complex conjugate solutions

$$\mathbf{z}(t) = e^{(-2+i)t}\begin{pmatrix} -5+i \\ 2 \\ 3-i \end{pmatrix} \quad \text{and}$$

$$\overline{\mathbf{z}}(t) = e^{(-2-i)t}\begin{pmatrix} -5-i \\ 2 \\ 3+i \end{pmatrix}.$$

Using Euler's formula, we find the real and imaginary parts of the solution $\mathbf{z}(t)$.

$$\begin{aligned} \mathbf{z}(t) &= e^{-2t}e^{it}\begin{pmatrix} -5+i \\ 2 \\ 3-i \end{pmatrix} \\ &= e^{-2t}(\cos t + i\sin t)\left[\begin{pmatrix} -5 \\ 2 \\ 3 \end{pmatrix} + i\begin{pmatrix} 1 \\ 0 \\ -1 \end{pmatrix}\right] \\ &= e^{-2t}\begin{pmatrix} -5\cos t - \sin t \\ 2\cos t \\ 3\cos t + \sin t \end{pmatrix} \\ &\quad + ie^{-2t}\begin{pmatrix} \cos t - 5\sin t \\ 2\sin t \\ -\cos t + 3\sin t \end{pmatrix} \end{aligned}$$

Thus we have the solutions

$$\mathbf{y}_2(t) = \operatorname{Re}(\mathbf{z}(t)) = e^{-2t}\begin{pmatrix} -5\cos t - \sin t \\ 2\cos t \\ 3\cos t + \sin t \end{pmatrix}$$

and

$$\mathbf{y}_3(t) = \operatorname{Im}(\mathbf{z}(t)) = e^{-2t}\begin{pmatrix} \cos t - 5\sin t \\ 2\sin t \\ -\cos t + 3\sin t \end{pmatrix}.$$

The general solution is

$$\mathbf{y}(t) = C_1\mathbf{y}_1(t) + C_2\mathbf{y}_2(t) + C_3\mathbf{y}_3(t).$$

27. In Exercise 21, the general solution was

$$\begin{aligned} \begin{pmatrix} x(t) \\ y(t) \\ z(t) \end{pmatrix} &= C_1 e^{2t}\begin{pmatrix} 0 \\ -1 \\ 1 \end{pmatrix} \\ &\quad + C_2\begin{pmatrix} \cos 4t + \sin 4t \\ \cos 4t \\ 0 \end{pmatrix} \\ &\quad + C_3\begin{pmatrix} -\cos 4t + \sin 4t \\ \sin 4t \\ 0 \end{pmatrix}. \end{aligned}$$

If $x(0) = 1$, $y(0) = 0$, and $z(0) = 0$, then

$$\begin{pmatrix} 1 \\ 0 \\ 0 \end{pmatrix} = C_1\begin{pmatrix} 0 \\ -1 \\ 1 \end{pmatrix} + C_2\begin{pmatrix} 1 \\ 1 \\ 0 \end{pmatrix} + C_3\begin{pmatrix} -1 \\ 0 \\ 0 \end{pmatrix}.$$

The augmented matrix reduces.

$$\begin{pmatrix} 0 & 1 & -1 & 1 \\ -1 & 1 & 0 & 0 \\ 1 & 0 & 0 & 0 \end{pmatrix} \rightarrow \begin{pmatrix} 1 & 0 & 0 & 0 \\ 0 & 1 & 0 & 0 \\ 0 & 0 & 1 & -1 \end{pmatrix}$$

Thus, $C_1 = C_2 = 0$ and $C_3 = -1$, giving the particular solution

$$\begin{pmatrix} x(t) \\ y(t) \\ z(t) \end{pmatrix} = \begin{pmatrix} \cos 4t - \sin 4t \\ -\sin 4t \\ 0 \end{pmatrix}.$$

29. In Exercise 23, the general solution was

$$\begin{pmatrix} x(t) \\ y(t) \\ z(t) \end{pmatrix} = C_1 e^{-2t} \begin{pmatrix} 0 \\ 1 \\ 0 \end{pmatrix} + C_2 e^{2t} \begin{pmatrix} \cos 4t - \sin 4t \\ 2\cos 4t \\ 2\cos 4t \end{pmatrix} + C_3 e^{2t} \begin{pmatrix} \cos 4t + \sin 4t \\ 2\sin 4t \\ 2\sin 4t \end{pmatrix}.$$

If $x(0) = -2$, $y(0) = -1$, and $z(0) = 0$, then

$$\begin{pmatrix} -2 \\ -1 \\ 0 \end{pmatrix} = C_1 \begin{pmatrix} 0 \\ 1 \\ 0 \end{pmatrix} + C_2 \begin{pmatrix} 1 \\ 2 \\ 2 \end{pmatrix} + C_3 \begin{pmatrix} 1 \\ 0 \\ 0 \end{pmatrix}.$$

The augmented matrix reduces.

$$\begin{pmatrix} 0 & 1 & 1 & -2 \\ 1 & 2 & 0 & -1 \\ 0 & 2 & 0 & 0 \end{pmatrix} \rightarrow \begin{pmatrix} 1 & 0 & 0 & -1 \\ 0 & 1 & 0 & 0 \\ 0 & 0 & 1 & -2 \end{pmatrix}$$

Thus, $C_1 = -1$, $C_2 = 0$ and $C_3 = -2$, giving the particular solution

$$\begin{pmatrix} x(t) \\ y(t) \\ z(t) \end{pmatrix} = \begin{pmatrix} e^{2t}(-2\cos 4t - 2\sin 4t) \\ -e^{-2t} - 4e^{2t}\sin 4t \\ -4e^{2t}\sin 4t \end{pmatrix}.$$

31. In Exercise 25, the general solution was

$$\begin{pmatrix} x(t) \\ y(t) \\ z(t) \end{pmatrix} = C_1 e^{-2t} \begin{pmatrix} 0 \\ 0 \\ 1 \end{pmatrix} + C_2 e^{-2t} \begin{pmatrix} -5\cos t - \sin t \\ 2\cos t \\ 3\cos t + \sin t \end{pmatrix} + C_3 e^{-2t} \begin{pmatrix} \cos t - 5\sin t \\ 2\sin t \\ -\cos t + 3\sin t \end{pmatrix}.$$

If $\mathbf{y}(0) = (-1, 1, 1)^T$, then

$$\begin{pmatrix} -1 \\ 1 \\ 1 \end{pmatrix} = C_1 \begin{pmatrix} 0 \\ 0 \\ 1 \end{pmatrix} + C_2 \begin{pmatrix} -5 \\ 2 \\ 3 \end{pmatrix} + C_3 \begin{pmatrix} 1 \\ 0 \\ -1 \end{pmatrix}.$$

The augmented matrix reduces.

$$\begin{pmatrix} 0 & -5 & 1 & -1 \\ 0 & 2 & 0 & 1 \\ 1 & 3 & -1 & 1 \end{pmatrix} \rightarrow \begin{pmatrix} 1 & 0 & 0 & 1 \\ 0 & 1 & 0 & 1/2 \\ 0 & 0 & 1 & 3/2 \end{pmatrix}$$

Thus, $C_1 = 1$, $C_2 = 1/2$ and $C_3 = 3/2$, giving the particular solution

$$\mathbf{y}(t) = e^{-2t} \begin{pmatrix} -\cos t - \sin t \\ \cos t + 3\sin t \\ 1 + 5\sin t \end{pmatrix}.$$

33. In matrix form,

$$\begin{pmatrix} x \\ y \\ z \end{pmatrix}' = \begin{pmatrix} 1 & 0 & 0 \\ 1 & 1 & 0 \\ -10 & 8 & 5 \end{pmatrix} \begin{pmatrix} x \\ y \\ z \end{pmatrix}.$$

Using a computer, matrix

$$A = \begin{pmatrix} 1 & 0 & 0 \\ 1 & 1 & 0 \\ -10 & 8 & 5 \end{pmatrix}$$

has characteristic polynomial

$$p(\lambda) = (\lambda - 1)^2(\lambda - 5).$$

Thus, A has eigenvalues 1 and 5 with algebraic multiplicities 2 and 1, respectively. For the eigenvalue 1, we look for a vector in the nullspace (eigenspace) of

$$A - I = \begin{pmatrix} 0 & 0 & 0 \\ 1 & 0 & 0 \\ -10 & 8 & 4 \end{pmatrix} \rightarrow \begin{pmatrix} 1 & 0 & 0 \\ 0 & 1 & 1/2 \\ 0 & 0 & 0 \end{pmatrix}.$$

Note that there is one free variable and the eigenspace is generated by the single eigenvector $(0, 1, -2)^T$. Therefore, the eigenvalue 1 has geometric multiplicity 1. For the eigenvalue 5, we look for a vector in the nullspace (eigenspace) of

$$A - 5I = \begin{pmatrix} -4 & 0 & 0 \\ 1 & -4 & 0 \\ -10 & 8 & 0 \end{pmatrix} \to \begin{pmatrix} 1 & 0 & 0 \\ 0 & 1 & 0 \\ 0 & 0 & 0 \end{pmatrix}.$$

Note again that there is only one free variable and the eigenspace is generated by the single eigenvector $(0, 0, 1)^T$. Therefore, the eigenvalue 5 has geometric multiplicity 1. Consequently, there are not enough independent eigenvectors to form a fundamental solution set.

35. In matrix form,

$$\begin{pmatrix} x \\ y \\ z \end{pmatrix}' = \begin{pmatrix} 4 & 0 & 0 \\ -6 & -2 & 0 \\ 7 & 1 & -2 \end{pmatrix} \begin{pmatrix} x \\ y \\ z \end{pmatrix}.$$

Using a computer, matrix

$$A = \begin{pmatrix} 4 & 0 & 0 \\ -6 & -2 & 0 \\ 7 & 1 & -2 \end{pmatrix}$$

has characteristic polynomial

$$p(\lambda) = (\lambda - 4)(\lambda + 2)^2.$$

Thus, A has eigenvalues 4 and -2 with algebraic multiplicities 1 and 2, respectively. For the eigenvalue 4, we look for a vector in the nullspace (eigenspace) of

$$A - 4I = \begin{pmatrix} 0 & 0 & 0 \\ -6 & -6 & 0 \\ 7 & 1 & -6 \end{pmatrix} \to \begin{pmatrix} 1 & 0 & -1 \\ 0 & 1 & 1 \\ 0 & 0 & 0 \end{pmatrix}.$$

Note that there is one free variable and the eigenspace is generated by the single eigenvector $(1, -1, 1)^T$. Therefore, the eigenvalue 4 has geometric multiplicity 1. For the eigenvalue -2, we look for a vector in the nullspace (eigenspace) of

$$A + 2I = \begin{pmatrix} 6 & 0 & 0 \\ -6 & 0 & 0 \\ 7 & 1 & 0 \end{pmatrix} \to \begin{pmatrix} 1 & 0 & 0 \\ 0 & 1 & 0 \\ 0 & 0 & 0 \end{pmatrix}.$$

Note again that there is only one free variable and the eigenspace is generated by the single eigenvector $(0, 0, 1)^T$. Therefore, the eigenvalue -4 has geometric multiplicity 1. Consequently, there are not enough independent eigenvectors to form a fundamental solution set.

37. Using a computer, matrix

$$A = \begin{pmatrix} -6 & 2 & -3 \\ -1 & -1 & -1 \\ 4 & -2 & 1 \end{pmatrix}$$

has eigenvalue–eigenvector pairs

$$-2 \to \begin{pmatrix} 1 \\ -1 \\ -2 \end{pmatrix}, \quad -3 \to \begin{pmatrix} -1 \\ 0 \\ 1 \end{pmatrix},$$

and

$$-1 \to \begin{pmatrix} -1 \\ -1 \\ 1 \end{pmatrix}.$$

Therefore,

$$\mathbf{y}_1(t) = e^{-2t} \begin{pmatrix} 1 \\ -1 \\ -2 \end{pmatrix}, \mathbf{y}_2(t) = e^{-3t} \begin{pmatrix} -1 \\ 0 \\ 1 \end{pmatrix}, \quad \text{and}$$

$$\mathbf{y}_3(t) = e^{-t} \begin{pmatrix} -1 \\ -1 \\ 1 \end{pmatrix}$$

form a fundamental set of solutions.

39. Using a computer, matrix

$$A = \begin{pmatrix} 8 & 12 & -4 \\ -9 & -13 & 4 \\ -1 & -3 & 0 \end{pmatrix}$$

has eigenvalue–eigenvector pairs

$$-1 \to \begin{pmatrix} 0 \\ 1 \\ 3 \end{pmatrix}, \quad -2 + 2i \to \begin{pmatrix} -2 \\ 2 \\ 1 + i \end{pmatrix}, \quad \text{and}$$

$$-2 - 2i \to \begin{pmatrix} -2 \\ 2 \\ 1 - i \end{pmatrix}.$$

Therefore,

$$\mathbf{y}_1(t) = e^{-t}\begin{pmatrix}0\\1\\3\end{pmatrix}$$

is a solution. Because

$$\begin{aligned}\mathbf{z}(t) &= e^{(-2+2i)t}\begin{pmatrix}-2\\2\\1+i\end{pmatrix}\\ &= e^{-2t}(\cos 2t + i\sin 2t)\left[\begin{pmatrix}-2\\2\\1\end{pmatrix} + i\begin{pmatrix}0\\0\\1\end{pmatrix}\right]\\ &= e^{-2t}\begin{pmatrix}-2\cos 2t\\2\cos 2t\\\cos 2t - \sin 2t\end{pmatrix}\\ &\quad + ie^{-2t}\begin{pmatrix}-2\sin 2t\\2\sin 2t\\\cos 2t + \sin 2t\end{pmatrix},\end{aligned}$$

the set $\mathbf{y}_1(t) = e^{-t}(0, 1, 3)^T$,

$$\mathbf{y}_2(t) = e^{-2t}\begin{pmatrix}-2\cos 2t\\2\cos 2t\\\cos 2t - \sin 2t\end{pmatrix},$$

and

$$\mathbf{y}_3(t) = e^{-2t}\begin{pmatrix}-2\sin 2t\\2\sin 2t\\\cos 2t + \sin 2t\end{pmatrix}$$

forms a fundamental set of solutions.

41. Using a computer, matrix

$$A = \begin{pmatrix}-18 & -18 & 10\\18 & 17 & -10\\10 & 10 & -7\end{pmatrix}$$

has eigenvalue–eigenvector pairs

$$-2 \to \begin{pmatrix}1\\-2\\-2\end{pmatrix}, \quad -3+2i \to \begin{pmatrix}-6+2i\\8-i\\5\end{pmatrix},$$

$$-3-2i \to \begin{pmatrix}-6-2i\\8+i\\5\end{pmatrix}.$$

Therefore,

$$\mathbf{y}_1(t) = e^{-2t}\begin{pmatrix}1\\-2\\-2\end{pmatrix}$$

is a solution. Because,

$$\begin{aligned}\mathbf{z}(t) &= e^{(-3+2i)t}\begin{pmatrix}-6+2i\\8-i\\5\end{pmatrix}\\ &= e^{-3t}(\cos 2t + i\sin 2t)\\ &\quad\times\left[\begin{pmatrix}-6\\8\\5\end{pmatrix} + i\begin{pmatrix}2\\-1\\0\end{pmatrix}\right]\\ &= e^{-3t}\begin{pmatrix}-6\cos 2t - 2\sin 2t\\8\cos 2t + \sin 2t\\5\cos 2t\end{pmatrix}\\ &\quad + ie^{-3t}\begin{pmatrix}2\cos 2t - 6\sin 2t\\-\cos 2t + 8\sin 2t\\5\sin 2t\end{pmatrix}.\end{aligned}$$

the set

$$\mathbf{y}_1(t) = e^{-2t}\begin{pmatrix}1\\-2\\-2\end{pmatrix},$$

$$\mathbf{y}_2(t) = e^{-3t}\begin{pmatrix}-6\cos 2t - 2\sin 2t\\8\cos 2t + \sin 2t\\5\cos 2t\end{pmatrix}, \quad \text{and}$$

$$\mathbf{y}_3(t) = e^{-3t}\begin{pmatrix}2\cos 2t - 6\sin 2t\\-\cos 2t + 8\sin 2t\\5\sin 2t\end{pmatrix}$$

forms a fundamental set of solutions.

43. The matrix

$$A = \begin{pmatrix}1 & 4 & 1 & -5\\-6 & -10 & -2 & 10\\3 & 4 & -1 & -5\\-3 & -4 & -1 & 3\end{pmatrix}$$

has characteristic polynomial

$$p(\lambda) = (\lambda+1)(\lambda+2)^3,$$

indicating eigenvalues -1 and -2, with algebraic multiplicities 1 and 3, respectively. The matrix

$$A+I=\begin{pmatrix}2&4&1&-5\\-6&-9&-2&10\\3&4&0&-5\\-3&-4&-1&4\end{pmatrix}\rightarrow\begin{pmatrix}1&0&0&1\\0&1&0&-2\\0&0&1&1\\0&0&0&0\end{pmatrix}$$

has one free variable, generating a single eigenvector and the solution

$$\mathbf{y}_1(t)=e^{-t}\begin{pmatrix}-1\\2\\-1\\1\end{pmatrix}.$$

The matrix

$$A+2I=\begin{pmatrix}3&4&1&-5\\-6&-8&-2&10\\3&4&1&-5\\-3&-4&-1&5\end{pmatrix}\rightarrow\begin{pmatrix}1&4/3&1/3&-5/3\\0&0&0&0\\0&0&0&0\\0&0&0&0\end{pmatrix}$$

has three free variables. A basis for the nullspace (eigenspace) of $A+2I$ contains the vectors

$$\begin{pmatrix}4\\-3\\0\\0\end{pmatrix},\quad\begin{pmatrix}1\\0\\-3\\0\end{pmatrix},\quad\text{and}\quad\begin{pmatrix}5\\0\\0\\3\end{pmatrix},$$

so, together with $\mathbf{y}_1(t)=e^{-t}(-1,2,-1,1)^T$,

$$\mathbf{y}_2(t)=e^{-2t}\begin{pmatrix}4\\-3\\0\\0\end{pmatrix},\quad\mathbf{y}_3(t)=e^{-2t}\begin{pmatrix}1\\0\\-3\\0\end{pmatrix},$$

$$\mathbf{y}_4(t)=e^{-2t}\begin{pmatrix}5\\0\\0\\3\end{pmatrix},$$

complete a fundamental set of solutions.

45. In Exercise 37, the fundamental set of solutions found there lead to the general solution

$$\mathbf{y}(t)=C_1e^{-2t}\begin{pmatrix}1\\-1\\-2\end{pmatrix}+C_2e^{-3t}\begin{pmatrix}-1\\0\\1\end{pmatrix}+C_3e^{-t}\begin{pmatrix}-1\\-1\\1\end{pmatrix}.$$

The initial condition $\mathbf{y}(0)=(-6,2,9)^T$ provides

$$\begin{pmatrix}-6\\2\\9\end{pmatrix}=C_1\begin{pmatrix}1\\-1\\-2\end{pmatrix}+C_2\begin{pmatrix}-1\\0\\1\end{pmatrix}+C_3\begin{pmatrix}-1\\-1\\1\end{pmatrix}.$$

The augmented matrix reduces.

$$\begin{pmatrix}1&-1&-1&-6\\-1&0&-1&2\\-2&1&1&9\end{pmatrix}\rightarrow\begin{pmatrix}1&0&0&-3\\0&1&0&2\\0&0&1&1\end{pmatrix}$$

Thus, $C_1=-3$, $C_2=2$ and $C_3=1$, leading to

$$\mathbf{y}(t)=\begin{pmatrix}-3e^{-2t}-2e^{-3t}-e^{-t}\\3e^{-2t}-e^{-t}\\6e^{-2t}+2e^{-3t}+e^{-t}\end{pmatrix}.$$

47. In Exercise 39, the fundamental set of solutions found there lead to the general solution

$$\mathbf{y}(t)=C_1e^{-t}\begin{pmatrix}0\\1\\3\end{pmatrix}+C_2e^{-2t}\begin{pmatrix}-2\cos 2t\\2\cos 2t\\\cos 2t-\sin 2t\end{pmatrix}+C_3e^{-2t}\begin{pmatrix}-2\sin 2t\\2\sin 2t\\\cos 2t+\sin 2t\end{pmatrix}.$$

The initial condition $\mathbf{y}(0)=(0,8,5)^T$ provides

$$\begin{pmatrix}0\\8\\5\end{pmatrix}=C_1\begin{pmatrix}0\\1\\3\end{pmatrix}+C_2\begin{pmatrix}-2\\2\\1\end{pmatrix}+C_3\begin{pmatrix}0\\0\\1\end{pmatrix}.$$

The augmented matrix reduces.

$$\begin{pmatrix}0&-2&0&0\\1&2&0&8\\3&1&1&5\end{pmatrix}\rightarrow\begin{pmatrix}1&0&0&8\\0&1&0&0\\0&0&1&-19\end{pmatrix}$$

Thus, $C_1 = 8$, $C_2 = 0$ and $C_3 = -19$, leading to

$$\mathbf{y}(t) = \begin{pmatrix} 38e^{-2t}\sin 2t \\ 8e^{-t} - 38e^{-2t}\sin 2t \\ 24e^{-t} - 19e^{-2t}\cos 2t - 19e^{-2t}\sin 2t \end{pmatrix}.$$

49. In Exercise 41, the fundamental set of solutions found there lead to the general solution

$$\mathbf{y}(t) = C_1 e^{-2t}\begin{pmatrix} 1 \\ -2 \\ -2 \end{pmatrix} + C_2 e^{-3t}\begin{pmatrix} -6\cos 2t - 2\sin 2t \\ 8\cos 2t + \sin 2t \\ 5\cos 2t \end{pmatrix} + C_3 e^{-3t}\begin{pmatrix} 2\cos 2t - 6\sin 2t \\ -\cos 2t + 8\sin 2t \\ 5\sin 2t \end{pmatrix}.$$

The initial condition $\mathbf{y}(0) = (-1, 7, 3)^T$ provides

$$\begin{pmatrix} -1 \\ 7 \\ 3 \end{pmatrix} = C_1\begin{pmatrix} 1 \\ -2 \\ -2 \end{pmatrix} + C_2\begin{pmatrix} -6 \\ 8 \\ 5 \end{pmatrix} + C_3\begin{pmatrix} 2 \\ -1 \\ 0 \end{pmatrix}.$$

The augmented matrix reduces.

$$\begin{pmatrix} 1 & -6 & 2 & -1 \\ -2 & 8 & -1 & 7 \\ -2 & 5 & 0 & 3 \end{pmatrix} \rightarrow \begin{pmatrix} 1 & 0 & 0 & 7 \\ 0 & 1 & 0 & 17/5 \\ 0 & 0 & 1 & 31/5 \end{pmatrix}$$

Thus, $C_1 = 7$, $C_2 = 17/5$ and $C_3 = 31/5$, leading to

$$\mathbf{y}(t) = \begin{pmatrix} 7e^{-2t} - e^{-3t}(8\cos 2t - 44\sin 2t) \\ -14e^{-2t} + e^{-3t}(21\cos 2t + 53\sin 2t) \\ -14e^{-2t} + e^{-3t}(17\cos 2t + 31\sin 2t) \end{pmatrix}.$$

51. In Exercise 43, the fundamental set of solutions found there lead to the general solution

$$\mathbf{y}(t) = C_1 e^{-t}\begin{pmatrix} -1 \\ 2 \\ -1 \\ 1 \end{pmatrix} + C_2 e^{-2t}\begin{pmatrix} 4 \\ -3 \\ 0 \\ 0 \end{pmatrix} + C_3 e^{-2t}\begin{pmatrix} 1 \\ 0 \\ -3 \\ 0 \end{pmatrix} + C_4 e^{-2t}\begin{pmatrix} 5 \\ 0 \\ 0 \\ 3 \end{pmatrix}.$$

The initial condition $\mathbf{y}(0) = (-1, 5, 2, 4)^T$ provides

$$\begin{pmatrix} -1 \\ 5 \\ 2 \\ 4 \end{pmatrix} = C_1\begin{pmatrix} -1 \\ 2 \\ -1 \\ 1 \end{pmatrix} + C_2\begin{pmatrix} 4 \\ -3 \\ 0 \\ 0 \end{pmatrix} + C_3\begin{pmatrix} 1 \\ 0 \\ -3 \\ 0 \end{pmatrix} + C_4\begin{pmatrix} 5 \\ 0 \\ 0 \\ 3 \end{pmatrix}.$$

The augmented matrix reduces.

$$\begin{pmatrix} -1 & 4 & 1 & 5 & -1 \\ 2 & -3 & 0 & 0 & 5 \\ -1 & 0 & -3 & 0 & 2 \\ 1 & 0 & 0 & 3 & 4 \end{pmatrix} \rightarrow \begin{pmatrix} 1 & 0 & 0 & 0 & 1 \\ 0 & 1 & 0 & 0 & -1 \\ 0 & 0 & 1 & 0 & -1 \\ 0 & 0 & 0 & 1 & 1 \end{pmatrix}$$

Thus, $C_1 = 1$, $C_2 = -1$, $C_3 = -1$ and $C_4 = 1$, leading to

$$\mathbf{y}(t) = \begin{pmatrix} -e^{-t} \\ 2e^{-t} + 3e^{-2t} \\ -e^{-t} + 3e^{-2t} \\ e^{-t} + 3e^{-2t} \end{pmatrix}.$$

53. The rate at which the salt content is changing in the topmost tank is

$$\frac{dx_1}{dt} = \text{Rate in} - \text{Rate out} = -\frac{x_1}{20}.$$

The rate at which the salt content is changing in the second tank is

$$\frac{dx_2}{dt} = \text{Rate in} - \text{Rate out} = \frac{x_1}{20} - \frac{x_2}{15}.$$

The rate at which the salt content is changing in the bottom tank is

$$\frac{dx_3}{dt} = \text{Rate in} - \text{Rate out} = \frac{x_2}{15} - \frac{x_3}{10}.$$

In matrix-vector form, the system becomes

$$\begin{pmatrix} x_1 \\ x_2 \\ x_3 \end{pmatrix}' = \begin{pmatrix} -1/20 & 0 & 0 \\ 1/20 & -1/15 & 0 \\ 0 & 1/15 & -1/10 \end{pmatrix} \begin{pmatrix} x_1 \\ x_2 \\ x_3 \end{pmatrix}.$$

As the coefficient matrix is lower triangular, the eigenvalues lie on the diagonal: $-1/20$, $-1/15$, and $-1/10$. With $\lambda = -1/20$,

$$A - \lambda I = A + \frac{1}{25} I = \begin{pmatrix} 0 & 0 & 0 \\ 1/20 & -1/60 & 0 \\ 0 & 1/15 & -1/20 \end{pmatrix}$$

$$\xrightarrow{\text{rref}} \begin{pmatrix} 1 & 0 & -1/4 \\ 0 & 1 & -3/4 \\ 0 & 0 & 0 \end{pmatrix}.$$

Thus, $(1/4, 3/4, 1)^T$ is an eigenvector, but we will use a multiple of this to form the solution $\mathbf{x}_1(t) = e^{-t/20}(1, 3, 4)^T$. In similar fashion, we can use the remaining eigenvalues and eigenvectors to form the general solution

$$\mathbf{x}(t) = C_1 e^{-t/20} \begin{pmatrix} 1 \\ 3 \\ 4 \end{pmatrix} + C_2 e^{-t/15} \begin{pmatrix} 0 \\ 1 \\ 2 \end{pmatrix} + C_3 e^{-t/10} \begin{pmatrix} 0 \\ 0 \\ 1 \end{pmatrix}.$$

Now, with the initial condition, we get

$$\begin{pmatrix} 4 \\ 2 \\ 1 \end{pmatrix} = \mathbf{x}(0) = C_1 \begin{pmatrix} 1 \\ 3 \\ 4 \end{pmatrix} + C_2 \begin{pmatrix} 0 \\ 1 \\ 2 \end{pmatrix} + C_3 \begin{pmatrix} 0 \\ 0 \\ 1 \end{pmatrix},$$

with solutions $C_1 = 4$, $C_2 = -10$, and $C_3 = 5$. Thus, the solution is:

$$\mathbf{x}(t) = 4e^{-t/20} \begin{pmatrix} 1 \\ 3 \\ 4 \end{pmatrix} - 10e^{-t/15} \begin{pmatrix} 0 \\ 1 \\ 2 \end{pmatrix} + 5e^{-t/10} \begin{pmatrix} 0 \\ 0 \\ 1 \end{pmatrix}$$

The components are

$$x_1 = 4e^{-t/20}$$
$$x_2 = 12e^{-t/20} - 10e^{-t/15}$$
$$x_3 = 16e^{-t/20} - 20e^{-t/15} + 5e^{-t/10}.$$

and the plot of each component is shown below.

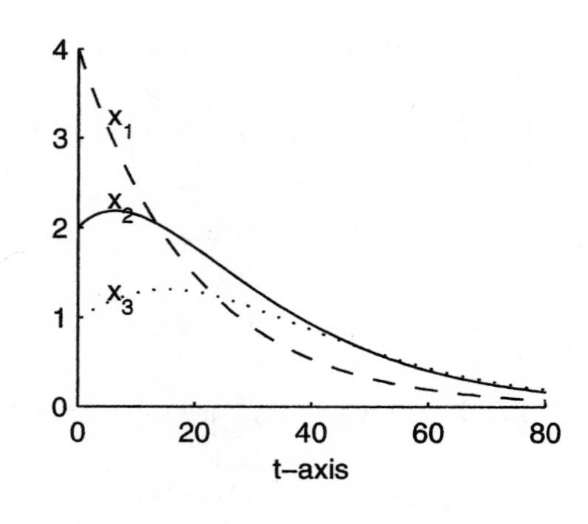

Section 9.6. The Exponential of a Matrix

1. It is easily checked that

$$A^2 = \begin{pmatrix} 0 & 0 \\ 0 & 0 \end{pmatrix}.$$

Therefore, the series

$$e^A = I + A + \frac{1}{2!}A^2 + \cdots$$

truncates and

$$e^A = I + A = \begin{pmatrix} 1 & 0 \\ 0 & 1 \end{pmatrix} + \begin{pmatrix} -2 & -4 \\ 1 & 2 \end{pmatrix}$$
$$= \begin{pmatrix} -1 & -4 \\ 1 & 3 \end{pmatrix}.$$

3. It is easily checked that

$$A^2 = \begin{pmatrix} 0 & 0 & 0 \\ 0 & 0 & 0 \\ 0 & 0 & 0 \end{pmatrix}.$$

Therefore, the series

$$e^A = I + A + \frac{1}{2!}A^2 + \cdots$$

truncates and

$$e^A = I + A = \begin{pmatrix} 1 & 0 & 0 \\ 0 & 1 & 0 \\ 0 & 0 & 1 \end{pmatrix} + \begin{pmatrix} 1 & -1 & 0 \\ 1 & -1 & 0 \\ 0 & 0 & 0 \end{pmatrix}$$
$$= \begin{pmatrix} 2 & -1 & 0 \\ 1 & 0 & 0 \\ 0 & 0 & 1 \end{pmatrix}.$$

5. (a) If $A^2 = \alpha A$, $\alpha \neq 0$, then

$$A^3 = AA^2 = A(\alpha A) = \alpha A^2 = \alpha(\alpha A)$$
$$= \alpha^2 A.$$

Similarly,

$$A^4 = AA^3 = A(\alpha^2 A) = \alpha^2 A^2$$
$$= \alpha^2(\alpha A) = \alpha^3 A.$$

Proceeding inductively,

$$A^k = \alpha^{k-1} A.$$

Now,

$$e^{tA} = I + tA + \frac{t^2}{2!}A^2 + \frac{t^3}{3!}A^3 + \cdots$$
$$= I + tA + \frac{t^2}{2!}(\alpha A) + \frac{t^3}{3!}(\alpha^2 A) + \cdots$$
$$= I + \left(t + \frac{\alpha t^2}{2!} + \frac{\alpha^2 t^3}{3!} + \cdots\right) A$$
$$= I + \frac{e^{\alpha t} - 1}{\alpha} A.$$

(b) One can easily show that

$$A^2 = \begin{pmatrix} 1 & 1 & 1 \\ 1 & 1 & 1 \\ 1 & 1 & 1 \end{pmatrix}^2 = \begin{pmatrix} 3 & 3 & 3 \\ 3 & 3 & 3 \\ 3 & 3 & 3 \end{pmatrix} = 3A.$$

Thus, we can apply the formula developed in part (a). With $\alpha = 3$:

$$e^{tA}$$
$$= I + \frac{e^{3t} - 1}{3} A$$
$$= \begin{pmatrix} 1 & 0 & 0 \\ 0 & 1 & 0 \\ 0 & 0 & 1 \end{pmatrix} + \frac{e^{3t} - 1}{3} \begin{pmatrix} 1 & 1 & 1 \\ 1 & 1 & 1 \\ 1 & 1 & 1 \end{pmatrix}$$
$$= \begin{pmatrix} (e^{3t} + 2)/3 & (e^{3t} - 1)/3 & (e^{3t} - 1)/3 \\ (e^{3t} - 1)/3 & (e^{3t} + 2/3 & (e^{3t} - 1)/3 \\ (e^{3t} - 1)/3 & (e^{3t} - 1)/3 & (e^{3t} + 2/3 \end{pmatrix}$$

7. Note that

$$A = \begin{pmatrix} a & -b \\ b & a \end{pmatrix} = \begin{pmatrix} a & 0 \\ 0 & a \end{pmatrix} + \begin{pmatrix} 0 & -b \\ b & 0 \end{pmatrix}$$
$$= aI + b\begin{pmatrix} 0 & -1 \\ 1 & 0 \end{pmatrix}.$$

Thus, by the result shown in Exercise 6,

$$e^{tA} = e^{atI + bt\left(\begin{smallmatrix} 0 & -1 \\ 1 & 0 \end{smallmatrix}\right)}$$
$$= e^{at} I e^{bt\left(\begin{smallmatrix} 0 & -1 \\ 1 & 0 \end{smallmatrix}\right)}$$
$$= e^{at} \begin{pmatrix} \cos bt & -\sin bt \\ \sin bt & \cos bt \end{pmatrix}.$$

9. (a) On the one hand

$$AB = \begin{pmatrix} -4 & 0 \\ 0 & 0 \end{pmatrix},$$

but

$$BA = \begin{pmatrix} 0 & 0 \\ 0 & -4 \end{pmatrix}.$$

(b) Note that if $t = 1$, the result from Exercise 7 becomes

$$e^{\begin{pmatrix} a & -b \\ b & a \end{pmatrix}} = e^a \begin{pmatrix} \cos b & -\sin b \\ \sin b & \cos b \end{pmatrix}.$$

Thus,

$$\begin{aligned} e^{A+B} &= e^{\begin{pmatrix} 0 & -2 \\ 0 & 0 \end{pmatrix} + \begin{pmatrix} 0 & 0 \\ 2 & 0 \end{pmatrix}} \\ &= e^{\begin{pmatrix} 0 & -2 \\ 2 & 0 \end{pmatrix}} \\ &= \begin{pmatrix} \cos 2 & -\sin 2 \\ \sin 2 & \cos 2 \end{pmatrix}. \end{aligned}$$

(c) Both A^2 and B^2 equal the zero matrix, so the series expansions for e^A and e^B truncate.

$$\begin{aligned} e^A &= I + A = \begin{pmatrix} 1 & 0 \\ 0 & 1 \end{pmatrix} + \begin{pmatrix} 0 & -2 \\ 0 & 0 \end{pmatrix} \\ &= \begin{pmatrix} 1 & -2 \\ 0 & 1 \end{pmatrix} \\ e^B &= I + B = \begin{pmatrix} 1 & 0 \\ 0 & 1 \end{pmatrix} + \begin{pmatrix} 0 & 0 \\ 2 & 0 \end{pmatrix} \\ &= \begin{pmatrix} 1 & 0 \\ 2 & 1 \end{pmatrix} \end{aligned}$$

Thus,

$$e^A e^B = \begin{pmatrix} 1 & -2 \\ 0 & 1 \end{pmatrix} \begin{pmatrix} 1 & 0 \\ 2 & 1 \end{pmatrix} = \begin{pmatrix} -3 & -2 \\ 2 & 1 \end{pmatrix},$$

which is not the same as e^{A+B} calculated in part (b). The problem arises because $AB \neq BA$, as was shown in part (a).

11. If

$$A = \begin{pmatrix} 2 & 6 \\ 0 & -1 \end{pmatrix},$$

then the characteristic polynomial is $p(\lambda) = (\lambda - 2)(\lambda + 1)$, giving eigenvalues $\lambda_1 = 2$ and $\lambda_2 = -1$. Set

$$D = \begin{pmatrix} 2 & 0 \\ 0 & -1 \end{pmatrix}.$$

The nullspace (eigenspace) of

$$A - 2I = \begin{pmatrix} 0 & 6 \\ 0 & -3 \end{pmatrix}$$

is generated by the single eigenvector $\mathbf{v}_1 = (1, 0)^T$. The nullspace of

$$A + I = \begin{pmatrix} 3 & 6 \\ 0 & 0 \end{pmatrix}$$

is generated by the single eigenvector $\mathbf{v}_1 = (2, -1)^T$. Set

$$P = \begin{pmatrix} 1 & 2 \\ 0 & -1 \end{pmatrix}.$$

It is easily checked that

$$P^{-1} = \begin{pmatrix} 1 & 2 \\ 0 & -1 \end{pmatrix} \quad \text{and} \quad A = PDP^{-1}.$$

Now,

$$\begin{aligned} e^{tA} &= Pe^{tD}P^{-1} \\ &= \begin{pmatrix} 1 & 2 \\ 0 & -1 \end{pmatrix} e^{\begin{pmatrix} 2t & 0 \\ 0 & -t \end{pmatrix}} \begin{pmatrix} 1 & 2 \\ 0 & -1 \end{pmatrix} \\ &= \begin{pmatrix} 1 & 2 \\ 0 & -1 \end{pmatrix} \begin{pmatrix} e^{2t} & 0 \\ 0 & e^{-t} \end{pmatrix} \begin{pmatrix} 1 & 2 \\ 0 & -1 \end{pmatrix} \\ &= \begin{pmatrix} 1 & 2 \\ 0 & -1 \end{pmatrix} \begin{pmatrix} e^{2t} & 2e^{2t} \\ 0 & -e^{-t} \end{pmatrix} \\ &= \begin{pmatrix} e^{2t} & 2e^{2t} - 2e^{-t} \\ 0 & e^{-t} \end{pmatrix}. \end{aligned}$$

13. If a 2×2 matrix A has a single eigenvalue λ_1, then the characteristic polynomial must be $p(\lambda) = (\lambda - \lambda_1)^2$. According to Cayley's Theorem, any matrix must satisfy its characteristic equation, so $(A - \lambda_1 I)^2 = 0$. Further, we know that We know that

$$\begin{aligned} e^{At} &= e^{\lambda_1 It + (A - \lambda_1 I)t} \\ &= e^{\lambda_1 It} e^{(A - \lambda_1)t} \\ &= e^{\lambda_1 t} I \Big[I + (A - \lambda_1 I)t \\ &\qquad + \frac{1}{2}(A - \lambda_1)^2 t^2 + \cdots \Big]. \end{aligned}$$

But, from the above, $(A - \lambda_1 I)^k = 0$, for $k \geq 2$. Hence,

$$e^{At} = e^{\lambda_1 t} \left[I + (A - \lambda_1 I)t \right].$$

15. Matrix

$$A = \begin{pmatrix} -1 & 0 \\ 1 & -1 \end{pmatrix}$$

has characteristic polynomial $p(\lambda) = (\lambda + 1)^2$ and repeated eigenvalue $\lambda = -1$. We can write

$$\begin{aligned} e^{tA} &= e^{t(-I+(A+I))} \\ &= e^{-tI}e^{t(A+I)} \\ &= e^{-t}\left(I + t(A+I) + \frac{t^2}{2!}(A+I)^2 + \cdots\right). \end{aligned}$$

Matrix A must satisfy its characteristic polynomial, so $(A+I)^2 = 0$ and $(A+I)^k = 0$ for $k \geq 2$. Thus, the series truncates.

$$\begin{aligned} e^{tA} &= e^{-t}(I + t(A+I)) \\ &= e^{-t}\left(\begin{pmatrix} 1 & 0 \\ 0 & 1 \end{pmatrix} + t\begin{pmatrix} 0 & 0 \\ 1 & 0 \end{pmatrix}\right) \\ &= e^{-t}\begin{pmatrix} 1 & 0 \\ t & 1 \end{pmatrix}. \end{aligned}$$

17. Matrix

$$A = \begin{pmatrix} -3 & -1 \\ 4 & 1 \end{pmatrix}$$

has characteristic polynomial $p(\lambda) = (\lambda + 1)^2$ and repeated eigenvalue $\lambda = -1$. We can write

$$\begin{aligned} e^{tA} &= e^{t(-I+(A+I))} \\ &= e^{-tI}e^{t(A+I)} \\ &= e^{-t}\left(I + t(A+I) + \frac{t^2}{2!}(A+I)^2 + \cdots\right). \end{aligned}$$

Matrix A must satisfy its characteristic polynomial, so $(A+I)^2 = 0$ and $(A+I)^k = 0$ for $k \geq 2$. Thus the series truncates.

$$\begin{aligned} e^{tA} &= e^{-t}(I + t(A+I)) \\ &= e^{-t}\left(\begin{pmatrix} 1 & 0 \\ 0 & 1 \end{pmatrix} + t\begin{pmatrix} -2 & -1 \\ 4 & 2 \end{pmatrix}\right) \\ &= e^{-t}\begin{pmatrix} 1-2t & -t \\ 4t & 1+2t \end{pmatrix}. \end{aligned}$$

19. Using a computer we find that A has eigenvalue -1 with algebraic multiplicity 3. We also find that

$$A + I = \begin{pmatrix} 0 & -1 & 0 \\ -1 & 1 & -1 \\ -1 & 2 & -1 \end{pmatrix},$$

$$(A+I)^2 = \begin{pmatrix} 1 & -1 & 1 \\ 0 & 0 & 0 \\ -1 & 1 & -1 \end{pmatrix},$$

and

$$(A+I)^3 = \begin{pmatrix} 0 & 0 & 0 \\ 0 & 0 & 0 \\ 0 & 0 & 0 \end{pmatrix}.$$

Thus

$$\begin{aligned} &e^{tA} \\ &= e^{\lambda t}e^{t(A-\lambda I)} \\ &= e^{-t}e^{t(A+I)} \\ &= e^{-t}\left[I + t(A+I) + \frac{t^2}{2}(A+I)^2\right] \\ &= e^{-t}\left[\begin{pmatrix} 1 & 0 & 0 \\ 0 & 1 & 0 \\ 0 & 0 & 1 \end{pmatrix} + t\begin{pmatrix} 0 & -1 & 0 \\ -1 & 1 & -1 \\ -1 & 2 & -1 \end{pmatrix}\right. \\ &\quad \left. + \frac{t^2}{2}\begin{pmatrix} 1 & -1 & 1 \\ 0 & 0 & 0 \\ -1 & 1 & -1 \end{pmatrix}\right] \\ &= e^{-t}\begin{pmatrix} 1+t^2/2 & -t-t^2/2 & t^2/2 \\ -t & 1+t & -t \\ -t-t^2/2 & 2t+t^2/2 & 1-t-t^2/2 \end{pmatrix} \end{aligned}$$

21. Using a computer, matrix

$$A = \begin{pmatrix} -2 & 0 & 0 \\ 0 & -2 & 0 \\ -1 & 1 & -2 \end{pmatrix}$$

has characteristic polynomial $p(\lambda) = (\lambda + 2)^3$ and repeated eigenvalue -2. We can write

$$\begin{aligned} &e^{tA} \\ &= e^{t(-2I+(A+2I))} \\ &= e^{-2tI}e^{t(A+2I)} \\ &= e^{-2t}\left(I + t(A+2I) + \frac{t^2}{2!}(A+2I)^2 + \cdots\right) \end{aligned}$$

Matrix A must satisfy its characteristic polynomial, so $(A+2I)^3 = 0$, but

$$A + 2I = \begin{pmatrix} 0 & 0 & 0 \\ 0 & 0 & 0 \\ -1 & 1 & 0 \end{pmatrix} \quad \text{and}$$

$$(A+2I)^2 = \begin{pmatrix} 0 & 0 & 0 \\ 0 & 0 & 0 \\ 0 & 0 & 0 \end{pmatrix},$$

so $(A+2I)^2 = 0$ for $k \geq 2$ and the series truncates at this point.

$$\begin{aligned} e^{tA} &= e^{-2t}(I + t(A+2I)) \\ &= e^{-2t}\left(\begin{pmatrix} 1 & 0 & 0 \\ 0 & 1 & 0 \\ 0 & 0 & 1 \end{pmatrix} + t\begin{pmatrix} 0 & 0 & 0 \\ 0 & 0 & 0 \\ -1 & 1 & 0 \end{pmatrix}\right) \\ &= e^{-2t}\begin{pmatrix} 1 & 0 & 0 \\ 0 & 1 & 0 \\ -t & t & 1 \end{pmatrix}. \end{aligned}$$

23. Using a computer, matrix

$$A = \begin{pmatrix} -5 & 0 & -1 & 4 \\ -4 & 0 & 1 & 5 \\ 4 & -4 & -5 & -4 \\ 0 & -1 & -1 & -2 \end{pmatrix}$$

has characteristic polynomial $p(\lambda) = (\lambda+3)^4$ and repeated eigenvalue $\lambda = -3$. We can write

$$\begin{aligned} &e^{tA} \\ &= e^{t(-3I+(A+3I))} \\ &= e^{-3tI}e^{t(A+3I)} \\ &= e^{-3t}\left(I + t(A+3I) + \frac{t^2}{2!}(A+3I)^2 + \cdots\right) \end{aligned}$$

Matrix A must satisfy its characteristic polynomial, so $(A+3I)^4 = 0$, but

$$A + 3I = \begin{pmatrix} -2 & 0 & -1 & 4 \\ -4 & 3 & 1 & 5 \\ 4 & -4 & -2 & -4 \\ 0 & -1 & -1 & 1 \end{pmatrix} \quad \text{and}$$

$$(A+3I)^2 = \begin{pmatrix} 0 & 0 & 0 & 0 \\ 0 & 0 & 0 & 0 \\ 0 & 0 & 0 & 0 \\ 0 & 0 & 0 & 0 \end{pmatrix},$$

So $(A+3I)^k = 0$ for $k \geq 2$ and the series truncates at this point.

$$\begin{aligned} e^{tA} &= e^{-3t}(I + t(A+3I)) \\ &= e^{-3t}\begin{pmatrix} 1 & 0 & 0 & 0 \\ 0 & 1 & 0 & 0 \\ 0 & 0 & 1 & 0 \\ 0 & 0 & 0 & 1 \end{pmatrix} \\ &\quad + te^{-3t}\begin{pmatrix} -2 & 0 & -1 & 4 \\ -4 & 3 & 1 & 5 \\ 4 & -4 & -2 & -4 \\ 0 & -1 & -1 & 1 \end{pmatrix} \\ &= e^{-3t}\begin{pmatrix} 1-2t & 0 & -t & 4t \\ -4t & 1+3t & t & 5t \\ 4t & -4t & 1-2t & -4t \\ 0 & -t & -t & 1+t \end{pmatrix}. \end{aligned}$$

25. Using a computer, matrix

$$A = \begin{pmatrix} 1 & 0 & 0 & 0 \\ -9 & 4 & 1 & 4 \\ 13 & -3 & -1 & -5 \\ 2 & -1 & 0 & 0 \end{pmatrix}$$

has characteristic polynomial $p(\lambda) = (\lambda-1)^4$ and repeated eigenvalue $\lambda = 1$. We can write

$$e^{tA} = e^{t(I+(A-I))} = e^{tI}e^{t(A-I)} = e^t\left(A + t(A-I) + \frac{t^2}{2!}(A-I)^2 + \cdots\right).$$

Matrix A must satisfy its characteristic polynomial, so $(A - I)^4 = 0$. But

$$A - I = \begin{pmatrix} 0 & 0 & 0 & 0 \\ -9 & 3 & 1 & 4 \\ 13 & -3 & -2 & -5 \\ 2 & -1 & 0 & -1 \end{pmatrix}, \quad (A - I)^2 = \begin{pmatrix} 0 & 0 & 0 & 0 \\ -6 & 2 & 1 & 3 \\ -9 & 2 & 1 & 3 \\ 7 & -2 & -1 & -3 \end{pmatrix}, \quad \text{and}$$

$$(A - I)^3 = \begin{pmatrix} 0 & 0 & 0 & 0 \\ 1 & 0 & 0 & 0 \\ 1 & 0 & 0 & 0 \\ -1 & 0 & 0 & 0 \end{pmatrix},$$

so $(A - I)^k = 0$ for $k \geq 4$ and the series truncates at this point.

$$\begin{aligned} e^{tA} &= \left(I + t(A - I) + \frac{t^2}{2!}(A - I)^2 + \frac{t^3}{3!}(A - I)^3\right) \\ &= e^t \left[\begin{pmatrix} 1 & 0 & 0 & 0 \\ 0 & 1 & 0 & 0 \\ 0 & 0 & 1 & 0 \\ 0 & 0 & 0 & 1 \end{pmatrix} + t \begin{pmatrix} 0 & 0 & 0 & 0 \\ -9 & 3 & 1 & 4 \\ 13 & -3 & -2 & -5 \\ 2 & -1 & 0 & -1 \end{pmatrix} \right. \\ &\qquad \left. + \frac{t^2}{2} \begin{pmatrix} 0 & 0 & 0 & 0 \\ -6 & 2 & 1 & 3 \\ -9 & 2 & 1 & 3 \\ 7 & -2 & -1 & -3 \end{pmatrix} + \frac{t^3}{6} \begin{pmatrix} 0 & 0 & 0 & 0 \\ 1 & 0 & 0 & 0 \\ 1 & 0 & 0 & 0 \\ -1 & 0 & 0 & 0 \end{pmatrix} \right] \\ &= e^t \begin{pmatrix} 1 & 0 & 0 & 0 \\ -9t - 3t^2 + t^3/6 & 1 + 3t + t^2 & t + t^2/2 & 4t + 3t^2/2 \\ 13t - 9t^2/2 + t^3/6 & -3t + t^2 & 1 - 2t + t^2/2 & -5t + 3t^2/2 \\ 2t + 7t^2/2 - t^3/6 & -t - t^2 & -t^2/2 & 1 - t - 3t^2/2 \end{pmatrix}. \end{aligned}$$

27. If

$$A = \begin{pmatrix} 1 & 0 & 1 \\ 2 & 2 & -2 \\ 0 & 0 & 2 \end{pmatrix},$$

then

$$\begin{aligned} p(\lambda) &= -\det(A - \lambda I) \\ &= -\begin{vmatrix} 1-\lambda & 0 & 1 \\ 2 & 2-\lambda & -2 \\ 0 & 0 & 2-\lambda \end{vmatrix}. \end{aligned}$$

Expanding across the third row,

$$p(\lambda) = (\lambda - 2)\begin{vmatrix} 1-\lambda & 0 \\ 2 & 2-\lambda \end{vmatrix} = (\lambda - 2)^2(\lambda - 1),$$

so the eigenvalues are 2 and 1, with algebraic multiplicities 2 and 1, respectively. For $\lambda_1 = 1$,

$$A - I = \begin{pmatrix} 0 & 0 & 1 \\ 2 & 1 & -2 \\ 0 & 0 & 1 \end{pmatrix} \xrightarrow{\text{rref}} \begin{pmatrix} 1 & 1/2 & 0 \\ 0 & 0 & 1 \\ 0 & 0 & 0 \end{pmatrix},$$

so the geometric multiplicity of λ_1 is 1 and an eigenvector is $\mathbf{v}_1 = (-1, 2, 0)^T$, providing exponential solution

$$\mathbf{y}_1(t) = e^{tA}\mathbf{v}_1 = e^t \begin{pmatrix} -1 \\ 2 \\ 0 \end{pmatrix}.$$

For $\lambda_2 = 2$,

$$A - 2I = \begin{pmatrix} -1 & 0 & 1 \\ 2 & 0 & -2 \\ 0 & 0 & 0 \end{pmatrix} \xrightarrow{\text{rref}} \begin{pmatrix} 1 & 0 & -1 \\ 0 & 0 & 0 \\ 0 & 0 & 0 \end{pmatrix},$$

so there are two free variables and the geometric multiplicity is 2. Thus, $\mathbf{v}_2 = (0, 1, 0)^T$ and $\mathbf{v}_3 = (1, 0, 1)^T$ are independent eigenvectors and

$$\mathbf{y}_2(t) = e^{tA}\mathbf{v}_2 = e^{2t}\begin{pmatrix}0\\1\\0\end{pmatrix} \quad \text{and}$$

$$\mathbf{y}_3(t) = e^{tA}\mathbf{v}_3 = e^{2t}\begin{pmatrix}1\\0\\1\end{pmatrix}$$

are independent solutions. Because

$$\det\left[\mathbf{y}_1(0), \mathbf{y}_2(0), \mathbf{y}_3(0)\right] = \begin{vmatrix}-1 & 0 & 1\\ 2 & 1 & 0\\ 0 & 0 & 1\end{vmatrix} = -1,$$

the solutions are independent for all t and form a fundamental set of solutions.

29. If

$$A = \begin{pmatrix}-1 & 0 & 0\\ 2 & -5 & -1\\ 0 & 4 & -1\end{pmatrix},$$

then

$$\begin{aligned} p(\lambda) &= -\det(A - \lambda I)\\ &= -\begin{vmatrix}-1-\lambda & 0 & 0\\ 2 & -5-\lambda & -1\\ 0 & 4 & -1-\lambda\end{vmatrix}.\end{aligned}$$

$$\begin{aligned} p(\lambda) &= (\lambda+1)\begin{vmatrix}-5-\lambda & -1\\ 4 & -1-\lambda\end{vmatrix}\\ &= (\lambda+1)(\lambda^2+6\lambda+9)\\ &= (\lambda+1)(\lambda+3)^2,\end{aligned}$$

providing eigenvalues $\lambda_1 = -1$ and $\lambda_2 = -3$, with algebraic multiplicities 1 and 2, respectively. Because

$$A + I = \begin{pmatrix}0 & 0 & 0\\ 2 & -4 & -1\\ 0 & 4 & 0\end{pmatrix} \xrightarrow{\text{rref}} \begin{pmatrix}1 & 0 & -1/2\\ 0 & 1 & 0\\ 0 & 0 & 0\end{pmatrix},$$

the geometric multiplicity of $\lambda_1 = -1$ is 1, and an eigenvector is $\mathbf{v}_1 = (1, 0, 2)^T$, providing the exponential solution

$$\mathbf{y}_1(t) = e^{tA}\mathbf{v}_1 = e^{-t}\begin{pmatrix}1\\0\\2\end{pmatrix}.$$

For $\lambda_2 = -3$

$$A + 3I = \begin{pmatrix}2 & 0 & 0\\ 2 & -2 & -1\\ 0 & 4 & 2\end{pmatrix} \xrightarrow{\text{rref}} \begin{pmatrix}1 & 0 & 0\\ 0 & 1 & 1/2\\ 0 & 0 & 0\end{pmatrix}$$

has one free variable, so the geometric multiplicity of $\lambda_2 = -3$ is 1. An eigenvector is $\mathbf{v}_2 = (0, -1, 2)^T$, giving a second exponential solution,

$$\mathbf{y}_2(t) = e^{tA}\mathbf{v}_2 = e^{-3t}\begin{pmatrix}0\\-1\\2\end{pmatrix}.$$

Next,

$$(A + 3I)^2 = \begin{pmatrix}4 & 0 & 0\\ 0 & 0 & 0\\ 8 & 0 & 0\end{pmatrix} \xrightarrow{\text{rref}} \begin{pmatrix}1 & 0 & 0\\ 0 & 0 & 0\\ 0 & 0 & 0\end{pmatrix}$$

has dimension 2, equaling the algebraic multiplicity of λ_2. Thus, we can pick a vector in the nullspace of $(A + 3I)^2$ that is not in the nullspace of $A + 3I$. Choose $\mathbf{v}_3 = (0, 1, 0)^T$, which is not a multiple of $\mathbf{v}_2$, making the set $\{\mathbf{v}_2, \mathbf{v}_3\}$ independent and giving a third solution,

$$\begin{aligned}\mathbf{y}_3(t) &= e^{tA}\mathbf{v}_3\\ &= e^{-3t}\left[\mathbf{v}_3 + t(A+3I)\mathbf{v}_3\right]\\ &= e^{-3t}\left[\begin{pmatrix}0\\1\\0\end{pmatrix} + t\begin{pmatrix}2 & 0 & 0\\ 2 & -2 & -1\\ 0 & 4 & 2\end{pmatrix}\begin{pmatrix}0\\1\\0\end{pmatrix}\right]\\ &= e^{-3t}\left[\begin{pmatrix}0\\1\\0\end{pmatrix} + t\begin{pmatrix}0\\-2\\4\end{pmatrix}\right]\\ &= e^{-3t}\begin{pmatrix}0\\1-2t\\4t\end{pmatrix}.\end{aligned}$$

Because

$$\det\left[\mathbf{y}_1(0), \mathbf{y}_2(0), \mathbf{y}_3(0)\right] = \begin{vmatrix}1 & 0 & 0\\ 0 & -1 & 1\\ 2 & 2 & 0\end{vmatrix} = -2,$$

the solutions are independent for all t and form a fundamental set of solutions.

31. Using a computer, matrix

$$A = \begin{pmatrix} 18 & -7 & 24 & 24 \\ 15 & -8 & 20 & 16 \\ 0 & 0 & -1 & 0 \\ -12 & 4 & -15 & -17 \end{pmatrix}$$

has characteristic polynomial $p(\lambda) = (\lambda + 3)^2(\lambda + 1)^2$, providing eigenvalues $\lambda_1 = -3$ and $\lambda_2 = -1$, with algebraic multiplicities 2 and 2, respectively. Because

$$A + 3I = \begin{pmatrix} 21 & -7 & 24 & 24 \\ 15 & -5 & 20 & 16 \\ 0 & 0 & 2 & 0 \\ -12 & 4 & -15 & -14 \end{pmatrix} \xrightarrow{\text{rref}} \begin{pmatrix} 1 & -1/3 & 0 & 0 \\ 0 & 0 & 1 & 0 \\ 0 & 0 & 0 & 1 \\ 0 & 0 & 0 & 0 \end{pmatrix},$$

the geometric multiplicity of $\lambda_1 = -3$ is 1, and an eigenvector is $\mathbf{v}_1 = (1, 3, 0, 0)^T$, giving solution

$$\mathbf{y}_1(t) = e^{tA}\mathbf{v}_1 = e^{-3t}\begin{pmatrix} 1 \\ 3 \\ 0 \\ 0 \end{pmatrix}.$$

Next, the nullspace of

$$(A + 3I)^2 = \begin{pmatrix} 48 & -16 & 52 & 56 \\ 48 & -16 & 60 & 56 \\ 0 & 0 & 4 & 0 \\ -24 & 8 & -28 & -28 \end{pmatrix} \xrightarrow{\text{rref}} \begin{pmatrix} 1 & -1/3 & 0 & 7/6 \\ 0 & 0 & 1 & 0 \\ 0 & 0 & 0 & 0 \\ 0 & 0 & 0 & 0 \end{pmatrix}$$

has dimension 2, equalling the algebraic multiplicity of λ_1. Thus, we can pick a vector in the nullspace of $(A + 3I)^2$ that is not in the nullspace of $A + 3I$. Choose $\mathbf{v}_2 = (-7, 0, 0, 6)^T$, which is not a multiple of $\mathbf{v}_1$., making the set $\{\mathbf{v}_1, \mathbf{v}_2\}$ independent, and giving a second solution

$$\begin{aligned} \mathbf{y}_2(t) &= e^{tA}\mathbf{v}_2 \\ &= e^{-3t}\left[\mathbf{v}_2 + t(A + 3I)\mathbf{v}_2\right] \\ &= e^{-3t}\left[\begin{pmatrix} -7 \\ 0 \\ 0 \\ 6 \end{pmatrix} + t\begin{pmatrix} -3 \\ -9 \\ 0 \\ 0 \end{pmatrix}\right] \\ &= e^{-3t}\begin{pmatrix} -7 - 3t \\ -9t \\ 0 \\ 6 \end{pmatrix}. \end{aligned}$$

For $\lambda_2 = -1$,

$$A + I = \begin{pmatrix} 19 & -7 & 24 & 24 \\ 15 & -7 & 20 & 16 \\ 0 & 0 & 0 & 0 \\ -12 & 4 & -15 & -16 \end{pmatrix} \xrightarrow{\text{rref}} \begin{pmatrix} 1 & 0 & 0 & 2 \\ 0 & 1 & 0 & 2 \\ 0 & 0 & 1 & 0 \\ 0 & 0 & 0 & 0 \end{pmatrix},$$

the geometric multiplicity of $\lambda_2 = -1$ is 1, and an eigenvector is $\mathbf{v}_3 = (-2, -2, 0, 1)^T$, giving the exponential solution

$$\mathbf{y}_3(t) = e^{tA}\mathbf{v}_3 = e^{-t}\begin{pmatrix} -2 \\ -2 \\ 0 \\ 1 \end{pmatrix}.$$

Next, the nullspace of

$$(A + I)^2 = \begin{pmatrix} -32 & 12 & -44 & -40 \\ -12 & 8 & -20 & -8 \\ 0 & 0 & 0 & 0 \\ 24 & -8 & 32 & 32 \end{pmatrix} \xrightarrow{\text{rref}} \begin{pmatrix} 1 & 0 & 1 & 2 \\ 0 & 1 & -1 & 2 \\ 0 & 0 & 0 & 0 \\ 0 & 0 & 0 & 0 \end{pmatrix}$$

has dimension 2, equaling the algebraic multiplicity of λ_2, so we can pick a vector in the nullspace of $(A + I)^2$ that is not in the nullspace of $A + I$. Choose $\mathbf{v}_4 = (-1, 1, 1, 0)^T$, which is not a multiple of $\mathbf{v}_3$, making the set $\{\mathbf{v}_3, \mathbf{v}_4\}$ independent, and giving a fourth solution.

$$\begin{aligned}
\mathbf{y}_4(t) &= e^{tA}\mathbf{v}_4 \\
&= e^{-t}\left[\mathbf{v}_4 + t(A+I)\mathbf{v}_4\right] \\
&= e^{-t}\left[\begin{pmatrix}-1\\1\\1\\0\end{pmatrix} + t\begin{pmatrix}-2\\-2\\0\\1\end{pmatrix}\right] \\
&= e^{-t}\begin{pmatrix}-1-2t\\1-2t\\1\\t\end{pmatrix}.
\end{aligned}$$

Because

$$\begin{aligned}
\det&\left[\mathbf{y}_1(0), \mathbf{y}_2(0), \mathbf{y}_3(0), \mathbf{y}_4(0)\right] \\
&= \begin{vmatrix}1 & -7 & -2 & -1\\3 & 0 & -2 & 1\\0 & 0 & 0 & 1\\0 & 6 & 1 & 0\end{vmatrix} \\
&= 3,
\end{aligned}$$

the solutions are independent for all t and form a fundamental set of solutions.

33. Using a computer, matrix

$$A = \begin{pmatrix}2 & 0 & 0 & 0 & 0 & 1\\-14 & -2 & -7 & 11 & -9 & -8\\-9 & -3 & -3 & 7 & -6 & -4\\-19 & -5 & -9 & 17 & -12 & -9\\-29 & -7 & -13 & 23 & -16 & -15\\19 & 5 & 9 & -15 & 12 & 11\end{pmatrix}$$

has characteristic polynomial

$$p(\lambda) = (\lambda-1)^3(\lambda-2)^3,$$

providing eigenvalues $\lambda_1 = 1$ and $\lambda_2 = 2$, with algebraic multiplicities 3 and 3, respectively. Using a computer, the nullspace of $A - I$ has dimension one, as $A - I$ reduces to

$$A - I = \begin{pmatrix}1 & 0 & 0 & 0 & 0 & 1\\0 & 1 & 0 & 0 & 0 & 0\\0 & 0 & 1 & 0 & 0 & 2\\0 & 0 & 0 & 1 & 0 & 1\\0 & 0 & 0 & 0 & 1 & -1\\0 & 0 & 0 & 0 & 0 & 0\end{pmatrix}.$$

Thus, $\lambda_1 = 1$ has geometric multiplicity 1. Using a computer to reduce $(A-I)^2$, you can check that the nullspace of $(A-I)^2$ has dimension 2. The key here is that $(A-I)^3$ reduces to

$$(A-I)^3 \xrightarrow{\text{rref}} \begin{pmatrix}1 & 0 & 1 & 0 & 1 & 2\\0 & 1 & -2 & 0 & -3/2 & -5/2\\0 & 0 & 0 & 1 & 0 & 1\\0 & 0 & 0 & 0 & 0 & 0\\0 & 0 & 0 & 0 & 0 & 0\\0 & 0 & 0 & 0 & 0 & 0\end{pmatrix}.$$

A basis for the nullspace is provided by the vectors.

$$\mathbf{v}_1 = \begin{pmatrix}-1\\2\\1\\0\\0\\0\end{pmatrix}, \quad \mathbf{v}_2 = \begin{pmatrix}-1\\3/2\\0\\0\\1\\0\end{pmatrix}, \quad \text{and}$$

$$\mathbf{v}_3 = \begin{pmatrix}-2\\5/2\\0\\-1\\0\\1\end{pmatrix}$$

Because $\mathbf{v}_1$, $\mathbf{v}_2$, and $\mathbf{v}_3$ are in the nullspace of $(A-I)^3$ we know that

$$\mathbf{y}(t) = e^{At}\mathbf{v} = \left[\mathbf{v} + t(A-I)\mathbf{v} + \frac{t^2}{2}(A-I)\mathbf{v}\right]$$

for each $\mathbf{v} = \mathbf{v}_1$, $\mathbf{v}_2$, and $\mathbf{v}_3$. This fact, and a computer, provide the following solutions.

$$\mathbf{y}_1(t) = e^{tA}\mathbf{v}_1 = e^t\begin{pmatrix}-1-t-t^2/2\\2+t\\1-t-t^2\\-t^2/2\\2t+t^2/2\\t^2/2\end{pmatrix},$$

$$\mathbf{y}_2(t) = e^{tA}\mathbf{v}_2 = e^t\begin{pmatrix}-1-t-t^2/4\\3/2+t/2\\-3t/2-t^2/2\\-t/2-t^2/4\\1+3t/2+t^2/4\\t/2+t^2/4\end{pmatrix},$$

and

$$\mathbf{y}_3(t) = e^{tA}\mathbf{v}_3 = e^t \begin{pmatrix} -2 - t - 3t^2/4 \\ 5/2 + 3t/2 \\ -t/2 - 3t^2/2 \\ -1 + t/2 - 3t^2/4 \\ 10t/4 + 3t^2/4 \\ 1 - t/2 + 3t^2/4 \end{pmatrix}.$$

On the other hand, $A - 2I$ reduces to

$$A - 2I \xrightarrow{\text{rref}} \begin{pmatrix} 1 & 0 & 1/6 & -5/6 & 1/2 & 0 \\ 0 & 1 & 7/6 & 1/6 & 1/2 & 0 \\ 0 & 0 & 0 & 0 & 0 & 1 \\ 0 & 0 & 0 & 0 & 0 & 0 \\ 0 & 0 & 0 & 0 & 0 & 0 \\ 0 & 0 & 0 & 0 & 0 & 0 \end{pmatrix},$$

so the nullspace of $A - 2I$ has dimension 3, and the geometric multiplicity of $\lambda_2 = 2$ is 3. A basis for the eigenspace contains the vectors

$$\mathbf{v}_4 = \begin{pmatrix} -1/6 \\ -7/6 \\ 1 \\ 0 \\ 0 \\ 0 \end{pmatrix}, \quad \mathbf{v}_5 = \begin{pmatrix} 5/6 \\ -1/6 \\ 0 \\ 1 \\ 0 \\ 0 \end{pmatrix}, \quad \text{and}$$

$$\mathbf{v}_6 = \begin{pmatrix} -1/2 \\ -1/2 \\ 0 \\ 0 \\ 1 \\ 0 \end{pmatrix}.$$

The corresponding solutions are

$$\mathbf{y}_4(t) = e^{tA}\mathbf{v}_4 = e^{2t} \begin{pmatrix} -1/6 \\ -7/6 \\ 1 \\ 0 \\ 0 \\ 0 \end{pmatrix},$$

$$\mathbf{y}_5(t) = e^{tA}\mathbf{v}_5 = e^{2t} \begin{pmatrix} 5/6 \\ -1/6 \\ 0 \\ 1 \\ 0 \\ 0 \end{pmatrix},$$

and

$$\mathbf{y}_6(t) = e^{tA}\mathbf{v}_6 = e^{2t} \begin{pmatrix} -1/2 \\ -1/2 \\ 0 \\ 0 \\ 1 \\ 0 \end{pmatrix}.$$

Because

$$\begin{aligned} &\det\left[\mathbf{y}_1(0), \mathbf{y}_2(0), \mathbf{y}_3(0), \mathbf{y}_4(0), \mathbf{y}_5(0), \mathbf{y}_6(0)\right] \\ &= \begin{vmatrix} -1 & -1 & -2 & -1/6 & 5/6 & -1/2 \\ 2 & 3/2 & 5/2 & -7/6 & -1/6 & -1/2 \\ 1 & 0 & 0 & 1 & 0 & 0 \\ 0 & 0 & -1 & 0 & 1 & 0 \\ 0 & 1 & 0 & 0 & 0 & 1 \\ 0 & 0 & 1 & 0 & 0 & 0 \end{vmatrix} \\ &= \frac{1}{12}, \end{aligned}$$

the solutions $\mathbf{y}_1(t)$, $\mathbf{y}_2(t)$, $\mathbf{y}_3(t)$, $\mathbf{y}_4(t)$, $\mathbf{y}_5(t)$, and $\mathbf{y}_6(t)$ are independent for all t and form a fundamental set of solutions.

35. If

$$A = \begin{pmatrix} 6 & 0 & -4 \\ -2 & 4 & 5 \\ 1 & 0 & 2 \end{pmatrix},$$

then the characteristic polynomial is found with the following computation

$$\begin{aligned} p(\lambda) &= -\det(A - \lambda I) \\ &= -\begin{vmatrix} 6-\lambda & 0 & -4 \\ -2 & 4-\lambda & 5 \\ 1 & 0 & 2-\lambda \end{vmatrix}. \end{aligned}$$

Expanding down the second column,

$$\begin{aligned} p(\lambda) &= (\lambda - 4)\begin{vmatrix} 6-\lambda & -4 \\ 1 & 2-\lambda \end{vmatrix} \\ &= (\lambda - 4)(\lambda^2 - 8\lambda + 16) \\ &= (\lambda - 4)^3. \end{aligned}$$

Because a matrix must satisfy it's characteristic, the series

$$\begin{aligned} e^{tA} &= e^{t[4I + (A-4I)]} \\ &= e^{4tI}e^{t(A-4I)} \\ &= e^{4t}\left[I + t(A - 4I) + \frac{t^2}{2!}(A - 4I)^2 + \cdots\right] \end{aligned}$$

truncates, with $(A - 4I)^k = 0$ for $k = 3, 4, \ldots$. Thus,

$$e^{tA} = e^{4t}\left[\begin{pmatrix}1 & 0 & 0\\ 0 & 1 & 0\\ 0 & 0 & 1\end{pmatrix} + t\begin{pmatrix}2 & 0 & -4\\ -2 & 0 & 5\\ 1 & 0 & -2\end{pmatrix} + \frac{t^2}{2!}\begin{pmatrix}0 & 0 & 0\\ 1 & 0 & -2\\ 0 & 0 & 0\end{pmatrix}\right]$$

$$= e^{4t}\begin{pmatrix}1+2t & 0 & -4t\\ -2t+t^2 & 1 & 5t-t^2\\ t & 0 & 1-2t\end{pmatrix}.$$

Choose

$$\mathbf{e}_1 = \begin{pmatrix}1\\0\\0\end{pmatrix}, \quad \mathbf{e}_2 = \begin{pmatrix}0\\1\\0\end{pmatrix}, \quad \text{and} \quad \mathbf{e}_3 = \begin{pmatrix}0\\0\\1\end{pmatrix}.$$

Then

$$\mathbf{y}_1(t) = e^{tA}\mathbf{e}_1 = e^{4t}\begin{pmatrix}1+2t\\ -2t+t^2/2\\ t\end{pmatrix},$$

$$\mathbf{y}_2(t) = e^{tA}\mathbf{e}_2 = e^{4t}\begin{pmatrix}0\\1\\0\end{pmatrix}, \quad \text{and}$$

$$\mathbf{y}_3(t) = e^{tA}\mathbf{e}_3 = e^{4t}\begin{pmatrix}-4t\\ 5t-t^2\\ 1-2t\end{pmatrix}$$

form a fundamental set of solutions.

37. In matrix form,

$$\begin{pmatrix}x\\y\\z\end{pmatrix}' = \begin{pmatrix}-2 & -4 & 13\\ 0 & 5 & -4\\ 0 & 1 & 1\end{pmatrix}\begin{pmatrix}x\\y\\z\end{pmatrix},$$

which leads to the characteristic polynomial

$$p(\lambda) = -\det(A - \lambda I)$$
$$= -\begin{vmatrix}-2-\lambda & -4 & 13\\ 0 & 5-\lambda & -4\\ 0 & 1 & 1-\lambda\end{vmatrix}.$$

Expanding down the first column,

$$p(\lambda) = (\lambda+2)\begin{vmatrix}5-\lambda & -4\\ 1 & 1-\lambda\end{vmatrix}$$
$$= (\lambda+2)(\lambda^2 - 6\lambda + 9)$$
$$= (\lambda+2)(\lambda-3)^2.$$

for $\lambda_1 = -2$,

$$A + 2I = \begin{pmatrix}0 & -4 & 13\\ 0 & 7 & -4\\ 0 & 1 & 3\end{pmatrix} \xrightarrow{\text{rref}} \begin{pmatrix}0 & 1 & 0\\ 0 & 0 & 1\\ 0 & 0 & 0\end{pmatrix}$$

and the eigenvector $\mathbf{v}_1 = (1, 0, 0)^T$ provides the solution

$$\mathbf{y}_1(t) = e^{tA}\mathbf{v}_1 = e^{-2t}\begin{pmatrix}1\\0\\0\end{pmatrix}.$$

For $\lambda_2 = 3$,

$$A - 3I = -\begin{pmatrix}-5 & -4 & 13\\ 0 & 2 & -4\\ 0 & 1 & -2\end{pmatrix} \xrightarrow{\text{rref}} \begin{pmatrix}1 & 0 & -1\\ 0 & 1 & -2\\ 0 & 0 & 0\end{pmatrix}$$

and the eigenvector $\mathbf{v}_2 = (1, 2, 1)^T$ provides the solution

$$\mathbf{y}_2(t) = e^{tA}\mathbf{v}_2 = e^{3t}\begin{pmatrix}1\\2\\1\end{pmatrix}.$$

Because

$$(A - 3I)^2 = \begin{pmatrix}25 & 25 & -75\\ 0 & 0 & 0\\ 0 & 0 & 0\end{pmatrix} \xrightarrow{\text{rref}} \begin{pmatrix}1 & 1 & -3\\ 0 & 0 & 0\\ 0 & 0 & 0\end{pmatrix}$$

the nullspace of $(A - 3I)^2$ has dimension 2. Pick $\mathbf{v}_3 = (3, 0, 1)^T$ in the nullspace of $(A - 3I)^2$. Note that $\mathbf{v}_2$ and $\mathbf{v}_3$ are independent. This gives a third solution,

$$\begin{aligned}\mathbf{y}_3(t) &= e^{tA}\mathbf{v}_3\\ &= e^{3t}\left[\mathbf{v}_3 + t(A-3I)\mathbf{v}_3\right]\\ &= e^{3t}\left[\begin{pmatrix}3\\0\\1\end{pmatrix} + t\begin{pmatrix}-5 & -4 & 13\\ 0 & 2 & -4\\ 0 & 1 & -2\end{pmatrix}\begin{pmatrix}3\\0\\1\end{pmatrix}\right]\\ &= e^{3t}\left[\begin{pmatrix}3\\0\\1\end{pmatrix} + t\begin{pmatrix}-2\\-4\\-2\end{pmatrix}\right]\\ &= e^{3t}\begin{pmatrix}3-2t\\-4t\\1-2t\end{pmatrix}.\end{aligned}$$

Because

$$\det\left[\mathbf{y}_1(0), \mathbf{y}_2(0), \mathbf{y}_3(0)\right] = \begin{vmatrix}1 & 1 & 3\\ 0 & 2 & 0\\ 0 & 1 & 1\end{vmatrix} = 2,$$

the solutions $\mathbf{y}_1(t)$, $\mathbf{y}_2(t)$, and $\mathbf{y}_3(t)$ are independent for all g and forma fundamental set of solutions.

39. If

$$A = \begin{pmatrix}5 & -1 & 0 & 2\\ 0 & 3 & 0 & 4\\ 1 & 1 & -1 & -3\\ 0 & -1 & 0 & 7\end{pmatrix},$$

a computer reveals the characteristic equation

$$p(\lambda) = (\lambda+1)(\lambda-5)^3,$$

so $\lambda_1 = 1$ and $\lambda_2 = 5$ are eigenvalues, with algebraic multiplicities 1 and 3 respectively. Because

$$A+I = \begin{pmatrix}6 & -1 & 0 & 2\\ 0 & 4 & 0 & 4\\ 1 & 1 & 0 & -3\\ 0 & -1 & 0 & 8\end{pmatrix} \xrightarrow{\text{rref}} \begin{pmatrix}1 & 0 & 0 & 0\\ 0 & 1 & 0 & 0\\ 0 & 0 & 0 & 1\\ 0 & 0 & 0 & 0\end{pmatrix},$$

the eigenvector $\mathbf{v}_1 = (0, 0, 1, 0)^T$ provides the solution

$$\mathbf{y}_1(t) = e^{At}\mathbf{v}_1 = e^{-t}\begin{pmatrix}0\\0\\1\\0\end{pmatrix}.$$

Because

$$A-5I = \begin{pmatrix}0 & -1 & 0 & 2\\ 0 & -2 & 0 & 4\\ 1 & 1 & -6 & -3\\ 0 & -1 & 0 & 2\end{pmatrix} \xrightarrow{\text{rref}} \begin{pmatrix}1 & 0 & -6 & -1\\ 0 & 1 & 0 & -2\\ 0 & 0 & 0 & 0\\ 0 & 0 & 0 & 0\end{pmatrix},$$

the nullspace of $A-5I$ has dimension 2 and the eigenvector $\mathbf{v}_2 = (6, 0, 1, 0)^T$ and $\mathbf{v}_3 = (1, 2, 0, 1)^T$ provide two more solutions.

$$\mathbf{y}_2(t) = e^{At}\mathbf{v}_2 = e^{5t}\begin{pmatrix}6\\0\\1\\0\end{pmatrix} \quad \text{and}$$

$$\mathbf{y}_3(t) = e^{At}\mathbf{v}_3 = e^{5t}\begin{pmatrix}1\\2\\0\\1\end{pmatrix}.$$

Because

$$(A-5I)^2 = \begin{pmatrix}0 & 0 & 0 & 0\\ 0 & 0 & 0 & 0\\ -6 & -6 & 36 & 18\\ 0 & 0 & 0 & 0\end{pmatrix} \xrightarrow{\text{rref}} \begin{pmatrix}1 & 1 & -6 & -3\\ 0 & 0 & 0 & 0\\ 0 & 0 & 0 & 0\\ 0 & 0 & 0 & 0\end{pmatrix},$$

the nullspace of $(A-5I)^2$ has dimension 3 and we can pick $\mathbf{v}_4 = (3, 0, 0, 1)^T$ independent of $\mathbf{v}_2$ and $\mathbf{v}_3$. This gives solution

$$\begin{aligned}\mathbf{y}_4(t) &= e^{At}\mathbf{v}_4\\ &= e^{5t}\left[\mathbf{v}_4 + t(A-5I)\mathbf{v}_4\right]\\ &= e^{5t}\left[\begin{pmatrix}3\\0\\0\\1\end{pmatrix} + t\begin{pmatrix}2\\4\\0\\2\end{pmatrix}\right]\\ &= e^{5t}\begin{pmatrix}3+2t\\4t\\0\\1+2t\end{pmatrix}.\end{aligned}$$

Because

$$\det\left[\mathbf{y}_1(0), \mathbf{y}_2(0), \mathbf{y}_3(0), \mathbf{y}_4(0)\right] = \begin{vmatrix} 0 & 6 & 1 & 3 \\ 0 & 0 & 2 & 0 \\ 1 & 1 & 0 & 0 \\ 0 & 0 & 1 & 1 \end{vmatrix} = 12,$$

the solutions are independents for all t and form a fundamental set of solutions.

41. If

$$A = \begin{pmatrix} -1 & 0 & 0 & 2 \\ -6 & 13 & 0 & -42 \\ 0 & -6 & -2 & 13 \\ -2 & 5 & 0 & -16 \end{pmatrix},$$

a computer reveals the characteristic polynomial

$$p(\lambda) = (\lambda + 2)^2(\lambda^2 + 2\lambda + 5).$$

For $\lambda = -2$,

$$(A + 2I)^2 = \begin{pmatrix} -3 & 10 & 0 & -26 \\ -12 & 15 & 0 & -54 \\ 10 & -25 & 0 & 70 \\ -4 & 5 & 0 & -18 \end{pmatrix} \xrightarrow{\text{rref}} \begin{pmatrix} 1 & 0 & 0 & 2 \\ 0 & 1 & 0 & -2 \\ 0 & 0 & 0 & 0 \\ 0 & 0 & 0 & 0 \end{pmatrix},$$

so we can pick $\mathbf{v}_1 = (0, 0, 1, 0)^T$ and $\mathbf{v}_2 = (-2, 2, 0, 1)^T$. Thus,

$$\mathbf{y}_1(t) = e^{tA}\mathbf{v}_1 = e^{-2t}\left[\mathbf{v}_1 + t(A + 2I)\mathbf{v}_1\right] = e^{-2t}\begin{pmatrix} 0 \\ 0 \\ 1 \\ 0 \end{pmatrix}$$

$$\mathbf{y}_2(t) = e^{tA}\mathbf{v}_2 = e^{-2t}\left[\mathbf{v}_2 + t(A + 2I)\mathbf{v}_2\right] = e^{-2t}\begin{pmatrix} -2 \\ 2 \\ t \\ 1 \end{pmatrix}.$$

The remaining eigenvalues are $-1 \pm 2i$. A computer reveals an eigenvector $\mathbf{w} = (1, 3i, -2 - i, i)^T$ associated with $\lambda = -1 + 2i$. Thus,

$$\mathbf{z}(t) = e^{tA}\mathbf{w} = e^{(-1+2i)t}\mathbf{w} = e^{-t}e^{2it}\begin{pmatrix} 1 \\ 3i \\ -2 - i \\ i \end{pmatrix}.$$

Using Euler's identity,

$$\begin{aligned} \mathbf{z}(t) &= e^{-t}(\cos 2t + i \sin 2t) \times \left[\begin{pmatrix} 1 \\ 0 \\ -2 \\ 0 \end{pmatrix} + i\begin{pmatrix} 0 \\ 3 \\ -1 \\ 1 \end{pmatrix}\right] \\ &= e^{-t}\left(\cos 2t \begin{pmatrix} 1 \\ 0 \\ -2 \\ 0 \end{pmatrix} - \sin 2t \begin{pmatrix} 0 \\ 3 \\ -1 \\ 1 \end{pmatrix}\right) \\ &\quad + ie^{-t}\left(\cos 2t \begin{pmatrix} 0 \\ 3 \\ -1 \\ 1 \end{pmatrix} + \sin 2t \begin{pmatrix} 1 \\ 0 \\ -2 \\ 0 \end{pmatrix}\right). \end{aligned}$$

Thus,

$$\mathbf{y}_3(t) = e^{-t}\begin{pmatrix} \cos 2t \\ -3\sin 2t \\ -2\cos 2t + \sin 2t \\ -\sin 2t \end{pmatrix} \quad \text{and}$$

$$\mathbf{y}_4(t) = e^{-t}\begin{pmatrix} \sin 2t \\ 3\cos 2t \\ -\cos 2t - 2\sin 2t \\ \cos 2t \end{pmatrix}$$

are solutions. Because

$$\det\left[\mathbf{y}_1(0), \mathbf{y}_2(0), \mathbf{y}_3(0), \mathbf{y}_4(0)\right] = \begin{vmatrix} 0 & -2 & 1 & 0 \\ 0 & 2 & 0 & 3 \\ 1 & 0 & -2 & -1 \\ 0 & 1 & 0 & 1 \end{vmatrix} = 1,$$

the solutions $\mathbf{y}_1(t)$, $\mathbf{y}_2(t)$, $\mathbf{y}_3(t)$, and $\mathbf{y}_4(t)$ are independent for all t and form a fundamental set of solutions.

43. If

$$A = \begin{pmatrix} -2 & 2 & -2 & 0 & -3 \\ -1 & 0 & -1 & 0 & -3 \\ 15 & -16 & -1 & 10 & 33 \\ 12 & -13 & 1 & 6 & 26 \\ -5 & 5 & 0 & -3 & -12 \end{pmatrix},$$

a computer reveals the characteristic equation

$$p(\lambda) = (\lambda+1)(\lambda+2)^4.$$

The eigenvalue-eigenvector pair $\lambda_1 = -1$, $\mathbf{v}_1 = (-3, -2, -2, -2, -1)^T$ give the solution

$$\mathbf{y}_1(t) = e^{tA}\mathbf{v}_1 = e^{-t}\begin{pmatrix} -3 \\ -2 \\ -2 \\ -2 \\ 1 \end{pmatrix}.$$

Because $(A+2I)^4$ reduces

$$(A+2I)^4 \xrightarrow{\text{rref}} \begin{pmatrix} 1 & -4/3 & 1/3 & 2/3 & 8/3 \\ 0 & 0 & 0 & 0 & 0 \\ 0 & 0 & 0 & 0 & 0 \\ 0 & 0 & 0 & 0 & 0 \\ 0 & 0 & 0 & 0 & 0 \end{pmatrix},$$

the dimension of the nullspace of $(A+2I)^4$ is 4 and the collection

$$\mathbf{v}_2 = \begin{pmatrix} 4 \\ 3 \\ 0 \\ 0 \\ 0 \end{pmatrix}, \quad \mathbf{v}_3 = \begin{pmatrix} -1 \\ 0 \\ 3 \\ 0 \\ 0 \end{pmatrix},$$

$$\mathbf{v}_4 = \begin{pmatrix} -2 \\ 0 \\ 0 \\ 3 \\ 0 \end{pmatrix}, \quad \mathbf{v}_5 = \begin{pmatrix} -8 \\ 0 \\ 0 \\ 0 \\ 3 \end{pmatrix}$$

is a basis for the nullspace of $(A+2I)^4$. Moreover, for $i = 2, 3, 4$, and 5,

$$\begin{aligned} \mathbf{y}_i(t) &= e^{tA}\mathbf{v}_i \\ &= e^{-2t}\Big[\mathbf{v}_i + t(A+2I)\mathbf{v}_i + \frac{t^2}{2!}(A+2I)^2\mathbf{v}_i \\ &\quad + \frac{t^3}{3!}(A+2I)^3\mathbf{v}_i\Big]. \end{aligned}$$

Using this result and a computer,

$$\mathbf{y}_2(t) = e^{tA}\mathbf{v}_2 = \frac{1}{2}e^{-2t}\begin{pmatrix} 8+12t-5t^2+t^3 \\ 6+4t+t^2+t^3 \\ 24t-5t^2+t^3 \\ 18t \\ -10t+3t^2 \end{pmatrix}$$

$$\mathbf{y}_3(t) = e^{tA}\mathbf{v}_3 = -\frac{1}{2}e^{-2t}\begin{pmatrix} 2+12t-5t^2+t^3 \\ 4t+t^2+t^3 \\ -6+24t-5t^2+t^3 \\ 18t \\ 10t+3t^2 \end{pmatrix}$$

$$\mathbf{y}_4(t) = e^{tA}\mathbf{v}_4 = \frac{1}{2}e^{-2t}\begin{pmatrix} -4+t^2 \\ 4+t \\ t^2 \\ 6 \\ 2t \end{pmatrix}$$

$$\mathbf{y}_5(y) = e^{tA}\mathbf{v}_4 = -e^{-2t}\begin{pmatrix} 8+9t-5t^2+t^3 \\ t+t^2+t^3 \\ 21t-5t^2+t^3 \\ 18t \\ -3-10t+3t^2 \end{pmatrix}$$

Because

$$\begin{aligned} &\det\left[\mathbf{y}_1(t), \mathbf{y}_2(0), \mathbf{y}_3(t), \mathbf{y}_4(0), \mathbf{y}_5(0)\right] \\ &= \begin{vmatrix} -3 & 4 & -1 & -2 & -8 \\ -2 & 3 & 0 & 0 & 0 \\ -2 & 0 & 3 & 0 & 0 \\ -2 & 0 & 0 & 3 & 0 \\ 1 & 0 & 0 & 0 & 3 \end{vmatrix} \\ &= 27, \end{aligned}$$

the solutions are independent for all t and form a fundamental set of solutions.

45. In matrix form,

$$\begin{pmatrix} x_1 \\ x_2 \\ x_3 \\ x_4 \\ x_5 \end{pmatrix}' = \begin{pmatrix} 5 & 7 & 1 & 1 & 8 \\ 3 & 6 & 5 & 4 & 5 \\ -3 & -8 & -2 & -5 & -12 \\ 3 & 14 & 8 & 10 & 18 \\ -4 & -9 & -6 & -5 & -9 \end{pmatrix}\begin{pmatrix} x_1 \\ x_2 \\ x_3 \\ x_4 \\ x_5 \end{pmatrix}.$$

A computer reveals the characteristic equation

$$p(\lambda) = -(\lambda + 1)^2(\lambda - 4)^3.$$

For $\lambda = -1$, $(A + I)^2$ reduces

$$(A + I)^2 \xrightarrow{\text{rref}} \begin{pmatrix} 1 & 0 & 0 & -1 & -1 \\ 0 & 1 & 0 & 1 & 2 \\ 0 & 0 & 1 & 0 & -1 \\ 0 & 0 & 0 & 0 & 0 \\ 0 & 0 & 0 & 0 & 0 \end{pmatrix},$$

and $\mathbf{v}_1 = (1, -1, 0, 1, 0)^T$ and $\mathbf{v}_2 = (1, -2, 1, 0, 1)^T$ form a basis for the nullspace of $(A + I)^2$. Moreover,

$$\mathbf{y}_i(t) = e^{tA}\mathbf{v}_i = e^{-t}[\mathbf{v}_i + t(A + I)\mathbf{v}_i]$$

for $i = 1, 2$. Using this result and a computer,

$$\mathbf{y}_1(t) = e^{tA}\mathbf{v}_1 = e^{-t}\begin{pmatrix} 1 \\ -1 \\ 0 \\ 1 \\ 0 \end{pmatrix}$$

$$\mathbf{y}_2(t) = e^{tA}\mathbf{v}_2 = e^{-t}\begin{pmatrix} 1 + t \\ -2 - t \\ 1 \\ t \\ 1 \end{pmatrix}.$$

For $\lambda = 4$, $(A - 4I)^3$ reduces

$$(A - 4I)^3 \xrightarrow{\text{rref}} \begin{pmatrix} 1 & 0 & 1 & 1 & 1 \\ 0 & 1 & 0 & 0 & 1 \\ 0 & 0 & 0 & 0 & 0 \\ 0 & 0 & 0 & 0 & 0 \\ 0 & 0 & 0 & 0 & 0 \end{pmatrix},$$

and $\mathbf{v}_3 = (-1, 0, 1, 0, 0)^T$, $\mathbf{v}_4 = (-1, 0, 0, 1, 0)^T$, and $\mathbf{v}_5 = (-1, -1, 0, 0, 1)^T$ form a basis for the nullspace of $(A - 4I)^3$. Moreover,

$$\mathbf{y}_i(t) = e^{tA}\mathbf{v}_i = e^{4t}\left[\mathbf{v}_i + t(A - 4I)\mathbf{v}_i + \frac{t^2}{2!}(A - 4I)^2\mathbf{v}_i\right]$$

for $i = 3, 4$, and 5. Using this result and a computer,

$$\mathbf{y}_3(t) = e^{At}\mathbf{v}_3 = \frac{1}{2}e^{4t}\begin{pmatrix} -2 \\ 4t - t^2 \\ 2 - 6t + t^2 \\ 10t - 2t^2 \\ -4t + t^2 \end{pmatrix}$$

$$\mathbf{y}_4(t) = e^{At}\mathbf{v}_4 = \frac{1}{2}e^{4t}\begin{pmatrix} -2 \\ 2t - t^2 \\ -4t + t^2 \\ 2 + 6t - 2t^2 \\ -2t + t^2 \end{pmatrix}$$

$$\mathbf{y}_5(t) = e^{At}\mathbf{v}_5 = \frac{1}{2}e^{4t}\begin{pmatrix} -2 \\ -2 - t^2 \\ -2t + t^2 \\ 2t - 2t^2 \\ 2 + t^2 \end{pmatrix}$$

Because

$$\det[\mathbf{y}_1(0), \mathbf{y}_2(0), \mathbf{y}_3(0), \mathbf{y}_4(0), \mathbf{y}_5(0)] = \begin{vmatrix} 1 & 1 & -1 & -1 & -1 \\ -1 & -2 & 0 & 0 & -1 \\ 0 & 1 & 1 & 0 & 0 \\ 1 & 0 & 0 & 1 & 0 \\ 0 & 1 & 0 & 0 & 1 \end{vmatrix} = 1,$$

the solutions $\mathbf{y}_1(t)$, $\mathbf{y}_2(t)$, $\mathbf{y}_3(t)$, $\mathbf{y}_4(t)$, and $\mathbf{y}_5(t)$ are independent for all t and form a fundamental set of solutions.

47. The salt content in the topmost tank is changing at a rate

$$\frac{dx_1}{dt} = \text{Rate in} - \text{Rate out} = -\frac{x_1}{25}.$$

The salt content in the second tank is changing at a rate

$$\frac{dx_2}{dt} = \text{Rate in} - \text{Rate out} = \frac{x_1}{25} - \frac{x_2}{25}.$$

The salt content in the third tank is changing at a rate

$$\frac{dx_3}{dt} = \text{Rate in} - \text{Rate out} = \frac{x_2}{25} - \frac{x_3}{25}.$$

In matrix-vector form,

$$\begin{pmatrix} x_1 \\ x_2 \\ x_3 \end{pmatrix}' = \begin{pmatrix} -1/25 & 0 & 0 \\ 1/25 & -1/25 & 0 \\ 0 & 1/25 & -1/25 \end{pmatrix}\begin{pmatrix} x_1 \\ x_2 \\ x_3 \end{pmatrix}.$$

Initially, $\mathbf{x}(0) = (x_1(0), x_2(0), x_3(0))^T = (10, 8, 4)^T$. Because the coefficient matrix is lower triangular, the main diagonal contains the eigenvalues. Note that there is a single eigenvalue of algebraic multiplicity 3, namely, $\lambda = -1/25$. Hence,

$$e^{tA} = e^{-t/25}\left[I + t\left(A + \frac{1}{25}I\right) + \frac{t}{2}\left(A + \frac{1}{25}I\right)^2\right].$$

However,

$$A + \frac{1}{25}I = \begin{pmatrix} 0 & 0 & 0 \\ 1/25 & 0 & 0 \\ 0 & 1/25 & 0 \end{pmatrix} \quad \text{and}$$

$$\left(A + \frac{1}{25}I\right)^2 = \begin{pmatrix} 0 & 0 & 0 \\ 0 & 0 & 0 \\ 1/625 & 0 & 0 \end{pmatrix}.$$

Therefore,

$$e^{tA} = e^{-t/25}\begin{pmatrix} 1 & 0 & 0 \\ 0 & 1 & 0 \\ 0 & 0 & 1 \end{pmatrix} + te^{-t/25}\begin{pmatrix} 0 & 0 & 0 \\ 1/25 & 0 & 0 \\ 0 & 1/25 & 0 \end{pmatrix} + \frac{t^2}{2}e^{-t/25}\begin{pmatrix} 0 & 0 & 0 \\ 0 & 0 & 0 \\ 1/625 & 0 & 0 \end{pmatrix}.$$

Simplifying,

$$e^{tA} = e^{-t/25}\begin{pmatrix} 1 & 0 & 0 \\ t/25 & 1 & 0 \\ t^2/1250 & t/25 & 1 \end{pmatrix}.$$

Finally, the solution with initial condition $\mathbf{x}_0 = (10, 8, 4)^T$ is

$$\begin{aligned} \mathbf{x}(t) &= e^{tA}\mathbf{x}_0 \\ &= e^{-t/25}\begin{pmatrix} 1 & 0 & 0 \\ t/25 & 1 & 0 \\ t^2/1250 & t/25 & 1 \end{pmatrix}\begin{pmatrix} 10 \\ 8 \\ 4 \end{pmatrix}, \\ &= e^{-t/25}\begin{pmatrix} 10 \\ 2t/5 + 8 \\ t^2/125 + 8t/25 + 4 \end{pmatrix}. \end{aligned}$$

At $t = 8$ minutes, the amount of salt in each tank is

$$\begin{pmatrix} x_1(8) \\ x_2(8) \\ x_3(8) \end{pmatrix} = \mathbf{x}(8) = \begin{pmatrix} 7.2614\text{ lb} \\ 8.1328\text{ lb} \\ 5.1353\text{ lb} \end{pmatrix}.$$

The plot of the salt in each tank over time is shown in the following figure.

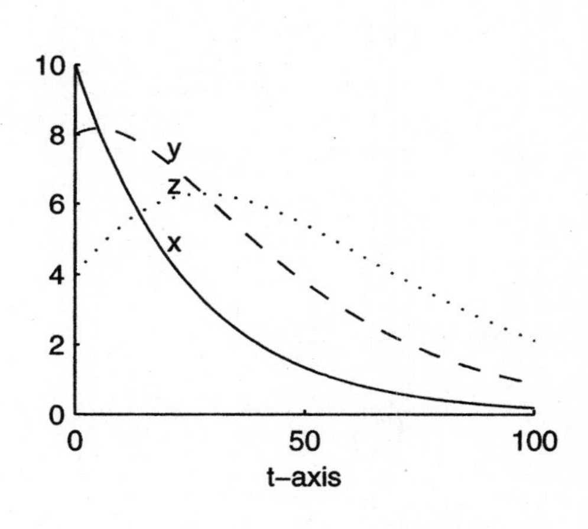

Section 9.7. Qualitative Analysis of Linear Systems

1. In matrix form,

$$\begin{pmatrix} x \\ y \end{pmatrix}' = \begin{pmatrix} -0.2 & 2.0 \\ -2.0 & -0.2 \end{pmatrix}\begin{pmatrix} x \\ y \end{pmatrix},$$

the coefficient matrix

$$A = \begin{pmatrix} -0.2 & 2.0 \\ -2.0 & -0.2 \end{pmatrix}$$

has characteristic polynomial

$$p(\lambda) = \lambda^2 + 0.4\lambda + 4.04,$$

producing eigenvalues $\lambda = -0.2 \pm 2i$. Because the real part of each eigenvalue is negative, the equilibrium point at the origin is asymptotically stable.

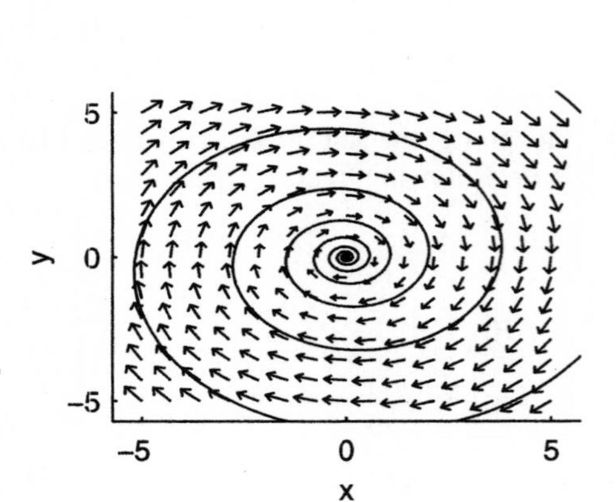

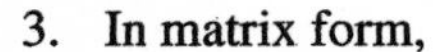

3. In matrix form,

$$\begin{pmatrix} x \\ y \end{pmatrix}' = \begin{pmatrix} -6 & -15 \\ 3 & 6 \end{pmatrix} \begin{pmatrix} x \\ y \end{pmatrix},$$

the coefficient matrix

$$A = \begin{pmatrix} -6 & -15 \\ 3 & 6 \end{pmatrix}$$

has characteristic polynomial

$$p(\lambda) = \lambda^2 + 9,$$

producing eigenvalues $\lambda = \pm 3i$. Therefore, the equilibrium point at the origin is a stable center.

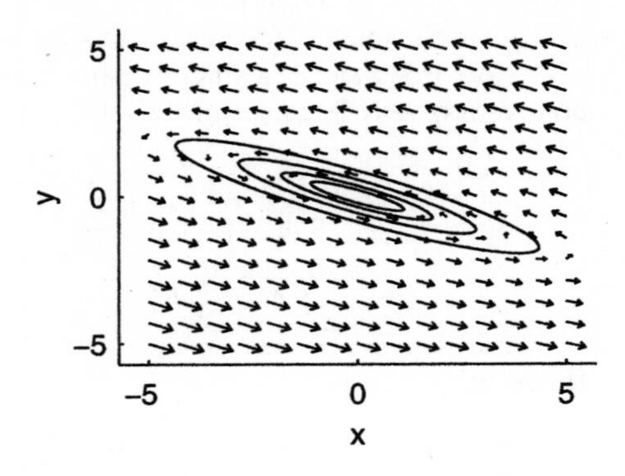

5. In system

$$\mathbf{y}' = \begin{pmatrix} 0.1 & 2.0 \\ -2.0 & 0.1 \end{pmatrix} \mathbf{y},$$

the coefficient matrix

$$A = \begin{pmatrix} 0.1 & 2.0 \\ -2.0 & 0.1 \end{pmatrix}$$

has characteristic polynomial

$$p(\lambda) = \lambda^2 - 0.2\lambda + 4.01,$$

producing eigenvalues $\lambda = 0.1 \pm 2i$. Therefore, the equilibrium point at the origin is a unstable. Indeed, the equilibrium point is a spiral source.

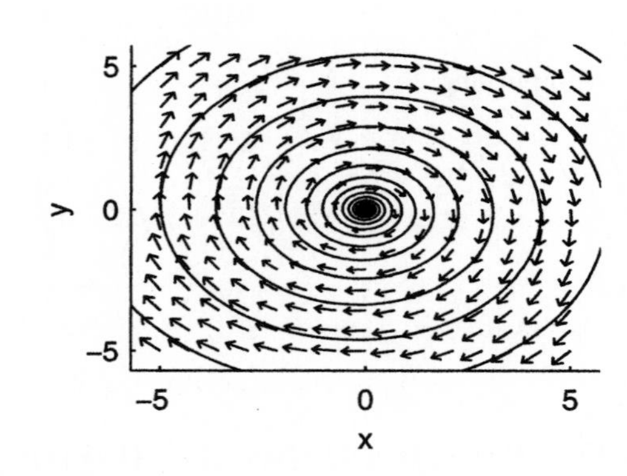

7. In system

$$\mathbf{y}' = \begin{pmatrix} 1 & -4 \\ 1 & -3 \end{pmatrix} \mathbf{y},$$

the coefficient matrix

$$A = \begin{pmatrix} 1 & -4 \\ 1 & -3 \end{pmatrix}$$

has characteristic polynomial

$$p(\lambda) = \lambda^2 + 2\lambda + 1,$$

producing the repeated eigenvalue $\lambda = -1$. Because the real part of every eigenvalue is negative, the equilibrium point at the origin is asymtotically stable. Indeed, the equilibrium point is a degenerate sink.

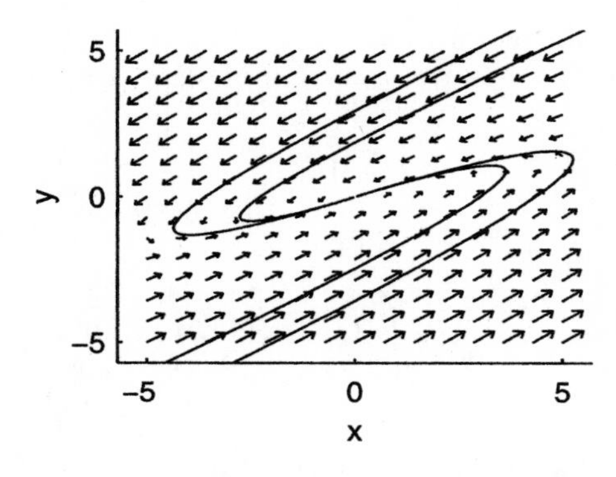

9. Consider the system

$$\mathbf{y}' = \begin{pmatrix} -3 & -4 & 2 \\ -2 & -7 & 4 \\ -3 & -8 & 4 \end{pmatrix} \mathbf{y}.$$

Using a computer, matrix

$$A = \begin{pmatrix} -3 & -4 & 2 \\ -2 & -7 & 4 \\ -3 & -8 & 4 \end{pmatrix}$$

has characteristic polynomial

$$p(\lambda) = \lambda^3 + 6\lambda^2 + 11\lambda + 6,$$

and eigenvalues $\lambda_1 = -3$, $\lambda_2 = -2$, and $\lambda_3 = -1$. Because the real parts of all eigenvalues are negative, the equilibrium point at the origin is asymptotically stable. One such solution, with initial condition $(1, 1, 1)^T$, is shown in the following figure.

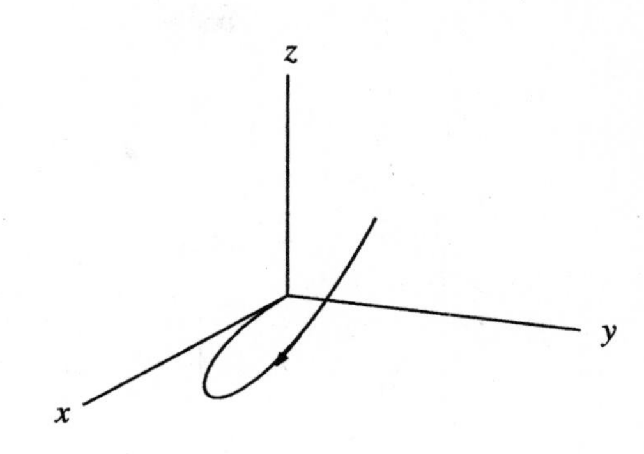

11. In matrix form,

$$\begin{pmatrix} x \\ y \\ z \end{pmatrix}' = \begin{pmatrix} -1 & 3 & 4 \\ 0 & 1 & 6 \\ 0 & -3 & -5 \end{pmatrix} \begin{pmatrix} x \\ y \\ x \end{pmatrix}.$$

Using a computer, matrix

$$A = \begin{pmatrix} -1 & 3 & 4 \\ 0 & 1 & 6 \\ 0 & -3 & -5 \end{pmatrix}$$

has characteristic polynomial

$$p(\lambda) = \lambda^3 + 5\lambda^2 + 1t\lambda + 13,$$

and eigenvalues $\lambda_1 = -1$, $\lambda_2 = -2 + 3i$, and $\lambda_3 = -2 - 3i$. Because all the real parts of the eigenvalues are negative, the equilibrium point at the origin is asymptotically stable. One such solution, with initial condition $(1, 1, 1)^T$, is shown in the following figure.

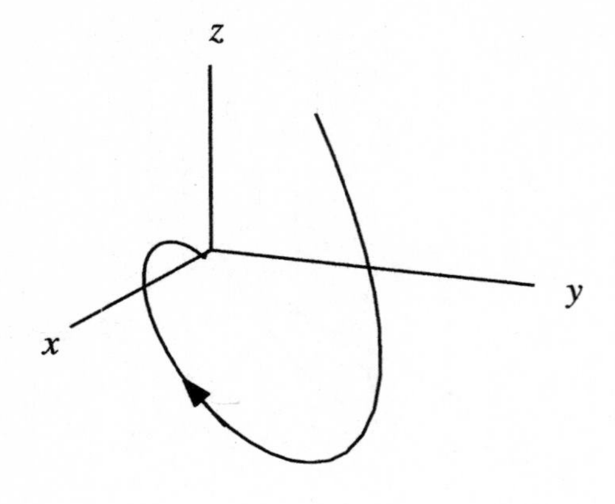

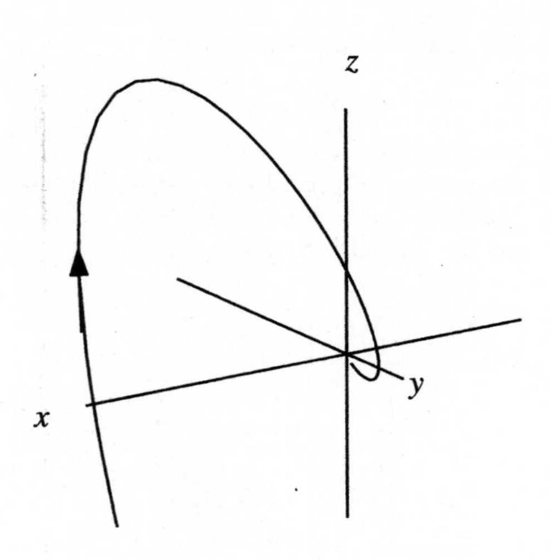

13. If

$$\mathbf{y}' = \begin{pmatrix} 0 & 0 & -1 \\ -1 & 0 & 0 \\ 4 & -2 & -3 \end{pmatrix} \mathbf{y},$$

then matrix

$$A = \begin{pmatrix} 0 & 0 & -1 \\ -1 & 0 & 0 \\ 4 & -2 & -3 \end{pmatrix}$$

has characteristic polynomial

$$p(\lambda) = (\lambda + 1)(\lambda^2 + 2\lambda + 2)$$

and eigenvalues -1, $-1+i$, and $-1-i$. Therefore, the real part of eigenvalue is negative, so the hypotheses of Theorem 7.4 are satisfied and the equilibrium point at the origin is asymptotically stable. One such solution, with initial condition $(2, -1, -2)^T$, is shown in the image that follows.

15. If

$$\mathbf{y}' = \begin{pmatrix} 3 & -2 & -5 & 3 \\ 16 & -6 & -17 & 9 \\ -14 & 5 & 15 & -8 \\ -19 & 8 & 23 & -13 \end{pmatrix} \mathbf{y},$$

then a computer reveals that matrix

$$A = \begin{pmatrix} 3 & -2 & -5 & 3 \\ 16 & -6 & -17 & 9 \\ -14 & 5 & 15 & -8 \\ -19 & 8 & 23 & -13 \end{pmatrix}$$

has characteristic equation

$$p(\lambda) = (\lambda - 2)(\lambda + 1)^3$$

and eigenvalues $\lambda_1 = 2$ and $\lambda_2 = -1$, the latter having algebraic multiplicity 3. Thus, one eigenvalue has positive real part and Theorem 7.4 predicts that the equilibrium point at the origin is unstable. One such solution, with initial conation $(0.1, 0.1, 0.1, 0.1)^T$, seems to approach the origin, only to veer away with the passage of time, much like a saddle point solution in the phase plane. This behavior is indicated in the following plot of each component of the solution versus time.

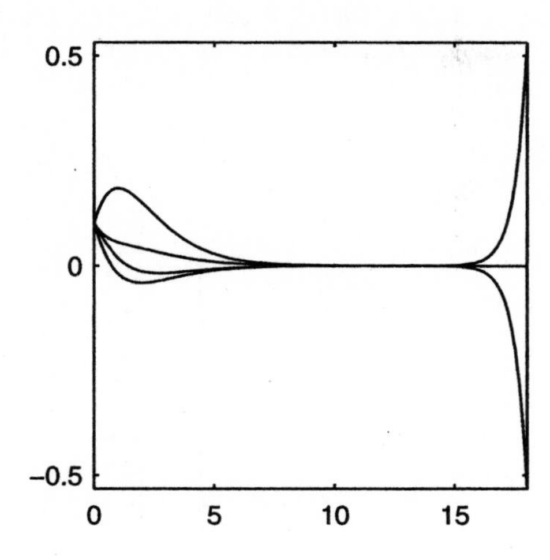

17. (a) In matrix form,

$$\begin{pmatrix} x \\ y \\ z \end{pmatrix}' = \begin{pmatrix} -3 & 0 & 0 \\ -2 & -1 & 0 \\ 0 & 0 & -2 \end{pmatrix} \begin{pmatrix} x \\ y \\ z \end{pmatrix}.$$

Using a computer, matrix

$$A = \begin{pmatrix} -3 & 0 & 0 \\ -2 & -1 & 0 \\ 0 & 0 & -2 \end{pmatrix}$$

has characteristic polynomial

$$p(\lambda) = (\lambda + 3)(\lambda + 2)(\lambda + 1)$$

and eigenvalues -3, -2 and -1. A computer also reveals the associated eigenvectors which lead to the following exponential solutions.

$$\mathbf{y}_1(t) = e^{-3t} \begin{pmatrix} 1 \\ 1 \\ 0 \end{pmatrix}, \quad \mathbf{y}_2(t) = e^{-2t} \begin{pmatrix} 0 \\ 0 \\ 1 \end{pmatrix}, \quad \text{and}$$

$$\mathbf{y}_3(t) = e^{-t} \begin{pmatrix} 0 \\ 1 \\ 0 \end{pmatrix}.$$

These exponential solutions generate the half-line solutions shown in the following figure. Each of the half-line solutions decay to the origin with the passage of time.

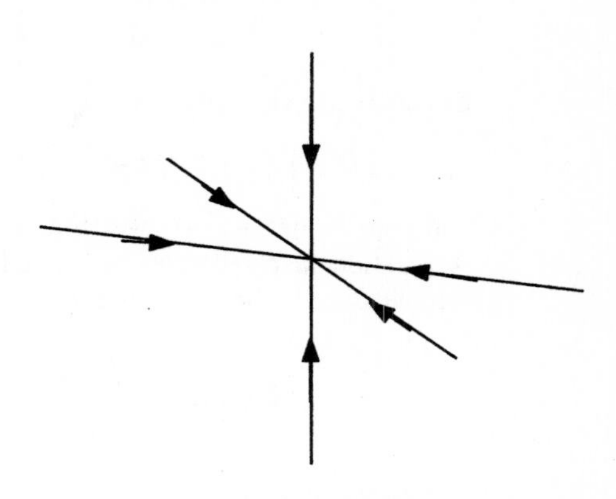

(b) We selected initial conditions $(1, 0, 1)^T$, $(-1, 0, 1)^T$, $(1/2, 1, 1)^T$, $(-1/2, -1, 1)^T$, $(1, 0, -1)^T$, $(-1, 0, -1)^T$, $(1/2, 1, -1)^T$, and $(-1/2, -1, -1)^T$ to craft the portrait in the following figure.

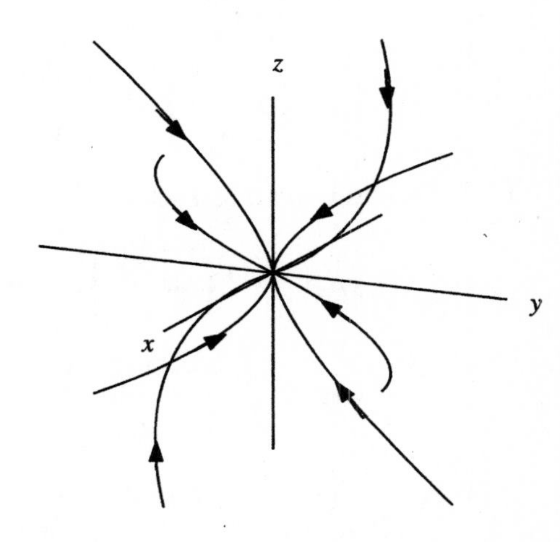

(c) Nodal sink

19. (a) If

$$\mathbf{y}' = \begin{pmatrix} -1 & -10 & 0 \\ 10 & -1 & 0 \\ 0 & 0 & -1 \end{pmatrix} \mathbf{y},$$

then, using a computer, matrix

$$A = \begin{pmatrix} -1 & -10 & 0 \\ 10 & -1 & 0 \\ 0 & 0 & -1 \end{pmatrix}$$

has characteristic polynomial

$$p(\lambda) = (\lambda + 1)(\lambda^2 + 2\lambda + 101)$$

and eigenvalues -1, $-1 + 10i$ and $-1 - 10i$. A computer also generates associated eigenvectors, leading to the real solution

$$\mathbf{y}_1(t) = e^{-t} \begin{pmatrix} 0 \\ 0 \\ 1 \end{pmatrix}$$

and the complex solution

$$\begin{aligned} \mathbf{z}(t) &= e^{(-1+10i)t} \begin{pmatrix} 1 \\ -i \\ 0 \end{pmatrix} \\ &= e^{-t}(\cos 10t + i \sin 10t) \\ &\quad \times \left[\begin{pmatrix} 1 \\ 0 \\ 0 \end{pmatrix} + i \begin{pmatrix} 0 \\ -1 \\ 0 \end{pmatrix} \right] \\ &= e^{-t} \begin{pmatrix} \cos 10t \\ \sin 10t \\ 0 \end{pmatrix} + i e^{-t} \begin{pmatrix} \sin 10t \\ -\cos 10t \\ 0 \end{pmatrix}. \end{aligned}$$

This leads to the real solutions

$$\mathbf{y}_2(t) = e^{-t} \begin{pmatrix} \cos 10t \\ \sin 10t \\ 0 \end{pmatrix}, \quad \text{and}$$

$$\mathbf{y}_3(t) = e^{-t} \begin{pmatrix} \sin 10t \\ -\cos 10t \\ 0 \end{pmatrix}.$$

(b) Any solution starting on the z-axis lies on the half-lines generated by the exponential solution

$$\mathbf{y}(t) = C_1 e^{-t} \begin{pmatrix} 0 \\ 0 \\ 1 \end{pmatrix}.$$

Thus, the solution will remain on the z-axis as it decays to the equilibrium point at the origin. In the following image, solutions with initial conditions $(0, 0, 1)^T$ and $(0, 0, -1)^T$ remain on the z-axis and decay to the origin.

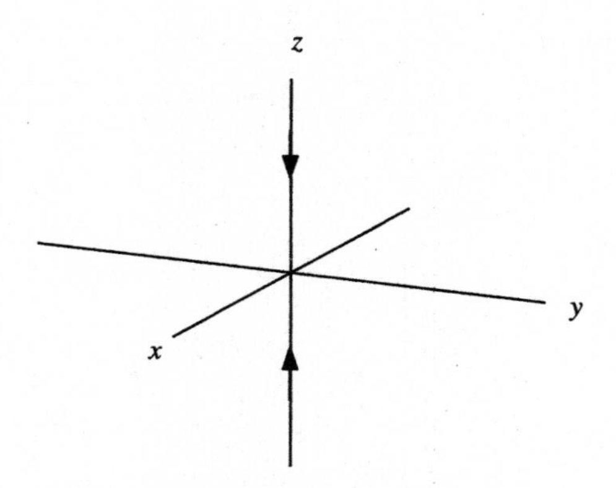

(c) The general solution is

$$\mathbf{y}(t) = C_1 e^{-t} \begin{pmatrix} 0 \\ 0 \\ 1 \end{pmatrix} + C_2 e^{-t} \begin{pmatrix} \cos 10t \\ \sin 10t \\ 0 \end{pmatrix} + C_3 e^{-t} \begin{pmatrix} \sin 10t \\ -\cos 10t \\ 0 \end{pmatrix}.$$

If a solution starts in the xy-plane with initial condition $\mathbf{y}(0) = (a, b, 0)^T$, then

$$\begin{pmatrix} a \\ b \\ 0 \end{pmatrix} = C_1 \begin{pmatrix} 0 \\ 0 \\ 1 \end{pmatrix} + C_2 \begin{pmatrix} 1 \\ 0 \\ 0 \end{pmatrix} + C_3 \begin{pmatrix} 0 \\ -1 \\ 0 \end{pmatrix},$$

leading to $C_1 = 0$, $C_2 = a$, and $C_3 = -b$. Thus, the particular solution is

$$\mathbf{y}(t) = a e^{-t} \begin{pmatrix} \cos 10t \\ \sin 10t \\ 0 \end{pmatrix} + b e^{-t} \begin{pmatrix} \sin 10t \\ -\cos 10t \\ 0 \end{pmatrix},$$

so these solutions will remain in the xy-plane and spiral inward to the equilibrium point at the origin. This is shown in the following figure, where we have plotted the solution with initial condition $(1, 1, 0)^T$.

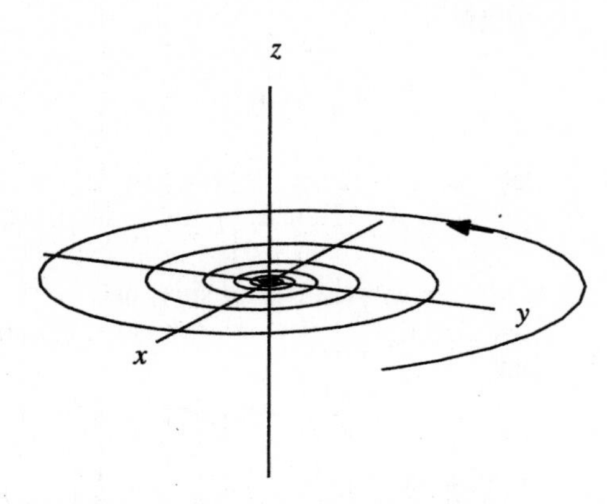

(d) A solution having initial condition $\mathbf{y}(0) = (a, b, c)^T$, where $c \neq 0$, would lead to

$$\begin{pmatrix} a \\ b \\ c \end{pmatrix} = C_1 \begin{pmatrix} 0 \\ 0 \\ 1 \end{pmatrix} + C_2 \begin{pmatrix} 1 \\ 0 \\ 0 \end{pmatrix} + C_3 \begin{pmatrix} 0 \\ -1 \\ 0 \end{pmatrix}$$

and $C_1 = c$, $C_2 = a$, and $C_3 = -b$. Thus, the particular solution is

$$\mathbf{y}(t) = \begin{pmatrix} e^{-t}(a \cos 10t - b \sin 10t) \\ e^{-t}(a \sin 10t + b \cos 10t) \\ ce^{-t} \end{pmatrix}.$$

We saw in part (c) that if $c = 0$, solutions spiral into the origin while remaining in the xy-plane. In this case, the z-coordinate decays to zero, so it is reasonable to assume that solutions will spiral into the origin, staying above or below the xy-plane. Solutions with initial conditions $(1, 1, 1)^T$ and $(-1, -1, -1)^T$ are shown in the following figure.

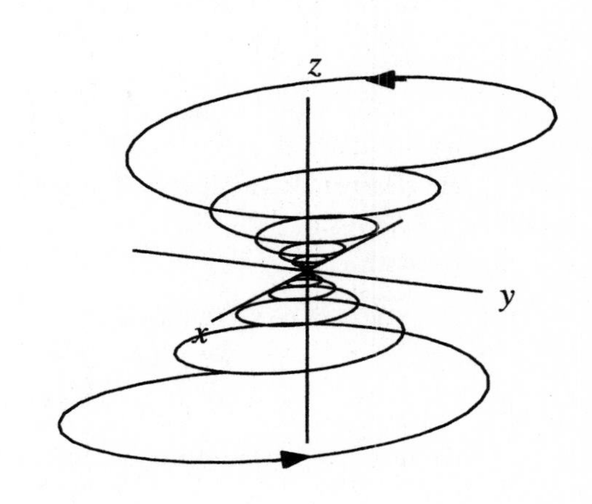

———×———

Section 9.8. Higher-Order Linear Systems

1. (a) If

$$\mathbf{x}_1(t) = \begin{pmatrix} e^{3t} \\ 3e^{3t} \end{pmatrix},$$

then

$$\mathbf{x}_1'(t) = \begin{pmatrix} 3e^{3t} \\ 9e^{3t} \end{pmatrix}$$

and

$$\begin{pmatrix} 0 & 1 \\ 3 & 2 \end{pmatrix} \mathbf{x}_1(t) = \begin{pmatrix} 0 & 1 \\ 3 & 2 \end{pmatrix} \begin{pmatrix} e^{3t} \\ 3e^{3t} \end{pmatrix} = \begin{pmatrix} 3e^{3t} \\ 9e^{3t} \end{pmatrix},$$

so $\mathbf{x}_1$ is a solution of

$$\mathbf{x}' = \begin{pmatrix} 0 & 1 \\ 3 & 2 \end{pmatrix} \mathbf{x}.$$

Similarly, if

$$\mathbf{x}_2(t) = \begin{pmatrix} e^{-t} \\ -e^{-t} \end{pmatrix},$$

then

$$\mathbf{x}_2'(t) = \begin{pmatrix} -e^{-t} \\ e^{-t} \end{pmatrix}$$

and

$$\begin{pmatrix} 0 & 1 \\ 3 & 2 \end{pmatrix} \mathbf{x}_2(t) = \begin{pmatrix} 0 & 1 \\ 3 & 2 \end{pmatrix} \begin{pmatrix} e^{-t} \\ -e^{-t} \end{pmatrix} = \begin{pmatrix} -e^{-t} \\ e^{-t} \end{pmatrix},$$

so $\mathbf{x}_2$ is a solution of

$$\mathbf{x}' = \begin{pmatrix} 0 & 1 \\ 3 & 2 \end{pmatrix} \mathbf{x}.$$

To show independence, we need only show that the functions are independent at one value of t. However,

$$\mathbf{x}_1(0) = \begin{pmatrix} 1 \\ 3 \end{pmatrix} \quad \text{and} \quad \mathbf{x}_2(0) = \begin{pmatrix} 1 \\ -1 \end{pmatrix}$$

are clearly independent ($\mathbf{x}_2(0)$ is not a multiple of $\mathbf{x}_1(0)$).

(b) Because

$$\begin{aligned} \mathbf{x}(t) &= C_1\mathbf{x}_1(t) + C_2\mathbf{x}_2(t) \\ &= C_1 \begin{pmatrix} e^{3t} \\ 3e^{3t} \end{pmatrix} + C_2 \begin{pmatrix} e^{-t} \\ -e^{-t} \end{pmatrix}, \end{aligned}$$

the first component of $\mathbf{x}(t)$ is $y(t) = C_1e^{3t} + C_2e^{-t}$. Thus,

$$\begin{aligned} y' &= 3C_1e^{3t} - C_2e^{-t} \\ y'' &= 9C_1e^{3t} + C_2e^{-t}, \end{aligned}$$

and

$$\begin{aligned} y'' - 2y' - 3y &= (9C_1e^{3t} + C_2e^{-t}) \\ &\quad - 2(3C_1e^{3t} - C_2e^{-t}) \\ &\quad - 3(C_1e^{3t} + C_2e^{-t}) \\ &= 0 \end{aligned}$$

3. If $y_1(t) = e^t$ and $y_2(t) = e^{2t}$, suppose that there exists constants c_1 and c_2 such that

$$c_1e^t + c_2e^{2t} = 0$$

for all t. Then,

$$\begin{aligned} t = 0 &\Rightarrow c_1 + c_2 = 0 \\ t = 1 &\Rightarrow c_1e + c_2e^2 = 0. \end{aligned}$$

Solving the first equation, $c_1 = -c_2$, and substituting into the second equation gives

$$\begin{aligned} -c_2e + c_2e^2 &= 0 \\ c_2(e^2 - e) &= 0. \end{aligned}$$

Because $e^2 - e \neq 0$, this give $c_2 = 0$, whence $c_1 = -c_2 = 0$. Hence, y_1 and y_2 are independent.

5. If $y_1(t) = \cos t$, $y_2(t) = \sin t$, and $y_3(t) = e^t$, suppose that there exists constants c_1, c_2, and c_3 such that

$$c_1 \cos t + c_2 \sin t + c_3e^t = 0$$

for all t. Then,

$$\begin{aligned} t = 0 &\Rightarrow c_1 + c_3 = 0 \\ t = \pi/2 &\Rightarrow c_2 + c_3e^{\pi/2} = 0 \\ t = \pi &\Rightarrow -c_1 + c_3e^{\pi}. \end{aligned}$$

Solving the first equation, $c_1 = -c_3$, and substituting this into the third equation give

$$0 = c_3 + c_3e^{\pi} = c_3(1 + e^{\pi}).$$

Because $e^{\pi} + 1 \neq 0$, this give $c_3 = 0$, whence $c_1 = -c_3 = 0$. Substituting $c_3 = 0$ into the second equation gives

$$0 = c_2 + 0e^{\pi} = c_2.$$

Therefore, y_1, y_2, and y_3 are linearly independent.

7. If $y_1(t) = \cos 3t$, then

$$\begin{aligned} y'(t) &= -3 \sin 3t \\ y''(t) &= -9 \cos 3t \end{aligned}$$

and

$$y_1'' + 9y_1 = -9 \cos 3t + 9 \cos 3t = 0.$$

Similarly, if $y_2(t) = \sin 3t$, then

$$\begin{aligned} y_2'(t) &= 3 \cos 3t \\ y_2''(t) &= -9 \sin 3t, \end{aligned}$$

and

$$y_2'' + 9y_2 = -9\sin 3t + 9\sin 3t = 0.$$

Thus, both y_1 and y_2 are solutions of $y'' + 9y = 0$. Finally, the Wronskian is

$$\begin{aligned} W(t) &= \det\begin{pmatrix} y_1 & y_2 \\ y_1' & y_2' \end{pmatrix} \\ &= \det\begin{pmatrix} \cos 3t & \sin 3t \\ -3\sin 3t & 3\cos 3t \end{pmatrix} \\ &= 3\cos^2 3tI + 3\sin^2 3t \\ &= 3, \end{aligned}$$

which is nonzero for all t. Hence, the solutions y_1 and y_2 are linearly independent.

9. If $y_1(t) = e^{2t}$, then

$$\begin{aligned} y'(t) &= 2e^{2t} \\ y''(t) &= 4e^{2t}. \end{aligned}$$

and

$$y_1'' - 4y_1' + 4y_1 = 4e^{2t} - 8e^{2t} + 4e^{2t} = 0.$$

Similarly, if $y_2(t) =$, then

$$\begin{aligned} y_2'(t) &= e^{2t}(2t+1) \\ y_2''(t) &= e^{2t}(4t+4), \end{aligned}$$

and

$$\begin{aligned} y_2'' - 4y_2' + 4y_2 &= e^{2t}(4t+4) - 4e^{2t}(2t+1) \\ &\quad + 4te^{2t} \\ &= e^{2t}(4t + 4 - 8t - 4 + 4t) \\ &= 0. \end{aligned}$$

Thus, both y_1 and y_2 are solutions of $y'' - 4y' + 4y = 0$. Finally, the Wronskian is

$$\begin{aligned} W(t) &= \det\begin{pmatrix} y_1 & y_2 \\ y_1' & y_2' \end{pmatrix} \\ &= \det\begin{pmatrix} e^{2t} & te^{2t} \\ 2e^{2t} & e^{2t}(2t+1) \end{pmatrix} \\ &= e^{4t}(2t+1) - 2te^{4t} \\ &= e^{4t}, \end{aligned}$$

which is nonzero for all t. Hence, the solutions y_1 and y_2 are linearly independent.

11. If $y_1(t) = e^t$, then

$$y_1''' - 3y_1'' + 3y_1' - y_1 = e^t - 3e^t + 3e^t - e^t = 0.$$

If $y_2(t) = te^t$, then

$$\begin{aligned} y_2' &= (t+1)e^t \\ y_2'' &= (t+2)e^t \\ y_2''' &= (t+3)e^t \end{aligned}$$

and

$$\begin{aligned} &y_2''' - 3y_2'' + 3y_2' - y_2 \\ &\quad = (t+3)e^t - 3(t+2)e^t \\ &\qquad 3(t+1)e^t - te^t \\ &\quad = e^t(t + 3 - 3t - 6 + 3t + 3 - t) \\ &\quad = 0. \end{aligned}$$

If $y_3(t) = t^2e^t$, then

$$\begin{aligned} y_3' &= (t^2 + 2t)e^t \\ y_3'' &= (t^2 + 4t + 2)e^t \\ y_3'' &= (t^2 + 6t + 6)e^t, \end{aligned}$$

and

$$\begin{aligned} &y_3''' - 3y_3'' + 3y_3' - y_3 \\ &\quad = (t^2 + 6t + 6)e^t - 3(t^2 + 4t + 2)e^t \\ &\qquad 3(t^2 + 2t)e^t - t^2e^t \\ &\quad = e^t(t^2 + 6t + 6 \\ &\qquad - 3t^2 - 12t - 6 + 3t^2 + 6t - t^2) \\ &\quad = 0. \end{aligned}$$

Thus, y_1, y_2, and y_3 are solutions of the equation $y''' - 3y'' + 3y' - y = 0$. Finally, the Wronskian is

$$\begin{aligned} W(t) &= \det\begin{pmatrix} y_1 & y_2 & y_3 \\ y_1' & y_2' & y_3' \\ y_1'' & y_2'' & y_3'' \end{pmatrix} \\ &= \det\begin{pmatrix} e^t & te^t & t^2e^t \\ e^t & (t+1)e^t & (t^2+2t)e^t \\ e^t & (t+2)e^t & (t^2+4t+2)e^t \end{pmatrix}. \end{aligned}$$

Using a computer, $W(t) = 2e^{3t}$, which is never zero. Therefore, the solutions y_1, y_2 and y_3 are linearly independent.

---

13. (a) If $y = e^{\lambda t}$, then

$$y' = \lambda e^{\lambda t}, \quad y'' = \lambda^2 e^{\lambda t}, \quad \text{and} \quad y''' = \lambda^3 e^{\lambda t}.$$

Subbing these results into $y''' + ay'' + by' + cy = 0$ gives

$$\begin{aligned} \lambda^3 e^{\lambda t} + a\lambda^2 e^{\lambda t} + b\lambda e^{\lambda t} + ce^{\lambda t} &= 0 \\ e^{\lambda t}(\lambda^3 + a\lambda^2 + b\lambda + c) &= 0. \end{aligned}$$

Because $e^{\lambda t}$ can never equal zero, we must have

$$\lambda^3 + a\lambda^2 + b\lambda + c = 0.$$

(b) If

$$y''' = -ay'' - by' - cy,$$

let $x_1 = y$, $x_2 = y'$, and $x_3 = y''$. Then

$$\begin{aligned} x_1' &= x_2 \\ x_2' &= x_3 \\ x_3' &= -ax_3 - bx_2 - cx_1. \end{aligned}$$

In matrix form,

$$\begin{pmatrix} x_1 \\ x_2 \\ x_3 \end{pmatrix}' = \begin{pmatrix} 0 & 1 & 0 \\ 0 & 0 & 1 \\ -c & -b & -a \end{pmatrix} \begin{pmatrix} x_1 \\ x_2 \\ x_3 \end{pmatrix},$$

and if

$$A = \begin{pmatrix} 0 & 1 & 0 \\ 0 & 0 & 1 \\ -c & -b & -a \end{pmatrix},$$

the characteristic polynomial is

$$p(\lambda) = -\det(A - \lambda I) = -\begin{vmatrix} -\lambda & 1 & 0 \\ 0 & -\lambda & 1 \\ -c & -b & -a - \lambda \end{vmatrix}.$$

Expanding across the first row,

$$\begin{aligned} p(\lambda) &= \lambda \begin{vmatrix} -\lambda & 1 \\ -b & -a - \lambda \end{vmatrix} + 1 \begin{vmatrix} 0 & 1 \\ -c & -a - \lambda \end{vmatrix} \\ &= \lambda(a\lambda + \lambda^2 + b) + 1(c) \\ &= \lambda^3 + a\lambda^2 + b\lambda + c. \end{aligned}$$

15. If $y''' - 3y'' - 4y' + 12y = 0$, then the characteristic equation factors

$$\lambda^3 - 3\lambda^2 - 4\lambda + 12 = 0$$
$$\lambda^2(\lambda - 3) - 4(\lambda - 3) = 0$$
$$(\lambda + 2)(\lambda - 2)(\lambda - 3) = 0.$$

Thus, the characteristic equation has roots -2, 2, and 3, leading to the general solution

$$y(t) = C_1e^{-2t} + C_2e^{2t} + C_3e^{3t}.$$

17. If $y^{(4)} - 13y'' + 36y = 0$, then the characteristic equation factors

$$\lambda^4 - 13\lambda^2 + 36 = 0$$
$$(\lambda^2 - 4)(\lambda^2 - 9) = 0$$
$$(\lambda + 2)(\lambda - 2)(\lambda + 3)(\lambda - 3) = 0.$$

Thus, the characteristic equation has roots -3, -2, 2, and 3, leading to the general solution

$$y(t) = C_1e^{-3t} + C_2e^{-2t} + C_3e^{2t} + C_4e^{3t}.$$

19. If $y''' - 4y'' - 11y' + 30y = 0$, then the characteristic equation is

$$\lambda^3 - 4\lambda^2 - 11\lambda + 30 = 0.$$

A plot of the characteristic equation (computer or calculator) reveals possible roots.

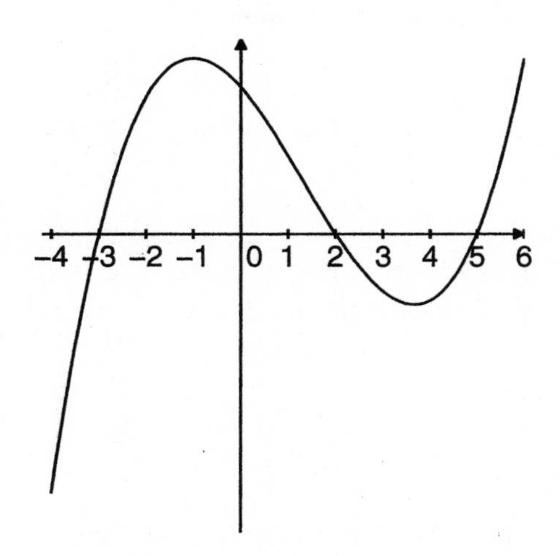

The plot suggests that -3 is a root, but division by $\lambda + 3$ guarantees that -3 is a root and $\lambda + 3$ is a factor.

$$(\lambda + 3)(\lambda^2 - 7\lambda + 10) = 0$$
$$(\lambda + 3)(\lambda - 2)(\lambda - 5) = 0$$

Thus, the roots are -3, 2, and 5, and the general solution is

$$y(t) = C_1e^{-3t} + C_2e^{2t} + C_3e^{5t}.$$

21. If $y^{(5)} - 4y^{(4)} - 13y''' + 52y'' + 36y' - 144y = 0$, then the characteristic equation is

$$\lambda^5 - 4\lambda^4 - 13\lambda^3 + 52\lambda^2 + 36\lambda - 144 = 0.$$

A plot of the characteristic equation (computer or calculator) reveals possible roots.

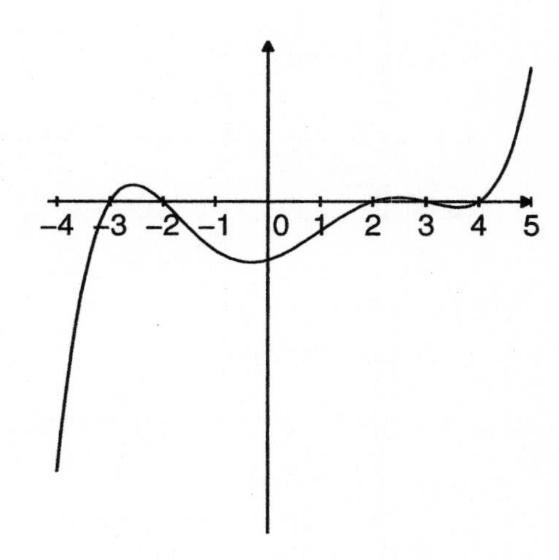

The plot suggests a root at -3. Long (or synthetic) division reveals

$$(\lambda + 3)(\lambda^4 - 7\lambda^3 + 8\lambda^2 + 28\lambda - 48) = 0.$$

The plot suggests a root at -2. Again, division reveals

$$(\lambda + 3)(\lambda + 2)(\lambda^3 - 9\lambda^2 + 26\lambda - 24) = 0.$$

The plot suggest a root at 2. Again, division reveals

$$(\lambda + 3)(\lambda + 2)(\lambda - 2)(\lambda^2 - 7\lambda + 12) = 0$$
$$(\lambda + 3)(\lambda + 2)(\lambda - 2)(\lambda - 3)(\lambda - 4) = 0.$$

Thus, the roots of the characteristic equation are -3, -2, 2, 3, and 4, and the general solution is

$$y(t) = C_1e^{-3t} + C_2e^{-2t} + C_3e^{2t} + C_4e^{3t} + C_5e^{4t}.$$

23. If $y''' + y'' - 8y' - 12y = 0$, then the characteristic equation is

$$\lambda^3 + \lambda^2 - 8\lambda - 12 = 0.$$

The plot of the characteristic equation

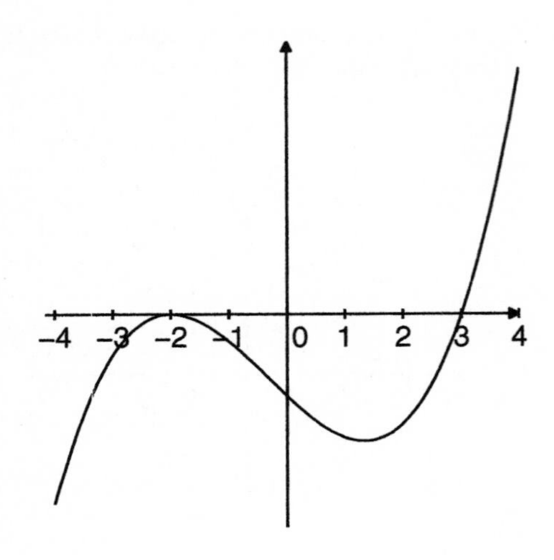

suggests a root at -2. Division shows that

$$(\lambda + 2)(\lambda^2 - \lambda - 6) = 0$$
$$(\lambda + 2)(\lambda + 2)(\lambda - 3) = 0.$$

Hence there are two roots, -2 and 3, with the former having algebraic multiplicity 2. Thus, the general solution is

$$y(t) = C_1e^{-2t} + C_2te^{-2t} + C_3e^{3t}.$$

25. If $y''' + 3y'' + 3y' + y = 0$, then the characteristic equation is

$$\lambda^3 + 3\lambda^2 + 3\lambda + 1 = 0.$$

The plot of the characteristic equation

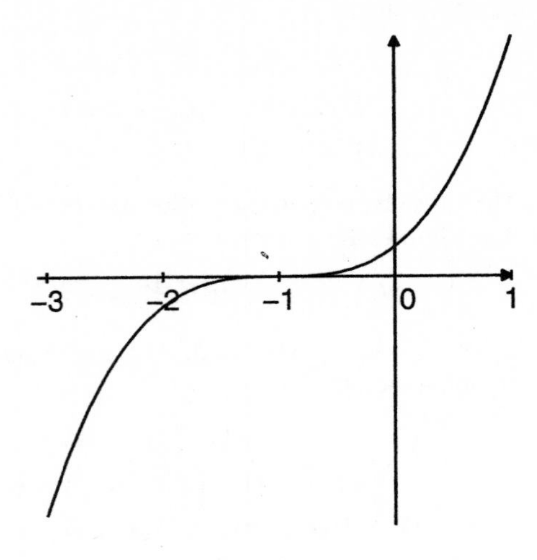

suggests a root at -1. Division show that

$$(\lambda + 1)(\lambda^2 + 2\lambda + 1) = 0$$
$$(\lambda + 1)^3 = 0.$$

Thus, -1 is a root of algebraic multiplicity 3. Therefore, the general solution is

$$y(t) = C_1e^{-t} + C_2te^{-t} + C_3t^2e^{-t}.$$

27. If $y^{(5)} - y^{(4)} - 6y''' + 14y'' - 11y' + 3y = 0$, then the characteristic equation is

$$\lambda^5 - \lambda^4 - 6\lambda^3 + 14\lambda^2 - 11\lambda + 3 = 0.$$

The plot of the characteristic equation

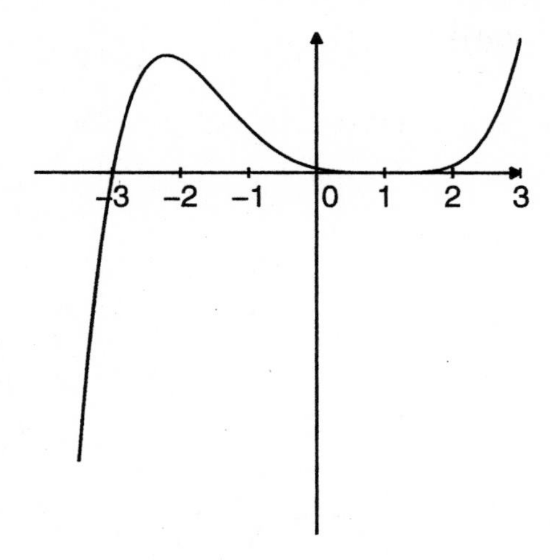

suggests a multiple root at 1. Repeated division by $\lambda - 1$ reveals that

$$(\lambda - 1)^4(\lambda + 3) = 0.$$

Thus, the roots are 1 and -3, with the former having algebraic multiplicity 4. Therefore, the general solution is

$$y(t) = C_1e^t + C_2te^t + C_3t^2e^t + C_4t^3e^t + C_5e^{-3t}.$$

29. If $y''' - y'' + 2y = 0$, then the characteristic equation is

$$\lambda^3 - \lambda^2 + 2 = 0.$$

A plot of the characteristic equation

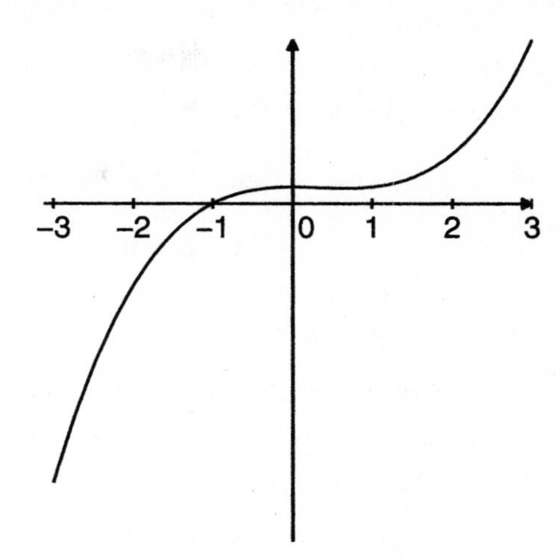

suggests a root at -1. Division reveals

$$(\lambda + 1)(\lambda^2 - 2\lambda + 2) = 0.$$

The quadratic formula provides the remaining roots, $1 \pm i$. Thus, the general solution is

$$y(t) = C_1e^{-t} + C_2e^t \cos t + C_3e^t \sin t.$$

31. If $y^{(4)} + 2y'' + y = 0$, then the characteristic equation is

$$\lambda^4 + 2\lambda^2 + 1 = 0,$$

which easily factors as

$$(\lambda^2 + 1)^2 = 0.$$

Thus, both i and $-i$ are roots of multiplicity 2. Therefore, the general solution is

$$y(t) = C_1 \cos t + C_2t \cos t + C_3 \sin t + C_4t \sin t.$$

33. If $y^{(6)} + 3y^{(4)} + 3y'' + y = 0$, then the characteristic equation is

$$\lambda^6 + 3\lambda^4 + 3\lambda^2 + 1 = 0.$$

The form of the characteristic equation suggests the binomial theorem and

$$(\lambda^2 + 1)^3 = 0.$$

Thus, i and $-i$ are roots, each having algebraic multiplicity 3. Therefore, the general solution is

$$y(t) = C_1 \cos t + C_2 t \cos t + C_3 t^2 \cos t + C_4 \sin t + C_5 t \sin t + C_6 t^2 \sin t.$$

35. If $y'' + 2y' + 5y = 0$, then the characteristic equation is

$$\lambda^2 + 2\lambda + 5 = 0.$$

The quadratic formula provides the roots, $-1 \pm 2i$. Thus, the general solution is

$$y(t) = C_1 e^{-t} \cos 2t + C_2 e^{-t} \sin 2t.$$

Substituting the initial condition $y(0) = 2$ provides $C_1 = 2$. The derivative of the general solution is

$$y'(t) = C_1 e^{-t}(-\cos 2t - 2\sin 2t) + C_2 e^{-t}(-\sin 2t + 2\cos 2t).$$

The initial condition $y'(0) = 0$ provides

$$0 = -C_1 + 2C_2,$$

which in turn, because $C_1 = 2$, generates $C_2 = 1$. Thus, the solution of the initial value problem is

$$y(t) = 2e^{-t} \cos 2t + e^{-t} \sin 2t.$$

37. If $y'' - 2y' + y = 0$, then the characteristic equation is

$$\lambda^2 - 2\lambda + 1 = (\lambda - 1)^2 = 0.$$

Thus, 1 is a single root of multiplicity 2, and the general solution is

$$y(t) = C_1 e^t + C_2 t e^t.$$

The initial condition $y(0) = 1$ provides $C_1 = 1$. The derivative of the general solution is

$$y'(t) = C_1 e^t + C_2(t + 1)e^t.$$

The initial condition $y'(0) = 0$ provides

$$0 = C_1 + C_2,$$

which in turn, because $C_1 = 1$, generates $C_2 = -1$. Therefore, the solution of the initial value problem is

$$y(t) = e^t - te^t.$$

39. If $y''' - 7y'' + 11y' - 5y = 0$, then the characteristic equation is

$$\lambda^3 - 7\lambda^2 + 11\lambda - 5 = 0.$$

The plot of the characteristic equation

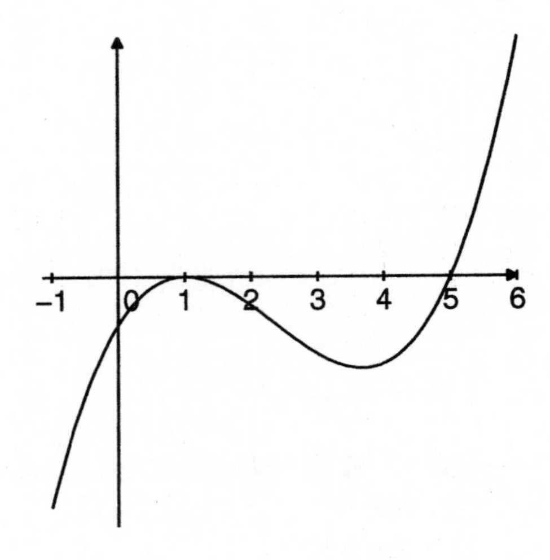

suggest a root at 1. Division reveals

$$(\lambda - 1)(\lambda^2 - 6\lambda + 5) = 0$$
$$(\lambda - 1)^2(\lambda - 5) = 0.$$

Thus, the roots are 1 and 5, the former having algebraic multiplicity 2. Thus, the general solution is

$$y(t) = C_1 e^t + C_2 t e^t + C_3 e^{5t}.$$

The initial condition $y(0) = -1$ provides

$$C_1 + C_3 = -1.$$

The derivative of the general solution is

$$y'(t) = C_1 e^t + C_2 e^t(t + 1) + 5C_3 e^{5t}.$$

The initial condition $y'(0) = 1$ provides

$$C_1 + C_2 + 5C_3 = 1.$$

The second derivative of the general solution is

$$y''(t) = C_1e^t + C_2e^t(t+2) + 25C_3e^{5t}.$$

The initial condition $y''(0) = 0$ provides

$$C_1 + 2C_2 + 25C_3 = 0.$$

The augmented matrix

$$\begin{pmatrix} 1 & 0 & 1 & -1 \\ 1 & 1 & 5 & 1 \\ 1 & 2 & 25 & 0 \end{pmatrix} \rightarrow \begin{pmatrix} 1 & 0 & 0 & -13/16 \\ 0 & 1 & 0 & 11/4 \\ 0 & 0 & 1 & -3/16 \end{pmatrix}$$

provides $C_1 = -13/16$, $C_2 = 11/4$, and $C_3 = -3/16$. Therefore, the solution of the initial value problem is

$$y(t) = -\frac{13}{16}e^t + \frac{11}{4}te^t - \frac{3}{16}e^{5t}.$$

41. If $y''' - 6y'' + 12y' - 8y = 0$, then the characteristic equation is

$$\lambda^3 - 6\lambda^2 + 12\lambda - 8 = 0.$$

The plot of the characteristic equation

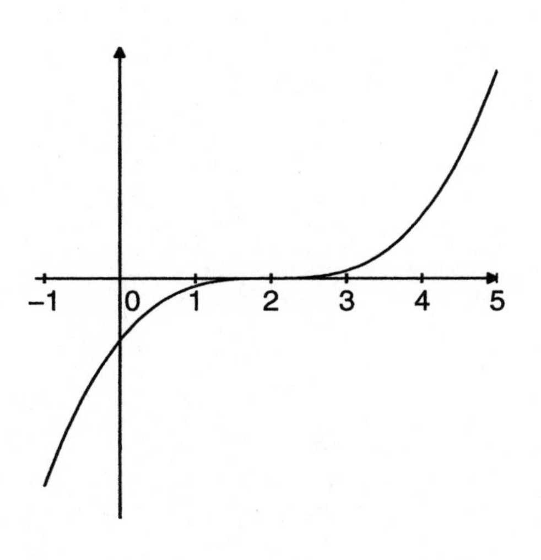

suggest a multiple root at 2. Repeated division by $\lambda - 2$ reveals

$$(\lambda - 2)^3 = 0.$$

Thus, the characteristic polynomial has a single root, 2, with algebraic multiplicity 3. Thus, the general solution is

$$y(t) = C_1e^{2t} + C_2te^{2t} + C_3t^2e^{2t}.$$

The initial condition $y(0) = -2$ provides $C_1 = -2$. The derivative of the general solution is

$$y'(t) = 2C_1e^{2t} + C_2e^{2t}(2t+1) + C_3e^{2t}(2t^2 + 2t).$$

The initial condition $y'(0) = 0$ provides

$$2C_1 + C_2 = 0,$$

which in turn, because $C_1 = -2$, provides $C_2 = 4$. The second derivative of the general solution is

$$y''(t) = 4C_1e^{2t} + C_2e^{2t}(4+4t) + C_3e^{2t}(4t^2+8t+2).$$

The initial condition $y''(0) = 2$ provides

$$4C_1 + 4C_2 + 2C_3 = 0,$$

which in turn, because $C_1 = -2$ and $C_2 = 4$, provides $C_3 = -3$. Therefore, the solution of the initial value problem is

$$y(t) = -2e^{2t} + 4te^{2t} - 3t^2e^{2t}.$$

43. If $y^{(4)} + 8y'' + 16y = 0$, then the characteristic equation is

$$\lambda^4 + 8\lambda^2 + 16 = (\lambda^2 + 4)^2 = 0.$$

Therefore, the roots are $\pm 2i$, each of which has algebraic multiplicity 2. Therefore, the general solution if

$$y(t) = C_1\cos 2t + C_2t\cos 2t + C_3\sin 2t + C_4t\sin 2t.$$

The initial condition $y(0) = 0$ provides $C_1 = 0$. The derivative of the general solution is

$$\begin{aligned} y'(t) &= -2C_1\sin 2t + C_2(\cos 2t - 2t\sin 2t) \\ &\quad + 2C_3\cos 2t + C_4(\sin 2t + 2t\cos 2t). \end{aligned}$$

The initial condition $y'(0) = -1$ generates $C_2 + 2C_3 = -1$. The second derivative of the general solution is

$$\begin{aligned} y''(t) &= -4C_1\cos 2t + C_2(-4\sin 2t - 4t\cos 2t) \\ &\quad - 4C_3\sin 2t + C_4(4\cos 2t - 4t\sin 2t). \end{aligned}$$

The initial condition $y''(0) = 2$ generates $-4C_1 + 4C_4 = 2$. The third derivative of the general solution is:

$$y'''(t) = 8C_1 \sin 2t + C_2(-12\cos 2t + 8t \sin 2t) - 8C_3 \cos 2t + C_4(-12 \sin 2t - 8t \cos 2t)$$

The initial condition $y'''(0) = 0$ generates $-12C_2 - 8C_3 = 0$. The augmented matrix

$$\begin{pmatrix} 1 & 0 & 0 & 0 & 0 \\ 0 & 1 & 2 & 0 & -1 \\ -4 & 0 & 0 & 4 & 2 \\ 0 & -12 & -8 & 0 & 0 \end{pmatrix}$$

reduces to

$$\begin{pmatrix} 1 & 0 & 0 & 0 & 0 \\ 0 & 1 & 0 & 0 & 1/2 \\ 0 & 0 & 1 & 0 & -3/4 \\ 0 & 0 & 0 & 1 & 1/2 \end{pmatrix}.$$

Thus, $C_1 = 0$, $C_2 = 1/2$, $C_3 = -3/4$, and $C_4 = 1/2$. Therefore, the solution of the initial value problem is

$$y(t) = \frac{1}{2}t \cos 2t - \frac{3}{4} \sin 2t + \frac{1}{2}t \sin 2t.$$

45. First, we must show that the set V is closed under addition. Let $\mathbf{a}$ and $\mathbf{b}$ be elements of $V \subset \mathbf{R}^q$. Then, $y_{\mathbf{a}}$ and $y_{\mathbf{b}}$ are solutions of

$$y^{(n)} + a_1 y^{(n-1)} + \cdots + a_{n-1}y' + a_n y = 0. \quad (*)$$

However, $y_{\mathbf{a}} + y_{\mathbf{b}}$, being a linear combination of solutions of $(*)$, is also a solution of $(*)$. However, by 8.31,

$$y_{\mathbf{a}} + y_{\mathbf{b}} = y_{\mathbf{a}+\mathbf{b}}.$$

Recall that the set $V \subset \mathbf{R}^q$ is defined

$$V = \{\mathbf{a} : \ y_{\mathbf{a}} \text{ is a solution of } (*)\}.$$

Therefore, $y_{\mathbf{a}+\mathbf{b}}$ is a solution of $(*)$ and $\mathbf{a} + \mathbf{b} \in V$. Therefore, V is closed under addition. Next, we must show that V is closed under scalar multiplication. Let $\mathbf{a} \in V$ and let $\alpha \in \mathbf{R}$ be a scalar. Then, by definition of V, $y_{\mathbf{a}}$ is a solution of $(*)$. However, $\alpha y_{\mathbf{a}}$, being a linear combination of solutions of $(*)$, is also a solution of $(*)$. By 8.31,

$$\alpha y_{\mathbf{a}} = y_{\alpha\mathbf{a}}.$$

Hence, $y_{\alpha\mathbf{a}}$ is a solution of $(*)$ and $\alpha\mathbf{a} \in V$. Therefore, V is closed under scalar multiplication and is a subspace of $\mathbf{R}^q$.

Section 9.9. Inhomogeneous Systems

1. If

$$A = \begin{pmatrix} 5 & 6 \\ -2 & -2 \end{pmatrix} \quad \text{and} \quad \mathbf{f} = \begin{pmatrix} e^t \\ e^t \end{pmatrix},$$

then the characteristic polynomial is

$$p(\lambda) = \lambda^2 - T\lambda + D = \lambda^2 - 3\lambda + 2 = (\lambda - 1)(\lambda - 2),$$

generating eigenvalues $\lambda_1 = 1$ and $\lambda_2 = 2$. The associated eigenvectors are

$$A - I = \begin{pmatrix} 4 & 6 \\ -2 & -3 \end{pmatrix} \Rightarrow \mathbf{v}_1 = \begin{pmatrix} 3 \\ -2 \end{pmatrix}, \text{ and}$$

$$A - 2I = \begin{pmatrix} 3 & 6 \\ -2 & -4 \end{pmatrix} \Rightarrow \mathbf{v}_2 = \begin{pmatrix} 2 \\ -1 \end{pmatrix}.$$

Thus, the homogeneous solution is $\mathbf{y}_h = C_1\mathbf{y}_1 +$

$C_2\mathbf{y}_2$, where

$$\mathbf{y}_1(t) = e^t\mathbf{v}_1 = \begin{pmatrix} 3e^t \\ -2e^t \end{pmatrix} \quad \text{and}$$
$$\mathbf{y}_2(t) = e^{2t}\mathbf{v}_2 = \begin{pmatrix} 2e^{2t} \\ -e^{2t} \end{pmatrix}.$$

The fundamental matrix is

$$Y(t) = [\mathbf{y}_1(t), \mathbf{y}_2(t)] = \begin{pmatrix} 3e^t & 2e^{2t} \\ -2e^t & -e^{2t} \end{pmatrix}.$$

The inverse[1] of $Y(t)$ is calculated

$$Y^{-1}(t) = \frac{1}{e^{3t}}\begin{pmatrix} -e^{2t} & -2e^{2t} \\ 2e^t & 3e^t \end{pmatrix} = \begin{pmatrix} -e^{-t} & -2e^{-t} \\ 2e^{-2t} & 3e^{-2t} \end{pmatrix}.$$

Hence,

$$Y^{-1}(t)\mathbf{f}(t) = \begin{pmatrix} -e^{-t} & -2e^{-t} \\ 2e^{-2t} & 3e^{-2t} \end{pmatrix}\begin{pmatrix} e^t \\ e^t \end{pmatrix} = \begin{pmatrix} -3 \\ 5e^{-t} \end{pmatrix},$$

and

$$\int Y^{-1}(t)\mathbf{f}(t)\,dt = \int \begin{pmatrix} -3 \\ 5e^{-t} \end{pmatrix} dt = \begin{pmatrix} -3t \\ -5e^{-t} \end{pmatrix}.$$

Thus,

$$\begin{aligned} \mathbf{y}_p &= Y(t)\int Y^{-1}\mathbf{f}(t)\,dt \\ &= \begin{pmatrix} 3e^t & 2e^{2t} \\ -2e^t & -e^{2t} \end{pmatrix}\begin{pmatrix} -3t \\ -5e^{-t} \end{pmatrix} \\ &= \begin{pmatrix} -9te^t - 10e^t \\ 6te^t + 5e^t \end{pmatrix}. \end{aligned}$$

Finally, the general solution is

$$\mathbf{y}(t) = C_1e^t\begin{pmatrix} 3 \\ -2 \end{pmatrix} + C_2e^{2t}\begin{pmatrix} 2 \\ -1 \end{pmatrix} + \begin{pmatrix} -9te^t - 10e^t \\ 6te^t + 5e^t \end{pmatrix}.$$

3. If

$$A = \begin{pmatrix} -3 & 6 \\ -2 & 4 \end{pmatrix} \quad \text{and} \quad \mathbf{f} = \begin{pmatrix} 3 \\ 4 \end{pmatrix},$$

then the characteristic polynomial is

$$p(\lambda) = \lambda^2 - T\lambda + D = \lambda^2 - \lambda = \lambda(\lambda - 1),$$

generating eigenvalues $\lambda_1 = 0$ and $\lambda_2 = 1$. The associated eigenvectors are

$$A - 0I = \begin{pmatrix} -3 & 6 \\ -2 & 4 \end{pmatrix} \Rightarrow \mathbf{v}_1 = \begin{pmatrix} 2 \\ 1 \end{pmatrix}, \quad \text{and}$$
$$A - I = \begin{pmatrix} -4 & 6 \\ -2 & 3 \end{pmatrix} \Rightarrow \mathbf{v}_2 = \begin{pmatrix} 3 \\ 2 \end{pmatrix}.$$

Thus, the homogeneous solution is $\mathbf{y}_h = C_1\mathbf{y}_1 + C_2\mathbf{y}_2$, where

$$\mathbf{y}_1(t) = e^{0t}\mathbf{v}_1 = \begin{pmatrix} 2 \\ 1 \end{pmatrix} \quad \text{and}$$
$$\mathbf{y}_2(t) = e^t\mathbf{v}_2 = \begin{pmatrix} 3e^t \\ 2e^t \end{pmatrix}.$$

The fundamental matrix is

$$Y(t) = [\mathbf{y}_1(t), \mathbf{y}_2(t)] = \begin{pmatrix} 2 & 3e^t \\ 1 & 2e^t \end{pmatrix}.$$

The inverse of $Y(t)$ is calculated

$$Y^{-1}(t) = \frac{1}{e^t}\begin{pmatrix} 2e^t & -3e^t \\ -1 & 2 \end{pmatrix} = \begin{pmatrix} 2 & -3 \\ -e^{-t} & 2e^{-t} \end{pmatrix}.$$

Hence,

$$\begin{aligned} Y^{-1}(t)\mathbf{f}(t) &= \begin{pmatrix} 2 & -3 \\ -e^{-t} & 2e^{-t} \end{pmatrix}\begin{pmatrix} 3 \\ 4 \end{pmatrix} \\ &= \begin{pmatrix} -6 \\ 5e^{-t} \end{pmatrix}, \end{aligned}$$

[1] Perhaps the easiest way to invert a 2×2 matrix is to use the following fact:

$$A = \begin{pmatrix} a & b \\ c & d \end{pmatrix} \Rightarrow A^{-1} = \frac{1}{\det(A)}\begin{pmatrix} d & -b \\ -c & a \end{pmatrix}.$$

and

$$\int Y^{-1}(t)\mathbf{f}(t)\,dt = \int \begin{pmatrix} -6 \\ 5e^{-t} \end{pmatrix} dt = \begin{pmatrix} -6t \\ -5e^{-t} \end{pmatrix}.$$

Thus,

$$\begin{aligned} \mathbf{y}_p &= Y(t) \int Y^{-1}\mathbf{f}(t)\,dt \\ &= \begin{pmatrix} 2 & 3e^t \\ 1 & 2e^t \end{pmatrix} \begin{pmatrix} -6t \\ -5e^{-t} \end{pmatrix} \\ &= \begin{pmatrix} -12t - 15 \\ -6t - 10 \end{pmatrix}. \end{aligned}$$

Finally, the general solution is

$$\mathbf{y}(t) = C_1 \begin{pmatrix} 2 \\ 1 \end{pmatrix} + C_2 \begin{pmatrix} 3e^t \\ 2e^t \end{pmatrix} + \begin{pmatrix} -12t - 15 \\ -6t - 10 \end{pmatrix}.$$

5. A has eigenvalues $2 \pm i$, and associated with the eigenvalue $2+i$ has eigenvector $\mathbf{w} = (-1-i, 1)^T$. Hence the homogenous equation has complex solution

$$\begin{aligned} \mathbf{z}(t) &= e^{(2+i)t} \begin{pmatrix} -1-i \\ 1 \end{pmatrix} \\ &= e^{2t}[\cos t + i \sin t]\left[\begin{pmatrix} -1 \\ 1 \end{pmatrix} + i \begin{pmatrix} -1 \\ 0 \end{pmatrix}\right] \\ &= e^{2t}\left\{\begin{pmatrix} \sin t - \cos t \\ \cos t \end{pmatrix} + i \begin{pmatrix} -\cos t - \sin t \\ \sin t \end{pmatrix}\right\} \end{aligned}$$

Thus the homogeneous equation has the real solutions

$$\mathbf{y}_1(t) = \operatorname{Re}\mathbf{z}(t) = e^{2t}\begin{pmatrix} \sin t - \cos t \\ \cos t \end{pmatrix} \quad \text{and}$$

$$\mathbf{y}_2(t) = \operatorname{Im}\mathbf{z}(t) = e^{2t}\begin{pmatrix} -\cos t - \sin t \\ \sin t \end{pmatrix}.$$

The fundamental matrix is

$$Y(t) = e^{2t}\begin{pmatrix} \sin t - \cos t & -\cos t - \sin t \\ \cos t & \sin t \end{pmatrix}.$$

Its inverse is

$$Y^{-1}(t) = e^{-2t}\begin{pmatrix} \sin t & \cos t + \sin t \\ -\cos t & \sin t - \cos t \end{pmatrix}.$$

Hence

$$\begin{aligned} Y^{-1}(t)\mathbf{f}(t) &= e^{-2t}\begin{pmatrix} \sin t & \cos t + \sin t \\ -\cos t & \sin t - \cos t \end{pmatrix}\begin{pmatrix} 0 \\ e^{2t} \end{pmatrix} \\ &= \begin{pmatrix} \sin t + \cos t \\ \sin t - \cos t \end{pmatrix}, \end{aligned}$$

and

$$\int Y^{-1}(t)\mathbf{f}(t)\,dt = \begin{pmatrix} \sin t - \cos t \\ -\cos t - \sin t \end{pmatrix}.$$

Then the particular solution is

$$\begin{aligned} \mathbf{y}_p(t) &= Y(t)\int Y^{-1}\mathbf{f}(t)\,dt \\ &= e^{2t}\begin{pmatrix} 2 \\ -1 \end{pmatrix}. \end{aligned}$$

The general solution is

$$\begin{aligned} \mathbf{y}(t) &= C_1 e^{2t}\begin{pmatrix} \sin t - \cos t \\ \cos t \end{pmatrix} \\ &\quad + C_2 e^{2t}\begin{pmatrix} -\cos t - \sin t \\ \sin t \end{pmatrix} + e^{2t}\begin{pmatrix} 2 \\ -1 \end{pmatrix}. \end{aligned}$$

7. A has eigenvalues 0, 2, and 1, with corresponding eigenvalues $(-1, 2, 0)^T$, $(1, 0, 1)^T$, and $(0, 3, 1)^T$. Thus

$$V = \begin{pmatrix} -1 & e^{2t} & 0 \\ 2 & 0 & 3e^t \\ 0 & e^{2t} & e^t \end{pmatrix}.$$

If we form the augmented matrix $[V, I]$ and use row operations to reduce to row echelon form $[I, V^{-1}]$, we discover that

$$V^{-1} = \begin{pmatrix} -3 & -1 & 3 \\ -2e^{-2t} & -e^{-2t} & 3e^{-2t} \\ 2e^{-t} & e^{-t} & -2e^{-t} \end{pmatrix}.$$

Then

$$V^{-1}\mathbf{f} = \begin{pmatrix} \sin t \\ 0 \\ 0 \end{pmatrix}.$$

Hence $\int V^{-1}\mathbf{f}\,dt = (-\cos t, 0, 0)^T$, and the particular solution is

$$\begin{aligned}\mathbf{y}_p(t) &= V\int V^{-1}\mathbf{f}\,dt\\ &= \begin{pmatrix}-1 & e^{2t} & 0\\ 2 & 0 & 3e^t\\ 0 & e^{2t} & e^t\end{pmatrix}\begin{pmatrix}-\cos t\\ 0\\ 0\end{pmatrix}\\ &= \begin{pmatrix}\cos t\\ -2\cos t\\ 0\end{pmatrix}.\end{aligned}$$

The general solution is

$$\mathbf{y}(t) = C_1\begin{pmatrix}-1\\ 2\\ 0\end{pmatrix} + C_2\begin{pmatrix}e^{2t}\\ 0\\ e^{2t}\end{pmatrix} + C_3\begin{pmatrix}0\\ 3e^t\\ e^t\end{pmatrix} + \begin{pmatrix}\cos t\\ -2\cos t\\ 0\end{pmatrix}.$$

9. A has eigenvalues 0, -2, and -1, with corresponding eigenvectors $(-1, 4, 1)^T$, $(-3, 2, 0)^T$, and $(0, 3, 1)^T$, so

$$V = \begin{pmatrix}-1 & -3e^{-2t} & 0\\ 4 & 2e^{-2t} & 3e^{-t}\\ 1 & 0 & e^{-t}\end{pmatrix}.$$

If we form the augmented matrix $[V, I]$ and use row operations to reduce to row echelon form $[I, V^{-1}]$, we discover that

$$V^{-1} = \begin{pmatrix}2 & 3 & -9\\ -e^{2t} & -e^{2t} & 3e^{2t}\\ -2e^t & -3e^t & 10e^t\end{pmatrix}.$$

Then $V^{-1}\mathbf{f} = (6, -2e^{2t}, -6e^t)^T$, and $\int V^{-1}\mathbf{f}\,dt = (6t, -e^{2t}, -6e^t)^T$. Hence the particular solution is

$$\begin{aligned}\mathbf{y}_p(t) &= V\int V^{-1}\mathbf{f}\,dt\\ &= \begin{pmatrix}-1 & -3e^{-2t} & 0\\ 4 & 2e^{-2t} & 3e^{-t}\\ 1 & 0 & e^{-t}\end{pmatrix}\begin{pmatrix}6t\\ -e^{2t}\\ -6e^t\end{pmatrix}\\ &= \begin{pmatrix}3-6t\\ 24t-20\\ 6t-6\end{pmatrix}.\end{aligned}$$

The general solution is

$$\mathbf{y}(t) = C_1\begin{pmatrix}-1\\ 4\\ 1\end{pmatrix} + C_2\begin{pmatrix}-3e^{-2t}\\ 2e^{-2t}\\ 0\end{pmatrix} + C_3\begin{pmatrix}0\\ 3e^{-t}\\ e^{-t}\end{pmatrix} + \begin{pmatrix}3-6t\\ 24t-20\\ 6t-6\end{pmatrix}.$$

11. The independent solutions $\mathbf{y}_1$ and $\mathbf{y}_2$ in Example 9.16 provide a fundamental matrix

$$\begin{aligned}Y &= [\mathbf{y}_1, \mathbf{y}_2]\\ &= \epsilon^{-2t/5}\begin{pmatrix}-5c(t) & -5s(t)\\ 3c(t)-4s(t) & 4c(t)+3s(t)\end{pmatrix}.\end{aligned}$$

where $c(t) = \cos(4t/5)$ ans $s(t) = \sin(4t/5)$. It's actually possible to invert this by hand if you use the fact that $(\alpha A)^{-1} = (1/\alpha)A^{-1}$ and

$$\begin{pmatrix}a & b\\ c & d\end{pmatrix}^{-1} = \frac{1}{ad-bc}\begin{pmatrix}d & -b\\ -c & a\end{pmatrix}.$$

Hence,

$$Y^{-1} = e^{2t/5}\frac{1}{-20}\begin{pmatrix}4c(t)+3s(t) & 5s(t)\\ -3c(t)+4s(t) & -5c(t)\end{pmatrix}.$$

Then,

$$Y^{-1}\mathbf{f} = e^{2t/5}\begin{pmatrix}-c(t)+(1/2)s(t)\\ -(1/2)c(t)-s(t)\end{pmatrix}.$$

At this point, we're facing a tricky integration by parts, so we will use a computer to integrate for us.

$$\mathbf{v} = \int Y^{-1}\mathbf{f} = e^{2t/5}\begin{pmatrix}-c(t)-(3/4)s(t)\\ (3/4)c(t)-s(t)\end{pmatrix}.$$

Finally, we will again let the computer do the multiplication in the final step,

$$\mathbf{y}_p = Y\mathbf{v} = \begin{pmatrix}5\\ 0\end{pmatrix},$$

which is the same particular solution found in Example 9.16.

13. The natural guess is $\mathbf{y}_p = \mathbf{a}$, where $\mathbf{a} = (a_1, a_2)^T$. thus, $\mathbf{y}_p' = \mathbf{0}$, so consequently, $A\mathbf{y}_p + \mathbf{f} = \mathbf{0}$. Thus,

$$\mathbf{y}_p = A^{-1}(-\mathbf{f}) = \begin{pmatrix} -6 & 8 \\ -4 & 6 \end{pmatrix}^{-1} \begin{pmatrix} -2 \\ -3 \end{pmatrix} = \begin{pmatrix} -3 \\ -5/2 \end{pmatrix}.$$

Thus, $\mathbf{y}_p = (-3, -5/2)^T$.

15. If $\mathbf{y}_p = \mathbf{a}\cos t + \mathbf{b}\sin t$, whee $\mathbf{a} = (a_1, a_2)^T$ and $\mathbf{b} = (b_1, b_2)^T$, then

$$\mathbf{y}_p' = -\mathbf{a}\sin t + \mathbf{b}\cos t = \begin{pmatrix} -a_1 \sin t + b_1 \cos t \\ -a_2 \sin t + b_2 \cos t \end{pmatrix}.$$

On the other hand,

$$\begin{aligned} A\mathbf{y}_p + \mathbf{f} &= A(\mathbf{a}\cos t + \mathbf{b}\sin t) + \mathbf{f} \\ &= \begin{pmatrix} -2 & 0 \\ 1 & -1 \end{pmatrix} \begin{pmatrix} a_1 \\ a_2 \end{pmatrix} \cos t \\ &\quad + \begin{pmatrix} -2 & 0 \\ 1 & -1 \end{pmatrix} \begin{pmatrix} b_1 \\ b_2 \end{pmatrix} \sin t + \begin{pmatrix} 0 \\ \sin t \end{pmatrix} \\ &= \begin{pmatrix} -2a_1 \cos t - 2b_1 \sin t \\ (a_1 - a_2)\cos t + (b_1 - b_2 + 1)\sin t \end{pmatrix}. \end{aligned}$$

Comparing coefficients of sine and cosine in the vector components, we have the following system of equations.

$$\begin{aligned} 2a_1 + b_1 &= 0 \\ a_1 - 2b_1 &= 0 \\ a_1 - a_2 - b_2 &= 0 \\ a_2 + b_1 - b_2 &= -1 \end{aligned}$$

This system is easily solved and $a_1 = 0$, $a_2 = -1/2$, $b_1 = 0$, and $b_2 = 1/2$. Thus,

$$\begin{aligned} \mathbf{y}_p &= \begin{pmatrix} 0 \\ -1/2 \end{pmatrix} \cos t + \begin{pmatrix} 0 \\ 1/2 \end{pmatrix} \sin t \\ &= \begin{pmatrix} 0 \\ (-1/2)\cos t + (1/2)\sin t \end{pmatrix}. \end{aligned}$$

17. If $\mathbf{y}_p = (\mathbf{a}t + \mathbf{b})e^{-t}$, then

$$\begin{aligned} \mathbf{y}_p' &= \mathbf{a}e^{-t} - (\mathbf{a}t + \mathbf{b})e^{-t} = [-\mathbf{a}t + (\mathbf{a} - \mathbf{b})]e^{-t} \\ &= \begin{pmatrix} -a_1 t + (a_1 - b_1) \\ -a_2 t + (a_2 - b_2) \end{pmatrix} e^{-t}. \end{aligned}$$

Next,

$$\begin{aligned} A\mathbf{y}_p + \mathbf{f} &= A(\mathbf{a}t + \mathbf{b})e^{-t} + \mathbf{f} \\ &= \begin{pmatrix} 1 & 2 \\ 2 & 1 \end{pmatrix} \begin{pmatrix} a_1 t + b_1 \\ a_2 t + b_2 \end{pmatrix} e^{-t} + \begin{pmatrix} 1 \\ 0 \end{pmatrix} e^{-t}, \\ &= \begin{pmatrix} (a_1 + 2a_2)t + (b_1 + 2b_2 + 1) \\ (2a_1 + a_2)t + (2b_1 + b_2) \end{pmatrix} e^{-t}. \end{aligned}$$

Thus,

$$\begin{aligned} &\begin{pmatrix} -a_1 t + (a_1 - b_1) \\ -a_2 t + (a_2 - b_2) \end{pmatrix} \\ &\quad = \begin{pmatrix} (a_1 + 2a_2)t + (b_1 + 2b_2 + 1) \\ (2a_1 + a_2)t + (2b_1 + b_2) \end{pmatrix}. \end{aligned}$$

Comparing coefficients of the polynomial entries in these vectors (e.g., $-a_1 = (a_1 + 2a_2)$) leads to the system

$$\begin{aligned} a_1 + a_2 &= 0 \\ a_1 - 2b_1 - 2b_2 &= 1 \\ a_2 - 2b_1 - 2b_2 &= 0. \end{aligned}$$

The solution of this system is $a_1 = 1/2$, $a_2 = -1/2$, $b_1 = -1/4 - b_2$, where b_2 is free. Choosing $b_2 = -1/8$ provides the solution $a_1 = 1/2$, $a_2 = -1/2$, $b_1 = -1/8$, and $b_2 = -1/8$, so

$$\mathbf{y}_p = (\mathbf{a}t + \mathbf{b})e^{-t} = \begin{pmatrix} (1/2)t - 1/8 \\ (-1/2)t - 1/8 \end{pmatrix} e^{-t},$$

which is identical to that found in Example 9.15.

19. (a) The current entering node "a" must equal the current coming out of node "a".

$$i_1 = i_2 + i_3 \tag{9.9}$$

Traversing (clockwise) the leftmost loop containing emf, resistor, inductor, and resistor, KVL provides

$$-E + 8i_1 + 2i_1' + 4i_3 = 0. \tag{9.10}$$

Traversing (clockwise) the rightmost loop containing inductor and resistor, KVL provides

$$i_2' - 4i_3 = 0. \tag{9.11}$$

Substitute (9.9) in both (9.10) and (9.11), then solve the resulting equations for i_1' and i_2'.

$$i_1' = -6i_1 + 2i_2 + \sin t \qquad (9.12)$$
$$i_2' = 4i_1 - 4i_2 \qquad (9.13)$$

Note that we've also substituted $E = 2\sin t$. Place the system in matrix form

$$\mathbf{y}' = A\mathbf{y} + \mathbf{f}, \qquad (9.14)$$

where

$$A = \begin{bmatrix} -6 & 2 \\ 4 & -4 \end{bmatrix} \quad \text{and} \quad \mathbf{f} = \begin{bmatrix} \sin t \\ 0 \end{bmatrix},$$

and $\mathbf{y} = (i_1, i_2)^T$. It is not difficult to find the eigenvalue-eigenvector pairs

$$-2 \longrightarrow \begin{bmatrix} 1 \\ 2 \end{bmatrix} \quad \text{and} \quad -8 \longrightarrow \begin{bmatrix} 1 \\ -1 \end{bmatrix},$$

which provide the homogeneous solution

$$\mathbf{y}_h = C_1 e^{-2t}\begin{bmatrix} 1 \\ 2 \end{bmatrix} + C_2 e^{-8t}\begin{bmatrix} 1 \\ -1 \end{bmatrix}. \qquad (9.15)$$

Because of the form of $\mathbf{f} = (\sin t, 0)^T$, we will try

$$\mathbf{y}_p = \sin t\begin{bmatrix} a \\ b \end{bmatrix} + \cos t\begin{bmatrix} c \\ d \end{bmatrix} \qquad (9.16)$$

as a particular solution. Substituting in the left hand side of equation (9.14),

$$\mathbf{y}_p' = \cos t\begin{bmatrix} a \\ b \end{bmatrix} - \sin t\begin{bmatrix} c \\ d \end{bmatrix}. \qquad (9.17)$$

Substituting in the right hand side of (9.14),

$$\begin{aligned} A\mathbf{y}_p + \mathbf{f} &= \begin{bmatrix} -6 & 2 \\ 4 & -4 \end{bmatrix}\left[\sin t\begin{bmatrix} a \\ b \end{bmatrix} + \cos t\begin{bmatrix} c \\ d \end{bmatrix}\right] + \begin{bmatrix} \sin t \\ 0 \end{bmatrix} \\ &= \cos t\begin{bmatrix} -6c + 2d \\ 4c - 4d \end{bmatrix} + \sin t\begin{bmatrix} -6a + 2b + 1 \\ 4a - 4b \end{bmatrix} \end{aligned} \qquad (9.18)$$

Comparing coefficients of $\cos t$ and $\sin t$ in equations (9.17) and (9.18) leads to the system of equations

$$\begin{aligned} a + 6c - 2d &= 0 \\ b - 4c + 4d &= 0 \\ 6a - 2b - c &= 1 \\ -4a + 4b - d &= 0. \end{aligned} \qquad (9.19)$$

A computer or calculator provides a solution, which when substituted in (9.16), provides the particular solution $\mathbf{y}_p$. Finally, the general solution is $\mathbf{y} = \mathbf{y}_h + \mathbf{y}_p$.

$$\begin{aligned} \mathbf{y} = {} & C_1 e^{-2t}\begin{bmatrix} 1 \\ 2 \end{bmatrix} + C_2 e^{-8t}\begin{bmatrix} 1 \\ -1 \end{bmatrix} \\ & + \sin t\begin{bmatrix} 14/65 \\ 12/65 \end{bmatrix} + \cos t\begin{bmatrix} -5/65 \\ -8/65 \end{bmatrix}. \end{aligned} \qquad (9.20)$$

The initial current through each inductor is zero, so the initial condition for our system is $\mathbf{y}(0) = (0, 0)^T$. In (9.20), this provides

$$\begin{bmatrix} 0 \\ 0 \end{bmatrix} = C_1\begin{bmatrix} 1 \\ 2 \end{bmatrix} + C_2\begin{bmatrix} 1 \\ -1 \end{bmatrix} + \begin{bmatrix} -5/65 \\ -8/65 \end{bmatrix}. \qquad (9.21)$$

It is not difficult to solve this equation for $C_1 = 1/15$ and $C_2 = 2/195$, providing the final solution

$$\begin{aligned} \mathbf{y} = {} & \frac{1}{15}e^{-2t}\begin{bmatrix} 1 \\ 2 \end{bmatrix} + \frac{2}{195}e^{-8t}\begin{bmatrix} 1 \\ -1 \end{bmatrix} \\ & + \sin t\begin{bmatrix} 14/65 \\ 12/65 \end{bmatrix} + \cos t\begin{bmatrix} -5/65 \\ -8/65 \end{bmatrix}. \end{aligned} \qquad (9.22)$$

Recalling that $\mathbf{y} = (i_1, i_2)^T$ and equating components of (9.22), we can find the current in each inductor as a function of time.

$$\begin{aligned} i_1 &= \frac{1}{15}e^{-2t} + \frac{2}{195}e^{-8t} + \frac{14}{65}\sin t - \frac{1}{13}\cos t \\ i_2 &= \frac{2}{15}e^{-2t} - \frac{2}{195}e^{-8t} + \frac{12}{65}\sin t - \frac{8}{65}\cos t. \end{aligned} \qquad (9.23)$$

(b) Note that the homogeneous solution

$$\mathbf{y}_h = C_1 e^{-2t}\begin{bmatrix}1\\2\end{bmatrix} + C_2 e^{-8t}\begin{bmatrix}1\\-1\end{bmatrix}. \quad (9.24)$$

can be written in the form

$$\mathbf{y}_h = Y\mathbf{c}, \quad (9.25)$$

where

$$Y(t) = \begin{bmatrix} e^{-2t} & e^{-8t} \\ 2e^{-2t} & -e^{-8t}\end{bmatrix} \quad \text{and} \quad \mathbf{c} = \begin{bmatrix}C_1\\C_2\end{bmatrix}.$$

We will seek a particular solution by replacing $\mathbf{c}$ in (9.25) with $\mathbf{v}(t)$,

$$\mathbf{y}_p(t) = Y(t)\mathbf{v}(t), \quad (9.26)$$

where $\mathbf{v}(t)$ is a vector of functions to be determined. Substituting equation (9.25) in

$$\begin{aligned} \mathbf{y} &= A\mathbf{y} + \mathbf{f} \\ (Y\mathbf{v})' &= A(Y\mathbf{v}) + \mathbf{f} \\ Y\mathbf{v}' + Y'\mathbf{v} &= AY\mathbf{v} + \mathbf{f} \end{aligned}$$

Using the fact that $Y' = AY$,

$$\begin{aligned} Y\mathbf{v}' &= \mathbf{f} \\ \mathbf{v}' &= Y^{-1}\mathbf{f}. \end{aligned}$$

To determine $\mathbf{v}$, we need to perform the calculation

$$\mathbf{v} = \int Y^{-1}\vec{f}. \quad (9.27)$$

To that end, we must first calculate Y^{-1}. We invoke the shortcut for finding the inverse of a 2×2 matrix.

$$\begin{aligned} Y^{-1} &= \frac{1}{\det(Y)}\begin{bmatrix} -e^{-8t} & -e^{-8t} \\ -2e^{-2t} & e^{-2t}\end{bmatrix} \\ &= -\frac{1}{3}e^{10t}\begin{bmatrix} -e^{-8t} & -e^{-8t} \\ -2e^{-2t} & e^{-2t}\end{bmatrix} \\ &= \begin{bmatrix} e^{2t}/3 & e^{2t}/3 \\ 2e^{8t}/3 & -e^{8t}/3 \end{bmatrix} \end{aligned}$$

Next,

$$\begin{aligned} Y^{-1}\mathbf{f} &= \begin{bmatrix} e^{2t}/3 & e^{2t}/3 \\ 2e^{8t}/3 & -e^{8t}/3 \end{bmatrix}\begin{bmatrix}\sin t\\0\end{bmatrix} \\ &= \begin{bmatrix}(e^{2t}\sin t)/3 \\ (2e^{8t}\sin t)/3\end{bmatrix}. \end{aligned}$$

Integrating by parts,

$$\begin{aligned} \mathbf{v} &= \int Y^{-1}\mathbf{f} \\ &= \begin{bmatrix} (-e^{2t}\cos t)/15 + (2e^{2t}\sin t)/15 \\ (-2e^{8t}\cos t)/195 + (16e^{8t}\sin t)/195 \end{bmatrix} \end{aligned}$$

Using (9.26),

$$\begin{aligned} \mathbf{y}_p &= Y\mathbf{v} \\ &= \begin{bmatrix} (-\cos t)/13 + (14\sin t)/65 \\ (-8\cos t)/65 + (12\sin t)/65 \end{bmatrix}. \end{aligned}$$

Finally, the general solution is $\mathbf{y} = \mathbf{y}_h + \mathbf{y}_p$.

$$\begin{aligned} \mathbf{y} = {} & C_1 e^{-2t}\begin{bmatrix}1\\2\end{bmatrix} + C_2 e^{-8t}\begin{bmatrix}1\\-1\end{bmatrix} \\ & + \begin{bmatrix} (-\cos t)/13 + (14\sin t)/65 \\ (-8\cos t)/65 + (12\sin t)/65 \end{bmatrix}. \end{aligned} \quad (9.28)$$

The initial current through each inductor is zero, so the initial condition for our system is $\mathbf{y}(0) = (0, 0)^T$. In (9.28), this provides

$$\begin{bmatrix}0\\0\end{bmatrix} = C_1\begin{bmatrix}1\\2\end{bmatrix} + C_2\begin{bmatrix}1\\-1\end{bmatrix} + \begin{bmatrix}-1/13\\-8/65\end{bmatrix}. \quad (9.29)$$

It is not difficult to solve this equation for $C_1 = 1/15$ and $C_2 = 2/195$, providing the final solution

$$\begin{aligned} \mathbf{y} = {} & \frac{1}{15}e^{-2t}\begin{bmatrix}1\\2\end{bmatrix} + \frac{2}{195}e^{-8t}\begin{bmatrix}1\\-1\end{bmatrix} \\ & + \begin{bmatrix} (-\cos t)/13 + (14\sin t)/65 \\ (-8\cos t)/65 + (12\sin t)/65 \end{bmatrix}. \end{aligned} \quad (9.30)$$

Recalling that $\mathbf{y} = (i_1, i_2)^T$ and equating components of (9.30), we can find the current in each inductor as a function of time.

$$\begin{aligned} i_1 &= \frac{1}{15}e^{-2t} + \frac{2}{195}e^{-8t} - \frac{1}{13}\cos t \\ &\quad + \frac{14}{65}\sin t \\ i_2 &= \frac{2}{15}e^{-2t} - \frac{2}{195}e^{-8t} - \frac{8}{65}\cos t \\ &\quad + \frac{12}{65}\sin t. \end{aligned} \tag{9.31}$$

21. A has eigenvalues 1 and 3, with corresponding eigenvectors $(2, 1)^T$ and $(1, 1)^T$. Thus

$$Y(t) = \begin{pmatrix} 2e^t & e^{3t} \\ e^t & e^{-t} \end{pmatrix}.$$

$$Y(0) = \begin{pmatrix} 2 & 1 \\ 1 & 1 \end{pmatrix} \quad \text{and} \quad Y(0)^{-1} = \begin{pmatrix} 1 & -1 \\ -1 & 2 \end{pmatrix},$$

so

$$e^{tA} = Y(t)Y(0)^{-1} = \begin{pmatrix} 2e^t - e^{3t} & 2e^{3t} - 2e^t \\ e^t - e^{3t} & 2e^{3t} - e^t \end{pmatrix}.$$

23. A has eigenvalues $1 \pm i$. An eigenvector corresponding to i is $(1 + 1, 1)^T$. Hence a complex solution is

$$\begin{aligned} \mathbf{z}(t) &= e^{(1+i)t}\mathbf{w} \\ &= e^t[\cos t + i\sin t]\left[\begin{pmatrix} 1 \\ 1 \end{pmatrix} + i\begin{pmatrix} 1 \\ 0 \end{pmatrix}\right] \\ &= e^t\left[\begin{pmatrix} \cos t - \sin t \\ \cos t \end{pmatrix} + i\begin{pmatrix} \cos t + \sin t \\ \sin t \end{pmatrix}\right]. \end{aligned}$$

The real and imaginary parts of $\mathbf{z}$ are solutions to the homogeneous equation, and so

$$V(t) = \begin{pmatrix} \cos t - \sin t & \cos t + \sin t \\ \cos t & \sin t \end{pmatrix}.$$

Thus

$$V(0) = \begin{pmatrix} 1 & 1 \\ 1 & 0 \end{pmatrix} \quad \text{and}$$

$$V(0)^{-1} = \begin{pmatrix} 0 & 1 \\ 1 & -1 \end{pmatrix},$$

and

$$\begin{aligned} e^{tA} &= V(t)V(0)^{-1} \\ &= \begin{pmatrix} \cos t + \sin t & -2\sin t \\ \sin t & \cos t - \sin t \end{pmatrix}. \end{aligned}$$

25. A has eigenvalue $\lambda = 2$, which has algebraic multiplicity 2 and geometric multiplicity 1. Thus

$$\begin{aligned} e^{tA} &= e^{\lambda t}e^{t(A-\lambda I)} \\ &= e^{2t}[I + t(A - 2I)] \\ &= e^{2t}\begin{pmatrix} 1 - 2t & t \\ -4t & 1 + 2t \end{pmatrix}. \end{aligned}$$

27. A has eigenvalues -4 and -1 with eigenvectors $(-1, 1)^T$ and $(-1, 2)^T$. Thus

$$Y(t) = \begin{pmatrix} -e^{-4t} & -e^{-t} \\ e^{-4t} & 2e^{-t} \end{pmatrix},$$

$$Y(0) = \begin{pmatrix} -1 & -1 \\ 1 & 2 \end{pmatrix}.$$

Thus

$$\begin{aligned} e^{tA} &= Y(t)Y(0)^{-1} \\ &= \begin{pmatrix} 2e^{-4t} - e^{-t} & e^{-4t} - e^{-t} \\ 2e^{-t} - 2e^{-4t} & 2e^{-t} - e^{-4t} \end{pmatrix}. \end{aligned}$$

The solution to the initial value problem is

$$\begin{aligned} \mathbf{y}(t) &= e^{tA}\mathbf{y}_0 \\ &= \begin{pmatrix} 2e^{-4t} - e^{-t} & e^{-4t} - e^{-t} \\ 2e^{-t} - 2e^{-4t} & 2e^{-t} - e^{-4t} \end{pmatrix}\begin{pmatrix} 1 \\ 0 \end{pmatrix} \\ &= \begin{pmatrix} 2e^{-4t} - e^{-t} \\ 2e^{-t} - 2e^{-4t} \end{pmatrix}. \end{aligned}$$

29. A has one eigenvalue, -3, with multiplicity 2. Hence

$$\begin{aligned} e^{tA} &= e^{-3t}e^{t(A+3I)} \\ &= e^{-3t}[I + t(A + 3I)] \\ &= e^{-3t}\begin{pmatrix} 1 + 2t & t \\ -4t & 1 - 2t \end{pmatrix}. \end{aligned}$$

The solution to the initial value problem is

$$\mathbf{y}(t) = e^{tA}\mathbf{y}_0 = e^{-3t}\begin{pmatrix} 1+2t & t \\ -4t & 1-2t \end{pmatrix}\begin{pmatrix} 2 \\ -1 \end{pmatrix} = e^{-3t}\begin{pmatrix} 2+3t \\ -1-6t \end{pmatrix}.$$

31. The matrix A has eigenvalues -1 and 5, with associated eigenvectors $(0, 1)^T$ and $(1, 1)^T$. Thus

$$Y(t) = \begin{pmatrix} 0 & e^{5t} \\ e^{-t} & e^{5t} \end{pmatrix},$$

$$Y(0) = \begin{pmatrix} 0 & 1 \\ 1 & 1 \end{pmatrix} \quad \text{and} \quad Y(0)^{-1} = \begin{pmatrix} -1 & 1 \\ 1 & 0 \end{pmatrix}.$$

Hence

$$e^{tA} = Y(t)Y(0)^{-1} = \begin{pmatrix} e^{5t} & 0 \\ e^{5t} - e^{-t} & e^{-t} \end{pmatrix}.$$

The solution to the initial value problem is

$$\mathbf{y}(t) = e^{(t-t_0)A}\mathbf{y}_0 = \begin{pmatrix} -e^{5t-5} \\ 4e^{1-t} - e^{5t-5} \end{pmatrix}.$$

33. A has eigenvalues $\pm 3i$. The eigenvalue $3i$ has associated eigenvector $(2+1, 5)^T$. The associated solution is

$$\mathbf{z}(t) = \left[\cos 3t \begin{pmatrix} 2 \\ 5 \end{pmatrix} - \sin 3t \begin{pmatrix} 1 \\ 0 \end{pmatrix}\right] \\ i\left[\cos 3t \begin{pmatrix} 1 \\ 0 \end{pmatrix} + \sin 3t \begin{pmatrix} 2 \\ 5 \end{pmatrix}\right].$$

Thus

$$Y(t) = \begin{pmatrix} 2\cos 3t - \sin 3t & \cos 3t + 2\sin 3t \\ 5\cos 3t & 5\sin 3t \end{pmatrix},$$

$$Y(0) = \begin{pmatrix} 2 & 1 \\ 5 & 0 \end{pmatrix} \quad \text{and} \quad Y(0)^{-1} = \frac{1}{5}\begin{pmatrix} 0 & 1 \\ 5 & -2 \end{pmatrix}.$$

Hence

$$e^{tA} = Y(t)Y(0)^{-1} = \begin{pmatrix} \cos 3t + 2\sin 3t & -\sin 3t \\ 5\sin 3t & \cos 3t - 2\sin 3t \end{pmatrix}.$$

The solution to the initial value problem is

$$\mathbf{y}(t) = e^{(t-t_0)A}\mathbf{y}_0 = \begin{pmatrix} -\cos 3(t-1) - 2\sin 3(t-1) \\ -5\sin 3(t-1) \end{pmatrix}.$$

35. The matrix A has eigenvalues -2, 3, and 1, with associated eigenvectors $(0, 1, 0)^T$, $(-2, 0, 1)^T$, and $(-1, 1, 1)^T$. Hence

$$Y(t) = \begin{pmatrix} 0 & -2e^{3t} & -e^t \\ e^{-2t} & 0 & e^t \\ 0 & e^{3t} & e^t \end{pmatrix},$$

$$Y(0) = \begin{pmatrix} 0 & -2 & -1 \\ 1 & 0 & 1 \\ 0 & 1 & 1 \end{pmatrix} \quad \text{and}$$

$$Y(0)^{-1} = \begin{pmatrix} -1 & 1 & -2 \\ -1 & 0 & -1 \\ 1 & 0 & 2 \end{pmatrix}.$$

Thus

$$e^{tA} = Y(t)Y(0)^{-1} = \begin{pmatrix} 2e^{3t} - e^t & 0 & 2e^{3t} - 2e^t \\ e^t - e^{-2t} & e^{-2t} & 2e^t - 2e^{-2t} \\ e^t - e^{3t} & 0 & 2e^t - e^{3t} \end{pmatrix}.$$

37. A has eigenvalues -2, -1, 1, and 2 with associated eigenvectors $(0, 1, 0, 1)^T$, $(-1, 1, 1, 0)^T$, $(0, 1, 0, 0)^T$, and $(1, 2, 0, 1)^T$. Thus we have the fundamental matrix

$$Y(t) = \begin{pmatrix} 0 & -e^{-t} & 0 & e^{2t} \\ e^{-2t} & e^{-t} & e^t & 2e^{2t} \\ 0 & e^{-t} & 0 & 0 \\ e^{-2t} & 0 & 0 & e^{2t} \end{pmatrix}.$$

We have

$$Y(0) = \begin{pmatrix} 0 & -1 & 0 & 1 \\ 1 & 1 & 1 & 2 \\ 0 & 1 & 0 & 0 \\ 1 & 0 & 0 & 1 \end{pmatrix} \quad \text{and} \quad Y(0)^{-1} = \begin{pmatrix} -1 & 0 & -1 & 1 \\ 0 & 0 & 1 & 0 \\ -1 & 1 & -2 & -1 \\ 1 & 0 & 1 & 0 \end{pmatrix}.$$

Finally,

$$e^{tA} = Y(t)Y(0)^{-1} = \begin{pmatrix} e^{2t} & 0 & e^{2t} - e^{-t} & 0 \\ 2e^{2t} - e^{t} - e^{-2t} & e^{t} & 2e^{2t} - 2e^{t} + e^{-t} - e^{-2t} & e^{-2t} - e^{t} \\ 0 & 0 & e^{-t} & 0 \\ e^{2t} - e^{-2t} & 0 & e^{2t} - e^{-2t} & e^{-2t} \end{pmatrix}.$$

39. We write

$$\mathbf{y}(t) = e^{tA}\left[\mathbf{y}_0 + \int_0^t e^{-sA}\mathbf{f}(s)\,ds\right].$$

We can now prove the result by direct substitution.

$$\mathbf{y}'(t) = A\mathbf{y}(t) + e^{tA}\left[e^{-tA}\mathbf{f}(t)\right] = A\mathbf{y}(t) + f(t).$$

$$\mathbf{y}(0) = e^{0A}\mathbf{y}_0 = \mathbf{y}_0.$$

Chapter 10. Nonlinear Systems

Section 10.1. The Linearization of a Nonlinear System

1. The x-nullcline consists of the two lines defined by $x = 0$ and $3x + 2y = 6$. It is shown dashed in the accompanying figure. The y-nullcline consists of the two lines defined by $y = 0$ and $x + y = 5$. It is shown dot-dashed in the accompanying figure. The equilibrium points are found by looking for the intersections. They are $(0, 0)$, $(0, 5)$, $(2, 0)$, and $(-4, 9)$.

The Jacobian is

$$J(x, y) = \begin{pmatrix} 6 - 6x - 2y & -2x \\ -y & 5 - x - 2y \end{pmatrix}.$$

At $(0, 0)$,

$$J(0, 0) = \begin{pmatrix} 6 & 0 \\ 0 & 5 \end{pmatrix}$$

has eigenvalues 6 and 5 so $(0, 0)$ is a nodal source. At $(0, 5)$,

$$J(0, 5) = \begin{pmatrix} -4 & 0 \\ -5 & -5 \end{pmatrix}$$

has eigenvalues -4 and -5 so $(0, 5)$ is a nodal sink. At $(2, 0)$,

$$J(2, 0) = \begin{pmatrix} -6 & -4 \\ 0 & 3 \end{pmatrix}$$

has eigenvalues -6 and 3 so $(2, 0)$ is a saddle. At $(-4, 9)$,

$$J(-4, 9) = \begin{pmatrix} 12 & 8 \\ -9 & -9 \end{pmatrix}$$

has determinant $D = -36$ so $(-4, 9)$ is a saddle.

3. The x-nullcline consists of the two lines defined by $x = 0$ and $2x + y = 2$. It is shown dashed in the accompanying figure. The y-nullcline consists of the two lines defined by $y = 0$ and $x + 2y = 2$. It is shown dot-dashed in the accompanying figure. The equilibrium points are found by looking for the intersections. They are $(0, 0)$, $(0, 1)$, $(1, 0)$, and $(2/3, 2/3)$.

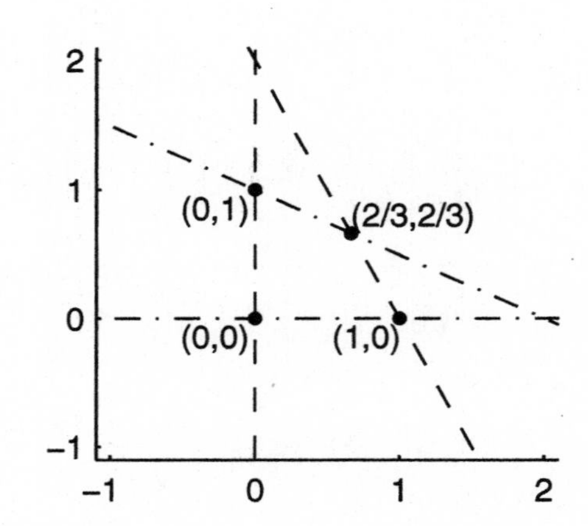

The Jacobian is

$$J(x, y) = \begin{pmatrix} 2 - 4x - y & -x \\ -y & 2 - x - 4y \end{pmatrix}.$$

At (0, 0),

$$J(0,0) = \begin{pmatrix} 2 & 0 \\ 0 & 2 \end{pmatrix}$$

has eigenvalues 2 and 2, so (0, 0) is a degenerate source. The type of equilibrium point for the nonlinear system is unknown. At (0, 1)

$$J(0,1) = \begin{pmatrix} 1 & 0 \\ -1 & -2 \end{pmatrix}$$

has eigenvalues 1 and −2, so (0, 1) is a saddle. At (1, 0)

$$J(1,0) = \begin{pmatrix} -2 & -1 \\ 0 & 1 \end{pmatrix}$$

has eigenvalues −2 and 1, so (1, 0) is a saddle. At (2/3, 2/3)

$$J(2/3, 2/3) = \begin{pmatrix} -4/3 & -2/3 \\ -2/3 & -4/3 \end{pmatrix}$$

has determinant $D = 4/3$, trace $T = -8/3$, and discriminant $T^2 - 4D = 16/9 > 0$, so (2/3, 2/3) is a nodal sink.

5. The x-nullcline consists of the two lines defined by $x = 0$ and $y = 5/4$. It is shown dashed in the accompanying figure. The y-nullcline consists of the two lines defined by $y = 0$ and $x = 3$. It is shown dot-dashed in the accompanying figure. The equilibrium points are found by looking for the intersections. They are (0, 0), and (3, 5/4).

The Jacobian is

$$J(x,y) = \begin{pmatrix} 4y - 5 & 4x \\ -y & 3 - x \end{pmatrix}.$$

At (0, 0)

$$J(0,0) = \begin{pmatrix} -5 & 0 \\ 0 & 3 \end{pmatrix}$$

has eigenvalues −5 and 3, so (0, 0) is a saddle. At (3, 5/4)

$$J(3, 5/4) = \begin{pmatrix} 0 & 12 \\ -5/4 & 0 \end{pmatrix}$$

has determinant $D = 15$ and trace $T = 0$, so (3, 5/4) is a linear center. Since a center is not generic, we can not classify the equilibrium point.

7. The x-nullcline consists of the line $y = 0$. It is shown dashed in the accompanying figure. The y-nullcline consists of the curve $y = -\sin x$. It is shown dot-dashed in the accompanying figure. The equilibrium points are where $0 = -\sin x$. Hence $x = k\pi$, where k is any integer. The equilibrium points are $(k\pi, 0)$, where k is any integer.

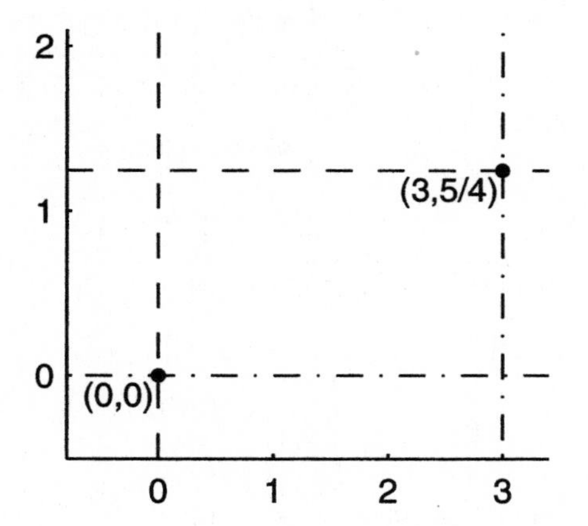

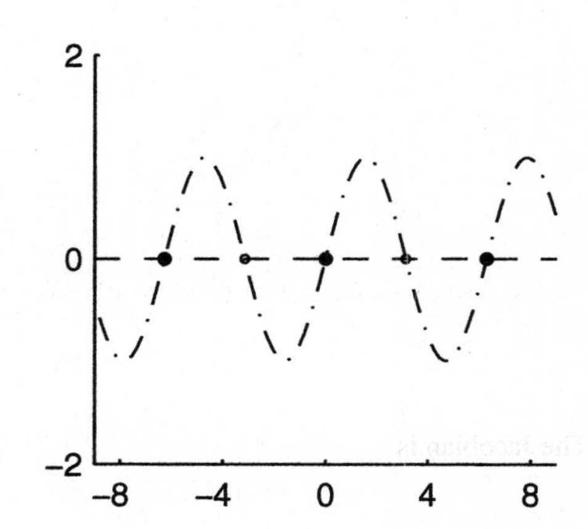

The Jacobian is

$$J(x, y) = \begin{pmatrix} 0 & 1 \\ -\cos x & -1 \end{pmatrix}.$$

If k is even,

$$J(k\pi, 0) = \begin{pmatrix} 0 & 1 \\ -1 & -1 \end{pmatrix}$$

has determinant $D = 1$, trace $T = -1$, and discriminant $T^2 - 4D = -3$, so $(k\pi, 0)$ is a spiral sink. These are shown by solid dots in the figure. If k is odd,

$$J(k\pi, 0) = \begin{pmatrix} 0 & 1 \\ 1 & -1 \end{pmatrix}$$

has determinant $D = -1$, so $(k\pi, 0)$ is a saddle. These are shown by open dots in the figure.

9. Near the point $(2, 0)$, the nonlinear system

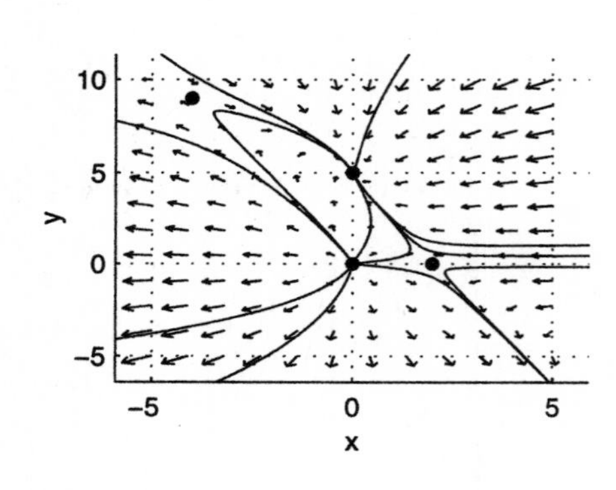

has linearization:

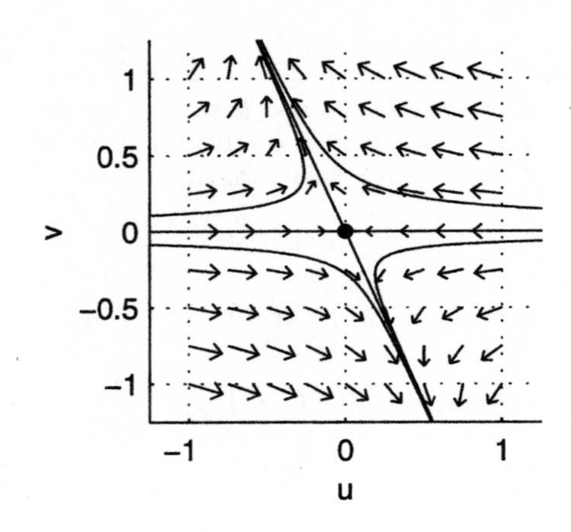

11. Near the point $(2/3, 2/3)$, the nonlinear system

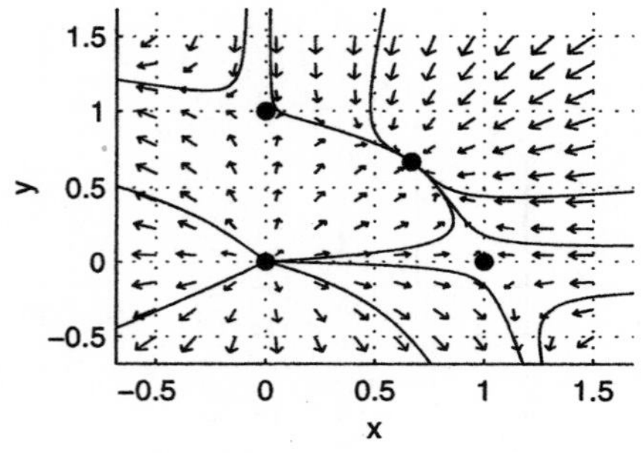

has linearization:

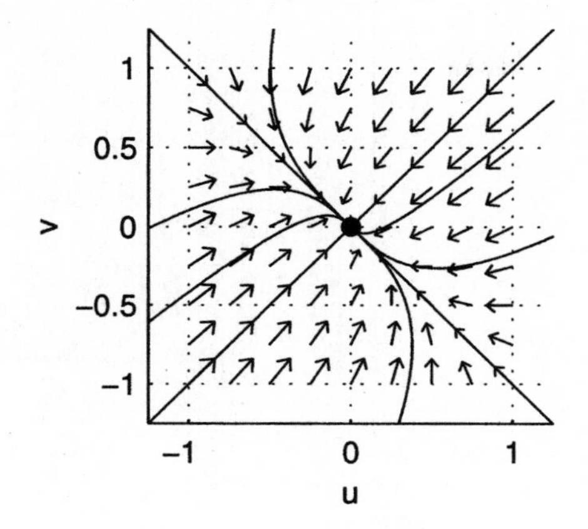

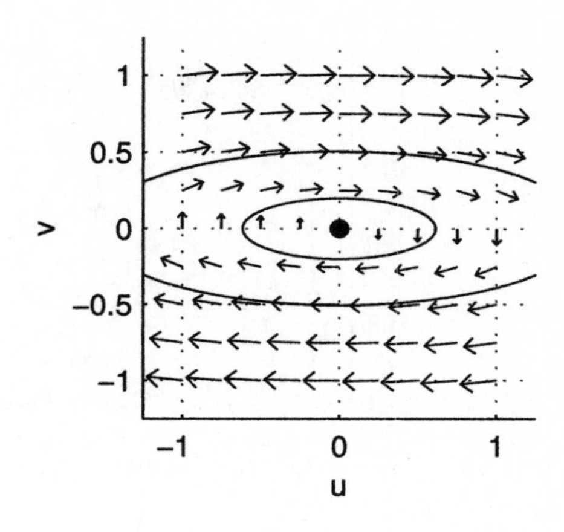

13. Near the point $(3, 5/4)$, the nonlinear system

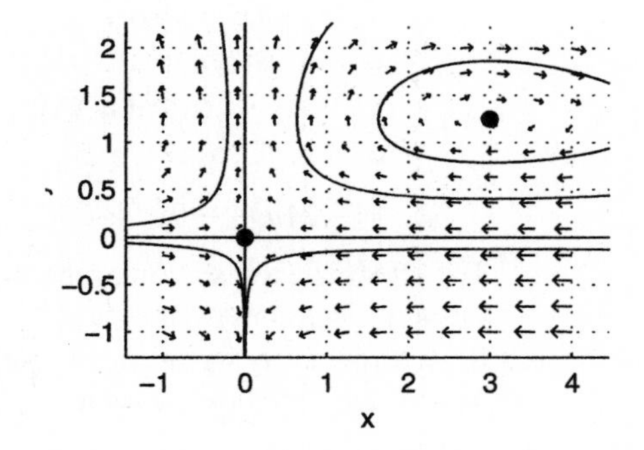

has linearization:

15. Near the point $(2\pi, 0)$, the nonlinear system

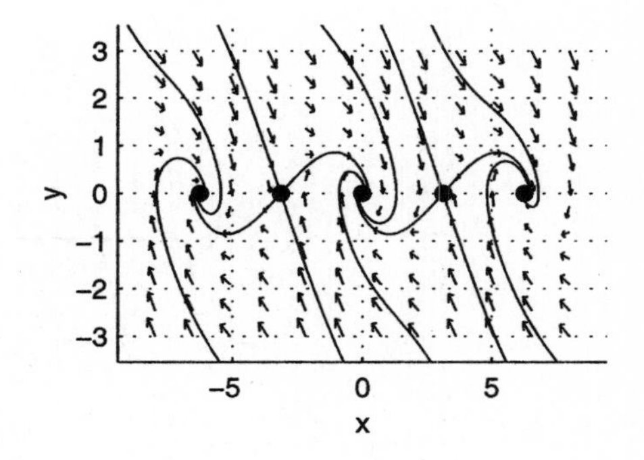

has linearization:

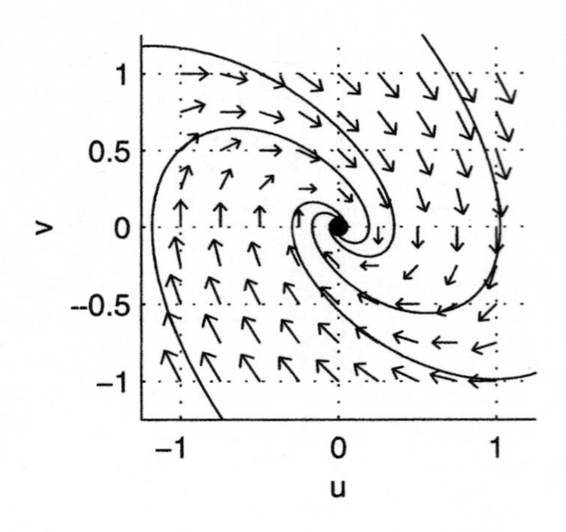

17. (a) Phase portrait and equilibrium points.

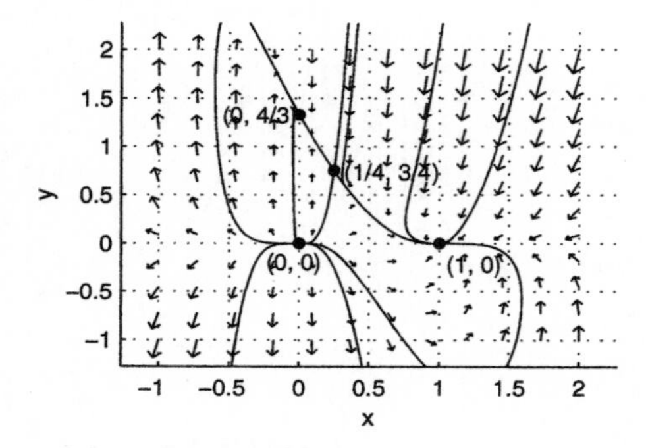

(b) Phase portrait and equilibrium points.

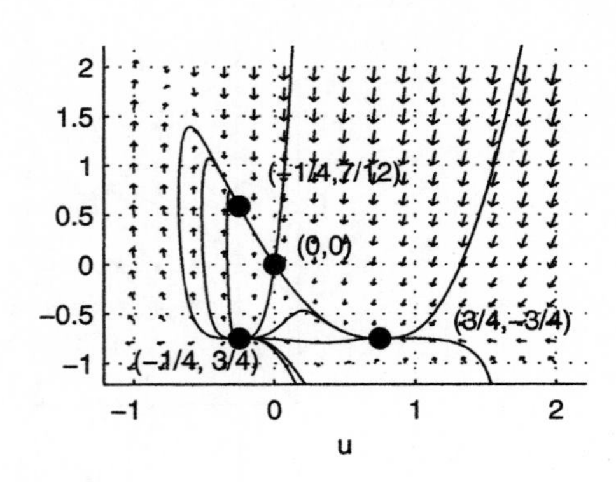

We have

$$\begin{aligned} u' &= (x - 1/4)' \\ &= x' \\ &= (1 - x - y)x \\ &= [1 - (1/4 + u) - (3/4 + v)](1/4 + u) \\ &= -(u + v)(1/4 + u) \\ &= -u/4 - v/4 - u^2 - uv. \end{aligned}$$

and

$$\begin{aligned} v' &= (y - 3/4)' \\ &= y' \\ &= (4 - 7x - 3y)y \\ &= [4 - 7(1/4 + u) \\ &\qquad - 3(3/4 + v)](3/4 + v) \\ &= -(7u + 3v)(3/4 + v) \\ &= -21u/4 - 9v/4 - 7uv - 3v^2 \end{aligned}$$

The effect is a translation of the figure, but the essential features are the same.

(c) If we throw away the terms having degree 2 or more we get the linearization.

19. (a) Differentiating both sides of $r^2 = x^2 + y^2$ we get

$$2r\frac{dr}{dt} = 2x\frac{dx}{dt} + 2y\frac{dy}{dt}.$$

Dividing by 2 we get the first result. Next differentiating both sides of $\tan\theta = y/x$ we get

$$\sec^2\theta\frac{d\theta}{dt} = \left(x\frac{dy}{dt} - y\frac{dx}{dt}\right)/x^2.$$

Since $\sec\theta = r/x$, multiplying by x^2/r^2 yields the second result.

(b) First we have

$$r\frac{dr}{dt} = x[y + \alpha x(x^2 + y^2)] + y[-x + \alpha y(x^2 + y^2)] = \alpha r^4.$$

Dividing by r we get the result. Next we have

$$r^2\frac{d\theta}{dt} = x[-x + \alpha y(x^2 + y^2)] - y[y + \alpha x(x^2 + y^2)] = -x^2 - y^2.$$

Dividing by r^2 we get the result.

(c) Integrating we get $\theta(t) = -t + C$. Hence θ is decreasing at a uniform rate. Hence the solution curve is spiraling clockwise. Since $r' = \alpha r^3$, we see that if $\alpha < 0$, r is decreasing, and the curve is spiraling into the origin. If $\alpha > 0$, r is increasing and the curve is spiraling out of the origin.

21. We have the model

$$x_1' = (a_1 - b_1x_1 + c_1x_2)x_1$$
$$x_2' = (a_2 - b_2x_2 + c_2x_1)x_2$$

where $a_1 < 0$, $b_1 = 0$, $c_1 > 0$, $a_2 > 0$, $b_2 > 0$, and $c_2 > 0$.

23. We have the model

$$x_1' = (a_1 - b_1x_1 + c_1x_2)x_1$$
$$x_2' = (a_2 - b_2x_2 + c_2x_1)x_2$$

where $a_1 < 0$, $b_1 = 0$, $c_1 > 0$, $a_2 < 0$, $b_2 = 0$, and $c_2 > 0$.

25. We have the model

$$x_1' = (a_1 - b_1x_1 + c_1x_2)x_1$$
$$x_2' = (a_2 - b_2x_2 + c_2x_1)x_2$$

where $a_1 > 0$, $b_1 > 0$, $c_1 > 0$, $a_2 > 0$, $b_2 > 0$, and $c_2 > 0$.

27. The model is

$$x_1' = (a_1 + b_{11}x_1 + b_{12}x_2 + b_{13}x_3)x_1$$
$$x_2' = (a_2 + b_{21}x_1 + b_{22}x_2 + b_{23}x_3)x_2$$
$$x_3' = (a_3 + b_{31}x_1 + b_{32}x_3 + b_{33}x_3)x_3.$$

The constant a_j is the reproductive rate of the population x_j in the absence of the others. The constant b_{jj} is negative if there is a logistic limit, and zero if not. For $i \neq j$ the constant b_{ij} measures the effect of x_j on the reproduction of x_i. If $b_{ij} > 0$ the effect is to increase the reproductive rate, and if $b_{ij} < 0$ the reproductive rate is diminished.

Section 10.2. Long-Term Behavior of Solutions

1. The equilibrium points are $(0, 0)$, $(1, 1)$, and $(-1, 1)$. The Jacobian is

$$J(x, y) = \begin{pmatrix} 1-y & -x \\ 2x & -1 \end{pmatrix}.$$

Since

$$J(0, 0) = \begin{pmatrix} 1 & 0 \\ 0 & -1 \end{pmatrix},$$

$(0, 0)$ is a saddle point. Since

$$J(1, 1) = \begin{pmatrix} 0 & -1 \\ 2 & -1 \end{pmatrix}$$

has determinant $D = 2$, trace $T = -1$, and discriminant $T^2 - 4D < 0$, $(1, 1)$ is a spiral sink. $J(-1, 1)$ has the same trace and determinant so it is also a spiral sink.

3. The equilibrium points are $(0,0)$ and $(-5,1)$. The Jacobian is

$$J(x,y) = \begin{pmatrix} 1-y & 9(1-y)-(x+9y) \\ -1 & -5 \end{pmatrix}.$$

Since

$$J(0,0) = \begin{pmatrix} 1 & 9 \\ -1 & -5 \end{pmatrix}$$

has determinant $D = 4$, trace $T = -4$, and discriminant $T^2 - 4D = 0$, the origin is a nongeneric sink. Since

$$J(-5,1) = \begin{pmatrix} 0 & -4 \\ -1 & -5 \end{pmatrix}$$

has determinant $D = -4 < 0$, $(-5,1)$ is a saddle.

5. The equilibrium points are $(0,0,0)$ and $(2/3, 4/9, 2/9)$. The Jacobian is

$$J(x,y,z) = \begin{pmatrix} -1 & 0 & 3 \\ 0 & -1 & 2 \\ 2x & 0 & -2 \end{pmatrix}.$$

Since

$$J(0,0,0) = \begin{pmatrix} -1 & 0 & 3 \\ 0 & -1 & 2 \\ 0 & 0 & -2 \end{pmatrix},$$

the origin is a sink (the eigenvalues are -1, -1, and -2). Since

$$J(2/3,4/9,2/9) = \begin{pmatrix} -1 & 0 & 3 \\ 0 & -1 & 2 \\ 4/3 & 0 & -2 \end{pmatrix}$$

the characteristic polynomial is $p(\lambda) = (\lambda+1)(\lambda^2 + 3\lambda - 2)$, and the eigenvalues are -1, $(-3 \pm \sqrt{17})/2$. Since $(-3+\sqrt{17})/2 = 0.5616 > 0$, $(2/3, 4/9, 2/9)$ is unstable.

7. The equilibrium points are $(0,0,0)$ and $(1,1,1)$. The Jacobian is

$$J(x,y,z) = \begin{pmatrix} 0 & 1 & -1 \\ 1 & 0 & -1 \\ 2x & 2y & -2 \end{pmatrix}.$$

Since

$$J(0,0,0) = \begin{pmatrix} 0 & 1 & -1 \\ 1 & 0 & -1 \\ 0 & 0 & -2 \end{pmatrix},$$

the characteristic polynomial is $p(\lambda) = (\lambda+1)(\lambda+2)(\lambda-1)$. Since 1 is an eigenvalue, $(0,0,0)$ is unstable. Since

$$J(1,1,1) = \begin{pmatrix} 0 & 1 & -1 \\ 1 & 0 & -1 \\ 2 & 2 & -2 \end{pmatrix},$$

the characteristic polynomial is $p(\lambda) = (\lambda^3 + 2\lambda^2 + 3\lambda + 2)$. The roots are $\lambda = -1$ and $\lambda = (-1 \pm \sqrt{7}i)/2$. Since all have negative real part, $(1,1,1)$ is asymptotically stable.

9. The Jacobian at the origin is

$$J = \begin{pmatrix} -1 & 4 & 4 \\ 0 & -3 & -2 \\ 0 & 4 & 3 \end{pmatrix}.$$

The characteristic polynomial is $p(\lambda) = (\lambda+1)(\lambda^2 - 1)$. Since $\lambda = 1$ is a root, the origin is unstable.

11. The Jacobian at the origin is

$$J = \begin{pmatrix} -5 & 6 & 0 \\ -3 & 1 & 1 \\ -3 & 2 & -1 \end{pmatrix}.$$

Using a computer we find that the eigenvalues are -2.3971 and $-1.3015 \pm 2.6583i$. Since all have negative real part, the origin is asymptotically stable.

13. The Jacobian at the origin is

$$J = \begin{pmatrix} 6 & -4 & 0 \\ 12 & -8 & 0 \\ 9 & -5 & -3 \end{pmatrix}.$$

The characteristic polynomial is $p(\lambda) = \lambda(\lambda+2)(\lambda+3)$. Although there are two negative roots, 0 is also an eigenvalue, so it is not possible to classify the equilibrium point using the Jacobian.

15. The Jacobian at the origin is

$$J = \begin{pmatrix} 1 & 3 & -3 & 0 \\ 0 & -2 & 1 & 1 \\ 0 & 0 & -1 & 0 \\ -9 & -9 & 11 & -2 \end{pmatrix}.$$

The characteristic polynomial is $p(\lambda) = (\lambda+1)(\lambda+2)(\lambda^2+\lambda+7)$. Since all roots have negative real part, the origin is asymptotically stable.

17. The equilibrium points are $\mathbf{0}$, $\mathbf{c}^+ = (4, 4, 6)^T$, and $\mathbf{c}^- = (-4, -4, 6)^T$. Since $r > 1$ we know that $\mathbf{0}$ is unstable.

$$J(\vec{c}^{\pm}) = \begin{pmatrix} -10 & 10 & 0 \\ 1 & -1 & \mp 4 \\ \pm 4 & \pm 4 & -8/3 \end{pmatrix}.$$

Both matrices have eigenvalues -12.1055, and $-0.7806 \pm 5.0818i$. Hence $\mathbf{c}^+$ and $\mathbf{c}^-$ are asymptotically stable. In the figures below we computed and plotted the solutions starting at $(2, 1, 1)$ and $(-2, -1, 1)$. Notice how they converge to $\mathbf{c}^+$ and $\mathbf{c}^-$, respectively.

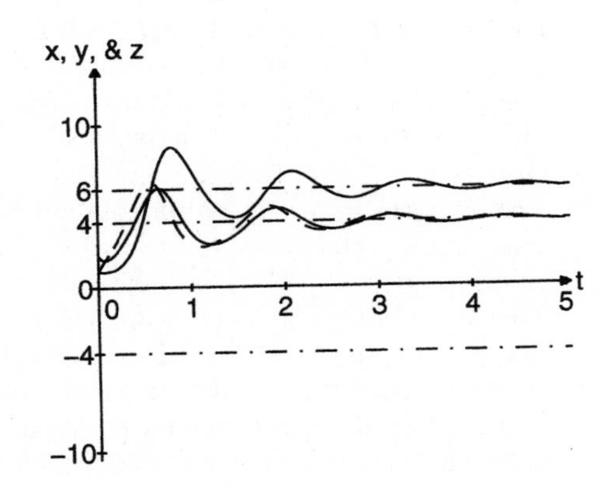

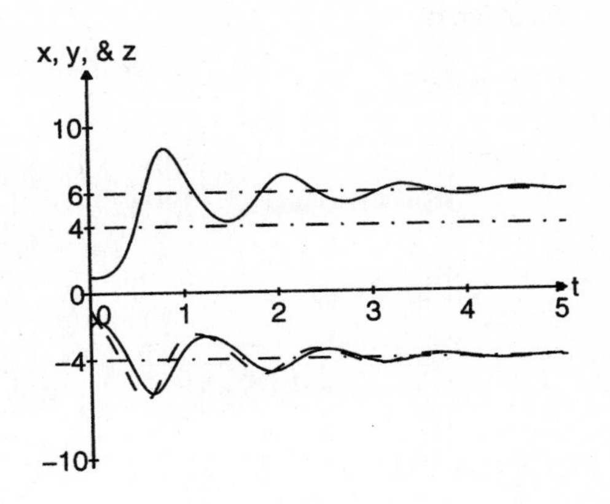

19. The equilibrium points are $\mathbf{0}$, $\mathbf{c}^+ = (x_0, x_0, z_0)^T$, and $\mathbf{c}^- = (x_0, x_0, z_0)^T$, where $x_0 = \sqrt{b(r-1)} \approx 8.4853$, and $z_0 = r - 1 = 27$. Since $r > 1$ we know that $\mathbf{0}$ is unstable.

$$J(\mathbf{c}^{\pm}) = \begin{pmatrix} -10 & 10 & 0 \\ 1 & -1 & \mp 8.4853 \\ \pm 8.4853 & \pm 8.4853 & -8/3 \end{pmatrix}.$$

Both matrices have eigenvalues -13.8546, and $0.094 \pm 10.1945i$. Since the last two have positive real part, $\mathbf{c}^+$ and $\mathbf{c}^-$ are unstable. In the figures below we computed and plotted the solutions starting at $(8, 8, 27)^T$ and $(-8, -8, 27)^T$. Notice that these starting positions are very close to $\mathbf{c}^+$ and $\mathbf{c}^-$. The solutions move away and then oscillate around $\mathbf{c}^+$ and $\mathbf{c}^-$ in turn.

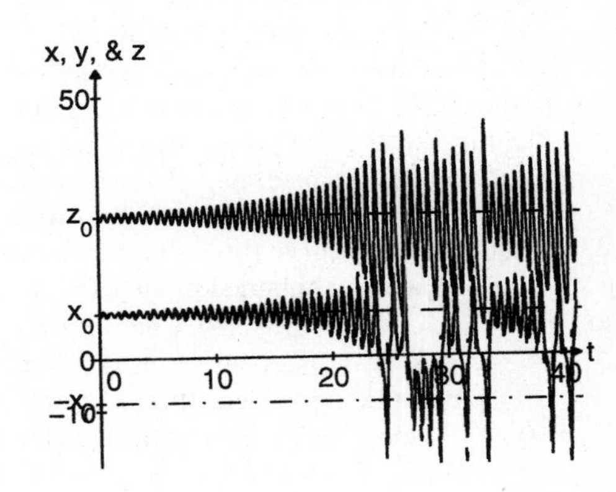

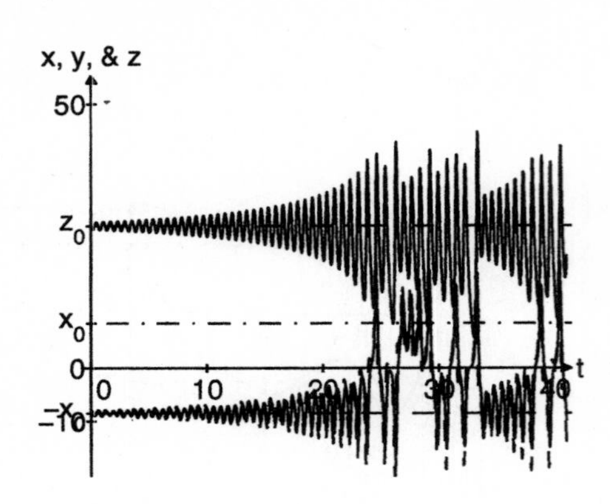

21. (a) The system displays all of the interaction among the three species.

(b) If all populations are positive, then we must have the three equations

$$2x_2 = 1$$
$$-2x_1 + 4x_3 = 0.5$$
$$x_2 + 2x_3 = 2$$

The unique solution is $\mathbf{x} = (5/4, 1/2, 3/4)^T$. The Jacobian J is:

$$\begin{pmatrix} -1+2x_2 & 2x_1 & 0 \\ -2x_2 & -1/2 - 2x_1 + 4x_3 & 4x_2 \\ 0 & -x_3 & 2 - x_2 - 4x_3 \end{pmatrix}$$

At the equilibrium point this is

$$J = \begin{pmatrix} 0 & 5/2 & 0 \\ -1 & 0 & 2 \\ 0 & -3/4 & -3/2 \end{pmatrix}.$$

The eigenvalues are -1.0610 and $-0.2195 \pm 1.8671i$. Since all have negative real part, the equilibrium point is asymptotically stable.

———————×———————

Section 10.3. Invariant Sets and the Use of Nullclines

1. The x-axis is defined by $y = 0$. Notice that if $x(t)$ solves the logistic equation $x' = (2 - x)x$ and $y(t) = 0$ then x and y are a solution to the system in the exercise. Every point in the x-axis is contained in such a solution curve, which stays in the x-axis. Hence the x-axis is invariant. Similarly the functions $x(t) = 0$ and $y(t)$ defined by $y' = (3 - y)y$ are solutions to the system in the exercise, and their solution curves exhaust the y-axis, making it invariant. A solution curve starting in one of the four quadrants must stay in that quadrant, because to get out it has to cross one of the axes. It cannot do so because the unique solution curve through any point in the axis must be entirely contained in the axis.

3. The x-axis is defined by $y = 0$. Notice that if $x(t)$ solves the logistic equation $x' = (1 - x)x$ and $y(t) = 0$ then x and y are a solution to the system in the exercise. Every point in the x-axis is contained in such a solution curve, which stays in the x-axis. Hence the x-axis is invariant. Similarly the functions $x(t) = 0$ and $y(t)$ defined by $y' = (6 - 3y)y$ are solutions to the system in the exercise, and their solution curves exhaust the y-axis, making it invariant. A solution curve starting in one of the four quadrants must stay in that quadrant, because to get out it has to cross one of the axes. It cannot do so because the unique solution curve through any point in the axis must be entirely contained in the axis.

5. The axes are invariant, so no solution curves can cross them. Along the line $x = 3$ we have $x' = (2 - x - y)x = -3(1 + y) < 0$, so x is decreasing. Thus the solution curve is moving into S. Along the line $y = 3$ we have $y' = (3 - 3x - y)y = -9x < 0$, so y is decreasing. Again the solution curves are moving into S. Consequently all solution curves cross the boundary of S going into S. Hence S is

invariant.

7. The axes are invariant, so no solution curves can cross them. Along the line $x = 4$ we have $x' = (1 - x - y)x = -4(3 + y) < 0$, so x is decreasing. Thus the solution curve is moving into S. Along the line $y = 3$ we have $y' = (6 - 2x - 3y)y = -3(3 + 2x) < 0$, so y is decreasing. Again the solution curves are moving into S. Consequently no solution curves cross the boundary of S going out of S. Hence S is invariant.

9. The x-nullcline is the union of the two lines $x = 0$ and $x + y = 2$, shown dashed.

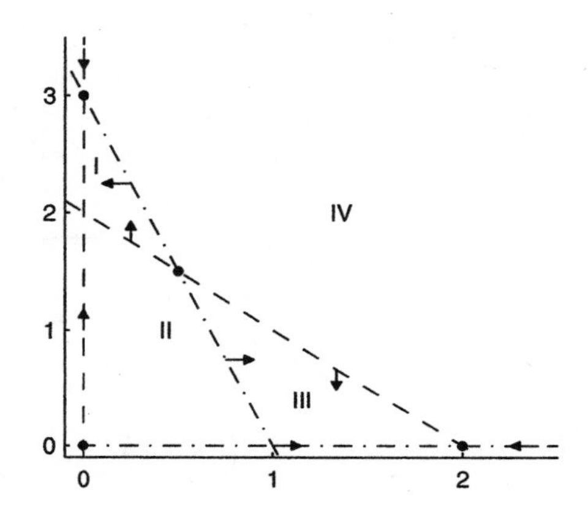

The y-nullcline is the union of the two lines $y = 0$ and $3x + y = 3$, shown dot-dashed. These intersect in the four equilibrium points $(0, 0)$, $(2, 0)$, $(0, 3)$, and $(1/2, 3/2)$. The Jacobian is

$$J(x, y) = \begin{pmatrix} 2 - 2x - y & -x \\ -3y & 3 - 3x - 2y \end{pmatrix}.$$

At $(0, 0)$,

$$J(0, 0) = \begin{pmatrix} 2 & 0 \\ 0 & 3 \end{pmatrix},$$

has eigenvalues 2 and 3, so the origin is a nodal source. At $(2, 0)$,

$$J(2, 0) = \begin{pmatrix} -2 & -2 \\ 0 & -3 \end{pmatrix},$$

has eigenvalues -2 and -3, so $(2, 0)$ is a nodal sink. At $(0, 3)$,

$$J(0, 3) = \begin{pmatrix} -1 & 0 \\ -9 & -3 \end{pmatrix},$$

has eigenvalues -1 and -3, so $(0, 3)$ is also a nodal sink. Finally,

$$J(1/2, 3/2) = \begin{pmatrix} -1/2 & -1/2 \\ -9/2 & -3/2 \end{pmatrix}$$

has determinant $D = -3/2 < 0$, so $(1/2, 3/2)$ is a saddle. The flow of the solutions along the nullclines is shown by the arrows. This information shows that regions I and III are invariant. The solution curves flow from region II into regions I and III, except the one stable solution curve for the saddle at $(1/2, 3/2)$. The solution curves in region IV can flow directly to the sinks, or into regions I and III, with the exception of the one stable solution curve for the saddle at $(1/2, 3/2)$. Finally solution curves in the invariant region I flow to the sink at $(0, 3)$ and the solution curves in the invariant region III flow to the sink at $(2, 0)$.

11. The x-nullcline is the union of the two lines $x = 0$ and $x + y = 1$, shown dashed.

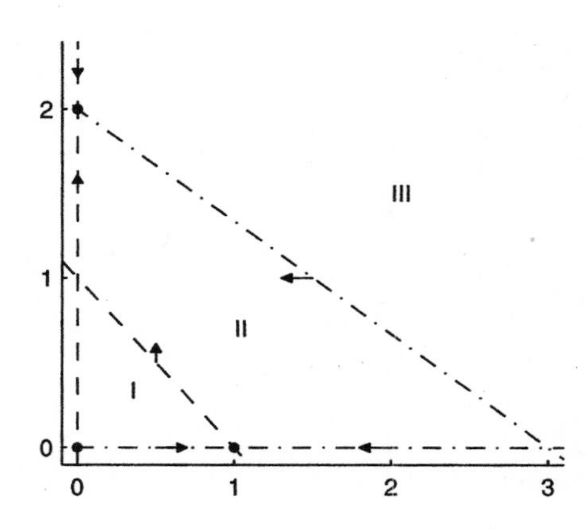

The y-nullcline is the union of the two lines $y = 0$ and $2x + 3y = 6$, shown dot-dashed. These intersect in the four equilibrium points $(0, 0)$, $(1, 0)$, $(0, 2)$, and $(-3, 4)$. Since we are only interested in the positive quadrant, the last of these is not shown. The Jacobian is

$$J(x, y) = \begin{pmatrix} 1 - 2x - y & -x \\ -2y & 6 - 2x - 6y \end{pmatrix}.$$

Since

$$J(0, 0) = \begin{pmatrix} 1 & 0 \\ 0 & 6 \end{pmatrix}$$

has eigenvalues 1 and 6, the origin is a nodal source. Since

$$J(1, 0) = \begin{pmatrix} -1 & -1 \\ 0 & 4 \end{pmatrix}$$

has eigenvalues -1 and 4, $(1, 0)$ is a saddle. Since

$$J(0, 2) = \begin{pmatrix} -1 & 0 \\ -4 & -2 \end{pmatrix}$$

has eigenvalues -1 and -2, $(0, 2)$ is a nodal sink. The flow of the solutions along the nullclines is shown by the arrows. This information shows that the solution curves flow from regions I and III into region II. Region II is invariant, and all solution curves flow to the sink at $(0, 2)$. Thus all solution curves in the positive quadrant flow to the sink at $(0, 2)$.

13. The x-nullcline is the line $y = 1$ shown dashed. The y-nullcline is the parabola $y = x^2$ shown dot-dashed.

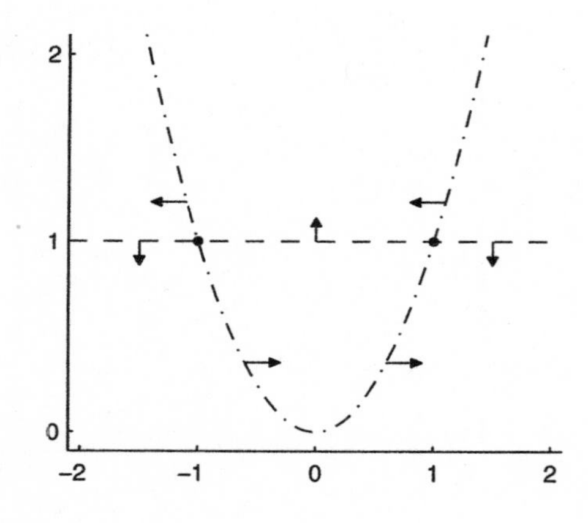

These intersect in the equilibrium points $(\pm 1, 1)$. The Jacobian is

$$J(x, y) = \begin{pmatrix} 0 & -1 \\ -2x & 1 \end{pmatrix}.$$

Since

$$J(1, 1) = \begin{pmatrix} 0 & -1 \\ -2 & 1 \end{pmatrix}$$

has determinant $D = -2$, so $(1, 1)$ is a saddle. Since

$$J(-1, 1) = \begin{pmatrix} 0 & -1 \\ 2 & 1 \end{pmatrix}$$

has determinant $D = 2$, trace $T = 1$, and discriminant $T^2 - 4D < 0$, $(-1, 1)$ is a spiral source. The flow of the solutions along the nullclines is shown by the arrows.

15. The x-nullcline is the cubic curve $y = x^3$ shown dashed. The y-nullcline is the line $x = y$ shown dot-dashed.

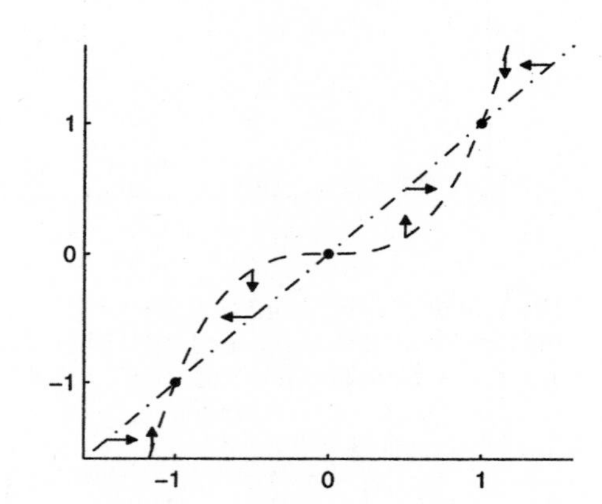

These intersect in three equilibrium points $(0, 0)$, $(1, 1)$ and $(-1, -1)$. The Jacobian is

$$J(x, y) = \begin{pmatrix} -3x^2 & 1 \\ 1 & -1 \end{pmatrix}.$$

Since

$$J(0, 0) = \begin{pmatrix} 0 & 1 \\ 1 & -1 \end{pmatrix}$$

has determinant $D = -1$, the origin is a saddle. At the other two equilibrium points

$$J = \begin{pmatrix} -3 & 1 \\ 1 & -1 \end{pmatrix},$$

which has determinant $D = 2$, trace $T = -4$, and discriminant $T^2 - 4D > 0$. Hence $(1, 1)$ and $(-1, -1)$ are nodal sinks. The flow of the solutions along the nullclines is shown by the arrows. The nullclines split all of $\mathbf{R}^2$ into six regions. Three of them are invariant. The flow of all solutions can be followed as they converge to one of the two sinks, except for the two stable solutions for the saddle point.

17. (a) If

$$u' = u(1 - u + av)$$
$$v' = rv(1 - v + bu),$$

the u-nullclines are

$$u = 0 \quad \text{and} \quad 1 - u + av = 0,$$

and the v-nullclines are

$$v = 0 \quad \text{and} \quad 1 - v + bu = 0.$$

The u-nullcline $u = 0$ will intersect the v-nullclines $v = 0$ and $1 - v + bu = 0$ at $(0, 0)$ and $(0, 1)$, respectively. The u-nullcline $1 - u + av = 0$ intersects the v-nullcline $v = 0$ at $(1, 0)$. What remains is to find where the u-nullcline $1 - u + av = 0$ meets the v-nullcline $1 - v + bu = 0$. Solving the first for u,

$$u = 1 + av,$$

and substituting into the v-nullcline,

$$1 - v + b(1 + av) = 0$$
$$v = \frac{1 + b}{1 - ab}$$

Substituting this result in $u = 1 + av$,

$$u = 1 + \frac{a(1 + b)}{1 - ab}$$
$$u = \frac{1 + a}{1 - ab}.$$

(b) In the case that $ab > 1$, we have that $1 - ab < 0$. This implies that the equilibrium point

$$\tilde{u} = \frac{1 + a}{1 - ab} \quad \text{and} \quad \tilde{v} = \frac{1 + b}{1 - ab}$$

lies in the third quadrant and will be ignored. The Jacobian is

$$J(u, v) = \begin{pmatrix} 1 - 2u + av & au \\ brv & r - 2rv + bru \end{pmatrix}.$$

Thus,

$$J(0, 0) = \begin{pmatrix} 1 & 0 \\ 0 & r \end{pmatrix}$$

and the origin is a source ($\lambda_1 = 1$, $\lambda_2 = r$). At $(1, 0)$,

$$J(1, 0) = \begin{pmatrix} -1 & a \\ 0 & r(1 + b) \end{pmatrix}$$

and $(1, 0)$ is a saddle ($\lambda_1 = -1$, $\lambda_2 = r(1 + b)$). Finally, at $(0, 1)$,

$$J(0, 1) = \begin{pmatrix} 1 + a & 0 \\ br & -r \end{pmatrix}$$

and $(1, 0)$ is also a saddle ($\lambda_1 = 1 + a$, $\lambda_2 = -r$). Because we restrict the analysis to the first quadrant, we can assume u and v are greater than zero. Thus, $u' = u(1 - u + av)$ is positive precisely when $1 - u + av$ is greater than zero, which occurs above the nullcline $1 - u + av = 0$. Thus, motion along the v-nullcline, $1 - v + bu = 0$, is strictly to the right. A similar analysis shows that motion across the u-nullcline, $1 - u + av = 0$ is strictly upward (at least in the first quadrant).

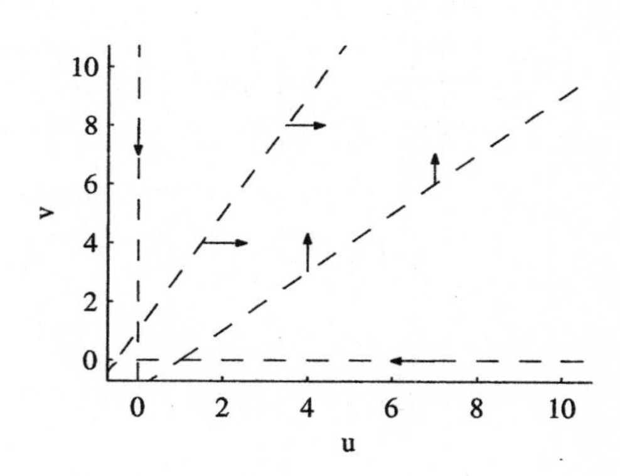

There is now enough information to draw the phase portrait.

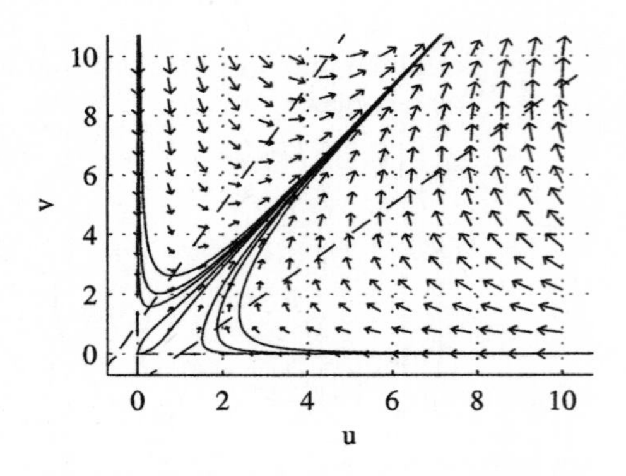

Note that both populations grow without bound if we start with any two initial nonzero populations.

(c) In the case that $ab < 1$, the Jacobian analysis at the points $(0, 0)$, $(1, 0)$, and $(0, 1)$ remains the same. However, this time the equilibrium point

$$\tilde{u} = \frac{1+a}{1-ab} \quad \text{and} \quad \tilde{v} = \frac{1+b}{1-ab}$$

lies in the first quadrant. Noting that $1 - \tilde{u} + a\tilde{v} = 0$ and $1 - \tilde{v} + b\tilde{u} = 0$,

$$J(\tilde{u}, \tilde{v}) = \begin{pmatrix} 1 - 2\tilde{u} + a\tilde{v} & a\tilde{u} \\ br\tilde{v} & r(1 - 2\tilde{v} + b\tilde{u}) \end{pmatrix}$$
$$= \begin{pmatrix} -\tilde{u} & a\tilde{u} \\ br\tilde{v} & -r\tilde{v} \end{pmatrix}.$$

The trace $T = -\tilde{u} - r\tilde{v} < 0$. The determinant $D = r\tilde{u}\tilde{v}(1 - ab) > 0$, because of the stipulation that $ab < 1$. Further,

$$\begin{aligned} T^2 - 4D &= (-\tilde{u} - r\tilde{v})^2 - 4r\tilde{u}\tilde{v}(1 - ab) \\ &= (\tilde{u} - r\tilde{v})^2 + 4abr\tilde{u}\tilde{v} \\ &> 0. \end{aligned}$$

Hence, the equilibrium point is a nodal sink. As in part (b), $u' > 0$ when $1 - u + av > 0$, which occurs above and to the left of the u-nullcline $1 - u + av = 0$. In a similar manner, $v' = 0$ when $1 - v + bu > 0$, which occurs below the v-nullcline $1 - v + bu = 0$.

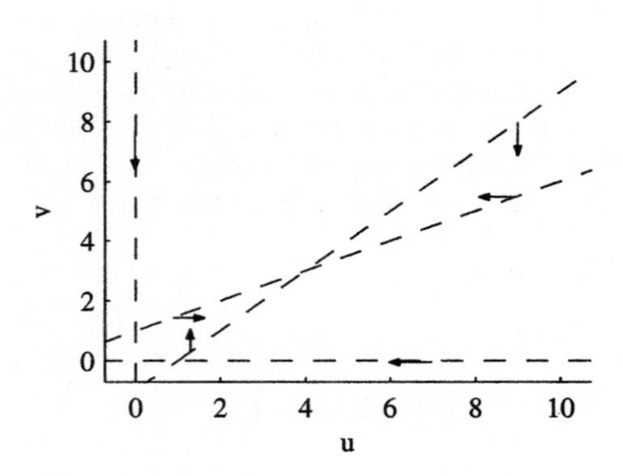

This information makes the phase portrait an easy task.

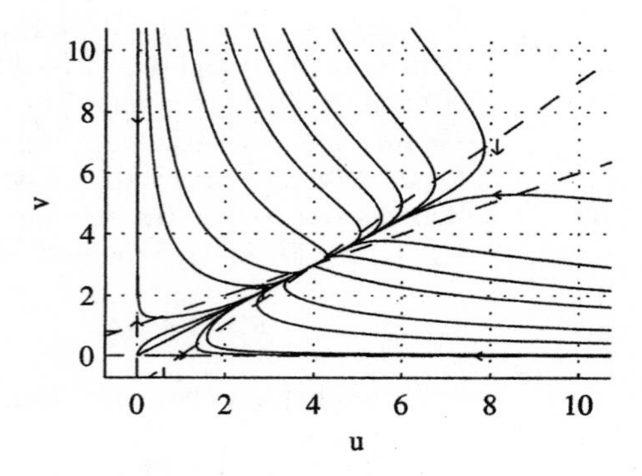

Note that any two positive initial conditions lead to a solution that flows into the sink at $(\tilde{u}, \tilde{v})$.

Section 10.4. Long-Term Behavior of Solutions to Planar Systems

1. For $r^2 = x^2 + y^2$ we have

$$\begin{aligned} rr' &= xx' + yy' \\ &= x[y + x(1 - r)] + y[-x + y(1 - r)] \\ &= r^2(1 - r). \end{aligned}$$

Thus $r' = r(1 - r)$. For this differential equation $r = 1$ is a stable equilibrium point. Hence $r = 1$ is an attracting limit cycle.

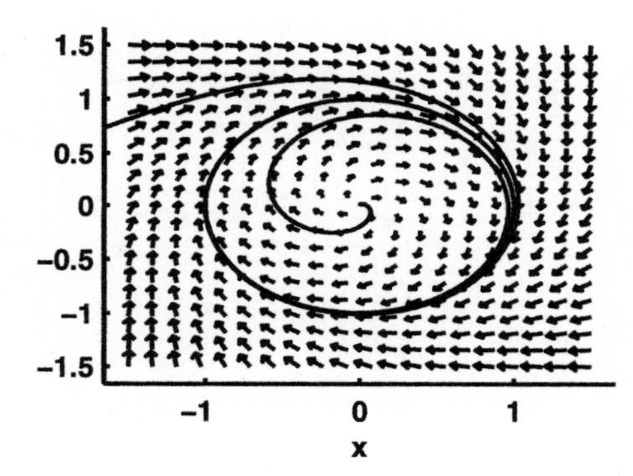

3. For $r^2 = x^2 + y^2$ we have

$$\begin{aligned} rr' &= xx' + yy' \\ &= x[-3y + x(4 - r^2)] + y[3x + y(4 - r^2)] \\ &= r^2(4 - r^2). \end{aligned}$$

Thus $r' = r(4 - r^2)$. For this differential equation $r = 2$ is a stable equilibrium point. Hence $r = 2$ is an attracting limit cycle.

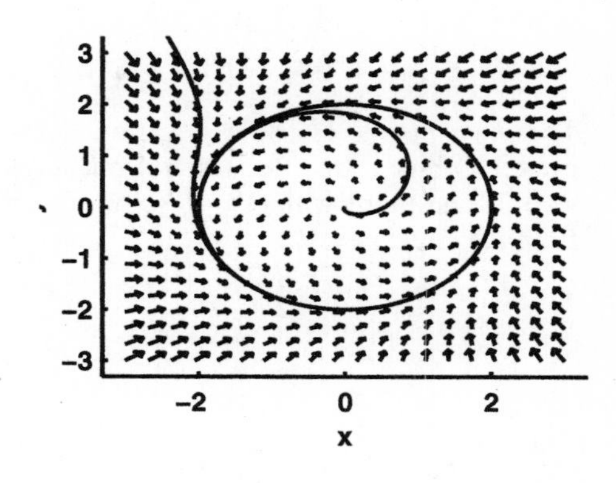

5. For $r^2 = x^2 + y^2$ we have

$$\begin{aligned} rr' &= xx' + yy' \\ &= x[-y + x(r - 3 + 2/r)] \\ &\quad + y[x + y(r - 3 + 2/r)] \\ &= r^3 - 3r^2 + 2r. \end{aligned}$$

Thus $r' = (r - 2)(r - 1)$, so $r = 1$ is an attracting limit cycle, while $r = 2$ is a repelling limit cycle.

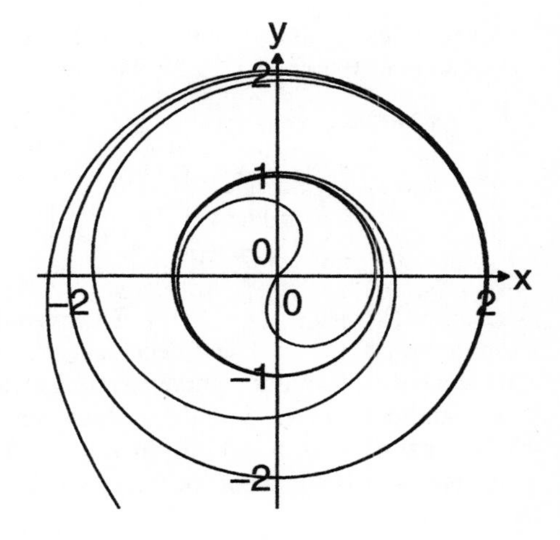

7. For $r^2 = x^2 + y^2$ we have

$$\begin{aligned} rr' &= xx' + yy' \\ &= x[y + x(r^2 - 1)^2] \\ &\quad + y[-x + y(r^2 - 1)^2] \\ &= r^2(r^2 - 1)^2. \end{aligned}$$

Thus $r' = r(r^2 - 1)^2$, so $r = 1$ is an unstable limit cycle. Orbits with $r < 1$ spiral out to the circle $r = 1$, and those with $r > 1$ spiral out and away from the circle $r = 1$.

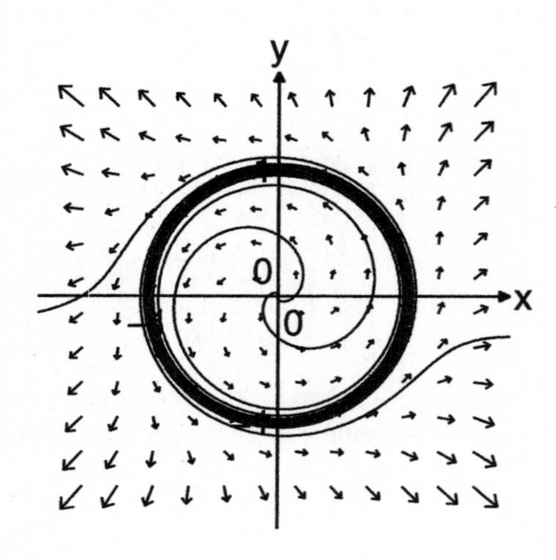

9. We verify, using the Jacobian, that the origin is a spiral source. For $r^2 = x^2 + y^2$ we have

$$\begin{aligned} rr' &= xx' + yy' \\ &= x[x + y - x(x^2 + 3y^2)] \\ &\quad + y[-x + 2y - 2y^3] \\ &= (x^2 + 2y^2)(1 - r^2). \end{aligned}$$

Thus $r' = (x^2 + 2y^2)(1 - r^2)/r$. This means that r is increasing for $r < 1$ and decreasing for $r > 1$. Consequently all solution curves starting away from the origin tend to the unit circle, which is invariant. Notice that $r' = 0$ if and only if $r = 1$. Thus all equilibrium points, except for the origin, are on the unit circle.

We also have

$$\begin{aligned} r^2\theta' &= xy' - yx' \\ &= -(x^2 - xy - y^2) + xyr^2 \end{aligned}$$

On the unit circle $r = 1$ we have $\theta' = -(x - y)^2$. Hence $\theta' = 0$ only if $x = y$. Thus the equilibrium points on the unit circle are at $(1/\sqrt{2}, 1/\sqrt{2})$ and $(-1/\sqrt{2}, -1/\sqrt{2})$. The Jacobian has determinant 0 at these points. Experimentally we see that there are solutions starting close to these points that move away. These equilibrium points are unstable. Nevertheless, every solution curve not through the origin is attracted to them!

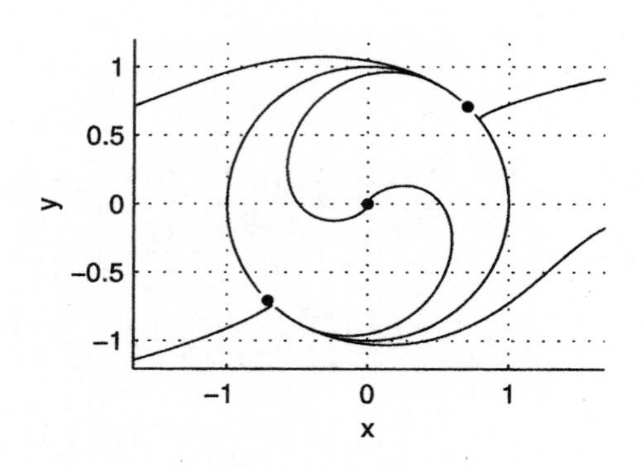

11. (a) For $r^2 = x^2 + y^2$,

$$\begin{aligned} rr' &= xx' + yy' \\ &= x[y] + y[-x + y(1 - 3x^2 - 2y^2)] \\ &= y^2(1 - 3x^2 - 2y^2). \end{aligned}$$

(b) We have

$$2r^2 = 2x^2 + 2y^2 \le 3x^2 + 2y^2,$$

and

$$3r^2 = 3x^2 + 3y^2 \ge 3x^2 + 2y^2.$$

On the inner boundary of R we have $r = 1/2$ so $(1/2)r' = y^2(1 - 3x^2 - 2y^2) \ge y^2(1 - 3r^2) \ge$

$y^2/4 \geq 0$. Thus $r \geq 0$ and r is not decreasing and so the solution curve stays in R. On the outer boundary of R we have $r = 1$ and $1r' = y^2(1 - 3x^2 - 2y^2) \leq y^2(1 - 2r^2) \leq -y^2 \leq 0$. Thus, $r' \leq 0$ and r is not increasing there and the solution curve stays in R.

(c) The only equilibrium points of the system is $(0, 0)$, and therefore there are none in R. By the Poincaré-Bendixson theorem (Theorem 4.10) there must be closed solution curve in R.

(d) A limit cycle.

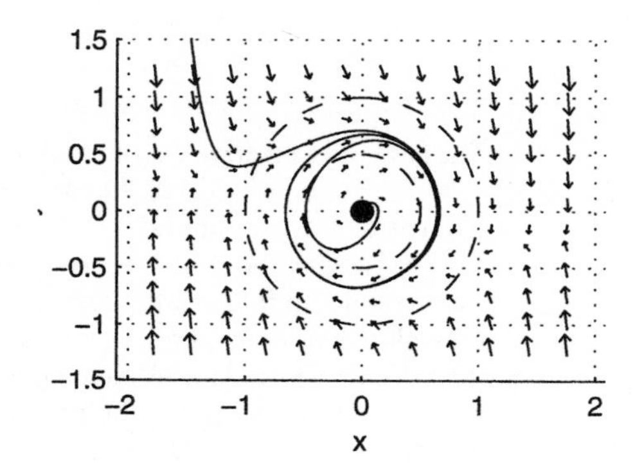

13. For $r^2 = x^2 + y^2$,

$$\begin{aligned} rr' &= xx' + yy' \\ &= x[x - y - x(3x^2 + y^2)] \\ &\qquad + y[x + y - y(2x^2 + y^2)] \\ &= r^2 - [3x^4 + 3x^2y^2 + y^4]. \end{aligned}$$

Notice that $r^4 \leq 3x^4 + 3x^2y^2 + y^4 \leq 3r^4$. Thus we have $r^2 - 3r^4 \leq rr' \leq r^2 - r^4$, or $r(1 - 3r^2) \leq r' \leq r(1 - r^2)$. From this we conclude that the annulus R defined by $1/3 \leq r \leq 1$ is invariant. since we are given that it contains no equilibrium points, it must contain a closed solution curve, which must be a limit cycle.

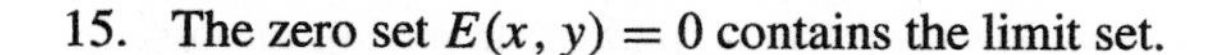

15. The zero set $E(x, y) = 0$ contains the limit set.

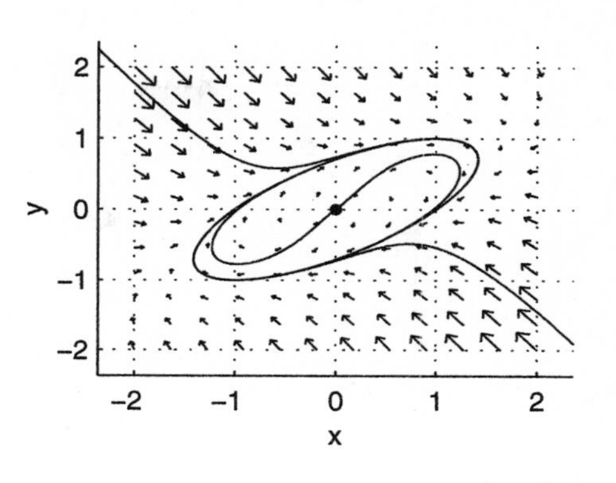

17. The zero set $E(x, y) = 0$ contains the limit set.

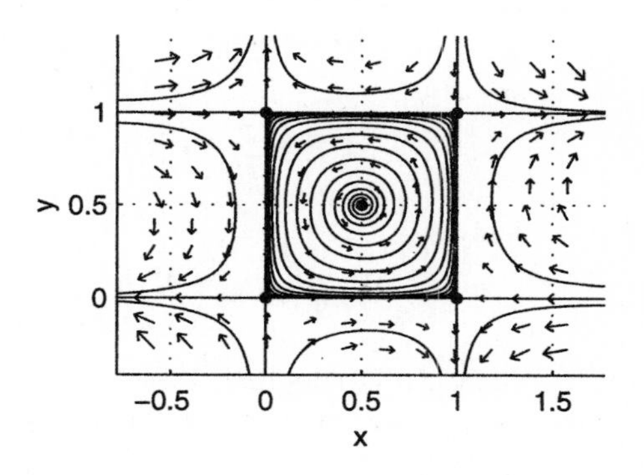

19. The zero set $E(x, y) = 0$ contains the limit set.

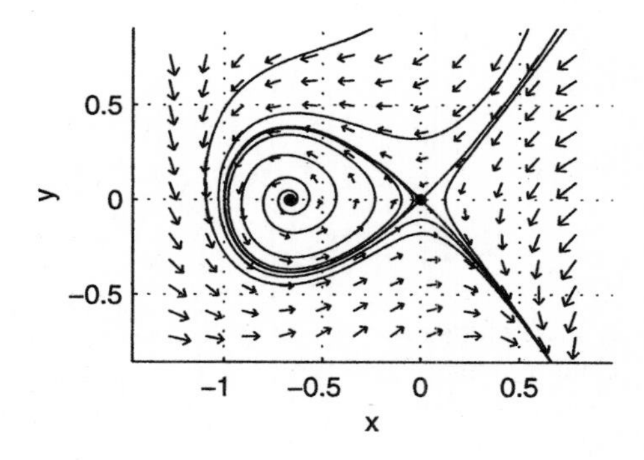

21. The zero set $E(x, y) = 0$ contains the limit set.

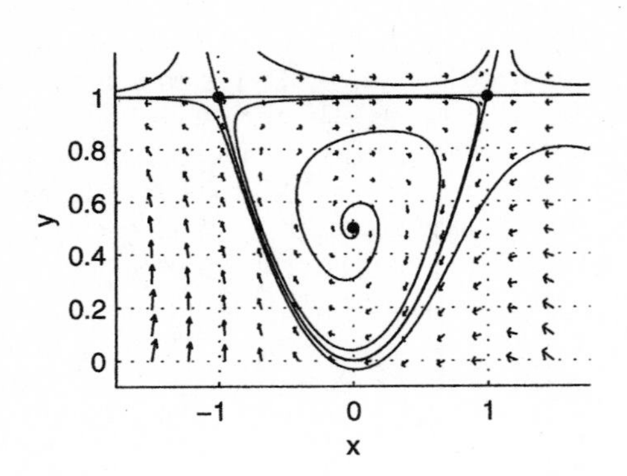

23. (a) The first equation means that $y = 0$ at any equilibrium point, and the second equation becomes $x = 0$. Thus the origin is the only equilibrium point. The Jacobian is

$$J = \begin{pmatrix} 0 & 1 \\ -1 & 1 - 3y^2 \end{pmatrix},$$

and

$$J(0,0) = \begin{pmatrix} 0 & 1 \\ -1 & 1 \end{pmatrix}.$$

Since $J(0,0)$ has trace $T = 1$ and determinant $D = 1$, we conclude that the origin is a spiral source.

(b) Does the vector field curve inward?

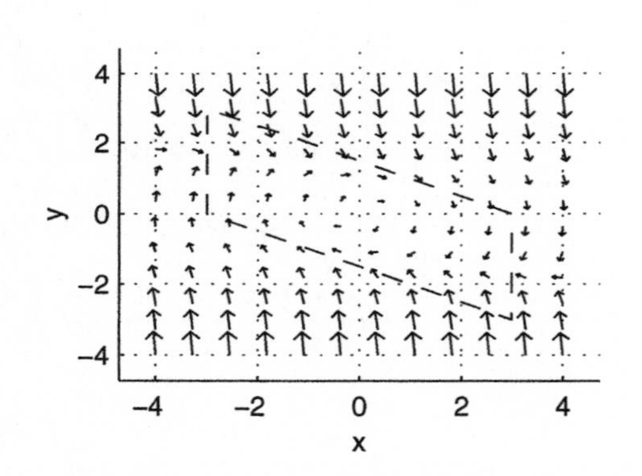

(c) The sides of the parallelogram are straight line segments defined by

$$\begin{aligned} x &= 3, && \text{for} \quad -3 \le y \le 0 \\ x &= -3, && \text{for} \quad 0 \le y \le 3 \\ x + 2y &= 3, && \text{for} \quad 0 \le y \le 3 \\ x + 2y &= -3, && \text{for} \quad 0 \le y \le 3 \end{aligned}$$

On the line $x = 3$ we have $x' = y \le 0$, so x is decreasing. On the line $x = -3$ we have $x' = y \ge 0$, so x is increasing. On the line $x + 2y = 3$ we have $(x + 2y)' = x' + 2y' = 3y - 2x - 2y^3$. We substitute $x = 3 - 2y$ to get $(x+2y)' = 7y - 7 - 2y^3$. We need this quantity to be negative for $0 \le y \le 3$. Using calculus we find that its maximum is at $y = \sqrt{7/3}$, and its value there is -2.4358. Similarly, on the line $x + 2y = -3$, we find that $(x + 2y)' \ge 2.4358$. Thus the parallelogram is invariant.

(d) The region is invariant, so it contains the limit set of any solution curve. the only equilibrium point is at the origin, and it is a spiral source, so it cannot be the limit set. According to the Bendixson alternatives, the only possible limit set which does not contain an equilibrium point is a closed solution curve. Starting with a solution curve on the boundary of the parallelogram which is moving into the parallelogram, we see that the limit set must be a limit cycle.

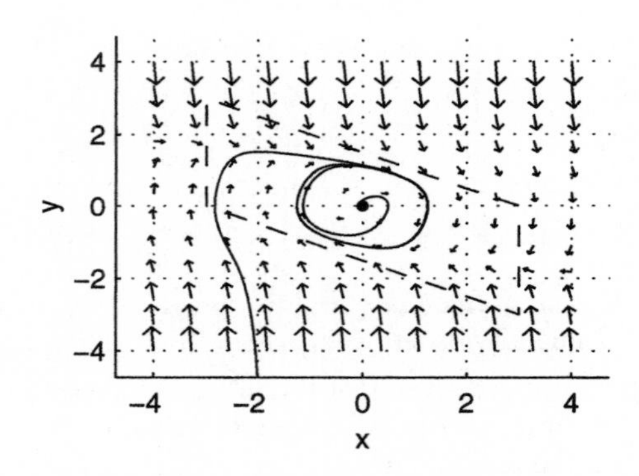

25. First we notice that the origin is the only equilibrium point, and this is a spiral source. We can use the

parallelogram with vertices $(3,4)$, $(3,0)$, $(-3,-4)$, and $(-3,0)$.

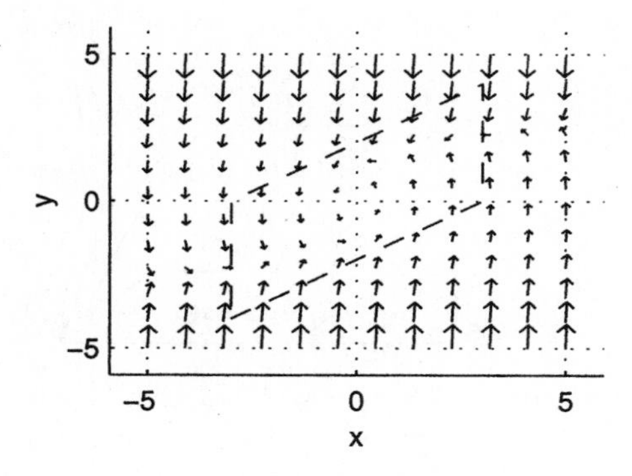

The sides are defined by the equations

$$\begin{aligned} x &= 3, \quad \text{for} \quad 0 \le y \le 4 \\ x &= -3, \quad \text{for} \quad -4 \le y \le 0 \\ 3y - 2x &= 3, \quad \text{for} \quad 0 \le y \le 4 \\ 3y - 2x &= -3, \quad \text{for} \quad -4 \le y \le 0 \end{aligned}$$

On the line $x = 3$ we have $x' = -y \le 0$, so x is decreasing. On the line $x = -3$ we have $x' = -y \ge 0$, so x is increasing. On the line $3y - 2x = 3$, we have

$$\begin{aligned} (3y - 2x)' &= 3y' - 2x' \\ &= 3[y + 3x - y^3] + 2y \\ &= 5y + 3x - 3y^3. \end{aligned}$$

On the line we have $3y - 2x = 3$, or $x = 3(y-1)/2$. making this substitution we have

$$\begin{aligned} (3y - 2x)' &= 5y + 3[3(y-1)/2] - 3y^3 \\ &= 19y/2 - 9/2 - 3y^3. \end{aligned}$$

Using calculus we find that the maximum of this quantity is at $y = \sqrt{(19/6)}$ and its value there is -4.5. Thus along this line $(3y - 2x)' < 0$. Similarly, on the line $3y - 2x = -3$ we find that $(3y - 2x)' > 4.5 > 0$. Thus the parallelogram is invariant. Since this is so, it contains the limit set of any solution curve. the only equilibrium point is at the origin, and it is a spiral source, so it cannot be the limit set. According to the Bendixson alternatives, the only possible limit set which does not contain an equilibrium point is a closed solution curve. Starting with a solution curve on the boundary of the parallelogram which is moving into the parallelogram, we see that the limit set must be a limit cycle.

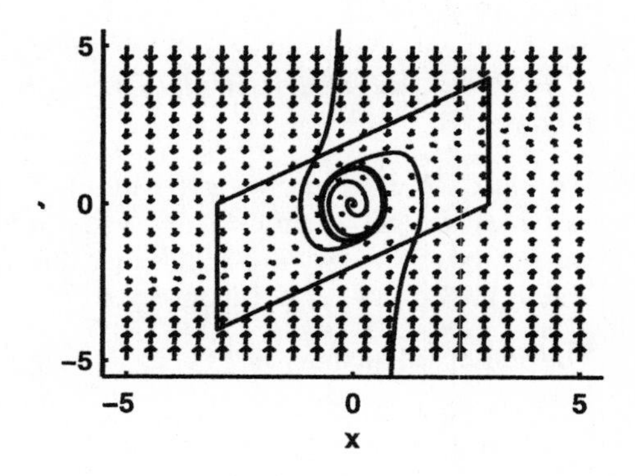

27. Any equilibrium point contained in limiting planar graph is vertex of the graph and must be the meeting place of two edges. Each of these edges must be a solution curve. Furthermore, since the edges are contained in the limit of a solution curve, one of the edges must be a stable solution curve for (x_0, y_0), and the other must be an unstable solution curve for (x_0, y_0). The only type of equilibrium point with a nonsingular determinant is a saddle. Clearly one of the edges is a stable separatrix and the other is an unstable separatrix.

Section 10.5. Conserved Quantities

1. We must solve $dy/dv = -v/(2y)$. Separating the variables we get $2y\,dy = -v\,dv$. Integrating we get $y^2 = -v^2/2 + C$. Hence the quantity $E(y, v) = y^2 + v^2/2$ is conserved.

3. We must solve $dy/dv = v/(y^2 - y)$. Separating variables we get $(y^2 - y)\,dy = v\,dv$. Integrating we get $y^3/3 - y^2/2 = v^2/2 + C$. Hence the quantity $E(y, v) = y^3/3 - y^2/2 - v^2/2$ is conserved.

5. We must solve $dy/dv = v/(y - y^3)$. Separating variables we get $(y - y^3)\,dy = v\,dv$. Integrating we get $y^2/2 - y^4/4 = v^2/2 + C$. Hence the quantity $E(y, v) = y^2/2 - y^4/4 - v^2/2$ is conserved.

7. We must solve $dx/dy = e^y/e^x$. Separating variables we get $e^x\,dx = e^y\,dy$. Integrating we get $e^x = e^y + C$. Hence the quantity $E(x, y) = e^x - e^y$ is conserved.

9. We must solve $dx/dy = y/(-\sin x)$. Separating variables we get $-\sin x\,dx = y\,dy$. Integrating we get $\cos x = y^2/2 + C$. Hence the quantity $E(x, y) = \cos x - y^2/2$ is conserved.

11. The conserved quantity is $E(y, v) = y^2 + v^2/2$.

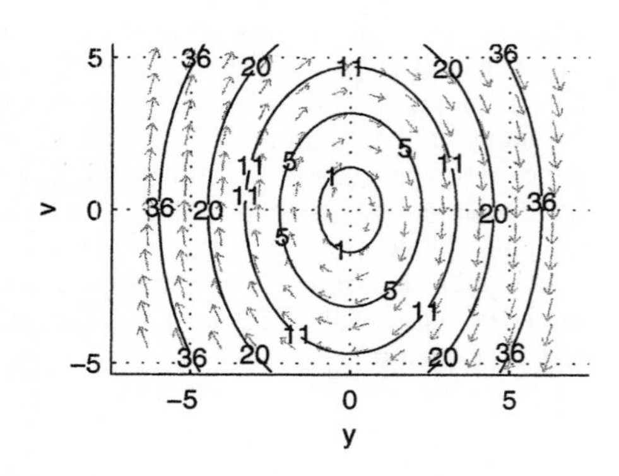

13. The conserved quantity is $E(y, v) = y^3/3 - y^2/2 - v^2/2$.

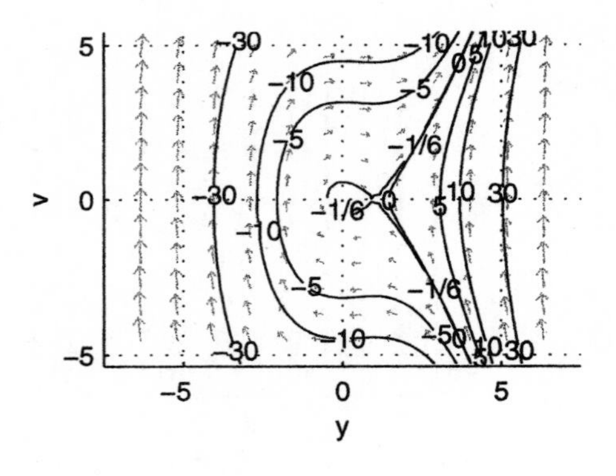

15. The conserved quantity is $E(y, v) = y^2/2 - y^4/4 - v^2/2$.

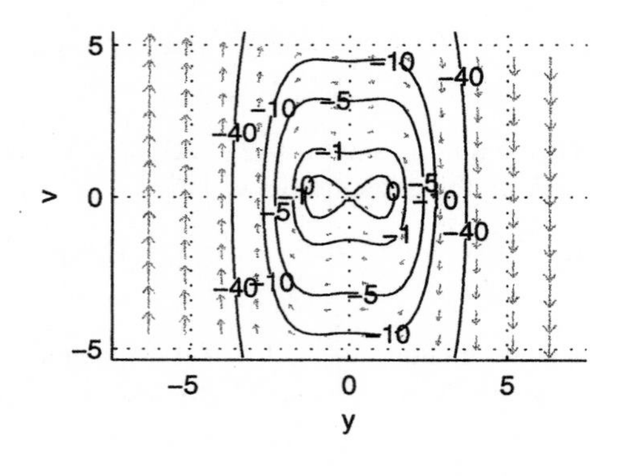

17. The conserved quantity is $E(x, y) = e^x - e^y$.

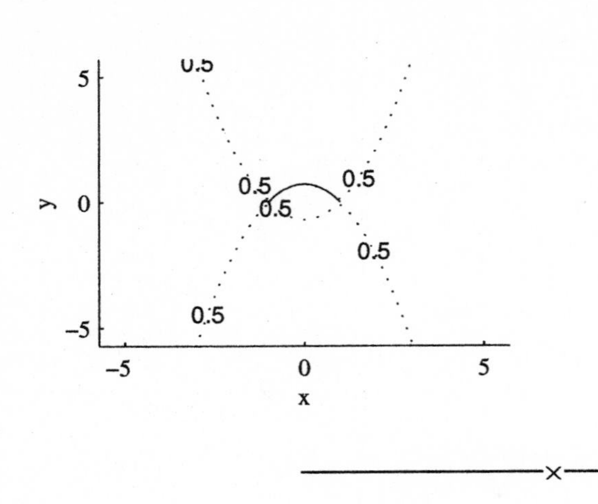

A similar strategy can be employed to start solution trajectories on the remaining pieces of the level curve $H(x, y) = 1/2$.

Section 10.6. Nonlinear Mechanics

1. The potential function has a minimum at $y = 0$. Hence, the phase portrait will have an equilibrium point at $(0, 0)$ that will be a center.

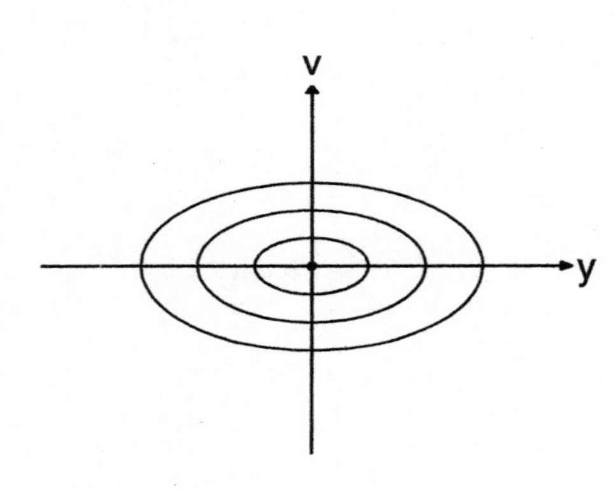

3. The potential curve has a minimum somewhere to the left of the v-axis, say at y_1. Consequently, the phase portrait will show an equilibrium point at $(y_1, 0)$ that is a center. Returning to the potential function, you can see a maximum to the right of the v-axis, say at y_2. Thus, the phase portrait will show a saddle at $(y_2, 0)$.

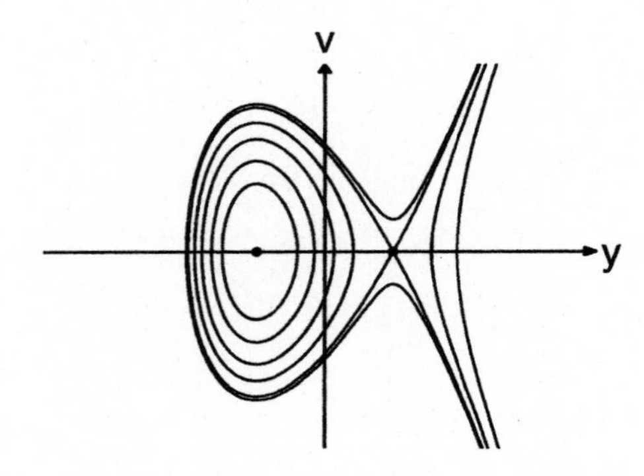

5. The potential curve has maxima both to the right and left of the v-axis, say at y_1, and by symmetry, at about $-y_1$. Therefore, the phase portrait will show saddle equilibrium points at $(y_1, 0)$ and $(-y_1, 0)$. Returning to the potential function, you can see a minimum at the origin. Hence, there will be a center in the phase portrait at the origin as well.

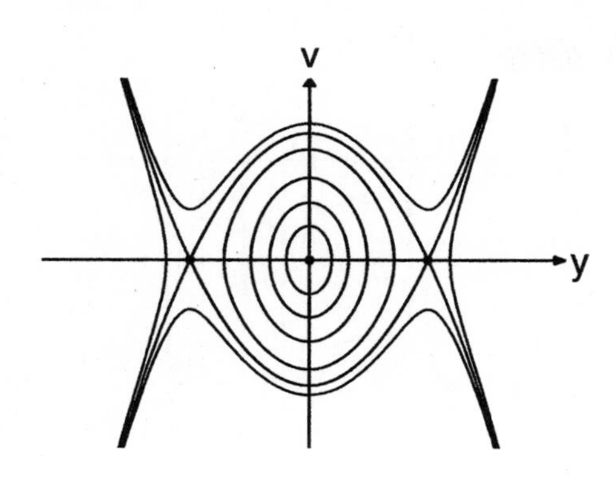

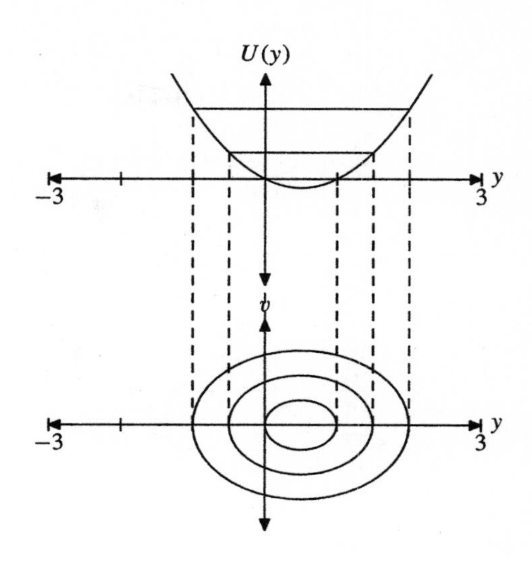

7. The potential function has derivative $U'(y) = -f(y)$. Thus,

$$\begin{aligned} U'(y) &= -(1-2y) \\ U(y) &= -y + y^2. \end{aligned}$$

The extrema of $U(y)$ occur when $U'(y) = 0$. Hence,

$$\begin{aligned} U'(y) &= 0 \\ -(1-2y) &= 0 \\ y &= \frac{1}{2}. \end{aligned}$$

As indicated in the plot that follows, the potential function has a minimum at $y = 1/2$, which leads to a center in the phase plane.

9. The potential function has derivative $U'(y) = -f(y)$. Thus,

$$\begin{aligned} U'(y) &= -(4y - y^2) \\ U(y) &= -2y^2 + \frac{1}{3}y^3. \end{aligned}$$

The extrema of $U(y)$ occur when $U'(y) = 0$. Hence,

$$\begin{aligned} U'(y) &= 0 \\ -(4y - y^2) &= 0 \\ y(y-4) &= 0 \\ y &= 0, 4. \end{aligned}$$

As indicated in the plot that follows, the potential function has a minimum at $y = 4$, which leads to a center in the phase plane. The potential also has a maximum at $y = 0$, which leads to a saddle in the phase plane.

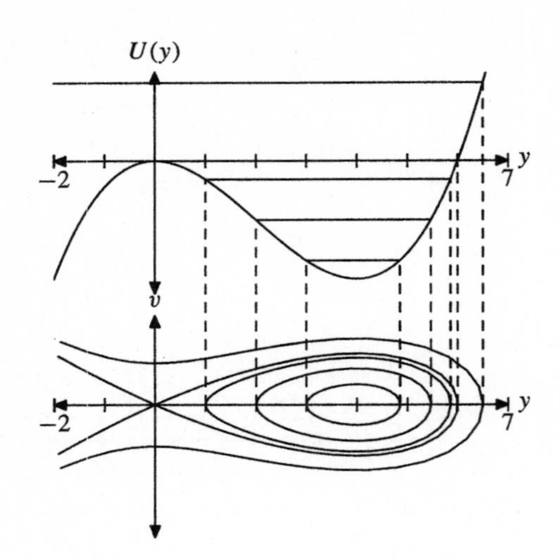

11. The potential function has derivative $U'(y) = -f(y)$. Thus,

$$\begin{aligned} U'(y) &= -\sin y \\ U(y) &= \cos y. \end{aligned}$$

The extrema of $U(y)$ occur when $U'(y) = 0$. Hence,

$$\begin{aligned} U'(y) &= 0 \\ -\sin y &= 0 \\ y &= -\pi, 0, \pi. \end{aligned}$$

As indicated in the plot that follows, the potential function has minima at $y = -\pi$ and $y = \pi$, which leads to centers in the phase plane. The potential also has a maximum at $y = -2\pi, 0, 2\pi$, which leads to saddles in the phase plane. Of course, this discussion can be extended beyond the interval $[-2\pi, 2\pi]$. In that case, the minima would occur at $y = k\pi$, where k is an odd integer, and the maxima would occur at $y = k\pi$, where k is an even integer. These would lead to centers at $(k\pi, 0)$, k odd, and saddles at $(k\pi, 0)$, k even.

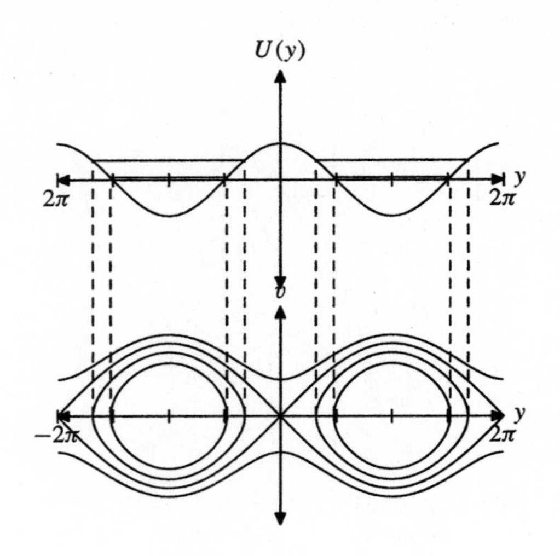

13. The phase portrait

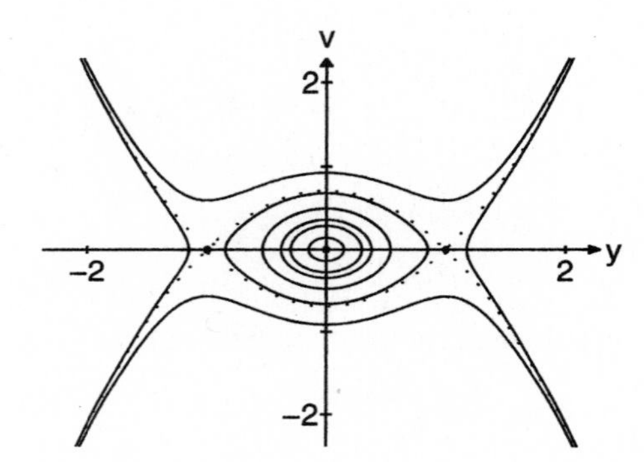

indicates the presence of saddles at $(-1, 0)$ and $(1, 0)$. In addition, there is a center at $(0, 0)$. Because the potential

$$\begin{aligned} U(y) &= -\int f(y)\,dy \\ &= -\int (y^3 - y)\,dy \\ &= -\frac{1}{4}y^4 + \frac{1}{2}y^2, \end{aligned}$$

the energy is given by

$$\frac{1}{2}v^2 + U(y) = E$$
$$\frac{1}{2}v^2 - \frac{1}{4}y^4 + \frac{1}{2}y^2 = E.$$

The presence of a saddle at $y = -1$ tells us that the potential is maximized at this value. But,

$$U(-1) = -\frac{1}{4}(-1)^4 + \frac{1}{2}(-1)^2 = \frac{1}{4}.$$

Thus, the equation of the separatrix is

$$\frac{1}{2}v^2 - \frac{1}{4}y^4 + \frac{1}{2}y^2 = \frac{1}{4}.$$

You will need an implicit function plotter to plot the separatrix as we have done on our figure (dotted lines).

15. The phase portrait

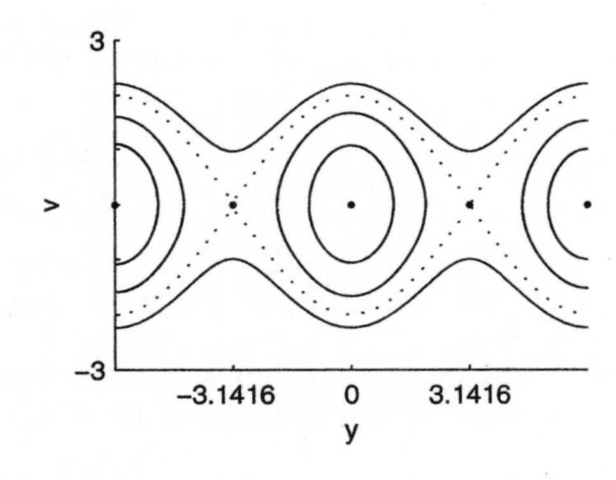

indicates the presence of saddles at $(-\pi, 0)$ and $(\pi, 0)$. In addition, there are centers at $(-2\pi, 0)$, $(0, 0)$, and $(2\pi, 0)$. Because the potential

$$U(y) = -\int f(y)\,dy = \int \sin y\,dy = -\cos y,$$

the energy is given by

$$\frac{1}{2}v^2 + U(y) = E$$
$$\frac{1}{2}v^2 - \cos y = E.$$

The presence of a saddle at $y = -\pi$ tells us that the potential is maximized at this value. But,

$$U(-\pi) = -\cos\pi = 1.$$

Thus, the equation of the separatrix is

$$\frac{1}{2}v^2 - \cos y = 1.$$

You can solve this equation for v, but we have chosen to use our implicit function plotter to plot the separatrix in our phase portrait (dotted lines). Finally, this discussion has focused on the interval $[-2\pi, 2\pi]$, but can easily be extended to a larger interval.

17. The linear spring defined by $f(y) = -\alpha y$, $\alpha > 0$, is shown in the following graph.

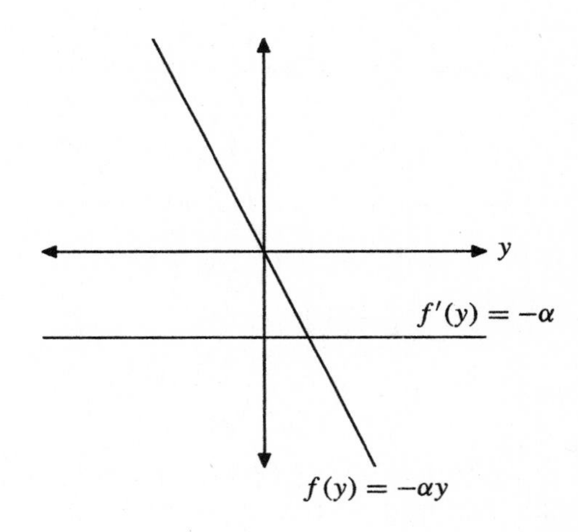

- Note that the force is negative when the displacement is positive (and positive when the displacement is negative).
- The magnitude of the force increases with the displacement. The further the spring is displaced from equilibrium, the more force is required to displace the spring an additional amount.

- The derivative of the force, $f'(y)$, is constant. Thus, the *stiffness* of the spring remains constant, regardless of the displacement.

If we add a term to the forcing function, as in $f(y) = -(\alpha y + \beta y^3)$, where $\beta > 0$, as y increases, it is intuitive that the function will decrease quicker. This is easily seen in the graph of f.

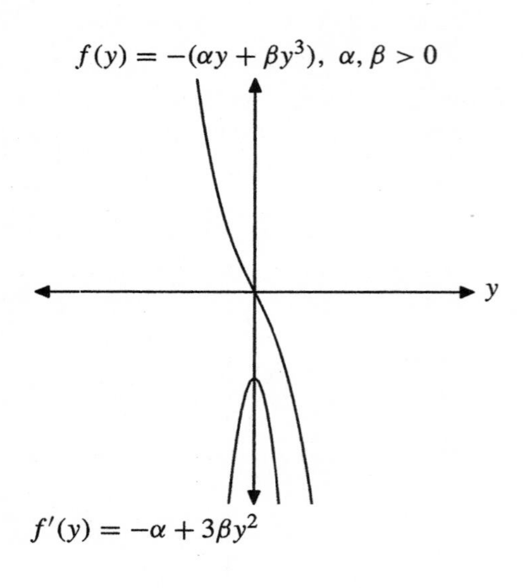

Note that the derivative, the stiffness, is no longer constant. Indeed, as the spring is displaced in the positive direction, the *magnitude* of the stiffness increases (the derivative is becoming more negative, but it is increasing in magnitude). This means that we have a *hard* spring, as the restoring force increases at a greater rate with displacement than that of the linear spring. This is easily seen when you graph the hard spring and linear spring in the same plot.

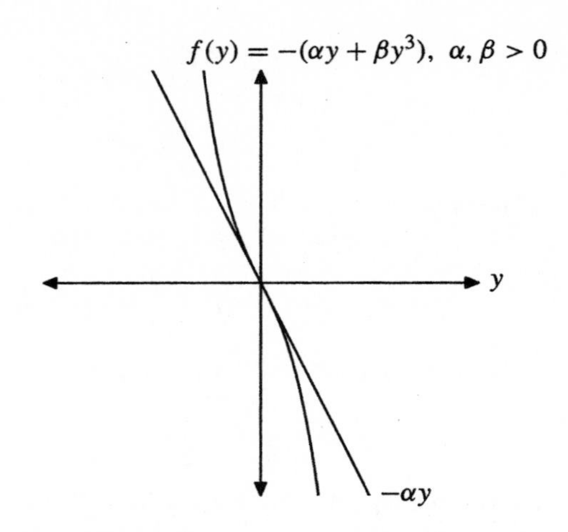

Similar comments are in order for the soft spring, which occurs when $f(y) = -(\alpha y + \beta y^3)$, $\alpha > 0$, $\beta < 0$.

19. (a) If absolute value is not used, then the damping term becomes $-b\omega^2$. Note that in this case the damping is always negative, and not opposite the motion. However, if the damping term $-b\omega|\omega|$ is used, then

$$-b\omega|\omega| = \begin{cases} -b\omega^2, & \text{if } \omega \geq 0, \\ b\omega^2, & \text{if } \omega < 0. \end{cases}$$

Therefore, if the damping is define as $-b\omega|\omega|$, then the damping is always opposite the motion.

To find the equilibrium points, set both right-hand sides of the system equal to zero.

$$\omega = 0$$
$$-a \sin\theta - b\omega|\omega| = 0$$

Thus, $\omega = 0$ and the second equation becomes

$$-a \sin\theta = 0,$$

which leads to equilibrium points at $(k\pi, 0)$, where k is an integer.

(b) In the image that follows, only the separatrices have been drawn. This is enough to indicate that spiral sinks occur at $(k\pi, 0)$, where k is an even integer, and saddles occur at $(k\pi, 0)$, where k is an odd integer.

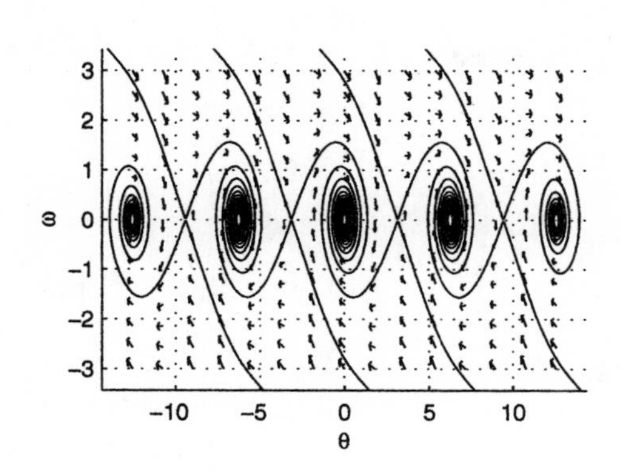

21. If the quantity H is conserved along solutions of the system

$$x' = ax + by$$
$$y' = cx + dy,$$

then

$$\frac{dH}{dt} = 0$$
$$\frac{\partial H}{\partial x}x' + \frac{\partial H}{\partial y}y' = 0.$$

Then,

$$\begin{aligned} 0 &= (2Ax + By)(ax + by) \\ &\quad + (Bx + 2Cy)(cx + dy) \\ 0 &= (2aA + cB)x^2 + (2bA + (a + d)B \\ &\quad + 2cC)xy + (bB + 2dC)y^2. \end{aligned}$$

If this is true for nontrivial solutions $(x(t), y(t))$, then we get the following system of equations by equating coefficients.

$$\begin{aligned} 2aA + cB &= 0 \\ 2bA + (a + d)B + 2cC &= 0 \\ bB + 2dC &= 0 \end{aligned}$$

If this system has a nontrivial solution for A, B, and C, then the determinant of the coefficient matrix is zero. Thus,

$$\begin{vmatrix} 2a & c & 0 \\ 2b & a+d & 2c \\ 0 & b & 2d \end{vmatrix} = 0$$

$$\begin{aligned} 2a\begin{vmatrix} a+d & 2c \\ b & 2d \end{vmatrix} - c\begin{vmatrix} 2b & 2c \\ 0 & 2d \end{vmatrix} &= 0 \\ 2a(2ad + 2d^2 - 2bc) - c(4bd) &= 0 \\ a^2d + ad^2 - abc - bcd &= 0 \\ (ad - bc)(a + d) &= 0. \end{aligned}$$

Thus, the determinant of the matrix of the system either has determinant or trace equal to zero.

23. If

$$x' = f(x, y) = -2x - 3y^2,$$
$$y' = g(x, y) = -3x^2 + 2y,$$

then

$$\frac{\partial f}{\partial x} = \frac{\partial}{\partial x}(-2x - 3y^2) = -2, \text{ and}$$
$$-\frac{\partial g}{\partial y} = -\frac{\partial}{\partial y}(-3x^2 + 2y) = -2.$$

Therefore, the system is Hamiltonian. Next,

$$\frac{\partial H}{\partial y} = f(x, y) = -2x - 3y^2, \text{ so}$$
$$H(x, y) = -2xy - y^3 + \phi(x).$$

But,

$$\frac{\partial H}{\partial x} = -g(x, y) = -(-3x^2 + 2y), \text{ so}$$
$$-2y + \phi'(x) = 3x^2 - 2y$$
$$\phi'(x) = 3x^2.$$

Thus, $\phi(x) = x^3$ and

$$H(x, y) = -2xy - y^3 + x^3.$$

25. If

$$x' = f(x, y) = 3y^2,$$
$$y' = g(x, y) = -3x^2$$

then

$$\frac{\partial f}{\partial x} = \frac{\partial}{\partial x}(3y^2) = 0, \text{ and}$$
$$-\frac{\partial g}{\partial y} = -\frac{\partial}{\partial y}(-3x^2) = 0.$$

Therefore, the system is Hamiltonian. Next,

$$\frac{\partial H}{\partial y} = f(x, y) = 3y^2, \text{ so}$$
$$H(x, y) = y^3 + \phi(x).$$

But,

$$\frac{\partial H}{\partial x} = -g(x, y) = -(-3x^2), \text{ so}$$
$$\phi'(x) = 3x^2.$$

Thus, $\phi(x) = x^3$ and

$$H(x, y) = y^3 + x^3.$$

27. If

$$x' = f(x, y) = -x + 2y,$$
$$y' = g(x, y) = -2x + y$$

then

$$\frac{\partial f}{\partial x} = \frac{\partial}{\partial x}(-x + 2y) = -1, \text{ and}$$
$$-\frac{\partial g}{\partial y} = -\frac{\partial}{\partial y}(-2x + y) = -1.$$

Therefore, the system is Hamiltonian. Next,

$$\frac{\partial H}{\partial y} = f(x, y) = -x + 2y, \text{ so}$$
$$H(x, y) = -xy + y^2 + \phi(x).$$

But,

$$\frac{\partial H}{\partial x} = -g(x, y) = -(-2x + y), \text{ so}$$
$$-y + \phi'(x) = 2x - y$$
$$\phi'(x) = 2x.$$

Thus, $\phi(x) = x^2$ and

$$H(x, y) = -xy + y^2 + x^2.$$

29. If

$$x' = f(x, y) = \cos x,$$
$$y' = g(x, y) = -y \sin x + 2x$$

then

$$\frac{\partial f}{\partial x} = \frac{\partial}{\partial x}(\cos x) = -\sin x, \text{ and}$$
$$-\frac{\partial g}{\partial y} = -\frac{\partial}{\partial y}(-y \sin x + 2x) = \sin x.$$

Therefore, the system is not Hamiltonian.

31. We compute that

$$\begin{aligned}\dot{E} &= \frac{\partial E}{\partial y_1}y_1' + \frac{\partial E}{\partial y_2}y_2' + \frac{\partial E}{\partial v_1}v_1' + \frac{\partial E}{\partial v_2}v_2' \\ &= -f_1 v_1 - f_2 v_2 + v_1 f_1 + v_2 f_2 \\ &= 0.\end{aligned}$$

Hence E is a conserved quantity.

Section 10.7. The Method of Lyapunov

1. If $V(x, y) = x^2 + 2y^2$, then $V(0, 0) = 0$, but $V(x, y) > 0$ for all $(x, y) \in \mathbf{R}^2$ that are different from $(0, 0)$. Thus, V is positive definite.

3. If $V(x, y) = x + y^2$, then $V(0, 0) = 0$. However, V is both positive and negative in any neighborhood containing $(0, 0)$. Thus, V is neither positive definite or negative definite, nor is it positive semidefinite or negative semidefinite.

5. If $V(x, y) = -2y^2$, then $V(0, 0) = 0$. Furthermore, $V(x, y) \le 0$ for all $(x, y) \in \mathbf{R}^2$. Note that any open neighborhood of $(0, 0)$ must contain a portion of the y-axis, where V is identically zero. Thus, V is negative semidefinite.

7. If $V(x, y) = x^2 + y$, then $V(0, 0) = 0$. However,

V is both positive and negative in any neighborhood containing $(0,0)$. Thus, V is neither positive definite or negative definite, nor is it positive semidefinite or negative semidefinite.

9. Note that the surface determined by $V(x,y) = x^2 - 2xy + 3y^2$ appears to lie completely above the xy-plane.

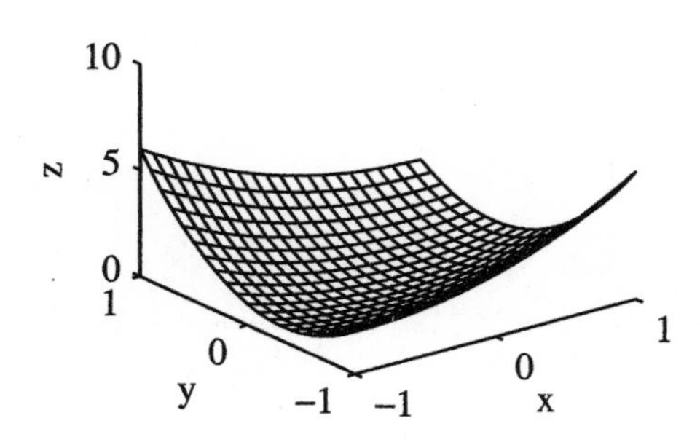

Indeed, $V(0,0) = 0$ and

$$\begin{aligned} V(x,y) &= x^2 - 2xy + 3y^2 \\ &= (x^2 - 2xy + y^2) + 2y^2 \\ &= (x-y)^2 + 2y^2. \end{aligned}$$

In order that $V(x,y) = 0$, both $x - y$ and y must simultaneously equal zero. This can only happen at $(0,0)$. Thus, $V(x,y) > 0$ for all $(x,y) \in \mathbf{R}^2$ different from $(0,0)$ and V is positive definite.

11. A standard test from multivariable calculus requires two conditions in order that $V(x,y) = ax^2 + 2bxy + cy^2$ have a minimum at $(0,0)$.

   - The second derivative, $V_{xx}(0,0) = 2a > 0$, which means that a must be greater than zero.
   - The second derivative test, $V_{xx}(0,0)V_{yy}(0,0) - V_{xy}(0,0) = (2a)(2c) - (2b)^2 > 0$. This requires that $ac - b^2 > 0$.

   You can also use linear algebra. Note that

$$V(x,y) = ax^2 + 2bxy + cy^2 = \begin{pmatrix} x & y \end{pmatrix} \begin{pmatrix} a & b \\ b & c \end{pmatrix} \begin{pmatrix} x \\ y \end{pmatrix}.$$

   Thus, $V(x,y) = \mathbf{x}^T A \mathbf{x}$, where

$$A = \begin{pmatrix} a & b \\ b & c \end{pmatrix} \quad \text{and} \quad \mathbf{x} = \begin{pmatrix} x \\ y \end{pmatrix}.$$

   Therefore, $V(x,y) > 0$ for all $(x,y) \neq (0,0)$ is equivalent to saying that $\mathbf{x}^T A\mathbf{x} > 0$ for all $\mathbf{x} \neq \vec{0}$. Note that if $\mathbf{x} = (1,0)^T$, then

$$\mathbf{x}^T A\mathbf{x} > 0 \quad \Rightarrow \quad a > 0.$$

   Further, if λ is an eigenvalue with associated eigenvector $\mathbf{x}$, then

$$0 < \mathbf{x}^T A\mathbf{x} = \mathbf{x}^T \lambda \mathbf{x} = \lambda \left\|\mathbf{x}\right\|.$$

   Because $\left\|\mathbf{x}\right\| \neq 0$, this implies that all eigenvalues of the matrix A are positive. Because the determinant of matrix A equals the product of its eigenvalues, and all the eigenvalues of matrix A are positive, this in turn implies that the determinant of matrix A is positive. Thus,

$$\begin{vmatrix} a & b \\ b & c \end{vmatrix} > 0 \quad \text{or} \quad ac - b^2 > 0.$$

13. If

$$\begin{aligned} \frac{dx}{dt} &= y \\ \frac{dy}{dt} &= -x - y^3 \end{aligned}$$

   and $V(x,y) = x^2 + y^2$, then

$$\begin{aligned} \frac{dV}{dt} &= \frac{\partial V}{\partial x} \cdot \frac{dx}{dt} + \frac{\partial V}{\partial y} \cdot \frac{dy}{dt} \\ &= 2x(y) + 2y(-x - y^3) \\ &= -2y^4. \end{aligned}$$

15. If

$$\begin{aligned} \frac{dx}{dt} &= y \\ \frac{dy}{dt} &= -x - y^3 \end{aligned}$$

   and $V(x,y) = x^2 + xy - y^2$, then

$$\begin{aligned} \frac{dV}{dt} &= \frac{\partial V}{\partial x} \cdot \frac{dx}{dt} + \frac{\partial V}{\partial y} \cdot \frac{dy}{dt} \\ &= (2x + y)(y) + (x - 2y)(-x - y^3) \\ &= 2y^4 - xy^3 + y^2 + 4xy - x^2 \end{aligned}$$

17. (a) If

$$\frac{dx}{dt} = -y - x^3$$
$$\frac{dy}{dt} = x - y^3,$$

then the Jacobian is

$$J(x, y) = \begin{pmatrix} -3x^2 & -1 \\ 1 & -3y^2 \end{pmatrix}.$$

Evaluating at the origin,

$$J(0, 0) = \begin{pmatrix} 0 & -1 \\ 1 & 0 \end{pmatrix},$$

the trace is $T = 0$ and the determinant is $D = 1$. Therefore, the linearization predicts that the equilibrium point at the origin is a center.

(b) If $V(x, y) = x^2 + y^2$, then $V(0, 0) = 0$, but $V(x, y) > 0$ for all $(x, y) \in \mathbf{R}^2$ different from $(0, 0)$. Hence, V is positive definite on $\mathbf{R}^2$. Further,

$$\begin{aligned}\dot{V} &= \frac{\partial V}{\partial x} \cdot \frac{dx}{dt} + \frac{\partial V}{\partial y} \cdot \frac{dy}{dt} \\ &= 2x(-y - x^3) + 2y(x - y^3) \\ &= -2(x^4 + y^4).\end{aligned}$$

Note that $\dot{V}(0, 0) = 0$, but $\dot{V}(x, y) < 0$ for all $(x, y) \in \mathbf{R}^2$ different from $(0, 0)$. Hence, $\dot{V}$ is negative definite, and the equilibrium point at the origin is asymptotically stable.

(c) Note that the equilibrium point is a spiral sink. Note further that our solver stopped short of the sink at the origin, but the theory of part (b) assures us that solutions should continue to spiral into the origin, provided we let enough time pass in our solver.

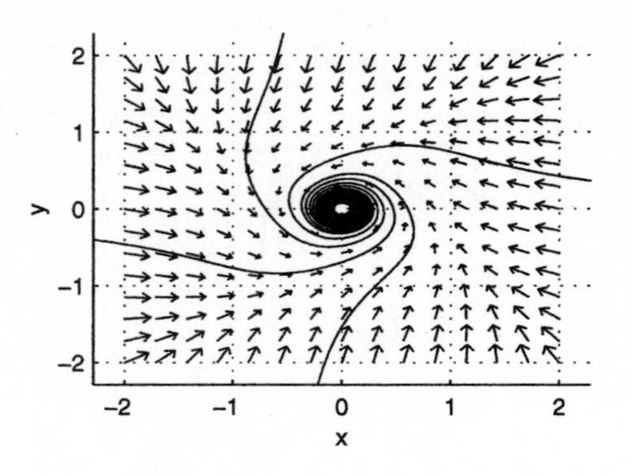

19. If

$$x' = -x + y$$
$$y' = -x - y,$$

and $V(x, y) = x^2 + y^2$, then V is positive definite with minimum at $(0, 0)$. Further,

$$\begin{aligned}\dot{V} &= \frac{\partial V}{\partial x} \cdot x' + \frac{\partial V}{\partial y} \cdot y' \\ &= 2x(-x + y) + 2y(-x - y) \\ &= -2(x^2 + y^2),\end{aligned}$$

so $\dot{V}$ is negative definite on $\mathbf{R}^2$. Thus, the origin is asymptotically stable as shown in the following phase portrait.

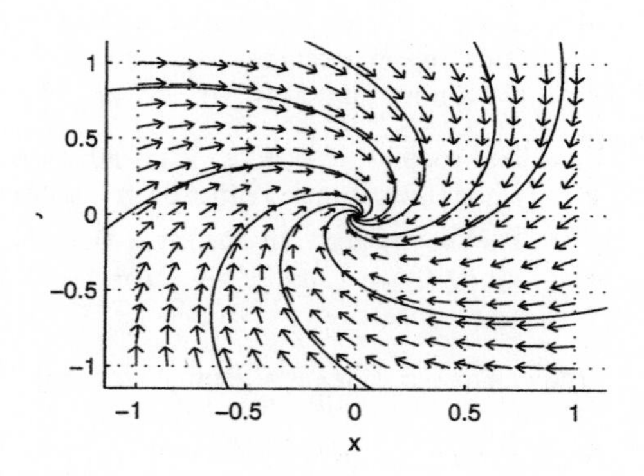

21. If

$$x' = -x + xy$$
$$y' = -y + xy,$$

and V is the positive definite function defined by $V(x, y) = x^2 + y^2$, then

$$\begin{aligned}\dot{V} &= \frac{\partial V}{\partial x} \cdot x' + \frac{\partial V}{\partial y} \cdot y' \\ &= 2x(-x + xy) + 2y(-y + xy) \\ &= -2(x^2 - x^2 y + y^2 - xy^2).\end{aligned}$$

A plot of the surface represented by $\dot{V}$ is revealing in that it appears to lie below the xy-plane on the domain $D = \{(x, y) : x, y < 1\}$.

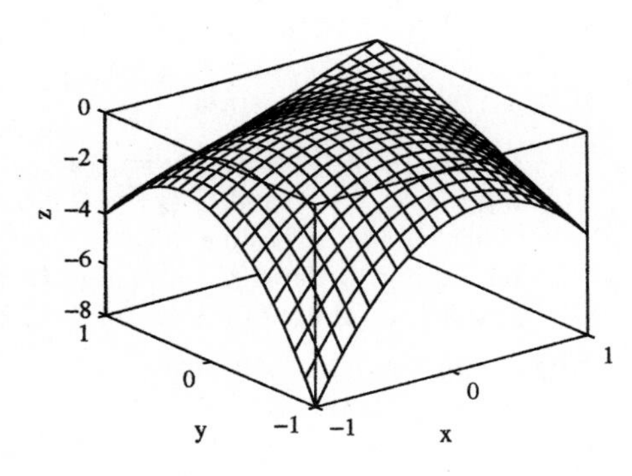

Indeed, we can write

$$\dot{V}(x, y) = -2(x^2(1 - y) + y^2(1 - x)),$$

so, if we restrict x and y so that both are less than 1, then $\dot{V}$ is negative definite on this restricted domain. Thus, the equilibrium point at the origin is asymptotically stable, as shown in the following phase portrait.

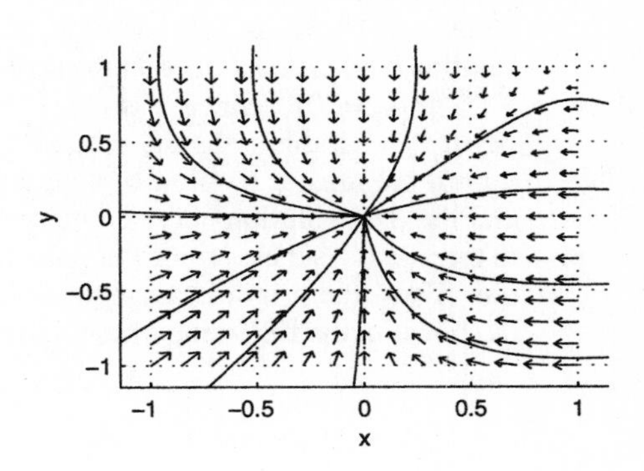

23. (a) If

$$x'' = -x' + \frac{1}{3}(x')^3 - x,$$

and we let $x_1 = x$ and $x_2 = x'$, then

$$x_1' = x_2$$
$$x_2' = -x_2 + \frac{1}{3}x_2^3 - x_1.$$

The function $V(x_1, x_2) = x_1^2 + x_2^2$ is positive definite with minimum at $(0, 0)$. Further,

$$\begin{aligned}\dot{V} &= \frac{\partial V}{\partial x_1} \cdot x_1' + \frac{\partial V}{\partial x_2} \cdot x_2' \\ &= 2x_1(x_2) + 2x_2(-x_2 + \frac{1}{3}x_2^3 - x_1) \\ &= -2x_2^2 + \frac{2}{3}x_2^4 \\ &= -\frac{2}{3}x_2^2(3 - x_2^2).\end{aligned}$$

We can make $\dot{V}(x_1, x_2) \le 0$ by choosing

$$3 - x_2^2 > 0$$
$$x_2^2 < 3$$
$$|x_2| < \sqrt{3}.$$

Thus, provided we stay in the domain $\{(x_1, x_2) : |x_2| < \sqrt{3}\}$, then V is negative semi-definite and the equilibrium point at the origin is stable.

(b) In part (a), we saw that $V(x_1, x_2) = x_1^2 + x_2^2$ is positive definite on $\{(x_1, x_2) : |x_2| < \sqrt{3}\}$. According to Theorem 7.10, the limit set of every solution curve is contained in the set $\{(x_1, x_2) : \dot{V}(x_1, x_2) = 0\}$. Since $\dot{V}(x_1, x_2) = 0$ only when $x_2 = 0$, this set is the x_1 axis. However, a limit set must e a union of solution curves, and the only solution curve contained entirely in the x_1 axis is the equilibrium point at the origin. Thus, the origin is asymptotically stable, as shown in the following phase portrait.

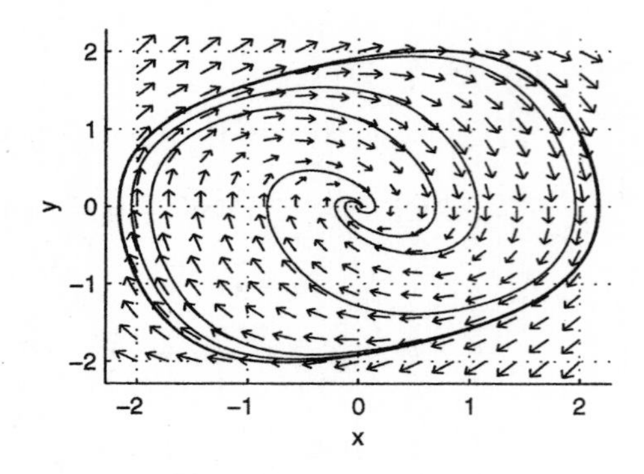

25. If

$$\ddot{x} + (\dot{x})^3 + x^3 = 0,$$

and we let $x_1 = x$ and $x_2 = \dot{x}$, then

$$\begin{aligned} \dot{x}_1 &= x_2 \\ \dot{x}_2 &= -x_2^3 - x_1^3. \end{aligned}$$

From here, we use $V(x_1, x_2) = (1/2)x_1^4 + x_2^2$ and argue as we did in Exercise 22 that the equilibrium point at the origin is asymptotically stable.

27. (a) If $xg(x) > 0$ for all x, this says that x and $g(x)$ must agree in sign. That is, when $x > 0$, so must $g(x) > 0$. Similarly, if $x < 0$, then $g(x) < 0$. One such example is show in the following graph.

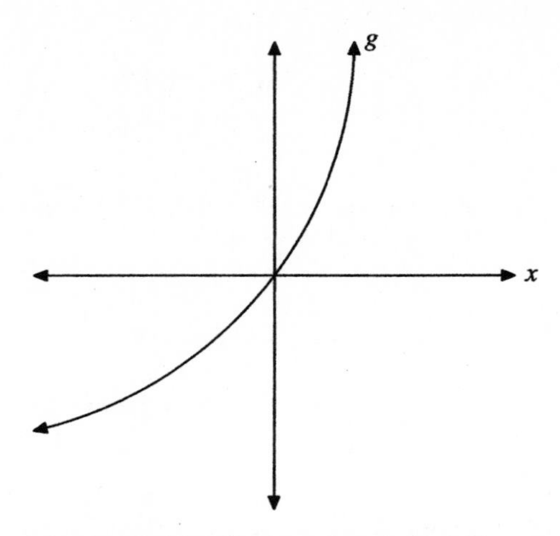

Another example is $g(x) = \sin x$, provided we restrict the domain to $|x| < \pi/2$.

Because the graph of g changes sign at $x = 0$ (negative to the left of the origin, positive to the right), the continuity of g requires that $g(0) = 0$. If we define

$$G(x) = \int_0^x g(u)\, du,$$

then $G(0) = 0$ and $G'(x) = g(x)$, so $G'(x) < 0$ if $-k < x < 0$ and $G'(x) > 0$ if $0 < x < k$. Thus, G is decreasing to the left of zero and increasing to the right, making $G(0) = 0$ an absolute minimum (at least on $(-k, k)$).

(b) If we let $y = x'$, then because $x'' = -f(x)x' - g(x)$, we can write

$$\begin{aligned} x' &= y \\ y' &= -f(x)y - g(x). \end{aligned}$$

If

$$V(x, y) = \frac{1}{2}y^2 + G(x),$$

then because we've shown that $G(0) = 0$ and G has an absolute minimum value of zero at $x = 0$ on $(-k, k)$, we can see that $V(0, 0) = 0$ and $V(x, y) > 0$ for all $(x, y) \neq (0, 0)$, provided that we restrict the neighborhood so that $-k < x < k$. Thus, V is positive definite with minimum at $(0, 0)$, provided $-k < x < k$.

(c) Furthermore,

$$\begin{aligned}\dot{V} &= \frac{\partial V}{\partial x}\cdot x' + \frac{\partial V}{\partial y}\cdot y' \\ &= G'(x)y + y(-f(x)y - g(x)) \\ &= g(x)y + y(-f(x)y - g(x)) \\ &= -f(x)y^2.\end{aligned}$$

Note that $\dot{V}(0,0) = 0$. Further, because it is given that $f(x) > 0$ for all x, $\dot{V}$ is negative semi-definite on $\{(x, y) : -k < x < k\}$. Thus, by Theorem 7.10, the forward limit set of all trajectories must lie in the set $\dot{V}(x, y) = 0$, which because $f(x) > 0$ for all x, must consist of that part of the x-axis between $-k$ and k. However, a limit set must be a union of solution curves, and the only solution curve contains in the x-axis is the equilibrium point at the origin. Therefore, the equilibrium point at the origin is asymptotically stable.

Section 10.8. Predator-Prey Systems

1. The system

$$\begin{aligned}x' &= x(a - ex - by) = ax - ex^2 - bxy \\ y' &= y(-c + dx) = -cy + dxy,\end{aligned}$$

has an equilibrium point which is the intersection of the nullclines

$$\begin{aligned}a - ex - by &= 0 \\ -c + dx &= 0.\end{aligned}$$

This equilibrium point has coordinates,

$$(x_0, y_0) = \left(\frac{c}{d}, \frac{ad - ec}{bd}\right).$$

The equilibrium point (x_0, y_0) lies in the first quadrant if $y_0 > 0$, which is true only if $ad - ec > 0$. The Jacobian of the system is

$$J(x, y) = \begin{pmatrix} a - 2ex - by & -bx \\ dy & -c + dx \end{pmatrix}.$$

If we evaluate the Jacobian at the equilibrium point (x_0, y_0), we have

$$\begin{aligned}J(x_0, y_0) &= \begin{pmatrix} a - 2ex_0 - by_0 & -bx_0 \\ dy_0 & -c + dx_0 \end{pmatrix} \\ &= \begin{pmatrix} (a - ex_0 - by_0) - ex_0 & -bx_0 \\ dy_0 & -c + dx_0 \end{pmatrix} \\ &= \begin{pmatrix} -ex_0 & -bx_0 \\ dy_0 & 0 \end{pmatrix}.\end{aligned}$$

This last result holds true because the equilibrium point is the intersection of the nullclines $a - ex - by = 0$ and $-c + dx = 0$. Consequently, $a - ex_0 - by_0 = 0$ and $-c + dx_0 = 0$. Now, the trace and determinant of $J(x_0, y_0)$ are easily determined. The trace is $T = -ex_0$, and because $x_0 > 0$, the trace is negative. The determinant is $D = bdx_0y_0$, and since both x_0 and y_0 lie in the positive quadrant, the determinant is also positive. Therefore, we have determined that the equilibrium point is some kind of sink.

To determine the kind of sink (nodal or spiral), we must examine $T^2 - 4D$.

3. In our derivation of the predator-prey equations, we assumed that the increase in predators would be proportional to the number of contacts between predators and prey, and therefore to the product xy. This lead us to the equation

$$y' = -sx + bxy,$$

where x is the prey and y the predators. Again, the term bxy assumes that the probability that a predator will attack is the same for every encounter. This may be reasonable for small prey populations, but as the prey increases, many encounters will occur when the predator has recently eaten and is not hungry. There are a number of ways to account for this, but assuming again that the predation is proportional to $xy/(x+D)$ for some constant D is a good way to satisfy this difficulty. Note that for large values of x, this quantity becomes approximately proportional to y, and not the larger quantity xy. Thus, we replace the term bxy with $cxy/(x+D)$ and write

$$y' = -sy + \frac{cxy}{x+D}$$

$$y' = -sy\left(1 - \frac{Cx}{x+D}\right).$$

5. When $u_0 < 0$, the parabolic portion of the u-nullcline devides the positive quadrant into two pieces. The direction of the nullclines show that the lower region is invariant, and that all solution curves in this region approach the equilibrium point at $(1, 0)$. All solution curves in the upper region move to the right and down, and are drawn into the lower region and then to $(1, 0)$, or are drawn directly to $(1, 0)$. Thus all solution curves end up at $(1, 0)$.

   When $u_0 > 1$, the parabolic portion of the u-nullcline and the line $u = u_0$ divide the positive quadrant into three pieces. Again the region below the parabola is invariant, and all solutions are drawn to the equilibrium point at $(1, 0)$. All solution curves starting to the left of the line $u = u_0$ move to the left and ultimately into the region between the line $u = u_0$ and the parabola. Once there they move to the left and down, either into the region below the parabola and then on to $u = u_0$, or they move directly to $u = u_0$. Again all solution curves approach the equilibrium point at $(1, 0)$.

Chapter 11. Series Solutions to Differential Equations

Section 11.1. Review of Power Series

1. If $\sum_{n=0}^{\infty} x^n/(n+1)$, then

$$\lim_{n\to\infty}\frac{|a_n|}{|a_{n+1}|}=\lim_{n\to\infty}\frac{\dfrac{1}{n+1}}{\dfrac{1}{n+2}}=\lim_{n\to\infty}\frac{1+2/n}{1+1/n}=1.$$

Thus, the radius of convergence is $R=1$.

3. If $\sum_{n=0}^{\infty} nx^n$, then

$$\lim_{n\to\infty}\frac{|a_n|}{|a_{n+1}|}=\lim_{n\to\infty}\frac{n}{n+1}=\lim_{n\to\infty}\frac{1}{1+1/n}=1.$$

Thus, the radius of convergence is $R=1$.

5. If $\sum_{n=0}^{\infty} x^{n+1}/n!$, then

$$\lim_{n\to\infty}\frac{|a_n|}{|a_{n+1}|}=\lim_{n\to\infty}\frac{\dfrac{1}{n!}}{\dfrac{1}{(n+1)!}}=\lim_{n\to\infty}(n+1)=\infty.$$

Thus, the radius of convergence is $R=\infty$; i.e., the series converges for all x.

7. If $\sum_{n=0}^{\infty} n!(x-1)^n$, then

$$\lim_{n\to\infty}\frac{|a_n|}{|a_{n+1}|}=\lim_{n\to\infty}\frac{n!}{(n+1)!}=\lim_{n\to\infty}\frac{1}{n+1}=0.$$

Therefore, the series diverges for all $x\neq 0$.

9. If $\sum_{n=2}^{\infty} x^n/\ln n$, then

$$\frac{|a_n|}{|a_{n+1}|}=\frac{\dfrac{1}{\ln n}}{\dfrac{1}{\ln(n+1)}}=\frac{\ln(n+1)}{\ln n}.$$

As $n\to\infty$, both numerator and denominator go to infinity, so we can apply L'Hôpital's rule.

$$\lim_{n\to\infty}\frac{|a_n|}{|a_{n+1}|}=\lim_{n\to\infty}\frac{\ln(n+1)}{\ln n}=\lim_{n\to\infty}\frac{\dfrac{1}{n+1}}{\dfrac{1}{n}}=\lim_{n\to\infty}\frac{1}{1+1/n}=1.$$

Thus, the radius of convergence is $R=1$.

11. If $\sum_{n=1}^{\infty} nx^n/(1\cdot 3\cdot 5\cdot\dots\cdot(2n-1))$, note that

$$\begin{aligned}a_n&=\frac{n}{1\cdot 3\cdot 5\cdot\dots\cdot(2n-1)}\\&=\frac{n(2\cdot 4\cdot 6\cdot\dots\cdot(2n-2))}{(2n-1)!}\\&=\frac{n\cdot 2^{n-1}(1\cdot 2\cdot 3\cdot\dots\cdot(n-1))}{(2n-1)!}\\&=\frac{2^{n-1}n!}{(2n-1)!}.\end{aligned}$$

Thus

$$\frac{|a_n|}{|a_{n+1}|}=\frac{2^{n-1}n!}{(2n-1)!}\cdot\frac{(2n+1)!}{2^n(n+1)!}=\frac{2n(2n+1)}{2(n+1)}=\frac{2n+1}{1+1/n}.$$

Thus, $\lim_{n\to\infty}|a_n|/|a_{n+1}|=\infty$ and the radius of convergence is $R=\infty$; i.e., the series converges for all x.

13. We know that

$$\cos t = 1 - \frac{t^2}{2!} + \frac{t^4}{4!} - \cdots + \frac{(-1)^n t^{2n}}{(2n)!} + \cdots$$

and the radius of convergence is $R = \infty$; i.e., the series converges for all x. Thus, letting $t = 2x$,

$$\cos 2x = 1 - \frac{(2x)^2}{2!} + \cdots + \frac{(-1)^n (2x)^{2n}}{(2n)!} + \cdots$$

and the series converges for all x.

15. Derivatives reveal the following details.

$$\begin{aligned} f(x) = \sin x &\Rightarrow f(\pi) = 0 \\ f'(x) = \cos x &\Rightarrow f'(\pi) = -1 \\ f''(x) = -\sin x &\Rightarrow f''(\pi) = 0 \\ f'''(x) = -\cos x &\Rightarrow f'''(\pi) = 1, \end{aligned}$$

after which this same pattern repeats itself in blocks of four. Therefore,

$$\begin{aligned} \sin x &= \sum_{n=0}^{\infty} \frac{f^{(n)}(\pi)}{n!}(x-\pi)^n \\ &= f(\pi) + f'(\pi)(x-\pi) + \frac{f''(\pi)}{2!}(x-\pi)^2 \\ &\quad + \frac{f'''(\pi)}{3!}(x-\pi)^3 + \cdots \\ &= -1(x-\pi) + \frac{1}{3!}(x-\pi)^3 - \frac{1}{5!}(x-\pi)^5 \\ &\quad + \frac{1}{7!}(x-\pi)^7 - \cdots \\ &\quad + \frac{(-1)^{n+1}}{(2n+1)!}(x-\pi)^{2n+1} + \cdots. \end{aligned}$$

Alternatively,

$$\begin{aligned} \sin x &= \sin(x - \pi + \pi) \\ &= \sin(x-\pi)\cos\pi + \sin\pi\cos(x-\pi) \\ &= -\sin(x-\pi). \end{aligned}$$

Now let $t = x - \pi$ in the series.

$$-\sin t = -t + \frac{t^3}{3!} - \frac{t^5}{5!} + \cdots + \frac{(-1)^{n+1} t^{2n+1}}{(2n+1)!} + \cdots.$$

Thus,

$$\begin{aligned} &\sin x \\ &= -\sin(x-\pi) \\ &= -(x-\pi) + \frac{1}{3!}(x-\pi)^3 - \frac{1}{5!}(x-\pi)^5 + \cdots \\ &\quad + \frac{(-1)^{n+1}}{(2n+1)!}(x-\pi)^{2n+1} + \cdots. \end{aligned}$$

The radius of convergence is $R = \infty$.

17. Derivatives reveal the following details.

$$\begin{aligned} f(x) = \frac{1}{x} &\Rightarrow f(3) = \frac{1}{3} \\ f'(x) = -\frac{1}{x^2} &\Rightarrow f'(3) = -\frac{1!}{3^2} \\ f''(x) = \frac{2}{x^3} &\Rightarrow f''(3) = \frac{2!}{3^3} \\ f'''(x) = -\frac{3 \cdot 2}{x^4} &\Rightarrow f'''(3) = -\frac{3!}{3^4} \\ f^{(4)}(x) = \frac{4 \cdot 3 \cdot 2}{x^5} &\Rightarrow f^{(4)}(3) = \frac{4!}{3^5}. \end{aligned}$$

We conjecture that

$$f^{(n)}(x) = \frac{(-1)^n n!}{x^{n+1}} \Rightarrow f^{(n)}(3) = \frac{(-1)^n n!}{3^{n+1}},$$

for $n \geq 0$. Thus,

$$\begin{aligned} \frac{1}{x} &= \sum_{n=0}^{\infty} \frac{f^{(n)}(3)}{n!}(x-3)^n \\ &= \sum_{n=0}^{\infty} \frac{(-1)^n}{3^{n+1}}(x-3)^n \\ &= \frac{1}{3} - \frac{1}{3^2}(x-3) + \frac{1}{3^3}(x-3)^3 - \frac{1}{3^4}(x-3)^4 \\ &\quad + \cdots + \frac{(-1)^n}{3^{n+1}}(x-3)^n + \cdots. \end{aligned}$$

The ratio test shows that the radius of convergence is $R = 1$.

19. We have

$$
\begin{aligned}
f(x) &= (1+x)\sum_{n=0}^{\infty} x^n \\
&= \sum_{n=0}^{\infty} x^n + x\sum_{n=0}^{\infty} x^n \\
&= \sum_{n=0}^{\infty} x^n + \sum_{n=0}^{\infty} x^{n+1} \\
&= \sum_{n=0}^{\infty} x^n + \sum_{n=1}^{\infty} x^n \\
&= 1 + 2\sum_{n=1}^{\infty} x^n.
\end{aligned}
$$

21.

$$
\begin{aligned}
f(x) &= \sum_{n=0}^{\infty} x^n - \sum_{n=1}^{\infty} \frac{x^{n-1}}{n} \\
&= \sum_{n=0}^{\infty} 0\frac{x^n}{n} - \sum_{n=0}^{\infty} \frac{x^n}{n+1} \\
&= \sum_{n=0}^{\infty} \left(1 - \frac{1}{n+1}\right) x^n \\
&= \sum_{n=0}^{\infty} \frac{n}{n+1} x^n
\end{aligned}
$$

23.

$$
\begin{aligned}
f(x) &= e^x - e^{-x} \\
&= \sum_{n=0}^{\infty} \frac{x^n}{n!} - \sum_{n=0}^{\infty} \frac{(-x)^n}{n!} \\
&= \sum_{n=0}^{\infty} [1 - (-1)^n] \frac{x^n}{n!} \\
&= 2\sum_{n=0}^{\infty} \frac{x^{2n}}{(2n)!}
\end{aligned}
$$

25. The derivatives reveal the following details.

$$
\begin{aligned}
f(x) &= \frac{1}{3+x} \Rightarrow f(0) = \frac{1}{3} \\
f'(x) &= \frac{-1}{(3+x)^2} \Rightarrow f'(0) = -\frac{1}{3^2} \\
f''(x) &= \frac{2}{(3+x)^3} \Rightarrow f''(0) = \frac{2!}{3^3} \\
f'''(x) &= \frac{-3\cdot 2}{(3+x)^4} \Rightarrow f'''(0) = \frac{-3!}{3^4}.
\end{aligned}
$$

We conjecture that

$$
f^{(n)}(x) = \frac{(-1)^n n!}{(3+x)^{n+1}} \Rightarrow f^{(n)}(0) = \frac{(-1)^n n!}{3^{n+1}}.
$$

Thus,

$$
\begin{aligned}
\frac{1}{3+x} &= \sum_{n=0}^{\infty} \frac{f^{(n)}(0)}{n!} x^n \\
&= \frac{1}{3} - \frac{1}{3^2}x + \frac{1}{3^3}x^2 - \frac{1}{3^4}x^3 + \cdots \\
&\quad + \frac{(-1)^n}{3^{n+1}} x^n + \cdots.
\end{aligned}
$$

Alternatively, we can expand in a geometric series.

$$
\begin{aligned}
\frac{1}{3+x} &= \frac{1}{3} \cdot \frac{1}{1-(-x/3)} \\
&= \frac{1}{3}\left[1 - \frac{x}{3} + \frac{x^2}{3^2} - \frac{x^3}{3^3} + \cdots \right. \\
&\quad \left. + \frac{(-1)^n x^n}{3^n} + \cdots\right] \\
&= \frac{1}{3} - \frac{x}{3^2} + \frac{x^2}{3^3} - \frac{x^3}{3^4} + \cdots \\
&\quad + \frac{(-1)^n x^n}{3^{n+1}} + \cdots
\end{aligned}
$$

The argument is valid provided $|x/3| < 1$, or, equivalently, $|x| < 3$. The radius of convergence is $R = 3$.

27. We expand in a geometric series.

$$\begin{aligned}\frac{1}{4+x^2} &= \frac{1}{4}\cdot\frac{1}{1+x^2/4}\\ &= \frac{1}{4}\left\{1-\frac{x^2}{4}+\frac{x^4}{4^2}-\frac{x^6}{4^3}+\cdots\right.\\ &\quad\left.+\frac{(-1)^n x^{2n}}{4^n}+\cdots\right\}\\ &= \frac{1}{4}-\frac{x^2}{4^2}+\frac{x^4}{4^3}-\frac{x^6}{4^4}+\cdots\\ &\quad+\frac{(-1)^n x^{2n}}{4^{n+1}}+\cdots,\end{aligned}$$

provided $|x| < 2$.

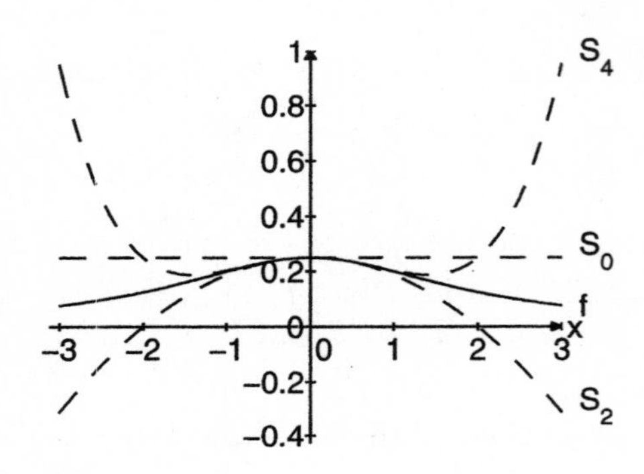

29. We know that

$$e^t = 1+t+\frac{t^2}{2!}+\frac{t^3}{3!}+\cdots+\frac{t^n}{n!}+\cdots.$$

Thus,

$$\begin{aligned}e^{-2x^2} &= 1-2x^2+\frac{2^2x^4}{2!}-\frac{2^3x^6}{3!}+\cdots\\ &\quad+\frac{(-1)^n 2^n x^{2n}}{n!}+\cdots,\end{aligned}$$

valid for all x.

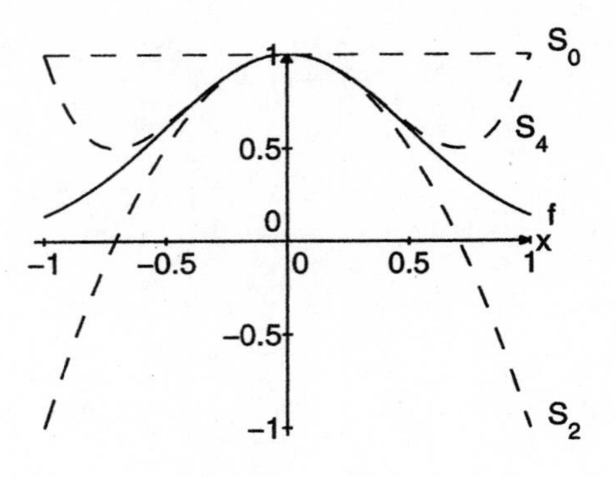

31. Derivatives reveal these details.

$$\begin{aligned}f(x) &= \frac{1}{x^2} \Rightarrow f(2) = \frac{1}{2^2}\\ f'(x) &= \frac{-2}{x^3} \Rightarrow f'(2) = \frac{-2}{2^3}\\ f''(x) &= \frac{3\cdot 2}{x^4} \Rightarrow f''(2) = \frac{3!}{2^4}\\ f'''(x) &= \frac{-4\cdot 3\cdot 2}{x^5} \Rightarrow f'''(2) = \frac{-4!}{2^5}\end{aligned}$$

We conjecture that

$$f^{(n)}(x) = \frac{(-1)^n(n+1)!}{x^{n+2}},$$

so

$$f^{(n)}(2) = \frac{(-1)^n(n+1)!}{2^{n+2}}.$$

Thus,

$$\begin{aligned}\frac{1}{x^2} &= \sum_{n=0}^{\infty}\frac{f^{(n)}(2)}{n!}(x-2)^n\\ &= \frac{1}{2^2}-\frac{2}{2^3}(x-2)+\frac{3}{2^4}(x-2)^2\\ &\quad-\cdots+\frac{(-1)^n(n+1)}{2^{n+2}}(x-2)^n+\cdots.\end{aligned}$$

Note that

$$\frac{|a_n|}{|a_{n+1}|} = \frac{2(n+1)}{n+2} \to 2$$

as $n \to \infty$. Thus, the radius of convergence is $R = 2$.

33. Start with the geometric series,

$$\frac{1}{1-x} = \sum_{n=0}^{\infty} x^n,$$

valid for $|x| < 1$. Differentiate.

$$\frac{1}{(1-x)^2} = \sum_{n=1}^{\infty} nx^{n-1}$$

Shift the index of the sum on the right.

$$\frac{1}{(1-x)^2} = \sum_{n=0}^{\infty}(n+1)x^n$$

$$\frac{1}{(1-x)^2} = \sum_{n=0}^{\infty} nx^n + \sum_{n=0}^{\infty} x^n$$

$$\frac{1}{(1-x)^2} = \sum_{n=0}^{\infty} nx^n + \frac{1}{1-x}.$$

Therefore,

$$\sum_{n=0}^{\infty} nx^n = \frac{1}{(1-x)^2} - \frac{1}{1-x}$$

$$\sum_{n=0}^{\infty} nx^n = \frac{1-(1-x)}{(1-x)^2}$$

$$\sum_{n=0}^{\infty} nx^n = \frac{x}{(1-x)^2},$$

provided $|x| < 1$.

35. If $f(x) = \tan x$, then

$$f'(x) = \sec^2 x,$$
$$f''(x) = 2\sec^2 x \tan x, \quad \text{and}$$
$$f'''(x) = 2\sec^2 x(1 + 3\tan^2 x).$$

Thus,

$$f(x) \approx f(0) + f'(0)x + \frac{f''(0)}{2!}x^2 + \frac{f'''(0)}{3!}x^3$$
$$\approx x + \frac{x^3}{3}.$$

37. We're given $f(x) = \sum_{n=0}^{\infty} a_n x^n$ and we wish to find

$$\frac{1}{f(x)} = \sum_{n=0}^{\infty} b_n x^n,$$

or equivalently

$$1 = f(x)\sum_{n=0}^{\infty} b_n x^n = \sum_{n=0}^{\infty} a_n x^n \sum_{n=0}^{\infty} b_n x^n$$
$$= \sum_{n=0}^{\infty} c_n x^n,$$

where

$$c_n = \sum_{k=0}^{n} a_{n-k} b_k.$$

Thus,

$$1 = a_0 b_0$$
$$b_0 = \frac{1}{a_0},$$

provided $a \neq 0$. The remaining coefficients on the right must equal zero. Thus,

$$0 = a_1 b_0 + a_0 b_1$$
$$b_1 = -\frac{a_1 b_0}{a_0}.$$

And, in general,

$$0 = a_n b_0 + a_{n-1} b_1 + \cdots + a_1 b_{n-1} + a_0 b_n$$
$$b_n = -\frac{a_1 b_{n-1} + \cdots + a_n b_0}{a_0},$$

for $n \geq 1$.

Section 11.2. Series Solutions near Ordinary Points

1. The equation $y' = 3y$ is separable with solution $y(x) = Ce^{3x}$. With $y(0) = a_0$, this becomes $y(x) = a_0e^{3x}$. We look for a solution of the form

$$y(x) = \sum_{n=0}^{\infty} a_n x^n.$$

Differentiating, then shifting the index,

$$y'(x) = \sum_{n=1}^{\infty} na_n x^{n-1} = \sum_{n=0}^{\infty}(n+1)a_{n+1}x^n.$$

Substituting,

$$\begin{aligned} 0 &= y' - 3y \\ &= \sum_{n=0}^{\infty}(n+1)a_{n+1}x^n - 3\sum_{n=0}^{\infty} a_n x^n \\ &= \sum_{n=0}^{\infty}\left[(n+1)a_{n+1} - 3a_n\right]x^n. \end{aligned}$$

Setting the coefficients equal to zero, we have the recurrence formula,

$$\begin{aligned} (n+1)a_{n+1} - 3a_n &= 0 \\ a_{n+1} &= \frac{3a_n}{n+1}, \quad n \geq 0. \end{aligned}$$

Thus, with $y(0) = a_0$,

$$\begin{aligned} a_1 &= 3a_0, \\ a_2 &= \frac{3}{2}a_1 = \frac{3 \cdot 3}{2}a_0, \\ a_3 &= \frac{3}{3}a_2 = \frac{3 \cdot 3 \cdot 3}{2 \cdot 3}a_0, \\ a_4 &= \frac{3}{4}a_3 = \frac{3 \cdot 3 \cdot 3 \cdot 3}{2 \cdot 3 \cdot 4}a_0. \end{aligned}$$

We conjecture that the general coefficient is

$$a_n = \frac{3^n}{n!}a_0.$$

Thus,

$$\begin{aligned} y(x) &= \sum_{n=0}^{\infty} a_n x^n = a_0 \sum_{n=0}^{\infty} \frac{3^n}{n!}x^n \\ &= a_0 \sum_{n=0}^{\infty} \frac{(3x)^n}{n!} = a_0 e^{3x}. \end{aligned}$$

This agrees nicely with the separable solution above.

3. The equation $y' = -x^3 y$ is separable.

$$\begin{aligned} \frac{dy}{y} &= -x^3\,dx \\ \ln y &= -\frac{1}{4}x^4 + C \\ y &= Ce^{-x^4/4}. \end{aligned}$$

With $y(0) = a_0$, this gives $y = a_0e^{-x^4/4}$. We look for a solution having form

$$y(x) = \sum_{n=0}^{\infty} a_n x^n.$$

Differentiating,

$$y'(x) = \sum_{n=1}^{\infty} na_n x^{n-1}.$$

Substituting,

$$\begin{aligned} 0 &= y' + x^3 y \\ &= \sum_{n=1}^{\infty} na_n x^{n-1} + x^3 \sum_{n=0}^{\infty} a_n x^n \\ &= \sum_{n=1}^{\infty} na_n x^{n-1} + \sum_{n=0}^{\infty} a_n x^{n+3}. \end{aligned}$$

To add these two series, we shift the index in the first one so that powers of x are the same.

$$\begin{aligned} \sum_{n=1}^{\infty} na_n x^{n-1} &= \sum_{n'=-3}^{\infty}(n'+4)a_{n'+4}x^{n'+3} \\ &= \sum_{n=-3}^{\infty}(n+4)a_{n+4}x^{n+3} \end{aligned}$$

Thus,

$$0 = \sum_{n=-3}^{\infty}(n+4)a_{n+4}x^{n+3} + \sum_{n=0}^{\infty} a_n x^{n+3}$$
$$= a_1 + 2a_2x + 3a_3x^2 + \sum_{n=0}^{\infty}(n+4)a_{n+4}x^{n+3}$$
$$+ \sum_{n=0}^{\infty} a_n x^{n+3}$$
$$= a_1 + 2a_2x + 3a_3x^2$$
$$+ \sum_{n=0}^{\infty}\left[(n+4)a_{n+4} + a_n\right]x^{n+3}.$$

Setting coefficients equal to zero, $a_1 = a_2 = a_3 = 0$ and the recurrence formula

$$(n+4)a_{n+4} + a_n = 0$$
$$a_{n+4} = \frac{-a_n}{n+4}, \quad n \geq 0.$$

Thus, with $y(0) = a_0$,

$$a_4 = -\frac{1}{4}a_0$$
$$a_5 = -\frac{1}{5}a_1 = 0$$
$$a_6 = -\frac{1}{6}a_2 = 0$$
$$a_7 = -\frac{1}{7}a_3 = 0$$
$$a_8 = -\frac{1}{8}a_4 = \frac{1}{4\cdot 8}a_0.$$

You can see that every fourth term is meaningful, with terms in between equaling zero. Thus,

$$a_{12} = -\frac{1}{12}a_8 = -\frac{1}{4\cdot 8\cdot 12}a_0$$
$$a_{16} = -\frac{1}{16}a_{12} = \frac{1}{4\cdot 8\cdot 12\cdot 16}a_0.$$

Note that

$$a_{16} = a_{4(4)} = \frac{1}{4^4\cdot 1\cdot 2\cdot 3\cdot 4}a_0.$$

We conjecture that

$$a_{4n} = \frac{(-1)^n}{4^n n!}a_0.$$

Because intervening terms are zero,

$$y(x) = \sum_{n=0}^{\infty} a_n x^n$$
$$= \sum_{n=0}^{\infty} a_{4n} x^{4n}$$
$$= \sum_{n=0}^{\infty} \frac{(-1)^n}{4^n n!} a_0 x^{4n}$$
$$= a_0 \sum_{n=0}^{\infty} \frac{(-x^4/4)^n}{n!}$$
$$= a_0 e^{-x^4/4}.$$

This agrees nicely with the separable solution found above.

5. The equation $(1-x)y' + y = 0$ is separable with solution

$$(1-x)\frac{dy}{dx} = -y$$
$$\frac{dy}{y} = \frac{dx}{x-1}$$
$$\ln y = \ln(x-1) + C$$
$$y = C(x-1).$$

With $y(0) = a_0$, $C = -a_0$ and $y = a_0(1-x)$. We seek a solution having form

$$y(x) = \sum_{n=0}^{\infty} a_n x^n$$

and derivative

$$y'(x) = \sum_{n=1}^{\infty} n a_n x^{n-1}.$$

Substituting,

$$0 = (1-x)y' + y$$
$$= (1-x)\sum_{n=1}^{\infty} n a_n x^{n-1} + \sum_{n=0}^{\infty} a_n x^n$$
$$= \sum_{n=1}^{\infty} n a_n x^{n-1} - \sum_{n=1}^{\infty} n a_n x^n + \sum_{n=0}^{\infty} a_n x^n.$$

Shifting the index on the first term and noting that the second term

$$\sum_{n=1}^{\infty} na_n x^n = \sum_{n=0}^{\infty} na_n x^n,$$

we write

$$\begin{aligned}0 &= \sum_{n=0}^{\infty}(n+1)a_{n+1}x^n - \sum_{n=0}^{\infty} na_n x^n + \sum_{n=0}^{\infty} a_n x^n \\ &= \sum_{n=0}^{\infty}\left[(n+1)a_{n+1} + (1-n)a_n\right]x^n.\end{aligned}$$

Setting the coefficients equal to zero,

$$\begin{aligned}(n+1)a_{n+1} + (1-n)a_n &= 0 \\ a_{n+1} &= \frac{n-1}{n+1}a_n, \quad n \ge 0.\end{aligned}$$

Thus, a_0 is free and

$$\begin{aligned}a_1 &= (-1)a_0 \\ a_2 &= 0a_1 = 0 \\ a_3 &= \frac{1}{3}a_2 = 0.\end{aligned}$$

Indeed, $a_n = 0$, $n \ge 2$. Thus,

$$y(x) = \sum_{n=0}^{\infty} a_n x^n = a_0 + a_1 x = a_0(1-x),$$

comparing nicely with the separable solution above.

7. The equation is separable.

$$\begin{aligned}(x-4)\frac{dy}{dx} &= -y \\ \frac{dy}{y} &= \frac{-dx}{x-4} \\ \ln y &= -\ln(x-4) + C \\ \ln y &= \ln\frac{D}{x-4} \\ y &= \frac{D}{x-4}\end{aligned}$$

With $y(0) = a_0$,

$$\begin{aligned}a_0 &= \frac{D}{0-4} \\ D &= -4a_0,\end{aligned}$$

and

$$y(x) = \frac{-4a_0}{x-4}.$$

We look for a solution having form

$$y(x) = \sum_{n=0}^{\infty} a_n x^n.$$

Differentiating,

$$y'(x) = \sum_{n=1}^{\infty} na_n x^{n-1}.$$

Substituting,

$$\begin{aligned}0 &= (x-4)y' + y \\ &= (x-4)\sum_{n=1}^{\infty} na_n x^{n-1} + \sum_{n=0}^{\infty} a_n x^n \\ &= \sum_{n=1}^{\infty} na_n x^n - \sum_{n=1}^{\infty} 4na_n x^{n-1} + \sum_{n=0}^{\infty} a_n x^n.\end{aligned}$$

Noting that the first term is

$$\sum_{n=1}^{\infty} na_n x^n = \sum_{n=0}^{\infty} na_n x^n,$$

and shifting the index of the second term, we write

$$\begin{aligned}0 &= \sum_{n=0}^{\infty} na_n x^n - \sum_{n=0}^{\infty} 4(n+1)a_{n+1}x^n + \sum_{n=0}^{\infty} a_n x^n \\ &= \sum_{n=0}^{\infty}\left[na_n - 4(n+1)a_{n+1} + a_n\right]x^n.\end{aligned}$$

Setting the coefficients equal to zero,

$$\begin{aligned}na_n - 4(n+1)a_{n+1} + a_n &= 0 \\ a_{n+1} &= \frac{1}{4}a_n, \quad n \ge 0.\end{aligned}$$

Thus,

$$a_1 = \frac{1}{4} a_0$$
$$a_2 = \frac{1}{4} a_1 = \frac{1}{4^2} a_0$$
$$a_3 = \frac{1}{4} a_2 = \frac{1}{4^3} a_0.$$

We conjecture that

$$a_n = \frac{1}{4^n} a_0.$$

Therefore,

$$y(x) = \sum_{n=0}^{\infty} a_n x^n = a_0 \sum_{n=0}^{\infty} \frac{x^n}{4^n}.$$

However, this can be written

$$\begin{aligned} y(x) &= a_0 \sum_{n=0}^{\infty} \left(\frac{x}{4}\right)^n \\ &= a_0 \left(\frac{1}{1 - x/4}\right) \\ &= \frac{-4a_0}{x-4}, \end{aligned}$$

which agrees nicely with our separable solution.

9. The equation is separable.

$$\begin{aligned} (2-x)\frac{dy}{dx} &= -2y \\ \frac{dy}{y} &= \frac{-2dx}{2-x} \\ \ln y &= 2\ln(2-x) + C \\ \ln y &= \ln D(2-x)^2 \\ y &= D(2-x)^2 \end{aligned}$$

With $y(0) = a_0$,

$$a_0 = D(2-0)^2 = 4D.$$

Thus, $D = a_0/4$ and

$$y(x) = \frac{1}{4} a_0 (2-x)^2.$$

We look for a solution having form

$$y(x) = \sum_{n=0}^{\infty} a_n x^n.$$

Differentiating,

$$y'(x) = \sum_{n=1}^{\infty} n a_n x^{n-1}.$$

Substituting,

$$\begin{aligned} 0 &= (2-x)y' + 2y \\ &= (2-x)\sum_{n=1}^{\infty} n a_n x^{n-1} + 2\sum_{n=1}^{\infty} a_n x^n \\ &= \sum_{n=1}^{\infty} 2n a_n x^{n-1} - \sum_{n=1}^{\infty} n a_n x^n + \sum_{n=0}^{\infty} 2a_n x^n. \end{aligned}$$

Noting that the second term

$$\sum_{n=1}^{\infty} n a_n x^n = \sum_{n=0}^{\infty} n a_n x^n,$$

and shifting the index of the first term,

$$\begin{aligned} 0 &= \sum_{n=0}^{\infty} 2(n+1)a_{n+1}x^n - \sum_{n=0}^{\infty} n a_n x^n + \sum_{n=0}^{\infty} 2a_n x^n \\ &= \sum_{n=0}^{\infty} \left[2(n+1)a_{n+1} - n a_n + 2a_n\right] x^n. \end{aligned}$$

Setting coefficients equal to zero,

$$0 = 2(n+1)a_{n+1} + (2-n)a_n$$
$$a_{n+1} = \frac{n-2}{2(n+1)} a_n, \quad n \geq 0.$$

Thus,

$$a_1 = -a_0$$
$$a_2 = \frac{-1}{4} a_1 = \frac{1}{4} a_0,$$
$$a_3 = 0,$$

and $a_n = 0$ for all $n \geq 3$. Thus,

$$\begin{aligned} y(x) &= \sum_{n=0}^{\infty} a_n x^n \\ &= a_0 \left[1 - x + \frac{1}{4} x^2\right]. \end{aligned}$$

However, this can be written

$$y(x) = \frac{1}{4}a_0\left[4 - 4x + x^2\right] = \frac{1}{4}a_0(2-x)^2,$$

which agrees nicely with the separable solution above.

11. The equation $y'' = y'$ has characteristic equation $\lambda^2 - \lambda = 0$ and roots $\lambda = 0, 1$. Thus, we have independent solutions $y_1(x) = 1$ and $y_2(x) = e^x$ and general solution $y(x) = a_0 + a_1e^x$ where $y(0) = a_0$ and $y'(0) = a_1$. We seek a solution having form

$$y(x) = \sum_{n=0}^{\infty} a_n x^n$$

and derivatives

$$y'(x) = \sum_{n=1}^{\infty} na_n x^{n-1}$$

$$y''(x) = \sum_{n=2}^{\infty} n(n-1)a_n x^{n-2} = \sum_{n=1}^{\infty}(n+1)na_{n+1}x^{n-1}.$$

Thus,

$$0 = y'' - y' = \sum_{n=1}^{\infty}(n+1)na_{n+1}x^{n-1} - \sum_{n=1}^{\infty} na_n x^{n-1} = \sum_{n=1}^{\infty}\left[n(n+1)a_{n+1} - na_n\right]x^{n-1}.$$

Setting coefficients equal to zero,

$$n\left[(n+1)a_{n+1} - a_n\right] = 0, \quad n \geq 1.$$

Thus,

$$a_{n+1} = \frac{a_n}{n+1}, \quad n \geq 1.$$

Thus, a_0 and a_1 are arbitrary, and

$$a_2 = \frac{1}{2}a_1$$

$$a_3 = \frac{1}{3}a_2 = \frac{1}{2\cdot 3}a_1$$

$$a_4 = \frac{1}{4}a_3 = \frac{1}{2\cdot 3\cdot 4}a_1.$$

Note that for $n \geq 2$,

$$a_n = \frac{1}{n!}a_1.$$

Thus, the general solution can be written

$$y(x) = a_0 + \sum_{n=1}^{\infty}\frac{a_1}{n!}x^n = a_0 + a_1\sum_{n=1}^{\infty}\frac{x^n}{n!} = a_0 + a_1e^x,$$

which agrees nicely with the sparable solution above.

---

13. The characteristic equation of $y'' + y = 0$ is $\lambda^2 + 1 = 0$ with zeros $\lambda = \pm i$. This yields independent solutions $y_1(x) = \cos x$ and $y_2(x) = \sin x$ and general solution $y(x) = C_1\cos x + C_2\sin x$. With $y(0) = a_0$, $a_0 = y(0) = C_1$. Differentiate.

$$y'(x) = -C_1\sin x + C_2\cos x$$

With $y'(0) = a_1$, $a_1 = y'(0) = C_2$. Thus,

$$y(x) = a_0\cos x + a_1\sin x.$$

We look for solutions having form $y(x) = \sum_{n=0}^{\infty} a_n x^n$, with second derivative

$$y''(x) = \sum_{n=2}^{\infty} n(n-1)a_n x^{n-2}.$$

Substituting,

$$0 = y'' + y = \sum_{n=2}^{\infty} n(n-1)a_n x^{n-2} + \sum_{n=0}^{\infty} a_n x^n.$$

Shift the index of the first term.

$$0 = \sum_{n=0}^{\infty}(n+2)(n+1)a_{n+2}x^n + \sum_{n=0}^{\infty} a_n x^n = \sum_{n=0}^{\infty}[(n+2)(n+1)a_{n+2} + a_n]x^n$$

Set the coefficients equal to zero.

$$(n+2)(n+1)a_{n+2} + a_n = 0 \quad \text{and} \quad a_{n+2} = \frac{-a_n}{(n+2)(n+1)}, \quad n \geq 0$$

Thus,

$$a_2 = \frac{-a_0}{2 \cdot 1}, \quad a_4 = \frac{-a_2}{4 \cdot 3} = \frac{a_0}{4 \cdot 3 \cdot 2 \cdot 1}, \quad \text{and} \quad a_6 = \frac{-a_4}{6 \cdot 5} = \frac{-a_0}{6 \cdot 5 \cdot 4 \cdot 3 \cdot 2 \cdot 1}$$

We conjecture that

$$a_{2n} = \frac{(-1)^n a_0}{(2n)!}.$$

Similarly,

$$a_3 = \frac{-a_1}{3 \cdot 2}, \quad a_5 = \frac{-a_3}{5 \cdot 4} = \frac{a_1}{5 \cdot 4 \cdot 3 \cdot 2}, \quad \text{and} \quad a_7 = \frac{-a_5}{7 \cdot 6} = \frac{-a_1}{7 \cdot 6 \cdot 5 \cdot 4 \cdot 3 \cdot 2}.$$

We conjecture that

$$a_{2n+1} = \frac{(-1)^n a_1}{(2n+1)!}.$$

Thus,

$$\begin{aligned} y(x) &= a_0\left[1 - \frac{1}{2!}x^2 + \frac{1}{4!}x^4 - \cdots + \frac{(-1)^n}{(2n)!}x^{2n} + \cdots\right] \\ &\quad + a_1\left[x - \frac{1}{3!}x^3 + \frac{1}{5!}x^5 - \cdots + \frac{(-1)^n}{(2n+1)!}x^{2n+1} + \cdots\right] \\ &= a_0 \cos x + a_1 \sin x, \end{aligned}$$

which agrees nicely with the result above.

15. In the equation $y'' + x^2 y = 0$, the coefficient of y is a polynomial, clearly analytic at $x = 0$. Thus, $x = 0$ is an ordinary point. According to Theorem 2.29, the solutions will have an infinite radius of convergence. We look for a solution having the form $y(x) = \sum_{n=0}^{\infty} a_n x^n$ with derivatives

$$y'(x) = \sum_{n=1}^{\infty} n a_n x^{n-1} \quad \text{and} \quad y''(x) = \sum_{n=2}^{\infty} n(n-1)a_n x^{n-2}.$$

Substituting,

$$0 = y'' + x^2 y = \sum_{n=2}^{\infty} n(n-1)a_n x^{n-2} + x^2 \sum_{n=0}^{\infty} a_n x^n = \sum_{n=2}^{\infty} n(n-1)a_n x^{n-2} + \sum_{n=0}^{\infty} a_n x^{n+2}.$$

Shifting the index of the first term,

$$0 = \sum_{n=-2}^{\infty} (n+4)(n+3)a_{n+4}x^{n+2} + \sum_{n=0}^{\infty} a_n x^{n+2} = 2a_2 + 6a_3 x + \sum_{n=0}^{\infty} \big[(n+4)(n+3)a_{n+4} + a_n\big] x^{n+2}.$$

Setting the coefficients equal to zero, $a_2 = a_3 = 0$ and

$$(n+4)(n+3)a_{n+4} + a_n = 0 \quad \text{or} \quad a_{n+4} = \frac{-a_n}{(n+4)(n+3)}.$$

Thus,

$$a_4 = \frac{-a_0}{4 \cdot 3}, \quad a_8 = \frac{-a_4}{8 \cdot 7} = \frac{a_0}{4 \cdot 8 \cdot 3 \cdot 7}, \quad \text{and} \quad a_{12} = \frac{-a_8}{12 \cdot 11} = \frac{-a_0}{4 \cdot 8 \cdot 12 \cdot 3 \cdot 7 \cdot 11}.$$

Note that

$$a_{16} = \frac{-a_{12}}{16 \cdot 15} \cdot = \frac{a_0}{4 \cdot 8 \cdot 12 \cdot 16 \cdot 3 \cdot 7 \cdot 11 \cdot 15} = \frac{a_0}{4^4(1 \cdot 2 \cdot 3 \cdot 4) \cdot [3 \cdot 7 \cdot 11 \cdot 15]}.$$

We conjecture that

$$a_{4n} = \frac{(-1)^n a_0}{4^n \cdot n![3 \cdot 7 \cdot \dots \cdot (4n-1)]}, \qquad n \ge 1.$$

Similarly,

$$a_5 = \frac{-a_1}{5 \cdot 4}, \quad a_9 = \frac{-a_5}{9 \cdot 8} = \frac{a_1}{5 \cdot 9 \cdot 4 \cdot 8}, \quad \text{and} \quad a_{13} = \frac{-a_9}{13 \cdot 12} = \frac{-a_1}{5 \cdot 9 \cdot 13 \cdot 4 \cdot 8 \cdot 12}.$$

Note that

$$a_{17} = \frac{-a_{13}}{17 \cdot 16} = \frac{a_1}{5 \cdot 9 \cdot 13 \cdot 17 \cdot 4 \cdot 8 \cdot 12 \cdot 16} = \frac{a_1}{4^4(1 \cdot 2 \cdot 3 \cdot 4)[5 \cdot 9 \cdot 13 \cdot 17]}.$$

We conjecture

$$a_{4n+1} = \frac{(-1)^n a_1}{4^n n![5 \cdot 9 \cdot 13 \cdot \dots \cdot (4n+1)]}.$$

Finally, since $a_2 = a_3 = 0$, it is easy to see that $a_6 = a_{10} = a_{14} = \cdots = 0$, and $a_7 = a_{11} = \cdots = 0$. Thus,

$$y(x) = a_0 \left[1 - \frac{1}{4 \cdot 3}x^4 + \frac{1}{4 \cdot 8 \cdot 3 \cdot 7}x^8 - \cdots\right] + a_1 \left[x - \frac{1}{5 \cdot 4}x^5 + \frac{1}{5 \cdot 9 \cdot 4 \cdot 8}x^9 - \cdots\right].$$

Choose $y_1(x)$ with $1 = y_1(0) = a_0$ and $0 = y_1'(0) = a_1$.

$$y_1(x) = 1 - \frac{1}{4 \cdot 3}x^4 + \frac{1}{4 \cdot 8 \cdot 3 \cdot 7}x^8 - \cdots = \sum_{n=0}^{\infty} \frac{(-1)^n x^{4n}}{4^n n![3 \cdot 7 \cdots (4n-1)]}.$$

Choose $y_2(x)$ with $0 = y_2(0) = a_0$ and $1 = y_2'(0) = a_1$.

$$y_2(x) = x - \frac{1}{5 \cdot 4}x^5 + \frac{1}{5 \cdot 9 \cdot 4 \cdot 8}x^9 - \cdots = \sum_{n=0}^{\infty} \frac{(-1)^n x^{4n+1}}{4^n n![5 \cdot 9 \cdot \dots \cdot (4n+1)]}.$$

Note that the solutions are chosen so that

$$W(y_1, y_2)(0) = \begin{vmatrix} y_1(0) & y_2(0) \\ y_1'(0) & y_2'(0) \end{vmatrix} = \begin{vmatrix} 1 & 0 \\ 0 & 1 \end{vmatrix} = 1 \neq 0,$$

so the solutions are independent.

17. The coefficients of y' and y are $p(x) = 2x$ and $q(x) = -1$, both polynomials are analytic at $x = 0$. Thus, $x = 0$ is an ordinary point. According to Theorem 2.29, the solutions will have an infinite radius of convergence. We seek a solution of the form $y(x) = \sum_{n=0}^{\infty} a_n x^n$ with derivatives

$$y'(x) = \sum_{n=1}^{\infty} n a_n x^{n-1} \quad \text{and} \quad y''(x) = \sum_{n=2}^{\infty} n(n-1) a_n x^{n-2}.$$

Substituting,

$$\begin{aligned} 0 = y'' + 2xy' - y &= \sum_{n=2}^{\infty} n(n-1) a_n x^{n-2} + 2x \sum_{n=1}^{\infty} n a_n x^{n-1} - \sum_{n=0}^{\infty} a_n x^n \\ &= \sum_{n=2}^{\infty} n(n-1) a_n x^{n-2} + \sum_{n=1}^{\infty} 2n a_n x^n - \sum_{n=0}^{\infty} a_n x^n. \end{aligned}$$

Shifting the index of the first term and noting that the second term

$$\sum_{n=1}^{\infty} 2n a_n x^n = \sum_{n=0}^{\infty} 2n a_n x^n,$$

we write

$$0 = \sum_{n=0}^{\infty} (n+2)(n+1) a_{n+2} x^n + \sum_{n=0}^{\infty} 2n a_n x^n - \sum_{n=0}^{\infty} a_n x^n = \sum_{n=0}^{\infty} \left[(n+2)(n+1) a_{n+2} + (2n-1) a_n\right] x^n.$$

Setting coefficients equal to zero,

$$(n+2)(n+1) a_{n+2} + (2n-1) a_n = 0 \quad \text{or} \quad a_{n+2} = \frac{1-2n}{(n+2)(n+1)} a_n, \quad n \geq 0.$$

Thus,

$$a_2 = \frac{1}{2 \cdot 1} a_0, \quad a_4 = \frac{-3}{4 \cdot 3} a_2 = \frac{-3}{4 \cdot 3 \cdot 2 \cdot 1} a_0, \quad \text{and} \quad a_6 = \frac{-7}{6 \cdot 5} a_4 = \frac{7 \cdot 3}{6!} a_0.$$

Note that

$$a_8 = \frac{-11}{8 \cdot 7} a_6 = \frac{-11 \cdot 7 \cdot 3}{8!} a_0.$$

We conjecture that

$$a_{2n} = \frac{(-1)^{n+1}(3 \cdot 7 \cdot 11 \cdot \cdots \cdot (4n-5))}{(2n)!} a_0, \quad n > 1.$$

Similarly,

$$a_3 = \frac{-1}{3 \cdot 2} a_1, \quad a_5 = \frac{-5}{5 \cdot 4} a_3 = \frac{5}{5 \cdot 4 \cdot 3 \cdot 2} a_1, \quad \text{and} \quad a_7 = \frac{-9}{7 \cdot 6} a_5 = \frac{-9 \cdot 5}{7 \cdot 6 \cdot 5 \cdot 4 \cdot 3 \cdot 2} a_1.$$

Note that

$$a_9 = \frac{-13}{9 \cdot 8} a_7 = \frac{13 \cdot 9 \cdot 5}{9!} a_1.$$

We conjecture that

$$a_{2n+1} = \frac{(-1)^n (5 \cdot 9 \cdot 13 \cdot \cdots \cdot (4n-3))}{(2n+1)!} a_1, \quad n > 1.$$

Thus,

$$y(x) = a_0 \left[1 + \frac{1}{2 \cdot 1} x^2 - \frac{3}{4 \cdot 3 \cdot 2 \cdot 1} x^3 + \cdots \right] + a_1 \left[x - \frac{1}{3 \cdot 2} x^3 + \frac{5}{5 \cdot 4 \cdot 3 \cdot 2} x^5 - \cdots \right].$$

Choose $y_1(x)$ with $1 = y_1(0) = a_0$ and $0 = y_1'(0) = a_1$.

$$y_1(x) = 1 + \frac{1}{2 \cdot 1} x^2 - \frac{3}{4 \cdot 3 \cdot 2 \cdot 1} x^4 + \cdots = 1 + \frac{1}{2} x^2 + \sum_{n=2}^{\infty} \frac{(-1)^{n+1} (3 \cdot 7 \cdot 11 \cdot \cdots \cdot (4n-5))}{(2n)!} x^{2n}$$

Choose $y_2(x)$ with $0 = y_2(0) = a_0$ and $1 = y_2'(0) = a_1$.

$$y_2(x) = x - \frac{1}{3 \cdot 2} x^3 + \frac{5}{5 \cdot 4 \cdot 3 \cdot 2} x^5 - \cdots = x - \frac{1}{6} x^3 + \sum_{n=2}^{\infty} \frac{(-1)^n (5 \cdot 9 \cdot 12 \cdot \cdots \cdot (4n-3))}{(2n+1)!} x^{2n+1}.$$

Note that the solutions are chosen so that

$$W(y_1, y_2)(0) = \begin{vmatrix} y_1(0) & y_2(0) \\ y_1'(0) & y_2'(0) \end{vmatrix} = \begin{vmatrix} 1 & 0 \\ 0 & 1 \end{vmatrix} = 1 \neq 0,$$

so the solutions are independent.

19. In the equation $y'' + xy' - y$, the coefficient of y' is $p(x) = x$. The coefficient of y is $q(x) = -1$. Both p and q are analytic at $x = 0$, so $x = 0$ is an ordinary point. According to Theorem 2.29, the solutions will have an infinite radius of convergence. We look for a solution of the form $y(x) = \sum_{n=0}^{\infty} a_n x^n$, with derivatives

$$y'(x) = \sum_{n=1}^{\infty} n a_n x^{n-1} \quad \text{and} \quad y''(x) = \sum_{n=2}^{\infty} n(n-1) x_n x^{n-2}.$$

Substituting,

$$\begin{aligned} 0 = y'' + xy' - y &= \sum_{n=2}^{\infty} n(n-1) a_n x^{n-2} + x \sum_{n=1}^{\infty} n a_n x^{n-1} - \sum_{n=0}^{\infty} a_n x^n \\ &= \sum_{n=2}^{\infty} n(n-1) a_n x^{n-2} + \sum_{n=1}^{\infty} n a_n x^n - \sum_{n=0}^{\infty} a_n x^n. \end{aligned}$$

Note that the second term is

$$\sum_{n=1}^{\infty} n a_n x^n = \sum_{n=0}^{\infty} n a_n x^n$$

and shift the index of the first term.

$$0 = \sum_{n=0}^{\infty}(n+2)(n+1)a_{n+2}x^n + \sum_{n=0}^{\infty} na_nx^n - \sum_{n=0}^{\infty} a_nx^n = \sum_{n=0}^{\infty}\left[(n+2)(n+1)a_{n+2} + (n-1)a_n\right]x^n.$$

Setting the coefficients equal to zero,

$$(n+2)(n+1)a_{n+2} + (n-1)a_n = 0 \quad \text{or} \quad a_{n+2} = \frac{1-n}{(n+2)(n+1)}a_n.$$

Thus,

$$a_2 = \frac{1}{2\cdot 1}a_0, \quad a_4 = \frac{-1}{4\cdot 3}a_2 = \frac{-1}{4\cdot 3\cdot 2\cdot 1}a_0, \quad \text{and} \quad a_6 = \frac{-3}{6\cdot 5}a_4 = \frac{1\cdot 3}{6!}a_0.$$

Note that

$$a_8 = \frac{-5}{8\cdot 7}a_6 = \frac{-1\cdot 3\cdot 5}{8!}a_0.$$

We conjecture that

$$a_{2n} = \frac{(-1)^{n+1}(1\cdot 3\cdot 5\cdot\dots\cdot(2n-3))}{(2n)!}a_0, \quad n \geq 2.$$

Similarly,

$$a_3 = 0a_1 = 0$$

and $a_{2n+1} = 0$ for $n \geq 1$. Thus,

$$y(x) = a_0\left[1 + \frac{1}{1\cdot 2}x^2 - \frac{1}{4\cdot 3\cdot 2\cdot 1}x^4 + \cdots\right] + a_1x.$$

Choose $y_1(x)$ with $a_0 = y_1(0) = 1$ and $a_1 = y_1'(0) = 0$. Then

$$y_1(x) = 1 + \frac{1}{2\cdot 1}x^2 - \frac{1}{4\cdot 3\cdot 2\cdot 1}x^4 + \cdots = 1 + \frac{1}{2}x^2 + \sum_{n=2}^{\infty}\frac{(-1)^{n+1}(1\cdot 3\cdot 5\cdot\dots\cdot(2n-3))}{(2n)!}x^{2n}.$$

Choose $y_2(x)$ with $a_0 = y_2(0) = 0$ and $a_1 = y_2'(0) = 1$. Then

$$y_2(x) = x.$$

Note that the solutions were chosen so that

$$W(y_1, y_2)(0) = \begin{vmatrix} y_1(0) & y_2(0) \\ y_1'(0) & y_2'(0) \end{vmatrix} = \begin{vmatrix} 1 & 0 \\ 0 & 1 \end{vmatrix} = 1 \neq 0,$$

so the solutions are independent.

21. When we put the equation in the form $y'' + y/(1+x) = 0$, we see that the coefficient $1/(1+x)$ is analytic at $x_0 = 0$ and has radius of convergence 1. Then $x_0 = 0$ is an ordinary point, and according to Theorem 2.29, all solutions have radius of convergence $R \geq 1$. We compute that a solution $y(x) = \sum_{n=0}^{\infty} a_nx^n$ must satisfy the recurrence formula

$$a_{n+2} = -\frac{n(n+1)a_{n+1} + a_n}{(n+1)(n+2)}, \quad \text{for } n \geq 0.$$

For the first solution we choose $a_0 = y_1(0) = 1$ and $a_1 = y_1'(0) = 0$. Then $a_2 = -1/2$, $a_3 = 1/6$, and $a_4 = 1/72$, so

$$y_1(x) = 1 - \frac{x^2}{2} + \frac{x^3}{6} + \frac{1}{72}x^4 + \cdots.$$

For the second solution we choose $a_0 = y_1(0) = 0$ and $a_1 = y_1'(0) = 1$. Then $a_2 = 0$, $a_3 = -1/6$, and $a_4 = 1/24$, so

$$y_2(x) = x - \frac{x^3}{6} + \frac{1}{24}x^4 + \cdots.$$

These solutions are linearly independent by Proposition 2.14.

23. The power series for the coefficient $\cos x = \sum_{n=0}^{\infty}(-1)^n x^{2n}/(2n)!$ converges for all x. Then $x_0 = 0$ is an ordinary point, and according to Theorem 2.29, all solutions have an infinite radius of convergence. If $y(x) = \sum_{n=0}^{\infty} a_n x^n$ is a solution, then

$$\begin{aligned} 0 &= y'' - (\cos x)y \\ &= \sum_{n=0}^{\infty}(n+2)(n+1)a_{n+2}x^n \\ &\quad - \left(1 - x^2/2 + \cdots\right)\sum_{n=0}^{\infty} a_n x^n \\ &= [2a_2 - a_0] + [6a_3 - a_1]x \\ &\quad + [12a_4 - a_2 + a_0/2]x^2 + \cdots. \end{aligned}$$

Setting the coefficients equal to 0, we see that $a_2 = a_0/2$, $a_3 = a_1/6$, and $a_4 = (a_2 - a_0/2)/12$. For the first solution we choose $a_0 = y_1(0) = 1$ and $a_1 = y_1'(0) = 0$. Then $a_2 = 1/2$, $a_3 = 0$, and $a_4 = 0$ so

$$S_4(x) = 1 + \frac{x^2}{2} \quad \text{and} \quad y_1(x) = 1 + \frac{x^2}{2} + \cdots.$$

For the second solution we choose $a_0 = y_1(0) = 0$ and $a_1 = y_1'(0) = 1$. Then $a_2 = 0$, $a_3 = 1/6$, and $a_4 = 0$, so

$$S_4(x) = x + \frac{x^3}{6} \quad \text{and} \quad y_2(x) = x + \frac{x^3}{6} + \cdots.$$

These solutions are linearly independent by Proposition 2.14.

25. The first figure contains the plot of y_1 and three partial sums, the second the plot of y_2 and three partial sums.

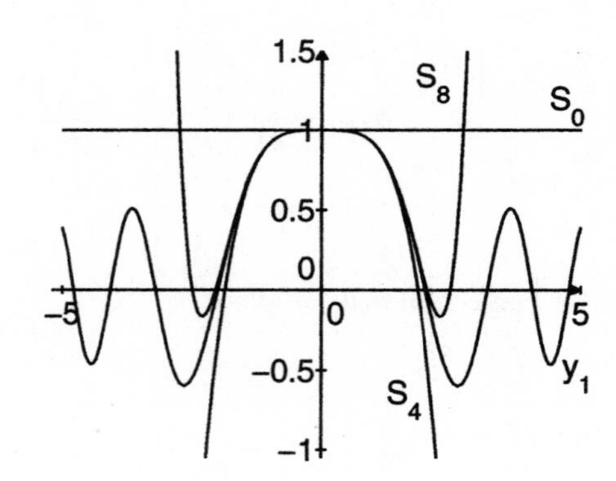

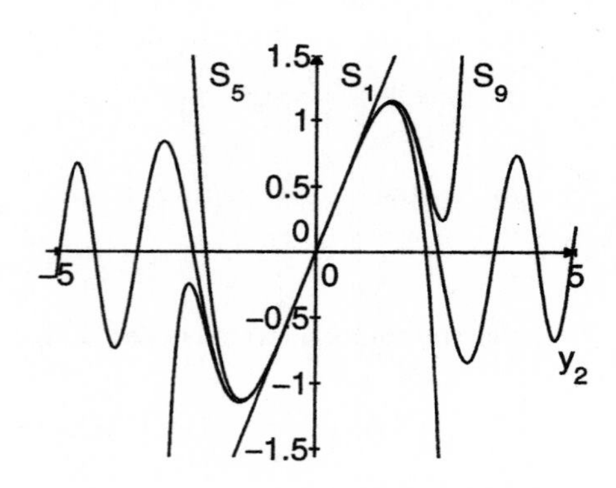

27. The first figure contains the plot of y_1 and three partial sums, the second the plot of y_2 and three partial sums.

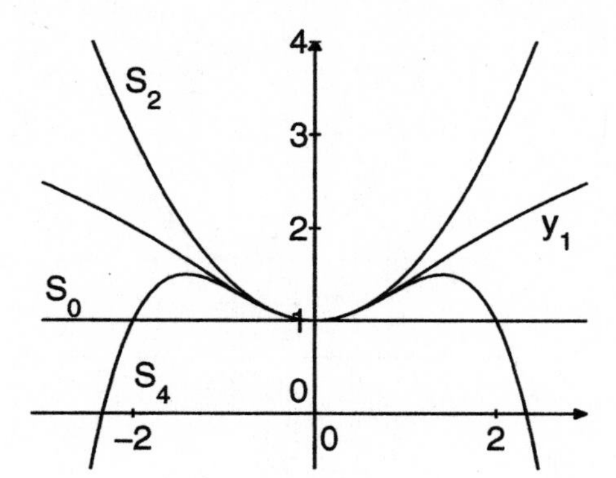

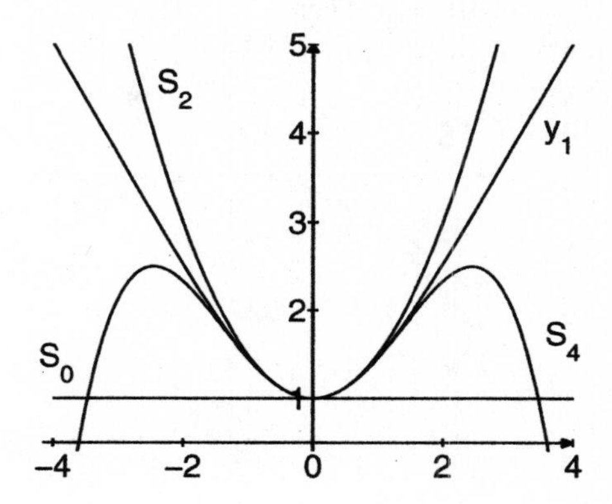

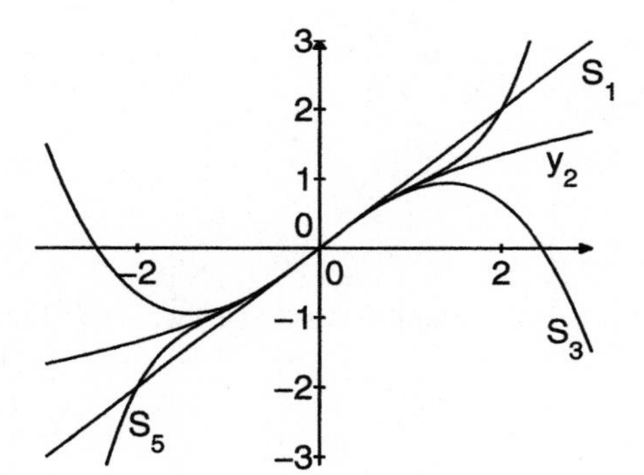

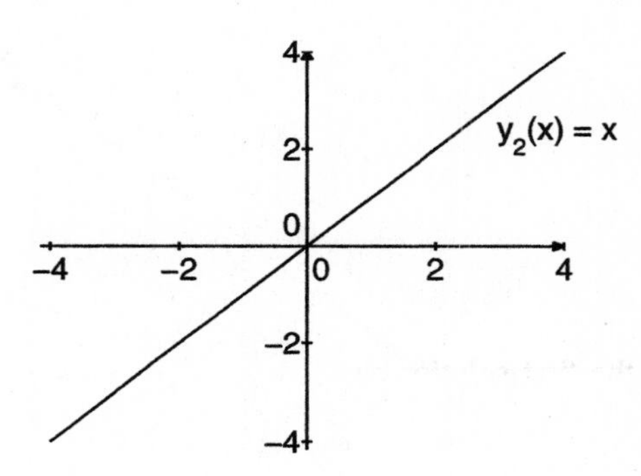

29. The first figure contains the plot of y_1 and three partial sums, the second the plot of y_2 and three partial sums.

31. The first figure contains the plot of y_1 and three partial sums, the second the plot of y_2 and three partial sums.

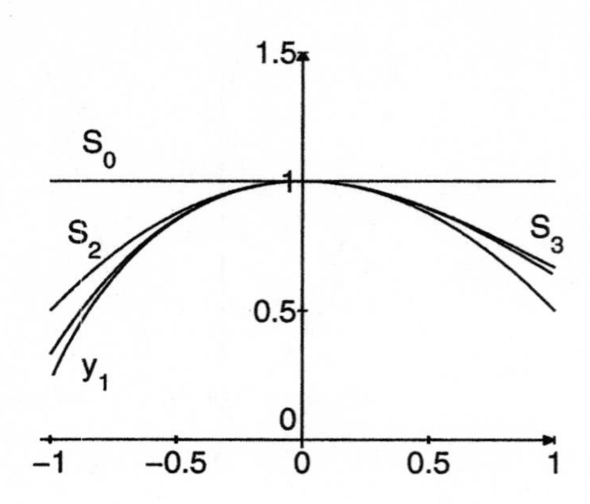

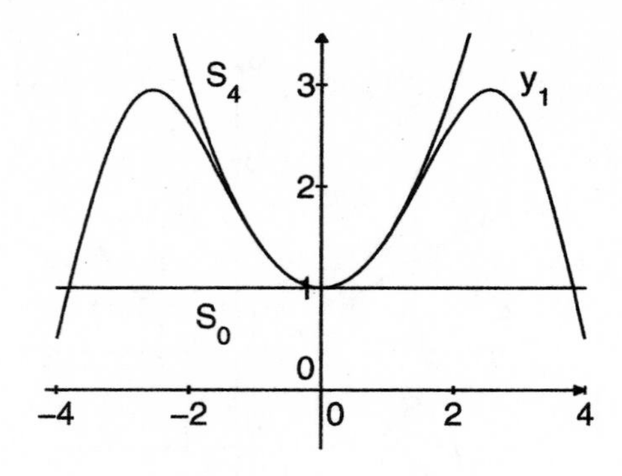

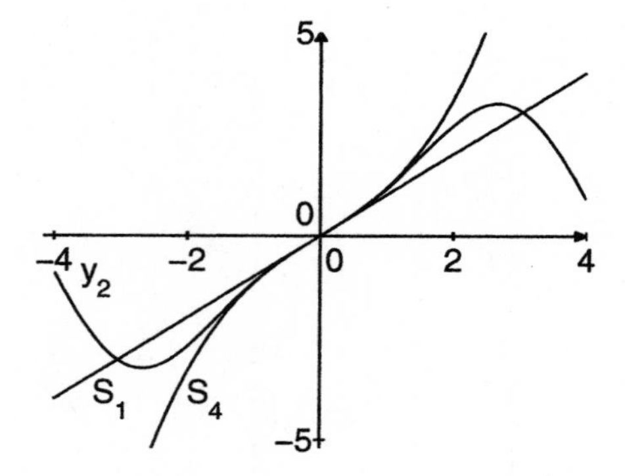

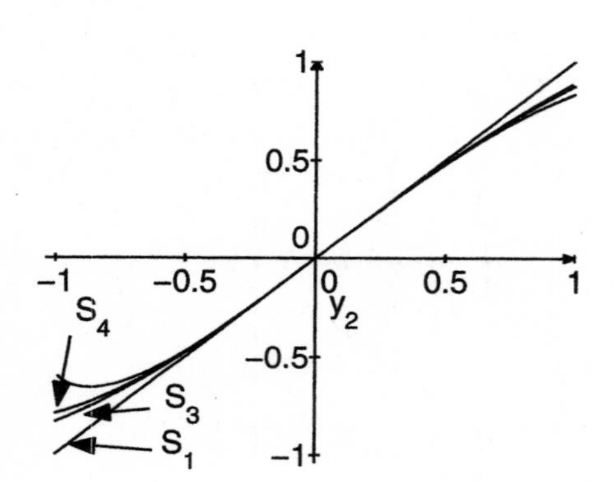

33. The first figure contains the plot of y_1 and three partial sums, the second the plot of y_2 and three partial sums.

35. Certainly, if $y(x) = (1+x)^p$, $y(0) = 1$. Further, if $p \neq 0$ (the equation is trivially satisfied if $p = 0$), then

$$\begin{aligned}
&(1+x)y' - py \\
&\quad = (1+x)p(1+x)^{p-1} - p(1+x)^p \\
&\quad = p(1+x)^p - p(1+x)^p \\
&\quad = 0,
\end{aligned}$$

showing that $y(x) = (1+x)^p$ is the unique solution of $(1+x)y' = py$, $y(0) = 1$. We now look for a solution having form

$$y = \sum_{n=0}^{\infty} a_n x^n.$$

Differentiating,

$$y' = \sum_{n=1}^{\infty} n a_n x^{n-1}.$$

Substituting,

$$\begin{aligned} 0 &= (1+x)y' - py \\ &= (1+x)\sum_{n=1}^{\infty} n a_n x^{n-1} - p\sum_{n=0}^{\infty} a_n x^n \\ &= \sum_{n=1}^{\infty} n a_n x^{n-1} + \sum_{n=1}^{\infty} n a_n x^n - \sum_{n=0}^{\infty} p a_n x^n. \end{aligned}$$

Note that the second term

$$\sum_{n=1}^{\infty} n a_n x^n = \sum_{n=0}^{\infty} n a_n x^n,$$

and shift the index of the first term.

$$\begin{aligned} 0 &= \sum_{n=0}^{\infty} (n+1)a_{n+1}x^n + \sum_{n=0}^{\infty} n a_n x^n - \sum_{n=0}^{\infty} p a_n x^n \\ &= \sum_{n=0}^{\infty} \left[(n+1)a_{n+1} + n a_n - p a_n\right] x^n. \end{aligned}$$

Setting coefficients equal to zero,

$$(n+1)a_{n+1} + (n-p)a_n = 0,$$

or

$$a_{n+1} = \frac{p-n}{n+1} a_n, \quad n \geq 0.$$

Thus, with $a_0 = y(0) = 1$,

$$\begin{aligned} a_1 &= p a_0 = p, \\ a_2 &= \frac{p-1}{2} a_1 = \frac{p(p-1)}{2}, \\ a_3 &= \frac{p-2}{3} a_2 = \frac{p(p-2)(p-2)}{3 \cdot 2}. \end{aligned}$$

We conjecture that

$$a_n = \frac{p(p-1)(p-2)\cdots(p-(n-1))}{n!}.$$

Thus,

$$\begin{aligned} y(x) &= 1 + px + \frac{p(p-1)}{2}x^2 + \cdots \\ &\quad + \frac{p(p-1)(p-2)\cdots(p-n+1)}{n!}x^n + \cdots. \end{aligned}$$

The radius of convergence is

$$\begin{aligned} &\lim_{n\to\infty} \left|\frac{a_n}{a_{n+1}}\right| \\ &= \lim_{n\to\infty} \left|\frac{p(p-1)\cdots(p-n+1)}{n!} \cdot \frac{(n+1)!}{p(p-1)\cdots(p-n)}\right| \\ &= \lim_{n\to\infty} \left|\frac{n+1}{p-n}\right| \\ &= \lim_{n\to\infty} \left|\frac{1+1/n}{p/n-1}\right| \\ &= 1 \end{aligned}$$

Thus, the series converges for $|x| < 1$.

Section 11.3. Legendre's Equation

1. (a) Note that

$$\begin{aligned}\frac{1}{\sqrt{1-z}} &= (1-z)^{-1/2} \\ &= 1 + \frac{1}{2}z + \frac{1\cdot 3}{2^2 2!}z^2 + \frac{1\cdot 3\cdot 5}{2^3 3!}z^3 + \cdots + \frac{1\cdot 3\cdot 5\cdots(2n-1)}{2^n n!}z^n + \cdots\end{aligned}$$

provided $|z| < 1$. Thus,

$$\begin{aligned}\frac{1}{\sqrt{1-2xt+t^2}} &= \frac{1}{\sqrt{1+t(2x-t)}} \\ &= 1 + \frac{1}{2}t(2x-t) + \frac{1\cdot 3}{2^2\cdot 2!}t^2(2x-t)^2 + \frac{1\cdot 3\cdot 5}{2^3\cdot 3!}t^3(2x-t)^3 + \\ &\quad \cdots + \frac{1\cdot 3\cdot 5\cdots(2n-1)}{2^n n!}t^n(2x-t)^n + \cdots\end{aligned}$$

Now, the idea is to gather together like powers of t. For example,

$$\begin{aligned}\frac{1}{\sqrt{1-2xt+t^2}} &= 1 + xt + \left(-\frac{1}{2} + \frac{1\cdot 3}{2^2\cdot 2!}(2^2x^2)\right)t^2 + \left(\frac{1\cdot 3}{2^2\cdot 2!}(-4x) + \frac{1\cdot 3\cdot 5}{2^3\cdot 3!}(8x^3)\right)t^3 + \cdots \\ &= 1 + xt + \left(-\frac{1}{2} + \frac{3}{2}x^2\right)t^2 + \left(-\frac{3}{2}x + \frac{5}{2}x^3\right)t^3 + \cdots.\end{aligned}$$

Note that the coefficients of t are the Legendre polynomials.

(b) If $g(t,x) = (1-2xt+t^2)^{-1/2}$, then

$$\frac{\partial g}{\partial x} = t(1-2xt+t^2)^{-3/2}.$$

Then,

$$\begin{aligned}\frac{\partial}{\partial x}\left[(1-x^2)\frac{\partial g}{\partial x}\right] &= \frac{\partial}{\partial x}t(1-x^2)(1-2xt+t^2)^{-3/2} \\ &= 3t^2(1-x^2)(1-2xt+t^2)^{-5/2} - 2tx(1-2xt+t^2)^{-3/2} \\ &= t(1-2xt+t^2)^{-5/2}(3t(1-x^2) - 2x(1-2xt+t^2)) \\ &= t(1-2xt+t^2)^{-5/2}(3t - 2x + tx^2 - 2t^2x).\end{aligned}$$

Next,

$$\begin{aligned}\frac{\partial}{\partial t}[tg] &= \frac{\partial}{\partial t}t(1-2xt+t^2)^{-1/2} \\ &= (1-2xt+t^2)^{-1/2} + t(x-t)(1-2xt+^2)^{-3/2}.\end{aligned}$$

Now,

$$\frac{\partial^2}{\partial t^2}[tg] = -\frac{1}{2}(1-2xt+t^2)^{-3/2}(-2x+2t) + t(x-t)\left[-\frac{3}{2}(1-2xt+t^2)^{-5/2}(-2x+2t)\right]$$
$$+ (x-2t)(1-2xt+t^2)^{-3/2}$$
$$= (x-t)(1-2xt+t^2)^{-3/2} + 3t(x-t)^2(1-2xt+t^2)^{-5/2} + (x-2t)(1-2xt+t^2)^{-3/2}$$
$$= (2x-3t)(1-2xt+t^2)^{-3/2} + 3t(x-t)^2(1-2xt+t^2)^{-5/2}$$
$$= (1-2xt+t^2)^{-5/2}[3t(x-t)^2 + (2x-3t)(1-2xt+t^2)]$$
$$= (1-2xt+t^2)^{-5/2}[-3t+2x-tx^2+2t^2x].$$

Thus,

$$\frac{\partial}{\partial x}\left[(1-x^2)\frac{\partial g}{\partial x}\right] + t\frac{\partial^2}{\partial t^2}[tg] = 0.$$

(c) If $g(t,x) = \sum_{n=0}^{\infty} Q_n(x)t^n$, then

$$\frac{\partial g}{\partial x} = \frac{\partial}{\partial x}\sum_{n=0}^{\infty} Q_n(x)t^n = \sum_{n=0}^{\infty} Q_n'(x)t^n$$
$$(1-x^2)\frac{\partial g}{\partial x} = \sum_{n=0}^{\infty}(1-x^2)Q_n'(x)t^n.$$

Taking another derivative,

$$\frac{\partial}{\partial x}\left[(1-x^2)\frac{\partial g}{\partial x}\right] = \sum_{n=0}^{\infty}\frac{\partial}{\partial x}\left[(1-x^2)Q_n'(x)\right]t^n = \sum_{n=0}^{\infty}\left[(1-x^2)Q_n'(x)\right]'t^n.$$

On the other hand,

$$\frac{\partial}{\partial t}tg = \frac{\partial}{\partial t}\sum_{n=0}^{\infty} Q_n(x)t^{n+1} = \sum_{n=0}^{\infty}(n+1)Q_n(x)t^n.$$

Taking another derivative,

$$\frac{\partial^2}{\partial t^2}tg = \sum_{n=0}^{\infty} n(n+1)Q_n(x)t^{n-1}.$$

Thus,

$$t\frac{\partial^2}{\partial t^2}tg = \sum_{n=0}^{\infty} n(n+1)Q_n(x)t^n$$

and

$$0 = \frac{\partial}{\partial x}\left[(1-x^2)\frac{\partial g}{\partial x}\right] + t\frac{\partial^2}{\partial t^2}[tg] = \sum_{n=0}^{\infty}[(1-x^2)Q_n'(x)]'t^n + \sum_{n=0}^{\infty} n(n+1)Q_n(x)t^n$$
$$= \sum_{n=0}^{\infty}\left\{\left[(1-x^2)Q_n'(x)\right]' + n(n+1)Q_n(x)\right\}t^n.$$

Thus, for each $n \geq 0$,

$$\left[(1-x^2)Q_n'(x)\right]' + n(n+1)Q_n(x) = 0$$
$$(1-x^2)Q_n''(x) - 2xQ_n'(x) + n(n+1)Q_n(x) = 0.$$

(d) Therefore, $Q_n(x)$ is a solution of

$$(1-x^2)y'' - 2xy' + n(n+1)y = 0$$

of order n.

(e) When $x = 1$,

$$g(t,1) = \frac{1}{1-t} = \sum_{n=0}^{\infty} t^n.$$

Hence, $Q_n(1) = 1$. Thus, $Q_n(x)$ is a polynomial solution of Legendre's equation of order n with $Q_n(1) = 1$. Therefore, $Q_n(x) = P_n(x)$, the Legendre polynomial of order n.

3. Given the recurrence formula

$$nP_n(x) = (2n-1)xP_{n-1}(x) - (n-1)P_{n-2}(x),$$

for $n \geq 2$, letting $x = 0$ produces the relation

$$nP_n(0) = -(n-1)P_{n-2}(0), \quad n \geq 2.$$

Replacing n with $2n$ provides

$$2nP_{2n}(0) = (1-2n)P_{2n-2}(0)$$
$$P_{2n}(0) = \frac{1-2n}{2n}P_{2n-2}(0), \quad n \geq 1.$$

Thus,

$$P_2(0) = \frac{1-2}{2}P_0(0), \quad P_4(0) = \frac{1-4}{4}P_2(0), \quad \text{and} \quad P_6(0) = \frac{1-6}{6}P_4(0),$$

etc., so that

$$P_2(0)P_4(0)P_6(0)\ldots P_{2n}(0) = \frac{1-2}{2}\cdot\frac{1-4}{4}\cdot\frac{1-6}{6}\cdots\frac{1-2n}{2n}P_0(0)P_2(0)P_4(0)\cdots P_{2n-2}(0).$$

Cancelling,

$$P_{2n}(0) = \frac{(-1)^n 1\cdot 3\cdot 5\cdots(2n-1)}{2^n n!}P_0(0).$$

Of course, $P_0(0) = 1$, and

$$1\cdot 3\cdot 5\cdots(2n-1) = \frac{1\cdot 2\cdot 3\cdot 4\cdot 5\cdots(2n-1)(2n)}{2\cdot 4\cdot 6\cdots(2n)} = \frac{(2n)!}{2^n n!}.$$

Thus,

$$P_{2n}(0) = \frac{(-1)^n\frac{(2n)!}{2^n n!}}{2^n n!} = (-1)^n\frac{(2n)!}{2^{2n}(n!)^2}.$$

5. We seek a solution having from $\sum_{n=0}^{\infty} a_n x^n$, with derivatives

$$y'(x) = \sum_{n=1}^{\infty} n a_n x^{n-1} \quad \text{and} \quad y''(x) = \sum_{n=2}^{\infty} n(n-1)a_n x^{n-2}.$$

Substituting,

$$0 = y'' + xy = \sum_{n=2}^{\infty} n(n-1)a_n x^{n-2} + x\sum_{n=0}^{\infty} a_n x^n = \sum_{n=2}^{\infty} n(n-1)a_n x^{n-2} + \sum_{n=0}^{\infty} a_n x^{n+1}.$$

We shift the index of each term to obtain a common power of x.

$$0 = \sum_{n=0}^{\infty}(n+2)(n+1)a_{n+2}x^n + \sum_{n=1}^{\infty} a_{n-1}x^n$$

Thus,

$$0 = 2a_2 + \sum_{n=1}^{\infty}\left[(n+2)(n+1)a_{n+2} + a_{n-1}\right]x^n.$$

Setting coefficients equal to zero, $a_2 = 0$ and

$$0 = (n+2)(n+1)a_{n+2} + a_{n-1} \quad \text{or} \quad a_{n+2} = \frac{-a_{n-1}}{(n+2)(n+1)}, \qquad n \geq 1.$$

Thus,

$$a_3 = \frac{-a_0}{3\cdot 2}, \quad a_6 = \frac{-a_3}{6\cdot 5} = \frac{a_0}{6\cdot 5\cdot 3\cdot 2}, \quad \text{and} \quad a_9 = \frac{-a_6}{9\cdot 8} = \frac{-a_0}{9\cdot 8\cdot 6\cdot 5\cdot 3\cdot 2}.$$

We conjecture that

$$a_{3n} = \frac{(-1)^n a_0}{(2\cdot 3)\cdot(5\cdot 6)\cdot\dots\cdot(3n-1)(3n)}, \qquad n \geq 1.$$

Similarly,

$$a_4 = \frac{-a_1}{4\cdot 3}, \quad a_7 = \frac{-a_4}{7\cdot 6} = \frac{a_1}{3\cdot 4\cdot 6\cdot 7}, \quad \text{and} \quad a_{10} = \frac{-a_7}{10\cdot 9} = \frac{-a_1}{3\cdot 4\cdot 6\cdot 7\cdot 9\cdot 10}.$$

We conjecture

$$a_{3n+1} = \frac{(-1)^n a_1}{(3\cdot 4)\cdot(6\cdot 7)\cdot\dots\cdot(3n)(3n+1)}, \qquad n \geq 1.$$

Finally, because $a_2 = 0$, $a_{3n+2} = 0$, for $n \geq 1$. Thus,

$$y(x) = a_0\left[1 - \frac{1}{2\cdot 3}x^3 + \frac{1}{2\cdot 3\cdot 5\cdot 6}x^6 - \cdots\right] + a_1\left[x - \frac{1}{3\cdot 4}x^4 + \frac{1}{3\cdot 4\cdot 6\cdot 7}x^7 - \cdots\right]$$

Choose $y_1(x)$ so that $a_0 = y_1(0) = 1$ and $a_1 = y_1'(0) = 0$.

$$y_1(x) = 1 - \frac{1}{2\cdot 3}x^3 + \frac{1}{2\cdot 3\cdot 5\cdot 6}x^6 - \cdots = 1 + \sum_{n=1}^{\infty}\frac{(-1)^n x^{3n}}{(2\cdot 3)\cdot(5\cdot 6)\cdot\dots\cdot(3n-1)(3n)}.$$

Choose $y_2(x)$ so that $a_0 = y_2(0) = 0$ and $a_1 = y_2'(0) = 1$.

$$y_2(x) = x - \frac{1}{3 \cdot 4}x^4 + \frac{1}{3 \cdot 4 \cdot 6 \cdot 7}x^7 - \cdots = x + \sum_{n=1}^{\infty} \frac{(-1)^n x^{3n+1}}{(3 \cdot 4) \cdot (6 \cdot 7) \cdot \cdots \cdot (3n)(3n+1)}.$$

In $y'' + xy = 0$, the coefficients of y is $Q(x) = x$. The function Q is analytic for all x. By Theorem 2.29, the series converges for all x. The solutions and partial sums are plotted below.

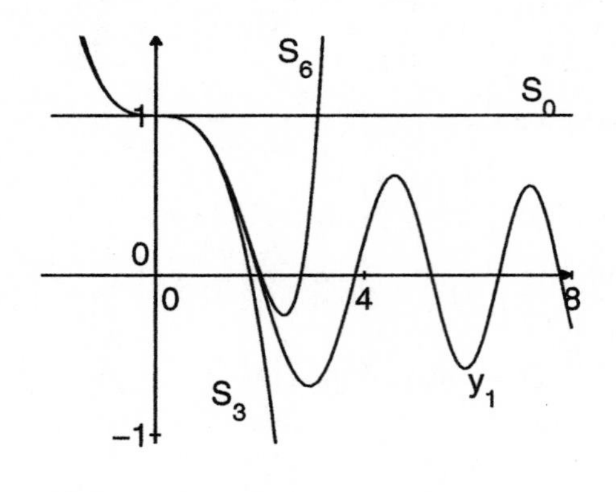

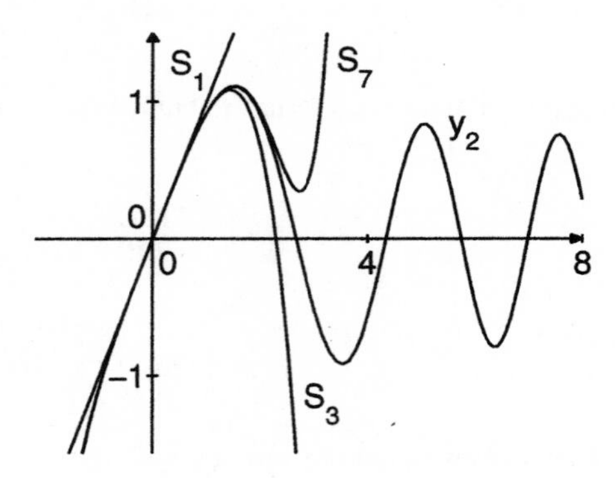

7. (a) Consider $y'' - 2xy' + py = 0$. Set

$$y = \sum_{n=0}^{\infty} a_n x^n, \quad y' = \sum_{n=1}^{\infty} n a_n x^{n-1}, \quad \text{and} \quad y'' = \sum_{n=2}^{\infty} n(n-1) a_n x^{n-2}.$$

Substitute,

$$\begin{aligned} 0 = y'' - 2xy' + py &= \sum_{n=2}^{\infty} n(n-1)a_n x^{n-2} - 2x \sum_{n=1}^{\infty} n a_n x^{n-1} + p \sum_{n=0}^{\infty} a_n x^n \\ &= \sum_{n=2}^{\infty} n(n-1)a_n x^{n-2} - \sum_{n=1}^{\infty} 2n a_n x^n + \sum_{n=0}^{\infty} p a_n x^n. \end{aligned}$$

We will shift indicies to the lowest common power of x.

$$\begin{aligned} 0 &= \sum_{n=2}^{\infty} n(n-1)a_n x^{n-2} - \sum_{n=3}^{\infty} 2(n-2)a_{n-2} x^{n-2} + \sum_{n=2}^{\infty} p a_{n-2} x^{n-2} \\ &= (2a_2 + pa_0)x^0 + \sum_{n=3}^{\infty} \{n(n-1)a_n - (2n - 4 - p)a_{n-2}\} x^{n-2} \end{aligned}$$

Thus,

$$2a_2 + pa_0 = 0 \quad \text{or} \quad a_2 = -\frac{p}{2} a_0,$$

and

$$n(n-1)a_n - (2n - p - 4)a_{n-2} = 0, \quad n \geq 3.$$

Equivalently,

$$a_n = \frac{2n - p - 4}{n(n-1)} a_{n-2}, \quad n \geq 3.$$

Thus,

$$a_4 = \frac{4-p}{4 \cdot 3} a_2, \quad a_6 = \frac{8-p}{6 \cdot 5} a_4, \quad \dots \quad , a_{2n} = \frac{4n-p-4}{2n(2n-1)} a_{2n-2}.$$

Thus,

$$a_2 a_4 a_6 \cdots a_{2n} = -\frac{p(4-p)(8-p)\cdots(4n-p-4)}{2 \cdot 3 \cdot 4 \cdot 5 \cdot 6 \cdots (2n-1)(2n)} a_0 a_2 a_4 \cdots a_{2n-2},$$

and

$$a_{2n} = -\frac{p(4-p)(8-p)\cdots(4n-p-4)}{(2n)!} a_0.$$

In a similar manner,

$$a_3 = \frac{2-p}{3 \cdot 2} a_1, \quad a_5 = \frac{6-p}{5 \cdot 4} a_3, \quad \dots \quad , a_{2n+1} = \frac{4n-p-2}{(2n+1)(2n)} a_{2n-1}.$$

Thus,

$$a_{2n+1} = \frac{(2-p)(6-p)\cdots(4n-p-2)}{(2n+1)!} a_1.$$

Hence,

$$y(x) = a_0 y_1(x) + a_1 y_2(x),$$

where

$$y_1(x) = 1 - \sum_{n=1}^{\infty} \frac{p(4-p)(8-p)\cdots(4n-p-4)}{(2n)!} x^{2n} \tag{3.1}$$

and

$$y_2(x) = x + \sum_{n=1}^{\infty} \frac{(2-p)(6-p)\cdots(4n-p-2)}{(2n+1)!} x^{2n+1}. \tag{3.2}$$

Using the ratio test on $y_1(x)$,

$$\begin{aligned}\left|\frac{A_{n+1}}{A_n}\right| &= \frac{p(4-p)\cdots(4n-p)}{(2n+2)!} \cdot \frac{(2n)!}{p(4-p)\cdots(4n-p-4)} x^2 \\ &= \frac{4n-p}{(2n+2)(2n+1)} x^2 \to 0 \quad \text{as} \quad n \to \infty.\end{aligned}$$

Therefore the series y_1 converges for all x. For $y_2(x)$,

$$\begin{aligned}\left|\frac{A_{n+1}}{A_n}\right| &= \frac{(2-p)\dots(4n-p+2)}{(2n+3)!} \cdot \frac{(2n+1)!}{(2-p)\dots(4n-p-2)} x^2 \\ &= \frac{4n-p+2}{(2n+3)(2n+2)} x^2 \to 0 \quad \text{as} \quad n \to \infty.\end{aligned}$$

Therefore the series y_2 converges for all x.

(b) We now consider the case when p is a nonnegative integer. In the case that p is odd, neither

$$\frac{-p(4-p)(8-p)\cdots(4n-p-4)}{(2n)!}$$

nor

$$\frac{(2-p)(6-p)\cdots(4n-p-2)}{(2n+1)!}$$

can equal zero. Let's examine the case where p is even. There are two possibilities. Either $p = 2, 6, 10, \ldots$ or $p = 4, 8, 12, \ldots$, etc. We examine each case separately. Consider the case where $p = 2 + 4l$, l a nonnegative integer. Then

$$\begin{aligned} 4n - p - 2 &= 0 \\ 4n - (2 + 4l) - 2 &= 0 \\ 4n &= 4l + 4 \\ n &= l + 1. \end{aligned}$$

Thus, for $n \geq l + 1$,

$$a_{2n+1} = \frac{(2-p)(6-p)\cdots(4n-p-2)}{(2n+1)!} = 0$$

and the series $y_2(x)$ terminates in a polynomial having degree

$$2n + 1 = 2l + 1 = p/2,$$

which is odd degree. It is easy to show that $y_1(x)$ does not terminate in this case. In the case where $p = 4 + 4l$, l a nonnegative integer,

$$\begin{aligned} 4n - p - 4 &= 0 \\ 4n - (4 + 4l) - 4 &= 0 \\ 4n &= 4l + 8 \\ n &= l + 2. \end{aligned}$$

Thus, for $n \geq l + 2$,

$$a_{2n} = -\frac{p(4-p)\cdots(4n-p-4)}{(2n)!} = 0$$

and the series $y_1(x)$ terminates in a polynomial having degree

$$2n = 2(l + 1) = 2l + 2 = p/2,$$

which is even degree.

(c) Using equations (3.1) and (3.2), we arrive at

$$\begin{aligned} p = 0 &\Rightarrow 1 \\ p = 2 &\Rightarrow x \\ p = 4 &\Rightarrow 1 - 2x^2 \\ p = 6 &\Rightarrow x - \frac{2}{3}x^3 \\ p = 8 &\Rightarrow 1 - 4x^2 + \frac{4}{3}x^4 \\ p = 10 &\Rightarrow x - \frac{4}{3}x^3 + \frac{4}{15}x^5. \end{aligned}$$

Each of these must be multiplied by an appropriate constant so that the coefficient of x^p is 2^p.

$$H_0(x) = 1$$
$$H_1(x) = 2x$$
$$H_2(x) = 4x^2 - 2$$
$$H_3(x) = 8x^3 - 12x$$
$$H_4(x) = 16x^4 - 48x^2 + 12$$
$$H_5(x) = 32x^5 - 160x^3 + 120x$$

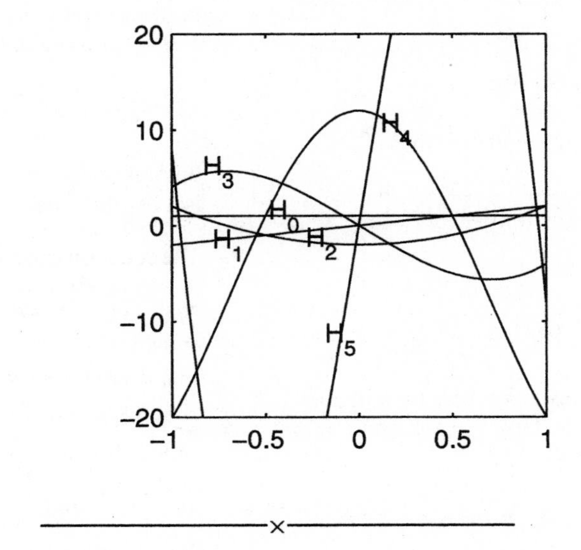

Section 11.4. Types of Singular Points—Euler's Equation

1. Divide by the leading coefficient.

$$x^2y'' + 4xy' - 2xy = 0$$
$$y'' + \frac{4}{x}y' - \frac{2}{x}y = 0$$

Hence, $x = 0$ is a singular point. Thus,

$$P(x) = \frac{4}{x} \quad \text{and} \quad Q(x) = -\frac{2}{x},$$

so

$$xP(x) = 4 \quad \text{and} \quad x^2Q(x) = -2x,$$

both of which are analytic at $x = 0$. Thus, the point $x = 0$ is a regular singular point.

3. Divide by the leading coefficient.

$$y'' + \frac{4}{x(2-x)^2}y' + \frac{5}{x^2(2-x)^2}y = 0.$$

The functions

$$P(x) = \frac{4}{x(2-x)^2} \quad \text{and} \quad Q(x) = \frac{5}{x^2(2-x)^2}$$

are analytic everywhere, except where their denominators equal zero. Thus, $x = 0$ and $x = 2$ are singular points. For $x = 0$,

$$xP(x) = \frac{4}{(2-x)^2} \quad \text{and} \quad x^2 Q(x) = \frac{5}{(2-x)^2}$$

are both analytic near $x = 0$. Thus, $x = 0$ is a regular singular point. For $x = 2$,

$$(x-2)P(x) = \frac{4}{x(2-x)}$$

is not analytic near $x = 2$, so $x = 2$ is an irregular singular point.

5. Divide by the leading coefficient.

$$(t^2 - t - 6)y'' + (t^2 - 9)y' + 2y = 0$$

$$y'' + \frac{t+3}{t+2}y' + \frac{2}{(t-3)(t+2)}y = 0$$

Thus,

$$P(t) = \frac{t+3}{t+2} \quad \text{and} \quad Q(t) = \frac{2}{(t-3)(t+2)}$$

are analytic everywhere, except where their denominators are equal to zero. Thus, $t = -2$ and $t = 3$ are singular points. For $t = -2$,

$$(t+2)P(t) = t+3 \quad \text{and} \quad (t+2)^2 Q(t) = \frac{2(t+2)}{t-3},$$

both of which are analytic near $t = -2$. Thus, $t = -2$ is a regular singular point. For $t = 3$,

$$(t-3)P(t) = \frac{(t+3)(t-3)}{t+2}$$

and

$$(t-3)^2 Q(t) = \frac{2(t-3)}{t+2},$$

both of which are analytic near $t = 3$. Thus, $t = 3$ is a regular singular point.

7. Divide by the leading coefficient.

$$(t^2 - 1)y'' + (t-2)y' + y = 0$$

$$y'' + \frac{t-2}{(t+1)(t-1)}y' + \frac{1}{(t+1)(t-1)}y = 0$$

Thus,

$$P(t) = \frac{t-2}{(t+1)(t-1)} \text{ and } Q(t) = \frac{1}{(t+1)(t-1)}$$

are analytic everywhere, except where their denominators equal zero. Thus, $t = -1$ and $t = 1$ are singular points. For $t = -1$,

$$(t+1)P(t) = \frac{t-2}{t-1} \quad \text{and} \quad (t+1)^2 Q(t) = \frac{t+1}{t-1},$$

both of which are analytic near $t = -1$. Thus, $t = -1$ is a regular singular point. For $t = 1$,

$$(t-1)P(t) = \frac{t-2}{t+1} \quad \text{and} \quad (t-1)^2 Q(t) = \frac{t-1}{t+1},$$

both of which are analytic near $t = 1$. Thus, $t = 1$ is a regular singular point.

9. The equation can be written as $y'' + P(t)y' + Q(t)y = 0$, where $P(t) = t/\sin^2 t$ and $Q(t) = -3/\sin^2 t$. Both P and Q are analytic except where $\sin t = 0$, or where $t = k\pi$ for some integer k. Hence the singular points are $t = k\pi$.

First let $t_0 = 0$. Then we have

$$tP(t) = \left(\frac{t}{\sin t}\right)^2, \quad \text{and} \quad t^2 Q(t) = -3\left(\frac{t}{\sin t}\right)^2.$$

Since $t/\sin t \to 1$ as $t \to t_0 = 0$, both functions are analytic at $t_0 = 0$. Hence 0 is a regular singular point.

Now suppose that $t_0 = k\pi$, where k is a nonzero integer. Then

$$(t - k\pi)P(t) = \frac{t(t-k\pi)}{\sin^2 t} = t \cdot \frac{t-k\pi}{\sin t} \cdot \frac{1}{\sin t}.$$

Since, by l'Hopital's rule,

$$\lim_{t \to k\pi} \frac{t-k\pi}{\sin t} = \lim_{t \to k\pi} \frac{1}{\cos t} = \frac{1}{\cos k\pi} = (-1)^k,$$

the first two factors in are analytic at $t_0 = k\pi$ and have nonzero values there. However, the third factor has no limit as $t \to k\pi$, so $(t - k\pi)P(t)$ is not analytic at $t_0 = k\pi$. Therefore $k\pi$ is an irregular singular point if k is a nonzero integer.

11. Consider $y = x^s$ and its derivatives.

$$y' = sx^{s-1} \quad \text{and} \quad y'' = s(s-1)x^{s-2}$$

Substitute.

$$\begin{aligned} 0 &= x^2y'' + 5xy' + 3y \\ &= x^2 s(s-1)x^{s-2} + 5xsx^{s-1} + 3x^s \\ &= [s(s-1) + 5s + 3]x^s. \end{aligned}$$

The indicial equation

$$0 = I(s) = s(s-1) + 5s + 3 = (s+3)(s+1)$$

has roots $s = -3$ and $s = -1$. These provide independent solutions $y_1(x) = x^{-3}$ and $y_2(x) = x^{-1}$ and the general solution

$$y(x) = C_1x^{-3} + C_2x^{-1}.$$

13. Consider $y = x^s$ and its derivatives.

$$y' = sx^{s-1} \quad \text{and} \quad y'' = s(s-1)x^{s-2}$$

Negate both sides of the equation.

$$x^2y'' - xy' + y = 0$$

Substitute.

$$\begin{aligned} 0 &= x^2 s(s-1)x^{s-2} - xsx^{s-1} + x^s \\ &= [s(s-1) - s + 1]x^s \end{aligned}$$

the indicial equation

$$0 = I(s) = s(s-1) - s + 1 = (s-1)^2$$

has repeated root $s = 1$. This provides independent solutions $y_1(x) = x$ and $y_2(x) = x \ln x$ and general solution

$$y(x) = C_1x + C_2x \ln x.$$

15. The equation $x^2y'' + 9xy' + 16y = 0$ has indicial equation

$$0 = I(s) = s(s-1) + 9s + 16 = (s+4)^2.$$

This equation has repeated root $s = -4$. This provides independent solutions $y_1(x) = x^{-4}$ and $y_2(x) = x^{-4} \ln x$ and general solution

$$y(x) = C_1x^{-4} + C_2x^{-4} \ln x.$$

17. The equation

$$x^2y'' - 5xy' + 9y = 0$$

has indicial equation

$$0 = I(s) = s(s-1) - 5s + 9 = (s-3)^2,$$

which has repeated zero $s = 3$. This provides independent solutions $y_1(x) = x^3$ and $y_2(x) = x^3 \ln x$ and general solution

$$y(x) = C_1x^3 + C_2x^3 \ln x.$$

19. Consider $y = x^s$ and its derivatives

$$y' = sx^{s-1} \quad \text{and} \quad y'' = s(s-1)x^{s-2}.$$

Substitute.

$$\begin{aligned} 0 &= x^2y'' + 2xy' - 6y \\ &= x^2 s(s-1)x^{s-2} + 2xsx^{s-1} - 6x^s \\ &= [s(s-1) + 2s - 6]x^s \end{aligned}$$

The indicial equation

$$0 = I(s) = s(s-1) + 2s - 6 = (s+3)(s-2)$$

has roots $s = -3$ and $s = 2$. These provide independent solutions $y_1(x) = x^{-3}$ and $y_2(x) = x^2$ and general solution

$$y(x) = C_1x^{-3} + C_2x^2.$$

The initial condition $y(1) = 3$ provides

$$3 = C_1 + C_2.$$

Differentiate.

$$y'(x) = -3C_1x^{-3} + 2C_2x$$

The initial condition $y'(1) = 1$ provides

$$1 = -3C_1 + 2C_2.$$

Solving, $C_1 = 1$ and $C_2 = 2$. Thus,

$$y(x) = x^{-3} + 2x^2.$$

21. Consider $y = (x-3)^s$ and its derivatives

$$y' = s(x-3)^{s-1} \quad \text{and} \quad y'' = s(s-1)(x-3)^{s-2}.$$

Substitute.

$$\begin{aligned} 0 &= (x-3)^2 y'' + 5(x-3)y' + 4y \\ &= (x-3)^2 s(s-1)(x-3)^{s-2} \\ &\quad + 5(x-3)s(s-3)^{s-1} + 4(x-3)^s \\ &= [s(s-1) + 5s + 4](x-3)^s \end{aligned}$$

The indicial equation

$$0 = I(s) = s(s-1) + 5s + 4 = (s+2)^2$$

has repeated zero $s = -2$. Using Theorem 4.21, this provides independent solutions $y_1(x) = |x-3|^{-2}$ and $y_2(x) = |x-3|^{-2}\ln|x-3|$. The general solution is

$$y(x) = C_1|x-3|^{-2} + C_2|x-3|^{-2}\ln|x-3|.$$

No solution can contain the point at which $x = 3$. Because the initial conditions use $x = 4$, the solution of interest is to the right of $x = 3$ and we can write

$$y(x) = C_1(x-3)^{-2} + C_2(x-3)^{-2}\ln(x-3).$$

The initial condition $y(4) = 1$ provides $C_1 = 1$. Differentiate.

$$\begin{aligned} y'(x) &= -2C_1(x-3)^{-3} + C_2(x-3)^{-3} \\ &\quad - 2C_2(x-3)^{-3}\ln(x-3) \end{aligned}$$

The initial condition $y'(4) = 1$ provides

$$1 = -2C_1 + C_2.$$

Solving, $C_1 = 1$, $C_2 = 3$, and

$$y(x) = (x-3)^{-2} + 3(x-3)^{-2}\ln(x-3).$$

23. (a) Substitute $s = \alpha + i\beta$ in the indicial equation $I(s) = s(s-1) + ps + q$.

$$\begin{aligned} 0 &= I(\alpha + i\beta) \\ &= (\alpha + i\beta)(\alpha + i\beta - 1) + p(\alpha + i\beta) + q \\ &= (\alpha^2 - \alpha - \beta^2 + \alpha p + q) \\ &\quad + i(2\alpha\beta - \beta + p\beta) \end{aligned}$$

Thus,

$$\begin{aligned} \alpha p + q &= \beta^2 + \alpha - \alpha^2, \quad \text{and} \\ p\beta &= \beta - 2\alpha\beta. \end{aligned}$$

If $\beta = 0$, then the root $s = \alpha + i\beta$ is real and handled by Theorem 4.21. Assuming $\beta \neq 0$, the second equation becomes

$$p = 1 - 2\alpha.$$

Substitute this result in the first equation.

$$\begin{aligned} \alpha(1-2\alpha) + q &= \beta^2 + \alpha - \alpha^2 \\ \alpha - 2\alpha^2 + q &= \beta^2 + \alpha - \alpha^2 \\ q &= \alpha^2 + \beta^2. \end{aligned}$$

(b) With $x > 0$,

$$y_1(x) = x^\alpha \cos(\beta \ln x).$$

The proof with $x < 0$ is similar. Differentiate.

$$y_1'(x) = \alpha x^{\alpha-1}\cos(\beta\ln x) - \beta x^{\alpha-1}\sin(\beta\ln x)$$

Differentiate a second time.

$$\begin{aligned} y_1''(x) &= [\alpha(\alpha-1) - \beta^2]x^{\alpha-2}\cos(\beta\ln x) \\ &\quad - [\alpha\beta + \beta(\alpha-1)]x^{\alpha-2}\sin(\beta\ln x) \end{aligned}$$

Substitute in $x^2y'' + pxy' + qy = 0$.

$$\begin{aligned} 0 &= x^2\Big\{[\alpha(\alpha-1) - \beta^2]x^{\alpha-2}\cos(\beta\ln x) \\ &\qquad - [\alpha\beta + \beta(\alpha-1)]x^{\alpha-2}\sin(\beta\ln x)\Big\} \\ &\quad + px[\alpha x^{\alpha-1}\cos(\beta\ln x) \\ &\quad - \beta x^{\alpha-1}\sin(\beta\ln x)] + qx^\alpha\cos(\beta\ln x) \\ &= \{(\alpha^2 - \alpha - \beta^2 + p\alpha + q)\cos(\beta\ln x) \\ &\quad - (2\alpha\beta - \beta + \beta p)\sin(\beta\ln x)\}x^\alpha \\ &= 0, \end{aligned}$$

by the development shown in part (a). In a similar manner, one can show that

$$y_2 = x^\alpha \sin(\beta\ln x)$$

is also a solution of $x^2y'' + pxy' + qy = 0$.

25. The equation $3x^2y'' + y = 0$ has indicial equation

$$0 = I(s) = 3s(s-1) + 1 = 3s^2 - 3s + 1.$$

This equation has complex zeros $s = 1/2 \pm \sqrt{3}i/6$. These provide two independent solutions $y_1(x) = x^{1/2}\cos(\sqrt{3}\ln x/6)$ and $y_2(x) = x^{1/2}\sin(\sqrt{3}\ln x/6)$ on $x > 0$. The general solution on $x > 0$ is

$$y(x) = C_1x^{1/2}\cos\left(\frac{\sqrt{3}\ln x}{6}\right) + C_2x^{1/2}\sin\left(\frac{\sqrt{3}\ln x}{6}\right).$$

27. The equation $x^2y''+xy'+4y = 0$ has indicial equation

$$0 = I(s) = s(s-1) + s + 4 = s^2 + 4.$$

This equation has complex zeros $s = \pm 2i$. These provide independent solutions $y_1(x) = \cos(2\ln x)$ and $y_2(x) = \sin(2\ln x)$ on $x > 0$. The general solution is

$$y(x) = C_1\cos(2\ln x) + C_2\sin(2\ln x).$$

29. The equation $x^2y'' + 3xy' + 2y = 0$ has indicial equation

$$0 = I(s) = s(s-1) + 3s + 2 = s^2 + 2s + 2.$$

This equation has complex zeros $s = -1 \pm i$. Thus, the general solution on $x > 0$ is

$$y(x) = C_1x^{-1}\cos(\ln x) + C_2x^{-1}\sin(\ln x).$$

The initial condition $y(1) = 3$ provides $C_1 = 3$. Differentiate.

$$\begin{aligned} y' &= -C_1x^{-2}\cos(\ln x) - C_1x^{-2}\sin(\ln x) \\ &\quad - C_2x^{-2}\sin(\ln x) + C_2x^{-2}\cos(\ln x) \end{aligned}$$

The initial condition $y'(1) = -2$ leads to

$$-2 = -C_1 + C_2.$$

Thus, $C_2 = 1$ and the solution on $x > 0$ is

$$y(x) = 3x^{-1}\cos(\ln x) + x^{-1}\sin(\ln x).$$

31. The equation $x^2y''-xy'+6y = 0$ has indicial equation

$$0 = I(s) = s(s-1) - s + 6 = s^2 - 2s + 6.$$

This equation has complex roots $s = 1 \pm \sqrt{5}i$. Thus, the general solution is

$$y(x) = C_1x\cos(\sqrt{5}\ln x) + C_2x\sin(\sqrt{5}\ln x).$$

The initial condition $y(1) = 2$ provides $C_1 = 2$. Differentiate.

$$\begin{aligned} y'(x) &= C_1\cos(\sqrt{5}\ln x) - \sqrt{5}C_1\sin(\sqrt{5}\ln x) \\ &\quad + C_2\sin(\sqrt{5}\ln x) + \sqrt{5}C_2\cos(\sqrt{5}\ln x) \end{aligned}$$

The initial condition $y'(1) = 2 - 3\sqrt{5}$ provides

$$2 - 3\sqrt{5} = C_1 + \sqrt{5}C_2.$$

Thus, $C_2 = -3$ and

$$y(x) = 2x\cos(\sqrt{5}\ln x) - 3x\sin(\sqrt{5}\ln x).$$

33. (a) Consider $x^2y''+pxy'+qy = 0$ and let $x = e^t$ and $Y(t) = y(e^t)$. Note that $t = \ln x$ and

$$dt = \frac{1}{x}\,dx.$$

Thus,

$$\frac{d}{dt} = x\frac{d}{dx} \quad \text{and} \quad \frac{d}{dx} = \frac{1}{x}\frac{d}{dt}.$$

Note further that $y(x) = y(e^t) = Y(t) = Y(\ln x)$. Assume that $y(x)$ is a solution; i.e.,

$$x^2y''(x) + pxy'(x) + qy(x) = 0.$$

Now,

$$Y'(t) = \frac{d}{dt}Y(t) = \frac{d}{dt}y(x) = x\frac{d}{dx}y(x).$$

Thus, $Y'(t) = xy'(x)$ and a second derivative is

$$Y''(t) = \frac{d}{dt}Y'(t) = \frac{d}{dt}xy'(x) = x\frac{d}{dx}xy'(x).$$

Thus,

$$Y''(t) = x(xy''(x) + y'(x)) \\ = x^2y''(x) + xy'(x).$$

Therefore,

$$Y''(t) + (p-1)Y'(t) + qY(t) \\ = [x^2y''(x) + xy'(x)] + (p-1)xy'(x) \\ + qy(x) \\ = x^2y''(x) + pxy'(x) + qy(x) \\ = 0.$$

Conversely, assume that $Y(t)$ is a solution of

$$Y''(t) + (p-1)Y'(t) + qY(t) = 0.$$

But,

$$y'(x) = \frac{d}{dx}y(x) = \frac{d}{dx}Y(t) = \frac{1}{x}\frac{d}{dt}Y(t).$$

Thus, $y'(x) = (1/x)\,Y'(t)$ and a second derivative is

$$y''(x) = \frac{d}{dx}y'(x) \\ = \frac{d}{dx}\left(\frac{1}{x}Y'(t)\right) \\ = \frac{1}{x}\frac{d}{dx}Y'(t) - \frac{1}{x^2}Y'(t) \\ = \frac{1}{x}\left(\frac{1}{x}\frac{d}{dt}Y'(t)\right) - \frac{1}{x^2}Y'(t) \\ = \frac{1}{x^2}Y''(t) - \frac{1}{x^2}Y'(t).$$

Thus,

$$x^2y''(x) + pxy'(x) + qy(x) \\ = x^2\left[\frac{1}{x^2}Y''(t) - \frac{1}{x^2}Y'(t)\right] + px\left[\frac{1}{x}Y'(t)\right] \\ + qY(t) \\ = Y''(t) - Y'(T) + pY'(t) + qY(t) \\ = Y''(t) + (p-1)Y'(t) + qY(t) \\ = 0.$$

Hence, $y(x)$ is a solution of $x^2y''+pxy'+qy = 0$ if and only if $Y(t)$ [or $Y(\ln x)$] is a solution of $Y''(t) + (p-1)Y'(T) + qY(t) = 0$.

(b) The characteristic polynomial of $Y'' + (p-1)Y' + qY = 0$ is found by substituting $Y(t) = e^{\lambda t}$.

$$e^{\lambda t}(\lambda^2 + (p-1)\lambda + q) = 0$$

Thus, the characteristic equation is

$$\lambda^2 + (p-1)\lambda + q = 0.$$

To find the indicial equation of $x^2y'' + pxy' + qy = 0$, substitute $y = x^s$.

$$x^2s(s-1)x^{s-2} + pxsx^{s-1} + qx^s = 0 \\ x^s[s(s-1) + ps + q] = 0$$

Thus, the indicial equation is

$$s(s-1) + ps + q = 0 \\ s^2 + (p-1)s + q = 0,$$

which matches the characteristic equation of $Y'' + (p-1)Y' + qY = 0$.

(c) If $Y''(t)+(p-1)Y'(t)+qY(t) = 0$ has solution

$$Y(t) = C_1e^{s_1t} + C_2e^{s_2t},$$

then $x^2y'' + pxy' + qy = 0$ has solution

$$y(x) = Y(\ln x) = C_1e^{s_1\ln x} + C_2e^{s_2\ln x} \\ = C_1x^{s_1} + C_2x^{s_2}.$$

(d) If $Y''(t)+(p-1)Y'(t)+qY(t) = 0$ has solution

$$Y(t) = C_1e^{s_0t} + C_2te^{s_0t}$$

then $x^2y'' + pxy' + qy = 0$ has solution

$$y(x) = Y(\ln x) = C_1e^{s_0\ln x} + C_2(\ln x)e^{s_0\ln x} \\ = C_1x^{s_0} + C_2x^{s_0}\ln x.$$

(e) If $Y''(t)+(p-1)Y'(t)+qY(t) = 0$ has solution

$$Y(t) = e^{\alpha t}(C_1\cos\beta t + C_2\sin\beta t),$$

then $x^2y'' + pxy' + qy = 0$ has solution

$$y(x) \\ = Y(\ln x) \\ = e^{\alpha\ln x}(C_1\cos(\beta\ln x) + C_2\sin(\beta\ln x)) \\ = x^{\alpha}(C_1\cos(\beta\ln x) + C_2\sin(\beta\ln x)).$$

Section 11.5. Series Solutions Near Regular Singular Points

1. To see if there is a nontrivial solution $y = \sum_n a_n x^n$, we differentiate y:

$$y = \sum_{n=0}^{\infty} a_n x^n, \quad y' = \sum_{n=1}^{\infty} n a_n x^{n-1}, \quad \text{and} \quad y'' = \sum_{n=2}^{\infty} n(n-1) a_n x^{n-2},$$

and then insert these expressions into the given differential equation:

$$\begin{aligned} 0 = x^2 y'' + 4xy' + 2y &= x^2 \sum_{n=2}^{\infty} n(n-1) a_n x^{n-2} + 4x \sum_{n=1}^{\infty} n a_n x^{n-1} + 2 \sum_{n=0}^{\infty} a_n x^n \\ &= 2a_0 + 6a_1 x + \sum_{n=2}^{\infty} (n(n-1) + 4n + 2) a_n x^n. \end{aligned}$$

Equating the coefficient of each power of x on the right with zero (since the left side is zero), we obtain $2a_0 = 04$, $6a_1 = 0$, and

$$(n(n-1) + 4n + 2) a_n = 0, \qquad n \geq 2.$$

Clearly, a_0 and a_1 must be zero. Since $n(n-1)+4n+2 = n^2+3n+2$ is never zero for $n \geq 2$, we conclude that all the $a_n = 0$, for $n \geq 2$, as well. Thus the only power series solution $\sum_{n=0} a_n x^n$ to this differential equation is the trivial one where all the a_n equal zero.

3. Compare $x^2y'' + 4xy' + 2y = 0$ with $x^2y'' + xp(x)y' + q(x) = 0$ and note that $p(x) = 4$ and $q(x) = 2$ are both analytic at $x = 0$. Hence $x = 0$ is a regular singular point. Further, note that $p_0 = p(0) = 4$ and $q_0 = q(0) = 2$, so the indicial polynomial should be

$$I(s) = s^2 + (p_0 - 1)s + q_0 = s^2 + 3s + 2.$$

Alternatively, let

$$y(x) = x^s \sum_{n=0}^{\infty} a_n x^n = \sum_{n=0}^{\infty} a_n x^{s+n}.$$

Then,

$$x^2 y'' = \sum_{n=0}^{\infty} (s+n)(s+n-1) a_n x^{s+n} \quad \text{and} \quad 4xy' = 4 \sum_{n=0}^{\infty} (s+n) a_n x^{s+n} = \sum_{n=0}^{\infty} 4(s+n) a_n x^{s+n}.$$

Substituting,

$$\begin{aligned} 0 = x^2y'' + 4xy' + 2y &= \sum_{n=0}^{\infty} (s+n)(s+n-1) a_n x^{s+n} + \sum_{n=0}^{\infty} 4(s+n) a_n x^{s+n} + \sum_{n=0}^{\infty} 2a_n x^{s+n} \\ &= [s(s-1) + 4s + 2]\, a_0 x^s + \sum_{n=1}^{\infty} \big[(s+n)(s+n-1) + 4(s+n) + 2\big] a_n x^{s+n} \\ &= (s^2 + 3s + 2) a_0 x^s + \sum_{n=1}^{\infty} \big[(s+n)^2 + 3(s+n) + 2\big]\, a_n x^{s+n}. \end{aligned}$$

Again, this gives the indicial polynomial $I(s) = s^2 + 3s + 2$.

5. We have to write the equation in the form in 5.2 with $x_0 = 0$. To do so we multiply by $x/\sin x$, to obtain

$$x^2 y'' - \frac{3x}{\sin x} y = 0.$$

Thus, the coefficients are $p(x) = 0$ and $q(x) = -3x/\sin x$. According to 6.8, the indicial equation is $I(s) = s^2 + (p_0 - 1)s + q_0 = 0$, where $p_0 = p(0) = 0$, and $q_0 = q(0)$. By l'Hôpital's rule, $q(0) = -3$. Hence the indicial equation is $s^2 - s - 3 = 0$.

7. Set

$$y(x) = x^2 \sum_{n=0}^{\infty} a_n x^n = \sum_{n=0}^{\infty} a_n x^{s+n}.$$

Then,

$$2x(x-1)y'' + 3(x-1)y' - y = [2x^2 y'' + 3xy' - y] + [-2xy'' - 3y'].$$

But,

$$2x^2 y'' + 3xy' - y = \sum_{n=0}^{\infty} [2(s+n)(s+n-1) + 3(s+n) - 1] a_n x^{s+n}.$$

Next,

$$\begin{aligned} -2xy'' - 3y' &= -2x \sum_{n=0}^{\infty} (s+n)(s+n-1) a_n x^{s+n-2} - 3 \sum_{n=0}^{\infty} (s+n) a_n x^{s+n-1} \\ &= -\sum_{n=0}^{\infty} 2(s+n)(s+n-1) a_n x^{s+n-1} - \sum_{n=0}^{\infty} 3(s+n) a_n x^{s+n-1}. \end{aligned}$$

Thus,

$$\begin{aligned} 0 &= [2x^2 y'' + 3xy' - y] + [-2xy'' - 3y'] \\ &= \sum_{n=0}^{\infty} [2(s+n)(s+n-1) + 3(s+n) - 1] a_n x^{s+n} \\ &\quad - \sum_{n=0}^{\infty} [2(s+n)(s+n-1) + 3(s+n)] a_n x^{s+n-1} \\ &= \sum_{n=1}^{\infty} [2(s+n-1)(s+n-2) + 3(s+n-1) - 1] a_{n-1} x^{s+n-1} \\ &\quad - \sum_{n=0}^{\infty} [2(s+n)(s+n-1) + 3(s+n)] a_n x^{s+n-1} \\ &= -[2s(s-1) + 3s] a_0 x^{s-1} + \sum_{n=1}^{\infty} \{[2(s+n-1)(s+n-2) + 3(s+n-1) - 1] a_{n-1} \\ &\quad - [2(s+n)(s+n-1) + 3(s+n)] a_n\} x^{s+n-1}. \end{aligned}$$

Thus, the indicial equation is

$$I(s) = -[2s^2 + s] = -s(2s+1) = 0,$$

with roots $s_2 = -1/2$ and $s_1 = 0$. The remaining terms provide the recurrence relation

$$[2(s+n-1)(s+n-2)+3(s+n-1)-1]a_{n-1} = [2(s+n)(s+n-1)+3(s+n)]a_n, \quad n \geq 1.$$

With $s = s_1 = 0$, this becomes

$$\begin{aligned}
[2n(n-1)+3n]a_n &= [2(n-1)(n-2)+3(n-1)-1]a_{n-1} \\
(2n^2+n)a_n &= [2n^2-6n+4+3n-3-1]a_{n-1} \\
n(2n+1)a_n &= n(2n-3)a_{n-1} \\
a_n &= \frac{2n-3}{2n+1}a_{n-1},
\end{aligned}$$

for $n \geq 1$.

9. Multiply by x to obtain

$$x^2y'' + x(1-x)y' - xy = 0,$$

then compare with $x^2y'' + xp(x)y' + q(x)y = 0$ to get $p(x) = 1 - x$ and $q(x) = -x$, both analytic at $x = 0$. Hence, $x = 0$ is a regular singular point. Further, $p_0 = p(0) = 1$ and $q_0 = q(0) = 0$, so the indicial polynomial should be

$$I(s) = s^2 + (p_0 - 1)s + q_0 = s^2.$$

Let

$$y(x) = x^s \sum_{n=0}^{\infty} a_n x^n = \sum_{n=0}^{\infty} a_n x^{s+n}.$$

Then,

$$x^2y'' = \sum_{n=0}^{\infty} (s+n)(s+n-1)a_n x^{s+n},$$

and

$$x(1-x)y' = (1-x)\sum_{n=0}^{\infty}(s+n)a_n x^{s+n} = \sum_{n=0}^{\infty}(s+n)a_n x^{s+n} - \sum_{n=1}^{\infty}(s+n-1)a_{n-1}x^{s+n},$$

where we've shifted the index of summation. Finally,

$$-xy = -x\sum_{n=0}^{\infty} a_n x^{s+n} = -\sum_{n=1}^{\infty} a_{n-1}x^{s+n},$$

where we've again shifted the index of summation. Substituting,

$$\begin{aligned}
0 &= x^2y'' + x(1-x)y' - xy \\
&= \sum_{n=0}^{\infty}(s+n)(s+n-1)a_n x^{s+n} + \sum_{n=0}^{\infty}(s+n)a_n x^{s+n} - \sum_{n=1}^{\infty}(s+n-1)a_{n-1}x^{s+n} - \sum_{n=1}^{\infty}a_{n-1}x^{s+n} \\
&= [s(s-1)+s]\,a_0x^s + \sum_{n=1}^{\infty}\{[(s+n)(s+n-1)+(s+n)]\,a_n - [(s+n-1)+1]\,a_{n-1}\}\,x^{s+n} \\
&= s^2a_0x^s + \sum_{n=1}^{\infty}\left[(s+n)^2a_n - (s+n)a_{n-1}\right]x^{s+n}.
\end{aligned}$$

Thus, the indicial equation is $I(s) = s^2 = 0$, with repeated roots $s_2 = s_1 = 0$. The remaining terms provide the recurrence relation,

$$(s+n)^2 a_n - (s+n)a_{n-1} = 0,$$

which for $s = 0$ becomes

$$n^2 a_n = n a_{n-1} \quad \text{or} \quad a_n = \frac{a_{n-1}}{n}, \quad n \geq 1.$$

11. Multiplying both sides by $x/4$, we obtain

$$x^2 y'' + \frac{x}{2} y' + \frac{x}{4} y = 0.$$

Comparing with $x^2 y'' + xp(x)y' + q(x)y = 0$, we see that $p(x) = 1/2$ and $q(x) = x/4$, both of which are analytic near $x = 0$. We seek a solution having form

$$y = x^s \sum_{n=0}^{\infty} a_n x^n = \sum_{n=0}^{\infty} a_n x^{s+n},$$

$a_0 \neq 0$, with derivatives

$$y' = \sum_{n=0}^{\infty} a_n (s+n) x^{s+n-1} \quad \text{and} \quad y' = \sum_{n=0}^{\infty} a_n (s+n)(s+n-1) x^{s+n-2}.$$

Therefore,

$$4x^2 y'' = \sum_{n=0}^{\infty} 4a_n (s+n)(s+n-1) x^{s+n}, \quad 2xy' = \sum_{n=0}^{\infty} 2a_n (s+n) x^{s+n},$$

and

$$xy = \sum_{n=0}^{\infty} a_n x^{s+n-1} = \sum_{n=1}^{\infty} a_{n-1} x^{s+n}.$$

Substituting,

$$\begin{aligned}
0 &= 4x^2 y'' + 2xy' + xy \\
&= \sum_{n=0}^{\infty} 4a_n (s+n)(s+n-1) x^{s+n} + \sum_{n=0}^{\infty} 2a_n (s+n) x^{s+n} + \sum_{n=1}^{\infty} a_{n-1} x^{s+n} \\
&= [4s(s-1) + 2s] a_0 x^s + \sum_{n=1}^{\infty} \{[4(s+n)(s+n-1) + 2(s+n)] a_n + a_{n-1}\} x^{s+n}.
\end{aligned}$$

Setting coefficients equal to zero,

$$[4s(s-1) + 2s] a_0 = 0.$$

But $a_0 \neq 0$, so the indicial equation

$$4s(s-1) + 2s = 0 \quad \text{or} \quad 2s(2s-1) = 0$$

has roots $s = 0$ and $s = 1/2$. Further,

$$[4(s+n)(s+n-1) + 2(s+n)] a_n + a_{n-1} = 0 \quad \text{or} \quad a_n = \frac{-a_{n-1}}{(s+n)(4s+4n-2)},$$

for $n \geq 1$. With $s = 0$,

$$a_n = \frac{-a_{n-1}}{n(4n-2)} = \frac{-a_{n-1}}{2n(2n-1)}.$$

Indeed,

$$a_1 = \frac{-a_0}{2 \cdot 1}, \quad a_2 = \frac{-a_1}{4 \cdot 3} = \frac{a_0}{4!}, \quad a_3 = \frac{-a_2}{6 \cdot 5} = \frac{-a_0}{6!}, \quad \ldots \quad , a_n = \frac{a_0(-1)^n}{(2n)!},$$

for $n \geq 1$. Hence, with $a_0 = 1$, we have solution

$$y_1(x) = x^0 \left[1 + \sum_{n=1}^{\infty} \frac{(-1)^n x^n}{(2n)!} \right] = \sum_{n=0}^{\infty} \frac{(-1)^n x^n}{(2n)!}.$$

With $s = 1/2$,

$$a_n = \frac{-a_{n-1}}{(1/2+n)4n} = \frac{-a_{n-1}}{2n(2n+1)},$$

for $n \geq 1$. Hence,

$$a_1 = \frac{-a_0}{2 \cdot 3}, \quad a_2 = \frac{-a_1}{4 \cdot 5} = \frac{-a_0}{5!}, \quad a_3 = \frac{-a_2}{6 \cdot 7} = \frac{a_0}{7!}, \quad \ldots \quad , a_n = \frac{a_0(-1)^n}{(2n+1)!},$$

for $n \geq 1$. With $a_0 = 1$, we have solution

$$y_2(x) = x^{1/2} \left[1 + \sum_{n=1}^{\infty} \frac{(-1)^n x^n}{(2n+1)!} \right] = x^{1/2} \sum_{n=0}^{\infty} \frac{(-1)^n x^n}{(2n+1)!},$$

13. Dividing by 2 produces

$$x^2 y'' - \frac{3}{2} x y' + (1+x)y = 0,$$

which upon comparison with $x^2 y'' + x p(x) y' + q(x) y = 0$, reveal $p(x) = -3/2$ and $q(x) = 1 + x$, both of which are analytic at $x = 0$. We seek solution

$$y = x^s \sum_{n=0}^{\infty} a_n x^n = \sum_{n=0}^{\infty} a_n x^{s+n},$$

$a_0 \neq 0$, with derivatives

$$y' = \sum_{n=0}^{\infty} a_n (s+n) x^{s+n-1} \quad \text{and} \quad y'' = \sum_{n=0}^{\infty} a_n (s+n)(s+n-1) x^{s+n-2}.$$

Thus,

$$2x^2 y'' = \sum_{n=0}^{\infty} 2a_n (s+n)(s+n-1) x^{s+n}, \quad -3xy' = -\sum_{n=0}^{\infty} 3a_n (s+n) x^{s+n},$$

and

$$(2+2x)y = \sum_{n=0}^{\infty} 2a_n x^{s+n} + \sum_{n=0}^{\infty} 2a_n x^{s+n+1} = \sum_{n=0}^{\infty} 2a_n x^{s+n} + \sum_{n=1}^{\infty} 2a_{n-1} x^{s+n}.$$

Thus,

$$\begin{aligned}0 &= 2x^2y'' - 3xy' + (2+2x)y \\ &= \sum_{n=0}^{\infty} 2a_n(s+n)(s+n-1)x^{s+n} - \sum_{n=0}^{\infty} 3a_n(s+n)x^{s+n} + \sum_{n=0}^{\infty} 2a_n x^{s+n} + \sum_{n=1}^{\infty} 2a_{n-1}x^{s+n} \\ &= [2s(s-1) - 3s + 2]a_0x^s + \sum_{n=1}^{\infty}\{[2(s+n)(s+n-1) - 3(s+n) + 2]a_n + 2a_{n-1}\}x^{s+n}\end{aligned}$$

Setting coefficients equal to zero,

$$[2s(s-1) - 3s + 2]a_0 = 0.$$

But, $a_0 \neq 0$, so the indicial equation

$$0 = 2s(s-1) - 3s + 2 = (2s-1)(s-2)$$

has zeros $s = 1/2$ and $s = 2$. Further,

$$[2(s+n)(s+n-1) - 3(s+n) + 2]a_n + 2a_{n-1} = 0,$$

so

$$a_n = \frac{-2a_{n-1}}{2(s+n)(s+n-1) - 3(s+n) + 2}, \quad n \geq 1.$$

With $s = 2$,

$$a_n = \frac{-2a_{n-1}}{2(n+2)(n+1) - 3(n+2) + 2} = \frac{-2a_{n-1}}{n(2n+3)}, \quad n \geq 1.$$

Thus,

$$a_1 = \frac{-2a_0}{1\cdot 5}, \quad a_2 = \frac{-2a_1}{2\cdot 7} = \frac{2^2a_0}{(1\cdot 2)(5\cdot 7)}, \quad a_3 = \frac{-2a_2}{3\cdot 9} = \frac{-2^3a_0}{(1\cdot 2\cdot 3)(5\cdot 7\cdot 9)},$$

and in general,

$$a_n = \frac{(-1)^n 2^n a_0}{n!(5\cdot 7\cdot\dots\cdot(2n+3))}, \quad n \geq 1.$$

With $a_0 = 1$, we have solution

$$y_1(x) = x^2\left[1 + \sum_{n=1}^{\infty}\frac{(-1)^n 2^n x^n}{n!(5\cdot 7\cdot\dots\cdot(2n+3))}\right].$$

With $s = 1/2$,

$$a_n = \frac{-2a_{n-1}}{2(1/2+n)(n-1/2) - 3(1/2+n) + 2} = \frac{-2a_{n-1}}{n(2n-3)}.$$

Hence,

$$a_1 = \frac{-2a_0}{1\cdot(-1)}, \quad a_2 = \frac{-2a_1}{2\cdot 1} = \frac{2^2a_0}{(1\cdot 2)(-1\cdot 1)}, \quad a_3 = \frac{-2a_2}{3\cdot 3} = \frac{-2^3a_0}{(1\cdot 2\cdot 3)(-1\cdot 1\cdot 3)},$$

and in general,

$$a_n = \frac{(-1)^{n+1}2^n a_0}{n!(1\cdot 3\cdot\dots\cdot(2n-3))}, \quad n \geq 2.$$

With $a_0 = 1$, we have a second solution

$$y_2(x) = x^{1/2}\left[1 + 2x + \sum_{n=2}^{\infty} \frac{(-1)^{n+1}2^n x^n}{n!(1 \cdot 3 \cdot \cdots \cdot (2n-3))}\right].$$

15. Compare $x^2y'' + xy' + (x^2 - 1/16)y = 0$ with $x^2y'' + xp(x)y' + q(x)y = 0$ and note that $p(x) = 1$ and $q(x) = x^2 - 1/16$, both of which are analytic at $x = 0$. Thus, $x = 0$ is a regular singular point. Further, note that $p_0 = p(0) = 1$ and $q_0 = q(0) = -1/16$, so the indicial polynomial should be

$$I(s) = s^2 + (p_0 - 1)s + q_0 = s^2 - \frac{1}{16}.$$

Let

$$y(x) = x^s \sum_{n=0}^{\infty} a_n x^n = \sum_{n=0}^{\infty} a_n x^{s+n}.$$

Then,

$$x^2y'' = \sum_{n=0}^{\infty}(s+n)(s+n-1)a_n x^{s+n}, \quad xy' = \sum_{n=0}^{\infty}(s+n)a_n x^{s+n},$$

and

$$\left(x^2 - \frac{1}{16}\right)y = \left(x^2 - \frac{1}{16}\right)\sum_{n=0}^{\infty} a_n x^{s+n} = \sum_{n=2}^{\infty} a_{n-2}x^{s+n} - \frac{1}{16}\sum_{n=0}^{\infty} a_n x^{s+n},$$

where we've shifted the first index of summation. Substituting,

$$\begin{aligned}
0 &= x^2y'' + xy' + \left(x^2 - \frac{1}{16}\right)y \\
&= \sum_{n=0}^{\infty}(s+n)(s+n-1)a_n x^{s+n} + \sum_{n=0}^{\infty}(s+n)a_n x^{s+n} + \sum_{n=2}^{\infty} a_{n-2}x^{s+n} - \frac{1}{16}\sum_{n=0}^{\infty} a_n x^{s+n} \\
&= \left[s(s-1) + s - \frac{1}{16}\right]a_0 x^s + \left[(s+1)s + (s+1) - \frac{1}{16}\right]a_1 x^{s+1} \\
&\quad + \sum_{n=2}^{\infty}\left\{\left[(s+n)(s+n-1) + (s+n) - \frac{1}{16}\right]a_n + a_{n-2}\right\}x^{s+n}.
\end{aligned}$$

The first term gives the indicial equation

$$I(s) = s^2 - \frac{1}{16} = \left(s + \frac{1}{4}\right)\left(s - \frac{1}{4}\right),$$

with roots $s_2 = -1/4$ and $s_1 = 1/4$. Setting the second term equal to zero,

$$\left(s^2 + 2s + \frac{15}{16}\right)a_1 = 0,$$

provides $a_1 = 0$ as the coefficient of a_1 is not zero for either root of the indicial equation. The remaining terms provide the recurrence relation,

$$\left[(s+n)^2 - \frac{1}{16}\right]a_n + a_{n-2} = 0, \quad n \geq 2.$$

Thus,

$$a_n = \frac{-a_{n-2}}{(s+n+1/4)(s+n-1/4)} = \frac{-16a_{n-2}}{(4s+4n+1)(4s+4n-1)}$$

for $n \geq 2$. With $s_1 = 1/4$, this becomes

$$a_n = \frac{-16a_{n-2}}{(4n+2)(4n)} = \frac{-2a_{n-2}}{(2n+1)n}, \quad n \geq 2.$$

Thus,

$$a_2 = \frac{-2a_0}{5 \cdot 2}, \quad a_4 = \frac{-2a_2}{9 \cdot 4}, \quad \ldots \quad , a_{2n} = \frac{-2a_{2n-2}}{(4n+1)(2n)}.$$

Multiplying the left and right-hand sides of these equations,

$$a_2\, a_4 \cdots a_{2n} = \frac{(-2)^n}{[5 \cdot 9 \cdot \ldots \cdot (4n+1)][2 \cdot 4 \cdot \ldots \cdot (2n)]|} a_0\, a_2 \cdots a_{2n-2}.$$

Cancelling,

$$a_{2n} = \frac{(-1)^n}{[5 \cdot 9 \cdot \ldots \cdot (4n+1)][1 \cdot 2 \cdot \ldots \cdot n]} a_0 \quad \text{or} \quad a_{2n} = \frac{(-1)^n}{[5 \cdot 9 \cdot \ldots \cdot (4n+1)]n!} a_0, \quad n \geq 1.$$

Because $a_1 = 0$, it is easily seen that $a_{2n+1} = 0$, for $n \geq 0$. That is, all odd coefficients are zero. Thus, with $s_1 = 1/4$ and $a_0 = 1$, we have

$$y_1(x) = x^{1/4}\left[1 + \sum_{n=1}^{\infty} \frac{(-1)^n x^{2n}}{[5 \cdot 9 \cdots (4n+1)]n!}\right].$$

In similar fashion, with $s_2 = -1/4$,

$$a_n = \frac{-a_{n-2}}{n\left(n - \frac{1}{2}\right)} = \frac{-2a_{n-2}}{n(2n-1)}, \quad n \geq 2.$$

Again, because $a_1 = 0$, all odd coefficients equal zero, but

$$a_2 = \frac{-2a_0}{2 \cdot 3}, \quad a_4 = \frac{-2a_2}{4 \cdot 7}, \quad \ldots \quad , a_{2n} = \frac{-2a_{2n-2}}{2n(4n-1)}, \quad n \geq 1.$$

Multiplying left and right-hand sides,

$$a_2\, a_4 \cdots a_{2n} = \frac{(-2)^n}{[2 \cdot 4 \cdot \ldots \cdot 2n][3 \cdot 7 \cdot \ldots \cdot (4n-1)]} a_0\, a_2 \cdots a_{2n-2}.$$

Cancelling,

$$a_{2n} = \frac{(-1)^n}{[1 \cdot 2 \cdot \ldots \cdot n][3 \cdot 7 \cdot \ldots \cdot (4n-1)]} a_0 = \frac{(-1)^n}{n![3 \cdot 7 \cdot \ldots \cdot (4n-1)]} a_0.$$

Thus, with $s_2 = -1/4$ and $a_0 = 1$, we have

$$y_2(x) = x^{-1/4}\left[1 + \sum_{n=1}^{\infty} \frac{(-1)^n x^{2n}}{n![3 \cdot 7 \cdots (4n-1)]}\right].$$

17. Dividing by 2,

$$x^2y'' - \frac{x}{2}y' + \frac{x+1}{2}y = 0,$$

then comparing with $x^2y'' + xp(x)y' + q(x)y = 0$ reveals $p(x) = -1/2$ and $q(x) = (x+1)/2$, both of which are analytic at $x = 0$. Hence, $x = 0$ is a regular singular point. Further, note that $p_0 = p(0) = -1/2$ and $q_0 = q(0) = 1/2$, so the indicial polynomial should be

$$I(s) = s^2 + (p_0 - 1)s + q_0 = s^2 - \frac{3}{2}s + \frac{1}{2}.$$

Let

$$y(x) = x^s \sum_{n=0}^{\infty} a_n x^n = \sum_{n=0}^{\infty} a_n x^{s+n}.$$

We'll find the equation $2x^2y'' - xy' + (x+1)y = 0$ easier to deal with. Thus,

$$2x^2y'' = \sum_{n=0}^{\infty} 2(s+n)(s+n-1)a_n x^{s+n}, \quad -xy' = -\sum_{n=0}^{\infty}(s+n)a_n x^{s+n},$$

and

$$(x+1)y = (x+1)\sum_{n=0}^{\infty} a_n x^{s+n} = \sum_{n=1}^{\infty} a_{n-1}x^{s+n} + \sum_{n=0}^{\infty} a_n x^{s+n},$$

where we've shifted the index of the first summation. Substituting,

$$\begin{aligned}
0 &= 2x^2y'' - xy' + (x+1)y \\
&= \sum_{n=0}^{\infty} 2(s+n)(s+n-1)a_n x^{s+n} - \sum_{n=0}^{\infty}(s+n)a_n x^{s+n} + \sum_{n=1}^{\infty} a_{n-1}x^{s+n} + \sum_{n=0}^{\infty} a_n x^{s+n} \\
&= [2s(s-1) - s + 1]a_0 x^s + \sum_{n=1}^{\infty}\{[2(s+n)(s+n-1) - (s+n) + 1]a_n + a_{n-1}\}x^{s+n}.
\end{aligned}$$

Thus, the indicial equation is

$$I(s) = 2s^2 - 3s + 1 = (2s-1)(s-1) = 0,$$

with roots $s_1 = 1/2$ and $s_1 = 1$. The remaining terms provide the recurrence relation

$$(2s + 2n - 1)(s + n - 1)a_n + a_{n-1} = 0, \quad n \geq 1.$$

With $s_1 = 1$, this becomes

$$n(2n+1)a_n = -a_{n-1} \quad \text{or} \quad a_n = \frac{-a_{n-1}}{n(2n+1)}.$$

Thus,

$$a_1 = \frac{-a_0}{1 \cdot 3}, \quad a_2 = \frac{-a_1}{2 \cdot 5}, \quad \ldots \quad , a_n = \frac{-a_{n-1}}{n(2n+1)}.$$

Multiplying left and right-hand sides of these equations,

$$a_1\, a_2 \cdots a_n = \frac{(-1)^n}{[1 \cdot 2 \cdot \ldots \cdot n][3 \cdot 5 \cdot \ldots \cdot (2n+1)]} a_0\, a_1 \cdots a_{n-1}.$$

Cancelling,

$$a_n = \frac{(-1)^n}{n![3 \cdot 5 \cdot \ldots \cdot (2n+1)]} a_0.$$

We can simplify a bit further by noticing that

$$3 \cdot 5 \cdot \ldots \cdot (2n+1) = \frac{2 \cdot 3 \cdot 5 \ldots \cdot (2n+1)}{2 \cdot 4 \cdot \ldots \cdot (2n)} = \frac{(2n+1)!}{2^n \cdot n!}$$

Thus,

$$a_n = \frac{(-1)^n 2^n n!}{n!(2n+1)!} a_0 = \frac{(-1)^n 2^n}{(2n+1)!} a_0.$$

Thus, with $s_1 = 1$, and $a_0 = 1$, we have one solution

$$y_1(x) = x^1 \sum_{n=0}^{\infty} \frac{(-1)^n 2^n}{(2n+1)!} x^n.$$

With $s_2 = 1/2$, the recurrence relation becomes

$$2n\left(n - \frac{1}{2}\right) a_n + a_{n-1} = 0 \quad \text{or} \quad a_n = \frac{-a_{n-1}}{n(2n-1)}, \qquad n \geq 1.$$

Thus,

$$a_1 = \frac{-a_0}{1 \cdot 1}, \quad a_2 = \frac{-a_1}{2 \cdot 3}, \quad \ldots \quad a_n = \frac{-a_{n-1}}{n(2n-1)}.$$

Multiplying left and right-hand sides together, we get

$$a_1 \, a_2 \cdots a_n = \frac{(-1)^n}{[1 \cdot 2 \cdot \ldots \cdot n][1 \cdot 3 \cdot \ldots \cdot (2n-1)]} a_0 \, a_1 \cdots a_{n-1}.$$

Cancelling,

$$a_n = \frac{(-1)^n}{n![1 \cdot 3 \cdot \ldots \cdot (2n-1)]} a_0.$$

We can simplify a bit further by noticing that

$$1 \cdot 3 \cdot \ldots \cdot (2n-1) = \frac{1 \cdot 2 \cdot 3 \cdot 4 \cdot \ldots \cdot (2n-1)}{2 \cdot 4 \cdot \ldots \cdot (2n-2)} = \frac{(2n-1)!}{2^{n-1}[1 \cdot 2 \cdot \ldots \cdot (n-1)]} = \frac{(2n-1)!}{2^{n-1}(n-1)!}.$$

Thus,

$$a_n = \frac{(-1)^n 2^{n-1}(n-1)!}{n!(2n-1)!} a_0 = \frac{(-1)^n 2^{n-1}}{n(2n-1)!} a_0.$$

Thus, with $s_2 = 1/2$ and $a_0 = 1$, we have a second independent solution

$$y_2(x) = x^{1/2}\left[1 + \sum_{n=1}^{\infty} \frac{(-1)^n 2^{n-1}}{n(2n-1)!} x^n\right].$$

19. Multiply by $x/(2(1+x))$,

$$x^2y'' + \frac{x}{2(1+x)}y' - \frac{x}{2(1+x)}y = 0,$$

then compare with $x^2y'' + xp(x)y' + q(x)y = 0$ and note that $p(x) = 1/(2(1+x))$ and $q(x) = -x/(2(1+x))$ are both analytic at $x = 0$, making $x = 0$ a regular singular point. Further, $p_0 = p(0) = 1/2$ and $q_0 = q(0) = 0$, so the indicial polynomial should be

$$I(s) = s^2 + (p_0 - 1)s + q_0 = s^2 - \frac{1}{2}s.$$

Let

$$y(x) = x^s \sum_{n=0}^{\infty} a_n x^n = \sum_{n=0}^{\infty} a_n x^{s+n}.$$

We'll find the equation $2(x + x^2)y'' + y' - y = 0$ easier to deal with. Thus,

$$\begin{aligned}
0 &= 2(x+x^2)y'' + y' - y \\
&= \sum_{n=0}^{\infty} 2(s+n)(s+n-1)a_n x^{s+n-1} + \sum_{n=1}^{\infty} 2(s+n-1)(s+n-2)a_{n-1}x^{s+n-1} \\
&\quad + \sum_{n=0}^{\infty}(s+n)a_n x^{s+n-1} - \sum_{n=1}^{\infty} a_{n-1}x^{s+n-1},
\end{aligned}$$

where we've shifted the index of the second summation and fourth summations. Thus,

$$\begin{aligned}
0 &= [2s(s-1) + s]a_0 x^{s-1} \\
&\quad + \sum_{n=1}^{\infty}\{[2(s+n)(s+n-1) + (s+n)]a_n + [2(s+n-1)(s+n-2) - 1]a_{n-1}\}x^{s+n-1},
\end{aligned}$$

and the indicial equation is

$$I(s) = 2s^2 - s = s(2s-1) = 0,$$

with roots $s_2 = 0$ and $s_1 = 1/2$. The remaining terms provide the recurrence relation

$$(s+n)(2s+2n-1)a_n + [2(s+n-1)(s+n-2) - 1]a_{n-1} = 0,$$

for $n \geq 1$. With $s_1 = 1/2$, this becomes

$$a_n = -\frac{2n^2 - 4n + 1/2}{n(2n+1)}a_{n-1}, \quad n \geq 1.$$

Thus,

$$a_1 = -\frac{-3/2}{1 \cdot 3}a_0, \quad a_2 = -\frac{1/2}{2 \cdot 5}a_1, \quad a_3 = -\frac{13/2}{3 \cdot 7}a_2, \quad \ldots$$

etc. Thus,

$$a_1 a_2 a_3 \ldots a_n = (-1)^n \frac{\frac{(-3)}{2} \cdot \frac{1}{2} \cdot \frac{13}{2} \cdots \frac{4n^2 - 8n + 1}{2}}{n!(3 \cdot 5 \cdot 7 \cdots (2n+1))} a_0 a_1 a_2 \ldots a_{n-1}.$$

Cancelling,

$$a_n = (-1)^n \frac{(-3) \cdot 1 \cdot 13 \cdots (4n^2 - 8n + 1)}{2^n n!(3 \cdot 5 \cdot 7 \cdots (2n + 1))} a_0, \quad n \geq 1.$$

But,

$$3 \cdot 5 \cdot 7 \cdots (2n + 1) = \frac{2 \cdot 3 \cdot 4 \cdot 5 \cdot 6 \cdot 7 \cdots 2n(2n + 1)}{2 \cdot 4 \cdot 6 \cdots 2n} = \frac{(2n + 1)!}{2^n n!}.$$

Thus,

$$a_n = (-1)^n \frac{(-3) \cdot 1 \cdot 13 \cdots (4n^2 - 8n + 1)}{(2n + 1)!} a_0.$$

With $a_0 = 1$ and $s_1 = 1/2$,

$$y_1(x) = x^{1/2} \left[1 + \sum_{n=1}^{\infty} \frac{(-1)^n(-3) \cdot 1 \cdot 13 \cdots (4n^2 - 8n + 1)}{(2n + 1)!} x^n \right].$$

With $s_2 = 0$, the recurrence relation becomes

$$a_n = -\frac{2n^2 - 6n + 3}{n(2n - 1)}, \quad n \geq 1.$$

Thus,

$$a_1 = -\frac{-1}{1 \cdot 1} a_0, \quad a_2 = -\frac{-1}{2 \cdot 3} a_1, \quad a_3 = -\frac{3}{3 \cdot 5} a_2, \quad \ldots$$

etc. Thus,

$$a_1 \, a_2 \, a_3 \cdots a_n = (-1)^n \frac{(-1)(-1)(3) \cdots (2n^2 - 6n + 3)}{n!(1 \cdot 3 \cdot 5 \cdots (2n - 1))} a_0 \, a_1 \, a_2 \cdots a_{n-1}.$$

Cancelling,

$$a_n = (-1)^n \frac{(-1) \cdot (-1) \cdot 3 \cdot 11 \cdots (2n^2 - 6n + 3)}{n!(1 \cdot 3 \cdot 5 \cdots (2n - 1))} a_0.$$

But $1 \cdot 3 \cdot 5 \cdots (2n - 1) = (2n)!/(2^n n!)$, so

$$a_n = \frac{(-1)^n 2^n ((-1) \cdot (-1) \cdot 3 \cdot 11 \cdots (2n^2 - 6n + 3))}{(2n)!} a_0.$$

Thus, with $s_2 = 0$ and $a_0 = 1$,

$$y_2(x) = 1 + \sum_{n=1}^{\infty} \frac{(-1)^n 2^n \cdot (-1) \cdot (-1) \cdot 3 \cdot 11 \cdots (2n^2 - 6n + 3)}{(2n)!} x^n.$$

21. Multiply by $x/2$,

$$x^2 y'' + \frac{1}{2} x y' + \frac{1}{2} x y = 0,$$

then compare with $x^2 y'' + x p(x) y' + q(x) y = 0$. Note that $p(x) = 1/2$ and $q(x) = x/2$, both analytic at $x = 0$, so $x = 0$ is a regular singular point. Further, note that $p_0 = p(0) = 1/2$ and $q_0 = q(0) = 0$, so the indicial polynomial is

$$I(s) = s^2 + (p_0 - 1)s + q_0 = s^2 - \frac{1}{2} s.$$

Let

$$y(x) = x^s \sum_{n=0}^{\infty} a_n x^n = \sum_{n=0}^{\infty} a_n x^{s+n}.$$

We'll find the equation $2x^2y'' + xy' + xy = 0$ easier to deal with. Thus,

$$2x^2y'' = \sum_{n=0}^{\infty} 2(s+n)(s+n-1)a_n x^{s+n}, \quad xy' = \sum_{n=0}^{\infty} (s+n)a_n x^{s+n},$$

and

$$xy = x\sum_{n=0}^{\infty} a_n x^{s+n} = \sum_{n=1}^{\infty} a_{n-1} x^{s+n},$$

where we've shifted the index of the last summation. Substituting,

$$\begin{aligned}
0 &= 2x^2y'' + xy' + xy \\
&= \sum_{n=0}^{\infty} 2(s+n)(s+n-1)a_n x^{s+n} + \sum_{n=0}^{\infty} (s+n)a_n x^{s+n} + \sum_{n=1}^{\infty} a_{n-1} x^{s+n} \\
&= [2s(s-1)+s]a_0 x^s + \sum_{n=1}^{\infty} \{[2(s+n)(s+n-1)+(s+n)]a_n + a_{n-1}\} x^{s+n} \\
&= (2s^2 - s)a_0 x^s + \sum_{n=1}^{\infty} \{[2(s+n)^2 - (s+n)]a_n + a_{n-1}\} x^{s+n}.
\end{aligned}$$

Thus, the indicial equation is

$$I(s) = 2s^2 - s = s(2s-1) = 0$$

with roots $s_2 = 0$ and $s_1 = 1/2$. The remaining terms provide the recurrence relation

$$(s+n)(2s+2n-1)a_n + a_{n-1} = 0, \quad n \geq 1.$$

With $s_1 = 1/2$, this becomes

$$n(2n+1)a_n = -a_{n-1} \quad \text{or} \quad a_n = \frac{-a_{n-1}}{n(2n+1)}.$$

Thus,

$$a_1 = \frac{-a_0}{1 \cdot 3}, \quad a_2 = \frac{-a_1}{2 \cdot 5}, \quad \ldots \quad a_n = \frac{-a_{n-1}}{n(2n+1)}.$$

Multiplying left and right-hand sides of these equations,

$$a_1 a_2 \cdots a_n = \frac{(-1)^n a_0 a_1 \cdots a_{n-1}}{[1 \cdot 2 \cdot \ldots \cdot n][3 \cdot 5 \cdot \ldots \cdot (2n+1)]}.$$

Cancelling,

$$a_n = \frac{(-1)^n a_0}{n![3 \cdot 5 \cdot \ldots \cdot (2n+1)]}.$$

We can further simplify by noticing

$$3 \cdot 5 \cdot \ldots \cdot (2n+1) = \frac{2 \cdot 3 \cdot 4 \cdot 5 \cdot \ldots \cdot (2n+1)}{2 \cdot 4 \cdot \ldots \cdot (2n)} = \frac{(2n+1)!}{2^n n!}.$$

Thus,

$$a_n = \frac{(-1)^n n! 2^n a_0}{n!(2n+1)!} = \frac{(-1)^n 2^n a_0}{(2n+1)!}.$$

Thus, with $s_1 = 1/2$ and $a_0 = 1$, one solution is

$$y_1(x) = x^{1/2} \sum_{n=0}^{\infty} \frac{(-1)^n 2^n x^n}{(2n+1)!}.$$

Next, with $s_2 = 0$, the recurrence relation becomes

$$n(2n-1)a_n + a_{n-1} = 0 \quad \text{or} \quad a_n = \frac{-a_{n-1}}{n(2n-1)}, \quad n \geq 1.$$

Thus,

$$a_1 = \frac{-a_0}{1 \cdot 1}, \quad a_2 = \frac{-a_1}{2 \cdot 3}, \quad \ldots \quad , a_n = \frac{-a_{n-1}}{n(2n-1)}.$$

Multiply the left and right-hand sides of these equation,

$$a_1 a_2 \cdots a_n = \frac{(-1)^n a_0 a_1 \cdots a_{n-1}}{[1 \cdot 2 \cdot \ldots \cdot n][1 \cdot 3 \cdot \ldots \cdot (2n-1)]}.$$

Cancelling,

$$a_n = \frac{(-1)^n a_0}{n![1 \cdot 3 \cdot \ldots \cdot (2n-1)]}.$$

We can simplify further by noticing that

$$1 \cdot 3 \cdot \ldots \cdot (2n-1) = \frac{1 \cdot 2 \cdot 3 \cdot 4 \cdot \ldots \cdot (2n-1) \cdot 2n}{2 \cdot 4 \cdot \ldots \cdot 2n} = \frac{(2n)!}{2^n n!}.$$

Thus,

$$a_n = \frac{(-1)^n 2^n n! a_0}{n!(2n)!} = \frac{(-1)^n 2^n a_0}{(2n)!}.$$

Thus, with $s_1 = 0$ and $a_0 = 1$, we get a second solution

$$y_2(x) = \sum_{n=0}^{\infty} \frac{(-1)^n 2^n x^n}{(2n)!} = \sum_{n=0}^{\infty} \frac{(-2x)^n}{(2n)!}.$$

23. Divide by 2,

$$x^2 y'' - \frac{1}{2} x y' + \frac{1-x^2}{2} y = 0,$$

then compare with $x^2y'' + xp(x)y' + q(x)y = 0$. Note that $p(x) = -1/2$ and $q(x) = (1-x^2)/2$, both analytic at $x = 0$, so $x = 0$ is a regular singular point. Further, note that $p_0 = p(0) = -1/2$ and $q_0 = q(0) = 1/2$, so the indicial polynomial is

$$I(s) = s^2 + (p_0 - 1)s + q_0 = s^2 - \frac{3}{2}s + \frac{1}{2}.$$

Let

$$y(x) = x^s \sum_{n=0}^{\infty} a_n x^n = \sum_{n=0}^{\infty} a_n x^{s+n}.$$

We'll find the equation $2x^2y'' - xy' + (1-x^2)y = 0$ easier to deal with. Thus,

$$2x^2y'' = \sum_{n=0}^{\infty} 2(s+n)(s+n-1)a_n x^{s+n}, \quad -xy' = -\sum_{n=0}^{\infty}(s+n)a_n x^{s+n},$$

and

$$(1-x^2)y = (1-x^2)\sum_{n=0}^{\infty} a_n x^{s+n} = \sum_{n=0}^{\infty} a_n x^{s+n} - \sum_{n=2}^{\infty} a_{n-2} x^{s+n},$$

where we've shifted the index on the second summation. Substituting,

$$\begin{aligned}
0 &= 2x^2y'' - xy' + (1-x^2)y \\
&= \sum_{n=0}^{\infty} 2(s+n)(s+n-1)a_n x^{s+n} - \sum_{n=0}^{\infty}(s+n)a_n x^{s+n} + \sum_{n=0}^{\infty} a_n x^{s+n} - \sum_{n=2}^{\infty} a_{n-2} x^{s+n} \\
&= [2s(s-1) - s + 1]a_0 x^s + [2(s+1)s - (s+1) + 1]a_1 x^{s+1} \\
&\quad + \sum_{n=2}^{\infty}\{[2(s+n)(s+n-1) - (s+n) + 1]a_n - a_{n-2}\}x^{s+n}.
\end{aligned}$$

Thus, the indicial equation is

$$I(s) = 2s^2 - 3s + 1 = (2s-1)(s-1) = 0,$$

with roots $s_1 = 1/2$ and $s_1 = 1$. The remaining terms provide the recurrence relation

$$(2s + 2n - 1)(s + n - 1)a_n - a_{n-2} = 0, \quad n \ge 2.$$

With $s_1 = 1$, this becomes

$$(2n+1)na_n - a_{n-2} = 0 \quad \text{or} \quad a_n = \frac{a_{n-2}}{n(2n+1)}.$$

Thus,

$$a_2 = \frac{a_0}{2 \cdot 5}, \quad a_4 = \frac{a_2}{4 \cdot 9}, \quad \ldots \quad a_{2n} = \frac{a_{2n-2}}{2n(4n+1)}.$$

Multiplying left and right-hand sides of these equations,

$$a_2 a_4 \cdots a_{2n} = \frac{a_0 a_2 \cdots a_{2n-2}}{[2 \cdot 4 \cdot \ldots \cdot 2n][5 \cdot 9 \cdot \ldots \cdot (4n+1)]}.$$

Cancelling,

$$a_{2n} = \frac{a_0}{2^n n![5 \cdot \ldots \cdot (4n+1)]}.$$

The coefficient of $[2(s+1)s - (s+1) + 1]a_1$ must also equal zero, but with $s_1 = 1$, this becomes $3a_1 = 0$, so $a_1 = 0$. With the recurrence relation, this implies that $a_{2n+1} = 0$ for all n. Thus, with $s_1 = 1$ and $a_0 = 1$, one solution is

$$y_1(x) = x\left[1 + \sum_{n=1}^{\infty} \frac{x^{2n}}{2^n n![5 \cdot 9 \cdots (4n+1)]}\right].$$

Next, with $s_2 = 1/2$, the recurrence relation becomes

$$n(2n-1)a_n - a_{n-2} = 0 \quad \text{or} \quad a_n = \frac{a_{n-2}}{n(2n-1)}, \qquad n \geq 2.$$

Thus,

$$a_2 = \frac{a_0}{2 \cdot 3}, \qquad a_4 = \frac{a_2}{4 \cdot 7}, \qquad \ldots \qquad, a_{2n} = \frac{a_{2n-2}}{2n(4n-1)}.$$

Multiplying the left and right-hand sides of these equations,

$$a_2\, a_4 \cdots a_{2n} = \frac{a_0\, a_2 \cdots a_{2n-2}}{[2 \cdot 4 \cdot \ldots \cdot 2n][3 \cdot 7 \cdot \ldots \cdot (4n-1)]}.$$

Cancelling,

$$a_{2n} = \frac{a_0}{2^n n![3 \cdot 7 \cdot \ldots \cdot (4n-1)]}.$$

Again, the coefficient of $[2(s+1)s - (s+1) + 1]a_1$ must equal zero, but with $s_2 = 1/2$, this becomes $a_1 = 0$. With the recurrence relation, this implies that $a_{2n+1} = 0$ for all n. Thus, with $s_2 = 1/2$ and $a_0 = 1$, a second independent solution is

$$y_2(x) = x^{1/2}\left[1 + \sum_{n=1}^{\infty} \frac{x^{2n}}{2^n n![3 \cdot 7 \cdot \ldots \cdot (4n-1)]}\right].$$

25. Multiply by $x/2$,

$$x^2 y'' + \frac{1}{2}xy' + \frac{1}{2}x^3 y = 0,$$

then compare with $x^2y'' + xp(x)y' + q(x)y = 0$. Note that $p(x) = 1/2$ and $q(x) = x^3/2$, both analytic at $x = 0$, making $x = 0$ a regular singular point. Further, note that $p_0 = p(0) = 1/2$ and $q_0 = q(0) = 0$. Thus, the indicial polynomial is

$$I(s) = s^2 + (p_0 - 1)s + q_0 = s^2 - \frac{1}{2}s.$$

It will be easier to work with the equation $2x^2y'' + xy' + x^3y = 0$. Let

$$y(x) = x^s \sum_{n=0}^{\infty} a_n x^n = \sum_{n=0}^{\infty} a_n x^{s+n}.$$

Then,

$$2x^2y'' = \sum_{n=0}^{\infty} 2(s+n)(s+n-1)a_n x^{s+n}, \quad xy' = \sum_{n=0}^{\infty} (s+n)a_n x^{s+n},$$

and

$$x^3 y = x^3 \sum_{n=0}^{\infty} a_n x^{s+n} = \sum_{n=3}^{\infty} a_{n-3} x^{s+n},$$

where we've shifted the index of summation. Substituting,

$$\begin{aligned}
0 &= 2x^2 y' + xy' + x^3 y \\
&= \sum_{n=0}^{\infty} 2(s+n)(s+n-1) a_n x^{s+n} + \sum_{n=0}^{\infty} (s+n) a_n x^{s+n} + \sum_{n=3}^{\infty} a_{n-3} x^{s+n} \\
&= [2s(s-1) + s] a_0 x^s + [2(s+1)s + (s+1)] a_1 x^{s+1} + [2(s+2)(s+1) + (s+2)] a_2 x^{s+2} \\
&\quad + \sum_{n=3}^{\infty} \{[2(s+n)(s+n-1) + (s+n)] a_n + a_{n-3}\} x^{s+n}.
\end{aligned}$$

Thus, the indicial equation is

$$I(s) = 2s^2 - s = s(2s-1) = 0,$$

with roots $s_2 = 0$ and $s_1 = 1/2$. The remaining terms provide the recurrence relation

$$(s+n)(2s+2n-1) a_n + a_{n-3} = 0, \quad n \geq 3.$$

With $s_1 = 1/2$, this becomes

$$n(2n+1) a_n + a_{n-3} = 0 \quad \text{or} \quad a_n = \frac{-a_{n-3}}{n(2n+1)}.$$

Thus,

$$a_3 = \frac{-a_0}{3 \cdot 7}, \quad a_6 = \frac{-a_3}{6 \cdot 13}, \quad \ldots \quad , a_{3n} = \frac{-a_{3n-3}}{3n(6n+1)}.$$

Multiplying left and right-hand sides,

$$a_3 \, a_6 \cdots a_{3n} = \frac{(-1)^n a_0 \, a_3 \cdots a_{3n-3}}{[3 \cdot 6 \cdot \ldots \cdot 3n][7 \cdot 13 \cdot \ldots \cdot (6n+1)]}.$$

Cancelling,

$$a_{3n} = \frac{(-1)^n a_0}{3^n n! [7 \cdot 13 \cdot \ldots \cdot (6n+1)]}.$$

We also need $[2(s+1)s + (s+1)]a_1 = 0$, but with $s = 1/2$, this becomes $3a_1 = 0$ so $a_1 = 0$. Thus, coupled with the recurrence relation means that $a_{3n+1} = 0$ for all n.

Similarly, $[2(s+2)(s+1) + (s+2)]a_2 = 0$ becomes $10a_2 = 0$, so $a_2 = 0$. The recurrence relation then reveals that $a_{3n+2} = 0$ for all n. Thus, with $s_1 = 1/2$ and $a_0 = 1$, one solution is

$$y_1(x) = x^{1/2} \left[1 + \sum_{n=1}^{\infty} \frac{(-1)^n x^{3n}}{3^n n! [7 \cdot 13 \cdot \ldots \cdot (6n+1)]} \right].$$

Next, with $s_2 = 0$, the recurrence relation becomes

$$n(2n-1) a_n + a_{n-3} = 0 \quad \text{or} \quad a_n = \frac{-a_{n-3}}{n(2n-1)}, \quad n \geq 3.$$

Thus,

$$a_3 = \frac{-a_0}{3 \cdot 5}, \quad a_6 = \frac{-a_3}{6 \cdot 11}, \quad \dots \quad , a_{3n} = \frac{-a_{3n-3}}{3n(6n-1)}.$$

Multiplying left and right-hand sides,

$$a_3\, a_6 \cdots a_{3n} = \frac{(-1)^n a_0\, a_3 \cdots a_{3n-3}}{[3 \cdot 6 \cdot \ldots \cdot 3n][5 \cdot 11 \cdot \ldots \cdot (6n-1)]}.$$

Cancelling,

$$a_{3n} = \frac{(-1)^n a_0}{3^n n! [5 \cdot 11 \cdot \ldots \cdot (6n-1)]}.$$

With $s_2 = 0$, $[2(s+1)s + (s+1)]a_1 = 0$, becomes $a_1 = 0$. The recurrence relation then gives $a_{3n+1} = 0$ for all n. With $s_2 = 0$, $[2(s+2)(s+1) + (s+2)]a_2 = 0$ becomes $6a_2 = 0$. Thus, $a_2 = 0$ and the recurrence relation then shows that $a_{3n+2} = 0$ for all n. Hence, with $s_2 = 0$ and $a_0 = 1$, a second independent solution is

$$y_2(x) = 1 + \sum_{n=1}^{\infty} \frac{(-1)^n x^{3n}}{3^n n! [5 \cdot 11 \cdot \ldots \cdot (6n-1)]}.$$

27. Compare

$$x^2 y'' + xy' + x^2 y = 0$$

with $x^2 y'' + xp(x)y' + q(x)y = 0$ and note that $p(x) = 1$ and $q(x) = x^2$, both of which are analytic at $x = 0$. Hence, $x = 0$ is a regular singular point. Further, $p_0 = p(0) = 1$ and $q_0 = q(0) = 0$, so the indicial polynomial is

$$I(s) = s^2 + (p_0 - 1)s + q_0 = s^2.$$

Let

$$y(x) = x^s \sum_{n=0}^{\infty} a_n x^n = \sum_{n=0}^{\infty} a_n x^{s+n}.$$

Then,

$$x^2 y'' = \sum_{n=0}^{\infty} (s+n)(s+n-1) a_n x^{s+n}, \quad xy' = \sum_{n=0}^{\infty} (s+n) a_n x^{s+n},$$

and

$$x^2 y = x^2 \sum_{n=0}^{\infty} a_n x^{s+n} = \sum_{n=2}^{\infty} a_{n-2} x^{s+n},$$

where we've shifted the index of summation. Substituting,

$$\begin{aligned}
0 &= x^2 y'' + xy' + x^2 y \\
&= \sum_{n=0}^{\infty} (s+n)(s+n-1) a_n x^{s+n} + \sum_{n=0}^{\infty} (s+n) a_n x^{s+n} + \sum_{n=2}^{\infty} a_{n-2} x^{s+n} \\
&= [s(s-1) + s]a_0 x^s + [(s+1)s + (s+1)]a_1 x^{s+1} + \sum_{n=2}^{\infty} \{[(s+n)(s+n-1) + (s+n)]a_n + a_{n-2}\} x^{s+n}.
\end{aligned}$$

Thus, the indicial equation is

$$I(s) = s^2 = 0,$$

with roots $s_2 = s_1 = 0$. We also need $[(s+1)s + (s+1)]a_1 = 0$, but with $s = 0$, this becomes $a_1 = 0$. The remaining terms provide the recurrence relation

$$(s+n)^2 a_n + a_{n-2} = 0, \quad n \geq 2.$$

With $s = 0$, this becomes

$$n^2 a_n + a_{n-2} = 0 \quad \text{or} \quad a_n = -\frac{a_{n-2}}{n^2}, \quad n \geq 2.$$

Thus,

$$a_2 = \frac{-a_0}{2^2}, \quad a_4 = \frac{-a_2}{4^2}, \quad \ldots \quad , a_{2n} = \frac{-a_{2n-2}}{(2n)^2}$$

Multiplying left and right-hand sides of these equations,

$$a_2\, a_4 \cdots a_{2n} = \frac{(-1)^n a_0\, a_2 \cdots a_{2n-2}}{2^2 \cdot 4^2 \cdot \ldots \cdot (2n)^2}.$$

Cancelling,

$$a_{2n} = \frac{(-1)^n a_0}{(2 \cdot 4 \cdot \ldots \cdot 2n)^2} = \frac{(-1)^n a_0}{[2^n(1 \cdot 2 \cdot \ldots \cdot n)]^2} = \frac{(-1)^n a_0}{2^{2n}(n!)^2}.$$

Finally, because $a_1 = 0$, the recurrence relation gives that $a_{2n+1} = 0$ for all n. Thus, with $s_1 = 0$ and $a_0 = 1$, we have one solution,

$$y_1(x) = \sum_{n=0}^{\infty} \frac{(-1)^n x^{2n}}{2^{2n}(n!)^2}.$$

29. Collecting the fundamental expressions, we can write the equation as

$$[x^2 y'' + xy - y/4] + x^2 y = 0.$$

Using the series for the fundamental expressions in (5.4) and (5.8), this becomes

$$0 = \sum_{n=0}^{\infty} [(n+s)(n+s-1) + (n+s) - 1/4] a_n x^{s+n} + \sum_{n=0}^{\infty} a_n x^{s+n+2}.$$

Shifting the index in the second series and isolating the extra terms in the first series, we get

$$0 = (s^2 - 1/4) a_0 x^s + [(s+1)^2 - 1/4] a_1 x^{s+1} + \sum_{n=2}^{\infty} \left\{[(s+n)^2 - 1/4] a_n + a_{n-2}\right\} x^{s+n}.$$

Since $a_0 \neq 0$, the first term gives the indicial equation $s^2 - 1/4 = 0$. Hence the roots are $s_1 = 1/2$ and $s_2 = -1/2$. Setting $s = s_1 = 1/2$, the second term becomes $2a_1 = 0$, so $a_1 = 0$. The coefficent of the general term gives the recurrence formula

$$[(n+1/2)^2 - 1/4] a_n = -a_{n-2} \quad \text{or} \quad n(n+1) a_n = -a_{n-2}$$

Since $a_1 = 0$, we conclude that $a_{2n+1} = 0$ for all n. For even subscripts, the recurrence formula becomes

$$a_{2n} = -\frac{a_{2(n-1)}}{2n(2n+1)}.$$

From this we conjecture that

$$a_{2n} = \frac{(-1)^n}{(2n+1)!}a_0.$$

Since $s_1 = 1/2$, the Frobenius solution is

$$y(x) = x^{1/2}\sum_{n=0}^{\infty}\frac{(-1)^n}{(2n+1)!}x^{2n}.$$

Next look at $s = s_2 = -1/2$. The second term in the series is $0 \cdot a_1 = 0$, so a_1 can be given any value. The coefficient of the general term gives us the recurrence formula $n(n-1)a_n = -a_{n-2}$. Since $n(n-1) \neq 0$ for $n \geq 2$, we can solve for all n. Hence there is a second Frobenius solution.

31. Multiply by x,

$$x^2y'' + x(1-x)y' + pxy = 0,$$

then comparison with $x^2y'' + xp(x)y' + q(x)y = 0$ reveals that $p(x) = 1 - x$ and $q(x) = px$ are both analytic at $x = 0$. Further, $p_0 = p(0) = 1$ and $q_0 = q(0) = 0$, so the indicial polynomial should be

$$I(s) = s^2 + (p_0 - 1)s + q_0 = s^2.$$

Let

$$y = x^s\sum_{n=0}^{\infty}a_nx^n = \sum_{n=0}^{\infty}a_nx^{s+n}.$$

Then,

$$x^2y'' = \sum_{n=0}^{\infty}(s+n)(s+n-1)a_nx^{s+n},$$

and

$$x(1-x)y' = (1-x)\sum_{n=0}^{\infty}(s+n)a_nx^{s+n} = \sum_{n=0}^{\infty}(s+n)a_nx^{s+n} - \sum_{n=1}^{\infty}(s+n-1)a_{n-1}x^{s+n},$$

where we've shifted the index of summation. Finally,

$$pxy = px\sum_{n=0}^{\infty}a_nx^{s+n} = \sum_{n=1}^{\infty}pa_{n-1}x^{s+n}.$$

Substituting,

$$\begin{aligned}
0 &= x^2y'' + x(1-x)y' + pxy \\
&= \sum_{n=0}^{\infty}(s+n)(s+n-1)a_nx^{s+n} + \sum_{n=0}^{\infty}(s+n)a_nx^{s+n} - \sum_{n=1}^{\infty}(s+n-1)a_{n-1}x^{s+n} + \sum_{n=1}^{\infty}pa_{n-1}x^{s+n} \\
&= [s(s-1)+s]a_0x^s + \sum_{n=1}^{\infty}\{[(s+n)(s+n-1)+(s+n)]a_n - [(s+n-1)-p]a_{n-1}\}x^{s+n} \\
&= s^2a_0 + \sum_{n=1}^{\infty}\left[(s+n)^2a_n - (s+n-1-p)a_{n-1}\right]x^{s+n}.
\end{aligned}$$

Thus, the indicial equation is $I(s) = s^2$, with equal roots $s_1 = s_2 = 0$. The remaining terms provide the recurrence relation

$$(s+n)^2 a_n - (s+n-1-p)a_{n-1} = 0, \quad n \geq 1.$$

With $s = 0$, this becomes

$$n^2 a_n - (n-1-p)a_{n-1} = 0 \quad \text{or} \quad a_n = \frac{n-1-p}{n^2} a_{n-1}, \quad n \geq 1.$$

Now, if p is a positive integer, the numerator of this recurrence relation will equal zero when

$$n - 1 - p = 0 \quad \text{or} \quad n = p + 1.$$

Thus, if $N = p+1$, then $a_n = 0$ for all $n \geq N$. That is, the Frobenius series terminates and becomes a polynomial of degree $N - 1$.

33. (a) If we multiply the equation by $x/(1-x)$ we get

$$x^2 y'' + x\frac{\gamma - (1+\alpha+\beta)x}{1-x} y' + \frac{-\alpha\beta x}{1-x} y = 0.$$

This equation is in the form of(6.1), with $p(x) = [\gamma - (1+\alpha+\beta)x]/(1-x)$ and $q(x) = -\alpha\beta x/(1-x)$. Both of these are analytic except at $x = 1$, so $x = 0$ is a regular singular point. The indicial equation is

$$s^2 + [p(0) - 1]s + q(0) = s^2 + [\gamma - 1]s = 0.$$

This equation has roots 0 and $1 = \gamma$.

(b) If we multiply the equation by $(1-x)/x$ we get

$$(x-1)^2 y'' - \frac{\gamma - (1+\alpha+\beta)x}{x}(x-1)y' - \alpha\beta(1-x)xy = 0.$$

This equation has the form of(6.1), with x replaced by $x - 1$, and coefficients $p(x) = -[\gamma - (1+\alpha+\beta)x]/x$, and $q(x) = -\alpha\beta(1-x)x$. Both are analytic except at $x = 0$, so $x = 1$ is a regualr singular point. The indicial polynomial is

$$s^2 + [p(1) - 1]s + q(1) = s^2 + [\alpha + \beta - \gamma]s = 0.$$

This equation has roots 0 and $\gamma - \alpha - \beta$.

(c) Multiplying the equation by x and separating the powers of x times fundamental expressions, we get

$$\begin{aligned} 0 &= x^2(1-x)y'' + x[\gamma - (1+\alpha+\beta)x]y' - \alpha\beta xy \\ &= [x^2 y'' + \gamma x y'] - x[x^2 y'' + (1+\alpha+\beta)xy' + \alpha\beta y]. \end{aligned}$$

Substituting $y(x) = \sum_{n=0}^{\infty} a_n x^{s+n}$, and similar formulas for the derivatives of y and using a little algebra, this becomes

$$\sum_{n=0}^{\infty}(n+s)(n+s+\gamma-1)a_n x^{s+n} - x\sum_{n=0}^{\infty}(n+s+\alpha)(n+s+\beta)a_n x^{s+n} = 0.$$

Shifting the index in the second sum, and isolating the first term we get

$$0 = s(s+\gamma-1)a_0 x^s + \sum_{n=1}^{\infty}[(n+s)(n+s+\gamma-1)a_n - (n+s+\alpha-1)(n+s+\beta-1)a_{n-1}]x^{s+n}.$$

The first term gives us the indicial equation and its roots $s = 0$ and $s = 1 - \gamma$. If $1 - \gamma$ is not a positive integer, then Theorem ?? says that we get a solution corresponding to $s = 0$. The general term of the series gives us the recurrence formula

$$a_n = \frac{(n + s + \alpha - 1)(n + s + \beta - 1)}{(n + s)(n + s + \gamma - 1)} a_{n-1} \quad \text{for } n \geq 1.$$

For $s = 0$ this becomes

$$a_n = \frac{(n + \alpha - 1)(n + \beta - 1)}{n(n + \gamma - 1)} a_{n-1} \quad \text{for } n \geq 1.$$

For $n = 1$, this becomes $a_1 = \alpha \cdot \beta / a \cdot \gamma$, and for $n = 2$,

$$a_2 = \frac{(\alpha + 1)(\beta + 1)}{2(\gamma + 1)} a_1 = \frac{\alpha \cdot (\alpha + 1) \cdot \beta \cdot (\beta + 1)}{2! \cdot \gamma \cdot (\gamma + 1)}.$$

(d) If $1 - \gamma$ is not an integer, then the two roots, 0 and $1 - \gamma$ do not differ by an integer, so, by Theorem ??, we get a Frobenius solution for each roots. If we set $s = 1 - \gamma$, the recurrence formula becomes

$$a_n = \frac{(n + \alpha - \gamma)(n + \beta - \gamma)}{(n + 1 - \gamma)n} a_{n-1} \quad \text{for } n \geq 1.$$

For $n = 1$ we get

$$a_1 = \frac{(1 + \alpha - \gamma)(1 + \beta - \gamma)}{(2 - \gamma) \cdot 1} a_0,$$

and for $n = 2$,

$$a_2 = \frac{(2 + \alpha - \gamma)(2 + \beta - \gamma)}{(3 - \gamma) \cdot 2} a_1 = \frac{(1 + \alpha - \gamma)(2 + \alpha - \gamma)(1 + \beta - \gamma)(2 + \beta - \gamma)}{(2 - \gamma)(3 - \gamma) \cdot 2!} a_0$$

———————×———————

Section 11.6. Series Solutions Near Regular Singular Points — The General Case

1. Compare $x^2y'' + 3xy' + (1 - 2x)y = 0$ with $x^2y'' + xp(x)y' + q(x)y = 0$ and note that $p(x) = 3$ and $q(x) = 1 - 2x$ are both analytic at $x = 0$, making this point a regular singular point. Further, $p_0 = p(0) = 3$ and $q_0 = q(0) = 1$, so the indicial equation should be

$$I(s) = s^2 + (p_0 - 1)s + q_0 = s^2 + 2s + 1 = (s + 1)^2.$$

Thus, we seek a solution having form

$$y(s, x) = x^s \sum_{n=0}^{\infty} a_n(s)x^n.$$

Hence,

$$x^2y'' = \sum_{n=0}^{\infty}(s+n)(s+n-1)a_n(s)x^{s+n}, \quad 3xy' = \sum_{n=0}^{\infty} 3(s+n)a_n(s)x^{s+n},$$

and

$$(1-2x)y = (1-2x)\sum_{n=0}^{\infty} a_n(s)x^{s+n} = \sum_{n=0}^{\infty} a_n(s)x^{s+n} - \sum_{n=1}^{\infty} 2a_{n-1}(s)x^{s+n},$$

where we've shifted the index of the second summation. Substituting, we get

$$0 = x^2y'' + 3xy' + (1-2x)y = (s+1)^2a_0x^s + \sum_{n=1}^{\infty}\left[(s+n+1)^2a_n(s) - 2a_{n-1}(s)\right]x^{s+n}.$$

Thus, the indicial equation is $I(s) = (s+1)^2 = 0$ with equal roots $s_2 = s_1 = -1$. The remaining terms provide the recurrence relation

$$a_n(s) = \frac{2a_{n-1}(s)}{(s+n+1)^2}, \quad n \geq 1.$$

Since the roots of the indicial equation are equal, we solve the recurrence relation for $a_n(s)$ as a function of s.

$$a_1(s) = \frac{2a_0}{(s+2)^2}, \quad a_2(s) = \frac{2a_1(s)}{(s+3)^2}, \quad \ldots \quad , a_n(s) = \frac{2a_{n-1}(s)}{(s+n+1)^2}.$$

Multiplying left and right-hand sides of these equations,

$$a_1(s)\, a_2(s) \cdots a_n(s) = \frac{2^n}{(s+2)^2(s+2)^2 \cdot \ldots \cdot (s+n+1)^2}\, a_0\, a_1(s) \cdots a_{n-1}(s).$$

Cancelling,

$$a_n(s) = \frac{2^n}{(s+2)^2(s+3)^2 \cdot \ldots \cdot (s+n+1)^2}\, a_0.$$

Evaluating at $s = -1$, we get

$$a_n(-1) = \frac{2^n}{1^2 \cdot 2^2 \cdot \ldots \cdot n^2}\, a_0.$$

With $a_0 = 1$, this becomes

$$a_n(-1) = \frac{2^n}{(n!)^2},$$

giving us the solution

$$y_1(x) = y(-1, x) = x^{-1}\sum_{n=0}^{\infty} a_n(-1)x^n = \sum_{n=0}^{\infty} \frac{2^n}{(n!)^2}\, x^{n-1}.$$

To find a second solution, we use logarithmic differentiation and write

$$\ln|a_n(s)| = n\ln 2 + \ln|a_0| - 2\sum_{k=2}^{n+1} \ln(s+k).$$

Thus,

$$\frac{1}{a_n(s)}\, a_n'(s) = -2\sum_{k=2}^{n+1} \frac{1}{s+k}.$$

Evaluating at $s = -1$,

$$a_n'(-1) = -2a_n(-1)\sum_{k=2}^{n+1}\frac{1}{k-1} = -\frac{2^{n+1}}{(n!)^2}\sum_{k=1}^{n}\frac{1}{k},$$

where we've shifted the index of summation. Hence, our second fundamental solution is

$$y_2(x) = y_1(x)\ln x + x^{-1}\sum_{n=1}^{\infty} a_n'(-1)x^n = y_1(x)\ln x - \sum_{n=1}^{\infty}\frac{2^{n+1}H(n)}{(n!)^2}x^{n-1},$$

where we've let $H(n) = \sum_{k=1}^{n} 1/k$.

3. Multiply by x.

$$x^2y'' + x(1-x)y' - xy = 0,$$

then comparison with $x^2y'' + xp(x)y' + q(x)y = 0$ reveals that $p(x) = 1 - x$ and $q(x) = -x$ are both analytic at $x = 0$. Thus, $x = 0$ is a regular singular point. Note that $p_0 = p(0) = 1$ and $q_0 = q(0) = 0$, so the indicial equation should be

$$I(s) = s^2 + (p_0 - 1)s + q_0 = s^2.$$

Thus, we seek a solution of the form

$$y(s, x) = x^s\sum_{n=0}^{\infty} a_n(s)x^n = \sum_{n=0}^{\infty} a_n(s)x^{s+n}.$$

Thus,

$$x^2y'' = \sum_{n=0}^{\infty}(s+n)(s+n-1)a_n(s)x^{s+n},$$

and

$$x(1-x)y' = (1-x)\sum_{n=0}^{\infty}(s+n)a_n(s)x^{s+n} = \sum_{n=0}^{\infty}(s+n)a_n(s)x^{s+n} - \sum_{n=1}^{\infty}(s+n-1)a_{n-1}(s)x^{s+n}$$

where we've shifted the index of the second summation. Finally,

$$-xy = -x\sum_{n=0}^{\infty} a_n(s)x^{s+n} = -\sum_{n=1}^{\infty} a_{n-1}(s)x^{s+n},$$

where again we have shifted the index of summation. Substituting, we get

$$0 = x^2y'' + x(1-x)y' - xy = s^2a_0x^n + \sum_{n=1}^{\infty}\left[(s+n)^2a_n(s) - (s+n)a_{n-1}(s)\right]x^{s+n}.$$

Thus, the indicial equation is $I(s) = s^2 = 0$ with equal roots $s_2 = s_1 = 0$. The remaining terms provide the recurrence relation

$$(s+n)^2a_n(s) = (s+n)a_{n-1}(s), \quad n \ge 1.$$

Since the roots of the indicial equation are equal, we solve the recurrence relation for $a_n(s)$ as a function of s. Thus,

$$a_1(s) = \frac{a_0}{s+1}, \quad a_2(s) = \frac{a_1(s)}{s+2}, \quad \ldots \quad , a_n(s) = \frac{a_{n-1}(s)}{s+n}.$$

Multiplying left and right-hand sides of these equations, we have

$$a_1(s)a_2(s)\cdots a_n(s) = \frac{1}{(s+1)(s+2)\cdots(s+n)} a_0 a_1(s)\cdots a_{n-1}(s).$$

Cancelling,

$$a_n(s) = \frac{1}{(s+1)(s+2)\cdots(s+n)} a_0.$$

Evaluating at $s = 0$, we get

$$a_n(0) = \frac{1}{n!} a_0.$$

Choosing $a_0 = 1$, our first solution is

$$y_1(x) = y(0,x) = x^0 \sum_{n=0}^{\infty} \frac{x^n}{n!} = e^x.$$

To find our second solution, we must differentiate $a_n(s)$. We'll use logarithmic differentiation and write

$$\ln|a_n(s)| = \ln|a_0| - \sum_{k=1}^{n} \ln|s+k|.$$

Thus,

$$\frac{1}{a_n(s)} a_n'(s) = -\sum_{k=1}^{n} \frac{1}{s+k}.$$

Evaluating at $s = 0$,

$$a_n'(0) = -a_n(0) \sum_{k=1}^{n} \frac{1}{k} = -\frac{a_0}{n!} \sum_{k=1}^{n} \frac{1}{k}.$$

We set $a_0 = 1$ and let $H(n) = \sum_{k=1}^{n}(1/k)$. Then,

$$a_n'(0) = -\frac{1}{n!} H(n).$$

Thus,

$$y_2(x) = y_1(x)\ln x + x^0 \sum_{n=0}^{\infty} a_n'(0)x^n = e^x \ln x - \sum_{n=1}^{\infty} \frac{H(n)x^n}{n!}$$

5. Multiply by x,

$$x^2y'' - x(4+x)y' + 2xy = 0,$$

then compare this with $x^2y'' + xp(x)y' + q(x)y = 0$ and note that $p(x) = -4 - x$ and $q(x) = 2x$ are both analytic at $x = 0$, making this a regular singular point. Further, $p_0 = p(0) = -4$ and $q_0 = q(0) = 0$, so the indicial equation should be

$$I(s) = s^2 + (p_0 - 1)s + q_0 = s^2 - 5s.$$

We seek a solution having form

$$y(s,x) = x^s \sum_{n=0}^{\infty} a_n(s)x^n.$$

Thus,

$$x^2y'' = \sum_{n=0}^{\infty}(s+n)(s+n-1)a_n(s)x^{s+n},$$

and

$$-x(4+x)y' = -(4+x)\sum_{n=0}^{\infty}(s+n)a_n(s)x^{s+n} = -\sum_{n=0}^{\infty}4(s+n)a_n(s)x^{s+n} - \sum_{n=1}^{\infty}(s+n-1)a_{n-1}(s)x^{s+n},$$

where we've shifted the index of the second summation. Finally,

$$2xy = 2x\sum_{n=0}^{\infty}a_n(s)s^{s+n} = \sum_{n=1}^{\infty}2a_{n-1}(s)x^{s+n},$$

where we've shifted the index of summation. Substituting, we get

$$\begin{aligned}0 &= x^2y'' - x(4+x)y' + 2xy \\ &= (s^2-5s)a_0x^s + \sum_{n=1}^{\infty}\left\{\left[(s+n)^2 - 5(s+n)\right]a_n(s) - (s+n-3)a_{n-1}(s)\right\}x^{s+n}.\end{aligned}$$

Thus, the indicial equation is $I(s) = s(s-5)$ with roots $s_1 = 5$ and $s_2 = 0$. The remaining terms provide the recurrence relation

$$(s+n)(s+n-5)a_n(s) = (s+n-3)a_{n-1}(s), \quad n \geq 1.$$

Because the roots differ by a positive integer, we use the smaller root, $s_2 = 0$, first. We get

$$n(n-5)a_n(0) = (n-3)a_{n-1}(0), \quad n \geq 1.$$

Hence

$$\begin{aligned}n=1: &\quad -4a_1(0) = -2a_0 \Rightarrow a_1(0) = \frac{1}{2}a_0 \\ n=2: &\quad -6a_2(0) = -1a_1(0) \Rightarrow a_2(0) = \frac{1}{6}a_1(0) = \frac{1}{12}a_0 \\ n=3: &\quad -6a_3(0) = 0\cdot a_2(0) \Rightarrow a_3(0) = 0 \\ n=4: &\quad -4a_4(0) = a_3(0) \Rightarrow a_4(0) = 0.\end{aligned}$$

Finally, if $n = 5$, then

$$0a_5(0) = 2a_4(0) = 0,$$

so $a_5(0) = a_5$ is arbitrary, and

$$a_n(0) = \frac{n-3}{n(n-5)}a_{n-1}(0), \quad n \geq 6.$$

Next,

$$a_6(0) = \frac{3}{6\cdot 1}a_5, \quad a_7(0) = \frac{4}{7\cdot 2}a_6(0), \quad \ldots \quad, a_n(0) = \frac{n-3}{n(n-5)}a_{n-1}(0).$$

Multiplying left and right-hand sides of the preceding equations,

$$a_6(0)\,a_7(0)\cdots a_n(0) = \frac{3\cdot 4\cdot\ldots\cdot(n-3)}{[6\cdot 7\cdot\ldots\cdot n][1\cdot 2\cdot\ldots\cdot(n-5)]}\,a_5\,a_6(0)\cdots a_{n-1}(0).$$

Cancelling, for $n \geq 6$,

$$a_n(0) = \frac{3\cdot 4\cdot\ldots\cdot(n-3)}{[6\cdot 7\cdot\ldots\cdot n](n-5)!}\,a_5 = \frac{3\cdot 4\cdot 5\,a_5}{(n-2)(n-1)n(n-5)!} = \frac{60a_5}{(n-2)(n-1)n(n-5)!}$$

Hence,

$$y = a_0\left(1+\frac{1}{2}x+\frac{1}{12}x^2\right) + a_5\left[x^5+\sum_{n=6}^{\infty}\frac{60x^n}{(n-2)(n-1)n(n-5)!}\right].$$

By letting $a_0 = 1$ and $a_5 = 0$, then $a_0 = 0$ and $a_5 = 1$, we obtain a fundamental set of solutions

$$y_1(x) = 1+\frac{1}{2}x+\frac{1}{12}x^2, \quad \text{and} \quad y_2(x) = x^5+\sum_{n=6}^{\infty}\frac{60x^n}{(n-2)(n-1)n(n-5)!}.$$

7. Multiply by x,

$$x^2y'' - x(3+x)y' + 2xy = 0,$$

then compare with $x^2y''+xp(x)y'+q(x)y = 0$ and note that $p(x) = -(3+x)$ and $q(x) = 2x$ are both analytic at $x = 0$, making $x = 0$ a regular singular point. Further, $p_0 = p(0) = -3$ and $q_0 = q(0) = 0$, so the indicial polynomial is

$$I(s) = s^2 + (p_0-1)s + q_0 = s^2 - 4s = s(s-4).$$

Let

$$y(s,x) = x^s\sum_{n=0}^{\infty}a_n(s)x^n = \sum_{n=0}^{\infty}a_n(s)x^{s+n}.$$

Substituting,

$$\begin{aligned} 0 &= x^2y'' - x(3+x)y' + 2xy \\ &= (s^2-4s)a_0x^s + \sum_{n=1}^{\infty}\left\{\left[(s+n)^2 - 4(s+n)\right]a_n(s) - (s+n-3)a_{n-1}(s)\right\}x^{s+n}. \end{aligned}$$

Thus, the indicial equation is $I(s) = s(s-4) = 0$ with roots $s_2 = 0$ and $s_1 = 4$. The remaining terms provide the recurrence relation

$$(s+n)(s+n-4)a_n(s) - (s+n-3)a_{n-1}(s) = 0, \quad n \geq 1.$$

Starting with the smaller root, $s_2 = 0$, the recurrence relation becomes

$$n(n-4)a_n(0) = (n-3)a_{n-1}(0), \quad n \geq 1.$$

For $n = 1$, 2, and 3, this becomes

$$a_1(0) = \frac{-2}{1\cdot(-3)}a_0 = \frac{2}{3}a_0, \quad a_2(0) = \frac{-1}{2(-2)}a_1(0) = \frac{1}{6}a_0, \quad \text{and} \quad a_3(0) = 0a_2(0) = 0.$$

For $n = 4$, the recurrence relation becomes

$$4 \cdot 0 \cdot a_4(0) - 1a_3(0) = 0 \quad \text{or} \quad 0a_4(0) = a_3(0) = 0.$$

Thus, we are free to choose $a_4(0) = a_4$, after which

$$a_n(0) = \frac{n-3}{n(n-4)} a_{n-1}(0), \quad n \geq 5.$$

Thus,

$$a_5(0) = \frac{2}{5 \cdot 1} a_4(0), \quad a_6(0) = \frac{3}{6 \cdot 2} a_5(0), \quad \ldots \quad , a_n(0) = \frac{n-3}{n(n-4)} a_{n-1}(0).$$

Multiplying left and right-hand sides of these equations, we get

$$a_5(0)\, a_6(0) \cdots a_n(0) = \frac{2 \cdot 3 \cdot \ldots \cdot (n-3) \cdot a_4(0)\, a_5(0) \cdots a_{n-1}(0)}{(5 \cdot 6 \cdot \ldots \cdot n)(1 \cdot 2 \cdot \ldots \cdot (n-4))}.$$

Cancelling,

$$a_n(0) = \frac{(n-3)a_4}{5 \cdot 6 \cdot \ldots \cdot n} = \frac{24(n-3)a_4}{n!}, \quad n \geq 5.$$

Thus,

$$y(x) = a_0 \left[1 + \frac{2}{3}x + \frac{1}{6}x^2 \right] + a_4 \left[x^4 + \sum_{n=5}^{\infty} \frac{24(n-3)}{n!} x^n \right].$$

By choosing $a_0 = 1$ and $a_4 = 0$, then $a_0 = 0$ and $a_4 = 1$, you get two independent solutions.

9. Multiply by x, $x^2y'' + xy = 0$, then compare with $x^2y'' + xp(x)y' + q(x)y = 0$ and note that $p(x) = 0$ and $q(x) = x$ are both analytic at $x = 0$, making $x = 0$ a regular singular point. Further, $p_0 = p(0) = 0$ and $q_0 = q(0) = 0$, so the indicial equation should be

$$I(s) = s^2 + (p_0 - 1)s + q_0 = s^2 - s = s(s-1) = 0.$$

Set

$$y(s, x) = x^s \sum_{n=0}^{\infty} a_n(s)x^n = \sum_{n=0}^{\infty} a_n(s)x^{s+n}.$$

Substituting,

$$0 = x^2y'' + xy = s(s-1)a_0x^s + \sum_{n=1}^{\infty} [(s+n)(s+n-1)a_n(s) + a_{n-1}(s)]x^{s+n}.$$

Thus, the indicial equation is

$$I(s) = s(s-1) = 0,$$

with roots $s_2 = 0$ and $s_1 = 1$. The remaining terms provide the recurrence relation

$$(s+n)(s+n-1)a_n(s) + a_{n-1}(s) = 0, \quad n \geq 1.$$

Working with the smaller root, $s = s_2 = 0$, this becomes

$$n(n-1)a_n(0) = -a_{n-1}(0), \quad n \geq 1.$$

However, for $N = s_1 - s_2 = 1 - 0 = 1$, this becomes

$$0a_1 = -a_0 \neq 0,$$

so we have a real degeneracy. Solve the recurrence relation for $a_n(s)$ as a function of s.

$$a_n(s) = \frac{-a_{n-1}(s)}{(s+n)(s+n-1)}, \quad n \geq 1.$$

Thus,

$$a_1(s) = \frac{-a_0}{(s+1)s}, \quad a_2(s) = \frac{-a_1(s)}{(s+2)(s+1)},$$

etc., and

$$a_1(s)\, a_2(s) \cdots a_n(s) = \frac{(-1)^n a_0\, a_1(s) \cdots a_{n-1}(s)}{[(s+1)(s+2)\cdots(s+n)][s(s+1)\cdots(s+n-1)]}.$$

Cancelling,

$$a_n(s) = \frac{(-1)^n a_0}{[(s+1)(s+2)\cdots(s+n)][s(s+1)\cdots(s+n-1)]}, \quad n \geq 1.$$

Because the smaller root is $s = s_2 = 0$, define

$$Y(s,x) = sy(s,x) = x^s \sum_{n=0}^{\infty} sa_n(s)x^n = x^s \sum_{n=0}^{\infty} b_n(s)x^n,$$

where $b_n(s) = sa_n(s)$. For example, $b_0(s) = sa_0$, $b_1(s) = sa_1(s) = -a_0/(s+1)$, and

$$b_n(s) = \frac{(-1)^n a_0}{[(s+1)(s+2)\cdots(s+n)][(s+1)(s+2)\cdots(s+n-1)]}, \quad n \geq 2.$$

Evaluating at $s = s_2 = 0$, $b_0(0) = 0$, $b_1(0) = -a_0$, and

$$b_n(0) = \frac{(-1)^n a_0}{(1\cdot 2\cdot \ldots \cdot n)(1\cdot 2\cdot \ldots \cdot (n-1))} = \frac{(-1)^n a_0}{n!(n-1)!},$$

for $n \geq 2$. Thus, remembering that

$$Y(s,x) = x^s \sum_{n=0}^{\infty} b_n(s)x^n,$$

and choosing $a_0 = 1$ with $s = s_2 = 0$,

$$y_1(x) = Y(0,x) = x^0 \sum_{n=0}^{\infty} b_n(0)x^n = -x + \sum_{n=2}^{\infty} \frac{(-1)^n x^n}{n!(n-1)!}.$$

We can absorb the first term into the summation and write

$$y_1(x) = \sum_{n=1}^{\infty} \frac{(-1)^n x^n}{n!(n-1)!}.$$

Take the derivative of

$$Y(s,x) = x^s \sum_{n=0}^{\infty} b_n(s)x^n$$

with respect to s.

$$\frac{\partial Y}{\partial s}(s,x) = x^s \ln x \sum_{n=0}^{\infty} b_n(s)x^n + x^s \sum_{n=0}^{\infty} b_n'(s)x^n$$

Then, a second solution is

$$\frac{\partial Y}{\partial s}(0,x) = x^0 \ln x \sum_{n=0}^{\infty} b_n(0)x^n + x^0 \sum_{n=0}^{\infty} b_n'(0)x^n = y_1(x)\ln x + \sum_{n=0}^{\infty} b_n'(0)x^n$$

Thus, we need to differentiate $b_n(s)$. First, $b_0'(s) = a_0$ and $b_1'(s) = a_0/(s+1)^2$, so $b_0'(0) = a_0$ and $b_1'(0) = a_0$. Next, using logarithmic differentiation, for $n \geq 2$, we can write

$$b_n'(s) = b_n(s)\left[-\frac{1}{s+1} - \frac{1}{s+2} - \cdots - \frac{1}{s+n} - \frac{1}{s+1} - \frac{1}{s+2} - \cdots - \frac{1}{s+n-1}\right],$$

and

$$b_n'(0) = -b_n(0)\left[1 + \frac{1}{2} + \cdots + \frac{1}{n} + 1 + \frac{1}{2} + \cdots + \frac{1}{n-1}\right] = \frac{(-1)^{n+1}a_0}{n!(n-1)!}[H(n) + H(n-1)],$$

where $H(n) = \sum_{k=1}^{n} 1/k$. Thus, a second independent solution is

$$y_2(x) = y_1(x)\ln x + \left[a_0 + a_0 x + \sum_{n=2}^{\infty} \frac{(-1)^{n+1}a_0 x^n}{n!(n-1)!}[H(n) + H(n-1)]\right].$$

With $a_0 = 1$, this becomes

$$y_2(x) = y_1(x)\ln x + 1 + x + \sum_{n=2}^{\infty} \frac{(-1)^{n+1}[H(n) + H(n-1)]}{n!(n-1)!} x^n.$$

11. Compare $x^2y'' - 3xy' + (3+4x)y = 0$ with $x^2y'' + xp(x)y' + q(x)y = 0$ and note that $p(x) = -3$ and $q(x) = 3 + 4x$ are both analytic at $x = 0$, making $x = 0$ a regular singular point. Further, $p_0 = p(0) = -3$ and $q_0 = q(0) = 3$, so the indicial equation should be

$$I(s) = s^2 + (p_0 - 1)s + q_0 = s^2 - 4s + 3 = (s-1)(s-3) = 0.$$

Set

$$y(s,x) = x^s \sum_{n=0}^{\infty} a_n(s)x^n = \sum_{n=0}^{\infty} a_n(s)x^{s+n}.$$

Substituting,

$$\begin{aligned} 0 &= x^2y'' - 3xy' + (3+4x)y \\ &= [s(s-1) - 3s + 3]a_0x^s + \sum_{n=1}^{\infty}\{[(s+n)(s+n-1) - 3(s+n) + 3]a_n(s) + 4a_{n-1}(s)\}x^{s+n}. \end{aligned}$$

Thus, the indicial equation is

$$I(s) = s^2 - 4s + 3 = (s-1)(s-3) = 0,$$

with roots $s_2 = 1$ and $s_1 = 3$. The remaining terms provide the recurrence relation

$$(s+n-1)(s+n-3)a_n(s) + 4a_{n-1}(s) = 0, \quad n \geq 1.$$

Working with the smaller root, $s = s_2 = 1$, this becomes

$$n(n-2)a_n(1) = -4a_{n-1}(1), \quad n \geq 1.$$

Thus, $-a_1(1) = -4a_0$ or $a_1(1) = 4a_0 \neq 0$. However, for $N = s_1 - s_2 = 3 - 1 = 2$, this becomes

$$0a_2(1) = -4a_1(1) \neq 0,$$

so we have a real degeneracy. Compute $a_n(s)$ as a function of s.

$$a_n(s) = \frac{-4a_{n-1}(s)}{(s+n-1)(s+n-3)}, \quad n \geq 1.$$

Thus,

$$a_1(s) = \frac{-4a_0}{s(s-2)}, \quad a_2(s) = \frac{-4a_1(s)}{(s+1)(s-1)},$$

etc., and

$$a_1(s)\,a_2(s)\cdots a_n(s) = \frac{(-4)^n a_0\, a_1(s)\cdots a_{n-1}(s)}{[s(s+1)\cdots(s+n-1)][(s-2)(s-1)\cdots(s+n-3)]}.$$

Cancelling,

$$a_n(s) = \frac{(-4)^n a_0}{[s(s+1)\cdots(s+n-1)][(s-2)(s-1)\cdots(s+n-3)]}$$

for $n \geq 1$. Because the smaller roots is $s = s_2 = 1$, define

$$Y(s,x) = (s-1)y(s,x) = x^s \sum_{n=0}^{\infty}(s-1)a_n(s)x^n = x^s \sum_{n=0}^{\infty} b_n(s)x^n,$$

where $b_n(s) = (s-1)a_n(s)$. For example, $b_0(s) = (s-1)a_0$, $b_1(s) = -4a_0(s-1)/(s(s-2))$, $b_2(s) = -4a_0/(s+1)$, and

$$b_n(s) = \frac{(-4)^n a_0}{[s(s+1)\cdots(s+n-1)][(s-2)s(s+1)\cdots(s+n-3)]},$$

for $n \geq 3$. Evaluating at $s = s_2 = 1$, $b_0(1) = 0$, $b_1(1) = 0$, $b_2(1) = -2a_0$, and

$$b_n(1) = \frac{(-4)^n a_0}{(1\cdot 2\cdot \ldots n)(-1)(1\cdot 2\cdot \ldots \cdot(n-2))} = \frac{(-1)^{n+1}4^n a_0}{n!(n-2)!}, \quad n \geq 3.$$

Remembering that

$$Y(s,x) = x^s \sum_{n=0}^{\infty} b_n(s)x^n,$$

then with $s = s_2 = 1$,

$$y_1(x) = x^1 \sum_{n=0}^{\infty} b_n(1)x^n = x\left[0 + 0x - 2a_0x^2 + \sum_{n=3}^{\infty} \frac{(-1)^{n+1}4^n a_0}{n!(n-2)!} x^n\right].$$

With $a_0 = -1$, this becomes

$$y_1(x) = 2x^3 + \sum_{n=3}^{\infty} \frac{(-1)^n 4^n}{n!(n-2)!} x^{n+1}.$$

A second solution is found by taking the derivative of $Y(s, x) = x^s \sum_{n=0}^{\infty} b_n(s)x^n$ with respect to s.

$$\frac{\partial Y}{\partial s} = x^s \ln x \sum_{n=0}^{\infty} b_n(s)x^n + x^s \sum_{n=0}^{\infty} b_n'(s)x^n = y_1(x)\ln x + x^s \sum_{n=0}^{\infty} b_n'(s)x^n$$

Thus, we need to differentiate $b_n(s)$. First, $b_0'(s) = a_0$,

$$b_1'(s) = b_1(s)\left[\frac{1}{s-1} - \frac{1}{s} - \frac{1}{s-2}\right] = \frac{-4a_0}{s(s-2)}\left[1 - \frac{s-1}{s} - \frac{s-1}{s-2}\right],$$

and

$$b_2'(s) = \frac{4a_0}{(s+1)^2}.$$

Thus, $b_0'(1) = a_0$, $b_1'(1) = 4a_0$, and $b_2'(1) = a_0$. Using more logarithmic differentiation, for $n \geq 3$, we have

$$b_n'(s) = b_n(s)\left[-\frac{1}{s} - \frac{1}{s+1} - \cdots - \frac{1}{s+n-1} - \frac{1}{s-2} - \frac{1}{s} - \frac{1}{s+1} - \cdots - \frac{1}{s+n-3}\right].$$

At $s = 1$,

$$b_n'(1) = -b_n(1)\left[1 + \frac{1}{2} + \cdots + \frac{1}{n} - 1 + 1 + \frac{1}{2} + \cdots + \frac{1}{n-2}\right] = \frac{(-1)^n 4^n a_0}{n!(n-2)!}[H(n) - 1 + H(n-2)],$$

for $n \geq 3$ and $H(n) = \sum_{k=1}^{n}(1/k)$. Hence,

$$\begin{aligned} y_2(x) &= y_1(x)\ln x + x^1 \sum_{n=0}^{\infty} b_n'(1)x^n \\ &= y_1(x)\ln x + x\left[a_0 + 4a_0x + +a_0x^2 + \sum_{n=3}^{\infty} \frac{(-1)^n 4^n a_0}{n!(n-2)!}[H(n) - 1 + H(n-2)]x^n\right]. \end{aligned}$$

With $a_0 = -1$,

$$y_2(x) = y_1(x)\ln x - x - x^2 - x^3 - \sum_{n=3}^{\infty} \frac{(-4)^n[H(n) - 1 + H(n-2)]}{n!(n-2)!} x^{n+1}.$$

13. Compare $x^2y'' + x(2+3x)y' - 2y = 0$ with $x^2y'' + xp(x)y' + q(x)y = 0$ and note that $p(x) = 2 + 3x$ and $q(x) = -2$ are both analytic at $x = 0$, making $x = 0$ a regular singular point. Further, $p_0 = p(0) = 2$ and $q_0 = q(0) = -2$, so the indicial equation should be

$$I(s) = s^2 + (p_0 - 1)s + q_0 = s^2 + s - 2 = (s+2)(s-1) = 0.$$

Set

$$y(s,x) = x^s \sum_{n=0}^{\infty} a_n(s)x^n = \sum_{n=0}^{\infty} a_n(s)x^{s+n}.$$

Substituting,

$$\begin{aligned}0 &= x^2y'' + x(2+3x)y' - 2y \\ &= [(s(s-1) + 2s - 2]a_0x^s + \sum_{n=1}^{\infty}\{[(s+n)(s+n-1) + 2(s+n) - 2]a_n(s) + 3(s+n-1)a_{n-1}(s)\}x^{s+n}.\end{aligned}$$

Thus, the indicial equation is

$$I(s) = s^2 + s - 2 = (s+2)(s-1) = 0,$$

with roots $s_2 = -2$ and $s_1 = 1$. The remaining terms provide the recurrence relation.

$$(s+n+2)(s+n-1)a_n(s) + 3(s+n-1)a_{n-1}(s) = 0, \quad n \geq 1.$$

Working with the smaller root, $s = s_2 = -2$, this becomes

$$n(n-3)a_n(-2) = -3(n-3)a_{n-1}(-2), \quad n \geq 1.$$

However, for $N = s_1 - s_2 = 1 - (-2) = 3$, we have $0a_3(-2) = 0$, so this is a false degeneracy. Providing $n \neq 3$, we have

$$a_n(-2) = \frac{-3a_{n-1}(-2)}{n}.$$

Thus,

$$a_1(-2) = -3a_0 \quad \text{and} \quad a_2(-2) = \frac{-3a_1(-2)}{2} = \frac{9}{2}a_0.$$

Letting a_3 be an arbitrary constant,

$$a_4(-2) = \frac{-3a_3}{4}, \quad a_5(-2) = \frac{-3a_4(-2)}{5},$$

etc., and

$$a_4(-2)\,a_5(-2)\ldots a_n(-2) = \frac{(-3)^{n-3}a_3\,a_4(-2)\cdots a_{n-1}(-2)}{4\cdot 5\cdot \ldots \cdot n}.$$

Cancelling,

$$a_n(-2) = \frac{6(-3)^{n-3}a_3}{n!}, \quad n \geq 4.$$

Thus, for $s = s_2 = -2$,

$$y(x) = x^s\sum_{n=0}^{\infty} a_n(s)x^n = x^{-2}\left[a_0 - 3a_0x + \frac{9}{2}a_0x^2\right] + x^{-2}\left[a_3x^3 + \sum_{n=4}^{\infty}\frac{6(-3)^{n-3}a_3}{n!}x^n\right].$$

With $a_0 = 1$ and $a_3 = 0$, one solution is

$$y_1(x) = x^{-2}\left(1 - 3x + \frac{9}{2}x^2\right) = x^{-2} - 3x^{-1} + \frac{9}{2}.$$

With $a_0 = 0$ and $a_3 = 1$, a second independent solution is

$$y_2(x) = x^{-2}\left[x^3 + \sum_{n=4}^{\infty} \frac{6(-3)^{n-3}}{n!} x^n\right] = x + 6\sum_{n=4}^{\infty} \frac{(-3)^{n-3}}{n!} x^{n-2}.$$

15. Multiply by x,

$$x^2y'' + x^2y' + xy = 0,$$

then compare with $x^2y'' + xp(x)y' + q(x)y = 0$ and note that $p(x) = x$ and $q(x) = x$, both analytic at $x = 0$, making $x = 0$ a regular singular point. Further, $p_0 = p(0) = 0$ and $q_0 = q(0) = 0$, so the indicial equation should be

$$I(s) = s^2 + (p_0 - 1)s + q_0 = s^2 - s = s(s-1).$$

Set

$$y(x) = x^s \sum_{n=0}^{\infty} a_n x^n = \sum_{n=0}^{\infty} a_n x^{s+n}.$$

Substituting,

$$\begin{aligned} 0 &= x^2y'' + x^2y' + xy \\ &= s(s-1)a_0x^s + \sum_{n=1}^{\infty}\{(s+n)(s+n-1)a_n + [(s+n-1)+1]a_{n-1}(s)\}x^{s+n}. \end{aligned}$$

Thus, the indicial equation is

$$I(s) = s(s-1) = 0,$$

with roots $s_2 = 0$ and $s_1 = 1$. The remaining terms provide the recurrence relation

$$(s+n)(s+n-1)a_n = -(s+n)a_{n-1}(s), \quad n \geq 1.$$

Working with the smaller root, $s = s_2 = 0$, this becomes

$$n(n-1)a_n = -na_{n-1}(s).$$

However, with $N = s_1 - s_2 = 1 - 0 = 1$,

$$0a_1 = -a_0.$$

Because $a_0 \neq 0$, this equation cannot be solved and we have a real degeneracy. We start by setting

$$y(s, x) = x^s \sum_{n=0}^{\infty} a_n(s)x^n = \sum_{n=0}^{\infty} a_n(s)x^{s+n},$$

and solve for a_n as a function of s.

$$a_n = \frac{-a_{n-1}(s)}{s+n-1}, \quad n \geq 1.$$

Thus,

$$a_1 = \frac{-a_0}{s}, \quad a_2 = \frac{-a_1}{s+1}, \quad \cdots, \quad a_n = \frac{-a_{n-1}(s)}{s+n-1}.$$

Multiplying these equation together,

$$a_1\, a_2 \cdots a_n = \frac{(-1)^n a_0\, a_1 \cdots a_{n-1}(s)}{s(s+1)\cdots(s+n-1)}.$$

Cancelling,

$$a_n(s) = \frac{(-1)^n a_0}{s(s+1)\cdots(s+n-1)}.$$

Define

$$\begin{aligned} Y(s,x) &= sy(s,x) \\ &= s\sum_{n=0}^{\infty} a_n(s)x^{s+n} \\ &= \sum_{n=0}^{\infty} b_n(s)x^{s+n}, \end{aligned}$$

where $b_n(s) = sa_n(s)$. Thus, for example,

$$b_0 = sa_0 \quad \text{and} \quad b_1(s) = -a_0.$$

For $n \geq 2$,

$$b_n(s) = sa_n(s) = \frac{(-1)^n a_0}{(s+1)\cdots(s+n-1)}.$$

Evaluating at $s = 0$,

$$b_0(0) = 0 \quad \text{and} \quad b_1(0) = -a_0.$$

Then,

$$b_n(0) = \frac{(-1)^n a_0}{(n-1)!},$$

for $n \geq 2$. Thus, remembering

$$Y(s,x) = x^s \sum_{n=0}^{\infty} b_n(s)x^n,$$

and choosing $a_0 = -1$ with $s = s_2 = 0$,

$$y_1(x) = Y(0,x) = x^0\left[x + \sum_{n=2}^{\infty} \frac{(-1)^{n+1}x^n}{(n-1)!}\right] = x + \sum_{n=2}^{\infty} \frac{(-1)^{n+1}}{(n-1)!}x^n.$$

A second solution is found by taking the derivative of

$$Y(s,x) = sy(s,x) = x^s \sum_{n=0}^{\infty} b_n(s)x^n$$

with respect to s.

$$\begin{aligned}\frac{\partial Y}{\partial s} &= x^s \ln x \sum_{n=0}^{\infty} b_n(s)x^n + x^s \sum_{n=0}^{\infty} b_n'(s)x^n \\ &= y_1(x)\ln x + x^s \sum_{n=0}^{\infty} b_n'(s)x^n\end{aligned}$$

Thus, we need to differentiate $b_n(s)$. First, $b_0'(s) = a_0$ and $b_1'(s) = 0$, so $b_0'(0) = a_0$ and $b_1'(0) = 0$. Next, using logarithmic differentiation,

$$b_n'(s) = b_n(s)\left[-\frac{1}{s+1} - \cdots - \frac{1}{s+n-1}\right]$$

and

$$b_n'(0) = -b_n(0)\sum_{k=1}^{n-1}\frac{1}{k} = \frac{(-1)^{n+1}a_0}{(n-1)!}H(n-1),$$

where $H(n) = \sum_{k=1}^{n} 1/k$. Choosing $a_0 = 1$, a second independent solution is, remembering that $s = s_2 = 0$,

$$\begin{aligned}y_2(x) &= y_1(x)\ln x + x^0 \sum_{n=0}^{\infty} b_n'(0)x^n \\ &= y_1(x)\ln x + 1 + \sum_{n=2}^{\infty}\frac{(-1)^{n+1}H(n-1)}{(n-1)!}x^n.\end{aligned}$$

17. Multiply by x.

$$x^2y'' + xy' - xy = 0,$$

then compare with $x^2y'' + xp(x)y' + q(x)y = 0$ and note that $p(x) = 1$ and $q(x) = -x$, both analytic at $x = 0$, making $x = 0$ a regular singular point. Further, note that $p_0 = p(0) = 1$ and $q_0 = q(0) = 0$, making the indicial equation

$$I(s) = s^2 + (p_0 - 1)s + q_0 = s^2.$$

Thus, we seek a solution of the form

$$y(x, s) = x^s \sum_{n=0}^{\infty} a_n(s)x^n = \sum_{n=0}^{\infty} a_n(s)x^{s+n}.$$

Substituting,

$$\begin{aligned}0 &= x^2y'' + xy' - xy \\ &= s^2a_0x^2 + \sum_{n=1}^{\infty}\left[(s+n)^2a_n - a_{n-1}(s)\right]x^{s+n}.\end{aligned}$$

Thus, the indicial equation is $I(s) = s^2 = 0$ with equal roots $s_2 = s_1 = 0$. The remaining terms provide the recurrence relation

$$\begin{aligned}(s+n)^2a_n &= a_{n-1}(s) \\ a_n &= \frac{a_{n-1}(s)}{(s+n)^2}, \quad n \geq 1.\end{aligned}$$

Since the roots of the indicial equation are equal, we solve the indicial equation for a_n as a function of s. Thus,

$$a_1 = \frac{a_0}{(s+1)^2}$$
$$a_2 = \frac{a_1}{(s+2)^2}$$
$$\vdots$$
$$a_n = \frac{a_{n-1}(s)}{(s+n)^2}.$$

Multiplying left and right-hand sides of these equations, we have

$$a_1\, a_2 \cdots a_n = \frac{1}{[(s+1)(s+2)\cdots(s+n)]^2}\, a_0\, a_1 \cdots a_{n-1}(s).$$

Cancelling,

$$a_n(s) = \frac{a_0}{[(s+1)(s+2)\cdots(s+n)]^2}.$$

Evaluating at $s = 0$,

$$a_n(0) = \frac{a_0}{(n!)^2},$$

and choosing $a_0 = 1$ gives

$$a_n(0) = \frac{1}{(n!)^2},$$

and a first solution

$$y_1 = y(0, x) = x^0 \sum_{n=0}^{\infty} \frac{x^n}{(n!)^2} = \sum_{n=0}^{\infty} \frac{x^n}{(n!)^2}.$$

We now use logarithmic differentiation to find $a_n'(s)$.

$$\ln|a_n(s)| = \ln|a_0| - 2\sum_{k=1}^{n} \ln|s+k|$$

$$\frac{1}{a_n(s)}\, a_n'(s) = -2\sum_{k=1}^{n} \frac{1}{s+k}$$

With $s = 0$,

$$a_n'(0) = -2a_n(0)\sum_{k=1}^{n} \frac{1}{k} = -\frac{2}{n!} H(n),$$

where we let $H(n) = \sum_{k=1}^{n} 1/k$. Thus, a second independent solution is

$$\begin{aligned} y_2 &= y_1 \ln x + x^0 \sum_{n=1}^{\infty} a_n'(0) x^n \\ &= y_1 \ln x - 2\sum_{n=1}^{\infty} \frac{H(n)}{n!} x^n. \end{aligned}$$

19. Compare $x^2y'' + xy' - (1 + x^2)y = 0$ with $x^2y'' + xp(x)y' + q(x)y = 0$ and note that $p(x) = 1$ and $q(x) = -(1 + x^2)$, both of which are analytic at $x = 0$, making $x = 0$ a regular singular point. Further, note that $p_0 = p(0) = 1$, and $q_0 = q(0) = -1$, so the indicial equation should be

$$I(s) = s^2 + (p_0 - 1)s + q_0 = s^2 - 1 = (s + 1)(s - 1) = 0.$$

Set

$$y(x) = x^s \sum_{n=0}^{\infty} a_n x^n = \sum_{n=0}^{\infty} a_n x^{s+n}.$$

Substituting,

$$\begin{aligned} 0 &= x^2y'' + xy' - (1 + x^2)y \\ &= [s(s-1) + s - 1]a_0 x^s + [(s+1)s + (s+1) - 1]a_1 x^{s+1} \\ &\quad + \sum_{n=2}^{\infty} \{[(s+n)(s+n-1) + (s+n) - 1]a_n - a_{n-2}\} x^{s+n}. \end{aligned}$$

Thus, the indicial equation is

$$I(s) = s^2 - 1 = (s + 1)(s - 1) = 0,$$

with roots $s_2 = -1$ and $s_1 = 1$. Setting the coefficient of the second term equal to zero,

$$s(s + 2)a_1 = 0,$$

and because the coefficient of a_1 does not equal zero for either $s = s_2 = -1$ or $s = s_1 = 1$, we must have $a_1 = 0$. The remaining terms provide the recurrence relation

$$(s + n + 1)(s + n - 1)a_n - a_{n-2} = 0, \quad n \geq 2.$$

Starting with the smaller root, $s = s_2 = -1$, this becomes

$$n(n - 2)a_n = a_{n-2}, \quad n \geq 2.$$

However, for $N = s_1 - s_2 = 1 - (-1) = 2$, this becomes

$$0a_2 = a_0 \neq 0,$$

so this equation has no solution and we have a real degeneracy. Set

$$y(s, x) = x^s \sum_{n=0}^{\infty} a_n(s) x^n = \sum_{n=0}^{\infty} a_n(s) x^{s+n}$$

and compute a_n as a function of s.

$$a_n = \frac{a_{n-2}}{(s + n + 1)(s + n - 1)}, \quad n \geq 2.$$

Because $a_1 = 0$, this gives us that $a_{2k+1}(s) = 0$ for all $k \geq 0$. Starting with $a_0 \neq 0$,

$$a_{2n} = \frac{a_{2n-2}}{(s + 2n + 1)(s + 2n - 1)}, \quad n \geq 1.$$

Thus,

$$a_2 = \frac{a_0}{(s+3)(s+1)}$$
$$a_4 = \frac{a_2}{(s+5)(s+3)},$$

etc., and

$$a_2 a_4 \cdots a_{2n} = \frac{a_0 a_2 \cdots a_{2n-2}}{[(s+3)(s+5)\cdots(s+2n+1)][(s+1)(s+3)\cdots(s+2n-1)]}.$$

Cancelling,

$$a_{2n}(s) = \frac{a_0}{[(s+3)(s+5)\cdots(s+2n+1)][(s+1)(s+3)\cdots(s+2n-1)]}.$$

As the smaller roots is $s = s_2 = -1$, define

$$Y(s,x) = (s+1)y(s,x) = x^s \sum_{n=0}^{\infty}(s+1)a_{2n}(s)x^{2n} = x^s \sum_{n=0}^{\infty} b_{2n}(s)x^{2n},$$

where $b_{2n}(s) = (s+1)a_{2n}(s)$. For example, $b_0(s) = (s+1)a_0$, $b_2(s) = a_0/(s+3)$, and

$$b_{2n}(s) = \frac{a_0}{[(s+3)(s+5)\cdots(s+2n+1)][(s+3)\cdots(s+2n-1)]}.$$

Evaluating at $s = s_2 = -1$, $b_0(-1) = 0$, $b_2(-1) = a_0/2$, and

$$b_{2n}(-1) = \frac{a_0}{[2\cdot 4\cdot \ldots \cdot 2n][2\cdot 4\cdot \ldots \cdot (2n-2)]}$$
$$= \frac{a_0}{[2^n n!][2^{n-1}(n-1)!]}$$
$$= \frac{a_0}{2^{2n-1}n!(n-1)!}.$$

Thus, remembering

$$Y(s,x) = x^s \sum_{n=0}^{\infty} b_{2n}(s)x^{2n},$$

and choosing $a_0 = 1$ with $s = s_2 = -1$,

$$y_1(x) = Y(-1,x) = x^{-1}\left[\frac{1}{2}x^2 + \sum_{n=2}^{\infty}\frac{x^{2n}}{2^{2n-1}n!(n-1)!}\right]$$
$$= \frac{1}{2}x + \sum_{n=2}^{\infty}\frac{x^{2n-1}}{2^{2n-1}n!(n-1)!}.$$

A second solution is found by taking the derivative of

$$Y(s,x) = (s+1)y(s,x) = x^s \sum_{n=0}^{\infty} b_{2n}(s)x^{2n}$$

with respect to s.

$$\frac{\partial Y}{\partial s} = x^s \ln x \sum_{n=0}^{\infty} b_{2n}(s)x^{2n} + x^s \sum_{n=0}^{\infty} b'_{2n}(s)x^{2n}$$
$$= y_1(x)\ln x + x^s \sum_{n=0}^{\infty} b'_{2n}(s)x^{2n}.$$

Thus, we need to differentiate $b_{2n}(s)$. First, $b'_0(s) = a_0$, $b'_2(s) = -a_0/(s+3)^2$, so $b'_0(-1) = a_0$ and $b'_2(-1) = -a_0/4$. Next, using logarithmic differentiation,

$$b'_{2n}(s) = b_{2n}(s)\left[-\frac{1}{s+3} - \frac{1}{s+5} - \cdots - \frac{1}{s+2n+1} - \frac{1}{s+3} - \frac{1}{s+5} - \cdots - \frac{1}{s+2n-1}\right],$$

and

$$\begin{aligned} b'_{2n}(-1) &= -b_{2n}(-1)\left[\frac{1}{2} + \frac{1}{4} + \cdots + \frac{1}{2n} + \frac{1}{2} + \frac{1}{4} + \cdots + \frac{1}{2n-2}\right] \\ &= -\frac{a_0}{2^{2n-1}n!(n-1)!}\left[1 + \frac{1}{2} + \frac{1}{3} + \cdots + \frac{1}{n-1} + \frac{1}{2n}\right] \\ &= -\frac{a_0}{2^{2n-1}n!(n-1)!}\left[H(n-1) + \frac{1}{2n}\right], \\ &= \frac{-a_0[2nH(n-1)+1]}{2^{2n}(n!)^2}, \end{aligned}$$

where $H(n) = \sum_{k=1}^{n} 1/k$. Remembering that $s = s_2 = -1$ and choosing $a_0 = 1$, a second independent solution is

$$\begin{aligned} y_2(x) &= y_1(x)\ln x + x^{-1}\sum_{n=0}^{\infty} b'_{2n}(-1)x^{2n} \\ &= y_1(x)\ln x + x^{-1}\left[1 - \frac{1}{4}x^2 - \sum_{n=2}^{\infty}\frac{[2nH(n-1)+1]}{2^{2n}(n!)^2}x^{2n}\right]. \end{aligned}$$

21. Compare $x^2y'' + 2x(x-2)y' + 2(2-3x)y = 0$ with $x^2y'' + xp(x)y' + q(x)y = 0$ and note that $p(x) = 2x - 4$ and $q(x) = 4 - 6x$ are both analytic at $x = 0$, making $x = 0$ a regular singular point. Further, $p_0 = p(0) = -4$ and $q_0 = q(0) = 4$, so the indicial equation should be

$$I(s) = x^2 + (p_0 - 1)s + q_0 = s^2 - 5s + 4 = (s-1)(s-4) = 0.$$

Set

$$y(x) = x^s \sum_{n=0}^{\infty} a_n x^n = \sum_{n=0}^{\infty} a_n x^{s+n}.$$

Substituting,

$$\begin{aligned} 0 &= x^2y'' + x(2x-4)y' + (4-6x)y \\ &= [s(s-1) - 4s + 4]a_0x^s \\ &\quad + \sum_{n=1}^{\infty}\{[(s+n)(s+n-1) - 4(s+n) + 4]a_n + [2(s+n-1) - 6]a_{n-1}(s)\}x^{s+n}. \end{aligned}$$

Thus, the indicial equation is

$$I(s) = s^2 - 5s + 4 = (s - 4)(s - 1) = 0,$$

with roots $s_2 = 1$ and $s_1 = 4$. The remaining terms provide the recurrence relation

$$(s + n - 4)(s + n - 1)a_n + 2(s + n - 4)a_{n-1}(s) = 0, \quad n \geq 1.$$

Working with the smaller root, $s = s_2 = 1$, this becomes

$$n(n - 3)a_n = -2(n - 3)a_{n-1}(s), \quad n \geq 1.$$

However, for $N = s_1 - s_2 = 4 - 1 = 3$, we have

$$0a_3 = 0,$$

so this is a false degeneracy. Providing $n \neq 3$, we have

$$a_n = \frac{-2a_{n-1}(s)}{n},$$

so

$$a_1 = -2a_0 \quad \text{and} \quad a_2 = \frac{-2a_1}{2} = 2a_0.$$

Letting a_3 be an arbitrary constant,

$$a_4 = \frac{-2a_3}{4},$$
$$a_5 = \frac{-2a_4}{5},$$

etc., and

$$a_4 a_5 \ldots a_n = \frac{(-2)^{n-3} a_3 a_4 \ldots a_{n-1}(s)}{4 \cdot 5 \cdot \ldots \cdot n}.$$

Cancelling,

$$a_n = \frac{(-2)^{n-3} a_3}{4 \cdot 5 \cdot \ldots \cdot n} = \frac{6(-2)^{n-3} a_3}{n!}, \quad n \geq 4.$$

Thus,

$$y(x) = x[a_0 - 2a_0 x + 2a_0 x^2] + x\left[a_3 x^3 + \sum_{n=4}^{\infty} \frac{6(-2)^{n-3}}{n!} x^n\right]$$
$$= a_0 x[1 - 2x + 2x^2] + a_3 x\left[x^3 + \sum_{n=4}^{\infty} \frac{6(-2)^{n-3}}{n!} x^n\right].$$

Setting $a_0 = 1$ and $a_3 = 0$ provides

$$y_1(x) = x - 2x^2 + 2x^3.$$

Setting $a_0 = 0$ and $a_3 = 1$ provides a second independent solution

$$y_2(x) = x^4 + \sum_{n=4}^{\infty} \frac{6(-2)^{n-3}}{n!} x^{n+1}.$$

23. (a) Divide by $1 - x^2$ to get the equation

$$y'' - \frac{2x}{1-x^2}y' + \frac{n(n+1)}{1-x^2}y = 0.$$

The coefficients are $P(x) = -2x/(1-x^2)$ and $Q(x) = n(n+1)/(1-x^2)$. We see that $(x-1)P(x) = 2x/(x+1)$ and $(x-1)^2Q(x) = -n(n+1)(x-1)/(x+1)$ are both analytic at $x = x_0 = 1$, so $x_0 = 1$ is a regular singular point. Similarly, $(x+1)P(x) = 2x/(x-1)$ and $(x+1)^2Q(x) = -n(n+1)(x+1)/(x-1)$ are both analytic at $x = x_0 = -1$, so $x_0 = -1$ is a regular singular point.

(b) We make the substitution $x = t + 1$, or $t = x - 1$, and the equation becomes

$$-t(2+t)y'' - 2(1+t)y' + n(n+1)y = 0.$$

We multiply by $-t$ so that the fundamental expressions are present, and the equation becomes

$$(2+t)t^2y'' + 2(1+t)ty' - n(n+1)ty = 0.$$

Separating powers of t times fundamental expressions, this becomes

$$[2t^2y'' + 2ty'] + t[t^2y'' + 2ty' - n(n+1)y] = 0.$$

Substituting the series for the fundamental expressions, we get

$$\sum_{n=0}^{\infty}[2(s+n)(s+n-1) + 2(s+n)]a_nt^{s+n}$$
$$+t\sum_{n=0}^{\infty}[(s+n)(s+n-1) + 2(s+n) - n(n+1)]a_nt^{s+n+1}$$
$$= 0.$$

The indicial polynomial is the coefficient of t^s, and it comes form the first term in the first sum. Thus $I(s) = 2s^2$. We have a double root at $s = 0$, so there is only one Frobenius solution. However, since $s = 0$, this solution has the form $y_1(t) = \sum_{n=0}^{\infty} a_nt^n$, where $a_0 \neq 0$. Hence y_1 is analytic at $t = 0$, and $y_1(0) \neq 0$. In terms of x we have the solution $Y_1(x) = y_1(x-1) = \sum_{n=0}^{\infty} a_n(x-1)^n$, so Y_1 is analytic at $x = x_0 = 1$, and therefore bounded there. By Theorem 6.46, the second solution has the form

$$Y_2(x) = Y_1(x)\ln|x-1| + \sum_{n=1}^{\infty} b_n(x-1)^n.$$

Since $\ln|x-1|$ approaches $-\infty$ as $x \to 1$ and $Y_1(1) \neq 0$, this solution is unbounded.

(c) We make the substitution $x = t - 1$, or $t = x + 1$, and the equation becomes

$$t(2-t)y'' - 2(t-1)y' + n(n+1)y = 0.$$

We multiply by t so that the fundamental expressions are present, and the equation becomes

$$(2-t)t^2y'' + 2(1-t)ty' + n(n+1)ty = 0.$$

Separating powers of t times fundamental expressions, this becomes

$$[2t^2y'' + 2ty'] + t[-t^2y'' - 2ty' + n(n+1)y] = 0.$$

Substituting the series for the fundamental expressions, we get

$$\sum_{n=0}^{\infty}[2(s+n)(s+n-1)+2(s+n)]a_n t^{s+n}$$
$$+t\sum_{n=0}^{\infty}[-(s+n)(s+n-1)-2(s+n)+n(n+1)]a_n t^{s+n+1}$$
$$=0.$$

The indicial polynomial is the coefficient of t^s, and it comes form the first term in the first sum. Thus $I(s)=2s^2$. We have a double root at $s=0$, so there is only one Frobenius solution. However, since $s=0$, this solution has the form $y_1(t)=\sum_{n=0}^{\infty}a_n t^n$, where $a_0\neq 0$. Hence y_1 is analytic at $t=0$, and $y_1(0)\neq 0$. In terms of x we have the solution $Y_1(x)=y_1(x+1)=\sum_{n=0}^{\infty}a_n(x+1)^n$, so Y_1 is analytic at $x=x_0=-1$, and therefore bounded there. By Theorem 6.46, the second solution has the form

$$Y_2(x)=Y_1(x)\ln|x+1|+\sum_{n=1}^{\infty}b_n(x+1)^n.$$

Since $\ln|x+1|$ approaches $-\infty$ as $x\to -1$ and $Y_1(-1)\neq 0$, this solution is unbounded.

———×———

Section 11.7. Bessel's Equation and Bessel Functions

1. Using

$$J_r(x)=\sum_{n=0}^{\infty}\frac{(-1)^n}{n!(r+n)!}\left(\frac{x}{2}\right)^{2n+r},$$

we can write

$$J_{1/2}(x)=\sum_{n=0}^{\infty}\frac{(-1)^n}{n!(1/2+n)!}\left(\frac{x}{2}\right)^{2n+1/2}=\left(\frac{x}{2}\right)^{-1/2}\sum_{n=0}^{\infty}\frac{(-1)^n}{n!(1/2+n)!}\left(\frac{x}{2}\right)^{2n+1}$$
$$=\sqrt{\frac{2}{x}}\sum_{n=0}^{\infty}\frac{(-1)^n x^{2n+1}}{2^n n! 2^{n+1}(1/2+n)!}$$

Now,

$$2^n n!=2^n n(n-1)\cdots 1=2n(2n-2)\cdots 2.$$

Also,

$$2^{n+1}\left(\frac{1}{2}+n\right)! = 2^{n+1}\Gamma\left(\frac{1}{2}+n+1\right)$$
$$= 2^{n+1}\left(\frac{1}{2}+n\right)\Gamma\left(\frac{1}{2}+n\right)$$
$$= 2^{n+1}\left(n+\frac{1}{2}\right)\left(n-\frac{1}{2}\right)\cdots\left(\frac{1}{2}\right)\Gamma\left(\frac{1}{2}\right)$$
$$= (2n+1)(2n-1)\cdots 1\cdot\sqrt{\pi}.$$

Thus, rearranging terms,

$$2^n n! 2^{n+1}\left(\frac{1}{2}+n\right)! = (2n+1)(2n)(2n-1)(2n-2)\cdots(2)(1)\sqrt{\pi} = (2n+1)!\sqrt{\pi}.$$

Therefore,

$$J_{1/2}(x) = \sqrt{\frac{2}{x}}\sum_{n=0}^{\infty}\frac{(-1)^n x^{2n+1}}{(2n+1)!\sqrt{\pi}} = \sqrt{\frac{2}{\pi x}}\sum_{n=0}^{\infty}\frac{(-1)^n x^{2n+1}}{(2n+1)!} = \sqrt{\frac{2}{\pi x}}\sin x.$$

In a similar fashion

$$J_{-1/2}(x) = \sum_{n=0}^{\infty}\frac{(-1)^n}{n!(n-1/2)!}\left(\frac{x}{2}\right)^{2n-1/2} = \left(\frac{x}{2}\right)^{-1/2}\sum_{n=0}^{\infty}\frac{(-1)^n x^{2n}}{2^n n! 2^n (n-1/2)!}.$$

Again,

$$2^n n! 2^n (n-1/2)! = 2^n n(n-1)\cdots 1\cdot 2^n\left(n-\frac{1}{2}\right)\left(n-\frac{3}{2}\right)\cdots\left(\frac{1}{2}\right)\Gamma\left(\frac{1}{2}\right)$$
$$= 2n(2n-2)\cdots 2\cdot(2n-1)(2n-3)\cdots 1\cdot\sqrt{\pi}$$
$$= (2n)!\sqrt{\pi}.$$

Thus,

$$J_{-1/2}(x) = \sqrt{\frac{2}{x}}\sum_{n=0}^{\infty}\frac{(-1)^n x^{2n}}{(2n)!\sqrt{\pi}} = \sqrt{\frac{2}{\pi x}}\sum_{n=0}^{\infty}\frac{(-1)^n x^{2n}}{(2n)!} = \sqrt{\frac{2}{\pi x}}\cos x.$$

3. By definition, we can write

$$x^{-p}J_p(x) = x^{-p}\sum_{n=0}^{\infty}\frac{(-1)^n}{n!(n+p)!}\left(\frac{x}{2}\right)^{2n+p} = x^{-p}\sum_{n=0}^{\infty}\frac{(-1)^n x^{2n+p}}{n!(n+p)!2^{2n+p}} = \sum_{n=0}^{\infty}\frac{(-1)^n x^{2n}}{n!(n+p)!2^{2n+p}}.$$

Thus,

$$[x^{-p}J_p(x)]' = \sum_{n=1}^{\infty}\frac{(-1)^n 2n x^{2n-1}}{n!(n+p)!2^{2n+p}} = \sum_{n=1}^{\infty}\frac{(-1)^n x^{2n-1}}{(n-1)!(n+p)!2^{2n+p-1}},$$

so

$$[x^{-p}J_p(x)]' = \sum_{n=0}^{\infty}\frac{(-1)^{n+1}x^{2n+1}}{n!(n+p+1)!2^{2n+p+1}} = -x^{-p}\sum_{n=0}^{\infty}\frac{(-1)^n x^{2n+p+1}}{n!(n+p+1)!2^{2n+p+1}} = -x^{-p}J_{p+1}(x).$$

5. In Exercise 4, we showed that

$$J_p' = J_{p-1} - \frac{p}{x} J_p$$
$$J_p' = -J_{p+1} + \frac{p}{x} J_p.$$

Thus,

$$J_{p-1} - \frac{p}{x} J_p = -J_{p+1} + \frac{p}{x} J_p$$
$$J_{p+1} = \frac{2p}{x} J_p - J_{p-1}.$$

Similarly, if we add the two original equations, we get

$$2J_p' = J_{p-1} - J_{p+1}.$$

7. Let a and b be consecutive zeros of J_0. That is, $J_0(a) = J_0(b) = 0$. By Rolle's theorem, there exists a $a < c < b$ so that $J_0'(c) = 0$. Use $J_p' = -J_{p+1} + (p/x)J_p$ with $p = 0$ to obtain

$$J_0' = -J_1.$$

Thus, $J_1(c) = -J_0'(c) = 0$. Therefore, there exists a zero of J_1 between any consecutive zeros of J_0.
On the other hand, let α and β be consecutive zeros of J_1; i.e., $J_1(\alpha) = J_1(\beta) = 0$. Then α and β are consecutive zeros of $xJ_1(x)$. Let $g(x) = xJ_1(x)$. By Rolle's Theorem, there exists $\alpha < \gamma < \beta$ so that $g'(\gamma) = 0$. However, using $J_p' = J_{p-1} - (p/x)J_p$ with $p = 1$,

$$J_1' = J_0 - \frac{1}{x} J_1$$
$$xJ_1' = xJ_0 - J_1$$
$$xJ_0 = xJ_1' + J_1$$
$$xJ_0 = \frac{d}{dx}\{xJ_1\}$$
$$xJ_0 = g'(x).$$

Hence, $xJ_0(x) = g'(x)$ and

$$\gamma J_0(\gamma) = g'(\gamma) = 0.$$

Because $\gamma > 0$, $J_0(\gamma) = 0$. Hence, between any two consecutive zeros of J_1 lies a zero of J_0.

9. Consider

$$x^2y'' + xy' + (\mu^2x^2 - p^2)y = 0,$$

then let $y(x) = u(\mu x)$. That is, set $t = \mu x$, then

$$y' = \frac{dy}{dx} = \frac{dy}{dt}\frac{dt}{dx} = \mu\frac{dy}{dt},$$

and

$$y'' = \frac{d}{dx}\left(\mu\frac{dy}{dt}\right) = \mu\frac{d^2y}{dt^2}\frac{dt}{dx} = \mu^2\frac{d^2y}{dt^2}.$$

Substituting,

$$\left(\frac{t^2}{\mu^2}\right)(\mu^2y'') + \left(\frac{t}{\mu}\right)(\mu y') + (t^2 - p^2)y = 0,$$

where y' now means differentiation with respect to t. Thus, our equation becomes

$$t^2y'' + ty' + (t^2 - p^2)y = 0,$$

which has general solution

$$y(t) = AJ_p(t) + BY_p(t).$$

Of course, with $t = \mu x$, this becomes,

$$y(x) = aJ_p(\mu x) + bY_p(\mu x),$$

where a and b are arbitrary constants.

11. With $t = ax^b$,

$$dt = abx^{b-1}dx \quad \text{and} \quad \frac{dx}{dt} = \frac{1}{ab}x^{1-b}.$$

Hence, we have the following operator.

$$\frac{d}{dt} = \frac{d}{dx}\frac{dx}{dt} = \frac{1}{ab}x^{1-b}\frac{d}{dx}.$$

Now, we can write

$$t^2\frac{d^2u}{dt^2} + t\frac{du}{dt} + (t^2 - r^2)u = 0$$

as

$$t\frac{d}{dt}\left(t\frac{du}{dt}\right) + (t^2 - r^2)u = 0.$$

Substituting $t = ax^b$ and using our operator,

$$ax^b\left(\frac{1}{ab}x^{1-b}\frac{d}{dx}\left(ax^b\left(\frac{1}{ab}x^{1-b}\frac{du}{dx}\right)\right)\right) + (a^2x^{2b} - r^2)u = 0$$

$$\frac{1}{b}x\frac{d}{dx}\left(\frac{1}{b}x\frac{du}{dx}\right) + (a^2x^{2b} - r^2)u = 0$$

$$x\frac{d}{dx}\left(x\frac{du}{dx}\right) + (a^2b^2x^{2b} - b^2r^2)u = 0.$$

Now, with $u = x^c y$,

$$x\frac{du}{dx} = x\left[x^c\frac{dy}{dx} + cx^{c-1}y\right] = x^{c+1}\frac{dy}{dx} + cx^c y.$$

Substituting these in our last result,

$$x\frac{d}{dx}\left[x^{c+1}\frac{dy}{dx} + cx^c y\right] + (a^2b^2x^{2b} - b^2r^2)x^c y = 0$$

$$x\left[x^{c+1}\frac{d^2y}{dx^2} + (c+1)x^c\frac{dy}{dx} + cx^c\frac{dy}{dx} + c^2x^{c-1}y\right] + (a^2b^2x^{2b} - b^2r^2)x^c y = 0$$

$$x^{c+2}\frac{d^2y}{dx^2} + x^{c+1}[2c+1]\frac{dy}{dx} + (a^2b^2x^{2b} - b^2r^2 + c^2)x^c y = 0.$$

Dividing by x^c,

$$x^2\frac{d^2y}{dx^2} + (2c+1)x\frac{dx}{dy} + (a^2b^2x^{2b} - b^2r^2 + c^2)y = 0.$$

Now, letting $\alpha = 2c+1$, $\beta = ab$, and $\gamma = c^2 - r^2b^2$, this last equation becomes

$$x^2\frac{d^2y}{dx^2} + \alpha x\frac{dy}{dx} + (\beta^2x^{2b} + \gamma)y = 0.$$

13. According to the text, the roots of the indicial equation for Bessel's equation of order 2 are $s = \pm 2$. All of the odd-numbered coefficients are equal to 0, and by 7.4, the recurrence formula for the even-numbered terms is

$$a_{2n} = -\frac{a_{2n-2}}{(s+2n-2)(s+2n+2)}.$$

Solving for the first two coefficients as functions of s, we get

$$a_2(s) = -\frac{a_0}{s(s+4)} \quad \text{and} \quad a_4(s) = -\frac{a_2(s)}{(s+2)(s+6)} = \frac{a_0}{s(s+2)(s+4)(s+6)}.$$

Proceeding inductively we find that

$$a_{2n}(s) = \frac{(-1)^n a_0}{s\cdot(s+2)\cdot(s+4)\cdot\ldots\cdot(s+2n-2)\cdot(s+4)\cdot\ldots\cdot(s+2n+2)}, \quad \text{for } n \geq 1.$$

Notice that a_2 is defined near $s = -2$, but for $n \geq 2$, $a_{2n}(s)$ contains the factor $s+2$ in the denominator, and so is undefined at $s = -1$, reflecting the fact that we have an exceptional case.

Following Section 6, we set $b_{2n}(s) = (s+2)a_{2n}(s)$. Then

$$b_0(s) = (s+2)a_0, b_2(s) = -\frac{(s+2)a_0}{s(s+4)}, \quad \text{and} \quad b_4(s) = \frac{a_0}{s(s+4)(s+6)}.$$

For $n \geq 3$

$$b_{2n}(s) = \frac{(-1)^n a_0}{s\cdot(s+4)\cdot\ldots\cdot(s+2n-2)\cdot(s+4)\cdot\ldots\cdot(s+2n+2)}.$$

We form

$$Y(s,x) = \sum_{n=0}^{\infty} b_{2n}(s)x^{s+2n}.$$

Our first solution is $Y(-2,x)$. Evaluating the coefficients at $s = -2$, we get $b_0(-2) = b_2(-2) = 0$, and

$$b_{2n}(-2) = \frac{(-1)^{n+1}a_0}{2^{2n-1}(n-2)!n!}, \quad \text{for } n \geq 2.$$

Thus our solution is

$$
\begin{aligned}
Y(-2,x) &= x^{-2}a_0\sum_{n=2}^{\infty}(-1)^{n+1}\frac{x^{2n}}{2^{2n-1}(n-2)!n!} \\
&= -a_0\sum_{n=0}^{\infty}(-1)^n\frac{x^{2n+2}}{2^{2n+3}n!(n+2)!} \\
&= \frac{-a_0}{2}\sum_{n=0}^{\infty}\frac{(-1)^n}{n!(n+2)!}\left(\frac{x}{2}\right)^{2n+2} \\
&= \frac{-a_0}{2}J_2(x).
\end{aligned}
$$

Our second solution will be

$$
\frac{\partial Y}{\partial s}(-2,x) = Y(-2,x)\ln x + \sum_{n=0}^{\infty} b_{2n}'(-2)x^{-1+2n}.
$$

We compute that $b_0' = a_0$ and $b_2'(-2) = a_0/4$. For $n \geq 2$ we have

$$
\ln|b_{2n}(s)| = \ln|a_0| - \sum_{k=1}^{n}\ln|s+2k+2| - \sum_{k=1}^{n-2}\ln|s+2k+2| - \ln|s|.
$$

Hence

$$
\begin{aligned}
b_{2n}'(-2) &= -b_{2n}(-2)\left[\sum_{k=1}^{n}\frac{1}{2k} + \sum_{k=1}^{n-2}\frac{1}{2k} - \frac{1}{2}\right] \\
&= \frac{(-1)^n a_0[\phi(n)+\phi(n-2)]}{2^{2n}(n-2)!n!} + \frac{1}{2}b_{2n}(-2)
\end{aligned}
$$

for $n \geq 2$. Hence, our second solution is

$$
\begin{aligned}
\frac{\partial Y}{\partial s}(-2,x) = \frac{-a_0}{2}J_2(x)\ln x + a_0x^{-2} + \frac{a_0}{4} + x^{-2}\sum_{n=2}^{\infty}\frac{(-1)^n a_0[\phi(n)+\phi(n-2)]}{2^{2n}(n-2)!n!}x^{2n} \\
+ \frac{1}{2}x^{-2}\sum_{n=0}^{\infty} b_{2n}(-2)x^{2n}
\end{aligned}
$$

The last term will be recognized as $Y(-2,x)/2 = -a_0J_2(x)/4$. If we choose $a_0 = -2$, our solutions are

$$
\begin{aligned}
y_1(x) &= Y(-2,x) = J_2(x) \quad \text{and} \\
y_2(x) &= \frac{\partial Y}{\partial s}(-2,x) \\
&= J_2(x)\ln x - \frac{2}{x^2} - \frac{1}{2} - \frac{1}{2}\sum_{n=0}^{\infty}\frac{(-1)^n[\phi(n)+\phi(n+2)]}{n!(n+2)!}\left(\frac{x}{2}\right)^{2n+2} + \frac{1}{2}J_2(x)
\end{aligned}
$$

Chapter 12. Fourier Series

Section 12.1. Computation of Fourier Series

1. The function $f(x) = |\sin x|$ is even on the interval $[-\pi, \pi]$. Hence, the Fourier expansion will contain only cosine terms. Its Fourier coefficients are

$$a_n = \frac{2}{\pi}\int_0^{\pi} |\sin x| \cos nx\, dx.$$

However, $\sin x \geq 0$ on $[0, \pi]$, so

$$a_n = \frac{2}{\pi}\int_0^{\pi} \sin x \cos nx\, dx.$$

Now, for $n = 0$,

$$a_0 = \frac{2}{\pi}\int_0^{\pi} \sin x dx = \frac{2}{\pi}[-\cos x]_0^{\pi} = \frac{4}{\pi}.$$

For $n = 1$,

$$a_1 = \frac{2}{\pi}\int_0^{\pi} \sin x \cos x dx = 0.$$

Now, for $n > 1$,

$$\begin{aligned}
a_n &= \frac{2}{\pi}\int_0^{\pi} \sin x \cos nx\, dx \\
&= \frac{1}{\pi}\int_0^{\pi} [\sin(1+n)x + \sin(1-n)x]\, dx \\
&= \frac{1}{\pi}\left[\frac{-1}{1+n}\cos(1+n)x - \frac{1}{1-n}\cos(1-n)x\right]_0^{\pi} \\
&= \frac{1}{\pi}\left[\frac{-1}{1+n}(-1)^{1+n} - \frac{1}{1-n}(-1)^{1-n}\right. \\
&\qquad \left. + \frac{1}{1+n} + \frac{1}{1-n}\right] \\
&= \frac{1}{\pi}\left\{\frac{1}{1+n}\left[1-(-1)^{1+n}\right]\right. \\
&\qquad \left. + \frac{1}{1-n}\left[1-(-1)^{1-n}\right]\right\}
\end{aligned}$$

Thus, if n is odd, $a_n = 0$. On the other hand, if n is even,

$$\begin{aligned}
a_n &= \frac{1}{\pi}\left[\frac{2}{1+n} + \frac{2}{1-n}\right] \\
&= \frac{1}{\pi}\left[\frac{2-2n+2+2n}{(1+n)(1-n)}\right] \\
&= \frac{4}{\pi(1-n^2)}.
\end{aligned}$$

Let us now write

$$a_{2n} = \frac{4}{\pi(1-4n^2)},$$

for $n \geq 1$. Thus, the Fourier representation is

$$f(x) \sim \sum_{n=1}^{\infty} \frac{4}{\pi(1-4n^2)} \cos 2nx.$$

The partial sum S_6 on $[-\pi, \pi]$ and $[-3\pi, 3\pi]$ follow.

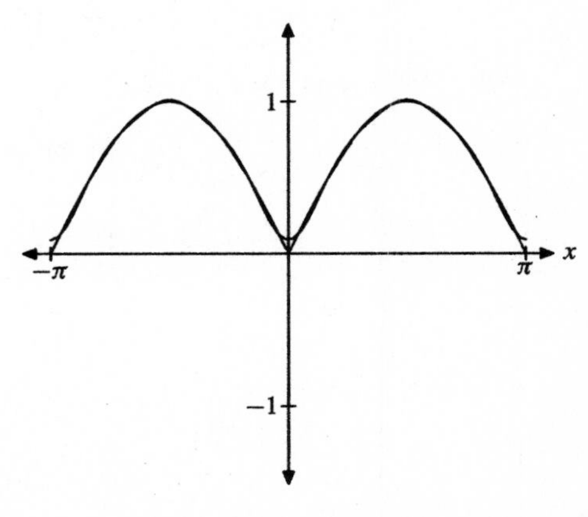

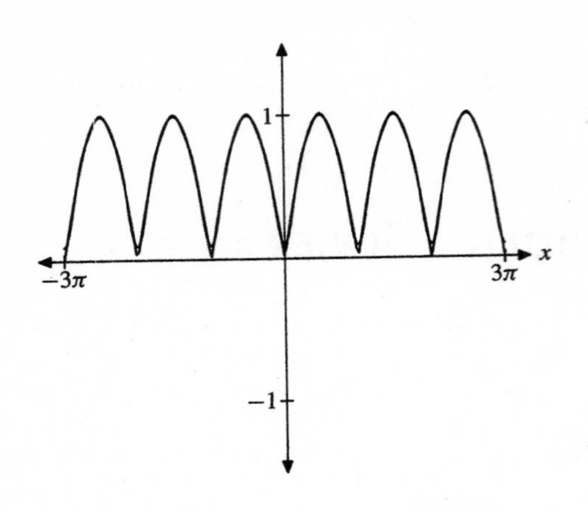

3. On the interval $[-\pi, \pi]$, the function

$$f(x) = \begin{cases} 0, & -\pi \le x < 0, \\ x, & 0 \le x \le \pi, \end{cases}$$

is neither even nor odd, so the Fourier series may have both sine and cosine terms. However, f is identically zero on $[-\pi, 0]$, so

$$a_n = \frac{1}{\pi}\int_0^{\pi} f(x)\cos nx\,dx = \frac{1}{\pi}\int_0^{\pi} x\cos nx\,dx.$$

Integration by parts provides

$$\begin{aligned} a_n &= \frac{1}{\pi}\left[\frac{x\sin nx}{n} + \frac{\cos nx}{n^2}\right]_0^{\pi} \\ &= \frac{1}{\pi}\left[\frac{\cos n\pi}{n^2} - \frac{1}{n^2}\right] \\ &= \frac{1}{\pi n^2}\left[(-1)^n - 1\right], \end{aligned}$$

provided $n \ne 0$. In the case that $n = 0$,

$$a_0 = \frac{1}{\pi}\int_0^{\pi} x\,dx = \frac{1}{\pi}\cdot\frac{1}{2}x^2\Big|_0^{\pi} = \frac{\pi}{2}.$$

Similarly,

$$\begin{aligned} b_n &= \frac{1}{\pi}\int_{-\pi}^{\pi} f(x)\sin nx\,dx \\ &= \frac{1}{\pi}\int_0^{\pi} x\sin nx\,dx. \end{aligned}$$

Integration by parts provides

$$\begin{aligned} b_n &= \frac{1}{\pi}\left[-\frac{x\cos nx}{n} + \frac{\sin nx}{n^2}\right]_0^{\pi} \\ &= \frac{1}{\pi}\left[\frac{-\pi\cos n\pi}{n}\right] \\ &= \frac{(-1)^{n+1}}{n}. \end{aligned}$$

Hence, $f(x)$ has Fourier series expansion

$$\begin{aligned} f(x) &= \frac{a_0}{2} + \sum_{n=1}^{\infty} a_n\cos nx + b_n\sin nx \\ &= \frac{\pi}{4} + \sum_{n=1}^{\infty}\left[\frac{[(-1)^n - 1]}{\pi n^2}\cos nx \right. \\ &\qquad \left. + \frac{(-1)^{n+1}}{n}\sin nx\right] \end{aligned}$$

The partial sum S_6 on $[-\pi, \pi]$ and $[-3\pi, 3\pi]$ follow.

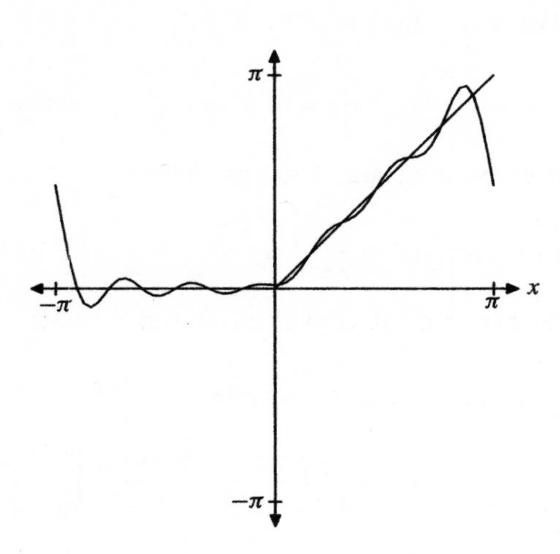

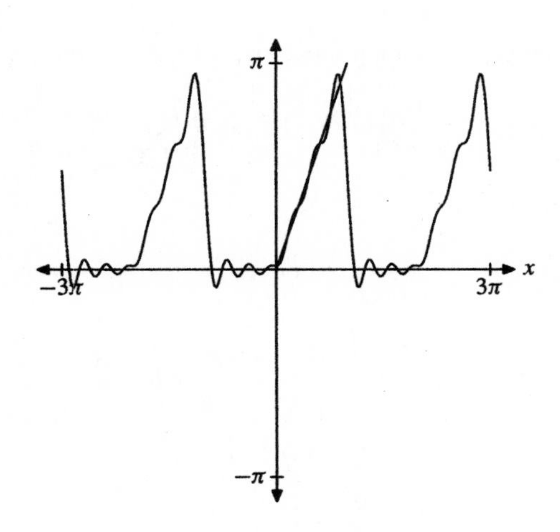

5. The function $f(x) = x\cos x$ is odd on the interval $[-\pi, \pi]$. Thus, the Fourier expansion will contain only sine terms.

$$b_n = \frac{1}{\pi}\int_{-\pi}^{\pi} f(x)\sin nx\,dx$$
$$= \frac{2}{\pi}\int_0^{\pi} f(x)\sin nx\,dx$$
$$= \frac{2}{\pi}\int_0^{\pi} x\cos x\sin nx\,dx.$$

Using a product to sum identity,

$$b_n = \frac{2}{\pi}\int_0^{\pi}\frac{x}{2}[\sin(n-1)x + \sin(n+1)x]\,dx$$
$$= \frac{1}{\pi}\int_0^{\pi} x\sin(n-1)x\,dx$$
$$+ \frac{1}{\pi}\int_0^{\pi} x\sin(n+1)x\,dx.$$

Using integration by parts, the first integral is

$$\frac{1}{\pi}\int_0^{\pi} x\sin(n-1)x\,dx$$
$$= \frac{1}{\pi}\left[\frac{-x\cos(n-1)x}{n-1} + \frac{\sin(n-1)x}{(n-1)^2}\right]_0^{\pi}$$
$$= \frac{1}{\pi}\left[\frac{-\pi\cos(n-1)\pi}{n-1}\right]$$
$$= \frac{-\cos(n-1)\pi}{n-1}$$
$$= \frac{(-1)^n}{n-1},$$

provided $n \neq 1$. Similarly, the second integral is

$$\frac{1}{\pi}\int_0^{\pi} x\sin(n+1)x\,dx = \frac{(-1)^n}{n+1}.$$

Thus,

$$b_n = (-1)^n\left[\frac{1}{n-1} + \frac{1}{n+1}\right] = (-1)^n\frac{2n}{n^2-1},$$

provided $n \neq 1$. In the case that $n = 1$,

$$b_1 = \frac{1}{\pi}\int_0^{\pi} x\cos x\sin x\,dx$$
$$= \frac{1}{2\pi}\int_0^{\pi} x\sin 2x\,dx.$$

Integrating by parts,

$$b_1 = \frac{1}{2\pi}\left[\frac{-x\cos 2x}{2} + \frac{\sin 2x}{4}\right]_0^{\pi}$$
$$= \frac{1}{2\pi}\left[\frac{-\pi}{2}\right]$$
$$= -\frac{1}{4}.$$

Thus, the Fourier series expansion is

$$f(x) = -\frac{1}{4}\sin x + \sum_{n=2}^{\infty}(-1)^n \frac{2n}{n^2-1}\sin nx.$$

The partial sum S_6 on $[-\pi, \pi]$ and $[-3\pi, 3\pi]$ follow.

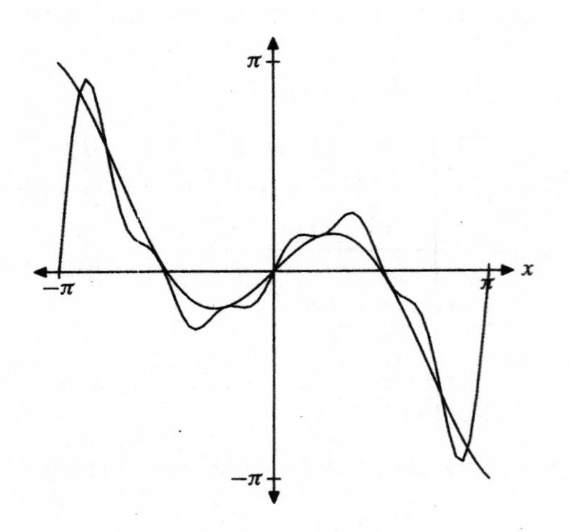

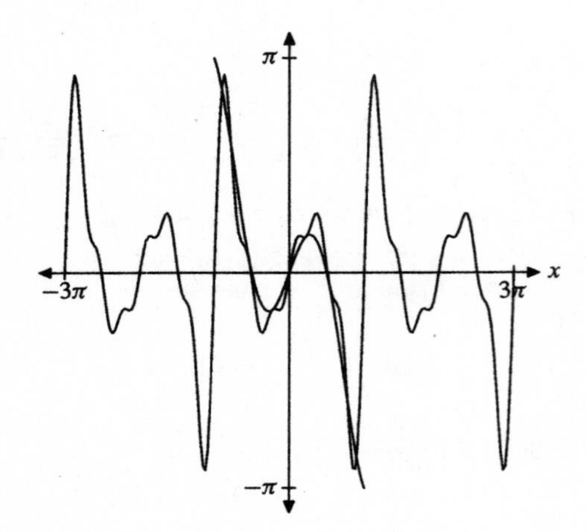

7. If

$$\begin{cases} 1+x, & -1 \le x \le 0, \\ 1, & 0 < x \le 1, \end{cases}$$

then on the interval $[-1, 1]$

$$\begin{aligned} a_n &= \int_{-1}^{1} f(x)\cos n\pi x dx \\ &= \int_{-1}^{0}(1+x)\cos n\pi x dx + \int_{0}^{1}\cos n\pi x dx. \end{aligned}$$

The second integral is straight forward.

$$\int_0^1 \cos n\pi x dx = \frac{1}{n\pi}\sin n\pi x\bigg|_0^1 = \frac{1}{n\pi}(0-0) = 0.$$

The first integral requires integration by parts.

$$\begin{aligned} &\int_{-1}^{0}(1+x)\cos n\pi x dx \\ &= \left[\frac{(1+x)\sin n\pi x}{n\pi} + \frac{\cos n\pi x}{n^2\pi^2}\right]_{-1}^{0} \\ &= \frac{1}{n^2\pi^2}[1-\cos n\pi] \end{aligned}$$

Hence, combining integrals, for $n \ge 1$,

$$a_n = \frac{1}{n^2\pi^2}[1-(-1)^n] = \begin{cases} 0, & n \text{ even}, \\ \frac{2}{n^2\pi^2}, & n \text{ odd}. \end{cases}$$

For $n = 0$,

$$\begin{aligned} a_0 &= \int_{-1}^{1} f(x)dx \\ &= \int_{-1}^{0}(1+x)dx + \int_0^1 dx \\ &= \left(x + \frac{1}{2}x^2\right)\bigg|_{-1}^{0} + 1 \\ &= \frac{3}{2}. \end{aligned}$$

Similarly,

$$\begin{aligned} b_n &= \int_{-1}^{1} f(x)\sin n\pi x dx \\ &= \int_{-1}^{0}(1+x)\sin n\pi x dx + \int_0^1 \sin n\pi x dx \\ &= -\frac{1}{n\pi} - \frac{1}{n\pi}[\cos n\pi - 1] \\ &= \frac{(-1)^{n+1}}{n\pi}. \end{aligned}$$

Thus,

$$f(x) \sim \frac{3}{4} + \frac{2}{\pi^2}\sum_{n=1}^{\infty}\frac{\cos((2n+1)\pi x)}{(2n+1)^2} - \frac{1}{\pi}\sum_{n=1}^{\infty}\frac{(-1)^n}{n}\sin n\pi x.$$

In the following figures, the first displays S_3, the second S_6.

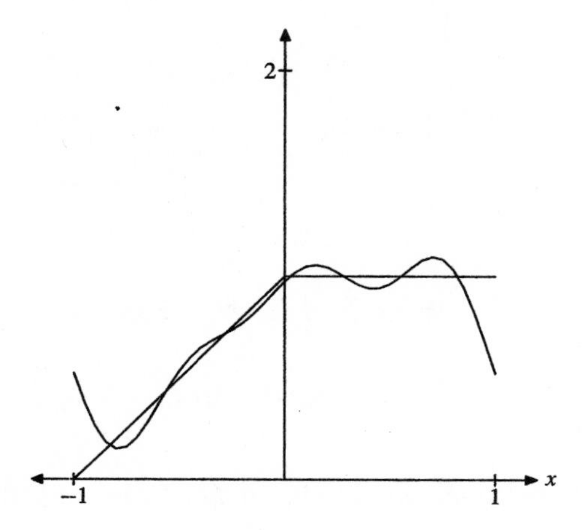

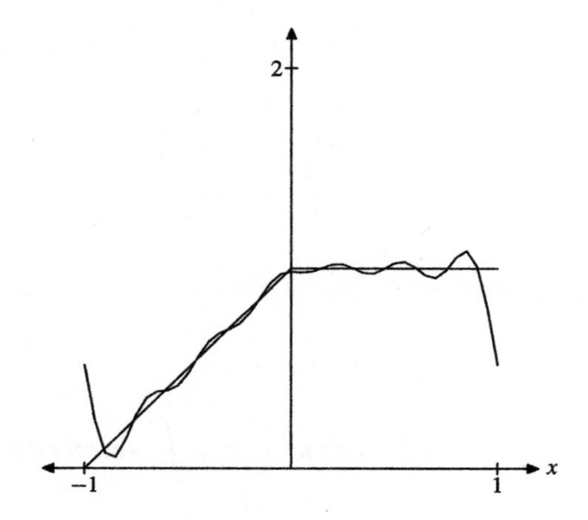

9. Note that $f(x) = x^3$ is odd on the interval $[-1, 1]$. Thus, the Fourier series representation will only contain sine terms with coefficient

$$\begin{aligned} b_n &= 2\int_0^1 f(x)\sin n\pi x\,dx \\ &= 2\int_0^1 x^3 \sin n\pi x\,dx. \end{aligned}$$

Several applications of integration by parts provides

$$\begin{aligned} b_n &= -\frac{2x^3\cos n\pi x}{n\pi} + \frac{6x^2\sin n\pi x}{n^2\pi^2} + \frac{12x\cos n\pi x}{n^3\pi^3} - \frac{12\sin n\pi x}{n^4\pi^4}\Bigg|_0^1 \\ &= -\frac{2\cos n\pi}{n\pi} + \frac{12\cos n\pi}{n^3\pi^3} \\ &= (-1)^n\frac{2(6-n^2\pi^2)}{n^3\pi^3}. \end{aligned}$$

Therefore, the Fourier representation is

$$f(x) \sim \sum_{n=1}^{\infty}(-1)^n\frac{2(6-n^2\pi^2)}{n^3\pi^3}\sin n\pi x.$$

In the following figures, the first displays S_3, the second right S_6.

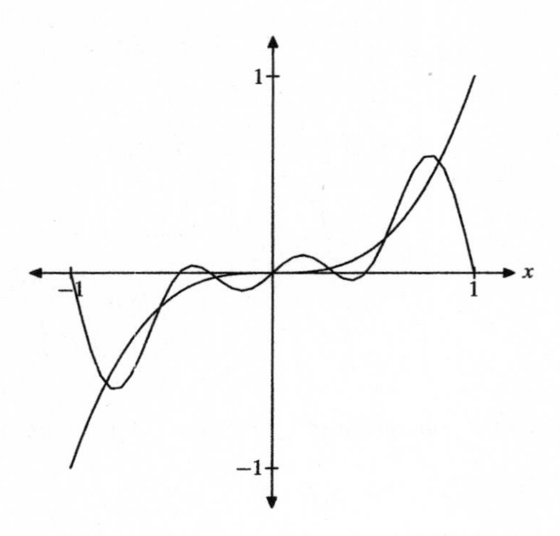

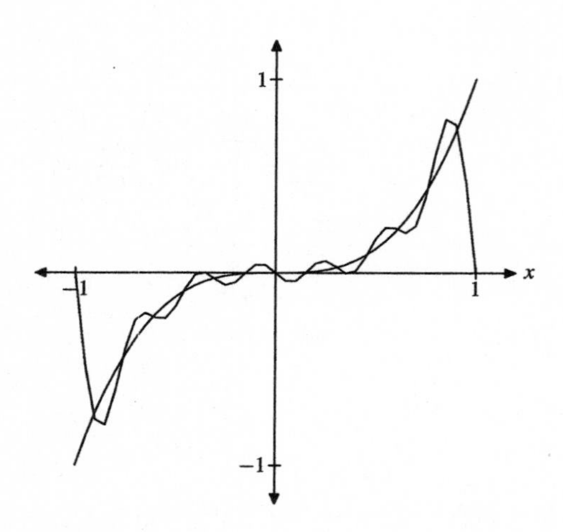

11. If

$$f(x) = \begin{cases} 0, & -1 \leq x \leq 0, \\ x^2, & 0 < x \leq 1, \end{cases}$$

then on the interval $[-1, 1]$

$$a_n = \int_{-1}^{1} f(x) \cos n\pi x dx$$
$$= \int_{0}^{1} x^2 \cos n\pi x dx.$$

With $n = 0$,

$$a_0 = \int_0^1 x^2 dx = \frac{1}{3}x^3\Big|_0^1 = \frac{1}{3}.$$

For $n \geq 1$, several applications of integration by parts provides

$$a_n = \frac{x^2 \sin n\pi x}{n\pi} + \frac{2x \cos n\pi x}{n^2\pi^2} - \frac{2 \sin n\pi x}{n^3\pi^3}\Big|_0^1$$
$$= \frac{2 \cos n\pi}{n^2\pi^2}$$
$$= (-1)^n \frac{2}{n^2\pi^2}.$$

Next, for $n \geq 1$

$$b_n = \int_{-1}^{1} f(x) \sin n\pi x dx$$
$$= \int_{0}^{1} x^2 \sin n\pi x dx.$$

Several applications of integration by parts provides

$$b_n = -\frac{x^2 \cos n\pi x}{n\pi} + \frac{2x \sin n\pi x}{n^2\pi^2} + \frac{2 \cos n\pi x}{n^3\pi^3}\Big|_0^1$$
$$= -\frac{\cos n\pi}{n\pi} + \frac{2 \cos n\pi}{n^3\pi^3} - \frac{2}{n^3\pi^3}$$
$$= \frac{1}{n^3\pi^3}\left[2(-1)^n - n^2\pi^2(-1)^n - 2\right].$$

Thus, the Fourier representation for the function is

$$f(x) \sim \frac{1}{6} + \sum_{n=1}^{\infty} \left[(-1)^n \frac{2}{n^2\pi^2} \cos n\pi x + \frac{2(-1)^n - n^2\pi^2(-1)^n - 2}{n^3\pi^3} \sin n\pi x \right].$$

In the following figures, the first displays S_3, the second S_6.

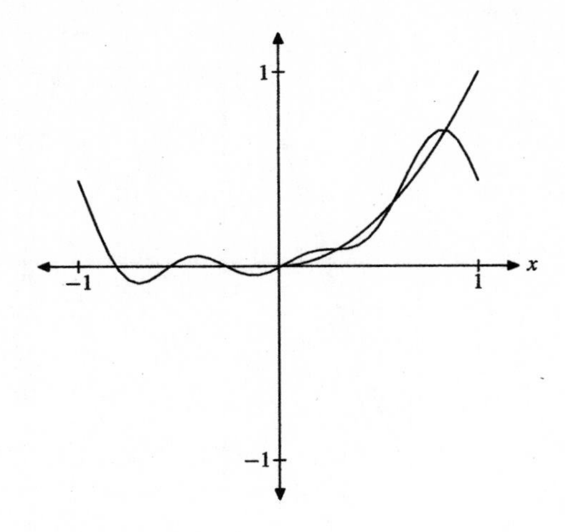

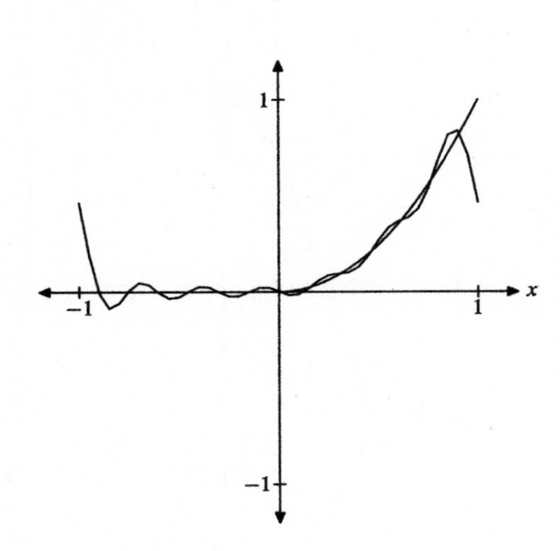

13. If

$$f(x) = \begin{cases} \cos \pi x, & -1 \le x \le 0, \\ 1, & 0 < x \le 1, \end{cases}$$

then on the interval $[-1, 1]$,

$$\begin{aligned} a_n &= \int_{-1}^{1} f(x) \cos n\pi x dx \\ &= \int_{-1}^{0} \cos \pi x \cos n\pi x dx + \int_{0}^{1} \cos n\pi x dx. \end{aligned}$$

For $n = 0$,

$$a_0 = \int_{-1}^{0} \cos \pi x dx + \int_{0}^{1} dx = 1.$$

For $n = 1$, the first integral becomes

$$\begin{aligned} \int_{-1}^{0} \cos^2 \pi x dx &= \int_{-1}^{0} \frac{1 + \cos 2\pi x}{2} dx \\ &= \frac{1}{2} \left[x + \frac{1}{2\pi} \sin 2\pi x \right]_{-1}^{0} \\ &= \frac{1}{2}. \end{aligned}$$

With $n > 1$,

$$\begin{aligned} &\int_{-1}^{0} \cos \pi x \cos n\pi x dx \\ &= \frac{1}{2} \int_{-1}^{0} [\cos(n-1)\pi x + \cos(n+1)\pi x] \, dx \\ &= \frac{1}{2} \left[\frac{\sin(n-1)\pi x}{(n-1)\pi} + \frac{\sin(n+1)\pi x}{(n+1)\pi} \right]_{-1}^{0} \\ &= 0. \end{aligned}$$

With $n \ge 1$, the second integral becomes

$$\int_{0}^{1} \cos n\pi x dx = \left. \frac{\sin n\pi x}{n\pi} \right|_{0}^{1} = 0.$$

In summary, $a_0 = 1$, $a_1 = 1/2$, and $a_n = 0$ for all $n > 1$. Next, with $n \ge 1$,

$$\begin{aligned} b_n &= \int_{-1}^{1} f(x) \sin n\pi x dx \\ &= \int_{-1}^{0} \cos \pi x \sin n\pi x dx + \int_{0}^{1} \sin n\pi x dx. \end{aligned}$$

With $n = 1$, the first integral becomes

$$\int_{-1}^{0} \cos \pi x \sin \pi x dx = \frac{1}{2} \int_{-1}^{0} \sin 2\pi x dx = 0.$$

With $n > 1$, the first integral becomes

$$\int_{-1}^{0} \cos \pi x \sin n\pi x dx$$
$$= \frac{1}{2}\int_{-1}^{0} [\sin(n-1)\pi x + \sin(n+1)\pi x] dx$$
$$= -\frac{1}{2}\left[\frac{\cos(n-1)\pi x}{(n-1)\pi} + \frac{\cos(n+1)\pi x}{(n+1)\pi}\right]_{-1}^{0}$$
$$= -\frac{1}{2}\left\{\left[\frac{1}{(n-1)\pi} + \frac{1}{(n+1)\pi}\right] - \left[\frac{(-1)^{n-1}}{(n-1)\pi} + \frac{(-1)^{n+1}}{(n+1)\pi}\right]\right\},$$

which, after some simplification, equals

$$\int_{-1}^{0} \cos \pi x \sin n\pi x dx = -n\left[\frac{1+(-1)^n}{(n-1)(n+1)\pi}\right].$$

The second integral is straight forward.

$$\int_0^1 \sin n\pi x dx = -\frac{\cos n\pi x}{n\pi}\Big|_0^1 = \frac{1}{n\pi}[1-(-1)^n].$$

In summary,

$$b_n = \begin{cases} \frac{2}{\pi}, & n = 1 \\ -n\left[\frac{1+(-1)^n}{(n-1)(n+1)\pi}\right] + \frac{1}{n\pi}[1-(-1)^n], & n > 1. \end{cases}$$

Thus, the Fourier representation of f is

$$f(x) \sim \frac{1}{2} + \frac{1}{2}\cos \pi x + \frac{2}{\pi}\sin \pi x + \sum_{n=2}^{\infty}\left\{-n\left[\frac{1+(-1)^n}{(n-1)(n+1)\pi}\right] + \frac{1}{n\pi}[1-(-1)^n]\right\}\sin n\pi x.$$

In the following figures, the first S_3, the second S_6.

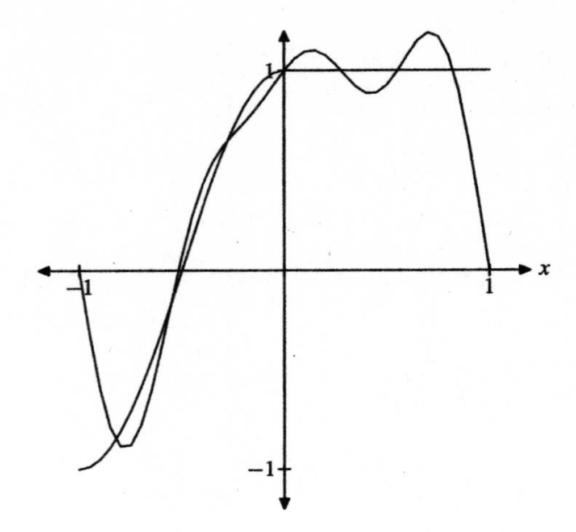

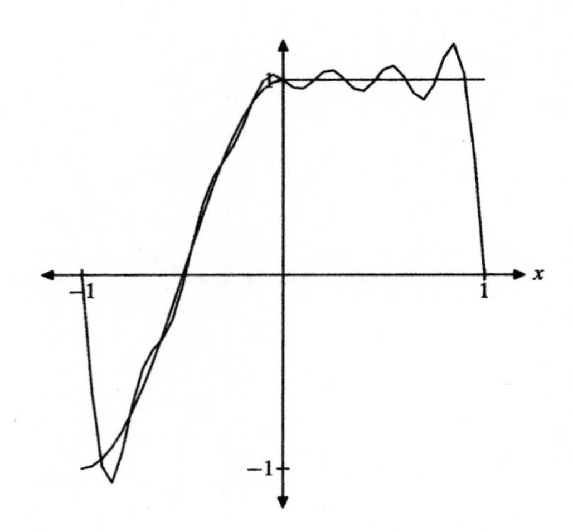

15. If

$$f(x) = \begin{cases} 2+x, & -2 \le x \le 0, \\ -2+x, & 0 < x \le 2, \end{cases}$$

then f is odd on the interval $[-2, 2]$ and we will need only calculate sine terms.

$$b_n = \frac{2}{2}\int_0^2 f(x)\sin\frac{n\pi x}{2}dx$$
$$= \int_0^2 (-2+x)\sin\frac{n\pi x}{2}dx$$

Integration by parts provides

$$b_n = \left[-\frac{2(x-2)\cos(n\pi x/2)}{n\pi} + \frac{4\sin(n\pi x/2)}{n^2\pi^2} \right]_0^2$$
$$= -\frac{4}{n\pi}.$$

Hence, the Fourier representation of the function f is

$$f(x) \sim \sum_{n=1}^{\infty} \frac{-4}{n\pi} \sin\frac{n\pi x}{2}.$$

In the following figures, the first displays S_3, the second S_6.

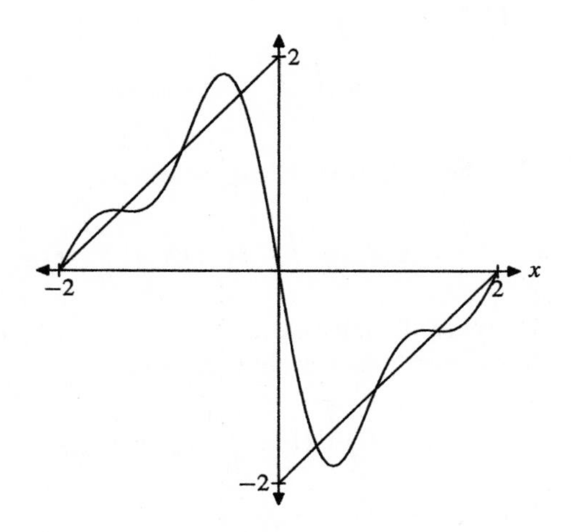

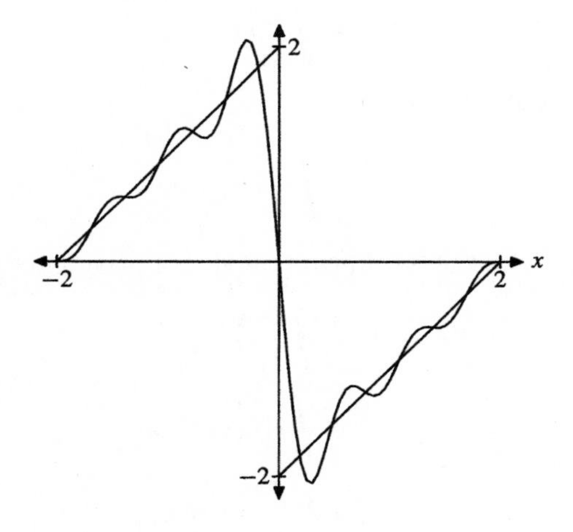

17. Note that $f(x) = x^2$ is even, so the Fourier series will have only cosine terms and

$$a_n = \frac{1}{\pi}\int_{-\pi}^{\pi} x^2 \cos nx\,dx = \frac{2}{\pi}\int_0^{\pi} x^2 \cos nx\,dx.$$

Two applications of integration by parts provides

$$a_n = \frac{2}{\pi}\left[\frac{x^2 \sin nx}{n} + \frac{2x\cos nx}{n^2} - \frac{2\sin nx}{n^3} \right]_0^{\pi}$$
$$= \frac{2}{\pi}\left[\frac{2\pi\cos n\pi}{n^2} \right]$$
$$= \frac{4\cos n\pi}{n^2}$$
$$= (-1)^n \frac{4}{n^2},$$

provided $n \neq 0$. In the case that $n = 0$,

$$a_0 = \frac{2}{\pi}\int_0^{\pi} x^2\,dx = \frac{2}{\pi}\cdot\frac{x^3}{3}\bigg|_0^{\pi} = \frac{2}{3}\pi^2.$$

Hence, $f(x) = x^2$ has Fourier series

$$f(x) = \frac{a_0}{2} + \sum_{n=1}^{\infty} a_n \cos nx$$
$$= \frac{\pi^2}{3} + 4\sum_{n=1}^{\infty} \frac{(-1)^n \cos nx}{n^2}.$$

The partial sum S_7 is displayed on $[-\pi, \pi]$ and $[-2\pi, 2\pi]$.

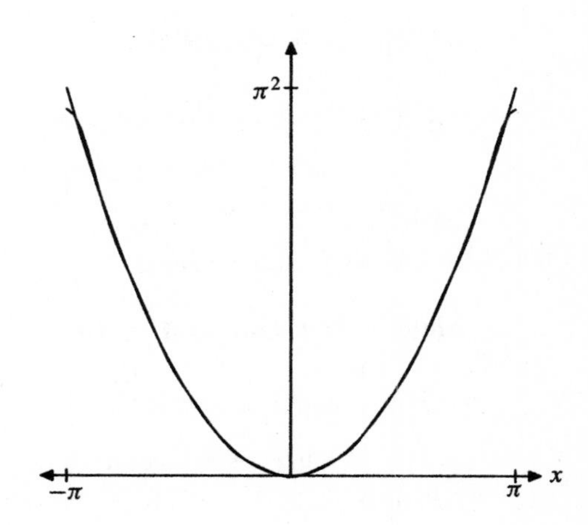

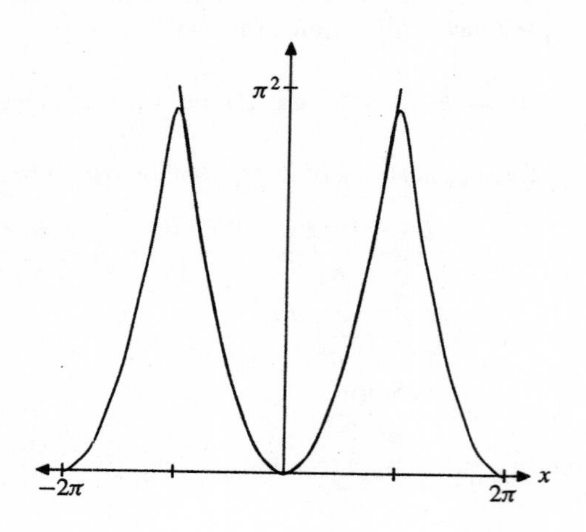

19. Since $\sin(-x) = -\sin(x)$, we have $|\sin(-x)| = |\sin(x)|$. Hence $f(x) = |\sin x|$ is even.

21. Since $f(-1) = e^{-1} = 1/e$, and $f(1) = e$, we see that f is neither even nor odd.

23. Using the cosine expansion,

$$\begin{aligned}&\frac{1}{2}[\cos(\alpha-\beta)+\cos(\alpha+\beta)]\\&=\frac{1}{2}[\cos\alpha\cos\beta+\sin\alpha\sin\beta\\&\qquad+\cos\alpha\cos\beta-\sin\alpha\sin\beta]\\&=\cos\alpha\cos\beta.\end{aligned}$$

Similarly,

$$\begin{aligned}&\frac{1}{2}[\cos(\alpha-\beta)-\cos(\alpha+\beta)]\\&=\frac{1}{2}[\cos\alpha\cos\beta+\sin\alpha\sin\beta\\&\qquad-\cos\alpha\cos\beta+\sin\alpha\sin\beta]\\&=\sin\alpha\sin\beta.\end{aligned}$$

Finally, the sine expansion delivers

$$\begin{aligned}&\frac{1}{2}[\sin(\alpha-\beta)+\sin(\alpha+\beta)]\\&=\frac{1}{2}[\sin\alpha\cos\beta-\cos\alpha\sin\beta\\&\qquad+\sin\alpha\cos\beta+\cos\alpha\sin\beta]\\&=\sin\alpha\cos\beta.\end{aligned}$$

25. We multiply both sides of

$$f(x)=\frac{a_0}{2}+\sum_{k=1}^{\infty}a_k\cos k\pi x+b_k\sin k\pi x$$

by $\sin nx$ and integrate. By the orthogonality relations in Lemma 1.2, all terms will equal zero, save one.

$$\int_{-\pi}^{\pi}f(x)\sin nx\,dx=b_k\int_{-\pi}^{\pi}\sin nx\sin nx\,dx$$
$$\int_{-\pi}^{\pi}f(x)\sin nx\,dx=b_k\pi$$

Hence,

$$b_k=\frac{1}{\pi}\int_{-\pi}^{\pi}f(x)\sin nx\,dx.$$

27. Whether f is even or odd, we can write

$$\int_{-L}^{L}f(x)\,dx=\int_{-L}^{0}f(x)\,dx+\int_{0}^{L}f(x)\,dx.$$

However, if we let $u=-x$, the $du=-dx$ and the first integral on the right becomes

$$\int_{-L}^{0}f(x)\,dx=\int_{L}^{0}f(-u)(-du).$$

If we reverse the bounds, then

$$=\int_{0}^{L}f(-u)\,du.$$

But f is even, so

$$\int_{0}^{L}f(u)\,du.$$

But u is just a dummy variable of integration, so

$$\begin{aligned}\int_{-L}^{L}f(x)\,dx&=\int_{-L}^{0}f(x)\,dx+\int_{0}^{L}f(x)\,dx\\&=\int_{0}^{L}f(x)\,dx+\int_{0}^{L}f(x)\,dx\\&=2\int_{0}^{L}f(x)\,dx.\end{aligned}$$

If f is odd, then with $u = -x$ and $du = -dx$,

$$\int_{-L}^{0} f(x)\,dx = \int_{L}^{0} f(-u)\,(-du).$$

Reversing the bounds,

$$= \int_{0}^{L} f(-u)\,du.$$

But f is odd, so

$$-\int_{0}^{L} f(u)\,du.$$

But u is just a dummy variable of integration, so

$$\begin{aligned}\int_{-L}^{L} f(x)\,dx &= \int_{-L}^{0} f(x)\,dx + \int_{0}^{L} f(x)\,dx\\ &= -\int_{0}^{L} f(x)\,dx + \int_{0}^{L} f(x)\,dx\\ &= 0.\end{aligned}$$

29. We start with f being odd on the interval $[-L, L]$. Then the Fourier representation contains only sine terms. Thus,

$$b_n = \frac{2}{L}\int_{0}^{L} f(x)\sin\frac{n\pi x}{L}\,dx.$$

Now, suppose we have the additional symmetry that f is symmetrical with respect to the line $x = L/2$. That is, $f(L-x) = f(x)$ for all $0 \le x \le L$. We can write

$$b_n = \frac{2}{L}\left[\int_{0}^{L/2} f(x)\sin\frac{n\pi x}{L}\,dx + \int_{L/2}^{L} f(x)\sin\frac{n\pi x}{L}\,dx\right].$$

Now, consider the second integral with the substitution

$$u = L - x \quad \text{and} \quad du = -dx.$$

Then,

$$\begin{aligned}&\int_{L/2}^{L} f(x)\sin\frac{n\pi x}{L}dx\\ &= \int_{L/2}^{0} f(L-u)\sin\frac{n\pi(L-u)}{L}(-du)\\ &= \int_{0}^{L/2} f(L-u)\sin\left(n\pi - \frac{n\pi u}{L}\right)du.\end{aligned}$$

But $f(L-u) = f(u)$, and the sine expansion provides

$$\begin{aligned}&\int_{0}^{L/2} f(u)\left[\sin n\pi\cos\frac{n\pi u}{L} - \sin\frac{n\pi u}{L}\cos n\pi\right]du\\ &= -(-1)^n\int_{0}^{L/2} f(u)\sin\frac{n\pi u}{L}\,du.\end{aligned}$$

Hence, if n is even,

$$\begin{aligned}b_n &= \frac{2}{L}\left[\int_{0}^{L/2} f(x)\sin\frac{n\pi x}{L}\,dx - \int_{0}^{L/2} f(u)\sin\frac{n\pi u}{L}\,du\right]\\ &= 0.\end{aligned}$$

As an example, consider

$$f(x) = \begin{cases} -x-2, & -2 \le x < -1\\ x, & -1 \le x < 1\\ -x+2, & 1 \le x \le 2\end{cases}.$$

Note that f is symmetric about $x = 1$ on $[0, 2]$. That is, $f(2-x) = f(x)$ for $0 \le x \le 2$.

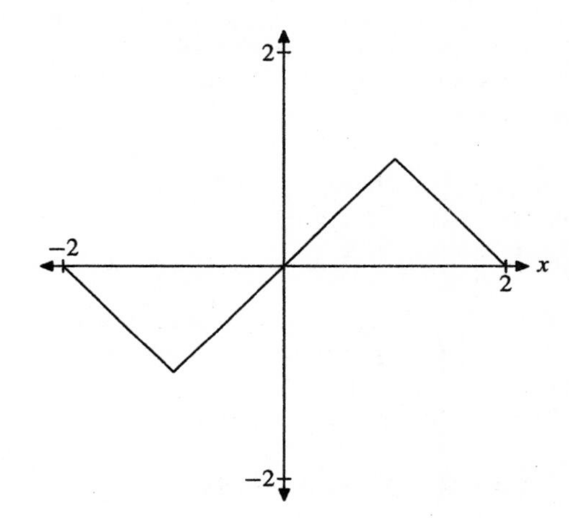

31. Set $f_{\text{odd}}(x) = [f(x) - f(-x)]/2$ and $f_{\text{even}}(x) = [f(x) + f(-x)]/2$. Then f_{odd} is off and f_{even} is even, and $f(x) = f_{\text{odd}}(x) + f_{\text{even}}(x)$ for all x.

Section 12.2. Convergence of Fourier Series

1. Since $\sin(x+\pi) = -\sin x$, $|\sin(x+\pi)| = |\sin x|$. Hence f is periodic with period π. For any $T < \pi$, we have $\sin T > 0 = \sin 0$. Hence f cannot be periodic with period T. Therefore the smallest period is π.

3. Since f is strictly increasing, $f(x) \neq f(y)$ as long as $x \neq y$. Hence f cannot be periodic.

5. Since $f(0) = 0$ and $f(x) \neq 0$ for $x \neq 0$, f cannot be periodic.

7. f_p has one-sided derivatives everywhere, and is continuous everywhere except at odd multiples of π. Therefore, the Fourier series converges to $f_p(x)$ except at the odd multiples of π. At the odd multiples of π the series converges to $[\pi + 0]/2 = \pi/2$.

9. f_p has one-sided derivatives everywhere, and is continuous everywhere except at the odd integers. Therefore, the Fourier series converges to $f_p(x)$ except at the odd integers. At the odd integers, the series converges to $[1 + 0]/2 = 1/2$.

11. f_p has one-sided derivatives everywhere, and is continuous everywhere except at the odd integers. Therefore, the Fourier series converges to $f_p(x)$ except at the odd integers. At the odd integers, the series converges to $[1 - 1]/2 = 0$.

13. f_p has one-sided derivatives everywhere, and is continuous everywhere, so the Fourier series converges to $f_p(x)$ everywhere.

15. The function $f(x) = |x|$ is even on the interval $[-\pi, \pi]$. Hence, the Fourier representation contains only cosine terms. Hence,

$$a_n = \frac{1}{\pi}\int_{-\pi}^{\pi} f(x)\cos nx\,dx = \frac{2}{\pi}\int_0^{\pi} x\cos nx\,dx,$$

since $|x| = x$ on $[0, \pi]$. For $n = 0$,

$$a_0 = \frac{2}{\pi}\int_0^{\pi} x\,dx = \frac{1}{\pi}x^2\Big|_0^{\pi} = \pi.$$

For $n \geq 1$, integration by parts provides

$$\begin{aligned} a_n &= \frac{2}{\pi}\left[\frac{x\sin nx}{n} + \frac{\cos nx}{n^2}\right]_0^{\pi} \\ &= \frac{2}{\pi}\left\{\frac{\cos n\pi}{n^2} - \frac{1}{n^2}\right\} \\ &= \frac{2}{\pi n^2}((-1)^n - 1). \end{aligned}$$

Thus, if n is even, $a_n = 0$. If n is odd,

$$a_n = -\frac{4}{\pi n^2}.$$

We can write

$$a_{2n+1} = -\frac{4}{\pi(2n+1)^2},$$

which is valid for $n \geq 1$. Thus, the Fourier representation of f is

$$f(x) = \frac{\pi}{2} - \frac{4}{\pi}\sum_{n=1}^{\infty}\frac{1}{(2n+1)^2}\cos((2n+1)x).$$

Using Corollary 2.4, the function $f(x) = |x|$ is continuous at $x = 0$ so the series converges to $f(0) = 0$ at $x = 0$. Thus,

$$0 = f(0) = \frac{\pi}{2} - \frac{4}{\pi}\sum_{n=1}^{\infty}\frac{1}{(2n+1)^2},$$

and

$$\sum_{n=1}^{\infty}\frac{1}{(2n+1)^2} = \frac{\pi^2}{8}.$$

17. The function $f(x) = x^4$ is even on the interval $[-\pi, \pi]$, so the Fourier representation has only cosine terms.

$$a_n = \frac{2}{\pi}\int_0^{\pi} x^4\cos nx\,dx$$

For $n = 0$,

$$a_0 = \frac{2}{\pi}\int_0^{\pi} x^4\,dx = \frac{2}{5\pi}x^5\Big|_0^{\pi} = \frac{2\pi^4}{5}.$$

For $n \geq 1$, several applications of integration by parts provides

$$a_n = \frac{2}{\pi}\left[\frac{x^4 \sin nx}{n} + \frac{4x^3 \cos nx}{n^2} - \frac{12x^2 \sin nx}{n^3} - \frac{24x \cos nx}{n^4} + \frac{24 \sin nx}{n^5}\right]_0^\pi$$

$$= \frac{2}{\pi}\left[\frac{4\pi^3 \cos n\pi}{n^2} - \frac{24\pi \cos n\pi}{n^4}\right].$$

Since $\cos n\pi = (-1)^n$,

$$a_n = 8(-1)^n\left[\frac{\pi^2}{n^2} - \frac{6}{n^4}\right].$$

Hence, the Fourier representation of $f(x) = x^4$ is

$$f(x) = \frac{\pi^4}{5} + 8\sum_{n=1}^{\infty}(-1)^n\left[\frac{\pi^2}{n^2} - \frac{6}{n^4}\right]\cos nx.$$

Because f is continuous at $x = 0$, the Fourier series converges to $f(0)$ at $x = 0$.

$$0 = f(0) = \frac{\pi^4}{5} + 8\sum_{n=1}^{\infty}(-1)^n\left[\frac{\pi^2}{n^2} - \frac{6}{n^4}\right]$$

$$-\frac{\pi^4}{40} = \pi^2\sum_{n=1}^{\infty}\frac{(-1)^n}{n^2} - 6\sum_{n=1}^{\infty}\frac{(-1)^n}{n^4}$$

In Exercise 16, we found that

$$\sum_{n=1}^{\infty}\frac{(-1)^{n+1}}{n^2} = \frac{\pi^2}{12}.$$

Substituting,

$$-\frac{\pi^4}{40} = \pi^2\left(-\frac{\pi^2}{12}\right) + 6\sum_{n=1}^{\infty}\frac{(-1)^{n+1}}{n^4}$$

$$6\sum_{n=1}^{\infty}\frac{(-1)^{n+1}}{n^4} = \frac{\pi^4}{12} - \frac{\pi^4}{40}$$

$$\sum_{n=1}^{\infty}\frac{(-1)^{n+1}}{n^4} = \frac{7\pi^4}{720}.$$

In similar fashion, $f(x) = x^4$ is continuous at $x = \pi$, so the series converges to $f(\pi)$ at $x = \pi$.

$$\pi^4 = f(\pi) = \frac{\pi^4}{5} + 8\sum_{n=1}^{\infty}(-1)^n\left[\frac{\pi^2}{n^2} - \frac{6}{n^4}\right]\cos n\pi$$

Since $\cos n\pi = (-1)^n$, we have $(-1)^n(-1)^n = 1$ and

$$\frac{4\pi^4}{5} = 8\sum_{n=1}^{\infty}\left[\frac{\pi^2}{n^2} - \frac{6}{n^4}\right]$$

$$\frac{\pi^4}{10} = \pi^2\sum_{n=1}^{\infty}\frac{1}{n^2} - 6\sum_{n=1}^{\infty}\frac{1}{n^4}.$$

From Exercise 16,

$$\sum_{n=1}^{\infty}\frac{1}{n^2} = \frac{\pi^2}{6},$$

so

$$\frac{\pi^4}{10} = \pi^2\left(\frac{\pi^2}{6}\right) - 6\sum_{n=1}^{\infty}\frac{1}{n^4}$$

$$6\sum_{n=1}^{\infty}\frac{1}{n^4} = \frac{\pi^4}{6} - \frac{\pi^4}{10} = \frac{5\pi^4 - 3\pi^4}{30} = \frac{\pi^4}{15}$$

$$\sum_{n=1}^{\infty}\frac{1}{n^4} = \frac{\pi^4}{90}.$$

19. Consider $f(x) = e^{rx}$ on the interval $[-\pi, \pi]$.

$$a_0 = \frac{1}{\pi}\int_{-\pi}^{\pi} e^{rx}\,dx = \frac{1}{\pi r}e^{rx}\Big|_{-\pi}^{\pi} = \frac{2}{\pi r}\sinh \pi r.$$

For $n \geq 1$,

$$a_n = \frac{1}{\pi}\int_{-\pi}^{\pi} e^{rx}\cos nx\,dx.$$

Two applications of integration by parts provides

$$a_n = \frac{1}{\pi}\cdot\frac{n^2}{n^2 + r^2}\left[\frac{e^{rx}\sin nx}{n} + \frac{re^{rx}\cos nx}{n^2}\right]_{-\pi}^{\pi}$$

$$= \frac{1}{\pi(n^2 + r^2)}\left[e^{rx}(n \sin nx + r\cos nx)\right]_{-\pi}^{\pi}$$

$$= \frac{1}{\pi(n^2 + r^2)}[e^{r\pi} r\cos n\pi - e^{-r\pi} r\cos n\pi]$$

$$= \frac{r(-1)^n}{\pi(n^2 + r^2)}[e^{r\pi} - e^{-r\pi}]$$

$$= \frac{2r(-1)^n}{\pi(n^2 + r^2)}\sinh \pi r.$$

Next,

$$b_n = \frac{1}{\pi}\int_{-\pi}^{\pi} e^{rx}\sin nx\,dx.$$

Again, two applications of integration by parts provides the answer.

$$\begin{aligned} b_n &= \frac{1}{\pi}\cdot\frac{n^2}{n^2+r^2}\left[-\frac{e^{rx}\cos nx}{n} + \frac{re^{rx}\sin nx}{n^2}\right]_{-\pi}^{\pi} \\ &= \frac{1}{\pi(n^2+r^2)}\left[e^{rx}(-n\cos nx + r\sin nx)\right]_{-\pi}^{\pi} \\ &= \frac{1}{\pi(n^2+r^2)}\left[e^{r\pi}(-n\cos n\pi)\right. \\ &\qquad \left. - e^{-r\pi}(-n\cos n\pi)\right] \\ &= \frac{n(-1)^{n+1}}{\pi(n^2+r^2)}\left[e^{r\pi} - e^{-r\pi}\right] \\ &= \frac{2n(-1)^{n+1}}{\pi(n^2+r^2)}\sinh r\pi. \end{aligned}$$

Thus, the Fourier representation is

$$\begin{aligned} e^{rx} &= \frac{\sinh \pi r}{\pi r} + \sum_{n=1}^{\infty}\frac{2r(-1)^n\sinh \pi r}{\pi(n^2+r^2)}\cos nx \\ &\quad - \sum_{n=1}^{\infty}\frac{2n(-1)^n\sinh \pi r}{\pi(n^2+r^2)}\sin nx. \end{aligned}$$

In the case that $r = 1/2$,

$$\begin{aligned} f(x) &= \frac{2}{\pi}\sinh\frac{\pi}{2} + \frac{\sinh(\pi/2)}{\pi}\sum_{n=1}^{\infty}\frac{(-1)^n}{n^2+1/4}\cos nx \\ &\quad - \frac{2\sinh(\pi/2)}{\pi}\sum_{n=1}^{\infty}\frac{(-1)^n n}{n^2+1/4}\sin nx. \end{aligned}$$

This is perhaps better written as

$$\begin{aligned} f(x) &= \frac{1}{\pi}\sinh\frac{\pi}{2}\left[2 + \sum_{n=1}^{\infty}\frac{(-1)^n}{n^2+1/4}\cos nx\right. \\ &\quad \left. - 2\sum_{n=1}^{\infty}\frac{(-1)^n n}{n^2+1/4}\sin nx\right]. \end{aligned}$$

The following plots use $N = 10$ terms.

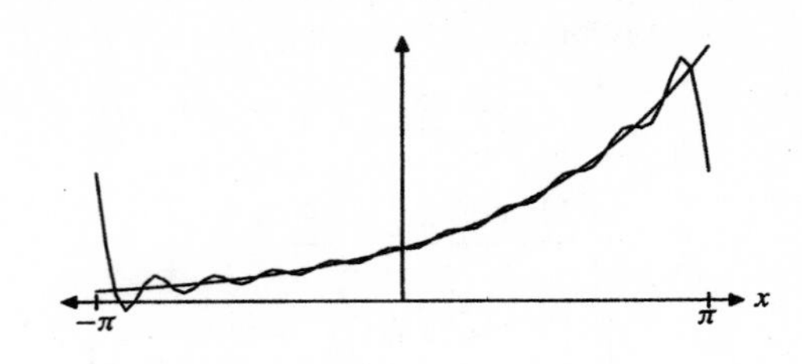

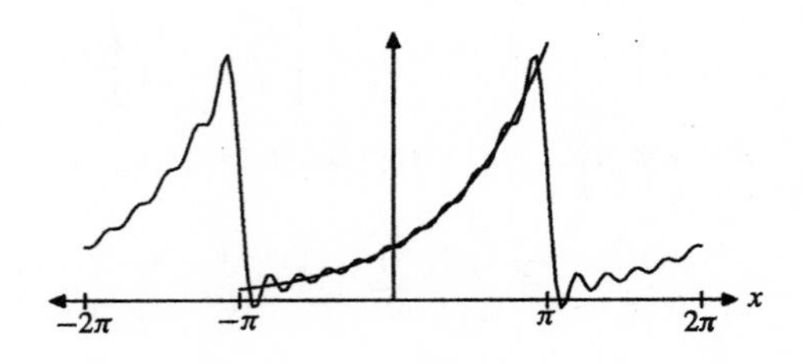

21. Consider

$$f(x) = \begin{cases} 0, & -1 < x \le -1/2, \\ 1, & -1/2 < x \le 1/2, \\ 0, & 1/2 < x \le 1. \end{cases}$$

The plot of $f_p(x)$ is shown in blue.

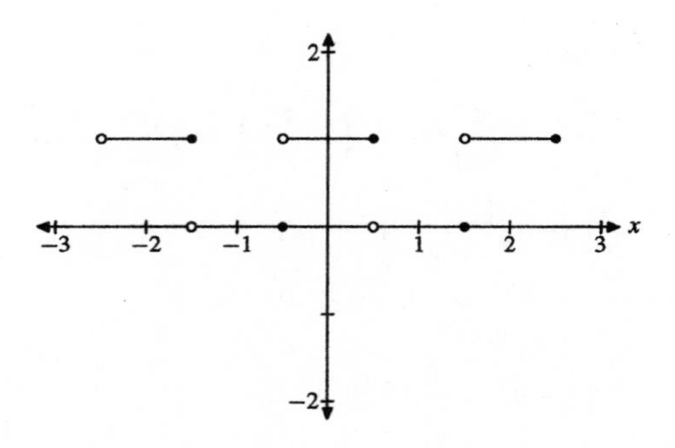

The Fourier series will converge to $f_p(x)$ whenever f_p is continuous at x. At each point of discontinuity we have a limit from one side equalling 1, while the limit coming in from the other side is 0. Thus, the

Fourier series should converge to 1/2 at these points, as is evident in the following image depicting $S_{13}(x)$.

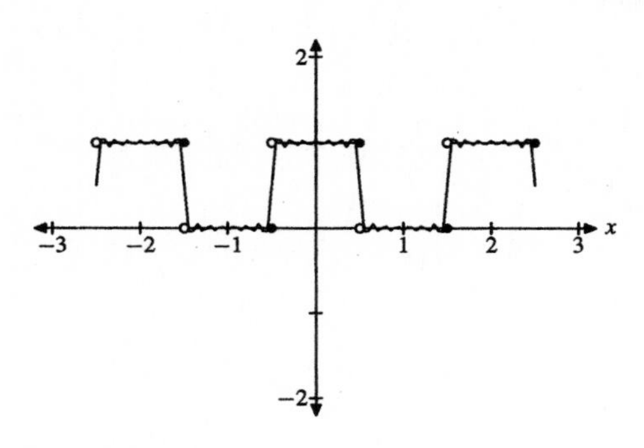

23. First notice that
$$\int_0^T f(x)\,dx = \int_0^a f(x)\,dx + \int_a^T f(x)\,dx.$$
For the first integral, we use the periodicity of f and the substitution $x = u + T$ to get
$$\int_0^a f(x)\,dx = \int_T^{a+T} f(u+T)\,du = \int_T^{a+T} f(u)\,du.$$

Since u is a dummy variable in the last integral, we can replace it by x. Then we get
$$\int_0^T f(x)\,dx = \int_T^{a+T} f(x)\,dx + \int_a^T f(x)\,dx$$
$$= \int_a^{a+T} f(x)\,dx.$$

Using this with a replaced by b, we get the final result
$$\int_b^{b+T} f(x)\,dx = \int_0^T f(x)\,dx = \int_a^{a+T} f(x)\,dx$$

If f is periodic with period L, then $F(x) = f(x)\cos(n\pi x/L)$ is also periodic with period L. Hence, for any c,
$$a_n = \frac{1}{L}\int_{-L}^{L} F(x)\,dx = \frac{1}{L}\int_c^{c+2L} F(x)\,dx.$$

The same argument works for b_n.

Section 12.3. Fourier Cosine and Sine Series

1. If $f(x) = 1 - x$ on $[0, 2]$, then
$$f_o(x) = \begin{cases} -(1+x), & -2 \le x < 0, \\ 0, & x = 0, \\ 1 - x, & 0 < x \le 2. \end{cases}$$

The graphs of the odd extension f_o on $[-2, 2]$ and the odd periodic extension f_{op} on $[-6, 6]$ follow.

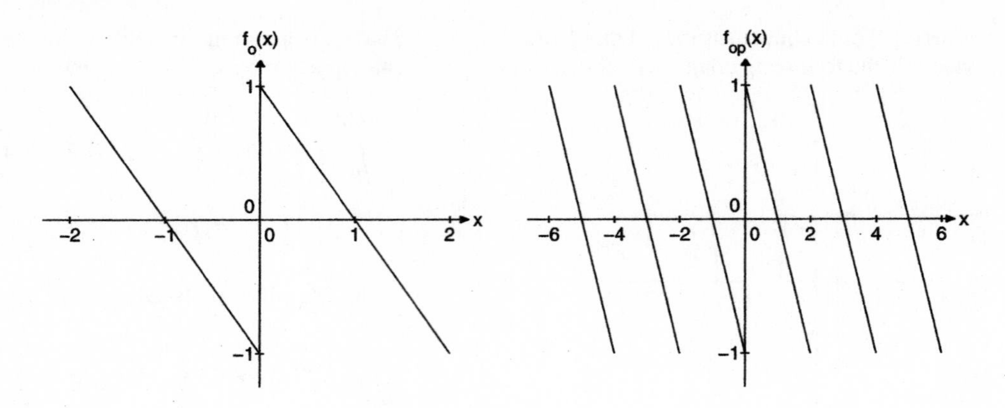

3. If $f(x) = x^2 - 1$ on $[0, 2]$, then

$$f_o(x) = \begin{cases} -(x^2-1), & -2 \le x < 0, \\ 0, & x = 0, \\ x^2 - 1, & 0 < x \le 2. \end{cases}$$

The graphs of the odd extension f_o on $[-2, 2]$ and the odd periodic extension f_{op} on $[-6, 6]$ follow.

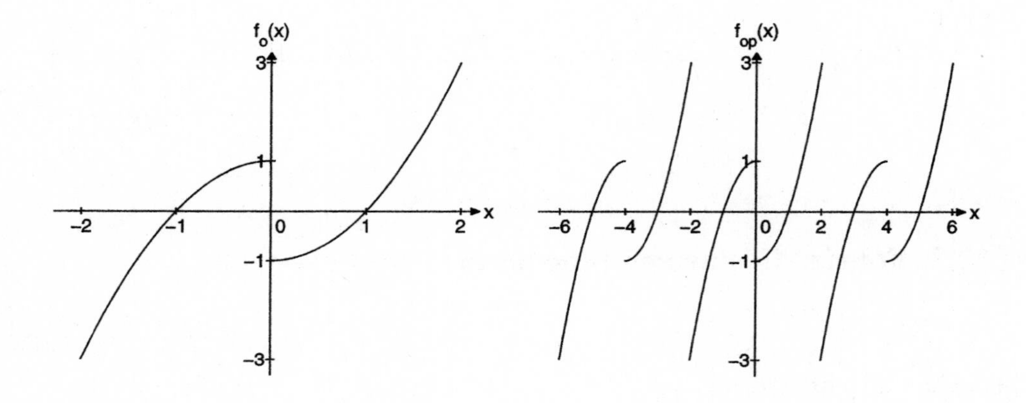

5. If $f(x) = 1 - x$ on $[0, 2]$, then

$$f_e(x) = \begin{cases} 1 + x, & -2 \le x < 0, \\ 1 - x, & 0 \le x \le 2. \end{cases}$$

The graphs of the even extension f_e on $[-2, 2]$ and the even periodic extension f_{ep} on $[-6, 6]$ follow.

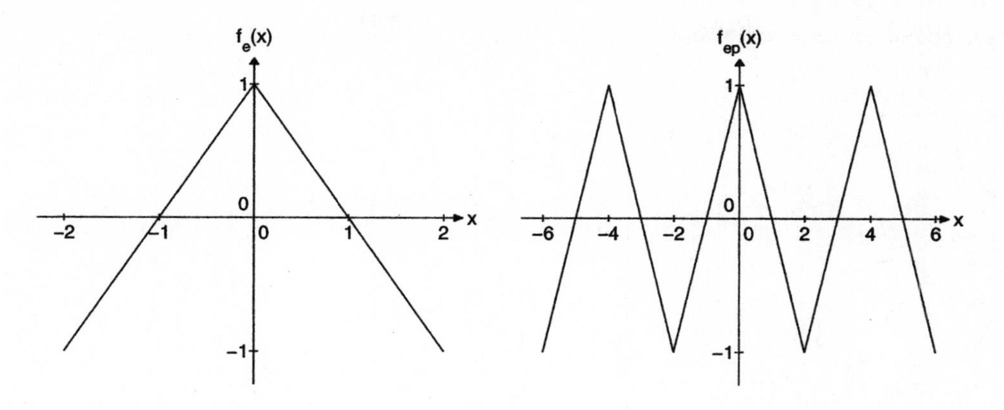

7. If $f(x) = x^2 - 1$ on $[0, 2]$, then

$$f_e(x) = \begin{cases} x^2 - 1, & -2 \le x < 0, \\ x^2 - 1, & 0 \le x \le 2. \end{cases}$$

The graphs of the even extension f_e on $[-2, 2]$ and the even periodic extension f_{ep} on $[-6, 6]$ follow.

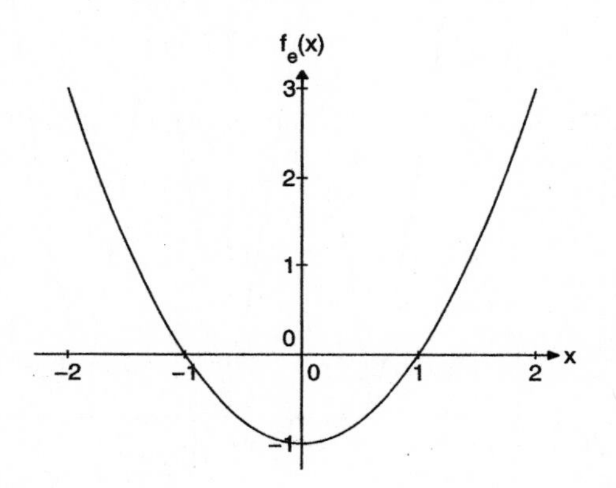

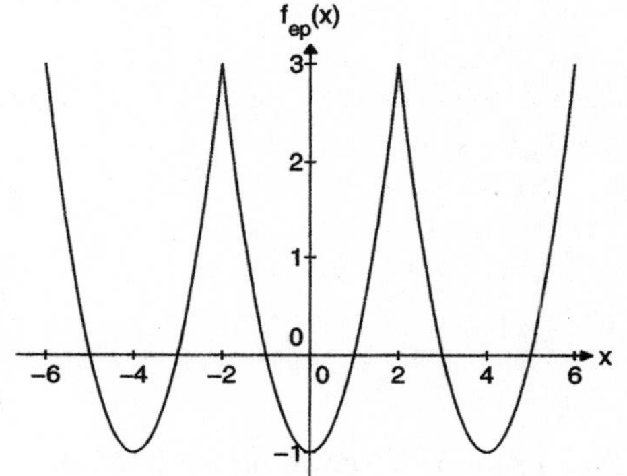

9. The function $f(x) = x$ on $[0, \pi]$ has even extension

$$f_e(x) = \begin{cases} -x, & -\pi \le x < 0, \\ x, & 0 \le x \le \pi, \end{cases}$$

on the interval $[-\pi, \pi]$. The Fourier expansion of $f_e(x)$ has only cosine terms and

$$\begin{aligned} a_n &= \frac{1}{\pi}\int_{-\pi}^{\pi} f_e(x)\cos nx\,dx \\ &= \frac{2}{\pi}\int_0^{\pi} f_e(x)\cos nx\,dx \\ &= \frac{2}{\pi}\int_0^{\pi} x\cos nx\,dx. \end{aligned}$$

Integrating by parts,

$$\begin{aligned} a_n &= \frac{2}{\pi}\left[\frac{x\sin nx}{n} + \frac{\cos nx}{n^2}\right]_0^{\pi} \\ &= \frac{2}{\pi}\left[\frac{\cos n\pi}{n^2} - \frac{1}{n^2}\right] \\ &= \frac{2}{\pi n^2}[(-1)^n - 1], \end{aligned}$$

provided $n \neq 0$. In the case $n = 0$,

$$a_0 = \frac{2}{\pi}\int_0^{\pi} x\,dx = \frac{2}{\pi}\left[\frac{1}{2}x^2\right]_0^{\pi} = \pi.$$

Thus, the Fourier series expansion is

$$f(x) = \frac{a_0}{2} + \sum_{n=1}^{\infty} a_n \cos nx$$
$$= \frac{\pi}{2} + \sum_{n=1}^{\infty} \frac{2}{\pi n^2}[(-1)^n - 1] \cos nx.$$

The partial sum S_6 is shown on the interval $[0, \pi]$ and $[-3\pi, 3\pi]$.

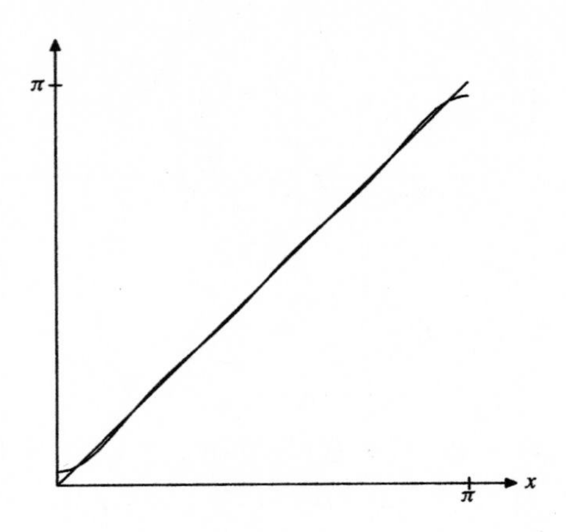

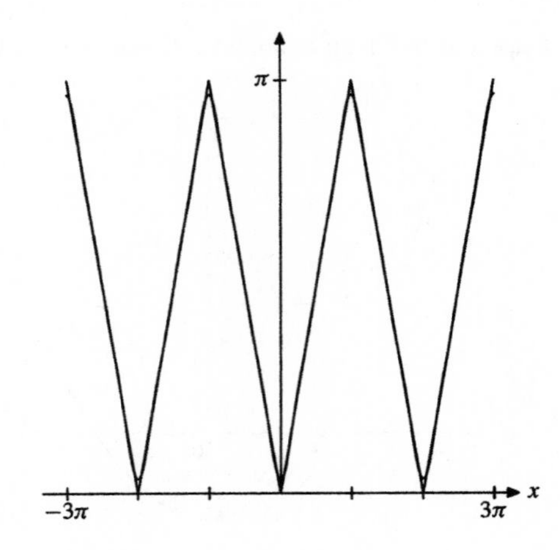

11. As $f(x) = \cos x$ is already a cosine series, any analysis we do should agree with this fact. The function has even extension

$$f_e(x) = \begin{cases} \cos(-x), & -\pi \le x < 0, \\ \cos x, & 0 \le x \le \pi, \end{cases}$$
$$= \begin{cases} \cos x, & -\pi \le x < 0, \\ \cos x, & 0 \le x \le \pi, \end{cases}$$

on the interval $[-\pi, \pi]$. Hence, the Fourier series expansion of $f_e(x)$ has only cosine terms and

$$a_n = \frac{1}{\pi} \int_{\pi}^{\pi} f_e(x) \cos nx \, dx$$
$$= \frac{2}{\pi} \int_0^{\pi} \cos x \cos nx \, dx.$$

Using a product to sum identity,

$$a_n = \frac{2}{\pi} \int_0^{\pi} \frac{1}{2}[\cos(1-n)x + \cos(1+n)x] \, dx$$
$$= \frac{1}{\pi} \left[\frac{\sin(n-1)x}{n-1} + \frac{\sin(n+1)x}{n+1} \right]_0^{\pi}$$
$$= 0,$$

provided $n \neq 1$. In the case that $n = 1$,

$$\begin{aligned} a_1 &= \frac{2}{\pi}\int_0^{\pi} \cos^2 x\,dx \\ &= \frac{2}{\pi}\int_0^{\pi} \frac{1+\cos 2x}{2}\,dx \\ &= \frac{1}{\pi}\left[x + \frac{1}{2}\sin 2x\right]_0^{\pi} \\ &= \frac{1}{\pi}[\pi] \\ &= 1. \end{aligned}$$

Hence, the Fourier series expansion is

$$f(x) = \cos x.$$

No surprises here! The partial sum S_6 is shown on the interval $[0, \pi]$ and $[-3\pi, 3\pi]$.

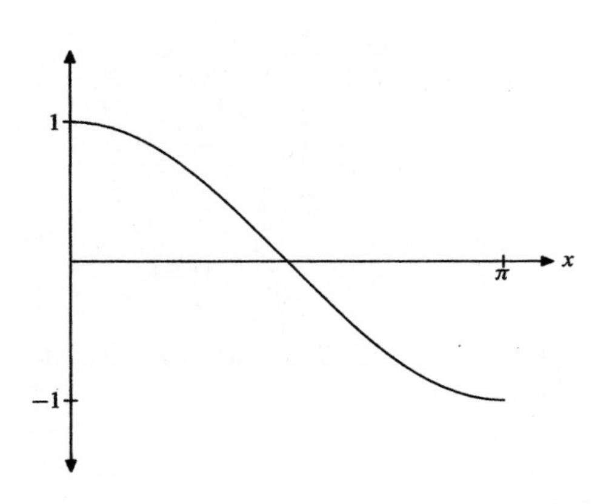

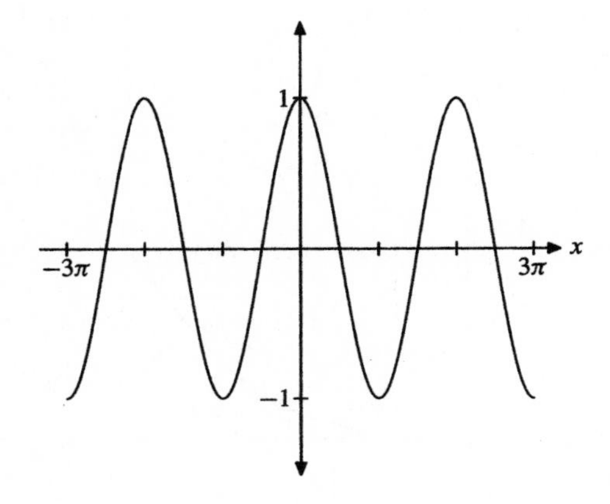

13. The function $f(x) = \pi - x$ on $[0, \pi]$ has even extension

$$f_e(x) = \begin{cases} \pi + x, & -\pi \le x < 0, \\ \pi - x, & 0 \le x \le \pi, \end{cases}$$

on the interval $[-\pi, \pi]$. The Fourier series expansion of $f_e(x)$ has only cosine terms and

$$\begin{aligned} a_n &= \frac{1}{\pi}\int_{-\pi}^{\pi} f_e(x)\cos nx\,dx \\ &= \frac{2}{\pi}\int_0^{\pi} (\pi - x)\cos nx\,dx. \end{aligned}$$

Integrating by parts,

$$\begin{aligned} a_n &= \frac{2}{\pi}\left[\frac{(\pi - x)\sin nx}{n} - \frac{\cos nx}{n^2}\right]_0^{\pi} \\ &= \frac{2}{\pi}\left[-\frac{\cos n\pi}{n^2} + \frac{1}{n^2}\right] \\ &= \frac{2}{\pi n^2}[1 - (-1)^n], \end{aligned}$$

provided $n \neq 0$. In the case that $n = 0$,

$$\begin{aligned} a_0 &= \frac{2}{\pi}\int_0^{\pi} (\pi - x)dx \\ &= \frac{2}{\pi}\left[\pi x - \frac{1}{2}x^2\right]_0^{\pi} \\ &= \pi. \end{aligned}$$

Thus, the Fourier series expansion is

$$\begin{aligned} f(x) &= \frac{a_0}{2} + \sum_{n=1}^{\infty} a_n \cos nx \\ &= \frac{\pi}{2} + \sum_{n=1}^{\infty} \frac{2}{\pi n^2}[1 - (-1)^n]\cos nx. \end{aligned}$$

The partial sum S_6 is shown on the interval $[0, \pi]$ and $[-3\pi, 3\pi]$.

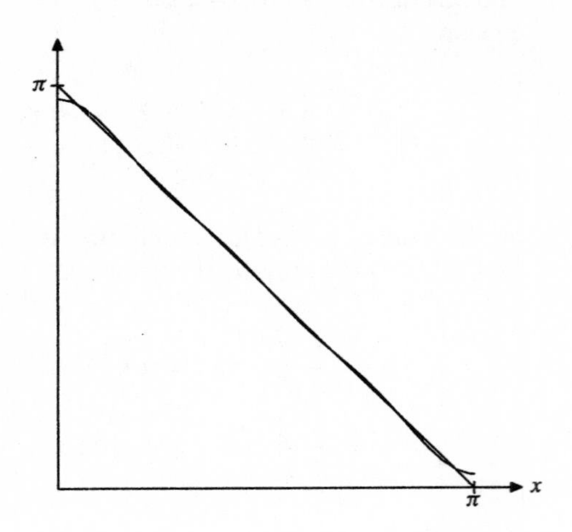

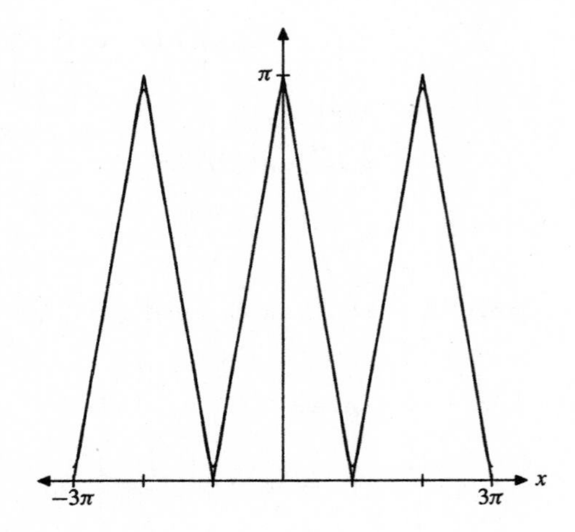

15. The function $f(x) = x^3$ on $[0, \pi]$ has even extension

$$f_e(x) = \begin{cases} -x^3, & -\pi \le x < 0, \\ x^3, & 0 \le x \le \pi, \end{cases}$$

on the interval $[-\pi, \pi]$. The Fourier expansion of $f_e(x)$ has only cosine terms and

$$\begin{aligned} a_n &= \frac{1}{\pi}\int_{-\pi}^{\pi} f_e(x)\cos nx\,dx \\ &= \frac{2}{\pi}\int_0^{\pi} x^3 \cos nx\,dx. \end{aligned}$$

Integrating by parts,

$$\begin{aligned} a_n &= \frac{2}{\pi}\left[\frac{x^3 \sin nx}{n} + \frac{3x^2\cos nx}{n^2} \right. \\ &\qquad \left. - \frac{6x\sin nx}{n^3} - \frac{6\cos nx}{n^4}\right]_0^{\pi} \\ &= \frac{2}{\pi}\left[\frac{3\pi^2\cos n\pi}{n^2} - \frac{6\cos n\pi}{n^4} + \frac{6}{n^4}\right] \\ &= \frac{6\pi(-1)^n}{n^2} - \frac{12}{\pi n^4}\left((-1)^n - 1\right), \end{aligned}$$

provided $n \neq 0$. In the case that $n = 0$,

$$a_0 = \frac{2}{\pi}\int_0^{\pi} x^3\,dx = \frac{2}{\pi}\left[\frac{1}{4}x^4\right]_0^{\pi} = \frac{\pi^3}{2}.$$

Hence, the Fourier series expansion is

$$\begin{aligned} f(x) &= \frac{a_0}{2} + \sum_{n=1}^{\infty} a_n \cos nx \\ &= \frac{\pi^3}{4} + \sum_{n=1}^{\infty}\left[\frac{6\pi(-1)^n}{n^2} \right. \\ &\qquad \left. - \frac{12}{\pi n^4}\left((-1)^n - 1\right)\right]\cos nx. \end{aligned}$$

The partial sum S_6 is shown on the interval $[0, \pi]$ and $[-3\pi, 3\pi]$.

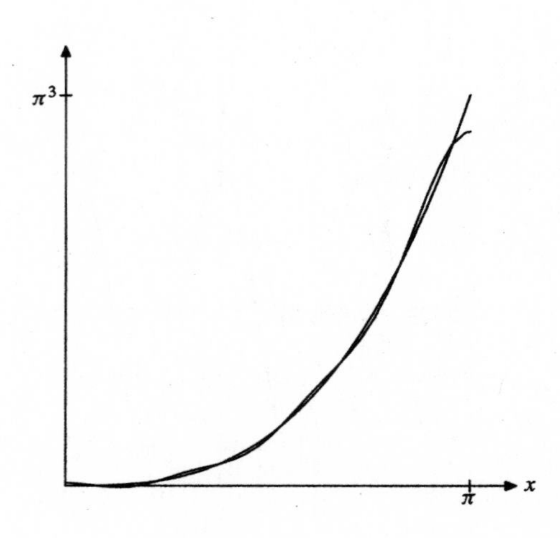

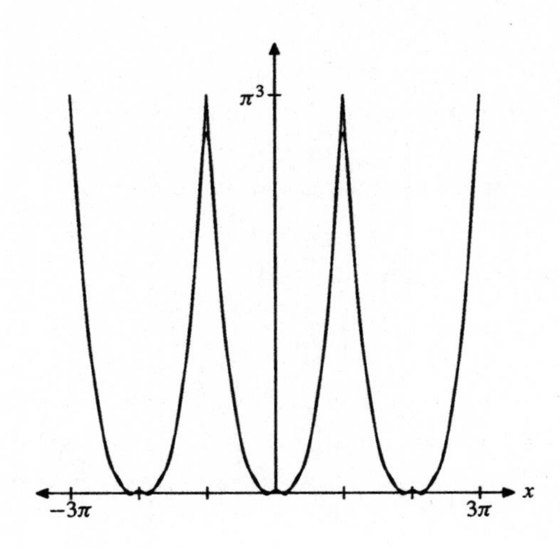

17. The function

$$f(x) = \begin{cases} 1, & 0 \le x < \pi/2, \\ 0, & \pi/2 \le x \le \pi, \end{cases}$$

has even extension

$$f_e(x) = \begin{cases} 0, & -\pi \le x < -\pi/2, \\ 1, & -\pi/2 \le x < \pi/2, \\ 0, & \pi/2 \le x < \pi, \end{cases}$$

on the interval $[-\pi, \pi]$. The Fourier expansion of $f_e(x)$ has only cosine terms and

$$\begin{aligned} a_n &= \frac{1}{\pi} \int_{-\pi}^{\pi} f_e(x) \cos nx \, dx \\ &= \frac{2}{\pi} \int_{0}^{\pi} f_e(x) \cos nx \, dx. \end{aligned}$$

However, $f_e(x) = 0$ on $[\pi/2, \pi]$, so this last integral becomes

$$\begin{aligned} a_n &= \frac{2}{\pi} \int_{0}^{\pi/2} \cos nx \, dx = \frac{2}{\pi} \left[\frac{\sin nx}{n} \right]_0^{\pi/2} \\ &= \frac{2}{n\pi} \sin \frac{n\pi}{2}, \end{aligned}$$

provided $n \ne 0$. In the case that $n = 0$,

$$a_0 = \frac{2}{\pi} \int_{0}^{\pi/2} dx = 1.$$

Thus, the Fourier series expansion is

$$\begin{aligned} f(x) &= \frac{a_0}{2} + \sum_{n=1}^{\infty} a_n \cos nx \\ &= \frac{1}{2} + \sum_{n=1}^{\infty} \frac{2}{n\pi} \sin \frac{n\pi}{2} \cos nx. \end{aligned}$$

The partial sum S_6 is shown on the interval $[0, \pi]$ and $[-3\pi, 3\pi]$.

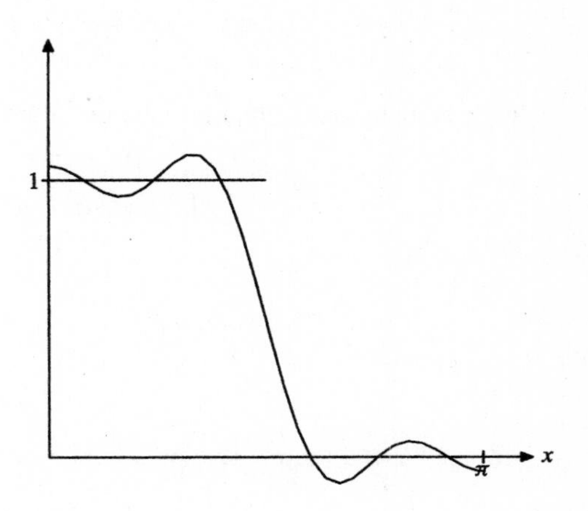

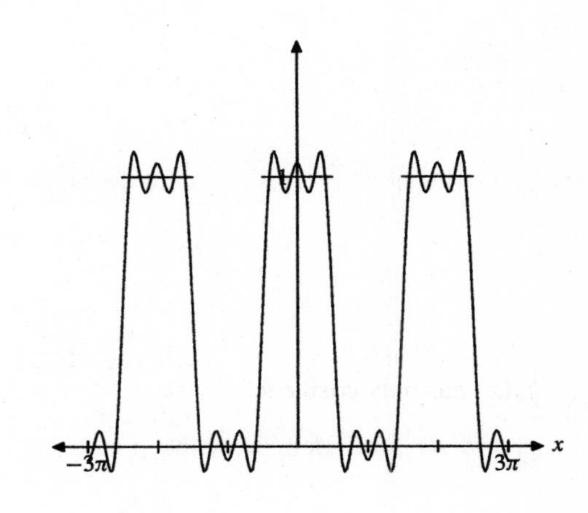

19. The function $f(x) = x\cos x$ on $[0, \pi]$ has even extension

$$f_e(x) = \begin{cases} -x\cos x, & -\pi \le x < 0, \\ x\cos x, & 0 \le x < \pi, \end{cases}$$

on the interval $[-\pi, \pi]$. The Fourier expansion of $f_e(x)$ has only cosine terms and

$$a_n = \frac{1}{\pi}\int_{-\pi}^{\pi} f_e(x)\cos nx\,dx = \frac{2}{\pi}\int_0^{\pi} x\cos x\cos nx\,dx.$$

Using a product to sum identity,

$$a_n = \frac{2}{\pi}\int_0^{\pi} \frac{x}{2}[\cos(1-n)x + \cos(1+n)x]\,dx = \frac{1}{\pi}\int_0^{\pi} x\cos(n-1)x\,dx + \frac{1}{\pi}\int_0^{\pi} x\cos(n+1)x\,dx.$$

Integrating by parts, the first integral becomes

$$\begin{aligned} \frac{1}{\pi}\int_0^{\pi} x\cos(n-1)x\,dx &= \frac{1}{\pi}\left[\frac{x\sin(n-1)x}{n-1} + \frac{\cos(n-1)x}{(n-1)^2}\right]_0^{\pi} \\ &= \frac{1}{\pi}\left[\frac{\cos(n-1)\pi}{(n-1)^2} - \frac{1}{(n-1)^2}\right] \\ &= \frac{1}{\pi(n-1)^2}\left[(-1)^{n-1} - 1\right], \end{aligned}$$

provided $n \ne 1$. Similarly, the second integral is

$$\frac{1}{\pi}\int_0^{\pi} x\cos(n+1)x\,dx = \frac{1}{\pi(n+1)^2}\left[(-1)^{n+1} - 1\right].$$

Adding,

$$\begin{aligned} a_n &= \frac{1}{\pi(n-1)^2}\left[(-1)^{n-1} - 1\right] + \frac{1}{\pi(n+1)^2}\left[(-1)^{n+1} - 1\right] \\ &= \frac{(n+1)^2\left((-1)^{n-1} - 1\right) + (n-1)^2\left((-1)^{n+1} - 1\right)}{\pi(n-1)^2(n+1)^2} \\ &= \frac{\left((-1)^{n-1} - 1\right)\left((n+1)^2 + (n-1)^2\right)}{\pi(n-1)^2(n+1)^2} \\ &= \frac{2\left((-1)^{n-1} - 1\right)(n^2+1)}{\pi(n-1)^2(n+1)^2}, \end{aligned}$$

provided $n \ne 1$. In the case that $n = 1$,

$$a_1 = \frac{2}{\pi}\int_0^{\pi} x\cos^2 x\,dx.$$

Using the half angle identity,

$$\begin{aligned}
a_1 &= \frac{2}{\pi}\int_0^{\pi} x\left(\frac{1+\cos 2x}{2}\right)dx \\
&= \frac{1}{\pi}\int_0^{\pi}(x + x\cos 2x)\,dx \\
&= \frac{1}{\pi}\left[\frac{1}{2}x^2 + \frac{1}{2}x\sin 2x + \frac{1}{4}\cos 2x\right]_0^{\pi} \\
&= \frac{1}{\pi}\left[\frac{\pi^2}{2}\right] \\
&= \frac{\pi}{2}.
\end{aligned}$$

Also, note that with computation above,

$$a_0 = \frac{2(-1-1)(1)}{\pi \cdot 1 \cdot 1} = -\frac{4}{\pi}.$$

Hence, the Fourier expansion is

$$\begin{aligned}
f(x) &= \frac{a_0}{2} + a_1\cos x + \sum_{n=2}^{\infty} a_n \cos nx \\
&= -\frac{2}{\pi} + \frac{\pi}{2}\cos x + \sum_{n=2}^{\infty}\frac{2\left((-1)^{n-1}-1\right)(n^2+1)}{\pi(n-1)^2(n+1)^2}\cos nx.
\end{aligned}$$

The partial sum S_6 is shown on the interval $[0, \pi]$ and $[-3\pi, 3\pi]$.

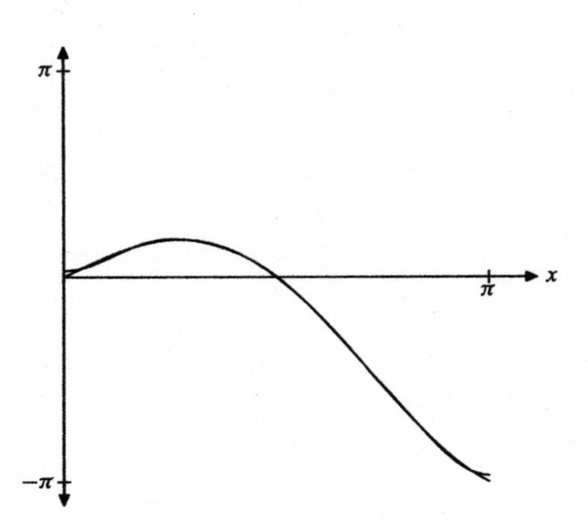

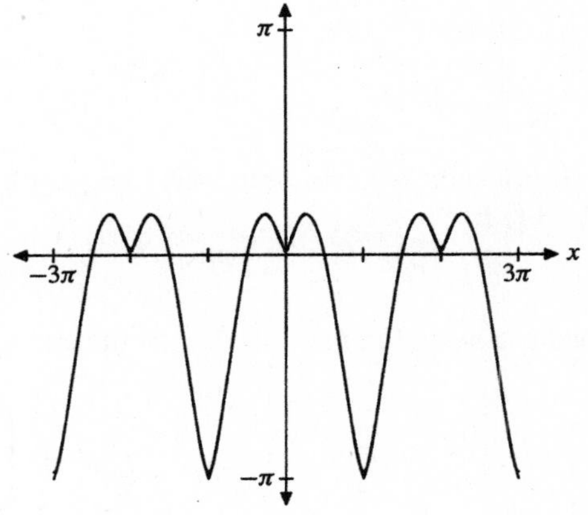

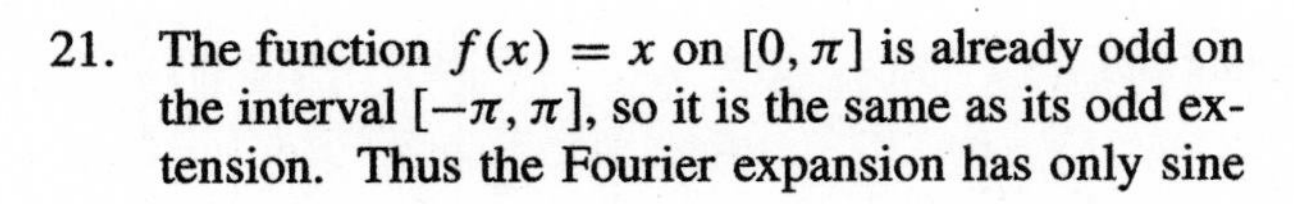
21. The function $f(x) = x$ on $[0, \pi]$ is already odd on the interval $[-\pi, \pi]$, so it is the same as its odd extension. Thus the Fourier expansion has only sine terms and

$$\begin{aligned}
b_n &= \frac{1}{\pi}\int_{-\pi}^{\pi} f(x)\sin nx\,dx \\
&= \frac{2}{\pi}\int_0^{\pi} x\sin nx\,dx.
\end{aligned}$$

Integrating by parts,

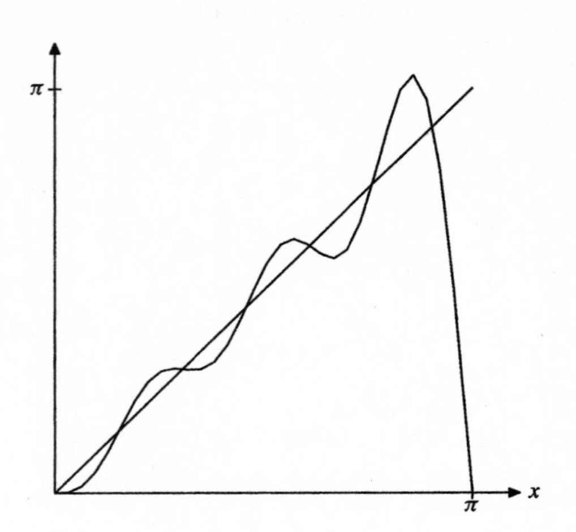

$$b_n = \frac{2}{\pi}\left[\frac{-x\cos nx}{n} + \frac{\sin nx}{n^2}\right]_0^\pi$$
$$= \frac{2}{\pi}\left[\frac{-\pi\cos n\pi}{n}\right]$$
$$= 2\left[\frac{-(-1)^n}{n}\right]$$
$$= (-1)^{n+1}\frac{2}{n},$$

for $n \geq 1$. Thus, the Fourier series expansion is

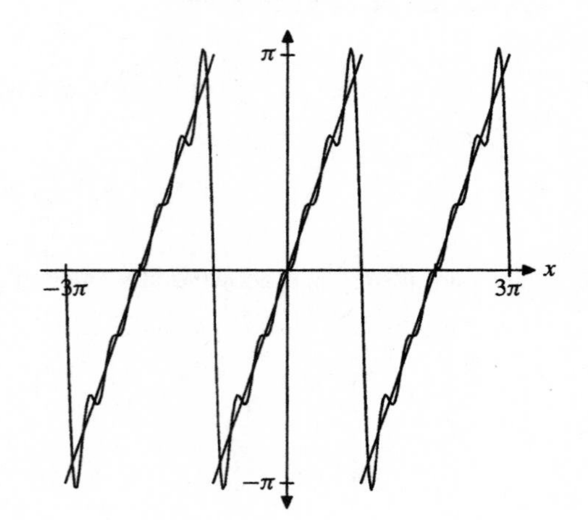

$$f(x) = \sum_{n=1}^{\infty} b_n \sin nx = \sum_{n=1}^{\infty}(-1)^{n+1}\frac{2}{n}\sin nx.$$

The partial sum S_6 is shown on the interval $[0, \pi]$ and $[-3\pi, 3\pi]$.

23. An odd extension for $f(x) = \cos x$ on the interval $[-\pi, \pi]$ is

$$f_o(x) = \begin{cases} -\cos x, & -\pi \leq x < 0, \\ \cos x, & 0 \leq x \leq \pi. \end{cases}$$

The Fourier expansion of $f_o(x)$ contains only sine terms and

$$b_n = \frac{1}{\pi}\int_{-\pi}^{\pi} f_o(x)\sin nx\, dx = \frac{2}{\pi}\int_0^{\pi} \cos x \sin nx\, dx.$$

Using a product to sum identity,

$$\begin{aligned} b_n &= \frac{2}{\pi}\int_0^{\pi} \frac{1}{2}\left[\sin(n-1)x + \sin(n+1)x\right] dx \\ &= \frac{1}{\pi}\left[-\frac{\cos(n-1)x}{n-1} - \frac{\cos(n+1)x}{n+1}\right]_0^{\pi} \\ &= -\frac{1}{\pi}\left\{\frac{(-1)^{n-1}}{n-1} + \frac{(-1)^{n+1}}{n+1} - \frac{1}{n-1} - \frac{1}{n+1}\right\} \\ &= -\frac{1}{\pi}\left\{\frac{(-1)^{n-1}(n+1) + (-1)^{n+1}(n-1) - (n+1) - (n-1)}{n^2-1}\right\}. \end{aligned}$$

Because $(-1)^{n+1} = (-1)^{n-1}$,

$$b_n = -\frac{1}{\pi}\left\{\frac{(-1)^{n-1}[(n+1)+(n-1)] - 2n}{n^2-1}\right\} = \frac{2n}{\pi}\cdot\frac{(-1)^n+1}{n^2-1},$$

provided $n \neq 1$. In the case that $n = 1$,

$$b_1 = \frac{2}{\pi}\int_0^{\pi} \cos x \sin x \, dx = \frac{1}{\pi}\int_0^{\pi} \sin 2x \, dx = \frac{1}{\pi}\left[-\frac{1}{2}\cos 2x\right]_0^{\pi} = 0.$$

Thus, the Fourier expansion is

$$f(x) = \sum_{n=1}^{\infty} b_n \sin nx = \sum_{n=2}^{\infty} \frac{2\left((-1)^n+1\right)n}{\pi(n^2-1)} \sin nx.$$

The partial sum S_6 is shown on the interval $[0, \pi]$ and $[-3\pi, 3\pi]$.

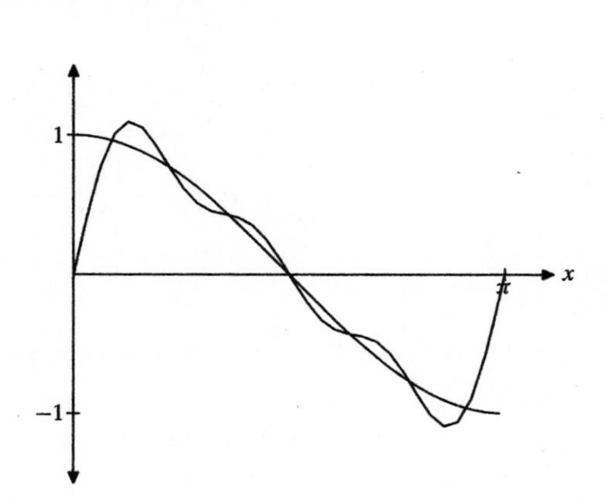

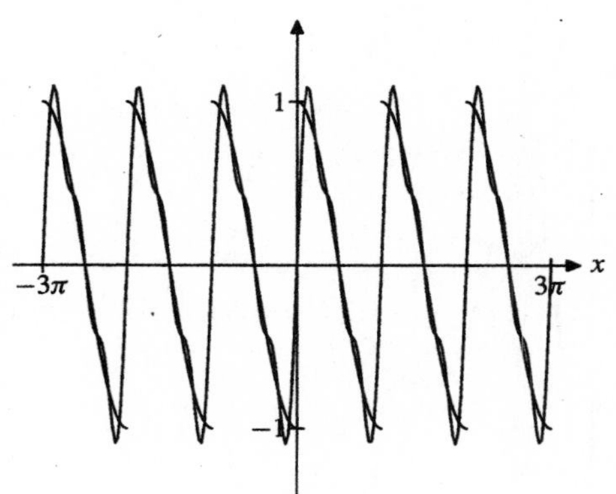

25. An odd extension of $f(x) = \pi - x$ on the interval $[-\pi, \pi]$ is

$$f_o(x) = \begin{cases} -\pi - x, & -\pi \le x < 0, \\ \pi - x, & 0 \le x \le \pi. \end{cases}$$

The Fourier expansion contains only sine terms with

$$b_n = \frac{1}{\pi}\int_{-\pi}^{\pi} f_o(x)\sin nx\,dx$$
$$= \frac{2}{\pi}\int_0^{\pi} (\pi - x)\sin nx\,dx.$$

Integrating by parts,

$$b_n = \frac{2}{\pi}\left[\frac{-(\pi - x)\cos nx}{n} - \frac{\sin nx}{n^2}\right]_0^{\pi}$$
$$= -\frac{2}{\pi n^2}\left[n(\pi - x)\cos nx + \sin nx\right]_0^{\pi}$$
$$= -\frac{2}{\pi n^2}(-n\pi)$$
$$= \frac{2}{n}.$$

Hence, the Fourier expansion is

$$f(x) = \sum_{n=1}^{\infty} b_n \sin nx = \sum_{n=1}^{\infty} \frac{2}{n}\sin nx.$$

The partial sum S_6 is shown on the interval $[0, \pi]$ and $[-3\pi, 3\pi]$.

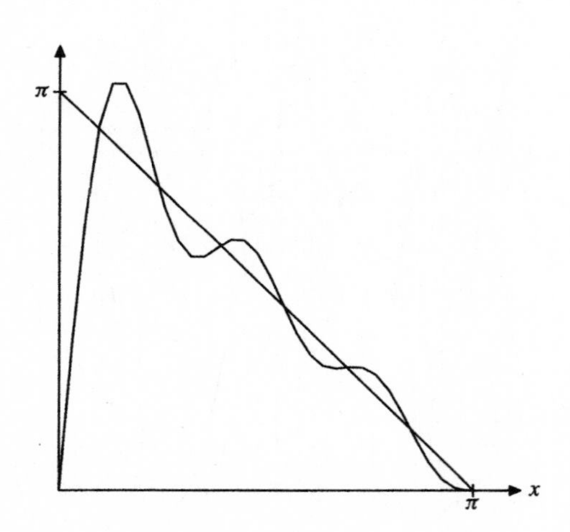

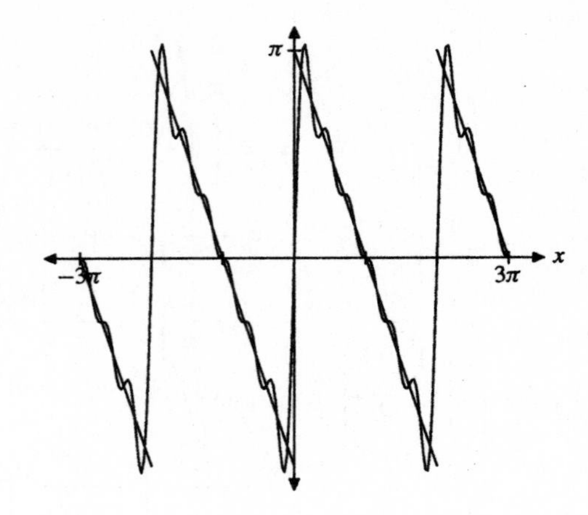

27. The function $f(x) = x^3$ is already odd on the interval $[-\pi, \pi]$, so its odd extension is itself and its Fourier expansion contains only sine terms.

$$b_n = \frac{1}{\pi}\int_{-\pi}^{\pi} f(x)\sin nx\,dx = \frac{2}{\pi}\int_0^{\pi} x^3 \sin nx\,dx.$$

Integrating by parts,

$$b_n = \frac{2}{\pi}\left[-\frac{x^3\cos nx}{n} + \frac{3x^2\sin nx}{n^2} + \frac{6x\cos nx}{n^3} - \frac{6\sin nx}{n^4}\right]_0^{\pi}$$
$$= \frac{2}{\pi}\left[-\frac{\pi^3\cos n\pi}{n} + \frac{6\pi\cos n\pi}{n^3}\right]$$
$$= \frac{2}{\pi}\left[\frac{-\pi^3 n^2(-1)^n + 6\pi(-1)^n}{n^3}\right]$$
$$= \frac{2(-1)^n(6 - n^2\pi^2)}{n^3}$$

Thus, the Fourier expansion is

$$f(x) = \sum_{n=1}^{\infty} b_n \sin nx$$
$$= \sum_{n=1}^{\infty} \frac{2(-1)^n(6 - n^2\pi^2)}{n^3}\sin nx.$$

The partial sum S_6 is shown on the interval $[0, \pi]$ and $[-3\pi, 3\pi]$.

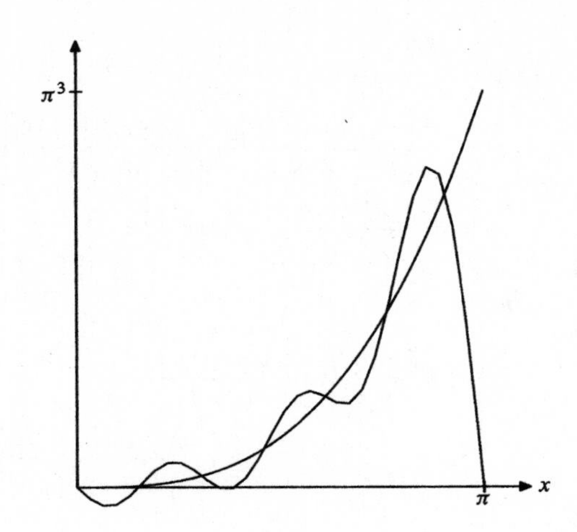

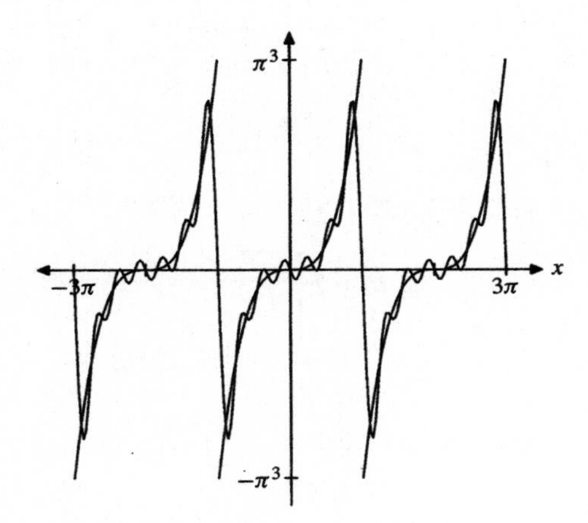

29. The function

$$f(x) = \begin{cases} 1, & 0 \le x < \pi/2, \\ 0, & \pi/2 \le x \le \pi, \end{cases}$$

has odd extension

$$f_o(x) = \begin{cases} 0, & -\pi \le x < -\pi/2, \\ -1, & -\pi/2 \le x < 0, \\ 1, & 0 \le x < \pi/2, \\ 0, & \pi/2 \le x \le \pi, \end{cases}$$

on the interval $[-\pi, \pi]$. The Fourier expansion of $f_o(x)$ has only sine terms with

$$\begin{aligned} b_n &= \frac{1}{\pi}\int_{-\pi}^{\pi} f_o(x)\sin nx\,dx \\ &= \frac{2}{\pi}\int_{0}^{\pi} f_o(x)\sin nx\,dx \\ &= \frac{2}{\pi}\int_{0}^{\pi/2} \sin nx\,dx \\ &= \frac{2}{\pi}\left[-\frac{\cos nx}{n}\right]_0^{\pi/2} \\ &= -\frac{2}{\pi n}\left[\cos\frac{n\pi}{2} - 1\right]. \end{aligned}$$

Hence, the Fourier expansion is

$$\begin{aligned} f(x) &= \sum_{n=1}^{\infty} b_n \sin nx \\ &= \sum_{n=1}^{\infty} \frac{2}{\pi n}\left[1 - \cos\frac{n\pi}{2}\right]\sin nx. \end{aligned}$$

The partial sum S_6 is shown on the interval $[0, \pi]$ and $[-3\pi, 3\pi]$.

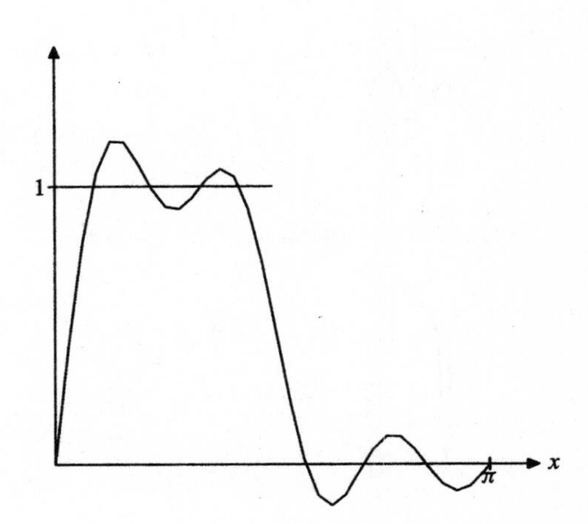

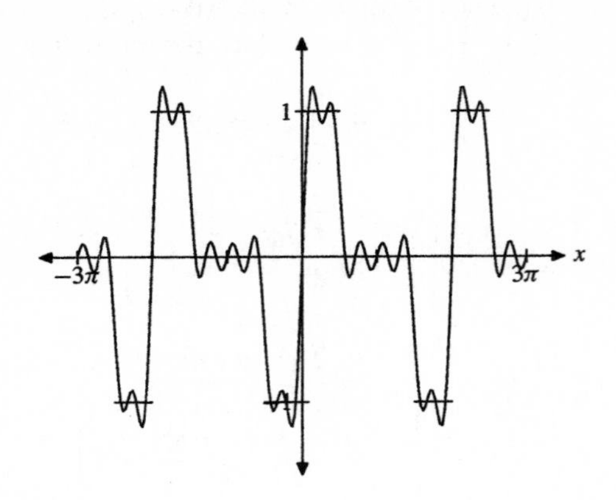

31. The function $f(x) = x\cos x$ is already odd on the interval $[-\pi, \pi]$, so the Fourier expansion contains only sine terms and

$$b_n = \frac{1}{\pi}\int_{-\pi}^{\pi} f(x)\sin nx\, dx = \frac{2}{\pi}\int_0^{\pi} x\cos x \sin nx\, dx.$$

Using a product to sum identity,

$$b_n = \frac{2}{\pi}\int_0^{\pi} x\left\{\frac{1}{2}[\sin(n-1)x + \sin(n+1)x]\right\} dx = \frac{1}{\pi}\left\{\int_0^{\pi} x\sin(n-1)x\, dx + \int_0^{\pi} x\sin(n+1)x\, dx\right\}.$$

Integrating by parts, the first integral is

$$\int_0^{\pi} x\sin(n-1)x\, dx = \left[-\frac{x\cos(n-1)x}{n-1} + \frac{\sin(n-1)x}{(n-1)^2}\right]_0^{\pi} = \frac{-\pi\cos(n-1)\pi}{n-1}.$$

Integrating by parts, the second integral is

$$\int_0^{\pi} x\sin(n+1)x\, dx = \left[\frac{-x\cos(n+1)x}{n+1} + \frac{\sin(n+1)x}{(n+1)^2}\right]_0^{\pi} = \frac{-\pi\cos(n+1)\pi}{n+1}.$$

Hence,

$$b_n = \frac{1}{\pi}\left\{\frac{-\pi\cos(n-1)\pi}{n-1} - \frac{\pi\cos(n+1)\pi}{n+1}\right\} = \frac{-(-1)^{n-1}}{n-1} - \frac{(-1)^{n+1}}{n+1} = \frac{(-1)^n}{n-1} + \frac{(-1)^{n+2}}{n+1} = (-1)^n\frac{2n}{n^2-1},$$

provided $n \neq 1$. In the case that $n = 1$,

$$b_1 = \frac{2}{\pi}\int_0^{\pi} x\cos x\sin x\, dx = \frac{1}{\pi}\int_0^{\pi} x\sin 2x\, dx.$$

Integrating by parts,

$$b_1 = \frac{1}{\pi}\left[\frac{-x\cos 2x}{2} + \frac{\sin 2x}{4}\right]_0^{\pi} = -\frac{1}{2}.$$

Hence, the Fourier expansion is

$$f(x) = \sum_{n=1}^{\infty} b_n \sin nx = -\frac{1}{2}\sin x + \sum_{n=2}^{\infty}\frac{(-1)^n 2n}{n^2-1}\sin nx.$$

The partial sum S_6 is shown on the interval $[0, \pi]$ and $[-3\pi, 3\pi]$.

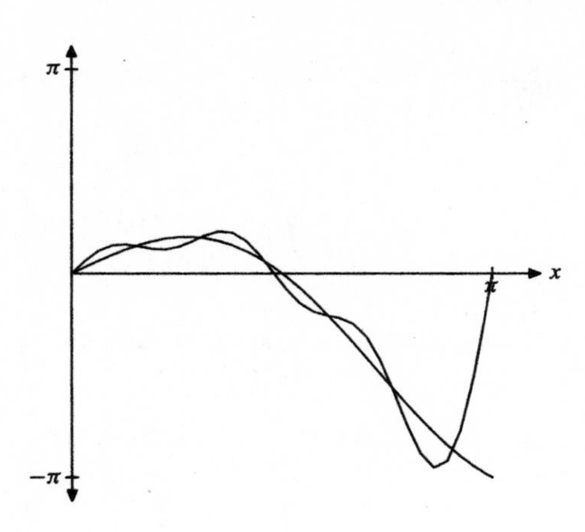

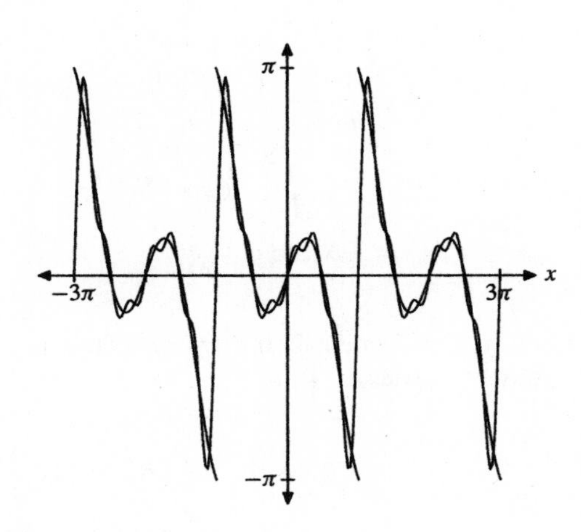

33. Use the trig identity $\cos\alpha\cos\beta = 1/2[\cos(\alpha-\beta)+\cos(\alpha+\beta)]$ to write (if $n \neq p$),

$$\int_0^L \cos\frac{n\pi x}{L}\cos\frac{p\pi x}{L}dx$$
$$= \frac{1}{2}\int_0^L \left[\cos\frac{(n-p)\pi x}{L} + \cos\frac{(n+p)\pi x}{L}\right]dx$$
$$= \frac{1}{2}\left[\frac{L}{(n-p)\pi}\sin\frac{(n-p)\pi x}{L} + \frac{L}{(n+p)\pi}\sin\frac{(n+p)\pi x}{L}\right]_0^L$$
$$= \frac{1}{2}\left[\frac{L}{(n-p)\pi}\sin(n-p)\pi + \frac{L}{(n+p)\pi}\sin(n+p)\pi\right]$$

Because $n-p$ and $n+p$ are integers, $\sin(n-p)\pi = 0$ and $\sin(n+p)\pi = 0$. Thus, if $n \neq p$, then

$$\int_0^L \cos\frac{n\pi x}{L}\cos\frac{p\pi x}{L}dx = 0.$$

35. Use the trig identity $\sin\alpha\cos\beta = 1/2[\sin(\alpha-\beta)+\sin(\alpha+\beta)]$ to write

$$\int_0^1 \cos(2n\pi x)\sin(2k\pi x)dx$$
$$= \frac{1}{2}\int_0^1 [\sin 2(k-n)\pi x + \sin 2(k+n)\pi x]\,dx$$
$$= \frac{1}{2}\left[-\frac{1}{2(k-n)\pi}\cos 2(k-n)\pi x - \frac{1}{2(k+n)\pi}\cos 2(k+n)\pi x\right]_0^1$$
$$= \frac{1}{2}\left[-\frac{1}{2(k-n)\pi}(\cos 2(k-n)\pi - 1) - \frac{1}{2(k+n)\pi}(\cos 2(k+n)\pi - 1)\right]$$

If k and n are different integers, $\cos 2(k-n)\pi = 1$ and $\cos 2(k+n)\pi = 1$. Thus, if $k \neq n$,

$$\int_0^1 \cos(2n\pi x)\sin(2k\pi x)dx = 0.$$

Now, if $k = n$,

$$\int_0^1 \cos(2n\pi x)\sin(2k\pi x)dx$$
$$= \int_0^1 \cos(2n\pi x)\sin(2n\pi x)dx$$
$$= \frac{1}{2}\int_0^1 \sin(4n\pi x)dx$$
$$= \frac{1}{2}\cdot\frac{-1}{4n\pi}\cos(4n\pi x)\Big|_0^1$$
$$= 0.$$

Section 12.4. The Complex Form of a Fourier Series

1. If f is real, we know that

$$\overline{\alpha_n} = \alpha_{-n}, \quad \text{for all } n.$$

However, if

$$\alpha_n = \frac{1}{L}\int_{-L}^{L} f(x)e^{-in\pi x/L}dx,$$

then

$$\alpha_{-n} = \frac{1}{L}\int_{-L}^{L} f(x)e^{in\pi x/L}dx.$$

Since f is even, $f(-x) = f(x)$ and

$$\alpha_{-n} = \frac{1}{L}\int_{-L}^{L} f(-x)e^{in\pi x/L}dx.$$

Now, let

$$u = -x \quad \text{and} \quad du = -dx,$$

and write

$$\begin{aligned}\alpha_{-n} &= \frac{1}{L}\int_{L}^{-L} f(u)e^{-in\pi u/L}(-du)\\ &= \frac{1}{L}\int_{-L}^{L} f(u)e^{-in\pi u/L}du.\end{aligned}$$

But this is precisely the definition for α_n (ignore the dummy variable u). Thus,

$$\overline{\alpha_n} = \alpha_{-n} = \alpha_n.$$

Because $\overline{\alpha_n} = \alpha_n$, it must be the case that α_n is real. Next, suppose that f is real and odd. Again, because f is real, we know that

$$\overline{\alpha_n} = \alpha_{-n}, \quad \text{for all } n.$$

The coefficient α_n is purely imaginary if and only if we can show that $\overline{\alpha_n} = -\alpha_n$ for all n. However,

$$\alpha_n = \frac{1}{L}\int_{-L}^{L} f(x)e^{-in\pi x/L}dx,$$

so again,

$$\alpha_{-n} = \frac{1}{L}\int_{-L}^{L} f(x)e^{in\pi x/L}dx.$$

Because f is odd, $f(-x) = -f(x)$ for all $x\epsilon[-L, L]$. We can write

$$\alpha_{-n} = \frac{1}{L}\int_{-L}^{L} -f(-x)e^{in\pi x/L}dx$$

and again make the change of variable

$$u = -x \quad \text{and} \quad du = -dx.$$

Thus,

$$\begin{aligned}\alpha_{-n} &= \frac{1}{L}\int_{L}^{-L} -f(u)e^{-in\pi u/L}(-du)\\ &= -\frac{1}{L}\int_{-L}^{L} f(u)e^{-in\pi u/L}du\\ &= -\alpha_n,\end{aligned}$$

the last statement being true because u is just a dummy variable. Thus,

$$\overline{\alpha_n} = \alpha_{-n} = -\alpha_n,$$

and thus α_n is purely imaginary for all n.

3. With $L = \pi$, the nth coefficient becomes

$$\begin{aligned}\alpha_n &= \frac{1}{2L}\int_{-L}^{L} f(x)e^{-in\pi x/L}\,dx\\ &= \frac{1}{2\pi}\int_{-\pi}^{\pi} |x|e^{-inx}\,dx\\ &= \frac{1}{2\pi}\int_{-\pi}^{0} (-x)e^{-inx}\,dx\\ &\quad + \frac{1}{2\pi}\int_{0}^{\pi} xe^{-inx}\,dx.\end{aligned}$$

Integrating by parts, the first integral becomes

$$\frac{1}{2\pi}\int_{-\pi}^{0}(-x)e^{-inx}\,dx$$
$$=\frac{1}{2\pi}\left[\frac{xe^{-inx}}{in}-\frac{e^{-inx}}{n^2}\right]_{-\pi}^{0}$$
$$=\frac{1}{2\pi}\left[e^{-inx}\left(\frac{x}{in}-\frac{1}{n^2}\right)\right]_{-\pi}^{0}$$
$$=\frac{1}{2\pi}\left[-\frac{1}{n^2}-e^{in\pi}\left(\frac{-\pi}{in}-\frac{1}{n^2}\right)\right]$$
$$=\frac{1}{2\pi}\left[-\frac{1}{n^2}-(-1)^n\left(-\frac{\pi}{in}-\frac{1}{n^2}\right)\right]$$
$$=\begin{cases}\frac{1}{2\pi}\left(\frac{\pi}{in}\right), & n \text{ even},\\ \frac{1}{2\pi}\left(-\frac{2}{n^2}-\frac{\pi}{in}\right), & n \text{ odd}\end{cases}$$
$$=\begin{cases}\frac{1}{2in}, & n \text{ even}\\ -\frac{1}{\pi n^2}-\frac{1}{2in}, & n \text{ odd}\end{cases}$$

The second integral is

$$\frac{1}{2\pi}\int_{0}^{\pi}xe^{-inx}\,dx$$
$$=\frac{1}{2\pi}\left[\frac{-xe^{-inx}}{in}+\frac{e^{-inx}}{n^2}\right]_{0}^{\pi}$$
$$=\frac{1}{2\pi}\left[e^{-inx}\left(\frac{1}{n^2}-\frac{x}{in}\right)\right]_{0}^{\pi}$$
$$=\frac{1}{2\pi}\left[e^{-in\pi}\left(\frac{1}{n^2}-\frac{\pi}{in}\right)-\frac{1}{n^2}\right]$$
$$=\frac{1}{2\pi}\left[(-1)^n\left(\frac{1}{n^2}-\frac{\pi}{in}\right)-\frac{1}{n^2}\right]$$
$$=\begin{cases}\frac{1}{2\pi}\left(-\frac{\pi}{in}\right), & n \text{ even},\\ \frac{1}{2\pi}\left(\frac{\pi}{in}-\frac{2}{n^2}\right), & n \text{ odd}\end{cases}$$
$$=\begin{cases}-\frac{1}{2in}, & n \text{ even},\\ \frac{1}{2in}-\frac{1}{\pi n^2}, & n \text{ odd}\end{cases}$$

Thus,

$$\alpha_n=\begin{cases}0, & n \text{ even},\\ -\frac{2}{\pi n^2}, & n \text{ odd},\end{cases}$$

provided $n \neq 0$. In the case that $n = 0$,

$$\alpha_0=\frac{1}{2\pi}\int_{-\pi}^{\pi}|x|\,dx=\frac{1}{2\pi}(\pi^2)=\frac{\pi}{2}.$$

Hence,

$$f(x)\sim\sum_{n=-\infty}^{\infty}\alpha_n e^{inx}$$
$$=\frac{\pi}{2}+\sum_{n \text{ odd}}\left(-\frac{2}{\pi n^2}\right)e^{inx}$$
$$=\frac{\pi}{2}-\frac{2}{\pi}\sum_{n \text{ odd}}\frac{1}{n^2}e^{inx}.$$

5. With $L = \pi$, the nth coefficient becomes

$$\alpha_n=\frac{1}{2\pi}\int_{-\pi}^{\pi}f(x)e^{-inx}\,dx$$
$$=\frac{1}{2\pi}\int_{0}^{\pi}e^{-inx}\,dx$$
$$=\frac{1}{2\pi}\left[\frac{e^{-inx}}{-in}\right]_{0}^{\pi}$$
$$=-\frac{1}{2\pi in}\left[e^{-in\pi}-1\right]$$
$$=\frac{1-(-1)^n}{2\pi in}$$
$$=\begin{cases}0, & n \text{ even},\\ \frac{1}{\pi in}, & n \text{ odd},\end{cases}$$

provided $n \neq 0$. In the case that $n = 0$,

$$\alpha_0=\frac{1}{2\pi}\int_{-\pi}^{\pi}f(x)\,dx=\frac{1}{2\pi}\int_{0}^{\pi}dx=\frac{1}{2}.$$

Thus,

$$f(x)\sim\sum_{n=-\infty}^{\infty}\alpha_n e^{inx}=\frac{1}{2}+\sum_{n \text{ odd}}\frac{1}{\pi in}e^{inx}.$$

7. With $L = \pi$, the nth coefficient becomes

$$\begin{aligned}\alpha_n &= \frac{1}{2\pi}\int_{-\pi}^{\pi} e^{bx}e^{-inx}\,dx \\ &= \frac{1}{2\pi}\int_{-\pi}^{\pi} e^{(b-in)x}\,dx \\ &= \frac{1}{2\pi}\frac{e^{(b-in)x}}{b-in}\bigg|_{-\pi}^{\pi} \\ &= \frac{1}{2\pi(b-in)}\left[e^{(b-in)\pi} - e^{(b-in)(-\pi)}\right] \\ &= \frac{1}{2\pi(b-in)}\left[e^{b\pi}e^{-in\pi} - e^{-b\pi}e^{in\pi}\right] \\ &= \frac{(-1)^n}{\pi(b-in)}\left[\frac{e^{b\pi}-e^{-b\pi}}{2}\right] \\ &= \frac{(-1)^n \sinh b\pi}{\pi(b-in)},\end{aligned}$$

provided, of course, that $b \neq in$. Hence,

$$f(x) \sim \sum_{n=-\infty}^{\infty} \alpha_n e^{inx} = \frac{\sinh b\pi}{\pi}\sum_{n=-\infty}^{\infty}\frac{(-1)^n}{b-in}e^{inx}$$

9. If $f(x) = \pi - x$ on the interval $[-\pi, \pi]$, then

$$\alpha_n = \frac{1}{2\pi}\int_{-\pi}^{\pi}(\pi - x)e^{-inx}dx.$$

Integration by parts provides

$$\begin{aligned}\alpha_n &= \frac{1}{2\pi}\left[-\frac{(\pi-x)e^{-inx}}{in} - \frac{e^{-inx}}{n^2}\right]_{-\pi}^{\pi} \\ &= \frac{1}{2\pi}\left[-e^{-inx}\left(\frac{\pi-x}{in}+\frac{1}{n^2}\right)\right]_{-\pi}^{\pi} \\ &= \frac{1}{2\pi}\left[-(-1)^n\cdot\frac{1}{n^2} + (-1)^n\left(\frac{2\pi}{in}+\frac{1}{n^2}\right)\right] \\ &= (-1)^{n+1}\frac{i}{n},\end{aligned}$$

provided $n \neq 0$. In the case that $n = 0$,

$$\begin{aligned}\alpha_0 &= \frac{1}{2\pi}\int_{-\pi}^{\pi}(\pi - x)dx \\ &= \frac{1}{2\pi}\left[\pi x - \frac{1}{2}x^2\right]_{-\pi}^{\pi} \\ &= \pi.\end{aligned}$$

Hence,

$$f(x) \sim \pi + i\sum_{n\neq 0}\frac{(-1)^{n+1}}{n}e^{inx}.$$

11. With $L = \pi$, the nth coefficient becomes

$$\begin{aligned}\alpha_n &= \frac{1}{2\pi}\int_{-\pi}^{\pi}|\sin x|e^{-inx}\,dx \\ &= \frac{1}{2\pi}\int_{-\pi}^{0}(-\sin x)e^{-inx}\,dx \\ &\quad + \frac{1}{2\pi}\int_{0}^{\pi}(\sin x)e^{-inx}\,dx.\end{aligned}$$

Integrating by parts,

$$\begin{aligned}&-\frac{1}{2\pi}\int_{-\pi}^{0}(\sin x)e^{-inx}\,dx \\ &= -\frac{1}{2\pi}\cdot\frac{n^2}{n^2-1}\left[\frac{-e^{-inx}\sin x}{in} + \frac{e^{-inx}\cos x}{n^2}\right]_{-\pi}^{0} \\ &= \frac{-n^2}{2\pi(n^2-1)}\left[e^{-inx}\left(\frac{\cos x}{n^2} - \frac{\sin x}{in}\right)\right]_{-\pi}^{0} \\ &= \frac{-n^2}{2\pi(n^2-1)}\left[\frac{1}{n^2} - e^{in\pi}\left(\frac{-1}{n^2}\right)\right] \\ &= \frac{-(-1)^n - 1}{2\pi(n^2-1)}.\end{aligned}$$

Similarly,

$$\frac{1}{2\pi}\int_0^{\pi}(\sin x)e^{-inx}\,dx = \frac{-1-(-1)^n}{2\pi(n^2-1)}.$$

Adding,

$$\begin{aligned}\alpha_n &= \frac{-2-2(-1)^n}{2\pi(n^2-1)} = \frac{-1-(-1)^n}{\pi(1-n^2)} \\ &= \begin{cases}2/(\pi(n^2-1)), & n \text{ even}, \\ 0, & n \text{ odd},\end{cases}\end{aligned}$$

providing $n \neq \pm 1, 0$. In the case that $n = -1$,

$$\begin{aligned}\alpha_{-1} &= \frac{1}{2\pi}\int_{-\pi}^{\pi} |\sin x| e^{ix}\,dx \\ &= \frac{1}{2\pi}\int_{-\pi}^{0} (-\sin x)e^{ix}\,dx \\ &\quad + \frac{1}{2\pi}\int_{0}^{\pi} (\sin x)e^{ix}\,dx.\end{aligned}$$

Note that

$$\begin{aligned}&\int e^{ix}\sin x dx \\ &= \int (\cos x + i\sin x)\sin x dx \\ &= \int \left(\frac{1}{2}\sin 2x + i\left(\frac{1-\cos 2x}{2}\right)\right)dx \\ &= -\frac{1}{4}\cos 2x + i\left(\frac{1}{2}x - \frac{1}{4}\sin 2x\right).\end{aligned}$$

Thus

$$\begin{aligned}&-\frac{1}{2\pi}\int_{-\pi}^{0} e^{ix}\sin x dx \\ &= -\frac{1}{2\pi}\left[-\frac{1}{4}\cos 2x + i\left(\frac{1}{2}x - \frac{1}{4}\sin 2x\right)\right]_{-\pi}^{0} \\ &= -\frac{1}{2\pi}\left[-\frac{1}{4} - \left(-\frac{1}{4} - \frac{\pi}{2}i\right)\right] \\ &= -\frac{1}{4}i.\end{aligned}$$

Similarly,

$$\frac{1}{2\pi}\int_{0}^{\pi} e^{ix}\sin x dx = \frac{1}{4}i.$$

Thus, $\alpha_1 = 0$. In a similar manner, $\alpha_{-1} = 0$. In the case that $n = 0$,

$$\alpha_0 = \frac{1}{2\pi}\int_{-\pi}^{\pi} |\sin x|\,dx = \frac{1}{\pi}\int_{0}^{\pi} \sin x dx = \frac{2}{\pi}.$$

Thus,

$$f(x) \sim \sum_{n=-\infty}^{\infty} \alpha_n e^{inx} = \frac{2}{\pi} + \sum_{\substack{n \text{ even} \\ n\neq 0}} \frac{2}{\pi(n^2-1)}e^{inx}$$

13. Begin with the assumption that

$$f(x) = \sum_{n=-\infty}^{\infty} \alpha_n e^{inx}, \quad -\pi \leq x \leq \pi.$$

Multiply both sides by $\overline{e^{ikx}} = e^{-ikx}$ and integrate.

$$\begin{aligned}f(x)e^{-ikx} &= \sum_{n=-\infty}^{\infty} \alpha_n e^{inx}e^{-ikx} \\ \int_{-\pi}^{\pi} f(x)e^{-ikx}dx &= \sum_{n=-\infty}^{\infty} \alpha_n \int_{-\pi}^{\pi} e^{inx}e^{-ikx}dx.\end{aligned}$$

In Exercise 12, we showed that e^{inx} and e^{ikx} are orthogonal, provided n and k are integers, $n \neq k$. Thus, the right side simplifies to

$$\int_{-\pi}^{\pi} f(x)e^{-ikx}dx = \alpha_k \int_{-\pi}^{\pi} e^{ikx}e^{-ikx}dx = \alpha_k(2\pi)$$

Therefore,

$$\alpha_k = \frac{1}{2\pi}\int_{-\pi}^{\pi} f(x)e^{-ikx}dx.$$

Now, k is a dummy variable, so just substitute n for k.

Section 12.5. The Discrete Fourier Transform and the FFT

1. The function f is plotted on the left below. Because of the terms involving $\cos 2x$ and $\sin 4x$, we would expect the coefficients for $n = 4$ to be the largest, followed by $n = 2$. The middle figure plots the discrete Fourier transform with N = 256, and it verifies the conjecture. The third figure is a plot of the partial sum of the Fourier series of order $n = 6$. The partial sum keeps the main features of f, but misses the high frequency vibrations in f.

2
0
-2
0
π
2π
t

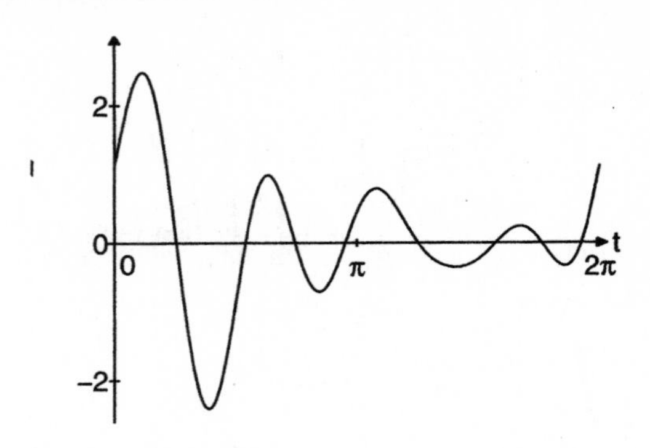

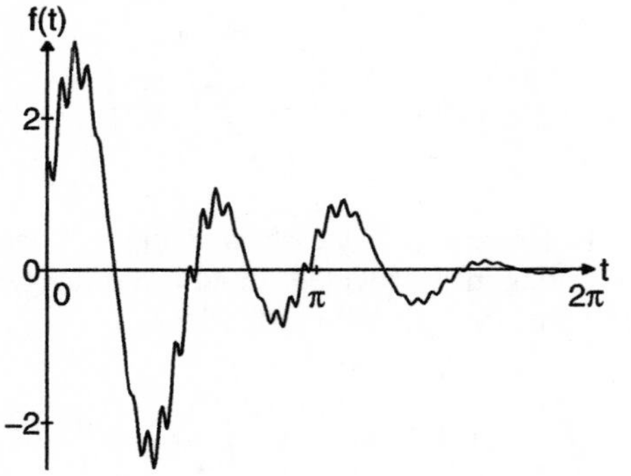

f(t)
2
0
-2
0
π
2π
t

3. By 5.5, $\widehat{y}_m = \sum_{j=0}^{N-1} y_j e^{(-2\pi imj/N)} = \sum_{j=0}^{N-1} y_j \overline{w}^{jm}$. Hence, using the facts that $\overline{y_j} = y_j$ and $w = \overline{w}^{-1}$, we have

$$\overline{\widehat{y}_m} = \overline{\sum_{j=0}^{N-1} y_j \overline{w}^{jm}} = \sum_{j=0}^{N-1} y_j w^{jm} = \sum_{j=0}^{N-1} y_j \overline{w}^{-jm}$$
$$= \widehat{y}_{-m} = \widehat{y}_{N-m}.$$

5. The graph of f and the FFT are shown in the previous exercise. The three figures shown below show the results of filtering with different tolerances. The first used tol $= 0.01$, and needed 71 nonzero coefficients. The second used tol $= 0.2$, and needed 15 nonzero coefficients. The second used tol $= 0.05$, and needed 27 nonzero coefficients.

100
0
0
128
255
k

1
0
-1
0
π
2π
t

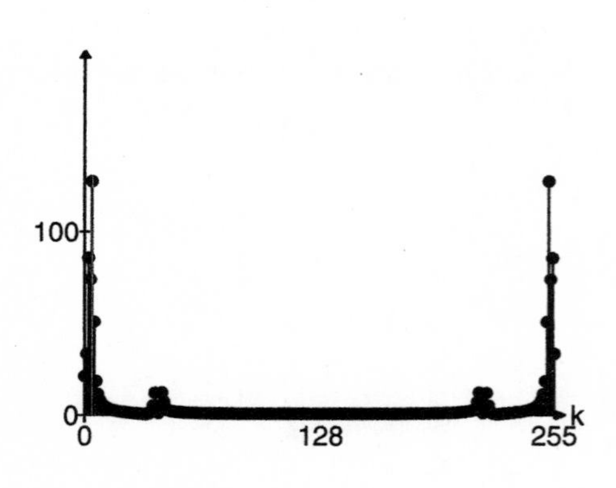

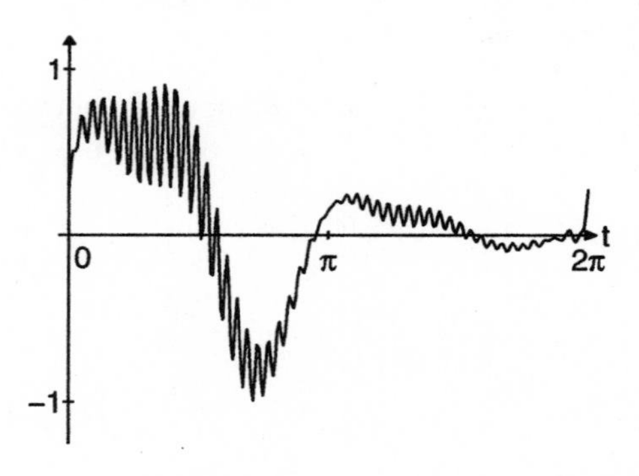

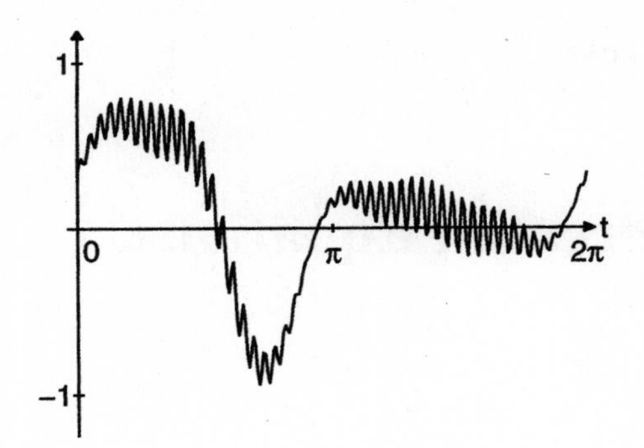
1
0
π
2π
t
−1

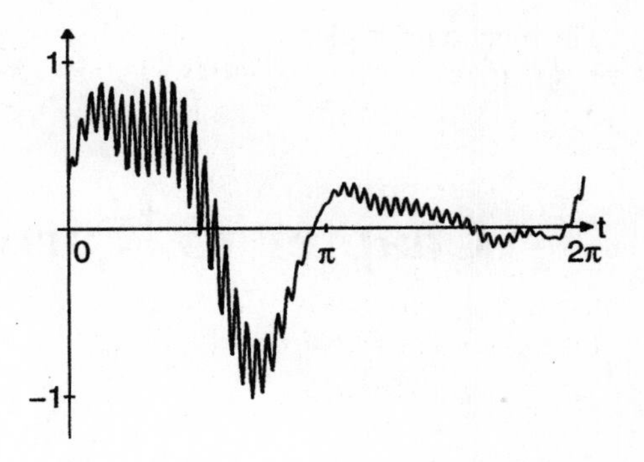
1
0
π
2π
t
−1

Chapter 13. Partial Differential Equations

Section 13.1. Derivation of the Heat Equation

1. The temperature satisfies Dirichlet conditions as given in (1.16). With the data as given the temperature must satisfy

$$\begin{aligned} &u_t(x,t) = ku_{xx}, \quad \text{for } 0 \le x \le L \text{ and } t > 0, \\ &u(0,t) = 5, \quad \text{and} \quad u(L,t) = 25, \quad \text{for } t > 0, \\ &u(x,0) = 15, \quad \text{for } 0 \le x \le L. \end{aligned}$$

3. By the equation just before (1.3),

$$\frac{dQ}{dt} = S\int_a^b \frac{\partial}{\partial t}[c\rho u]\,dx.$$

By (1.7)

$$\frac{dQ}{dt} = S\int_a^b \frac{\partial}{\partial x}\left(C\frac{\partial u}{\partial x}\right)\,dx.$$

Setting these equal, and using the argument leading to (1.9), we get the result.

5. Since α and β are constants, the linearity of differentiation implies that $w_t = \alpha u_t + \beta v_t$ and $w_{xx} = \alpha u_{xx} + \beta v_{xx}$. The result follows immediately.

7. From Table 1 we see that the thermal diffusivity of aluminum is $k = 0.86$. The boundary conditions are Dirichlet conditions so modifying (1.16) we get

$$\begin{aligned} &u_t(x,t) = 0.86u_{xx}, \quad \text{for } 0 < x < L \text{ and } t > 0, \\ &u(0,t) = 20, \quad \text{and} \quad u(L,t) = 35, \quad \text{for } t > 0, \\ &u(x,0) = 15, \quad \text{for } 0 \le x \le L. \end{aligned}$$

9. From Table 1 we see that the thermal diffusivity of silver is $k = 1.71$. We now have a Robin condition at the left-hand endpoint.

$$\begin{aligned} &u_t(x,t) = 1.71u_{xx}, \quad \text{for } 0 < x < L \text{ and } t > 0, \\ &u_x(0,t) = 0.0013(u(0,t) - 15), \quad \text{and} \\ &u(L,t) = 35, \quad \text{for } t > 0, \\ &u(x,0) = 15, \quad \text{for } 0 \le x \le L. \end{aligned}$$

———x———

Section 13.2. Separation of Variables for the Heat Equation

1. The thermal diffusivity of gold is $k = 1.18\,\text{cm}^2/\text{sec}$. We will let the unit of length be centimeters, so $L = 50$. The boundary conditions are $u(0,t) = 0$ and $u(50,t) = 0$, so the steady-state temperature is 0. Hence the solution is given by

$$u(x,t) = \sum_{n=1}^{\infty} b_n e^{-kn^2\pi^2 t/L^2}\sin(n\pi x/L),$$

where the coefficients are

$$\begin{aligned} b_n &= \frac{2}{L}\int_0^L 100\sin(n\pi x/L)\,dx \\ &= \frac{200}{n\pi}[1 - \cos(n\pi)] \\ &= \begin{cases} \frac{400}{n\pi} & \text{if } n \text{ is odd,} \\ 0 & \text{if } n \text{ is even.} \end{cases} \end{aligned}$$

Thus the solution is given by

$$u(x,t) = \sum_{p=0}^{\infty} \frac{400}{(2p+1)\pi} e^{-1.18\times(2p+1)^2\pi^2 t/2500} \times \sin\left(\frac{n\pi x}{50}\right).$$

For $t = 100$, the term for $p = 1$ is bounded by 0.65. Hence one term of the series will suffice to estimate the temperature within 1°. Using just this one term, we solve

$$\frac{400}{\pi} e^{-1.18\times\pi^2 t/2500} = 10$$

for $t = 546$ sec. Hence we see that it will take about 546 sec for the temperature to drop below 10°C. The temperature at 100 second intervals is plotted below.

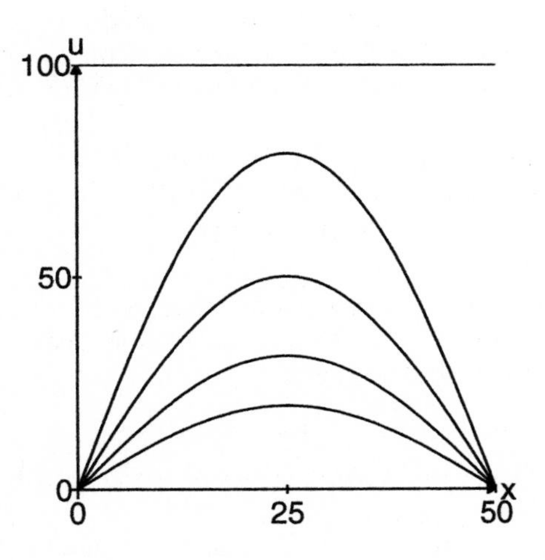

3. (a) For $0 \le x \le 10$, the steady-state temperature is given by $f(x) = 20 - x$.

   (b) Let $u(x,t)$ be the temperature. It must solve the initial/boundary value problem

$$u_t - ku_{xx} = 0, \quad \text{for } 0 < x < 10 \text{ and } t > 0,$$
$$u(0,t) = 20 \quad \text{and} \quad u(10,t) = -10,$$
$$u(x,0) = 20 - x.$$

   This solution will tend to a new steady-state temperature $u_{ss}(x)$ with $u_{ss}(0) = 20$ and $u_{ss}(10) = -10$. Hence $u_{ss}(x) = 20 - 3x$. The difference $v(x,t) = u(x,t) - u_{ss}(x)$ must solve

$$v_t - kv_{xx} = 0, \quad \text{for } 0 < x < 10 \text{ and } t > 0,$$
$$v(0,t) = 0 \quad \text{and} \quad v(10,t) = 0,$$
$$v(x,0) = (20 - x) - (20 - 3x) = 2x.$$

   The solution v is given by

$$v(x,t) = \sum_{n=1}^{\infty} b_n e^{-kn^2\pi^2 t/100} \sin\frac{n\pi x}{10},$$

   where $k = 0.0057$ cm^2/sec is the thermal diffusivity of brick, and

$$b_n = \frac{2}{10}\int_0^{10} 2x \sin\left(\frac{n\pi x}{10}\right) dx = (-1)^{n+1}\frac{40}{n\pi}.$$

   Hence the solution is

$$u(x,t) = (20 - 3x) + \sum_{n=1}^{\infty}(-1)^{n+1}\frac{40}{n\pi} e^{-0.0057\times n^2\pi^2 t/100} \times \sin\frac{n\pi x}{10}.$$

5. Since $T_0 = T_L = 0$, the steady-state temperature is 0°. The solution is given by

$$u(x,t) = \sum_{n=1}^{\infty} b_n e^{-4n^2\pi^2 t} \sin(n\pi x),$$

   where the coefficients are computed from

$$b_n = 2\int_0^1 x(1-x)\sin(n\pi x)\,dx.$$

   Computing this integral we see that $b_n = 0$ if n is even, and $b_{2n+1} = 4/((2n+1)^3\pi^3)$. Thus the solution is

$$u(x,t) = \frac{4}{\pi^3}\sum_{n=0}^{\infty}\frac{1}{(2n+1)^3} e^{-4(2n+1)^2\pi^2 t} \times \sin((2n+1)\pi x).$$

7. Since $T_0 = T_L = 0$, the steady-state temperature is $0°$. The solution is given by

$$u(x, t) = \sum_{n=1}^{\infty} b_n e^{-n^2 t} \sin(nx),$$

where the coefficients are

$$\begin{aligned} b_n &= \frac{2}{\pi} \int_0^{\pi} \sin^2 x \sin(nx)\, dx \\ &= \begin{cases} 0 & \text{if } n \text{ is even,} \\ -8/[\pi n(n^2 - 4)] & \text{if } n \text{ is odd.} \end{cases} \end{aligned}$$

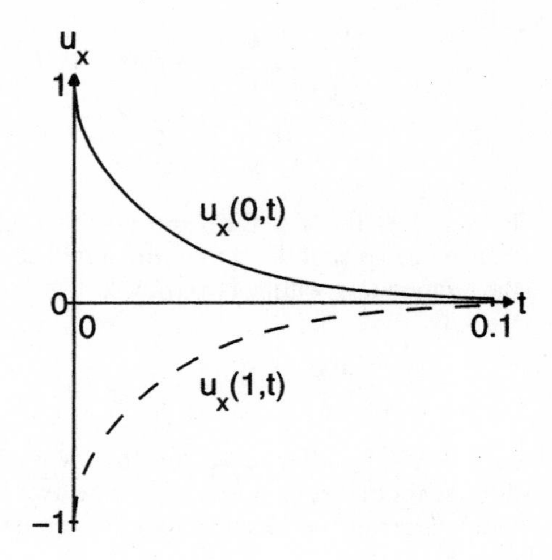

9. This solution is plotted in the first figure for $t = 0$ and then for values of t at intervals of 0.01. The temperature profiles show a steady decrease in temperature as it approaches steady-state. The partial u_x at the endpoints is plotted in the second figure. Notice that $u_x(0, t) > 0$ and $u_x(1, t) < 0$, reflecting the fact that heat is flowing out of the rod at each endpoint, leading to the decrease in the temperature in the rod.

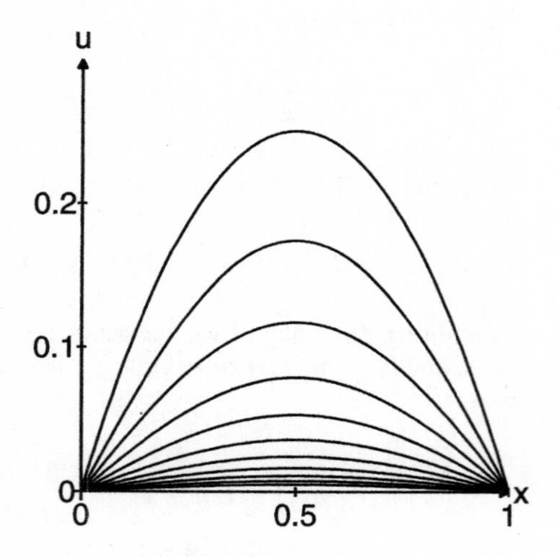

11. This solution is plotted in the first figure for $t = 0$ and then for values of t at intervals of 0.2. The temperature profiles show a steady decrease in temperature as it approaches steady-state. The partial u_x at the endpoints is plotted in the second figure. Notice that $u_x(0, t) > 0$ and $u_x(1, t) < 0$, reflecting the fact that heat is flowing out of the rod at each endpoint, leading to the decrease in the temperature in the rod. However, notice that there is an increase in the temperature near the endpoints for a short period of time. This is because heat flow faster from th center of the rod into these areas than it flows out through the endpoints.

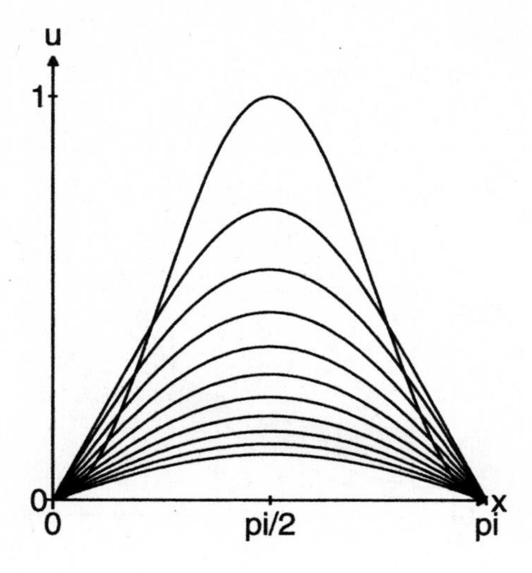

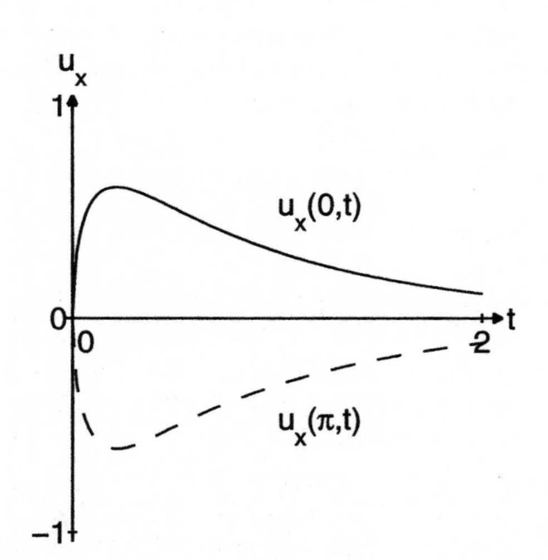

13. The solution is

$$u(x,t) = \frac{a_0}{2} + \sum_{n=1}^{\infty} a_n e^{-n^2\pi^2 t} \cos(n\pi x),$$

where the coefficients a_n are the Fourier cosine coefficients of f. After computing the coefficients, we get the series

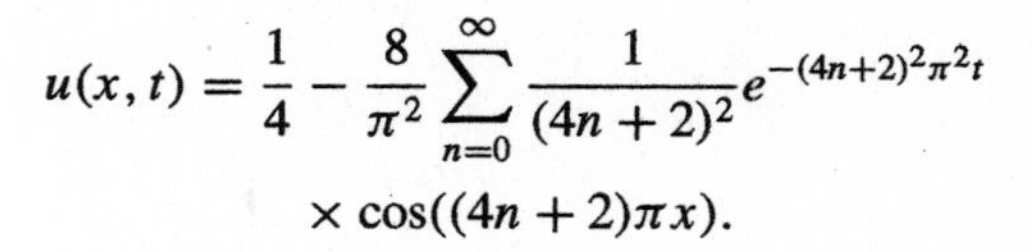

$$u(x,t) = \frac{1}{4} - \frac{8}{\pi^2} \sum_{n=0}^{\infty} \frac{1}{(4n+2)^2} e^{-(4n+2)^2\pi^2 t} \times \cos((4n+2)\pi x).$$

The solution is plotted in the next figure at $t = 0$ and then at intervals of 0.005. The temperature approaches the steady-state temperature of 1/4, which is the average temperature in the rod initially.

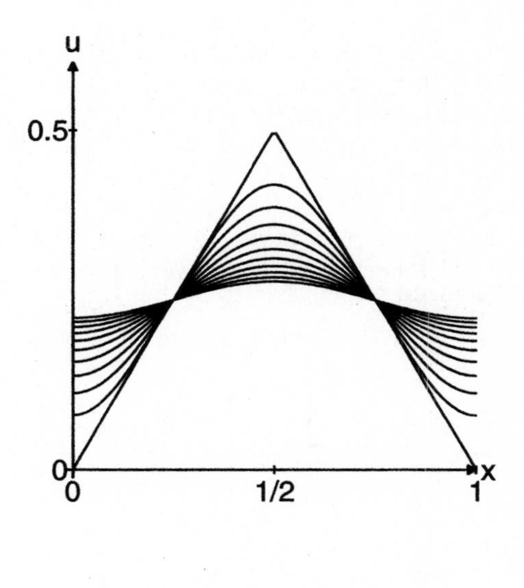

15. The solution is

$$u(x,t) = \frac{2}{\pi} - \frac{4}{\pi} \sum_{n=1}^{\infty} \frac{1}{4n^2 - 1} e^{-4n^2\pi^2 t} \cos(2n\pi x).$$

The solution is plotted in the next figure at $t = 0$ and then at intervals of 0.1. The temperature approaches the steady-state temperature of $2/\pi$, which is the average temperature in the rod initially.

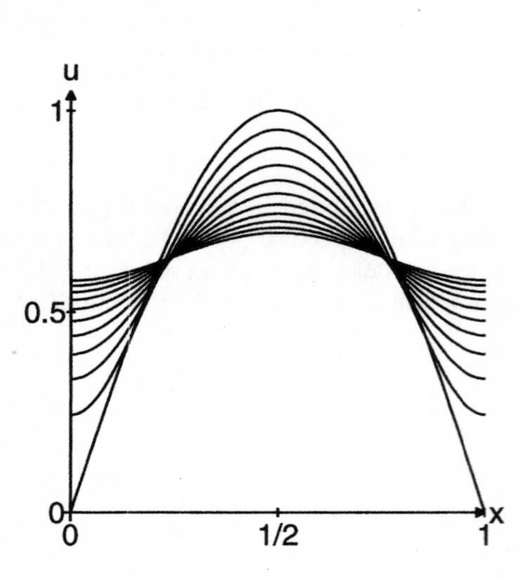

17. The solution is

$$u(x,t) = \frac{1}{4} + \sum_{n=1}^{\infty} a_n e^{-n^2\pi^2 t/4} \cos\left(\frac{n\pi x}{2}\right),$$

where for $n \geq 1$,

$$a_n = \frac{4}{n^2\pi^2} \times \begin{cases} 0, & \text{if } n = 4k, \\ -1, & \text{if } n = 4k \pm 1, \\ 2, & \text{if } n = 4k + 2. \end{cases}$$

The solution is plotted in the next figure at $t = 0$ and then at intervals of 0.1. The temperature approaches the steady-state temperature of 1/4, which

is the average temperature in the rod initially.

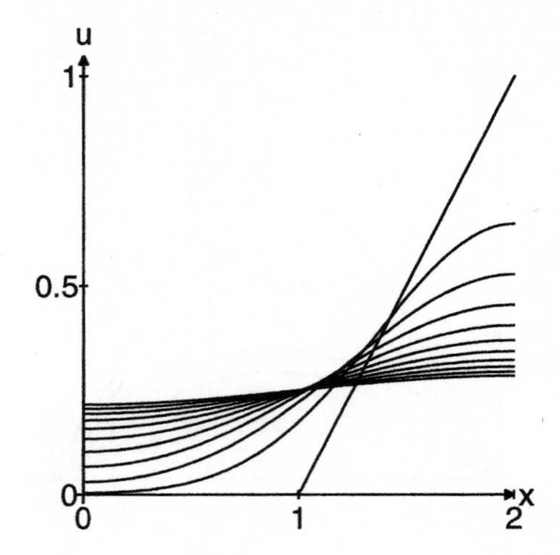

19. Since v does not depend on t, the partial differential equation reduces to $-kv'' = p/c\rho$, which is equivalent to what is to be shown. If $u(x,t) = u_h(x,t) + v(x)$, then

$$\frac{\partial u}{\partial t} = \frac{\partial u_h}{\partial t} \quad \text{and} \quad \frac{\partial^2 u}{\partial x^2} = \frac{\partial^2 u_h}{\partial x^2} + v'' = \frac{\partial^2 u_h}{\partial x^2} - \frac{P}{k}.$$

Hence

$$\frac{\partial u}{\partial t} - k\frac{\partial^2 u}{\partial x^2} = \frac{\partial u_h}{\partial t} - k\left(\frac{\partial^2 u_h}{\partial x^2} - \frac{P}{k}\right) = P.$$

In addition, $u(x,0) = u_h(x,0) + v(x) = f(x)$. Finally, $u(0,t) = u_h(0,t) + v(0) = A$ and $u(L,t) = u_h(L,t) + v(L) = B$.

21. Since $v'' = -P(x)/k = -e^{-x}$, we have $v(x) = -e^{-x} + \alpha x + \beta$, where α and β are constants of integration. The boundary conditions say that $1 = \beta$ and $-1/e = -1/e + \alpha + \beta$. Hence $\beta = 1$ and $\alpha = -1$, so $v(x) = 1 - x - e^{-x}$.

We have to solve the initial/boundary value problem for the heat equation with initial value $u_h(x,0) = f(x) - v(x) = \sin \pi x + e^{-x} - 1 + x$. The Fourier sine series for v is

$$v(x) = \sum_{n=1}^{\infty} a_n \sin n\pi x,$$

where

$$a_n = 2\int_0^1 v(x)\sin n\pi x\,dx$$
$$= 2\int_0^1 (1 - x - e^{-x})\sin n\pi x\,dx$$
$$= \frac{2}{n\pi} - \frac{2n\pi[1 - (-1)^n/e]}{1 + n^2\pi^2}.$$

Therefore

$$u_h(x,t) = e^{-\pi^2 t}\sin\pi x - \sum_{n=1}^{\infty} a_n e^{-n^2\pi^2 t}\sin n\pi x.$$

Finally,

$$u(x,t) = u_h(x,t) + v(x)$$
$$= 1 - x - e^{-x} + e^{-\pi^2 t}\sin\pi x$$
$$- \sum_{n=1}^{\infty} a_n e^{-n^2\pi^2 t}\sin n\pi x.$$

Section 13.3. The Wave Equation

1. The Fourier sine series for f is $\sum_{n=1}^{\infty} a_n \sin n\pi x$, where

$$a_n = 2\int_0^1 f(x)\sin n\pi x\,dx$$
$$= \frac{1}{2}\int_0^1 x(1-x)\sin n\pi x\,dx$$
$$= \begin{cases} 0, & \text{if } n \text{ is even,} \\ \dfrac{2}{n^3\pi^3}, & \text{if } n \text{ is odd.} \end{cases}$$

From (3.5), we conclude that

$$\sum_{k=0}^{\infty} \frac{2\sin((2k+1)\pi x)\cos((2k+1)\pi t)}{(2k+1)^3\pi^3}.$$

3. The Fourier sine series for g is $\sum_{n=1}^{\infty} b_n \sin n\pi x$, where

$$b_n = 2\int_0^1 \sin n\pi x\,dx$$
$$= \frac{2}{n\pi}[1 - \cos n\pi]$$
$$= \begin{cases} 0 & n \text{ even} \\ 4/n\pi & n \text{ odd.} \end{cases}$$

From (3.5) and (3.7), we conclude that

$$u(x,t) = 4\sum_{n=0}^{\infty} \frac{\sin(2n+1)\pi x \cdot \sin(2n+1)\pi t}{(2n+1)^2\pi^2}.$$

5. The Fourier sine series for g is $\sum_{n=1}^{\infty} b_n \sin n\pi x$, where

$$b_n = \frac{2}{3}\int_0^3 g(x)\sin\frac{n\pi x}{3}\,dx$$
$$= \frac{2}{3}\int_1^2 \sin\frac{n\pi x}{3}\,dx$$
$$= \frac{2[\cos(n\pi/3) - \cos(2n\pi/3)]}{n\pi}.$$

From (3.5) and (3.7), we conclude that

$$u(x,t) = \sum_{n=1}^{\infty} B_n \sin(n\pi x/3)\sin(n\pi t/3),$$

where

$$B_n = \frac{3b_n}{n\pi} = \frac{6[\cos(n\pi/3) - \cos(2n\pi/3)]}{n^2\pi^2}.$$

7. The odd periodic extension of $f(x) = \sin\pi x$ is

$f_{op}(x) = \sin \pi x$. Hence the solution is

$$\begin{aligned} u(x,t) &= [f_{op}(x+ct) + f_{op}(x-ct)]/2 \\ &= [\sin \pi(c+2t) + \sin \pi(x-2t)]/2 \\ &= \sin \pi x \cos 2\pi t. \end{aligned}$$

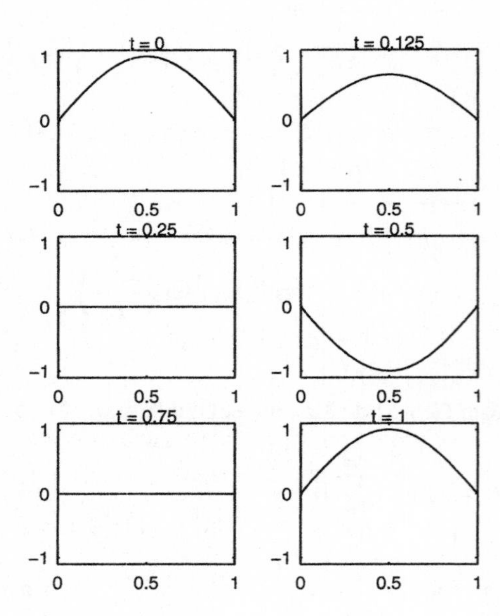

9. The Fourier sine series converges to $f_{op}(x)$ for all x, so

$$f_{op}(x) = \sum_{n=1}^{\infty} a_n \sin n\pi x, \quad \text{for all } x.$$

Using the trigonometric identity, we get

$$\begin{aligned} u_1(x,t) &= \sum_{n=1}^{\infty} a_n \sin n\pi x \cos n\pi t \\ &= \frac{1}{2}\sum_{n=1}^{\infty} a_n \sin n\pi(x+t) \\ &\quad + \frac{1}{2}\sum_{n=1}^{\infty} a_n \sin n\pi(x-t) \\ &= \frac{1}{2}\left[f_{op}(x+t) + f_{op}(x-t)\right] \\ &= u_2(x,t). \end{aligned}$$

11. We start with the d'Alembert solution

$$u(x,t) = F(x+ct) + G(x-ct).$$

The initial conditions are

$$\begin{aligned} 0 &= F(x) + G(x) \\ g(x) &= c[F'(x) - G'(x)]. \end{aligned}$$

By the first equation, $G = -F$, and then by the second equation

$$F'(x) = \frac{1}{2c} g(x) \quad \text{for } 0 \le x \le L.$$

The first boundary condition is

$$0 = u(0,t) = F(ct) - F(-ct),$$

from which we conclude that F is an even function. The second boundary condition is

$$0 = u(L,t) = F(L+ct) - F(L-ct).$$

If we set $ct = y + L$, and remember that F is even, this equation implies that

$$F(y+2L) = F(-y) = F(y).$$

Hence F is even and $2L$-periodic. Therefore, the derivative F' is odd and periodic. Since $F'(x) = g(x)/2c$ for $0 \le x \le L$, we set $F'(x) = g_{op}(x)$ for all x. Then by the fundamental theorem of calculus,

$$\begin{aligned} u(x,t) &= F(x+ct) - F(x-ct) \\ &= \int_{x-ct}^{x+ct} F'(s)\, ds \\ &= \frac{1}{2c}\int_{x-ct}^{x+ct} g_{op}(s)\, ds. \end{aligned}$$

13. Let u_1 be the solution with the same data, but with $g = 0$. According to (3.18)

$$u_1(x,t) = \frac{1}{2}\left[f_{op}(x+ct) - f_{op}(x-ct)\right].$$

Let u_2 be the solution with the same data, but with $f = 0$. According to Exercise 11

$$u_2(x,t) = \frac{1}{2c}\int_{x-ct}^{x+ct} g_{op}(s)\, ds.$$

Since the wave equation is linear, the function $u = u_1 + u_2$ is a solution, and satisfies the needed boundary and initial conditions. Hence

$$u(x,t) = \frac{1}{2}\left[f_{op}(x+ct) - f_{op}(x-ct)\right] + \frac{1}{2c}\int_{x-ct}^{x+ct} g_{op}(s)\,ds.$$

15. Proceeding according to the hint, we get

$$\begin{aligned} E'(t) &= \frac{1}{2}\int_0^L \left[\rho u_t^2 + T u_x^2\right]_t dx \\ &= \int_0^L [\rho u_t u_{tt} + T u_x u_{xt}]\,dx \\ &= T\int_0^L [u_t u_{xx} + u_x u_{xt}]\,dx, \end{aligned}$$

since $u_{tt} = c^2 u_{xx} = T u_{xx}/\rho$. It then follows that

$$\begin{aligned} E'(t) &= T\int_0^L [u_t u_x]_x\,dx \\ &= 0 \quad \text{since } u_t(0,t) = 0 = u_t(L,t). \end{aligned}$$

For the last step we notice that $u_t(0,t) = [u(0,t)]_t = 0$, and a similar calculation for the other endpoint.

17. (a) When we substitute $u(x,t) = X(x)T(t)$ into equation (3.22), and separate variables, we get

$$\frac{T'' + 2kT'}{c^2 T} = \frac{X''}{X}.$$

Since the left-hand side is a function of t and the right-hand side is a function of x, both must be equal to a constant, which we designate as $-\lambda$. Thus X must satisfy $X'' = -\lambda X$, with $X(0) = X(L) = 0$. This is the same boundary value problem as discussed in the text. Hence the solutions are

$$\lambda_n = \frac{n^2\pi^2}{L^2} \quad \text{and} \quad X_n(x) = \sin\frac{n\pi x}{L},$$

for $n = 1, 2, \ldots.$

The equation for T becomes

$$T'' + 2kT' + [n^2\pi^2c^2/L^2]T = 0.$$

Since $k < \pi c/L$, the functions $e^{-kt}\sin(\mu_n)$ and $e^{-kt}\cos(\mu_n)$, where

$$\mu_n = \sqrt{(n^2\pi^2c^2/L^2) - k^2},$$

form a fundamental set of solutions. Hence we have product solutions

$$\begin{aligned} u_n(x,t) &= e^{-kt}\cos(\mu_n t)\sin\left(\frac{n\pi x}{L}\right) \quad \text{and} \\ v_n(x,t) &= e^{-kt}\sin(\mu_n t)\sin\left(\frac{n\pi x}{L}\right) \end{aligned}$$

for $n = 1, 2, \ldots.$

(b) The general solution to equation (3.22) and the boundary conditions is the infinite series

$$u(x,t) = e^{-kt}\sum_{n=1}^{\infty}[a_n u_n(x,t) + b_n v_n(x,t)].$$

To satisfy the initial conditions we must have

$$\begin{aligned} f(x) = u(x,0) &= \sum_{n=1}^{\infty} a_n \sin\left(\frac{n\pi x}{L}\right) \quad \text{and} \\ g(x) = u_t(x,0) &= -\sum_{n=1}^{\infty} \mu_n b_n \sin\left(\frac{n\pi x}{L}\right). \end{aligned}$$

Hence we must have

$$\begin{aligned} a_n &= \frac{2}{L}\int_0^L f(x)\sin\left(\frac{n\pi x}{L}\right)dx \quad \text{and} \\ b_n &= -\frac{2}{L\mu_n}\int_0^L g(x)\sin\left(\frac{n\pi x}{L}\right)dx. \end{aligned}$$

Section 13.4. Laplace's Equation

1. Using Figure 1, we have $f(x) = 10$, and $g = h = k = 0$. According to (4.12), (4.13), and (4.14), the solution is given by

$$u(x, y) = \sum_{n=1}^{\infty} b_n \sinh[n\pi(y - 2)] \sin(n\pi x),$$

where

$$b_n = -\frac{2}{\sinh(2n\pi)} \int_0^1 10 \sin(n\pi x)\, dx = \begin{cases} 0 & \text{if } n \text{ is even,} \\ -40/n\pi \sinh(2n\pi) & \text{if } n \text{ is odd.} \end{cases}$$

Hence

$$u(x, y) = \sum_{k=0}^{\infty} \frac{40}{(2k+1)\pi} \frac{\sinh[(2k+1)\pi(2-y)]}{\sinh[(4k+2)\pi]} \sin(n\pi x).$$

3. The derivation is just like that in the text. In outline, we use separation of variables to find the product solutions

$$\sinh\frac{n\pi x}{b} \sin\frac{n\pi y}{b} \quad \text{and} \quad \sinh\frac{n\pi(x-a)}{b} \sin\frac{n\pi y}{b}.$$

The general solution therefore has the form

$$u(x, y) = \sum_{n=1}^{\infty} \left[a_n \sinh\frac{n\pi x}{b} + b_n \sinh\frac{n\pi(x-a)}{b} \right] \sin\frac{n\pi y}{b},$$

The constants are determined from the boundary conditions. At $x = 0$, we have

$$g(y) = u(0, y) = \sum_{n=1}^{\infty} -b_n \sinh\frac{n\pi a}{b} \sin\frac{n\pi y}{b},$$

and at $x = a$,

$$k(y) = u(a, y) = \sum_{n=1}^{\infty} a_n \sinh\frac{n\pi a}{b} \sin\frac{n\pi y}{b}.$$

These are the Fourier sine series for g and k, so the result follows.

5. Since $f = 0$ and $a = b = 1$, we see from (4.12) that the solution is given by

$$u(x, y) = \sum_{n=1}^{\infty} a_n \sinh(n\pi y) \sin(n\pi x),$$

where

$$a_n = \frac{2}{\sinh(n\pi)} \int_0^1 h(x) \sin(n\pi x)\, dx.$$

We find that $a_n = 0$ if n is even, and

$$a_{2n+1} = \frac{4}{\sinh((2n+1)\pi)} \cdot \frac{(-1)^n}{(2n+1)^2\pi^2}.$$

Hence the temperature is

$$u(x, y) = 4\sum_{n=0}^{\infty} \frac{(-1)^n \sinh((2n+1)\pi y)\sin((2n+1)\pi x)}{(2n+1)^2\pi^2 \sinh((2n+1)\pi)}.$$

7. The function $f(x) = \sin(2\pi x)$ is expressed in terms of its Fourier sine series, so we can immediately write down the temperature:

$$u(x, y) = \frac{-\sinh(2\pi(y-1))\sin(2\pi x)}{\sinh(2\pi)}.$$

9. We see from (4.12) that the solution is given by

$$u(x, y) = \sum_{n=1}^{\infty} [a_n \sinh(n\pi y) + b_n \sinh(n\pi(y-2)] \sin(n\pi x),$$

where

$$\begin{aligned} a_n = b_n &= \frac{2}{\sinh(2n\pi)} \int_0^1 \sin(n\pi x)\,dx \\ &= \frac{2}{\sinh(2n\pi)} \begin{cases} 0, & \text{if } n \text{ is even,} \\ \dfrac{2}{n\pi}, & \text{if } n \text{ is odd.} \end{cases} \end{aligned}$$

Therefore the temperature is

$$u(x, y) = \frac{4}{\pi}\sum_{n=0}^{\infty} \frac{\sinh((2n+1)\pi y) + \sinh((2n+1)\pi(y-2))}{(2n+1)\sinh((4n+2)\pi)} \sin((2n+1)\pi x).$$

11. (a) Using (4.12), we see that the temperature is given by

$$u(x, y) = -\sum_{n=0}^{\infty} B_n \frac{\sinh(n\pi(y-L))}{\sinh(n\pi L)} \sin(n\pi x).$$

(b) In part (a), the dependence on L comes from the factor

$$\frac{-\sinh(n\pi(y-L))}{\sinh(n\pi L)} = \frac{e^{n\pi(L-y)} - e^{n\pi(y-L)}}{e^{n\pi L} - e^{-n\pi L}}.$$

If we factor $e^{n\pi L}$ from both numerator and denominator, we get

$$\frac{e^{-n\pi y} - e^{n\pi(y-2L)}}{1 - e^{-2n\pi L}}.$$

When we let L increase to ∞, this converges to $e^{-n\pi y}$. Hence the solution over the finite rectangle converges to

$$u(x, y) = \sum_{n=1}^{\infty} B_n e^{-n\pi y} \sin(n\pi x).$$

Section 13.5. Laplace's Equation on a Disk

1. The computations of u_{xx} and u_{yy} proceed just as they do in polar coordinates, since the equations for x and y are are the same. Since the coordinate z is unchanged, u_{zz} is also unchanged. Hence

$$\nabla^2 = \frac{\partial^2}{\partial r^2} + \frac{1}{r}\frac{\partial}{\partial r} + \frac{1}{r^2}\frac{\partial^2}{\partial \theta^2} + \frac{\partial^2}{\partial z^2}.$$

3. The function $u(x, y) = 1 + y$ is a solution to Laplace's equation, and is equal to $f(x, y)$ on the surface of the can. Hence the temperature is $u(x, y) = 1 + y$.

5. Since the Fourier series for f is $f(\theta) = \sin^2\theta = [1 - \cos 2\theta]/2$, the temperature is $u(r, \theta) = [1 - r^2 \cos 2\theta]/2$.

7. The Fourier series for f is

$$f(\theta) = \frac{A_0}{2} + \sum_{n=1}^{\infty}[A_n \cos n\theta + B_n \sin n\theta].$$

First, we have

$$A_0 = \frac{1}{\pi}\int_0^{2\pi} f(\theta)\, d\theta = \frac{1}{\pi}\int_0^{2\pi} \theta(2\pi - \theta)\, d\theta = \frac{4\pi^2}{3},$$

Next, for $n \geq 1$, we integrate by parts twice to get

$$A_n = \frac{1}{\pi}\int_0^{2\pi} f(\theta)\cos n\theta\, d\theta = \frac{1}{\pi}\int_0^{2\pi} \pi(2\pi - \theta)\cos n\theta\, d\theta = \frac{-4}{n^2}.$$

Finally, since the periodic extension of f is even, the sine coefficients are $B_n = 0$. Thus the temperature is

$$u(r, \theta) = \frac{2\pi^2}{3} - 4\sum_{n=1}^{\infty}\frac{r^n \cos n\theta}{n^2}.$$

9. Let $u(r)$ denote the temperature at distance r from the center of the circles. Then by (5.2), $\nabla^2 u = u_{rr} + u_r/r = 0$. Thus u satisfies the Euler equation $ru_{rr} + u_r = 0$, and has the form $u(r) = A + B\ln r$. Using the boundary conditions to solve for the coefficients A and B we find that

$$u(r) = T_1 + (T_2 - T_1)\frac{\ln(r/a)}{\ln(b/a)}.$$

11. Since the origin is not in the ring-shaped plate, we cannot eliminate the second solution to Euler's equation in (5.9). Hence the general series solution has the form

$$u(r, \theta) = \frac{A_0 + C_0 \ln r}{2} + \sum_{n=1}^{\infty}\left[(A_n r^n + C_n r^{-n})\cos n\theta + (B_n r^n + D_n r^{-n} \sin n\theta\right]$$

Since $u(a, \theta) = 0$, all of the coefficients of the trigonometric functions must equal 0 at $r = a$. Using this to evaluate A_n and B_n we get

$$u(r, \theta) = \frac{C_0 \ln(r/a)}{2} + \sum_{n=1}^{\infty}\left[C_n(r^{-n} - a^{-2n}r^n)\cos n\theta + D_n(r^{-n} - a^{-2n}r^n)\sin n\theta\right]$$

We can then use $u(b, \theta) = f(\theta)$ to compute the upper case coefficients, since these reduce to the Fourier coefficients of f. We get

$$C_0 \ln(b/a) = \frac{1}{\pi} \int_{-\pi}^{\pi} f(\theta)\, d\theta$$

$$C_n(b^{-n} - a^{-2n} b^n) = \frac{1}{\pi} \int_{-\pi}^{\pi} f(\theta) \cos n\theta\, d\theta$$

$$D_n(b^{-n} - a^{-2n} b^n) = \frac{1}{\pi} \int_{-\pi}^{\pi} f(\theta) \sin n\theta\, d\theta$$

13. Separation of variables in polar coordinates proceeds as before, but we now have that $u(r, 0) = u(r, \theta_0) = 0$. Hence the Sturm-Liouville problem is

$$T'' + \lambda T = 0 \quad \text{with} \quad T(0) = T(\theta_0) = 0.$$

The eigenvalues and eigenfunctions are

$$\lambda_n = n^2/\theta_0^2 \quad \text{and} \quad T_n(\theta) = \sin(n\theta/\theta_0) \quad \text{for } n \geq 1.$$

Hence the product solutions are of the form $r^{n\pi/\theta_0} \sin(n\theta/\theta_0)$, and the general series solution is

$$u(r, \theta) = \sum_{n=1}^{\infty} B_n r^{n\pi/\theta_0} \sin(n\theta/\theta_0).$$

At $r = a$ we have $u(a, \theta) = f(\theta)$ so we get the Fourier sine series for f. Hence the coefficients are given by

$$B_n = \frac{2}{\theta_0 a^n} \int_0^{\theta_0} f(\theta) \sin(n\theta/\theta_0)\, d\theta.$$

———×———

Section 13.6. Sturm-Liouville Problems

1. The operator in (b) is $L\phi = x\phi'' + \phi' = (x\phi')'$. The operator in (e) is $L\phi = \sin x\phi'' + \cos x\phi' = (\sin x\phi')'$. The operator in (f) is $L\phi = (1 - x^2)\phi'' - 2x\phi' = [(1 - x^2)\phi']'$. These are therefore formally self-adjoint. The others are not.

3. We use Proposition 6.24 to conclude that all of the eigenvalues are positive. Setting $\lambda = \omega^2$, where $\omega > 0$, the differential equation becomes $\phi'' + \omega^2\phi = 0$. The general solution is $\phi(x) = A\cos\omega x + B\sin\omega x$. By the first boundary condition, $0 = \phi'(0) = \omega B$. Hence $B = 0$. Then, by the second boundary condition, $0 = \phi(1) = A\cos\omega$. Hence $\omega = \pi/2 + n\pi$ for a nonnegative integer n. We conclude that the eigenvalues and eigenfunctions are

$$\lambda_n = (\pi/2 + n\pi)^2 \quad \text{and}$$
$$\phi_n(x) = \cos(\pi/2 + n\pi)x$$

for $n = 0, 1, 2, 3, \ldots$.

5. We use Proposition 6.24 to conclude that all of the eigenvalues are positive. Setting $\lambda = \omega^2$, where $\omega > 0$, the differential equation becomes $\phi'' + \omega^2\phi = 0$. The general solution is $\phi(x) = A\cos\omega x + B\sin\omega x$. By the first boundary condition, $0 = \phi'(0) - \phi(0) =$

$\omega B - A$. Hence $A = \omega B$. so the solution becomes $\phi(x) = B(\omega\cos\omega x + \sin\omega x)$. Then, by the second boundary condition, $0 = \phi(1) = B[\omega\cos\omega + \sin\omega]$. Hence, we are looking for solutions to the equation $\tan\omega = -1/\omega$. For every $n \geq 1$, there is a solution ω_n satisfying $n\pi - \pi/2 < \omega_n < n\pi$. We conclude that the eigenvalues and eigenfunctions are

$$\lambda_n = \omega_n^2 \quad \text{and}$$
$$\phi_n(x) = \omega_n \cos\omega_n x + \sin\omega_n x,$$

for $n = 1, 2, 3, \ldots$.

7. We need to find functions p, q, and μ such that

$$\begin{aligned}\mu[2x\phi'' + \lambda\phi] &= (p\phi')' + \lambda q\phi \\ &= p\phi'' + p'\phi' + \lambda q\phi.\end{aligned}$$

Equating the coefficients of the derivatives of ϕ, we see that $p = 2x\mu$, $p' = 0$, and $q = \mu$. From the second equation we see that p is a constant, so we set $p = 1$. Then $q = \mu = 1/2x$, so the equation becomes $-\phi'' = \lambda(1/2x)\phi$.

9. We need to find functions p, q, and μ such that

$$\begin{aligned}\mu[x^2\phi'' - 2x\phi' + \lambda\phi] &= (p\phi')' + \lambda q\phi \\ &= p\phi'' + p'\phi' + \lambda q\phi.\end{aligned}$$

Equating the coefficients of the derivatives of ϕ, we see that $p = x^2\mu$, $p' = -2x\mu$, and $q = \mu$. From the first two equations we get the equation $p' = -2x^{-1}p$, which has solution $p(x) = x^{-2}$. Hence $\mu = x^{-4}$, and $q = \mu = x^{-4}$, so the equation becomes $-(x^{-2}\phi')' = \lambda x^{-4}\phi$.

11. (a) If $\phi \neq 0$ satisfies the boundary conditions we have $p\phi\phi'|_0^1 = \phi(1)\phi'(1) = a\phi(1)^2 \geq 0$.

(b) First notice that when $\lambda = 0$, the equation is $\phi'' = 0$, which has solution $\phi(x) = Ax + B$. The first boundary condition implies that $B = \phi(0) = 0$, so $\phi(x) = Ax$. The second condition then implies that $A(1 - a) = 0$. We get a nontrivial solution when $A \neq 0$, and therefore only when $a = 1$. Thus 0 is an eigenvalue only when $a = 1$. Next set $\lambda = -\nu^2$ where $\nu > 0$. The differential equation becomes $\phi'' = \nu^2\phi$, and has solution $\phi(x) = Ae^{\nu x} + Be^{-\nu x}$. The first boundary condition is $0 = \phi(0) = A + B$. Hence $B = -A$, and $\phi(x) = A(e^{\nu x} - e^{-\nu x}) = 2A\sinh\nu x$. The second boundary condition can be written as $\phi'(1) = a\phi(1)$, which becomes $2\nu A\cosh\nu = aA\sinh\nu$. This can be written as $\nu = a\tanh\nu$. To examine solutions, we graph the two functions ν and $a\tanh\nu$. Since $\tanh'(0) = 1$, we see that the slope of $a\tanh\nu$ at $\nu = 0$ is a. If $a < 1$, the graph of $a\tanh\nu$ lies below the graph of ν, so there is no solution, and therefore no negative eigenvalue. On the other hand if $a > 1$, the two graphs intersect precisely once, and the value of ν for which this occurs is the lone negative eigenvalue.

13. (a) Proposition 6.12 says that if L is a formally self-adjoint operator, then

$$\int_a^b Lf\cdot g\,dx = \int_a^b f\cdot Lg\,dx + p(fg' - f'g)\big|_a^b,$$

for any two functions f and g which have two continuous derivatives. If both f and g vanish at the endpoints, then $p(fg' - f'g)|_a^b = 0$, so the result follows.

(b) Direct computation shows that

$$\begin{aligned}\int_0^1 Lf\cdot g\,dx &= -\int_0^1 [x^2 + x^3 - 2x^4]\,dx \\ &= \frac{-13}{30},\end{aligned}$$

while

$$\begin{aligned}\int_a^b f\cdot Lg\,dx &= \int_0^1 [2x - 6x^2 + x^3 + 3x^4]\,dx \\ &= \frac{-3}{20}.\end{aligned}$$

15. Substituting $u(x,t) = X(x)T(t)$ into the differential equation, we get $XT_t = k[X_{xx} - qX]T$. Dividing by kXT, we get

$$\frac{T_t}{kT} = \frac{X_{xx} - qX}{X}.$$

Since the left-hand side is a function of t and the right-hand side is a function of x, both must be constant. Therefore, there is a constant λ such that

$$T' = \lambda T \quad \text{and} \quad -X'' + qX = \lambda X.$$

At $x = 0$ we have $0 = u(0,t) = X(x)T(t)$. Hence $X(0) = 0$. At $x = L$ the boundary condition is $0 = u_t(L,t) = X'(L)T(t)$, so $X'(L) = 0$.

Section 13.7. Orthogonality and Generalized Fourier Series

1. The eigenfunctions are $\phi_n(x) = \sin((2n+1)\pi x/2)$. We compute

$$(\phi_n, \phi_n) = \int_0^1 \sin^2 \frac{(2n+1)\pi x}{2}\,dx = \frac{1}{2}, \quad \text{and}$$
$$(f, \phi_n) = \int_0^1 \sin \frac{(2n+1)\pi x}{2}\,dx = \frac{2}{(2n+1)\pi}.$$

Consequently, the coefficients are

$$c_n = \frac{4}{(2n+1)\pi},$$

and the generalized Fourier series is

$$1 = \sum_{n=0}^{\infty} \frac{4}{(2n+1)\pi} \sin \frac{(2n+1)\pi x}{2}.$$

3. From Exercise 1 we know that $(\phi_n, \phi_n) = 1/2$. Using integration by parts, we compute that

$$\begin{aligned}(f, \phi_n) &= \int_0^1 (1-x) \cdot \sin \frac{(2n+1)\pi x}{2}\,dx \\ &= -\frac{2(1-x)}{(2n+1)\pi} \cos \frac{(2n+1)\pi x}{2}\bigg|_0^1 \\ &\quad - \frac{2}{(2n+1)\pi} \int_0^1 \cos \frac{(2n+1)\pi x}{2}\,dx \\ &= \frac{2}{(2n+1)\pi} + (-1)^{n+1} \frac{4}{(2n+1)^2\pi^2}.\end{aligned}$$

The generalized Fourier series is

$$\begin{aligned}f(x) &= 1 - x \\ &= \sum_{n=0}^{\infty} \left[\frac{4}{(2n+1)\pi} + (-1)^{n+1} \frac{8}{(2n+1)^2\pi^2}\right] \\ &\quad \times \sin \frac{(2n+1)\pi x}{2}.\end{aligned}$$

5. The eigenfunctions are $\phi_n(x) = \sin(\theta_n x)$, where θ_n is the nth root of the equation $\tan\theta = -\theta$. This equation can also be written as $\sin\theta = -\theta\cos\theta$. We compute

$$\begin{aligned}(\phi_n, \phi_n) &= \int_0^1 \sin^2 \theta_n x\,dx \\ &= \frac{1}{2}\int_0^1 [1 - \cos 2\theta_n x]\,dx \\ &= \frac{1}{2} - \frac{1}{4\theta_n} \sin 2\theta_n \\ &= \frac{1}{2}\left[1 - \frac{1}{\theta_n} \sin\theta_n \cos\theta_n\right] \\ &= \frac{1}{2}[1 + \cos^2\theta_n].\end{aligned}$$

We also compute

$$(f, \phi_n) = \int_0^1 \sin\theta_n x\,dx = \frac{1 - \cos\theta_n}{\theta_n}.$$

Thus the coefficients are

$$c_n = \frac{2(1 - \cos\theta_n)}{\theta_n(1 + \cos^2\theta_n)}.$$

We have

$$1 = f(x) = \sum_{n=1}^{\infty} \frac{2(1 - \cos\theta_n)}{\theta_n(1 + \cos^2\theta_n)} \sin(\theta_n x).$$

7. Integrating by parts, we get

$$\begin{aligned}(f, \phi_n) &= \int_0^1 (1-x)\sin\theta_n x\,dx \\ &= -\frac{1}{\theta_n}(1-x)\cos\theta_n x\bigg|_0^1 \\ &\quad + \frac{1}{\theta_n}\int_0^1 \cos\theta_n x\,dx \\ &= \frac{\theta_n - \sin\theta_n}{\theta_n^2}.\end{aligned}$$

Using the fact that $\sin\theta_n = -\theta_n\cos\theta_n$, we see that the coefficients are

$$c_n = \frac{2(\theta_n - \sin\theta_n)}{\theta_n^2(1 + \cos^2\theta_n)} = \frac{2(1 + \cos\theta_n)}{\theta_n(1 + \cos^2\theta_n)},$$

and

$$1 - x = \sum_{n=1}^{\infty} \frac{2(1+\cos\theta_n)}{\theta_n(1+\cos^2\theta_n)} \sin\theta_n x.$$

9. The solution in general is given in 7.22. For $f(x) = 1$ we computed the generalized Fourier coefficients in Exercise 1. The solution is

$$u(t,x) = \sum_{n=0}^{\infty} \frac{4}{(2n+1)\pi} e^{-(2n+1)^2\pi^2 t/4} \times \sin\frac{(2n+1)\pi x}{2}.$$

11. The solution in general is given in 7.22. For $f(x) = 1 - x$ we computed the generalized Fourier coefficients in Exercise 3. The solution is

$$u(t,x) = \sum_{n=0}^{\infty} \left[\frac{4}{(2n+1)\pi} + (-1)^{n+1}\frac{8}{(2n+1)^2\pi^2}\right] e^{-(2n+1)^2\pi^2 t/4} \sin\frac{(2n+1)\pi x}{2}.$$

13. The solution to the initial/boundary value problem is

$$u(x,t) = \sum_{n=1}^{\infty} c_n e^{-\lambda_n t} \sin\frac{\theta_n x}{L},$$

where $\lambda_n = \theta_n^2$. For $f(x) = 1$ we computed the generalized Fourier coefficients in Exercise 5. The solution is

$$u(t,x) = \sum_{n=1}^{\infty} \frac{2(1-\cos\theta_n)}{\theta_n(1+\cos^2\theta_n)} e^{-\theta_n^2 t} \sin\theta_n x.$$

15. The solution to the initial/boundary value problem is

$$u(x,t) = \sum_{n=1}^{\infty} c_n e^{-\lambda_n t} \sin\frac{\theta_n x}{L},$$

where $\lambda_n = \theta_n^2$. For $f(x) = 1 - x$ we computed the generalized Fourier coefficients in Exercise 7. The solution is

$$u(t,x) = \sum_{n=1}^{\infty} \frac{2(1+\cos\theta_n)}{\theta_n(1+\cos^2\theta_n)} e^{-\theta_n^2 t} \sin\theta_n x.$$

17. We look for product solutions of the form $u(x,y) = X(x)Y(y)$ which satisfy the boundary conditions in the variable x. Separating the variables in the differential equation

$$\nabla^2 u = X''Y + XY'' = 0,$$

we get the ordinary differential equations

$$-X'' = \lambda X \quad \text{and} \quad Y'' - \lambda Y = 0,$$

where λ is a constant. Invoking the boundary conditions, we see that X must be a solution to the Sturm Liouville problem

$$-X'' = \lambda X \quad \text{with} \quad X(0) = X'(1) = 0.$$

This is the Sturm Liouville problem in Example 6.26. The solutions are

$$\lambda_n = \frac{(2n+1)^2\pi^2}{4} \quad \text{and} \quad \phi_n(x) = \sin\frac{(2n+1)\pi x}{2} \quad \text{for } n = 0, 1, 2, \ldots.$$

In analogy with equation 4.11, for each positive integer n, we get two product solutions

$$u_n(x, y) = \sinh\left(\frac{(2n+1)\pi y}{2}\right)\sin\left(\frac{(2n+1)\pi x}{2}\right) \quad \text{and}$$

$$v_n(x, y) = \sinh\left(\frac{(2n+1)\pi(y-1)}{2}\right)\sin\left(\frac{(2n+1)\pi x}{2}\right)$$

Thus we get the general solution

$$\begin{aligned} u(x, y) &= \sum_{n=1}^{\infty} a_n u_n(x, y) + \sum_{n=1}^{\infty} b_n v_n(x, y) \\ &= \sum_{n=1}^{\infty} a_n \sinh\left(\frac{(2n+1)\pi y}{2}\right)\sin\left(\frac{(2n+1)\pi x}{2}\right) \\ &\quad + \sum_{n=1}^{\infty} b_n \sinh\left(\frac{(2n+1)\pi(y-1)}{2}\right)\sin\left(\frac{(2n+1)\pi x}{2}\right), \end{aligned}$$

For the boundary condition at $y = 0$ we have the generalized Fourier series

$$T_1 = u(x, 0) = -\sum_{n=1}^{\infty} b_n \sinh\left(\frac{(2n+1)\pi}{2}\right)\sin\left(\frac{(2n+1)\pi x}{2}\right).$$

The coefficients are given by

$$-b_n \sinh\left(\frac{(2n+1)\pi}{2}\right) = \frac{(T_1, X_n)}{(X_n, X_n)} = \frac{4T_1}{(2n+1)\pi},$$

or

$$b_n = \frac{-4T_1}{(2n+1)\pi\sinh((2n+1)\pi/2)}.$$

Similarly

$$a_n = \frac{4T_2}{(2n+1)\pi\sinh((2n+1)\pi/2)}.$$

Hence the temperature is

$$\begin{aligned} u(x, y) = \sum_{n=1}^{\infty} & \frac{4}{(2n+1)\pi\sinh((2n+1)\pi/2)}\sin\left(\frac{(2n+1)\pi x}{2}\right) \\ & \cdot \left[T_2\sinh\left(\frac{(2n+1)\pi y}{2}\right) - T_1\sinh\left(\frac{(2n+1)\pi(y-1)}{2}\right)\right]. \end{aligned}$$

Section 13.8. Temperatures in a Ball—Legendre Polynomials

1. We have the boundary temperature $f(z) = T$. If a is the radius of the sphere, we have $f(as) = T = T P_0(s)$. Hence $u(r, phi) = T P_0(\cos\phi) = T$.

3. We have $f(s) = s^3 = [2P_3(s) + 3P_1(s)]/5$. Therefore the temperature is

$$u(r,\phi) = \frac{2}{5}r^3 P_3(\cos\phi) + \frac{3}{5}r P_1(\cos\phi) = \frac{1}{5}r^3\left[5\cos^3\phi - 3\cos\phi\right] + \frac{3}{5}r\cos\phi.$$

When put into cartesian coordinates, this becomes

$$u(x,y,z) = z^3 + \frac{3}{5}z(1 - x^2 - y^2 - z^2).$$

5. We have

$$f(s) = \begin{cases} 10, & \text{for } s > 0, \text{ and} \\ 0, \text{ for } s \le 0. \end{cases}$$

The Legendre coefficients for f are given by

$$c_n = \frac{2n+1}{2}\int_0^1 P_n(s)\,ds.$$

Evaluating explicitly we get

$$c_0 = \frac{1}{2}\int_0^1 10\,ds = 5, \quad c_1 = \frac{3}{2}\int_0^1 10s\,ds = \frac{15}{2},$$
$$c_2 = \frac{5}{2}\int_0^1 10\frac{3s^2-1}{2}\,ds = 0, \quad c_3 = \frac{7}{2}\int_0^1 10\frac{5s^3-3s}{2}\,ds = -\frac{35}{8}.$$

Hence the temperature is given approximately by

$$\begin{aligned} u(r,\phi) &= \frac{1}{2} + \frac{3}{4}r\cos\phi - \frac{7}{16}r^3\frac{5\cos^3\phi - 3\cos\phi}{2} + \cdots \\ &= \frac{1}{2} + \frac{3}{4}z - \frac{35}{32}z^3 + \frac{21}{32}z(x^2+y^2+z^2) + \cdots. \end{aligned}$$

7. The Legendre coefficients for f are given by

$$c_n = \frac{2n+1}{2}\int_0^1 s P_n(s)\,ds.$$

Evaluating explicitly we get

$$c_0 = \frac{1}{2}\int_0^1 s\,ds = \frac{1}{4}, \quad c_1 = \frac{3}{2}\int_0^1 s\cdot s\,ds = \frac{1}{2},$$
$$c_2 = \frac{5}{2}\int_0^1 s\cdot\frac{3s^2-1}{2}\,ds = \frac{5}{16}, \quad c_3 = \frac{7}{2}\int_0^1 s\cdot\frac{5s^3-3s}{2}\,ds = 0.$$

9. Consider the temperature v of the ball with the botton half at 10° and the top half at 0°. Then $u + v = 100$ on the boundary, so $u + v = 100$ throughout the ball. In addition, by symmetry, $u = v$ at the center of the ball. Therefore $u = 5°$ at the center.

Section 13.9. The Heat and Wave Equations in Higher Dimension

1. If $u(t, x, y) = T(t)\phi(x, y)$, then the equation $u_t = \nabla^2 u$ becomes $T_t\phi = T\nabla^2\phi$. Separating variables we get

$$\frac{T_t}{T} = \frac{\nabla^2\phi}{\phi}.$$

Since the left-hand side depends only on t, while the right-hand side depends on (x, y), we conclude that both are equal to the same constant, which we write as $-\lambda$. Both differential equations follow. Since $0 = u(t, x, 0) = T(t)\phi(x, 0)$, we conclude that $\phi(x, 0) = 0$. The other boundary conditions follow in the same way.

3.

$$\begin{aligned}\int_D \phi_{p,q}(x, y)\,\phi_{p',q'}(x, y)\,dx\,dy \\ = \int_0^1 \sin\left(\frac{(2p+1)\pi x}{2}\right)\sin\left(\frac{(2p'+1)\pi x}{2}\right)dx \int_0^1 \sin(q\pi y)\sin(q'\pi y)\,dy\end{aligned}$$

The first integral is equal to 0 if $p' \neq p$ and equal to 1/2 if $p' = p$. The second integral is equal to 0 if $q' \neq q$ and and equal to 1/2 if $q' = q$. The result follows from this.

5. It is necessary to start by finding a steady-state temperatue u_s that solves the boundary value problem

$$\begin{aligned}\nabla^2 u_s = 0, \quad &\text{in } D,\\ u_s(x, 0) = u_s(x, 1) = T_1, \quad &\text{and} \quad u_s(0, y) = (u_s)_x(1, y) = 0.\end{aligned}$$

Then the function $v(t, x, y) = u(t, x, y) - u_s(x, y)$ must be found to solve the initial/boundary value problem

$$\begin{aligned}v_t &= \nabla^2 v,\\ v(t, x, 0) = v(t, x, 1) &= T_1, \quad \text{and} \quad v(t, 0, y) = v_x(t, 1, y) = 0\\ v(0, x, y) &= f(x, y) - u_s(x, y),\end{aligned}$$

which is a variant of the problem in these exercises. Then the solution is $u(t, x, y) = u_s(x, y) + v(t, x, y)$

7. (a) We check the four cases. Remember that $0 < x < \pi$ and $0 < y < \pi$.

$$0 < x < \min\{y, \pi - x, \pi - y\} \Rightarrow \begin{cases} -\pi < x - y < 0 & \text{so } F(x-y) = \pi - y + x \\ 0 < x < \pi/2 & \\ 0 < x + y < \pi & \text{so } F(x+y) = \pi - x - y \end{cases}$$

Therefore, $F(x - y) - F(x + y) = 2x$.

$$0 < y < \min\{x, \pi - x, \pi - y\} \Rightarrow \begin{cases} 0 < x - y < \pi & \text{so } F(x-y) = \pi - x + y \\ 0 < x + y < \pi & \text{so } F(x+y) = \pi - x - y \\ 0 < y < \pi/2 & \end{cases}$$

Therefore, $F(x - y) - F(x + y) = 2y$.

$$0 < \pi - x < \min\{x, y, \pi - y\} \Rightarrow \begin{cases} x > \pi/2 \\ x + y > \pi \\ -\pi < x - y < 0 \quad \text{so } F(x-y) = \pi - y + x \end{cases}$$

The second inequality and the periodicity of F implies that $F(x+y) = F(x+y-2\pi) = \pi-(2\pi-x-y) = x+y-\pi$. Therefore, $F(x-y) - F(x+y) = 2(\pi - x)$.

$$0 < \pi - y < \min\{x, y, \pi - x\} \Rightarrow \begin{cases} x + y > \pi \\ y > \pi/2 \\ -\pi < x - y < 0 \quad \text{so } F(x-y) = \pi - y + x \end{cases}$$

The first inequality and the periodicity of F implies that $F(x+y) = F(x+y-2\pi) = \pi - (2\pi - x - y) = x + y - \pi$. Therefore, $F(x-y) - F(x+y) = 2(\pi - y)$.

(b) F is an even function, so its Fourier sereis in the cosine series

$$F(z) = \frac{a_0}{2} + \sum_{n=1}^{\infty} a_n \cos nz,$$

where the coefficients are $a_0 = \pi$, and

$$a_n = \frac{2}{\pi}\int_0^{\pi} (\pi - z)\cos nz\, dz = \begin{cases} 0, & \text{if } n \text{ is even,} \\ 4/(n^2\pi), & \text{if } n \text{ is odd.} \end{cases}$$

Hence

$$F(z) = \frac{\pi}{2} + \frac{4}{\pi}\sum_{n=0}^{\infty} \frac{\cos(2n+1)z}{(2n+1)^2}.$$

(c) The addition formula for the cosine is $\cos(\alpha+\beta) = \cos(\alpha)\cos(\beta) - \sin(\alpha)\sin(\beta)$. Hence

$$F(x-y) = \frac{\pi}{2} + \frac{4}{\pi}\sum_{n=0}^{\infty} \frac{\cos(2n+1)x \cdot \cos(2n+1)y + \sin(2n+1)x \cdot \sin(2n+1)y}{(2n+1)^2}, \quad \text{and}$$

$$F(x+y) = \frac{\pi}{2} + \frac{4}{\pi}\sum_{n=0}^{\infty} \frac{\cos(2n+1)x \cdot \cos(2n+1)y - \sin(2n+1)x \cdot \sin(2n+1)y}{(2n+1)^2}.$$

Hence

$$\begin{aligned} f(x, y) &= [F(x-y) - F(x+y)]/2 \\ &= \frac{4}{\pi}\sum_{n=1}^{\infty} \frac{\sin(2n+1)x \cdot \sin(2n+1)y}{(2n+1)^2}. \end{aligned}$$

(d) Notice that only the diagonal terms $p = q = 2n+1$ occur. The eigenvalues are $\lambda_{2n+1,2n+1} = 2(2n+1)^2$. Hence,

$$u(t, x, y) = \frac{4}{\pi}\sum_{n=1}^{\infty} \frac{\sin(2n+1)x \cdot \sin(2n+1)y \cdot \cos\sqrt{2}c(2n+1)t}{(2n+1)^2}.$$

The time frequencies are $(2n+1)\cdot\sqrt{2}ct$. Since they are all integer multiples of the fundamental frequency $\sqrt{2}ct$, the vibration is periodic in time with period $2\pi/[\sqrt{2}ct]$.

Section 13.10. Domains with Circular Symmetry—Bessel Functions

1. Using polar coordinates, we have

$$\int_D \phi_{n,k}\phi_{n',k'}\,dx\,dy = \int_D \cos n\theta \cdot J_n(\alpha_{n,k}r/a) \cdot \cos n'\theta \cdot J_{n'}(\alpha_{n',k'}r/a)\,dx\,dy$$
$$= \int_0^{2\pi} \cos n\theta \cdot \cos n'\theta\,d\theta \cdot \int_0^a J_n(\alpha_{n,k}r/a) \cdot J_{n'}(\alpha_{n',k'}r/a)\,r\,dr.$$

If $n \neq n'$, the first integral is equal to 0. If $k \neq k'$, the second integral is equal to 0. If $n = n'$, the first integral is equal to π, and if $k = k'$ the second is equal to $a^2 J_{n+1}^2(\alpha_{n,k})/2$, so the first orthogonality relation is verified. Similarly,

$$\int_D \psi_{n,k}\psi_{n',k'}\,dx\,dy = \int_D \sin n\theta \cdot J_n(\alpha_{n,k}r/a) \cdot \sin n'\theta \cdot J_{n'}(\alpha_{n',k'}r/a)\,dx\,dy$$
$$= \int_0^{2\pi} \sin n\theta \cdot \sin n'\theta\,d\theta \cdot \int_0^a J_n(\alpha_{n,k}r/a) \cdot J_{n'}(\alpha_{n',k'}r/a)\,r\,dr.$$

Again, if $n \neq n'$, the first integral is equal to 0, and if $k \neq k'$, the second integral is equal to 0. If $n = n'$, the first integral is equal to π, and if $k = k'$ the second is equal to $a^2 J_{n+1}^2(\alpha_{n,k})/2$, so the second orthogonality relation is verified. Finally

$$\int_D \phi_{n,k}\psi_{n',k'}\,dx\,dy = \int_D \cos n\theta \cdot J_n(\alpha_{n,k}r/a) \cdot \sin n'\theta \cdot J_{n'}(\alpha_{n',k'}r/a)\,dx\,dy$$
$$= \int_0^{2\pi} \cos n\theta \cdot \sin n'\theta\,d\theta \cdot \int_0^a J_n(\alpha_{n,k}r/a) \cdot J_{n'}(\alpha_{n',k'}r/a)\,r\,dr.$$

The first integral is equal to 0, so we are through.

3.

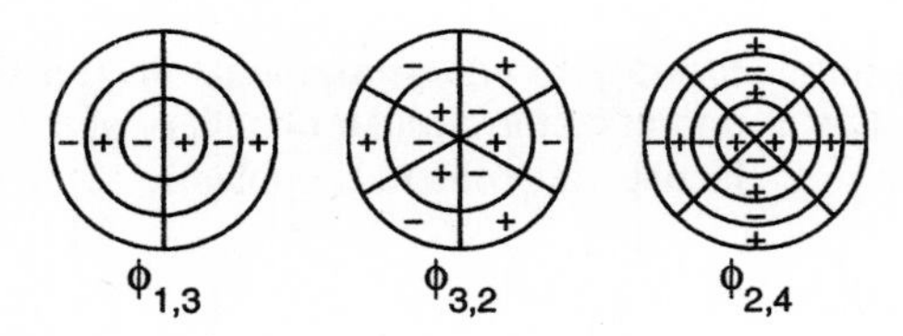

5. (a) The Laplacian of u in polar coordinates is $\nabla^2 u = u_{rr} + u_r/r + u_{\theta\theta}/r^2$. Since u does not depend on θ, this reduces to $\nabla^2 u = u_{rr} + u_r/r$.

   (b) If we substitute $u(t,r) = T(t)R(r)$ into the heat equation, we get $RT_t = [R_{rr} + R_r/r]T$. When we separate variables, we get the two equations

$$T' = -\lambda k T \quad \text{and} \quad -[R_{rr} + \frac{1}{r}R_r] = \lambda R.$$

   The first equation has solution $T(t) = e^{-\lambda k t}$. If we multiply the second equation by r it becomes $-(rR')' = \lambda r R$. Thus we are left with the Sturm Liouville problem in (10.7) with $n = 0$. The solutions are given in (10.8):

$$\lambda_p = \frac{\alpha_{0,p}^2}{a^2} \quad \text{and} \quad R_p(r) = J_0(\alpha_{0,p}r/a), \quad \text{for } p = 1, 2, \cdots.$$

Hence the product solutions are

$$e^{-k\alpha_{0,p}^2 t/a^2} J_0(\alpha_{0,p}r/a), \quad \text{for } p = 1, 2, \cdots.$$

(c) The solution is

$$u(t,r) = \sum_{p=1}^{\infty} c_p e^{-kt\alpha_{0,p}^2 t/a^2} J_0(\alpha_{0,p}r/a),$$

where

$$f(r) = u(0,r) = \sum_{p=1}^{\infty} c_p J_0(\alpha_{0,p}r/a)$$

is the Bessel series for f. The coefficients are computed in (10.10):

$$c_p = \frac{2}{a^2 J_1^2(\alpha_{0,p})} \int_0^a f(r) J_0(\alpha_{0,p}r/a)\, r\, dr.$$

7. (a) We look for product solutions $u(t,x,y) = T(t)\phi(x,y)$. When we substitute into the heat equation and separate variables, we see that there is a constant λ such that $T' = -k\lambda T$, and $-\nabla^2\phi = \lambda\phi$. Since

$$\frac{\partial u}{\partial \mathbf{n}}(t,x,y)T(t) = T(t)\frac{\partial \phi}{\partial \mathbf{n}}(x,y),$$

we conclude that ϕ must satisfy a Neumann condition. Hence the eigenvalue problem is

$$-\nabla^2\phi(x,y) = \lambda\phi(x,y), \quad \text{for } (x,y) \in D, \text{ and}$$
$$\frac{\partial \phi}{\partial \mathbf{n}}(x,y) = 0, \quad \text{for } (x,y) \in \partial D.$$

(b) Since we know the Laplacian in polar coordinates, it is only necessary to understand what $\partial\phi/\partial\mathbf{n}$ iis in polar coordinates. Since the normal direction to the boundary at a point on the circular boundary of the disk points in the direction of increasing r, we see that $\partial\phi/\partial\mathbf{n} = \phi_r$. Hence the problem becomes

$$-\left[\phi_{rr} + \frac{1}{r}\phi_r + \frac{1}{r^2}\phi_{\theta\theta}\right](r,\theta) = \lambda\phi(r,\theta), \quad \text{for } r < 1,$$
$$\phi_r(a,\theta) = 0 \quad \text{for } 0 \le \theta \le 2\pi.$$

(c) We look for product solutions $\phi(r,\theta) = R(r)U(\theta)$. Substituting and separating variables, we get the equations in refcdrum3. The Sturm Liouville problem for U is the same and has the solutions in (10.4). The Sturm Liouville problem for R, however, is different. It becomes

$$-(rR')' + \frac{n^2}{r}R = \lambda r R \quad \text{for } 0 < r < 1,$$
$$R \text{ and } R' \text{ are continuous at } r = 0,$$
$$R'(a) = 0.$$

Proposition 6.24 implies that all of the eigenvalues are nonnegative. If $\lambda = 0$ is an eigenvalue, the corresponding eigenfunction must be a constant. Notice that this occurs when $n = 0$. Thus $\lambda_{0,0} = 0$ as an eigenvalue, with eigenfunction $R_0 = 1$. In other cases the eigenvalues are positive, so we set $\lambda = \nu^2$. As before, the differential equation is transformed into Bessel's equation and has solutions $J_n(\nu r)$ which are bounded at $r = 0$. To satisfy the second boundary condition, we must have $J_n'(\nu) = 0$. Let $\beta_{n,k}$ be the sequence of zeros of J_n'. Thus the eigenvalues and eigenfunctions are

$$\lambda_k = \beta_{n,k}^2 \quad \text{and} \quad R_k(r) = J_n(\beta_{n,k} r/a) \quad \text{for } k = 1,\ 2,\ \cdots,$$

for $k \geq 1$, with the addition of $\lambda_{0,0} = 0$ and $R_0 = 1$ when $n = 0$. Consequently, the eigenvalues for the eigenvalue problem on the disk are $\lambda_{0,0} = 0$ and $\lambda_{n,k} = \beta_{n,k}^2$ for $n \geq 0$ and $k \geq 1$.

9. (a) We have $-\nabla^2 u = -\nabla^2 A \cdot B - A \cdot B_{zz} = \lambda AB$. If we divides by $u = AB$, this becomes

$$\frac{-\nabla^2 A}{A} + \frac{-B''}{B} = \lambda.$$

Since each of the summands depend on different variables, they must both be constant. Hence there exist constants μ and ν such that $\mu + \nu = \lambda$ for which

$$-\nabla^2 A = \mu A \quad \text{and} \quad -B'' = \nu B.$$

Since $u = AB$ vanishes on the boundary, we must have

$$A(a, \theta) = 0 \quad \text{for } (r, \theta) \in \partial D, \qquad \text{and} \quad B(0) = B(L) = 0.$$

(b) The eigenvalue problem for A is the same as that in (10.1). The solutions are in (10.11):

$$\mu_{0,k} = \frac{\alpha_{0,k}^2}{a^2} \quad \text{with} \quad \phi_{0,k}(r, \theta) = J_0(\alpha_{0,k} r/a)$$
for $n = 0$ and $k = 1,\ 2,\ 3,\ \ldots$

$$\mu_{n,k} = \frac{\alpha_{n,k}^2}{a^2} \quad \text{with} \quad \begin{cases} \phi_{n,k}(r, \theta) = \cos n\theta \cdot J_n(\alpha_{n,k} r/a) & \text{and} \\ \psi_{n,k}(r, \theta) = \sin n\theta \cdot J_n(\alpha_{n,k} r/a) \end{cases}$$
for $n = 1,\ 2,\ 3,\ \ldots$ and $k = 1,\ 2,\ 3,\ \ldots$.

We are well familiar with the Sturm Liouville problem for B. The solutions are

$$\nu_l = \frac{l^2\pi^2}{L^2} \quad \text{with} \quad Z_l(z) = \sin\frac{l\pi z}{L}.$$

Consequently, the eigenvalues for the cylinder are

$$\lambda_{n,k,l} = \mu_{nk} + \nu_l = \frac{\alpha_{n,k}^2}{a^2} + \frac{l^2\pi^2}{L^2},$$

for $n = 0, 1, 2, \ldots, k = 1, 2, \ldots$, and $l = 1, 2, \ldots$. The associated eigenfunctions are

$$u_{0,k,l} = \phi_{0,k} \cdot Z_l = J_0(\alpha_{0,k} r/a) \cdot \sin\frac{l\pi z}{L},$$

for $n = 0$, and

$$u_{n,k,l} = \phi_{n,k} \cdot Z_l = \cos n\theta \cdot J_n(\alpha_{n,k} r/a) \cdot \sin \frac{l\pi z}{L},$$

$$v_{n,k,l} = \psi_{n,k} \cdot Z_l = \sin n\theta \cdot J_n(\alpha_{n,k} r/a) \cdot \sin \frac{l\pi z}{L},$$

for $n \geq 1$.